CORAL REEFS OF THE WORLD

Volume 3: Central and Western Pacific

UNEP

The **United Nations Environment Programme (UNEP)** is a Secretariat within the United Nations which has been charged with the responsibility of working with Governments to catalyze the most sound forms of development, and to co-ordinate global action for development without destruction.

The **Regional Seas Programme** was initiated by UNEP in 1974. Since then the Governing Council of UNEP has repeatedly endorsed a regional approach to the control of marine pollution and the management of marine and coastal resources and has requested the development of regional action plans.

The Regional Seas Programme at present includes ten regions* and has over 120 coastal States participating in it. Each regional action plan is formulated according to the needs of the region as perceived by the Governments concerned, and is designed to link assessment of the quality of the marine environment, and of the causes of its deterioration, with activities for the management and development of the marine and coastal environment. The action plans promote the parallel development of regional legal agreements and of action-oriented programme activities**.

IUCN

The **International Union for Conservation of Nature and Natural Resources (IUCN)** is a membership organization comprising governments, non-governmental organizations (NGOs), research institutions, and conservation agencies, whose objective is to promote and encourage the protection and sustainable use of living resources.

Founded in 1948, IUCN has nearly 600 members representing 116 countries. Its six Commissions comprise a global network of experts on threatened species, protected areas, ecology, environmental planning, environmental law, and environmental education. Its thematic programmes include tropical forests, wetlands, marine ecosystems, plants, oceanic islands, the Sahel, Antarctica, and population and sustainable development.

The **Conservation Monitoring Centre (CMC)** is the division of IUCN that provides an information service to the Union, its members, and the conservation and development communities. CMC has developed an integrated and cross-referenced global database on animals, plants and habitats of conservation concern, on protected areas throughout the world, and on the international trade in wildlife species and products. CMC produces a wide variety of specialist outputs and reports based on the analysis of this data, including such major publications as the Red Data Books and Protected Areas Directories which are now recognized as the authoritative reference works in their field.

---

*Mediterranean Region, Kuwait Action Plan Region, West and Central African Region, Wider Caribbean Region, East Asian Seas Region, South-East Pacific Region, South Pacific Region, Red Sea and Gulf of Aden Region, Eastern African Region and South Asian Seas Region.

**UNEP: Achievements and planned development of UNEP's Regional Seas Programme and comparable programmes sponsored by other bodies. UNEP Regional Seas Reports and Studies No. 1 UNEP, 1982.

CORAL REEFS OF THE WORLD

Volume 3: Central and Western Pacific

Prepared by

The IUCN Conservation Monitoring Centre
Cambridge, U.K.

in collaboration with

The United Nations Environment Programme

with financial support from the World Wide Fund for Nature,
the United Nations Environment Stamp Conservation Fund,
and Exxon Corporation

The first draft of this volume was compiled by

Susan M. Wells

The final version was revised, updated and edited by

Susan M. Wells and **Martin D. Jenkins**
The IUCN Conservation Monitoring Centre, Cambridge, U.K.

The Hawaii section was compiled by
Dr R. Grigg and Dr P. Jokiel, Hawaii Institute of Marine Biology, Kaneohe, Hawaii
and
Dr J.E. Maragos, Environment Resource Planning Branch,
U.S. Army Corps of Engineers, Hawaii

The Hong Kong section was compiled by
Dr B. Morton, Dept of Zoology, University of Hong Kong

The accounts for Bikini and Kwajalein Atolls in the Marshall Islands were compiled by
Dr J.E. Maragos, Environment Resource Planning Branch,
U.S. Army Corps of Engineers, Hawaii

The first draft of the Nauru section was compiled by
P. Lili, Dept of Fisheries, Republic of Nauru

The Taiwan section was compiled by
Dr Lee-Shing Fang, National Sun Yat-sen University, Kaohsiung, Taiwan

United Nations Environment Programme
International Union for Conservation of Nature and Natural Resources
1988

Published jointly by UNEP, Nairobi, Kenya, and IUCN, Gland, Switzerland and Cambridge, U.K.

Citation: UNEP/IUCN (1988). *Coral Reefs of the World. Volume 3: Central and Western Pacific*. UNEP
                  Regional Seas Directories and Bibliographies. IUCN, Gland, Switzerland and Cambridge,
                  U.K./UNEP, Nairobi, Kenya. xlix + 329 pp., 30 maps.

ISBN: 2-88032-945-0
Printed by: Page Bros (Norwich) Ltd, U.K.
Cover design by: James Butler, Nagui Henein
Typeset by: IUCN Publications Services Unit
Cover photo: Reef scene with diver, Red Sea: David George

Available from: IUCN Publications Services,
                  219c Huntingdon Road, Cambridge CB3 0DL, U.K.

Trade enquiries: Belhaven Press (a division of Pinter Publishers),
                  25 Floral Street, London, WC2E 9DS, U.K.
                  or for North America:
                  Columbia University Press,
                  136 South Broadway, Irvington, NY 10533, U.S.A.

The designations of geographical entities in this book, and the presentation of the material, do not imply the expression of
any opinion whatsoever on the part of UNEP or IUCN concerning the legal status of any country, territory, or area, or of
its authorities, or concerning the delimitation of its frontiers or boundaries.

NOTE

This document is not an official publication but a compilation of information on coral reefs of international importance.
It is a contribution to UNEP sponsored regional action plans for the protection and development of the marine
environment and coastal areas in the central and western Pacific Ocean, specifically to the South Pacific Action Plan.

CONTENTS

MAPS

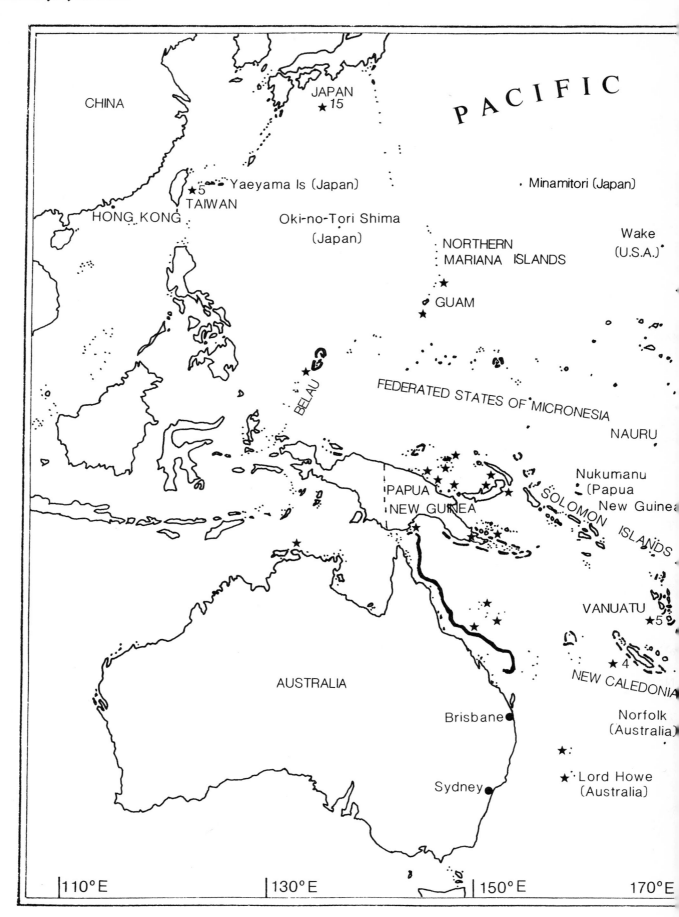

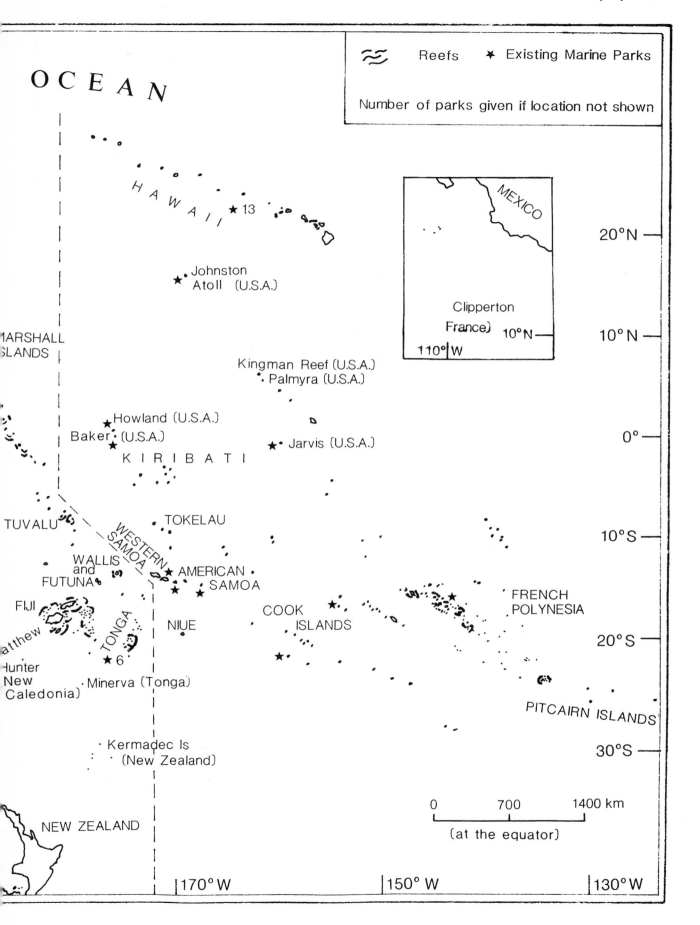

O C E A N

Reefs ✹ Existing Marine Parks

Number of parks given if location not shown

MEXICO

Clipperton
France) 10°N —

110° W

20°N —

H A W A I I ★ 13

Johnston
★ Atoll (U.S.A.)

MARSHALL
ISLANDS

10° N —

Kingman Reef (U.S.A.)
Palmyra (U.S.A.)

★ Howland (U.S.A.)
Baker (U.S.A.) ★

0° —

★ Jarvis (U.S.A.)

K I R I B A T I

TUVALU

WESTERN
SAMOA

TOKELAU

10°S —

WALLIS
and
FUTUNA

★ AMERICAN
★ SAMOA

FRENCH
POLYNESIA

FIJI

COOK
ISLANDS

NIUE

TONGA

20°S —

Matthew

★ 6

Hunter
New
Caledonia)

Minerva (Tonga)

PITCAIRN ISLANDS

Kermadec Is
(New Zealand)

30°S —

0   700   1400 km

(at the equator)

NEW ZEALAND

170° W   150° W   130° W

PREFACE

The last decade has seen growing concern for the future of coral reefs worldwide. Until comparatively recently, man used coral reefs for subsistence purposes only, as sources of food and craft materials. With the development of commercial fisheries, rapidly increasing populations still dependent on a subsistence lifestyle, growth of coastal ports and urban areas, increasing soil run-off due to deforestation and poor land use practices and most recently the exponential growth of coastal tourism, reefs have come under increasing pressure from both indirect and direct impacts. The theme 'The Reef and Man' was chosen for both the 4th and 5th International Coral Reef Congresses (1981 and 1985) when numerous papers were presented documenting the deterioration of reefs around the world. Reef protection and management were major issues at the 3rd World National Parks Congress, Bali in 1982 and at the XV Pacific Science Congress, Dunedin, New Zealand in 1983, and are now integrated into many environmental programmes and projects at international, regional and national levels.

Associated with this concern, there has been growing demand for information about coral reefs and their status. In response to this, the Coral Reef Working Group of the IUCN Commission on Ecology carried out two projects, one to document threats to reefs (see below) and a second to compile an inventory of parks and reserves containing coral reefs. The latter resulted in a preliminary list of coral reef protected areas, published in the IUCN Coral Reef Newsletters (Salvat, 1978, 1979, 1981 and 1982) which forms the basis of this Directory.

The aims of the Coral Reef Directory are:

1. to provide a broad survey of the reefs of the world giving sufficient detail to enable priorities for reef conservation to be established at both national and international levels.
2. to identify those areas where further research and conservation action is required.
3. to establish to what extent those reefs currently receiving or recommended for some form of protection are representative of the full range of reef types.
4. to promote more effective management of coral reefs by making basic information widely available.
5. to facilitate comparison between areas, thereby providing a working tool for those concerned with reef management.
6. to stimulate increased interest in coral reefs on the part of the general public, government officials, planners, scientists and students.

Many of the sections in this directory are incomplete and much of the information included will inevitably become rapidly out of date. However, the coral reef database at the IUCN Conservation Monitoring Centre in Cambridge will be continuously updated. Information should be sent to:

IUCN Conservation Monitoring Centre,
219c Huntingdon Road,
Cambridge, CB3 0DL, U.K.

ACKNOWLEDGEMENTS

We are very grateful to the many people and institutions who have given generous assistance in the preparation of this volume.

The staff of the IUCN Conservation Monitoring Centre gave invaluable support and assistance, as well as Simon Nash who helped to edit the first draft.

The maps were prepared by Francine Adams and Caroline Harcourt.

We would particularly like to thank the following who prepared drafts, reviewed sections and provided information for the country sections:

AMERICAN SAMOA
Dr C. Birkeland, Marine Laboratory, University of Guam
T. Buckley, Office of Marine and Wildlife Resources, Pago Pago, American Samoa
Dr A.L. Dahl, St-Pierre d'Albigny, France
B. Harry, USDI National Park Service, Pacific Area Office, Honolulu, Hawaii
D. Itano, Office of Marine and Wildlife Resources, Pago Pago, American Samoa
Dr A.E. Lamberts, Michigan, U.S.A.
Dr J.E. Maragos, Environment Resource Planning Branch, U.S. Army Corps of Engineers, Hawaii
Dr M. Pichon, Australian Institute of Marine Sciences, Townsville
W. Thomas, NOAA, USDI, Washington D.C., U.S.A.
Dr R.C. Wass, USDI, FWS, Honolulu, Hawaii

AUSTRALIA
Dr J.T. Baker, Australian Institute of Marine Sciences, Townsville
Dr J. Bunt, Australian Institute of Marine Sciences, Townsville
Dr W. Craik, Great Barrier Reef Marine Park Authority, Townsville
Dr Z. Dinesen, Queensland National Parks and Wildlife Service, Rockhampton
Dr H. Heatwole, University of New England, Armidale
Dr P. Hunnam, Queensland National Parks and Wildlife Service, Cairns
Dr R. Kenchington, Great Barrier Reef Marine Park Authority, Townsville
Dr D.A. Pollard, Fisheries Research Institute, Dept of Agriculture, New South Wales
M. Samoilys, Kilifi, Mombasa, Kenya
Dr J. Veron, Australian Institute of Marine Sciences, Townsville

BELAU
Dr A.L. Dahl, St-Pierre d'Albigny, France
Dr M. Gawel, Dept of Resources and Development, Kolonia, Pohnpei
B. Harry, USDI National Park Service, Pacific Area Office, Honolulu, Hawaii
Dr L. Eldredge, Marine Laboratory, University of Guam
Dr G. Heslinga, Micronesian Mariculture Demonstration Center, Koror, Republic of Palau
Dr J.E. Maragos, Environment Resource Planning Branch, U.S. Army Corps of Engineers, Hawaii
L.H. Strauss, Washington D.C., U.S.A.

CHINA
Dr P.O. Ang, Marine Sciences Center, University of the Philippines, Quezon City, Philippines
Dr Qian-Wen Lan, First Institute of Oceanography, Qingdao
Dr Bao-Ling Wu, First Institute of Oceanography, Qingdao
Dr Qiqan Wu, Third Institute of Oceanography, Xiamen
Dr Zhaoxuan Zeng, Dept of Geography, South China Normal University, Guangzhou
Dr Dakui Zhu, Dept of Geo and Ocean Sciences, Nanjing University
Dr Ren-lin Zou, South China Sea Institute of Oceanology, Academia Sinica, Guangzhou

COOK ISLANDS
Dr G. Andrews, Australian Institute of Marine Sciences, Townsville
Dr A.L. Dahl, St-Pierre d'Albigny, France
Dr G. Paulay, Dept of Zoology, University of Washington, Seattle, U.S.A.
N.A. Sims, Ministry of Marine Resources, Rarotonga, Cook Islands

FEDERATED STATES OF MICRONESIA
Dr A. Antonius, King Abdulaziz University, Saudi Arabia
Dr C. Birkeland, Marine Laboratory, University of Guam
Dr L. Eldredge, Marine Laboratory, University of Guam
Dr M. Gawel, Dept of Resources and Development, Kolonia, Pohnpei
B. Harry, USDI National Park Service, Pacific Area Office, Honolulu, Hawaii
Dr P. Holthus, SPREP, South Pacific Commission, Noumea, New Caledonia
Dr J.E. Maragos, Environment Resource Planning Branch, U.S. Army Corps of Engineers, Hawaii

FIJI
Dr T. Adams, Ministry of Primary Industries, Suva
Dr G.B.K. Baines, Development Unit, Western Province Government, Gizo, Solomon Islands
Dr J.E. Brodie, University of the South Pacific, Suva
Dr F. Clunie, Fiji Museum, Suva
Dr M. Guinea, Palmerston, Australia
Dr D. Kobluk, Earth and Planetary Science, University of Toronto, Ontario, Canada
Dr N. Penn, University College of Swansea, U.K.
Dr U. Raj, University of the South Pacific, Suva
Dr P. Rodda, Mineral Resources Dept, Government of Fiji, Suva
I. Rowles, National Trust for Fiji
Prof. J.S. Ryland, Dept of Zoology, University College of Swansea, U.K.
Dr B. Smith, Marine Laboratory, University of Guam, Guam

FRENCH POLYNESIA
Dr F.G. Bourrouilh-Le Jan, Laboratoire de géologie-sédimentologie comparée et appliquée, Université de Pau et des Pays de l'Adour, Pau, France
Dr C. Gabrié, Paris, France
Dr B. Salvat, Centre de Biology "Tropicale et Mediterranéene", Université de Perpignan, France
M. Hicks, Moorea

Dr P. Holthus, SPREP, Noumea, New Caledonia
Dr J.C. Thibault, Corsica, France
Dr J. de Vaugelas, Laboratoire de biologie et écologie marine, Université de Nice, France

KIRIBATI
Dr M. Garnett, Dolgellau, Gwynedd, U.K.
Dr P. Holthus, SPREP, South Pacific Commission, Noumea, New Caledonia
Dr D. Newill, Transport and Road Research Laboratory Dept of Transport, U.K.
Prof. J.S. Ryland, Dept of Zoology, University College of Swansea, U.K.
C. Stevens, Sir Alexander Gibb and Partners, Berks, U.K.

JAPAN
Dr O. Ikenouye, Marine Parks Center of Japan, Tokyo
Dr D.H.H. KÜhlmann, Berlin, F.R.G.
T. Milliken, TRAFFIC Japan, Tokyo
Dr J.T. Moyer, Tatsuo Tanaka Memorial Biological Station, Miyake-jima
Dr K. Muzik, Okinawa Expo Aquarium, Okinawa
Dr M. Nishihira, Dept of Biology, University of the Ryukyus, Okinawa
T. Obara, Nature Conservation Bureau, Environment Agency, Tokyo
Dr M. Ohta, UNEP Regional Office for Asia and the Pacific, Bangkok, Thailand
Dr G. Tribble, University of Hawaii at Manoa, Honolulu, Hawaii
Dr M. Yamaguchi, Dept of Marine Sciences, Univ. of the Ryukyus, Okinawa
M. Yasuko, WWF Japan, Tokyo

GUAM
Dr C. Birkeland, Marine Laboratory, University of Guam
Dr L. Eldredge, Marine Laboratory, University of Guam
Dr M. Gawel, Dept of Resources and Development, Kolonia, Pohnpei
D.T. Lotz, Dept of Parks and Recreation, Guam

MARSHALL ISLANDS
Dr L. Eldredge, Marine Laboratory, University of Guam
Dr M. Gawel, Dept of Resources and Development, Kolonia, Pohnpei
Dr J.E. Maragos, Environment Resource Planning Branch, U.S. Army Corps of Engineers, Hawaii

NAURU
Dr L. Eldredge, Marine Laboratory, University of Guam

NEW CALEDONIA
Dr P. Bouchet, Museum National d'Histoire Naturelle, Paris, France
Dr W. Bour, ORSTOM-Nouméa, Nouméa
Dr A.L. Dahl, St-Pierre d'Albigny, France

NEW ZEALAND
Dr M.P. Francis, Leigh Marine Lab., Northland

NIUE
Dr A.L. Dahl, St-Pierre d'Albigny, France
Dr G. Paulay, Dept of Zoology, University of Washington, Seattle, U.S.A.

NORTHERN MARIANA ISLANDS
Dr C. Birkeland, Marine Laboratory, University of Guam
Dr L. Eldredge, Marine Laboratory, University of Guam

Dr M. Gawel, Dept of Resources and Development, Kolonia, Pohnpei
B. Harry, USDI National Park Service, Pacific Area Office, Honolulu, Hawaii
Dr J.E. Maragos, Environment Resource Planning Branch, U.S. Army Corps of Engineers, Hawaii

PAPUA NEW GUINEA
F.B.S. Antram, TRAFFIC Oceania, Sydney, Australia
Dr M. Claereboudt, Université Libre de Bruxelles, Belgium
J. Genolagani, PNG University of Technology, Lae
Dr R.E. Johannes, CSIRO, Tasmania, Australia
Dr B. Kojis and Dr N. Quinn, Heron Island Research Station, Australia
K. Kisokau, Dept of Environment and Conservation, Boroko
Dr N. Kwapena, Wildlife Division, Dept of Environment and Conservation, Boroko
Dr E. Lindgren, Dept of Environment and Conservation, Boroko
Dr J.L. Munro, ICLARM, South Pacific Office, Honiara, Solomon Islands
Dr N. Polunin, Dept of Biology, University of Newcastle, Newcastle, U.K.
L.H. Strauss, Washington D.C., U.S.A.
A. Wright, Forum Fisheries Agency, Honiara, Solomon Islands

PITCAIRN
Dr N. Broodbakker, London, U.K.
S. Oldfield, Great Gransden, Cambridge, U.K.
A. Parkes, University of Hull, U.K.
Dr G. Paulay, Dept of Zoology, University of Washington, Seattle, U.S.A.
M. Richmond, Southampton, U.K.
B. Vittery, Nature Conservancy Council, Peterborough, U.K.

SOLOMON ISLANDS
Dr G.B.K. Baines, Development Unit, Western Province Government, Gizo
A. Worsnop, Honiara
S. Diake, Ministry of Natural Resources, Honiara

TAIWAN
Dr K.T. Shao, Inst. Zool. Acad. Sinica, Taipei
R. Quen-Jan, c/o Dept Biology, University of York, U.K.

TOKELAU
Dr R. Gillett, FAO/UNDP Regional Fishery Support Programme, Suva, Fiji
Dr A. Hooper, Univ. Auckland, New Zealand
I. Reti, SPREP, South Pacific Commission, Noumea, New Caledonia

TONGA
Dr R. Chesher, Neiafu, Vava'u
Dr A.L. Dahl, St-Pierre d'Albigny, France
U. Samani, Ministry of Lands, Survey and Natural Resources, Nuku'alofa

TUVALU
Dr A.L. Dahl, St-Pierre d'Albigny, France
B. Sloth, Ministry of the Environment, Copenhagen, Denmark
Dr L. Zann, Great Barrier Reef Marine Park Authority, Townsville, Australia

VANUATU

Dr G.B.K. Baines, Development Unit, Western Province
 Government, Gizo, Solomon Islands
Wycliff Bakeo, Fisheries Dept, Ministry of Agriculture,
 Fisheries and Forests, Luganville
Dr M.R. Chambers, Environmental Unit, Ministry of
 Lands, Energy and Water Supply, Port Vila
Dr J. Crossland, James Crossland and Associates,
 Auckland, New Zealand
Dr A.L. Dahl, St-Pierre d'Albigny, France
Dr D. Dickinson, Vanuatu Natural Science Society,
 Malapoe College, Port Vila
Dr R. Pickering, Vanuatu Natural Science Society,
 Port Vila
A. Power, Santo Dive Tours, Santo

WESTERN SAMOA

Dr A.L. Dahl, St-Pierre d'Albigny, France
Dr P. Holthus, SPREP, South Pacific Commission,
 Noumea, New Caledonia
T. Uli, Ministry of Agriculture, Forests and Fisheries,
 Apia
E. Bishop, Ministry of Agriculture, Forests and Fisheries,
 Apia

The following institutions and individuals have provided
valuable general assistance: B. Bishop, Pacific Science
Association, Honolulu, Hawaii; Dr D. Challinor,
Smithsonian Institution, Washington D.C., U.S.A.;
International Council for Bird Preservation, Cambridge;
N. Wendt and Dr P. Holthus, SPREP, Noumea, New
Caledonia

INTRODUCTION

METHODS AND FORMAT

Regional coverage

The Directory consists of three volumes covering 1) the Wider Caribbean, Atlantic and Eastern Pacific, 2) the Indian Ocean (up to and including South-east Asia), the Red Sea and the "Gulf"* and 3) the Pacific and Australasia. This volume (3) includes 29 countries in the region extending from Clipperton Island and the Pitcairn Islands in the East Pacific to the eastern seaboard of Asia, Papua New Guinea and Eastern Australia. Most of this area lies within the Oceanian Realm as defined by Udvardy (1975), but China, Taiwan and Hong Kong lie within the Indomalayan Realm, and Japan within the Palaearctic Realm. Western Australia, Indonesia and the Philippines, which have close biological, geographical or political affinities with some of the countries in this volume are covered in Volume 2. Of the regions defined by the UNEP Regional Seas Programme, only the South Pacific Region is covered in this volume. Eastern Australia, Hawaii, Hong Kong, Taiwan, Japan, China and New Zealand are not within this Region although some of them participate in the Regional Seas Programme.

Sources of information

Information has been obtained from a wide variety of sources, including published and unpublished material, and from a worldwide correspondence with coral reef researchers, conservationists and government officials. However, there are many gaps, either because the information is not available or because it has not been possible to contact the person who could provide it.

Accounts for protected areas which are being compiled for the IUCN Directory of Oceanian Protected Areas (IUCN, in prep.) have been rewritten in the appropriate format and additional information included where relevant; data on terrestrial aspects have generally been reduced. A recent source of information on marine parks is Silva *et al.* (1986) which provides an update on the list compiled by Björklund (1974). A complete and accurate listing of marine parks of the world is not yet available, partly because there are numerous instances where it is not known whether the park legislation covers marine (or submarine) habitat.

In order to keep the volume to a manageable size, coral reefs have had to be considered in isolation from other closely associated marine habitats, such as seagrass beds and mangroves. A directory of wetlands of international importance is currently being compiled for Asia by IWRB (International Waterfowl Research Bureau) with the support of IUCN and UNEP; this will provide a complementary volume for the Asian countries, with its emphasis on other coastal and marine habitats.

Information on reef-related species has been taken mainly from the IUCN Red Data Books, including Groombridge (1982) for turtles and crocodiles and Wells *et al.* (1983) for marine invertebrates. Information on seabirds is largely from papers in Croxall *et al.* (1984).

Maps showing the distribution of reefs and of proposed and established protected areas have been compiled for each country using, where possible, material sent in by correspondents and, where this was unavailable, British Admiralty and other charts and maps available at the University of Cambridge. In many instances, it was possible to give only a very rough approximation of true reef distribution and the extent of reef coverage is probably greatly under-represented. The majority of countries in this volume comprise groups of islands; information about individual islands is generally presented in tabular form, using as a basis the checklist provided by Douglas (1969), updated wherever possible. Orthography generally follows that of Motteler (1986).

Site descriptions

The format has been adapted from that used in other IUCN directories. For each country, an introductory section describes the distribution of reefs within the country, their status, relevant conservation issues and legislation. A comprehensive reference list, including scientific monographs and papers, popular books and articles, bibliographies, management plans and unpublished reports, is included. This section is followed by detailed accounts for reefs already protected in national parks and reserves, reefs proposed for protection, and reefs recommended by qualified experts as requiring protection or management on the basis of their scientific interest or economic importance. These accounts have the following format:

1. *Geographical Location*   province, region, geographical coordinates; where relevant the proximity to major towns, national borders, other protected areas and major features is noted.

2. *Area, Depth and Altitude* in hectares and metres: areas of both protected and unprotected areas; minimum/maximum depth of reef; altitude of associated cay/atoll/terrestrial ecosystem.

3. *Land Tenure* public (government-owned), freehold, private, etc., with percentages or hectarage where there is multiple ownership.

4. *Physical Features* topography, geology, climate, hydrology, and other physical features (e.g. salinity, water clarity, wave action, currents, water temperature), particularly as they affect management of the area; reef type e.g. barrier, fringing, atoll, patch.

5. *Reef Structure and Corals* reef zonation: coral morphology, diversity, per cent live cover and dominant species.

6. *Noteworthy Fauna and Flora* predominant algae and other vegetation; vertebrates and invertebrates which are of particular importance due to their dominance in the ecosystem, rarity, size of population, etc., in particular, species of possible economic importance (e.g. dugong, turtles, fish,

molluscs, crustaceans, echinoderms, etc.) and those included in the IUCN Red Data Books.

7. *Scientific Importance and Research* importance of the reef in terms of the scientific interest of its coral formations and fauna; major research conducted in the area; details of current projects and scientific facilities.

8. *Economic and Social Benefits* use or potential use for fisheries (commercial or subsistence), tourism, recreation, mariculture, education, harbour protection.

9. *Disturbance or Deficiencies* natural disturbances such as hurricanes and *Acanthaster* damage; siltation; pollution; damaging fishing methods (e.g. explosives, poisons, trawling); collection of corals for the curio trade, lime or building materials; overcollection of shells and other marine invertebrates; overfishing of reef fish for food or the aquarium trade; anchor damage and boat groundings; damage from tourists (e.g. trampling) and SCUBA divers; dredging, filling and other forms of coastal development.

10. *Legal Protection* degree of legal or special protection afforded to certain elements within the area.

11. *Management* local administrative entity for the area; presence of interpretative centres and wardens; degree of enforcement of legislation; system of zoning.

12. *Recommendations* legislation and management required; research priorities.

CORALS AND REEF DISTRIBUTION

Structure of Coral Reefs

Coral reefs rank as among the most biologically productive and diverse of all natural ecosystems, their high productivity stemming from efficient biological recycling, high retention of nutrients and a structure which provides habitat for a vast array of other organisms. They are tropical, shallow water ecosystems, largely restricted to the area between the latitudes 30°N and 30°S. The exact areal extent of coral reefs in the world is unknown and extremely difficult to estimate. However, Smith (1978) has produced a figure of 600 000 sq. km for reefs to a depth of 30 m. Under this analysis, 13%, or about 77 000 sq. km, lie within the South Pacific (including Eastern Australia), and 12%, or about 76 000 sq. km, lie within the North Pacific (including the Galapagos and west coast of North America). It must be emphasised that these figures are only a rough approximation; one estimate for the Great Barrier Reef alone gives a total of 348 000 sq. km, presumably including reefs to depths greater than 30 m.

General descriptions of coral reefs, their ecology and environmental requirements are given in Wood (1983), Salm and Clark (1984), Kenchington and Hudson (1984) and Snedaker and Getter (1985).

The true reef-building corals (hermatypic or stony corals) are animals (polyps) that collectively deposit calcium carbonate to build colonies. The coral polyps have symbiotic algae (zooxanthellae) within their tissues which process the polyp's waste products, thus retaining vital nutrients. The term "reef" is used in the directory for a population of stony corals which continues to build on products of its own making (Stoddart, 1969). Not all reefs are constructed predominantly of corals. In particular, several genera of red algae grow as heavily calcified encrustations which bind the reef framework together, forming structures such as algal ridges. In other cases, populations of corals exist, often in deeper, colder waters, which either do not build on themselves or are formed of ahermatypic, non-symbiotic corals which do not build reefs. Many of these have been included in the directory (like true reefs they may also have a very high productivity) and have been termed coral assemblages or communities.

Present day reefs fall into two basic categories: shelf reefs, which form on the continental shelf of large land masses; and oceanic reefs, which develop adjacent to deeper waters, often in association with oceanic islands. Within these two categories are a number of different reef types: fringing reefs which grow close to shore (i.e. most shelf reefs, although some develop around oceanic islands); patch reefs which form on irregularities on shallow parts of the sea bed; bank reefs which occur deeper than patch reefs, both on the continental shelf and in oceanic waters; barrier reefs which develop along the edge of the continental shelf or through land subsidence in deeper water and are separated from the mainland or island by a relatively deep, wide lagoon; and atolls, which are roughly circular reefs around a central lagoon and are typically found in oceanic waters, probably corresponding to the fringing reefs of long since submerged islands.

Marine and reef-related research in the Pacific Ocean, apart from in Eastern Australia, lags behind that of other parts of the world. This is largely because of the inaccessibility of many islands, and the vast distances that are involved. It is therefore striking that some of the earliest, and arguably the most important, reef work was carried out here. It was in the Tuamotu Archipelago and Society Islands of French Polynesia that Darwin began to develop his ideas on the orderly development of reefs on slowly sinking ocean volcanoes, on comparing the morphology of reef types with the theory of subsidence (Darwin, 1842). Until comparatively recently, research was restricted to expedition oriented work (Yonge, 1973). Early expeditions to the South Pacific included the Carnegie Institute expeditions to American Samoa (1917-1920), the Whitney South Seas Expedition (1922), the Great Barrier Reef Expedition (1928-29) and the Mangareva Expedition (1934). More recent ones include the Singer Polignac expeditions to the New Caledonia area (1960s), the Te Vega Expedition to the West Pacific from Stamford University (1965), the Royal Society Expedition to the Solomons (1965), the Cook Bicentenary Expedition (1969), the National Geographic Society-Oceanic Institute Expedition (1970), and the University of Hawaii expeditions to Tabuaeran (1970-1972). The Palau Tropical Biological Station, run by the Japanese until World War II, was one of the earliest permanent marine research bases.

Long-term reef projects are still comparatively rare in the region but reef research activities have increased enormously recently, through the activities of institutions such as the University of the South Pacific in Fiji, universities in Hawaii, Papua New Guinea, Guam and Australia, the Atoll Research Unit in Kiribati, the South Pacific Commission and ORSTOM in New Caledonia, and numerous cruises and expeditions such as the missions from the Museum National d'Histoire Naturelle in Paris to French Polynesia. The Atoll Research Bulletin provides a focus for much of the published work for this area. Eldredge (1987c) provides a compilation of references to Pacific marine ecosystems, and the 23rd South Pacific Conference in 1983 resulted in the compilation of a bibliography of marine related references (SPC, 1984).

Reef Distribution

Coral makes up a great proportion of Pacific islands, either by active growth of the reef itself or by accumulation of reef debris by mechanical forces such as waves and currents. Coral growth is optimal only within a fairly narrow range of water temperatures and salinities, and reef structure and development is influenced by other oceanogaphic factors such as currents and wave force. Compared to the Atlantic and Indian Oceans, the Pacific region is characterized by a lack of continental land masses, the absence of much continental shelf and the striking contrast between high volcanic oceanic islands and low coral atolls.

In the South Pacific, the continental shelf down to the 200 m isobath demarcates the continent proper to the west; most of the Arafura Sea is less than 100 m deep and the Torres Strait less than 20 m. The margin area of the continental shelf extends eastwards to the island arcs which form an almost continuous chain of islands or shallows stretching from New Zealand, through Tonga, the Solomon Islands, the Bismarck Archipelago, Belau and the Marianas to Japan; it includes the Tasman Sea, Coral Sea, Solomon Sea and Philippine Sea. Inside this margin area, seamounts support archipelagoes such as Fiji, Vanuatu, New Caledonia and the Loyalty Islands, interspersed with several very deep basin areas such as the Coral and Solomon Seas. To the east, the island arcs are bounded by deep trenches, the most distinctive of which are the Marianas Trench, the Palau Trench, the Bougainville Trench, the New Hebrides Trench, the Tonga Trench, and the Kermadec Trench, several of which are more than 10 000 m deep. Further east, depth averages more than 4000 m, and seamounts in the form of ocean ridges stretch over considerable distances, varying in width from 200 to 500 km. These are more or less parallel, running NW-SE and form the basement of atoll chains such as the Carolines (Federated States of Micronesia), Marshalls, Kiribati, Tuvalu, Tokelau, Cook Islands and Tuamotu Archipelago. Other less extensive relief features created by volcanic activity form the basements of more isolated island groups such as the Society Islands, Samoa, the Marquesas and Hawaii (Wauthy, 1986). The islands form three major groupings: Micronesia in the north-west, which has a minute proportion of emergent land in relation to sea area and runs from Ogasawara-shoto and Kazan-retto (the Bonin and Volcano Islands) to the Marianas and Marshalls; Polynesia, the group most remote from continental influence, which includes all the islands and atolls from Hawaii to Easter Island and west to Tuvalu and Tonga; and the larger islands of Melanesia which include Fiji, Papua New Guinea and New Caledonia, and which form part of the structural mass of the Australian Continent. The following table of distribution of reef types in the Pacific is taken from Dahl (1980):

	AR	WAR	LAR	BR	FR	LR	NGR	SR
New Guinea	X	X	X	X	X	X	X	X
Bismark Achipelago				X	X	X		
Solomon Islands	X			X	X	X		
New Caledonia-Loyalty	X	X	X	X	X	X		
New Hebrides (Vanuatu)		X	X		X	X		X
Norfolk-Lord Howe-Kermadec					X			
Fiji	X	X	X	X	X	X		
Tonga-Niue	X			X	X	X		
Samoa-Wallis	X	X	X	X	X	X	X	X
Tuvalu-Tokelau	X	X	X			X		X
Kiribati-Nauru		X	X		X	X		X
Mariana Islands				X	X	X	X	
Caroline Islands	X	X	X	X	X	X	X	X
Marshall Islands	X	X	X			X		
Phoenix-Line-Northern Cook	X	X	X			X		
Cook-Austral	X	X	X	X	X	X		
Society Islands		X	X	X	X	X		
Tuamotu Archipelago	X	X	X		X	X		X
Marquesas					X			
Pitcairn-Gambier-Rapa		X	X	X	X	X		

AR = algal reef	WAR= windward atoll reef	LAR = leeward atoll reef
BR = barrier reef	FR = fringing reef	LR = lagoon reef
NGR= non-growing reef	SR = submerged reef	

A brief summary of some of the oceanographic characteristics of the Pacific is given in Groves (1979) and a more detailed account is given in Wauthy (1986). The tropical Pacific is dominated by the North-easterly Trade Winds (strongest November - May) and the South-easterly Trade Winds (strongest June - October), while at higher latitudes, the winds are from the west. Wind waves are particularly important in the Pacific, having a major impact on islet (motu) construction and destruction on atolls. The large size of the Pacific means that almost everywhere there is appreciable wave energy at long wave periods and the trade winds cause almost constant shorter waves on eastern shores of islands. Hurricanes are comparatively rare in the eastern Pacific but common in the west. From October to January there are often westerly storms related to the monsoon winds of South-east Asia. Rainfall is very varied; it is highest in the region of Palmyra, (at about 7°N) under the intertropical convergence zone and smallest near the equator, for example in parts of Kiribati. Sea surface temperatures along the Equator are warmer in the west than in the east, and as the South Equatorial Current flows west it is heated several degrees before reaching the western Pacific. Furthermore, sea temperatures show much greater annual variation in the east than in the west. The El Niño phenomenon is a characteristic of the Pacific, and occurs at intervals of 2-10 years when anomalously warm water currents are triggered in the Eastern Pacific, and are accompanied by a rise in temperature of surface waters, high rainfall, and a sea level rise in the east and fall in the west (Cane, 1983). There are no major upwellings in the Pacific and no major input of nutrients from large land masses to favour primary productivity, which leads to a strong contrast between the productive coral reefs and the oligotrophic ocean.

Coral Diversity

The study of marine biogeography in this region is still in its infancy. Within the central Indo-Pacific, coral faunas are essentially homogenous at both generic and species level (Veron, 1985 and 1986). The most obvious trend is the decreasing diversity of scleractinian corals from the Western Pacific, adjacent to the Indo-Malayan centre of coral reef evolution and diversity, to the Central and Eastern Pacific; and from the equator north and south to more temperate waters. The east-west trend is probably a result of the prevailing westward currents preventing eastward dispersal of many corals. The main easterly tropical current flows north of the equator, missing most reef areas. The South Equatorial Current, the other weaker easterly current, may be the main vehicle for larval transport during El Niño disturbances. In the Tuamotu and Society Island groups, diversity is fairly high and there are a wide variety of reef types, but further east in the Marquesas reef diversity is poor. Glynn and Wellington (1984) discuss the relationship of these reefs with those of the Eastern Pacific, and the Galapagos in particular. The Line Islands (part of Kiribati) probably serve as a source for much of the coral reef biota of the Eastern Pacific. The reefs of the Central Pacific tend to be more strongly dominated by algae rather than corals as in the west. This is reflected on a small scale by the difference between the windward and leeward reefs of Pacific atolls. On eastern, wave-exposed reef slopes, melobesioid red-algal coral is very conspicuous.

The Ryukyu-retto (Ryukyu Archipelago) in Japan and the southern Great Barrier Reef are the northern and southern limits of the Indo-west Pacific centre of high diversity and have a high similarity between species. This is due to larval transport from the central region via the warm Kuroshio Current to the north and the East Australian Current to the south (Veron, 1985).

With the decrease in diversity away from the western Pacific centre of endemism there is also a noticeable increase in levels of endemism. Hawaii is very isolated from other reefs and has a marginal northern tropical location; it has an impoverished fauna but greater reef development and diversity than reefs in the Galapagos or eastern Polynesia (Glynn and Wellington, 1984). Kay (1987) discussed patterns of speciation in the Pacific basin with reference to molluscs, crustaceans and echinoderms and identifies two foci of endemism: Hawaii and south-east Polynesia. Other marginal areas for coral growth include the Kermadec Islands, Norfolk and Lord Howe Island, and the Pitcairn Group.

Zann and Bolton (1985) describe the distribution of the Blue Coral *Heliopora coerulea* throughout the Pacific, noting that it is more abundant in the equatorial Central Pacific such as the Marshall Islands, Tuvalu and the Gilbert Islands than in the Western Pacific. This is considered to be a result of competition with scleractinian corals. *Heliopora* is a "living fossil" and the only member of its family, and is possibly outcompeted by scleractinians which have a higher species diversity. Richmond (1985) discusses the distribution of *Pocillopora damicornis* across the Pacific.

ECONOMIC IMPORTANCE OF REEFS

The World Conservation Strategy (IUCN/UNEP/WWF, 1980) identifies coral reefs as one of the "essential life-support systems" necessary for food production, health and other aspects of human survival and sustainable development. Reefs protect the coastline against waves and storm surge, prevent erosion and contribute to the formation of sandy beaches and sheltered harbours. They are a source of raw materials such as corals and coral sand for building materials, black coral for jewellery, and stony coral and shells for ornamental objects. Increasing numbers of reef species are being found to contain compounds with medical properties. Reef resources are particularly important in the many island nations of the Pacific where lack of soil and water mean that agricultural potential is limited. A broader view of the marine economy and its importance in this region is given in Gopalakrishnan (1984). The Directory gives particular attention to the crucial role of reefs in fisheries and to their new role as a major focus of the tourist industry, owing to their aesthetic appeal and recreational value.

Fisheries

Reefs provide the fish, molluscs and crustaceans on which coastal communities in developing countries depend and, with other inshore habitats, provide nutrients and breeding grounds for many commercial species. Reefs are mainly important in small-scale traditional fisheries, particularly in island nations, the sea providing up to 90% of the animal protein of many Pacific islanders. Yields are often underestimated because of the large numbers of subsistence fishermen whose catches are not recorded, and are very variable (partly due to the varying conditions under which estimates have been made). It is thought that in general a sustainable yearly harvest of 15 t/sq. km can be obtained from coralline areas in depths of less than 30 m (Munro and Williams, 1985). In Belau a potential yield of 2000-11 000 tonnes of reef fish has been estimated, a range of the same magnitude as the tuna fishery, which seems to apply to Micronesia as a whole (Johannes, 1977). Excluding subsistence fishing, reef and lagoon fish may constitute 29% of the commercialized local fishery in the South Pacific, a yield of 100 000 tonnes a year (Salvat, 1980). Commercial reef fish include Trevally *Caranx* spp. and other large species caught in deep water passages in the reef, and smaller fish caught on the reef flat. The outer reef is occasionally used for fishing although it is normally too dangerous. With the increase in tuna fishing, a complementary bait fishery has developed around coral reefs for small fish such as anchovies *Stolepherus* spp., sprats *Spratelloides* spp. and sardines *Sardinella* spp. The Great Barrier Reef region off Australia is an exception in the region in that there is no subsistence harvesting on the reefs except in the very north in the Torres Straits. However, the area is very important for recreational and commercial fishing. There is also controlled exploitation of aquarium fish, a resource which is being increasingly exploited in other areas although it is still of minor importance in comparison with the South-east Asian trade (Wood, 1985; Randall, 1987). Additional information on fisheries in the Pacific region is provided in Mitchell (1975), Zann (1979), King and Stone (1979), Gopalakrishnan (1984) and Kearney (1985).

In addition to finfish, a wide range of other species are taken including molluscs, crustaceans, turtles, corals, shells and algae. The importance of invertebrate fisheries in Oceania, including those referred to as "miscellaneous marine products", particularly Giant Clams, trochus and bêche-de-mer, is described by Pearson (1980), Salvat (1980) and Carleton (1984a and b). At present exploitation of many of these resources is largely opportunistic and involves the export of raw or only simply processed products.

The productivity of certain mollusc species appears to be enormous in some circumstances; in French Polynesia, estimates for productivity range from 12 kg/ha for *Tridacna maxima* to 460 kg/ha for *Cardium fragum* in lagoons in the Tuamotu atolls (Richard, 1982). The Pacific countries contribute about 60% of the mother-of-pearl imported by Japan for the button industry, mainly trochus *Trochus niloticus* and pearl oysters *Pinctada margaritifera*. Exploitation has recently increased through the introduction of culture techniques, and trochus provides an important source of foreign exchange in many countries. Giant Clams are collected for local consumption, but are also fished commercially, largely by the Taiwanese for the lucrative Asian market; the meat is the chief product although there is also a market for the shells in Europe and the USA. The ornamental coral and shell trade is not as important as it is in South-east Asia but there is some exploitation, for example in Fiji, Papua New Guinea, and New Caledonia, and there is a small amount of regulated exploitation on the Great Barrier Reef. Black coral is exploited in a few countries such as Hawaii, Tonga and Papua New Guinea, and surveys suggest that it has potential value elsewhere provided its

vulnerability to over-exploitation is recognised (Carleton and Philipson, 1987). In many Pacific countries, reef algae such as *Caulerpa* are an important source of food (King and Stone, 1979); the commercial potential of marine algal resources in the area is reviewed by Furtado and Wereko-Brobby (1987).

Tourism

The tourist industry is developing rapidly and, although not yet as important as in the Caribbean and Indian Ocean, it is becoming a major economic factor. In island nations, the favourable climates, clear waters, beaches and reefs are major attractions. Pacific tourism started in the 1950s with the expansion of airlines and the development of cruises by shipping companies. Fiji and Hawaii are the main island tourist destinations, and in both countries the industry is focused on a few resort areas such as the 200 km Coral Coast on Viti Levu in Fiji. French Polynesia (mainly Tahiti and Moorea) and Vanuatu attract a medium number of tourists, but in many other countries tourism is at a very low level. In Papua New Guinea, the tourist industry is important but is still largely restricted to the interior; the potential for reef-based tourism in this country has yet to be fulfilled (Baines, 1980). The Queensland coast of Australia is probably the most important area economically for recreation and tourism, the primary attraction being the Great Barrier Reef. The total income from island resorts, camping, charterboats and reef fishing on the reef has been estimated at A$116.8 million, the rapid growth of the industry largely being a result of the new technology, such as high speed catamarans and semi-submersible viewing chambers, which make the reef easily accesible (Driml, 1987). A regional tourism programme has been set up for the South Pacific, supported by SPREP and the EEC, to promote tourism and to provide an information centre (Anon., 1986c).

VULNERABILITY OF REEFS

Biological research on reefs in the 1960s and early 1970s led to the view that reefs were fragile ecosystems, particularly vulnerable to human activities and slow to recover if damaged (Johannes, 1975). Subsequent work led to contrasting ideas, that reef communities are dynamic and unstable and that self-replacement and recovery from natural disturbance is normal and contributes to the maintenance of high diversity on the reef (Connell, 1978). These theories have been reviewed by Pearson (1981), Brown and Howard (1985), Brown (1987a) and Grigg and Dollar (in press), the consensus being that reefs are perhaps not as fragile as was previously thought.

However, corals generally have very specific requirements for light, temperature, water clarity, salinity and oxygen. Their lack of mobility makes them vulnerable to siltation, through smothering and oxygen depletion. Coral growth tends to be slower where sediments are regularly disturbed, and silted substrates inhibit larval settlement. Light penetration is decreased in turbid water, reducing photosynthesis by the symbiotic zooxanthellae; even in the clearest seas, reef-building corals are restricted to depths of less than 30 m and are generally found in much shallower waters. Many stony corals have slow growth rates, which may be slowed further by adverse environmental conditions; Davies (1983) provides figures for reef growth in the range of 0.38-12 m per thousand years. Brown and Howard (1985) review the responses of corals to stress, in terms of altered growth rate and metabolism (photosynthesis and respiration), loss of zooxanthellae, behavioural responses such as filament extrusion and mucus production, sediment shedding, altered reproductive biology and the appearance of disease.

Hurricanes, storms, diseases and sea-level changes show that reefs are well adapted to recovery from a variety of sources of natural stress (Pearson, 1981), although the manner and speed with which this occurs may be immensely variable. Recovery rates have been estimated at from 20 to 50 years for complex reefs to only a few years for simpler reefs at a subclimax state of succession. Shallow water monospecific thickets of, for example, *Acropora palmata* may recover even more quickly (Grigg and Dollar, in press). There is some concern that such phenomena, particularly outbreaks of coral predators (see below), are occurring at increasing frequencies, perhaps as a result of human activities, although at present there is no general consensus of opinion (Brown, 1987a). The observed increase in incidents may simply be an artefact of the rapid increase in reef studies over the last 20 years. Some of the more recent natural events having an impact on reefs in the Pacific Ocean region are summarized in Table 1.

Climatic, tidal and geological events

Storms and cyclones can reduce large areas of reef to rubble through freshwater inundation and the breaking of branching corals. Hurricanes are a frequent phenomenon in the Pacific; their impact on reefs has been documented in several countries including Fiji, French Polynesia and Tokelau. Stoddart (1985) summarizes current thinking on the impact of storm damage; most research and the only long-term monitoring studies (on the impact of hurricanes) have been carried out in the Caribbean. Geological and tectonic events, such as earthquakes and volcanic eruption, may affect reefs and have had a recorded impact in Fiji, the Northern Marianas and the Solomon Islands.

Cold temperatures periodically cause coral mortality in the more northerly and southerly parts of the region. Elevated temperatures may be equally damaging; the abnormally high sea-water temperatures which accompanied the severe 1982-83 El Niño event were probably responsible for the widespread "bleaching" (i.e. loss of symbiotic zooxanthellae) and death of corals in the tropical Pacific and western Atlantic (Glynn, 1984). It is thought that coral reef recovery may take many years or decades in the Eastern Pacific. In the Central and Western Pacific there have been no detailed studies but reefs have been seriously damaged in areas as widely separated as Japan and French Polynesia (Table 1). In some areas these events were accompanied by abnormally low tides and storms which exacerbated the damage.

Coral predators

Coral predators can have a significant impact on reef structure and development. This is particularly evident in the Indo-Pacific, where outbreaks of the Crown-of-Thorns Starfish, *Acanthaster planci*, have affected many reefs over the last 25-30 years (Table 1) (Endean, 1973; Moran, 1986). This phenomenon has been well studied in the Pacific, but remains a controversial subject. It is of greatest concern in Australia because of the economic impact it could potentially have on the Great Barrier Reef. Research has still not indicated whether outbreaks are simply natural recurring phenomena or whether they are the result of human activities; there is some evidence to suggest, for example, that increased soil-run off may aggravate outbreaks and in some instances cause them (Birkeland, 1982). A major study is currently under way in Australia which may provide new information to resolve these problems.

Whereas asteroids appear to be the most influential predators on Pacific reefs, echinoids seem to be more important in the Caribbean (Birkeland, 1987). Sea urchins can cause significant damage if they occur in high numbers. They graze on algal turf on coral rock and may weaken the reef structure through erosion of the rock surface (Hutchings, 1986). Urchin outbreaks have not yet been recorded in the Pacific, possible reasons for this being discussed in SPC/SPEC/ESCAP/UNEP (1985) and Birkeland (1987). One theory is that outbreaks are a result of human impact on the reefs, which is recognisably greater in the Caribbean than in the Pacific.

Disease

Two diseases, white band and black band, are widespread in the Caribbean but in the Pacific region have only been recorded from Japan. It is possible that the great size of the Pacific helps to buffer the spread of disease (Birkeland, 1987). Coral genera and species differ in their response but the general effect is a weakening of the reef framework. The diseases do not appear to be correlated with human activities and may be caused by bacteria (Antonius, 1981; Peters, 1983; Gladfelter, 1982).

HUMAN IMPACT ON REEFS

Although reefs may be constantly experiencing natural change, there is increasing evidence that human impact, combined with natural disturbance, may significantly slow the recovery rate of a reef, particularly since man-induced damage is often chronic rather than temporary. Where reefs are of economic importance, their long recovery time may become an important issue. Kenchington and Hudson (1984), Salm and Clark (1984), Sorensen *et al.* (1984), Brown and Howard (1985), Salvat (1987a) and Grigg and Dollar (in press) review the impact of human activities on coral reefs and Clark (1985) provides more detailed case studies.

Threats to coral reefs are intimately related to the high population densities of coastal areas. Although many of the Pacific nations have comparatively low population densities, the small size of many of the islands, and the fact that urban developments are often of necessity located on the coast, means that land-based activities have a very direct impact on the adjacent inshore waters. Fringing reefs, lying immediately off shore, are particularly vulnerable to pollutants and sediments washed off the land and may be affected by activities taking place many kilometres away, such as deforestation. They also suffer greatest damage from overexploitation and recreational use owing to their accessibility. Atoll and barrier reefs are less vulnerable but may be affected by pollutants carried on oceanic currents or released from ships and, in areas which are important to fisheries, may be vulnerable to overexploitation.

The lack of long-term studies and monitoring projects in the region has meant that threats to reefs have largely been inferred from known damaging activities. The Pacific however probably has proportionately more pristine and untouched reefs than other oceans (Dahl, 1985a). The reefs of New Caledonia and the Solomons are thought to be largely pristine, and those of Vanuatu, Tokelau and Niue have probably been little affected. Reefs in Melanesia may also be comparatively healthy because of the large area of reef resulting from the presence of the continental shelf and the fact that populations are less dense and less oriented towards the sea. In contrast, most accessible reefs in Polynesia are considered to be in various stages of decline, particularly in Fiji, Tonga, Western Samoa, American Samoa (except the remote Rose Atoll) and the Cook Islands (particularly Rarotonga).

In the early 1980s, three quarters of the countries in the South Pacific region reported reef pollution, nearly half noted damage from illegal fishing with explosives and poisons, and one third cited siltation from erosion and dredging as problems (Dahl and Baumgart, 1983). Reef fisheries have clearly declined in areas where the population has increased or traditional fishing techniques have been superceded by modern methods. Tsuda (1981) gave a brief overview of threats to reefs in Micronesia and concluded that in the late 1970s, dynamiting for fish, and dredging and construction were the most damaging activities. Overviews of environmental problems in the Pacific including those associated with reefs, are also provided by Dahl (1985a) and Gomez and Yap (1985). The main impacts on reefs in the region as identified in this volume are discussed below and are listed in Table 2.

Run-off from land clearance

Soil run-off is the most frequent source of increased sediment content in coastal waters, and is considered to be one of the most damaging impacts on corals in many parts of the world. In the Pacific, the problem is largely restricted to the high volcanic islands, such as the Society Islands in French Polynesia, and to continental countries such as Japan. It has been reported from about 11 countries in the region (Table 2). Deforestation through logging and agricultural malpractices such as slash-and-burn, are the main causes, as well as destruction of mangroves, which act as sediment traps. The impact is often compounded by the input of fertilizers, pesticides and other pollutants. The responses of corals to sediment are very variable, many species being able to withstand low levels and others having behavioural or physiological responses to remove sediment (Grigg and Dollar, in press). Nevertheless, there are certainly many instances where damage is severe. Furthermore, siltation of this kind is often difficult to control, since the source of sediment may be far from the site of damage and come under the control of different agencies and government authorities. Although soil run-off is not yet as serious in

the Pacific as in other regions, it could potentially cause major long term changes since, unlike reefs fringing continental land masses, most of those round oceanic islands have evolved in the absence of any significant terrestrial input (Birkeland, 1987).

Industrial, domestic and agricultural pollution

The impact of pollution in the South Pacific region is reviewed by Helfrich and Maragos (1979), Matsos (1981) and Brodie and Morrison (1984). It is not yet a major problem in the region except in urban areas, as there is very little industrial development, but is of considerable concern for the future, given the difficulty of disposal of wastes on small islands.

Sewage pollution has been reported from a number of countries (Table 2) and is a potential problem in many more. It causes eutrophication and ensuing accelerated algal growth which smothers corals; oxygen depletion and toxic contamination compound this (Johannes, 1975; Brown and Howard, 1985; Marszalek, 1987; Pastorok and Bilyard, 1985). It is of particular concern on small islands where urbanisation or tourism are developing rapidly, such as Fiji. Recovery of the reef may take place rapidly once the source of pollution is removed, and increasing efforts are now being made to place outfalls in water deeper than optimal for coral growth and in sites exposed to strong currents and unrestricted water circulation. Kaneohe Bay in Hawaii has become a classic example of this, having been studied both before and after the implementation of pollution controls.

The long-term impact of heavy metals and other similar pollutants on coral reefs is still far from clear (Howard and Brown, 1984; Brown, 1987b) but is of concern in countries where mining is a major activity on the coast. In the Pacific, minerals are extracted from some high islands, particularly Papua New Guinea, Fiji, New Caledonia, and to a lesser extent the Solomon Islands and Vanuatu. The only major resource on coral islands is phosphate but the exploitation of this has had a major impact on several small islands, particularly Nauru, Banaba (Kiribati) and Makatea (French Polynesia). The main consequences of mining in both these cases is sedimentation (Dupon, 1986; Anon., 1986d).

The impact of oil on coral reefs has been the subject of intensive research in the Caribbean and Red Sea. Of particular concern are oil terminals, tanker traffic, refineries and offshore oil reserves adjacent to reefs, all of which are potential sources of pollution. However, as yet it is not a major source of concern in the Pacific. The long-term effect of petroleum hydrocarbons on corals is poorly known but short-term sublethal effects may be felt (Knapp *et al.*, 1983), reef fish may be affected (Gettleson, 1980) and various other detrimental effects have been recorded (Loya and Rinkevich, 1987). Single-event episodes, such as oil spills, rarely seem to have detrimental effects but chronic oil pollution may, particularly in the intertidal zone where the reef surface may be exposed to the air. Major damage may also result from the clean-up operations which take place following spills, chemical detergents often being toxic.

Thermal pollution, from power plants and industrial complexes, affects reproduction and may cause expulsion of zooxanthellae (bleaching), a temperature rise of 4°C generally causing damage (Neudecker, 1987); the impact of thermal pollution has been investigated in Guam and Taiwan. The disposal of chemical and hazardous military wastes is another problem in the Pacific, particularly in the U.S. affiliated islands.

Coastal development

Activities such as filling to provide sites for industry, housing, recreation, and airports, and dredging to create, deepen or improve harbours, ports and marinas have major impacts, through increased turbidity and altered water circulation (White, 1987). In the Pacific, most settlement is on the coast and although the size of urban centres on many islands is comparatively small, the rate of urbanization is as large as elsewhere (Low, 1981). Causeways linking small islands, the construction of perimeter roads (Falanruw, 1983) and tourist developments are often the cause of damage. Helfrich (1979) discusses various aspects of such problems. Military activities have also caused significant damage, particularly in the U.S. affiliated islands.

Coral and sand mining and dredging

Large quantities of coral and sand are mined for use as lime, and for road and building materials, particularly on islands where terrestrial sources of such materials are limited. This causes beach erosion and transportation of sand to other sites as a result of altered water circulation. Sedimentation and pollution from the introduction of toxic substances in the mining process also affect adjacent coral reefs. These is a widespread problem in the Pacific (Table 2), its impact have been studied particularly in French Polynesia (Dubois and Towle, 1985; Salvat, 1987b).

Overexploitation

Exploitation of reef species is increasing in countries where cash economies have been introduced relatively recently, in those with high population growth rates, and where tourism has expanded rapidly. In most cases, the concern is not with the extinction of species but with lowered reef productivity. The impact of fisheries on reefs and lagoons in the Pacific is reviewed by Parrish (1980) and more generally by Munro *et al.* (1987). Fish catches have deteriorated in some reef areas and fishermen may be now obliged to travel further afield to find new, undamaged reefs with worthwhile fish stocks. Overfishing is reported in many countries in the region, but does not appear to be as significant as in other oceans. There is also less concern about the aquarium fish industry (Wood, 1985; Randall, 1987), as this is not yet a major issue in the Pacific.

Some reef invertebrates are however very vulnerable. Mother-of-pearl species have been overfished throughout much of the region and Giant Clams have been seriously depleted. The impact of collection of shells and coral for the marine curio trade does not yet seem to be serious in the Pacific but as the tourist industry increases this could lead to localised reef damage (Wells and Alcala, 1987; Wells, in press; Wood and Wells, in prep.; Munro, in press).

Deleterious fishing methods

Although illegal in most countries, fishing using dynamite, poisons or intoxicants still occurs, the use of dynamite having increased considerably after World War II when ammunition was readily available. These methods can be extremely damaging, destroying the reef, wasting fish and killing invertebrates (Alcala and Gomez, 1987; Eldredge, 1987b). Dynamite fishing has been reported recently in at least 11 countries (Table 2), nearly half the countries in the South Pacific region (Dahl and Baumgart, 1983), and it is particularly serious in Micronesia. Spearfishing is practiced by many subsistence fishermen and although of great concern in the Caribbean where recreational spearfishing has depleted some reef species, it does not yet seem to be an issue in the Pacific.

Intensive recreational use

Recreational activities are less widespread and therefore less damaging in the Pacific than in the Caribbean and Indian Ocean, but have been cited in the deterioration of reefs of some countries, such as French Polynesia, Hawaii, Japan and Taiwan (Table 2). Anchoring, boat groundings, trampling, littering, and increased exploitation of marine resources as a result of tourism cause localized reef damage (Gomez, 1983a; Tilmant, 1987). However, the main impact is probably not so much directly from these activities as from the indirect effect of tourist-generated pollution and coastal development. Tourist facilities are usually concentrated along narrow coastal strips and in semi-enclosed bays. Breakwaters and jetties are often constructed to protect beach fronts and this leads to the alteration of water circulation and erosion of beaches elsewhere. The impact of this in Oceania is discussed in Baines (1980) with particular reference to Fiji where this has been a significant issue.

Nuclear detonations and radioactivity

This is a subject of great political discussion and moral concern at present specific to the Pacific. The main issues are nuclear explosions in French Polynesia, nuclear missile testing in the U.S. affiliated Pacific states and proposals to include the Pacific Ocean in strategies for radioactive waste management. From 1946 to 1982, 203 explosions took place on Bikini, Enewetak, Johnston and Kiritimati under the U.S programme, on Kiritimati under the U.K. programme and on Moruroa and Fangataufa in the Tuamotu archipelago under the French programme. A Technical Group on Radioactivity in the South Pacific Region was created in 1982 and reviewed knowledge to that date (SPC/SPEC/ESCAP/UNEP, 1983b). Further information is given in Bablet and Perrault (1987), Seymour (1982), Gopalakrishnan (1984) and Bacon *et al.* (1985) but many of the data collected on the impact of testing have not been published. Nuclear explosions cause more direct harm in the immediate area concerned than most other human activities, but environmental damage in the wider context and in the long-term is not fully understood.

REEF MANAGEMENT

Reef management is intimately related to island management in many of the smaller island nations of Oceania. At a 1973 regional symposium on reefs and lagoons (SPC, 1973), several resolutions and recommendations were drawn up for the conservation and management of reefs in the area, covering the creation of marine protected areas, control of damaging fishing methods, particularly dynamiting, improvement of reef research and development of tourism. Early conservation efforts in the region are described in Costin and Groves (1973) and these issues have also been discussed at recent Pacific Science Congresses; papers concerning the marine environment from the 1983 congress in Dunedin, New Zealand are published in UNEP (1985). Dahl (1985d) provides a brief introduction to reef management in the region, indicating the numerous efforts now under way although many issues remain as relevant as they were fifteen years ago. Efforts under way in the U.S. affiliated countries (American Samoa, Guam, Northern Marianas, Marshall Islands, Federated States of Micronesia and Belau), following an assessment of the opportunities to improve sustainable renewable resource development and management by the U.S. Office of Technology Assessment, are described in Anon. (1987).

Traditional resource management

Oceania, unlike the Caribbean and Indian Ocean, is notable for the importance of the continuing existence of customary law or "traditional use rights in fisheries" (TURFs). The western concept of ownership tends to be too inflexible for islands where a small amount of land and few coastal resources must be shared among many multiple uses. Throughout much of the region, clans or large communal groups own and control the rights to fish on their adjacent reefs, this area generally extending from the shoreline across the reef flats and lagoon to the outer reef slope. Management may include limited entry to fishing grounds, self-imposed closed seasons and restrictions on the types of gear used (Eaton, 1985; Johannes, 1977, 1978, 1984 and in press; Nietschmann, 1984; Ruddle and Johannes, 1985; SPC/SPEC/ESCAP/UNEP, 1985). In many countries, particularly in Polynesia (Fiji, Tonga, Tokelau) and parts of Micronesia (F.S.M. (Yap), Belau), such systems appear to have evolved to ensure a sustainable yield of reef resources, and to minimize conflicts and distribute resources effectively. Elsewhere, notably in parts of Melanesia such as Papua New Guinea, they seem to have a political or social, rather than conservation, basis (Chapman, 1985 and 1987; Polunin, 1984). Increasingly, traditional rights are tending to disappear in the vicinity of urban centres as a result of the arrival of outsiders. In several countries, such as Hawaii, the Marianas, Pohnpei and American Samoa, these rights have almost completely disappeared. Although these systems are mainly found on oceanic islands, in Japan, traditional rights have until recently been controlled by traditional village leaders; these rights are now being transferred to village-based co-operatives but the basic system is being maintained, ensuring local involvement in the management of resources.

Traditional fishing rights and marine tenure, where they still exist, affect all aspects of current reef management efforts and offer new solutions, through the integration of traditional methods with modern strategies. Although systems are comparatively well understood in many parts of Oceania, information is noticeably lacking for China, Japan and Taiwan (Ruddle and Johannes, 1985). A high priority is to determine the boundaries of marine tenure systems throughout the region in order to assess their current applicability (Johannes, in press). With the arrival of the colonial powers in the Pacific, the western principle of all land and sea below the high water mark or mean high tide level belonging to the government or crown was generally adopted in the legislation. This often led to two systems of marine tenure existing for the same areas, and had to be resolved by government recognition of traditional fishing rights but not ownership of the marine areas involved. If new management systems can reinforce traditional ways or ideas they will have a better chance of success. The traditional tenure system involves enforcement of regulations by local people, whereas under the western system of management the government is required to provide and maintain an enforcement system. The appointment of local villagers as wardens of marine protected areas is one way of integrating traditional systems with modern requirements. Baines (1980) discusses the integration of tourism with traditional management.

It is now widely recognised that information on traditional resource use patterns should be assembled to provide the necessary background for modern management of subsistence fisheries (Fakalau and Shephard, 1986). Fisheries in the tropics involve far more species than in temperate regions, and it will be a long time before the scientific data are available to manage them in a western style; a short cut is provided by building on the traditional knowledge of local fishermen, which in such areas is usually considerable. Where such systems still operate effectively it is recommended that they should simply be left; in other cases it may be desirable to openly recognise them and to provide legal institutionalization for them provided that this does not impede their evolution as situations change. Johannes (in press) suggests that private enterprises such as seaweed and Giant Clam farming, pearl culture and trochus fisheries, could be integrated effectively into traditional resource ownership systems.

Protected areas

Reef management through a system of protected areas can help halt further degradation, facilitate the recovery of devastated areas, protect breeding stocks, improve recruitment in neighbouring areas and maintain the sustainable utilization of reef resources (Bakus, 1983; Salm and Clark, 1984; Lien and Graham, 1985).

The economic benefits of marine parks are being increasingly recognised (van't Hof, 1985; Salm and Clark, 1984; Lien and Graham, 1985). Income to a park accrues through entrance fees, concessions from commercial diving and boat operators, permits for particular activities, direct management of commercial activities by park staff, and the sale of souvenirs and educational materials. Against this is balanced the cost of staffing a park, the maintenance of facilities, the management of the environment where necessary, the provision of educational and recreational activities and in some cases the purchase of the site. Studies in the Caribbean have shown a benefit:cost ratio of 10:1 in some parks, particularly where SCUBA diving is an important activity (van't Hof, 1985). In the Pacific, with its low level of tourism, the economic value of marine parks has yet to be demonstrated in the smaller island nations but in Australia and Japan it is clearly significant.

The protected area system of Oceania has been reviewed in Anon. (1979 and 1985) and Dahl (1980, 1985b and 1986). At the Third South Pacific National Parks and Reserves Conference, sponsored by SPREP in co-operation with IUCN, an Action Strategy for Protected Areas in the South Pacific Region was produced (Anon., 1985) which incorporates the general goals for improvement of protected area coverage in this area suggested by Dahl (1985b).

Of the 29 countries included in this volume, 15 have a total of 71 protected areas which include coral reefs, but many of these are not yet properly implemented (Table 3). There are also 22 areas in which some form of fishing control operates. Several of the protected areas have been proposed for upgrading or improved management. About 60 sites adjacent to terrestrial protected areas are recommended for inclusion or could be included, and over 170 other sites have been recommended for protection. The main problem in most parks is poor management through lack of funding and trained personnel. Some countries, however, such as Australia, Guam, Hawaii, Japan and Tonga, now have well developed marine park programmes. In the Pacific, a large number of countries (F.S.M., Fiji, Hong Kong, Marshall Islands, Nauru, Niue, Kiribati, Pitcairn, Solomon Islands, Tokelau, Tuvalu, Wallis and Futuna and possibly China) still have no coral reef protected areas, although in several of these countries projects are under way.

In many countries, the continuing existence of traditional tenure and customary law means that the western concept of protected areas may not be appropriate, as discussed by Eaton (1985). In some countries, successful integration of the two systems is taking place, for example Fiji, and also Papua New Guinea which has a particular category of protected area (Wildlife Management Areas) in which local people are closely involved in the establishment and maintenance of management. Furthermore, the concept of coastal zone management has particular relevance in this area and has been adopted in several countries. Tallies of protected reef areas in this region therefore do not necessarily provide a good indication of successful protection and management of the reefs of a country.

The World Conservation Strategy emphasizes that any system of protected areas must aim to protect a representative selection of ecosystem types. This is still a problem as far as marine ecosystems are concerned, as there is no widely accepted classification. Problems in the classification of marine ecosystems are discussed by Ray (1975) and Salm (1984). Hayden *et al.* (1984) produced a preliminary classification based on attributes of the physical environment combined with faunal assemblage data.

Marine protected areas vary immensely in size, the area chosen often depending on practical rather than scientific considerations. The critical minimum core area for a coral reef protected area is considered by Salm (1984) to be the smallest area in which all coral species found in the overall area have a 100% chance of being found on all reefs of the same size. For example, in the Chagos Archipelago, core areas would correspond to at least 300 ha for each reef type (Salm, 1980). The remainder of the reserve (including reef-flats, land and intervening and surrounding waters) should function as a buffer and is zoned for different uses, permitting optimal use of the area by different interest groups with minimum conflict and maximum control. Reef flats and seagrass beds are often overlooked but are important in the recycling of nutrients in the reef ecosystem, and several fish have life cycles which involve two or more of these systems (Ogden and Gladfelter, 1983; SPC/SPEC/ESCAP/ UNEP, 1985). The creation of protected areas around and including entire island systems, and the extension of boundaries of terrestrial protected areas to include marine habitats are effective means of achieving protection of a group of interrelated ecosystems. Large multiple-use areas are therefore generally more practical than small reserves, although sanctuaries or strict reserves may still be required for critical habitat areas, such as nutrient sources, areas of high biological diversity and nesting, or to protect breeding stocks of important fish (Salm and Clark, 1984).

The Man and the Biosphere programme (MAB), established in 1971 by Unesco, is of particular relevance in this context. One aim of the programme is the establishment of an international network of "Biosphere Reserves". These sites, selected to provide representative examples of the world's major ecosystems, encompass multiple zones including one or more highly protected core areas (to protect natural ecosystems and genetic diversity), traditional use areas (to study and document traditional use patterns), experimental areas (for manipulative research on resource utilisation) and rehabilitation areas (to study techniques for the restoration of degraded ecosystems). They must also offer possibilities for sharing personnel and educational research and training facilities and have the potential for involving local communities in research and educational programmes.

The 1984 Action Plan for Biosphere Reserves emphasizes the need to improve representation of coastal and aquatic ecosystems. Of the 252 listed Biosphere sites, less than a dozen include or are adjacent to reefs, and only the W.A. Robinson Biosphere Reserve in French Polynesia lies within the region covered by this volume. An early MAB project involved an intensive study of some of the Fijian and French Polynesian islands but there was no follow-up in terms of creating biosphere reserves. Biosphere Reserves are particularly appropriate for the coastal zone owing to their emphasis on linkages between conservation and development and because of the difficulty of including complete marine and coastal ecosystems within traditional forms of protected areas. The diverse terrestrial and marine ecosystems, and varied cultural histories, traditions and traditional tenure systems in the Pacific provide good opportunites for applying this concept.

While Biosphere Reserves are aimed at protecting representative samples of ecosystems, the World Heritage Site system has been designed to conserve unique and outstanding examples of the world's natural heritage. These may be nominated by parties to the World Heritage Convention (Convention concerning the Protection of the World Cultural and Natural Heritage), concluded at Paris, 23 November 1972. Of the 62 listed natural sites, the following include or are adjacent to reefs: Aldabra, Great Barrier Reef (Australia), Sierra Nevada de Santa Marta (Colombia) and the Galápagos. There is clearly scope for listing other reefs on this convention.

In 1968 the International Biological Programme drew up a list of 39 "Islands for Science", with the aim of developing a convention to provide them with long-term protection for future research, the main criterion being that all were to be free from outside human impact (Douglas, 1969; Elliott, 1973). This proposal was never put into practice, but the islands on the list from the region covered by this volume are repeated here for interest:

- Japan: Minami-io-jima
- Cook Islands: Suwarrow Atoll
- Pitcairn: Ducie, Henderson, Oeno
- Kiribati: Birnie, Rawaki (Phoenix), Vostock, Malden
- American Samoa: Rose Atoll
- USA: Jarvis, Kingman Reef, Howland
- Hawaii: Pearl and Hermes Reef, Laysan, Gardner Pinnacles, Necker, Nihoa
- Northern Marianas: Farallon de Pajaros, Maug, Guguan and Farallon de Medinilla
- Belau: Helen
- F.S.M.: East Fayu
- Marshall Islands: Bokaak, Bikar

There is now renewed interest in establishing protected sites for long-term research work and monitoring of reefs and there may be a need to identify suitable reef areas for such activities (Birkeland, 1987).

Coastal zone management planning

Reef management has increasingly to be considered in the context of the entire coastal zone, and even catchment area; this is particularly important where siltation, caused by upstream sources, is a problem. The technology to avoid much human-induced damage to coral reefs is now available. For example, sewage outfalls can be placed below the level of coral growth, thermal effluent can be discharged in deep water, there are new methods for dispersing oil and alternative materials to coral are available for construction. This technology is often not applied owing to lack of technical expertise, funding or co-ordination between the relevant government authorities, but many countries are now moving in this direction.

In many Pacific countries, the concept of coastal zone management is particularly appropriate, and programmes are now under way in several countries, particularly the U.S. affiliated islands (Hawaii, Guam, American Samoa). Enforcement of regulations in countries consisting of scattered islands, as the majority are in the Pacific, is a particular problem; the introduction of broad management strategies with the involvement of local people is likely to be more effective in the long-term than attempting to enforce legislation controlling fisheries and protected areas, unless these are part of the broader plan (Baines, 1985).

Coastal zone management is discussed by Sorensen *et al.* (1984) and Snedaker and Getter (1985) who provide guidelines for the creation of national coastal resources management programmes; actions by international aid organizations; land use and coastal planning; and environmental impact assessment, many of which are specific to coral reefs. The International Affairs Office of the U.S. National Parks Service, in co-operation with the U.S. Agency for International Development and other organizations, has set up the C.A.M.P. (Coastal Area Management and Planning) Network to provide information and training opportunities. The Commonwealth Science Council (CSC) has initiated a South Pacific Coastal Zone Management Programme (SOPACOAST) which provides for comprehensive, community-based projects and other activities designed to improve knowledge and understanding of island coastal resource systems, to develop the capacity of communities with traditional resource rights to assess their resources and monitor change, and to impart training in coastal area management at all levels. There are two site-specific projects: a "high" island project in the Solomon Islands, and a "low" island project in the Cook Islands. As the projects progress, increasing attention will be paid to the application of the results in other parts of the region (Commonwealth Science Council, 1986).

The Unesco Coral Reef Management Handbook (Kenchington and Hudson, 1984) aims to provide political, administrative and technical decision-makers who have responsibility for coral reefs with the means to ensure that relevant issues are properly considered in the course of their work. UNEP is currently preparing a set of Coral Reef Management Guidelines in collaboration with the Australian Great Barrier Reef Marine Park Authority.

Reef fisheries management

The World Conservation Strategy recommends that the maintenance of coral reef fisheries at sustainable levels be considered a global priority. In addition to customary law, most countries in the region covered by this volume have national legislation to regulate fisheries (Table 4). Venkatesh and Vava'i (1983) and SPC/SPEC/ESCAP/ UNEP (1985) summarise marine resource legislation, and the latter includes a discussion of the applicability of western concepts of legislation to Oceania. A regional approach to fisheries management is proving to be advantageous. For example, in the Caribbean, a Fisheries Act has been developed under the auspices of the Organization of Eastern Caribbean States (OECS) and FAO which provides for the preparation of fisheries management and development plans, appointment of national fisheries advisory committees, regional co-operation, establishment of marine reserves and conservation programmes and prohibition of certain methods (Goodwin, 1985). In the Pacific, the Forum Fisheries Agency has been established under the South Pacific Forum Fisheries Agency Convention, for the conservation of fisheries resources and the promotion of optimal utilisation (a convention - the Fishing and Conservation of the Living Resources of the High Seas Convention of 1958 - also exists between Fiji and Tonga in 1958 (Venkatesh and Vava'i, 1983)). The Fisheries Forum Agency is active in encouraging improved fisheries management including reef fisheries.

Management of of reef and lagoon fisheries in the Pacific is discussed by Marshall (1980) and specifically in the U.S. affiliated islands in Callaghan (in press); general recommendations are given in Munro *et al.* (1987). Although some resources have been overexploited, landings could probably be improved through stock enhancement, such as use of underexploited non-reef species, artificial reefs and fish aggregating devices. In the U.S. affiliated islands, efforts are being made to direct fisheries away from the more fragile and heavily fished reef and lagoon resources to the comparatively underexploited pelagic and outer reef slope resources (Anon., 1987). The success of the Japanese in Micronesia during the mandate period (Smith, 1947) suggests that there is considerable potential for economic development of a variety of marine resources (Anon., 1987). Recommendations of improved management of the aquarium fish trade are given in Wood (1985) and Randall (1987).

Particular attention is being paid to reef invertebrates (Pearson, 1980; Smith, in press) and the Fisheries Forum Agency has initiated a programme to examine their marketing (Carleton, 1984a and b). New Caledonia and Australia are notable for having developed systems for the exploitation and management of invertebrates such as corals and shells. The ornamental coral and shell trades generally need better management, both nationally and internationally (Wells and Alcala, 1987; Wood and Wells, in prep.; Wells, in press). The shell trade potentially provides an important source of income as many molluscs can probably support fairly intense exploitation, provided damage to the reef environment and local overexploitation of the more popular species is avoided. Attempts have been made in Papua New Guinea and are being made in Fiji to develop the industry in a sustainable manner. Exploitation of stony corals is considered inadvisable because of the damage it causes to the reef (Gomez, 1983b) but there is potential for managed exploitation of the semi-precious corals. Carleton and Philipson (1987) provide recommendations for the management of black coral, suggesting that only small-scale, high craft content, processing enterprises should be encouraged; raw black coral should not be exported. A management plan drawn up for precious corals by the Western Pacific Regional Fishery Management Council (1980) could provide a model for other coral resources. Furtado and Wereko-Brobby (1987) recommend a survey of commercially valuable algal resources in Fiji, Kiribati, Papua New Guinea and the Solomon Islands prior to the development of management strategies for exploitation.

Little work has been carried out on artificial reefs and FADs (fish aggregating devices) in this region but projects are under way in some countries such as Japan and Western Samoa, and their potential is discussed in Gopalakrishnan (1984) and Anon. (1987). Compared with South-east Asia, aquaculture is a recent development in the Pacific (King and Stone, 1979). Rapid advances are being made, however, at the Micronesian Mariculture Demonstration Center in Belau, particularly for the commercially valuable but often overexploited molluscs: Giant Clams and trochus (Anon., 1987), on several of the Tuamotu atolls, where pearl oyster culture is proving to be a viable economic activity, and in the U.S. affiliated islands (Nelson, in press). A regional progamme for Giant Clam farming is under way under the auspices of ICLARM and includes Australia,

Papua New Guinea, Fiji, Solomon Islands, Belau and Western Samoa (Lucas and Munro, 1985; Munro, 1986 and in press). In addition to developing techniques for improving harvest yields, the trochus and Giant Clam programmes are also aiming at restocking depleted reefs, and in some cases introducing stock to new areas. Guidelines for such activities will need to be drawn up to ensure that in the long-term such introductions do not have deleterious effects.

The Convention on International Trade in Endangered Species of Wild Fauna and Flora (CITES), concluded in Washington D.C., 1973, provides a mechanism for controlling international trade in threatened species. However, although many reef species are involved in international trade, few fit the strict criteria required for listing on the Appendices. All turtles (Cheloniidae and Dermochelydae) and the Dugong *Dugong dugon* (except Australian populations) are listed in Appendix I which prohibits international trade between parties. Black coral (Antipatharia), the Giant Clams Tridacnidae and a number of stony coral genera are listed in Appendix II, which means that an export permit is required from the country of origin.

Research, training and education

Research into the coastal zone has increased dramatically in recent years, particularly in relation to the conflict between economic development and natural resource conservation, although, compared with the Caribbean and Australia, reef-related research in the Pacific Ocean is still fragmentary. UNEP/FAO (1985) lists marine environmental centres in the region and Eldredge (1987a) provides a directory of Pacific coral reef researchers; a world list of coral reef research institutes is currently being prepared (Eldredge and Potter, in prep.). Further information on marine research institutes and scientists is also available in Unesco/FAO (1983) and in the Coral Reef Newsletter produced by the Scientific Committee on Coral Reefs of the Pacific Science Association. A Pacific Islands Marine Resources Information Service (PIMRIS) is to be set up at the University of the South Pacific in Fiji (Eldredge, pers. comm., 1987).

The Division of Marine Sciences of Unesco has played a major role in stimulating research and in training specialists and technicians in marine sciences through the Major Inter-regional Project on Research and Training leading to Integrated Management of Coastal Systems (COMAR). Pilot projects are being drawn up to give each region the practical experience and background for decision-making in the field of coastal zone management (Unesco, 1986a). In Asia and the Pacific COMAR activities have focused on productivity and interactions between coral reefs and mangroves, traditional knowledge and the development of a network of coral reef institutions (Unesco, 1982). Regional workshops have been held on coral taxonomy (Unesco, 1985), reef survey methods, the assessment of damage caused by humans (Brown, 1986) and a comparison of tropical marine and coastal processes in the Atlantic and Pacific (Birkeland, 1987). The need for long-term research projects and permanently based laboratories which could carry out such work is recognised; it is recommended that a broad-scale programme to assess productivity gradients should be initiated in the Pacific as in the Caribbean (Birkeland, 1987).

International seminars are held in the U.S.A. for administrators, managers and professional personnel involved in coastal and marine protected areas policy, planning, design, management and operations (Anon., 1986a).

Monitoring changes on reefs is a high priority (Brown and Howard, 1985; Unesco, 1985). Simple, low-cost techniques for assessment have been designed with the Pacific in mind (Dahl, 1977 and 1981). At the more sophisticated level, remote sensing of coral reefs, although costly, is becoming an increasingly useful technique. A training course on the subject has recently been run under the auspices of the Unesco COMAR project (Unesco, 1986b) and an ESCAP/UNDP/ADAB/USP Pacific Island Regional Remote Sensing workshop examined the potential for resource mapping in the Pacific; a pilot project is now under way in Tonga (Chesher and Thaman, 1987). Extensive work on remote sensing applications to reef management is also under way in New Caledonia and French Polynesia. Marine resource atlases have been prepared for a number of Pacific countries (Hawaii, American Samoa, Pohnpei in the F.S.M.), the U.S. Army Corps of Engineers taking a lead role in this activity (Maragos and Elliott, 1985; Anon., 1987), and these provide a useful baseline for planners and resource managers.

Education programmes in schools and villages are also extremely important to help local people to understand the importance of reef resources and lead to improved enforcement of planning controls and regulations. Foreign visitors and tourists also need to understand the reef ecosystem, and marine parks are increasingly contributing to this through their provision of interpretative centres, underwater trails and guidebooks. SPREP (1984) describes activities under way at that time and the research, educational and monitoring programmes under way in different institutions within the region.

International and regional efforts

Multinational collaboration and regional strategies are particularly important in the conservation and management of marine resources which are so often shared by several countries. Forster (1985) and Pulea (1984 and 1985) discuss the conventions of particular relevance to the South Pacific region. Kenchington and Hudson (1984) summarize the international agencies and organizations which carry out activities relating to reef management.

The UNEP Regional Seas Programme was initiated in 1974. Since then the Governing Council of UNEP has repeatedly endorsed a regional approach to the control of marine pollution and the management of marine and coastal resources and has requested the development of regional action plans.

The Regional Seas Programme at present includes ten regions and has over 120 coastal States participating in it. It is conceived as an action-orientated programme having concern not only for the consequences but also for the causes of environmental degradation and encompassing a comprehensive approach to combating environmental problems through the management of marine and coastal areas. Each regional action plan is formulated according to the needs of the region as perceived by the Governments concerned. It is designed to link assessment of the quality of the marine environment and the causes of its deterioration with activities for the management and development of the marine and coastal environment. The action plans promote the parallel development of regional legal agreements and of action-orientated programme activities.

An important UNEP contribution to the concept of promoting regional cooperation in assessing and combating marine pollution together with resource conservation and management in oceans and coastal areas has been the preparation and publication of a series of regional directories and bibliographies.

UNEP's recognition of the environmental importance of coral reefs, as well as the tremendous pressures on and global exploitation of these fragile ecosystems, has been demonstrated in the inclusion and encouragement of various measures to protect reefs in the regional action plans. The publication of the Pacific Coral Reef Researchers in UNEP's Regional Seas Directories and Bibliographies (UNEP/PSA/SPREP/UG, 1984) is also an indication of this concern. In order to give coral reefs and associated problems a deservedly much greater exposure it was decided to publish the three volumes of directories that would cover the coral reefs globally. This volume includes a single region within the Programme: the South Pacific Region.

SPREP, the South Pacific Regional Environment Programme, is the central environmental organisation in the region. It was developed by the South Pacific Bureau for Economic Co-operation (SPEC), the Economic and Social Commission for Asia and the Pacific (ESCAP) and the South Pacific Commission (SPC). There are 22 participating countries, and the developed countries involved (Australia, France, New Zealand, U.K. and U.S.A) support the programme. The Action Plan for Managing the Natural Resources and Environment of the South Pacific Region (SPC/SPEC/SPREP/UNEP, 1983a) was adopted, with the South Pacific Declaration on Natural Resources and Environment of the South Pacific Region, at the 1982 Conference on the Human Environment in the South Pacific which was convened at Rarotonga in 1982 (SPREP, 1982; Carew-Reid and Sheppard, 1985; Dahl, 1985c).

The activities of Unesco and the Commonwealth Science Council in the region with respect to reefs are described above. IUCN/WWF activities include the development of marine parks and the general support of national park services through assistance provided by the IUCN Commission on National Parks and Protected Areas (CNPPA). The Conservation for Development Centre of IUCN is contributing to the development of National Conservation Strategies which will take reef ecosystems, where relevant, into account.

International conventions

Most international wildlife conventions make little reference to coral reefs. The UNEP Regional Seas Convention, two conventions relating to fisheries, CITES and the World Heritage Convention are mentioned above and a summary of environmental conventions of relevance in the Pacific region is given in Venkatesh and

Va'ai (1983). The following conventions also have, or could have, some bearing on reef management and conservation:

The Convention on the Protection and Development of the Natural Resources of the South Pacific Region (SPREP Convention) was approved in 1986; 10 countries have signed, of which two have ratified. It concerns the marine and coastal environment and emphasises pollution control. There are two protocols, one concerning prevention of pollution by dumping and the other concerning regional co-operation in combating pollution emergencies (Pulea, 1985).

The Convention for the Conservation of Nature in the South Pacific (Apia Convention) 1976 was sponsored by the SPC and IUCN and resulted from a regional symposium held in 1971 (Costin and Groves, 1973). It provides for the encouragement of the establishment of protected areas and the protection of species. Papua New Guinea, France (New Caledonia), Western Samoa and most recently the Cook Islands have ratified, and the convention can now come into force. Discussions are currently under way as to the best means of rationalising this with the Convention for the Protection and Development of the Natural Resources of the South Pacific Region which has a broader, more modern approach to conservation issues but which is limited to the marine and coastal zone (Forster, 1985).

The Convention on the Prevention of Marine Pollution by Dumping of Wastes and Other Matter, 1972 and the International Convention for the Prevention of Marine Pollution from Ships (MARPOL 73/78) have been drawn up by the International Maritime Organization to prevent marine pollution from ships which may often have a direct impact on reefs. They are discussed by Kenchington and Hudson (1984) and Hayes (1981).

The Convention on Wetlands of International Importance especially as Waterfowl Habitat, concluded at Ramsar, Iran, 2 February 1971, lists wetlands of international importance primarily to waterfowl, but sites may be selected on a variety of criteria. Wetlands are defined as areas of marsh, fen, peatland or water, fresh and marine, the depth of which at low tide does not exceed six metres; shallow coral reefs are therefore included. At present the only Ramsar site which includes reefs is the St Lucia System in South Africa. It is generally felt that, in its present form, the Convention is not appropriate for reefs since a) the emphasis is strongly on bird habitat, b) reefs generally extend to depths greater than 6 m, and c) reefs generally come under different national legislation from other wetland habitats, which would complicate the implementation of the Convention within a country (Anon., 1986b; Wells, 1984). However, efforts are currently under way to revise the convention to make it an important force for the protection of coastal wetlands essential for supporting fisheries.

REFERENCES

Alcala, A.C. and Gomez, E.D. (1987). Dynamiting coral reefs for fish: A resource-destructive fishing method. Chap. 4. In: Salvat, B. (Ed.). Pp. 51-60.
Anon. (1979). Second South Pacific Conference on National Parks and Reserves. Sydney, Australia. Proceedings Vols 1 and 2.
Anon. (1985). Summary record of proceedings of ministerial and technical sessions. *Report of the 3rd South Pacific National Parks and Reserves Conference, Apia 1.* 95 pp.
Anon. (1986a). *The Marine Connection* 1(1). (Joint newsletter of the Oceanic Symposia of the 4th World Wilderness Congress and the International Marine Protected Areas Network).
Anon. (1986b). Is the Ramsar Convention appropriate for the Caribbean? *Caribbean Wetlands Newsletter* 5: 1-3.
Anon. (1986c). TCSP Tourism Topics. *Newsletter of the Tourism Council of the South Pacific* 1. 2 pp.
Anon. (1986d). Mining activities in the South Pacific. *Environment Newsletter* 7: 12-17.
Anon. (1987). Integrated Renewable Resource Management for U.S. Insular Areas. Summary. Office of Technology Assessment, Congress of the United States, Washington D.C. 51 pp.
Antonius, A. (1981). The "band" diseases in coral reefs. *Proc. 4th Int. Coral Reef Symp.* 2: 7-14.
Bablet, J.-P. and Perrault, G.-H. (1987). Effects on a coral environment of a nuclear detonation. In: Salvat, B. (Ed.).
Bacon, M.P., Lambert, G., Rafter, T.A., Samisoni, J.I. and Stevens, D.J. (1985). Radioactivity in the South Pacific Region. In: UNEP (1985). Pp. 151-156.
Baines, G.B.K. (1980). Coastal tourism development in Oceania. In: *Marine and Coastal Processes in the Pacific: Ecological Aspects of Coastal Zone Management.* Unesco, ROSTEA, Jakarta. Pp. 191-207.
Baines, G. (1985). Coastal zone management and conservation in the South Pacific. *Report of the 3rd South Pacific National Parks and Reserves Conference, Apia* 2: 40-56.
Bakus, G. (1983). The selection and management of coral reef preserves. *Ocean Management* 8: 305-316.
Birkeland, C. (1982). Terrestrial runoff as a cause of outbreaks of *Acanthaster planci* (Echinodermata: Asteroidea). *Mar. Biol.* 69: 175-185.
Birkeland, C. (Ed.) (1987). Comparison between Atlantic and Pacific tropical marine coastal ecosystems: community structure, ecological processes, and productivity. *Unesco Rep. Mar. Sci.* 46. 262 pp.
Björklund, M.I. (1974). Achievements in marine conservation. 1. Marine Parks. *Env. Cons.* 1: 205-223.
Brodie, J.E. and Morrison, R.J. (1984). The management and disposal of hazardous wastes in the Pacific islands. *Ambio* 13(5-6): 331-333.
Brown, B.E. (Ed.) (1986). Human induced damage to coral reefs. *Unesco Rep. Mar. Sci.* 40. 180 pp.
Brown, B.E. (1987a). Worldwide death of corals: Natural cyclical events or man-made pollution? *Mar. Poll. Bull.* 18(1): 9-13.
Brown, B.E. (1987b). Heavy metals pollution on coral reefs. Chap. 10. In: Salvat, B. (Ed.). Pp. 119-134.
Brown, B.E. and Howard, L.S. (1985). Assessing the effects of "stress" on reef corals. *Adv. Mar. Biol.* 22: 1-63.
Callaghan, P. (in press). The development and management of nearshore fisheries in the U.S. Affiliated Pacific islands. In: Smith, B.D. (Ed.). *Topic Reviews in Insular Resource Development and Management in the Pacific U.S.-affiliated Islands.* Univ. Guam Mar. Lab. Tech. Rept 88.
Cane, M.A. (1983). Oceanographic events during El Niño. *Science* 222(4629): 1189-1194.
Carew-Reid, J. and Sheppard, D. (1985). Environmental management through regional co-operation: South Pacific Regional Environmental Programme. *Report of the 3rd*

South Pacific National Parks and Reserves Conference, Apia 2: 28-37.

Carleton, C. (1984a). Marketing studies on the miscellaneous marine resources of the South Pacific. *Infofish Marketing Digest* 5: 28-31.

Carleton, C. (1984b). Miscellaneous marine products in the South Pacific: A survey of the markets for specific groups of miscellaneous marine products. South Pacific Forum Fisheries Agency, Honiara, Solomon Islands. 147 pp.

Carleton, C.C. and Philipson, P.W. (1987). Report on a study of the marketing and processing of precious coral products in Taiwan, Japan and Hawaii. South Pacific Forum Fisheries Agency, Honiara, Solomon Islands.

Chapman, M. (1985). Environmental influences on the development of traditional conservation in the South Pacific region. *Env. Cons.* 12: 217-230.

Chapman, M.D. (1987). Traditional political structure and conservation in Oceania. *Ambio* 16(4): 201-205.

Chesher, R.H. and Thaman, R. (1987). Remote sensing in the Pacific islands. Report of the Pacific Island Regional Remote Sensing Workshop and Training Course on Resource Mapping. UNDP/ESCAP Regional Remote Sensing Programme, Bangkok, Thailand. 87 pp.

Clark J.R. (Ed.) (1985). *Coastal Resources Management: Development Case Studies.* Renewable Resources Information Series, Coastal Management Publication 3. Research Planning Institute, Inc., Columbia.

Commonwealth Science Council (1986). Environmental Planning Programme. Coastal Zone Management of Tropical Islands. South Pacific Coastal Zone Management Programme (SOPACOAST). Project Document. CSC Technical Publication Series 204. Commonwealth Science Council.

Connell, J. (1978). Diversity in tropical rainforests and coral reefs. *Science* 199: 1302-1310.

Costin, A.B. and Groves, R.H. (Eds) (1973). *Nature Conservation in the Pacific.* IUCN Pubs N.S. 25. 337 pp.

Croxall, J.P., Evans, P.G.H. and Schreiber, R.W. (1984). *Status and Conservation of the World's Seabirds.* ICBP Technical Publication 2, ICBP, Cambridge, U.K. 778 pp.

Dahl, A.L. (1977). Monitoring man's impact on Pacific island reefs. *Proc. 3rd Int. Coral Reef Symp., Miami*: 571-575.

Dahl, A.L. (1980). Regional Ecosystems Survey of the South Pacific Area. *SPC/IUCN Technical Paper* 179. South Pacific Commission, Noumea, New Caledonia. 99 pp.

Dahl, A.L. (1981). *Coral Reef Monitoring Handbook.* South Pacific Commission, Noumea, New Caledonia.

Dahl, A.L. (1985a). Status and conservation of South Pacific coral reefs. *Proc. 5th Int. Coral Reef Cong., Tahiti* 6: 509-513.

Dahl, A.L. (1985b). Adequacy of coverage of protected areas in Oceania. *Report of the 3rd South Pacific National Parks and Reserves Conference, Apia* 2: 2-8.

Dahl, A.L. (1985c). The South Pacific Regional Environment Programme. In: UNEP (1985). Pp. 3-6.

Dahl, A.L. (1985d). The challenge of conserving and managing coral reef ecosystems. In: UNEP (1985). Pp. 85-87.

Dahl, A.L. (1986). *Review of the Protected Areas System in Oceania.* IUCN/UNEP. 239 pp.

Dahl, A.L. and Baumgart, I.L. (1983). The state of the environment in the South Pacific. *UNEP Regional Seas Reports and Studies* 31. 25 pp.

Darwin, C.R. (1842). *The Structure and Distribution of Coral Reefs.* Smith, Elder and Co., London.

Davies, P.J. (1983). Reef growth. In: Barnes, D.J. (Ed.), *Perspectives on Coral Reefs.* AIMS, Townsville. Pp. 69-106.

Douglas, G. (1969). Checklist of Pacific Oceanic Islands. *Micronesica* 5(2): 327-463.

Driml, S. (1987). Economic Impacts of Activities on the Great Barrier Reef. Research Publication. Great Barrier Reef Marine Park Authority.

Dubois, R. and Towle, E.L. (1985). Coral harvesting and sand mining practices. Case Study 3. In: Clark, J.R. (Ed.). Pp. 203-289.

Dupon, J.F. (1986). The effects of mining on the environment of high islands: A case study of nickel mining in New Caledonia. Environmental Case Studies 1. SPREP, South Pacific Commission, Noumea, New Caledonia.

Eaton, P. (1985). Land tenure and conservation: Protected areas in the South Pacific. *SPREP Topic Review* 17. 103 pp.

Eldredge, L. (1987a). Coral Reef Researchers: Pacific. *UNEP Regional Seas Directories and Bibliographies.* FAO, Rome (in cooperation with PSA, SPREP and UOG).

Eldredge, L. G. (1987b). Poisons for fishing on coral reefs. Chap. 5. In: Salvat, B. (Ed.). Pp. 61-66.

Eldredge, L.G. (1987c). Bibliography of marine ecosystems: Pacific Islands. *UNEP Regional Seas Directories and Bibliographies.* FAO, Rome (in co-operation with SPREP and UOG). 72 pp.

Eldredge, L.G. and Potter, T.S. (in prep.). *Directory of Coral Reef Research Facilities of the World.* University of Guam.

Elliott, H. (1973). Pacific oceanic islands recommended for designation as Islands for Science. In: SPC (1973).

Endean, R. (1973). Population explosions of *Acanthaster planci* and associated destruction of hermatypic corals in the Indo-west Pacific region. In: Jones, O.A. and Endean, R. (Eds), *Biology and Geology of Coral Reefs.* Vol. 2. Biology 1. Academic Press, N.Y.

Fakalau, S. and Shepherd, M. (1986). Fisheries research needs in the South Pacific. South Pacific Fisheries Forum Agency, Honiara, Solomon Islands. 90 pp.

Falanruw, M.V.C. (1980). Marine environment impacts of land-based activities in the Trust Territory of the Pacific Islands. In: *Marine and Coastal Processes in the Pacific: Ecological Aspects of Coastal Zone Management.* Unesco, ROSTEA, Jakarta. Pp. 19-47.

Forster, M. (1985). Towards an effective international instrument for the conservation of nature in the South Pacific: Some options which may be considered. Paper prepared for Third South Pacific National Parks and Reserves Conference, Apia Western Samoa.

Furtado, J.I. and Wereko-Brobby, C.Y. (1987). Tropical marine algal resources of the Asia-Pacific Region: A status report. Commonwealth Science Council Tech. Pub. Series 181. 125 pp.

Gettleson, D.L. (1980). Effects of oil and gas drilling operations on the marine environment. In: Geyer, R.A. (Ed.), *Marine Environmental Operation.* Elsevier Scientific Publishing Co., New York.

Gladfelter, W. (1982). White-band disease in *Acropora palmata*: Implications for the structure and growth of shallow reefs. *Bull. Mar. Sci.* 32: 639-643.

Glynn, P.W. (1984). Widespread coral mortality and the 1982-83 El Niño warming event. *Env. Cons.* 11(2): 133-146.

Glynn, P.W. and Wellington, G.M. (1983). *Corals and Coral Reefs of the Galapagos Islands.* Univ. California Press, Berkeley/LA/London. 330 pp.

Gomez, E.D. (1983a). Direct and indirect impacts of tourism on the coastal environment. In: *Marine and Coastal Processes in the Pacific: Ecological Aspects of Coastal Zone Management*. Unesco, ROSTEA, Jakarta. Pp. 209-219.

Gomez, E.D. (1983b). Perspectives on coral reef research and management in the Pacific. *Ocean Management* 8: 281-295.

Gomez, E.D. and Yap, H.T. (1985). Coral reefs in the Pacific: Potentials and limitations. In: UNEP (1985). Pp. 89-106.

Gopalakrishnan, C. (1984). *The Emerging Marine Economy of the Pacific*. Butterworth Publishers, MA, USA. 256 pp.

Goodwin, M.H. (1985). *Characterization of Lesser Antillean Regional Fisheries*. Island Resources Foundation. 48 pp.

Grigg, R.W. and Dollar, S.J. (in press). Natural and anthropogenic disturbance on coral reefs. In: *Coral Reef Ecosystems*. Elsevier Press.

Groombridge, B. (1982). *IUCN Amphibia-Reptilia Red Data Book. Part 1. Testudines, Crocodilia and Rhynchocephalia*. IUCN, Gland, Switzerland. 426 pp.

Groves, G.W. (1979). Wind, waves and currents in the tropical Pacific. In: Helfrich, P. (Ed.).

Hayden, B.P., Ray, G.C. and Dolan, R. (1984). Classification of coastal and marine environments. *Env. Cons.* 11(3): 199-207.

Hayes, T.M. (1981). Activities of the Inter-governmental maritime consultative organization in the South Pacific relating to marine pollution prevention control and response. *SPREP Topic Review 12*. 6 pp.

Helfrich, P. (Ed.) (1979). *Utilization and Management of Inshore Marine Ecosystems of the Tropical Pacific Islands*. Seagrant Co-operative Report UNIHI-SEAGRANT-CR-82-01.

Helfrich, P. and Maragos, J.E. (1979). Industrial pollution on coral reefs: An overview. In: Helfrich, P. (Ed.).

Howard, L.S. and Brown, B.E. (1984). Heavy metals and reef corals. *Oceanogr. Mar. Biol. Ann. Rev.* 22: 195-210.

Hutchings, P.A. (1986). Biological destruction of coral reefs. *Coral Reefs* 4: 239-252.

IUCN (in prep.). *Directory of Oceanian Protected Areas*. IUCN, Gland, Switzerland and Cambridge, U.K.

IUCN/UNEP/WWF (1980). *World Conservation Strategy*. IUCN, Gland, Switzerland.

Johannes, R.E. (1975). Pollution and degradation of coral reef communities. In: Ferguson Wood, E.J. and Johannes, R.E. (Eds), *Tropical Marine Pollution*. Elsevier Scientific Publishing, Amsterdam. Pp. 13-50.

Johannes, R. (1977). Traditional law of the sea in Micronesia. *Micronesica* 13: 121-127.

Johannes, R. (1978). Traditional marine conservation methods in Oceania and their demise. *Ann. Rev. Ecol. Syst.* 9: 349-364.

Johannes, R.E. (1984). Traditional conservation methods and protected areas in Oceania. In: McNeely, J.A. and Miller, K.R. (Eds). *National Parks, Conservation and Development: The Role of Protected Areas in Sustaining Society*. Smithsonian Institution Press, Washington D.C. Pp. 344-347.

Johannes, R.E. (in press). The role of marine resource tenure systems (TURFs) in sustainable nearshore marine resource development and management in U.S.-affiliated tropical Pacific islands. In: Smith, B.D. (Ed.), *Topic Reviews in Insular Resource Development and Management in the Pacific U.S.-affiliated Islands*. Univ. Guam Mar. Lab. Tech. Rept 88.

Kay, E.A. (1987). Patterns of speciation in the Indo-West Pacific.

Kearney, R. (1985). Fishery potentials in the tropical central and western Pacific. In: UNEP (1985). Pp. 75-84.

Kenchington, R.A. and Hudson, B.E.T. (1984). *Coral Reef Management Handbook*. Unesco Regional Office of Science and Technology for South-East Asia, Jakarta. 281 pp.

King, M.G. and Stone, R.M. (1979). Commercial fisheries in Western Pacific islands. In: Helfrich, P. (Ed.).

Knapp, A.H., Sleeter, T.D., Dodge, R.E., Wyers, S.C., Frith, H.R. and Smith, S.R. (1983). The effects of oil spills and dispersant use on corals. *Oil and Petrochemical Pollution* 1(3): 157-169.

Lien, J. and Graham, R. (1985). *Marine Parks and Conservation: Challenge and Promise*. Vols 1 and 2. Henderson Park Book Series 10, The National Parks and Provincial Parks Association of Canada.

Low, J. (1981). Urbanization and its effects on the South Pacific environment. *SPREP Topic Review* 3. 18 pp.

Loya, Y. and Rinkevich, B. (1987). Effects of petroleum hydrocarbons on corals. Chap. 8. In: Salvat, B. (Ed.). Pp. 91-102.

Lucas, J.S. and Munro, J.L. (1985). International Giant Clam mariculture project. *Proc. 5th Int. Coral Reef Cong., Tahiti* 2: 229 (Abs).

Maragos, J.E. and Elliott, M.E. (1985). Coastal resouce inventories in Hawaii, Samoa and Micronesia. *Proc. 5th Int. Coral Reef Cong., Tahiti* 5: 577-582.

Marshall, N. (1980). Management of reef and lagoon fisheries. In: *Marine and Coastal Processes in the Pacific: Ecological Aspects of Coastal Zone Management*. Unesco, ROSTEA, Jakarta. Pp. 159-175.

Marszalek, D.S. (1987). Sewage and eutrophication. Chap. 7. In: Salvat, B. (Ed.). Pp. 77-90.

Matsos, C.A. (1981). Marine pollution in the South Pacific. *SPREP Topic Review* 11. 7 pp.

Mitchell, C.K. (1975). Food from the sea in Micronesia. *Micronesian Reporter* 23(4): 12-15.

Moran, P.J. (1986). The *Acanthaster* phenomenon. *Oceanogr. Mar. Biol. Ann. Rev.* 24: 398-480.

Motteler, L.S. (1986). Pacific Island Names. *B.P. Bishop Mus. Misc. Publ.* 34. 91 pp.

Munro, J.L. (1986). Report on progress of the International Giant Clam Mariculture Project and on the development of a pilot-scale Giant Calm hatchery. ICLARM. 5 pp.

Munro, J.L. (in press). Fisheries for giant clams (Tridacnidae: Bivalvia) and prospects for stock enhancement. In: Caddy, J.F. (Ed.). *Scientific Approaches to Management of Shellfish Resources*. John Wiley and Sons, New York.

Munro, J.L., Parrish, J.D. and Talbot, F.H. (1987). The biological effects of intensive fishing upon coral reef communities. Chap. 3. In: Salvat, B. (Ed.).

Munro, J.L. and Williams, D.McB. (1985). Assessment and Management of Coral Reef Fishes: Biological, environmental and socio-economic aspects. Seminar C. *Proc. 5th Int. Coral Reef Cong., Tahiti* 4: 545-578.

Nelson, S.G. (in press). Development of aquaculture in the U.S.-affiliated islands of Micronesia. In: Smith, B.D. (Ed.). *Topic Reviews in Insular Resource Development and Management in the Pacific U.S.-affiliated Islands*. Univ. Guam Mar. Lab. Tech. Rept 88.

Neudecker, S. (1987). Environmental effects of power plants on coral reefs and ways to minimize them. Chap. 9. In: Salvat, B. (Ed.). Pp. 103-118.

Nietschmann, B. (1984). Indigenous island peoples, living resources and protected areas. In: McNeely, J.A. and Miller, K.R. (Eds). *National Parks, Conservation and Development: The Role of Protected Areas in Sustaining Society*. Smithsonian Institution Press, Washington D.C. Pp: 333-343.

Ogden, J.C. and Gladfelter, E.H. (Eds) (1983). Coral reefs, seagrass beds and mangroves: Their interaction in the coastal zone of the Caribbean. *Unesco Reports in Marine Science* 23. 131 pp.

Parrish, J.D. (1980). Effects of exploitation upon reef and lagoon communities. In: *Marine and Coastal Processes in the Pacific: Ecological Aspects of Coastal Zone Management*. Unesco, ROSTEA, Jakarta. Pp. 85-121.

Pastorok, R.A. and Bilyard, G.R. (1985). Effects of sewage pollution on coral reef communities. *Mar. Ecol. Prog. Ser.* 21: 175-189.

Pearson, J.D. (1980). Assessment and management of fisheries for sessile invertebrates. In: *Marine and Coastal Processes in the Pacific: Ecological Aspects of Coastal Zone Management*. Unesco, ROSTEA, Jakarta. Pp. 123-157.

Pearson, R. (1981). Recovery and recolonisation of coral reefs. *Mar. Eco. Prog. Ser.* 4: 105-122.

Peters, E. (1983). Possible causal agent of "white-band" disease in Caribbean acroporid corals. *J. Invert. Path.* 41: 394-396.

Polunin, N. (1984). Do traditional marine "reserves" conserve? A view of Indonesian and Papua New Guinea evidence. *Senri Ethnology Studies* 17: 267-283.

Pulea, M. (1984). Environmental legislation in the Pacific Region. *Ambio* 13(5-6): 369-371.

Pulea, M. (1985). Legal measures for implementation of environmental policies in the Pacific Region. In: UNEP (1985). Pp. 157-162.

Randall, J.E. (1987). Collecting reef fishes for aquaria. Chap. 2. In: Salvat, B. (Ed.).

Ray, G.C. (1975). *A preliminary classification of coastal and marine environments*. IUCN Occ. Pap. 14, Morges, Switzerland.

Richard, G. (1982). Mollusques lagunaires et récifaux de Polynésie française: inventaire faunistique, bionomie, bilan quantitatif, croissance, production. Thèse de Doctorat d'Etat, Paris. 313 pp.

Richmond, R.H. (1985). Variations in the population biology of *Pocillopora damicornis* across the Pacific. *Proc. 5th Int. Coral Reef Cong., Tahiti* 6: 101-106.

Ruddle, K. and Johannes, R.E. (1985). *The Traditional Knowledge and Management of Coastal Systems in Asia and the Pacific*. Unesco, ROSTEA, Jakarta, Indonesia. 313 pp.

Salm, R.V. (1980). The genus-area relation of corals on reefs of the Chagos Archipelago, Indian Ocean. Ph.D. dissertation, Johns Hopkins University, Baltimore, Maryland, U.S.A.

Salm, R.V. (1984). Ecological boundaries for coral-reef reserves: Principles and guidelines. *Env. Cons.* 11(3): 209-215.

Salm, R.V. and Clark, J.R. (1984). *Marine and Coastal Protected Areas: A Guide for Planners and Managers*. IUCN, Gland, Switzerland.

Salvat, B. (Ed.) (1978, 1979, 1981, 1982). *Coral Reef Newsletters* 1,2,3,4. EPHE, Paris.

Salvat, B. (1980). The living marine resources of the South Pacific - past, present and future. In: Population-environment relations in tropical islands: The case of eastern Fiji. *Unesco MAB Technical Notes* 13: 131-148.

Salvat, B. (Ed.) (1987a). *Human Activities on Coral Reefs: Facts and Recommendations*. Antenne Museum EPHE, French Polynesia. 253 pp.

Salvat, B. (1987b). Dredging in coral reefs. Chap. 13. In: Salvat, B. (Ed.). Pp. 166-184.

Seymour, A.H. (1982). The impact on ocean ecosystems. *Ambio* 11(2-3): 132-137.

Silva, M.E., Gateley, E.M. and Desilvestre, I. (1986). *A Bibliographical listing of Coastal and Marine Protected Areas: A Global Survey*. Technical Report WHOI-86-11, Woods Hole Oceanographic Institution, Woods Hole, Massachusetts, U.S.A.

Smith, B.D. (in press). Development and management of nonfood marine resources in the U.S.-affiliated islands of the Pacific. In: Smith, B.D. (Ed.). *Topic Reviews in Insular Resource Development and Management in the Pacific U.S.-affiliated Islands*. Univ. Guam Mar. Lab. Tech. Rept 88.

Smith, R.O. (1947). Fishery resources of Micronesia. *Fishery leaflet* 239. Fish and Wildlife Service, U.S.D.I. 46 pp.

Smith, S.V. (1978). Coral reef area and contributions of reefs to processes and resources of the world's oceans. *Nature* 273(5659): 225.

Snedaker, S.C. and Getter, C.D. (1985). *Coastal Resources Management Guidelines*. Renewable Resources Information Series, Coastal Management Publication 2. Research Planning Institute Inc., Columbia, South Carolina. 205 pp.

Sorensen, J.C., McCreary, S.T. and Hershman, M.J. (1984). *Institutional Arrangements for Management of Coastal Resources*. Renewable Resources Information Series, Coastal Management Publication 1. Research Planning Institute Inc., Columbia, South Carolina. 165 pp.

SPC (1973). *Proceedings and Papers. Regional Symposium on Conservation of Nature - Reefs and Lagoons*. South Pacific Commission, Noumea, New Caledonia. 314 pp.

SPC (1984). Publications on Pacific maritime resources catalogued in SPC library, Jan. 1980 - July 1984. Unpub. ms.

SPC/SPEC/ESCAP/UNEP (1983a). Action Plan for managing the natural resources and environment of the South Pacific Region. *UNEP Regional Seas Reports and Studies* 29. 14 pp.

SPC/SPEC/ESCAP/UNEP (1983b). Radioactivity in the South Pacific. *SPREP Topic Review* 14. 211 pp.

SPC/SPEC/ESCAP/UNEP (1985). Ecological interactions between tropical coastal ecosystems. *UNEP Regional Seas Reports and Studies* 73. 71 pp.

SPREP (1982). Report of the Conference on the Human Environment in the South Pacific. Rarotonga, Cook Islands, 8-11 March 1982. SPREP/Conf. Human Environment/Report. 71 pp.

SPREP (1984). Report of the second consultative meeting of research and training institutions in the South Pacific Region. Port Moresby, Papua New Guinea, 23-27 Jan. 1984. South Pacific Commission, Noumea, New Caledonia.

Stoddart, D.R. (1969). Ecology and morphology of recent coral reefs. *Biol. Rev. Camb. Phil. Soc.* 44: 433-498.

Stoddart, D.R. (1985). Hurricane effects on coral reefs. Conclusion to Symposium 3. *Proc. 5th Int. Coral Reef Cong., Tahiti* 3: 349-350.

Tilmant, J.T. (1987). Impacts of recreational activities on coral reefs. Chap. 15. In: Salvat, B. (Ed.). Pp. 195-214.

Tsuda, R.T. (1981). Coral reefs in Micronesia. *Coral Reef Newsletter* 3: 25-27.

Udvardy, M.D.F. (1975). *A Classification of the Biogeographical Provinces of the World.* IUCN Occ. Pap. 18: 1-48.

UNEP (1985). Environment and resources in the Pacific. *UNEP Regional Seas Reports and Studies* 69. 294 pp.

UNEP/FAO (1985). Directory of marine environmental centres in South Pacific. *Regional Seas Directories and Bibliographies.* FAO, Rome. 147 pp.

Unesco (1982). Coral reef management in Asia and the Pacific: Some research and training priorities. *Unesco Rep. Mar. Sci.* 18. 22 pp.

Unesco (1984). Comparing coral reef survey methods. *Unesco Rep. Mar. Sci.* 21. 170 pp.

Unesco (1985). Coral Taxonomy. *Unesco Rep. Mar. Sci.* 33.

Unesco (1986a). Research on coastal marine systems. *Unesco Tech. Pap. Mar. Sci.* 47. 27 pp.

Unesco (1986b). The applications of digital remote sensing techniques in coral reef, oceanographic and estuarine studies. *Unesco Rep. Mar. Sci.* 42.

Unesco/FAO (1983). *International Directory of Marine Scientists.* FAO/IOC/UN/OETB.

van't Hof, T. (1985). The economic benefits of marine parks and protected areas in the Caribbean region. *Proc. 5th Int. Coral Reef Cong., Tahiti* 6: 551-556.

Venkatesh, S. and Va'ai, S. (1983). An overview of environmental protection legislation in the South Pacific countries. *SPREP Topic Review* 13. 63 pp.

Veron, J.E. (1985). Aspects of the biogeography of hermatypic corals. *Proc. 5th Int. Coral Reef Cong., Tahiti* 4: 83-88.

Veron, J.E.N. (1986). *Corals of Australia and the Indo-Pacific.* Angus and Robertson.

Wauthy, B. (1986). Physical ocean environment in the South Pacific Commission Area. *UNEP Regional Seas Reports and Studies* 83. 88 pp.

Wells, S.M. (1984). Coral reefs and the Ramsar Convention. *IUCN Bulletin* 15(4-6): 56-57.

Wells, S.M. (in press). Impacts of the precious shell harvest and trade: Conservation of rare or fragile resources. In: Caddy, J.F. (Ed.). *Scientific Approaches to Management of Shellfish Resources.* John Wiley and Sons, New York.

Wells, S.M. and Alcala, A.C. (1987). Collecting of corals and shells. In: Salvat, B. (Ed.).

Wells, S.M., Pyle, R.M. and Collins, N.M. (1983). *The IUCN Invertebrate Red Data Book.* IUCN, Gland, Switzerland and Cambridge, U.K.

Western Pacific Regional Fishery Management Council (1980). Fishery Management Plan and Regulations for the precious corals fishery in the Western Pacific Region. WPRFMC, Honolulu, Hawaii.

White, A. (1987). Effects of construction activity on coral reef and lagoon systems. Chap. 14. In: Salvat, B. (Ed.). Pp. 185-193.

Wood, E.M. (1983). *Corals of the World.* T.F.H. Publications, Inc. Ltd.

Wood, E. (1985). *Exploitation of Coral Reef Fishes for the Aquarium Trade.* Marine Conservation Society, Ross-on-Wye, U.K. 121 pp.

Wood, E.M. and Wells, S.M. (in prep.). The Marine Curio Trade: Conservation issues. Marine Conservation Society, Ross-in-Wye, U.K.

Yonge, M. (1973). The need for long-term research on coral reefs. In: Costin and Groves (Eds). Part 2: 7-18.

Zann, L.P. (1979). Changing technology in subsistence fisheries. In: Helfrich, P. (Ed.).

Zann, L.P. and Bolton, L. (1985). The distribution, abundance and ecology of the blue coral *Heliopora coerulea* (Pallas) in the Pacific. *Coral Reefs* 4: 125-134.

TABLE 1: DAMAGE TO REEFS DUE TO NATURAL EVENTS

	Climate, tides and tectonic events	Coral predators	Disease
American Samoa	storm (1987, Swains Atoll), heavy rain (1924, Tutuila), low tides (Tutuila), hurricanes (1966, 1987*)	*A. planci* (1978-79, Tutuila; 1987, Ofu)	
Australia, Eastern	floods (1974, Moreton Bay)	*A. planci* (1962-74, 1978, 1980/81, 1985-? mid part of Great Barrier Reef)	
Belau		*A. planci* (1930s, 1970s)	
China	hurricanes (1975, Xisha Qundao)		
Cook Is	hurricanes	*A. planci* (1971; 1987*)	
Federated States of Micronesia		*A. planci* (1970s; 1987*)	
Fiji	hurricanes (1965, Viti Levu; 1975, Lau Group; 1978, Taveuni; 1980, Viti Levu; 1983, Malololailai; 1985, Mamanuca Group), earthquakes (1953, 1979, Taveuni, Vanua Levu), heavy rainfall (1986, Viti Levu)	*A. planci* (1960s, 1970s*, 1978/80, 1984)	
French Polynesia	hurricanes (1903, Hao; 1906, Tuamotu Gp and Gambiers; 1980, 1981, Mururoa; 1982-83, Society Islands and Tuamotu Gp), sea level drop and El Niño (1983, Society Islands), low tides (1978, 1980, Mataiva)	*A. planci* (1981, Moorea, and others)	
Guam	hurricanes, sea-level changes	*A. planci* (1968-70) *Terpios*	
Hawaii	storms, tsunamis, hurricane (1982)	*A. planci* (1960s, Molokai)	
Hong Kong	cold temperatures, low tides		
Japan	coral bleaching and high water temperatures, (El Niño), low water temperatures, hurricanes	*A. planci* (1957, Miyako; 1960s-1980s, Ryukyus, Kyusho, Shikoku, Honshu), *Drupella*, (1976-80s, Miyake-jima, Kushimoto), *Terpios*	white and black band (Ishigaki)
Kiribati	high rainfall and high sea temperature (El Niño) (1983*)	*A. planci* (1970s, Tarawa, Abaiang; 1940s, Kiritimati)	
Marshall Is	hurricanes	*A. planci*	
Nauru	wave action		
New Caledonia			
Niue	hurricane (1979)		

Table 1 (Contd)

	Climate, tides and tectonic events	Coral predators	Disease
Northern Mariana Is	typhoons and wave action*, volcanic eruption	*A. planci* (1968, 1971)	
Papua New Guinea	storms (1969, Conflict Group, and others), earthquake (Madang)	*A. planci* (1983, Milne Bay)	
Pitcairn	low water temperature? (Ducie)	(single specimen *A. planci*, Ducie)	
Solomon Is	earthquake (1978), low tide/ El Niño (1983), cyclones	*A. planci* (1966, 1968, Malaita; 1981, Guadalcanal)	
Taiwan	cyclones	*A. planci*	
Tokelau	hurricane (1987), sea level drop/El Niño (1983)	*A. planci* (1960s*)	
Tonga	cyclone (1982), low tides and heavy rain/El Niño (1984, Hihifo peninsula)	*A. planci* (1969, Vava'u Group)	
Tuvalu	hurricane (1972, Funafuti)		
Vanuatu	hurricanes (1972, Reef Islands; 1985 Espiritu Santo; 1987 Efate), emergence	*A. planci* (1986)	
Wallis and Futuna			
Western Samoa		*A. planci* (1930s, 1960s-70s, Upolu and others)	

* = event occurred but damage not reported

TABLE 2: KNOWN THREATS TO REEFS

	Erosion/land clearance	Construction/ dredging/bombing	Pollution	Overcollection	Recreational use	Fishing activities
American Samoa	agriculture	dredging, blasting, roads, airports	sewage, oil, industrial, littering	Giant Clams, fish		dynamiting, poison
Australia, Eastern	agriculture	dredging, blasting,	pesticides, fertilizers, rubbish, sewage	coral, shells, fish	walking, anchor damage, propellers	
Belau		*port	coconut oil spill, *sewage	Giant Clams		dynamiting and poisons (1970s)
China		dredging, coral mining	*oil	corals, fish, shellfish, Giant Clams		
Cook Is	soil run-off	construction, reclamation, harbour, *phosphate mining	chemicals	pearl oysters, Giant Clams, fish		dynamiting
Federated States of Micronesia	soil run-off	dredging, reclamation, airports, roads, harbours, causeways	sewage, oil	Giant Clams, fish		dynamiting, poison
Fiji	deforestation, overgrazing	roads, hotels, *sand dredging	oil, sewage, *mining, *oil exploration	coral, Giant Clams, shellfish, shells, fish		dynamiting, poison
French Polynesia	soil run-off	dredging, sand mining, reclamation, phosphate mining, airport, hotels, nuclear testing	sewage, oil, insecticide, industry	fish	tourism, boat damage	

Table 2 (Contd)

	Erosion/land clearance	Construction/ dredging/bombing	Pollution	Overcollection	Recreational use	Fishing activities
Guam		dredging, hotels, reclamation, military, industry, shipwreck	thermal, oil, chemical	shellfish, fish, shells	*tourism	dynamiting, bleach
Hawaii	agriculture, overgrazing	dredging, military, airports, bombing	sewage, thermal, industrial, chemical	fish, black coral	tourism, fishing, anchor damage	
Hong Kong		reclamation	sewage, industrial	corals, shells, Giant Clams, fish	SCUBA diving	dynamiting
Japan	agriculture, deforestation	reclamation, dredging, roads, airports, underwater observatory, *military	oil, unspecified	boring clam	SCUBA diving, anchor damage,	reef gleaning
Kiribati		reclamation, causeway, airport, nuclear testing (1950/60s)	rubbish, *sewage	Giant Clams, coral, *fish		
Marshall Is		dredging, reclamation, military, roads, airport, nuclear tests	radiation			
Nauru			oil			
New Caledonia	mining, bush fires		sewage, industrial, *mining	fish, trochus, crustaceans, shells	fishing	
New Zealand				*fish		
Niue	siltation			*shells		

Table 2 (Contd)

	Erosion/land clearance	Construction/ dredging/bombing	Pollution	Overcollection	Recreational use	Fishing activities
Northern Mariana Is		*resorts, *military	thermal		tourism	dynamiting, poison
Papua New Guinea	*deforestation	dredging	oil, sewage, mining	shells, corals, holothurians		dynamiting
Pitcairn						
Solomon Is	*deforestation	*port	industry, *sewage	shells, *Giant Clams		dynamiting
Taiwan		dredging, construction, aquaculture	oil, unspecified, thermal	aquarium fish, corals, Giant Clams, shellfish	tourism, SCUBA diving, anchor damage	dynamiting, spearfishing, poison
Tokelau		channel blasting, shipwreck	pesticides	Giant Clams		
Tonga		causeways, sand mining, quarrying	sewage	Giant Clams, fish, Black Coral	tourism	"tu'afeo", gleaning, dynamite, poison
Tuvalu		channel blasting	unspecified, *oil	Giant Clams, fish		
Vanuatu		sand mining	unspecified	Black Coral, fish		spearfishing
Wallis and Futuna	*siltation	urban development	sewage	fish, trochus, echinoderms		"futu" poison
Western Samoa	agriculture, fuel-wood	sand mining, reclamation, blasting, roads	sewage, *pesticides	shells, Giant Clams, palolo worm, fish		dynamiting, coral breaking, poison, "faamo'a"

* = potential threat

TABLE 3: EXISTING, PROPOSED AND RECOMMENDED PROTECTED AREAS
ADJACENT TO OR INCLUDING REEFS

*	= party to World Heritage Convention
()	= marine habitats not included within protected area
F	= some controls over fishing/fishing reserve
rm	= resource management
I.	= Island
Is	= Islands

The IUCN Categories for Protected Areas have not been applied to the protected areas in this region.

	Established	Proposed/ Recommended
AMERICAN SAMOA		
Rose Atoll National Wildlife Refuge	+	
Fagatele Bay National Marine Sanctuary	+	
North Tutuila National Park (inc. Pola I.)		+
Papaloloa Pt, Ofu		+
South Ta'u National Park		+
Nu'utele Islet, Ofu		+
Goat I. Pt - Utulei Reef, Tutuila		+
Lepisi Pt, Tutuila		+
Ogegasa Pt, Tutuila		+
Pago Pago Bay and Pala Lagoon, Tutuila		+
Swains Atoll		+
AUSTRALIA, EASTERN		
Cobourg Peninsula Marine National Park and Sanctuary	+	
Coringa - Herald National Nature Reserve	+	
Elizabeth and Middleton Reefs MarineNational Nature Reserve	+	
Great Barrier Reef Marine Park	+	
Cairns Section		
Central Section		
Cormorant Pass Section		
Far Northern Section		
Mackay-Capricorn Section		
Lihou Reef National Nature Reserve	+	
Lord Howe Island Permanent Park Preserve	+	
Solitary Is Marine Reserve		+
BELAU		
Chelbacheb (Rock Is) National Park		+
Ngerukeuid Is Wildlife Preserve	+	
Ngerumekaol Channel	F	
Ikedelukes Reef trochus reserve	F	
Ngederrak Reef	F	
Helen Reef		+
CHINA		
Hainan Reefs		+
Xisha Qundao		+
COOK ISLANDS		
Suwarrow Atoll National Park	+	
Avatiu foreshore reserve	(+)	
Aitutaki trochus reserve	F	
Palmerston trochus reserve	F	
Manuae trochus reserve	F	
Takutea		(+)
Pukapuka		rm
Black Rocks (Tuoro), Rarotonga		+
Ngatangiia Harbour and Muri Lagoon, Rarotonga		+
Aitukaki lagoon and eastern motus		+

	Established	Proposed/ Recommended
FEDERATED STATES OF MICRONESIA		
Pohnpei		rm
Gaferut		+
Elato turtle reserve		+
Pikelot turtle reserve		+
West Fayu turtle reserve		+
Oroluk turtle reserve		+
FIJI		
(T = terrestrial, M = marine, P = park, R = reserve, (+) = seabird rookery or turtle reserve)		
Viti Levu, Lomaitivi, Kadavu and Yasawas		
Natadola Bay		TP
Makuluva and reefs		MR
Suva Barrier Reef and cays		MP, MR
Vuo	(+)	
Draunibota	(+)	
Labiko	(+)	
Bird I., Beqa Lagoon		(+)
Vatu-i-ra		(+)
Mabualau		(+)
Beqa Lagoon		MP, MR
Nadi Bay reefs (inc. Tai and Luvuka)		MP, MR
Coral Coast reefs		MP, MR
Makogai and reefs		TP, MR
Wakaya and reefs		MP
Cakau Momo (Horseshoe Reef)		MP
Great Astrolabe reef		MP, MR
North Astrolabe Reef		MR
Yabu		(+)
Taqa Rock		(+)
White Rock		(+)
Mamanuca Group		MR
Malamala and reefs		TP, TR, MP, MR
Vanua Levu and Taveuni		
Namenalala and reef		TR, MR
Nanuku islets		(+)
Great Sea Reef		MP, MR
Qelelevu Atoll		TR, MR
Rainbow Reef		MP, MR
Yadua Taba		+
Rotuma		TP, MP
Lau Group		
Wailagilala Atoll		TR, MR
Qilaqila (Bay of Islands)		TP, MP
Fulaga Bay of Islands		MP, PR
Sovu Is		(+)
Cakau Lekaleka Barrier Reef		MP, MR
Nukutolu islets and reefs		MR
Other areas recommended:		
Leleuvia		+
Ra coast		+
Mana		+
Yasawa-i-rara		+
Makodroga		+
Balolo Point, Ovalau Island		+
Cobia		+
Vetauua		(+)
Nukubasaga		(+)
Nukubalati		(+)

Table 3 (Contd)

	Established	Proposed/ Recommended
FRENCH POLYNESIA		
Austral Islands		
Gambier Islands		
Marquesas		
Hatutaa	(+)	
Eiao	(+)	
Mohotani	(+)	
Motu One	(+)	
Society Islands		
Manuae (Scilly) Reserve	+	
Huahine		rm
Tetiaroa		+
Moorea reefs - Temae, Tiahura		+
Raiatea		+
Tupai		+
Maupihaa		+
Motu One		+
Tuamotu Archipelago		
W.A. Robinson Integral Reserve	+	
Anuanuraro		+
Anuanurunga		+
Apataki		(+)
Hereheretue		+
Kauehi		(+)
Matureivavao		(+)
Napuka		(+)
Nukutipipi		+
Pukapuka		+
Tekokota		(+)
Toau		+
Motu Paio (Rangiroa)		(+)
Clipperton I. (France)		+
GUAM		
Anao Conservation Reserve	+	
Haputo Ecological Reserve Area	+	
Orote Peninsula Ecological Reserve Area	+	
Pati Point Natural Area	+	
War in the Pacific National Historical Park	+	
Guam Territorial Seashore Park	+	
Luminao Barrier Reef		+
HAWAII		
(MLCD = Marine Life Conservation District, MFMA = Marine Fisheries Management Area, SP = State Park, SNAR = State Natural Area Reserve, NWR = National Wildlife Refuge)		
Hawaii Island		
Kealakekua Bay MLCD	+	
Lapakahi State Historical Park MLCD	+	
Wailea Bay MLCD	+	
Hilo Bay MFMA	F	
Puako Bay and Reef MFMA	F	
Kailua Bay MFMA	F	
Kahoolawe Island		+
Kauai Island		
Waimea Bay and Recreational Pier MFMA	F	
Hanamaulu Bay and Ahukini Recreational Pier MFMA	F	
Milolii Reef SP	+	
Nualolo-Kai Reef SP	+	
Lanai Island		
Hulopoe Bay - Palawai, Manele Bay - Kamao MLCD	+	
Manele Boat Harbor MFMA	F	

Table 3 (Contd)

	Established	Proposed/ Recommended
HAWAII (Contd)		
Maui		
Honolua and Mokuleia Bay MLCD	+	
Kahului Harbor MFMAs	F	
Cape Kinau, Ahihi and La Perouse Bays SNAR	+	
Molokini Islet MLCD	+	
Hawaiian Islands NWR	+	
Kure MFMA	F	
Oahu Island		
Hanauma Bay MLCD	+	
Pupukea Beach Park MLCD	+	
Waikiki-Diamond Head MFMA	F	
Coconut Island Marine Refuge	+	
U.S. Dependencies:		
Baker Island NWR	+	
Howland Island NWR	+	
Jarvis Island NWR	+	
Johnston Island NWR	+	
Palmyra Atoll		+
HONG KONG		
Country Park 6	(+)	+
Country Park 7	(+)	+
Country Park 8	(+)	+
Country Park 18	(+)	+
Kat O Chau Special Area	(+)	
JAPAN		
(Q.N.P. = Quasi-National Park; NP = National Park)		
Minamiboso QNP	+	
Katsuura		
Yoshino-Kumano NP	+	
Kushimoto		
Muroto-Anan Kaigan QNP	+	
Awaoshima		
Awatakegashima		
Ashizuri-Uwakai NP	+	
Tatsukushi		
Okinoshima		
Kashinishi		
Uwakai		
Genkai QNP	+	
Genkai		
Saikai NP	+	
Wakamatsu		
Fukue		
Unzen-Amakusa NP	+	
Tomioka		
Amakusa		
Ushibuka		
Nippo Kaigan QNP	+	
Nanpokuura		
Kamae		
Nichinan Kaigan QNP	+	
Nichinan		
Kirishima-Yaku NP	+	
Sakurajima		
Sata Misaki		
Amami o-shima QNP	+	
Kasari Hanto-Higashi Kaigan		
Surikozaki		

Table 3 (Contd)

	Established	Proposed/ Recommended
JAPAN (Contd)		
Kametoku		
Setouchi		
Yoronto		
Okinawa Kaigan QNP	+	
Onna Kaigan		
Tokashiki		
Zamami		
Iriomote NP	+	
Yaeyama		
Sakiyama Bay Nature Conservation Area		
Fuji-Hakone Izu NP	+	
Miyake-jima		+
Ogasawara NP	+	
Ogaswara		
Kabira Bay marine sanctuary, Ishigakishima	F	
Nagura Bay marine sanctuary, Ishigakishima	F	
Shiraho Lagoon, Ishigakishima		+
KIRIBATI		
Kiritimati Wildlife Sanctuary	(+)	
Malden Wildlife Sanctuary	(+)	
Starbuck Wildlife Sanctuary	(+)	
Rawaki Wildlife Sanctuary	(+)	
McKean Wildlife Sanctuary	(+)	
Vostock Wildlife Sanctuary	(+)	
Birnie Wildlife Sanctuary	(+)	
Phoenix Island National Park		+
Enderbury		(+)
Caroline		(+)
Flint		(+)
Kanton		(+)
Butaritari		(+)
Abaiang		(+)
MARSHALL ISLANDS		
Kwajalein		rm
Jemo		+
Bikini		(+)
NAURU		
NEW CALEDONIA		
New Caledonia "protected zone"	F	
Maître Islet Nature Reserve	+	
Amédée Islet Nature Reserve	+	
Great Reef Rotating Reserve	+	
Yves Merlet Marine Reserve	+	
Isle of Pines		+
Ngo and Uie Bays		+
Mba and Mbo		+
Chesterfield		+
Hunter		+
Beautemps-Beaupré		+
Terrain Bas and La Foa		+
D'Entrecasteaux Reefs		+
NEW ZEALAND		
Kermadec Islands		+

	Established	Proposed/ Recommended
NIUE		
Fatiau Tuai		+
Makapu		(+)
Hio		(+)
Limu		(+)
Makatutaha		(+)
Tahileleka		(+)
Hikutavake Reef		+
Vaihoko		+
Tuo		+
Vaitafe		+
Motu		+
NORTHERN MARIANA ISLANDS		
Guguan	(+)	+
Maug	(+)	+
Farallon de Pajaros	(+)	+
Asuncion	(+)	+
Saipan (selected sites)		+
Managaha	+	
Rota (selected sites)		+
Tinian (selected sites)		+
PAPUA NEW GUINEA		
(WMA = Wildlife Management Area)		
East New Britain		
Talele I. Nature Reserve	+	
Nanuk I. Provincial Park	+	
West New Britain		
Garu WMA	+	
Hoskins Bay		+
Cape Anukur		+
Manus		
Lou I. WMA	+	
Ndrolowa WMA	+	
Wuvulu Is		+
Ninigo Group		+
Hermit Is		+
Western Is		+
Sabben Is		+
Alim Is		+
Milne Bay		
Sawataetae Wildlife Management Bay	+	
Baniara I. Protected Area	+	
Trobriand Is		+
Muyua		+
Goodenough		+
Fergusson		+
Normanby		+
Pocklington Reef		+
Misima		+
Yela Is		+
Calvados Chain		+
Conflict Group		+
Wari Is		+
Milne Bay Is		+
Madang		
Ranba (Long) I. WMA	+	
Macclay Park	+	
Manam I.		+
Hansa Bay (Laing Island)		+

Table 3 (Contd)

	Established	Proposed/ Recommended
PAPUA NEW GUINEA (Contd)		
Karkar		
Bagiai WMA	+	
Northern Province		
Mangrove I.		+
Cape Nelson		+
Central Province		
Horseshoe Reef Marine Park		+
Coutance Is		+
Motupore I. WMA		+
Idlers Bay		+
Papuan Barrier Reef		+
Western Province		
Maza WMA	+	
National Capital		
Taurama Beach Recreational Park		+
East Sepik		
Unei Island Village Reserve		+
Shouten Is		+
West Sepik		
Tumleo Ali, Seleo and Angel Island		+
Morobe		
Fly Island Marine Park		+
Tami Is		+
Salamaua Peninsula		+
New Ireland		
St Mathias Group		+
Islands between Lavongai and Kavieng		+
Djaul Is		+
Tabar Is		+
Lihir Group		+
Tanga Is		+
Feni Is		+
N. Solomons		
Pinipel-Nissau Group		+
Kulu, Manus, Passau		+
PITCAIRN		
Henderson I.		+
SOLOMON ISLANDS		
Tulagi	(+)	
Oema	(+)	
Mandoleana	(+)	
Dalakalau	(+)	
Dalakalonga	(+)	
Marovo Lagoon		rm
Arnavon Is, Manning Strait		+
Rennell I.		+
TAIWAN		
(CCZ = Coastal Conservation Zone; FCA = Fishery Conservation Area)		
North-east Coast CCZ	+	
Kenting National Park and CCZ	+	
Lang-Yang River Mouth CCZ	+	
Su-Hua CCZ	+	
Hua-Tung CCZ	+	
Hsiao-liu-chiu Reef FCA	F	
Lu-tao Reef FCA	F	
Pen-hu Reefs FCA	F	
Two FCAs near Kenting	F	

	Established	Proposed/ Recommended
TAIWAN (Contd)		
Lan-yu Reef		+
TOKELAU		
TONGA		
Fanga'uta and Fangakakau Lagoon Reserve	+	
Ha'atafu Beach Reserve	+	
Hakaumama'o Reef Reserve	+	
Malinoa Island Park and Reef Reserve	+	
Monuafe Island Park and Reef Reserve	+	
Pangaimotu Reef Reserve	+	
'Eua National Park		+
Nuapapu-Vaka'eita marine reserve		+
Muihopohoponga Beach Reserve		+
'Ata Island Biosphere Reserve		+
TUVALU		
Funafuti		F
Vaitupu		F
Kosciusko Bank		+
VANUATU		
President Coolidge and Million Dollar Point	+	
Whitesands Recreational Reserve	+	
Bucaro Recreational Reserve	+	
Site on Aore, Recreational Reserve	+	
Naomebaravu Recreational Reserve	+	
Reef Is		+
Anatom I.		+
Cook Reef		+
North west coasts of Makekula and Santo		+
WALLIS AND FUTUNA		
Alofi I.		(+)
WESTERN SAMOA		
Palolo Deep Marine Reserve, Upolu	+	
Aleipata and Nu'utele Is, Upolu		+
Satuimalufilufi/Fuailolo'o, Upolu		+
Fusi/Tafitoala, Upolu		+
Aganoa, Upolu		+
Nu'usaf'e I., Upolu		+
Salamumu, Upolu		+
Leanamoea, Savai'i		+
Cape Puava, Savai'i		+

TABLE 4: NATIONAL LEGISLATION RELATING TO CORAL REEFS

This list is known to be incomplete but has been included as a guide to the types of legislation which exist for the management of coral reefs and reef fisheries; customary law is not included. Information was not obtained for New Zealand. Additional information is available in Venkatesh and Va'ai (1983) and SPC/SPEC/ESCAP/ UNEP (1985).

AMERICAN SAMOA
U.S. Marine Protection, Research and Sanctuaries Act, 1972
- designates ocean waters as marine sanctuaries.
U.S. Endangered Species Act, 1973
- bans turtle collection.
U.S. Fisheries Conservation and Management Act, 1976
U.S. Clean Water Act, 1977
U.S. Coastal Zone Management Act, 1972
  Executive Order 3-80
- establishes the Coastal Management Program, and includes provision for establishment of Special Management Areas.
Public Law 16-58
- prohibits use of poison in territorial waters.
Territorial laws
- prohibits use of dynamite for marine harvesting.

AUSTRALIA, EASTERN
Great Barrier Reef Marine Park Act, 1975
- establishes Great Barrier Reef Marine Park Authority and provides for establishment, control, care and development of a marine park in the Great Barrier Reef region.
Great Barrier Reef Marine Park Regulations, 1983
- include prohibition on drilling for petroleum within Great Barrier Reef region.
Torres Strait Treaty, 1978, amended 1984, ratified 15.2.85 (bilateral agreement with Papua New Guinea)
- establishes protected zone in Torres Strait where traditional resource exploitation can continue.
Queensland legislation protects *Charonia tritonis* and bans spearfishing with SCUBA.

BELAU
N.B. Belau is the last remaining part of the Trust Territory of the Pacific Islands. It is expected shortly to become an independent nation in close association with the U.S.A. When this happens much existing legislation may no longer apply.
U.S. Clean Water Act, 1977
U.S. Endangered Species Act, 1973
Palau (Belau) Code (Chapter 2)
- establishes Ngerukeuid Islands Wildlife Preserve; provides for conservation of dugongs and other marine life; prohibits fishing in Ngerumekaol during grouper spawning season.
Trust Territory of the Pacific Islands Code Title 45
- provides for control of fishing with explosives and poisons; limits collection of turtles, sponges, trochus, pearl shell.
Republic of Palau Public Laws
- restrict location of trochus collection; prohibit export of clam meat; establish Palau Lagoon Monument.
Koror (Oreor) State Ordinances
- establish trochus sanctuaries and prohibit shelling.

CHINA
Regulations for the Propagation and Protection of Fishery Resources, 1979
Marine Environmental Protection Law, 1982
- covers wide range of issues; administered by the Environmental Protection Department.
Coastal Zone and Shallow Sea Areas Protection and Management Law (in preparation)

COOK ISLANDS
Conservation Act, 1975
- establishes Suwarrow Atoll National Park.
Conservation Act, 1987
- includes provision for establishment of marine parks.
Trochus Act, 1975
- restricts harvest of trochus within reserves.
Territorial Sea and Exclusive Economic Zone Act, 1979
- includes provision for protection and preservation of marine environment.
Legislation
- prohibits use of poisons and dynamite.

FEDERATED STATES OF MICRONESIA
Trust Territory of the Pacific Islands regulations no longer apply and F.S.M. environmental legislation has not yet been enacted.

FIJI
Fisheries Act, 1978
- Chap. 158 prohibits use of explosives and poisons for fishing; minimum size limits for turtles, Trochus, mother-of-pearl, *Scylla serrata*; protects turtles and eggs during nesting season, and *Charonia tritonis* and *Cassis cornuta*.
Town Planning Act, 1946
- Chap. 109 provides for preparation of Town Planning Schemes, including conservation of foreshores, harbours and other parts of the sea.
Harbour Act No. 3, 1974
- provides for penalties to be imposed for pollution of harbour and coastal waters (never been used).
Public Health Act
- Chap. 91 controls sewage discharge.
National Trust for Fiji Ordinance, 1970
- Chap. 265 provides for the permanent preservation of lands (including reefs) and for the development of parks and reserves.
Mining Act
- requires environmental impact assessment for mining activities.

FRENCH POLYNESIA
Fisheries legislation
- provides for protection and management of marine and coastal resources; controls exploitation of pearl oysters and trochus.
Marine pollution legislation

GUAM
U.S. Endangered Species Act, 1973
- bans turtles collection.
U.S. Coastal Zone Management Act
- provides for planning controls and regulations for coastal resources.
U.S. Clean Water Act, 1977

- Section 208 mandates Guam Water Quality Management Plan, designates conservation areas where pollutants banned.

Fisheries regulations
- prohibit poisons and explosives; limit exploitation of live coral, coconut crabs, trochus, spiny lobsters.

Law 12-209, 1975
- set up Guam Territorial Park System, creating Guam Territorial Seashore Park and Anao Conservation Area.

HAWAII (U.S.A.)

The Marine Protection, Research and Sanctuaries Act, 1972
- designates ocean waters as marine sanctuaries.

U.S. Endangered Species Act, 1973
- bans turtle collection.

Fishery Conservation and Management Act, 1977
- regulates fisheries within 200 mi. (320 km) territorial limit; provides for establishment of regional fishery management councils; regulates establishment of Habitat Areas of Particular Concern (HAPC); regulates take of coral; controls fishing; bans take of aquarium fish in protected areas.

HONG KONG

Fisheries Protection Ordinance Chapter 171, 1964
- prohibits use of explosive and poison in fishing.

Legislation
- regulates activities in coastal waters (details not known).

Country Parks Ordinance, 1976
- provides for establishment of country parks (includes coastline but not seabed).

JAPAN

Fisheries Law, 1949
- establishes framework for operation of Fisheries Cooperative Associations in management of local fisheries resources; places limits (including seasonal and size limits) and sometimes bans on some fisheries, including turtle eggs and corals in the orders Scleractinia, Gorgonacea and Stolonifera in Okinawa Prefecture.

National Parks Law 1931, revised 1957
- provides for creation of national, quasi-national and prefectural nature parks.

Environment Preservation Law, 1972
- designates the Environment Agency responsible for national parks.

Okinawan Prefectural Government regulations, 1973
- provide for conservation of natural environment, including establishment of conservation areas.

KIRIBATI

Fisheries Ordinance, 1957
- prohibits fishing without license, or with explosives, gas, poison; protects ancient and customary fishing grounds.

Wildlife Conservation Ordinance, 1975, amended 1979
- partial protection of Green Turtles; provides for establishment of wildlife sanctuaries.

MARSHALL ISLANDS

Trust Territory of the Pacific Islands regulations no longer apply and Marshall Is environmental legislation has not yet been enacted; U.S. laws apply to U.S. activities.

NAURU

Marine Resources Act, 1978
- provides for exploitation, conservation and management of fish and aquatic resources.

NEW CALEDONIA

Law 76-12222, 1976
- environmental legislation.

Laws 64-1331, 1964; 73-477, 1973; 79-5, 1979
- prohibit marine pollution by hydrocarbons.

Law 76-599, 1976
- prevents and controls pollution by dumping from ships and aircraft.

Délibération 245, 1981, modified by Délibération 510, 1982
- regulates fishing, including trochus fishing, and prohibits dynamite fishing.

Arrêté 83-002/CG, 1983
- requires permit for trochus fishing and establishes size limit.

Délibération 509, 1982, and Arrêté 85-321/CM, 1985
- authorizes controlled collection of coral on trial basis in Tetembia section of barrier reef.

Law 56-1106, 1956, Resolution 225, 1965, Délibération 108, 1980, Decree 1504, 1980
- establishes protected areas.

Water Resources and Pollution Law, Délibération 105, 1968
- controls or prohibits activities likely to endanger water quality; allows for establishment of protected zones.

Legislation:
- prohibits night fishing
- controls collection of turtles, aquarium fish, bryozoa and sponges.

NIUE

Fish Protection Ordinance, 1965

NORTHERN MARIANA ISLANDS

Public Law 1-8, Chapter 13
- empowers Dept of Natural Resources to protect and enhance natural resources.

U.S. Coastal Zone Management Act
- provides for Coastal Resources Management Program which regulates activities in territorial waters.

P.L.2-51 Fish, Game and Endangered Species Act, 1981

Emergency Regulations Protecting Fish and Wildlife, 1983
- establishes licensing requirements and controls for collecting trochus, corals, turtles and lobsters; prohibits use of explosives and poisons; designates sanctuaries.

Environmental Planning Act, 1978

Environmental Contaminants Act, 1978

PAPUA NEW GUINEA

Continental Shelf (Living and Natural Resources) Act (Chapter 210)
- controls the harvesting of sedentary organisms, including clams, trochus and Gold-lip Pearl shell.

Fisheries Act (Chapter 214)
- defines the powers of the Minister and officers of the Dept of Fisheries and Marine Resources in relation to licensing and fisheries management (does not cover whales and sedentary organisms).

Fauna (Protection and Control) Act, 1966
- provides for establishment of Wildlife Sanctuaries and Wildlife Management Areas.

Native Customs (Recognition) Act 1963 (Chapter 19, Section 5b)
- specifically recognizes the customary rights or ownership in connection with the sea, reef, seabed, rivers and lakes, including rights to fishing.

Torres Strait Treaty, 1978, amended 1984, ratified 15.2.85 (bilateral agreement with Australia)
- establishes protected zone in Torres Strait where traditional resource exploitation can continue.

Coral harvesting law

PITCAIRN ISLANDS
Fisheries Zone Ordinance
- provides for management of fisheries resources.

SOLOMON ISLANDS
Delimitation of Marine Waters Act, 1978
- provides for protection and preservation of marine environment.

National Parks Act, 1954
Fisheries Act, 1972/1977
- provides for licenses and permits for commercial fishing, enables limits on gear type and harvestable fish sizes to be set and closed areas to be declared.

Regulations limit size of harvestable trochus and turtles.

TAIWAN
National Park Law, 1972
- provides for establishment of natural parks.

Coastal Area Environment Protection Plan
- establishes coastal conservation zones.

TOKELAU
(Traditional conservation measures only)

TONGA
Parks and Reserves Act, 1976
- authorizes establishment of protected areas.

Bird and Fish Preservation Act, 1915, amended 1974
- partial or complete protection for some species, including turtles.

Environmental Protection and Fisheries Acts in preparation.

TUVALU
Fisheries Ordinance, 1977
- prohibits fishing with explosives and poisons.

Prohibited Areas Ordinance
- provides for establishment of wildlife reserves.

Wildlife Conservation Ordinance, 1975
Legislation:
- controls pollution and waste disposal.
- regulates sand and coral removal.

VANUATU
Fisheries Act No. 37, 1982
- provides for establishment of marine reserves.

Fisheries Regulations, 1983
- provides for issue of fishing licenses and conservation and regulation of fisheries, including size limits for spiny and slipper lobsters, coconut crabs, trochus, *Turbo* and *Charonia*; partially protects turtles; totally protects marine mammals; limits coral collection; requires permits for export of trochus, green snail, crustaceans, aquarium fish, coral and bêche-de-mer.

WALLIS AND FUTUNA
Fisheries Regulation Order No. 83, 1965
- prohibits use of explosives or poison for fishing.

WESTERN SAMOA
Fish Protection Act, 1972
- regulates exploitation of marine resources.

Fish Dynamiting Act, 1972
- prohibits use of dynamite for fishing.

Police Offences Act, Section 4(f), 1961
- prohibits use of *Derris* and *Barringtonia* plants and derivatives for fishing.

National Parks and Reserves Act, 1974
- enables establishment of marine parks and reserves.

Water Act, 1965
- controls discharge of pollutants into coastal waters.

AMERICAN SAMOA

INTRODUCTION

General Description

American Samoa, an unincorporated territory of the U.S.A., comprises the six eastern islands of the Samoan Archipelago as well as Swains Atoll, which is geographically part of the Tokelau group. The land area covers 76 sq. mi. The geology of the Samoan Islands is described by Stearns (1944). All the islands except Swains Atoll are aligned along the crest of a discontinuous submarine ridge which extends over 485 km and trends roughly north-west by south-east. Swains and Rose Atolls are limestone, the others are composed principally of volcanic rock and are typically steep-sided, with little in the way of coastal plains, and with lush vegetation. Terrestrial aspects of the ecology of the islands are described in some detail in Amerson et al. (1982) and Dahl (1970 and 1973) provides short ecological reports.

Tutuila is the largest island and is the top of a composite volcano rising approximately three miles (4.8 km) from the ocean floor, which results in deep water near shore. There are many small but no major streams and the island is almost bisected on the south-east by a deep natural harbour, Pago Pago Harbor. The coastline, except at the mouths of drowned alluvial valleys, is irregular, rocky and composed of steep cliffs of variable height.

American Samoa has a warm, humid tropical climate with average temperature of 70-90°F (21-32°C) and average humidity of 80%; mean rainfall is about 200 in. (5080 mm), the heaviest rains occurring from December to March. It is in the zone of the south-east trade winds which are moderate from May to November. During the remainder of the year winds are variable; the strongest occur during the winter months of June to August, and the weakest are from December to February. Major hurricanes are experienced about once every five years; these normally approach from the north but occasionally come from other directions. Tsunamis may also occur although only inner Pago Pago Harbor has experienced any sizeable run-up (USDC, 1984).

Table of Islands

Tutuila (X) 52 sq. mi. (135 sq. km) (32 x 4 km). volcanic, 2141 ft (653 m) with chain of mountains; fringing reef along eastern part of south coast as described below (see separate accounts for Fagatele Bay and Goat Island Pt - Utulei Reef).

Aunu'u (X) 1 sq. mi. (2.6 sq. km); volcanic islet with 200 ft (61 m) cone, 1.6 km off south-east coast of Tutuila; fringing reef.

Manu'a Group

Ofu (X) 3 sq. mi. (7.8 sq. km), 1621 ft (494 m) volcanic; undisturbed fringing reef; important reefs at Asaga Strait, Alaufau and Ofu; important structural reefs occur as offshore banks near Asaga Strait and Tumua'i Pt (Maragos, 1986); reef on west side at anchorage, 200 m off shore from Alaufau at 10-20 m depth, described by Dahl et al. (1974): water clear and warm with abundant fish; bottom has marked topographic relief, with small corals and red algae on top of elevations, dense coral cover; high species diversity on walls and white sand in troughs; (see account for Papaloloa Pt); brief descriptions of some sites in Itano (1987).

Olosega (X) 2 sq.mi. (5.2 sq. km); volcanic, 2095 ft (639 m); important reefs at Tamatupu Pt and Sili and important structural reef at Pouono Pt (Maragos, 1986); reef about 200 m off sheltered west side at 20-25 m depth described by Dahl et al. (1974): bottom rocky with huge blocks; coral cover more extensive and diversity higher than Ta'u, with huge colonies of table Acropora and Porites and abundant alcyonarians.

Ta'u (X) 17 sq. mi. (44 sq. km); 3170 ft (966 m) central peak; important reefs at Faleasao, Si'ufaga, Fusi and Saua (Maragos, 1986); reef on north-west coast described by Dahl et al. (1974); other reefs described by Itano (1987); see below for descriptions; fish recorded at various sites listed in Itano (1987).

Rose Atoll (see separate account).

Swains Atoll (X) (see separate account).

(X) = Inhabited

The islands are rich in fringing reefs although these are relatively narrow and lack good near-shore drop-offs. They typically consist of a shallow lagoon or moat (about 2 m deep), a shallower fore-reef, a reef crest (usually emergent at low tide), a surge zone (with spur and groove formation on the south-west windward side) and a sharp reef front dropping 5-10 m to a reef terrace and gradually descending to deep water. Most of the reefs have passes; the maximum width is 500 m and most are much narrower (Dahl, 1970 and 1973).

The reefs of American Samoa are among the best documented in the South Pacific region, those around Tutuila having been studied for nearly 70 years (Dahl, 1985). Lamberts (1983a) describes early scientific expeditions to the islands, the only extensive study of the reefs being the Carnegie Institution programme of 1917 to 1920. Early coral studies include those of Cary (1921 and 1931), Chamberlin (1921), Davis (1921), Helfrich et al. (1975), Hoffmeister (1925) and Maragos (1972). A resource survey of selected sites was carried out in the early 1970s by Randall and Devaney (1974). Monitoring surveys of reefs around Tutuila were made between 1970 and 1980 (Dahl, 1981). In 1979, a complete survey of the reefs was carried out for the U.S. Army Corps of Engineers and a reef inventory prepared for Tutuila, Aunu'u, Ofu, Olosega and Ta'u (AF and AECOS, 1980; Maragos and Elliott, 1985). Reef monitoring stations have been set up by Itano and Buckley of the Office of Marine and Wildlife Resources

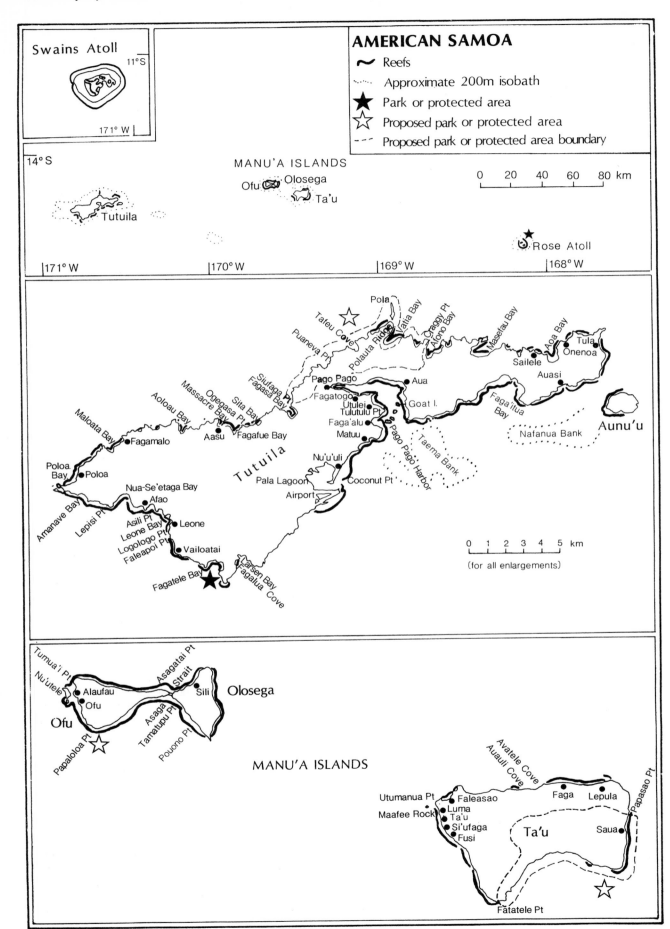

Swains Atoll

11°S

171° W

AMERICAN SAMOA

~ Reefs

⋯⋯ Approximate 200m isobath

★ Park or protected area

☆ Proposed park or protected area

--- Proposed park or protected area boundary

14°S

MANU'A ISLANDS

Ofu Olosega

Ta'u

Tutuila

0 20 40 60 80 km

Rose Atoll

171° W   170° W   169° W   168° W

Pola

Tafeu Cove

Puaneva Pt

Polauta Ridge

Vatia Bay

Craggy Pt

Afono Bay

Masefau Bay

Aoa Bay

Tula

Onenoa

Sailele

Auasi

Siufaga

Fagasa Bay

Sita Bay

Pago Pago

Aua

Fagatogo

Utulei

Goat I.

Faga'itua Bay

Aunu'u

Massacre Bay

Ogegasa Pt

Aoloau Bay

Tulutulu Pt

Faga'alu

Matuu

Nafanua Bank

Maloata Bay

Aasu

Fagafue Bay

Pago Pago Harbor

Taema Bank

Fagamalo

Tutuila

Nu'u'uli

Poloa
Bay

Poloa

Pala Lagoon

Coconut Pt

Nua-Se'etaga Bay

Afao

Airport

Amanave Bay

Lepisi Pt

Asili Pt

Leone Bay

Leone

Logologo Pt

Faleapoi Pt

Vailoatai

0 1 2 3 4 5 km

(for all enlargements)

Fagatele Bay

Larsen Bay

Fagalua Cove

Tumua'i Pt

Asagatai Pt

Asaga Strait

Nu'utele

Alaufau

Ofu

Sili

Olosega

Ofu

Asaga
Tamatupu Pt

Pouono Pt

Papaloloa Pt

MANU'A ISLANDS

Avatele Cove

Auauli Cove

Utumanua Pt

Faleasao

Faga

Lepula

Maafee Rock

Luma

Ta'u

Si'ufaga

Fusi

Ta'u

Saua

Papasao Pt

Fatatele Pt

of the American Samoa Government on all islands except Ta'u. Some of these transects duplicate areas previously studied by Wass (1978) and Dahl *et al.* (1974), others are new and were chosen because they are representative of a particular biotope. On Tutuila, transects have been set up at Whale Rock, Faga'alu, Taema Bank, Nafanua Bank, Faga'itua Bay, Tula, Aoa, Masefau, Puaneva Pt, Fagasa, Sita Bay, Poloa, Afao, Leone Bay and Coconut Pt. Periodic monitoring is carried out at the monitoring stations and transects for reef fish and corals are made. Corals are identified to species where possible and a coral reference collection is maintained and added to (Itano *in litt.*, 29.7.87). Lamberts (1983a) lists 174 scleractinian corals in 48 genera and subgenera for American Samoa. Surveys by the Office of Marine and Wildlife Resources have recorded Blue Coral *Heliopora caerulea* on Ofu, Olosega, Swains Island, and Taema Bank off Tutuila, with highest abundance on Ofu (Itano *in litt.*, 29.7.87); Dahl *et al.* (1974) also recorded it at Leone Bay, Tutuila. Reef algae were studied by Dahl (1972).

Two thirds of the coastline of Tutuila (about 55 km), particularly to the south, is bordered by narrow fringing reefs which are partially exposed at low tide. There are no barrier reefs and only a single well-developed lagoon (Pala Lagoon). Beyond the breakers on the seaward margin of the reef flat, the bottom slopes rapidly to very deep water, except south-west of Aunu'u Island where depths are less because of the drowned barrier reef Nafanua Bank. Because of rapid submergence during the last period of Pleistocene sea-level rise, the limited areas of fringing reefs are discontinuous and consist primarily of bedded calcareous sand and silt rather than coral reef framework (Stearns, 1944).

The most important reefs, identified in the course of compiling the coral reef inventory, are in Pago Pago Harbor, Utulei, Aua, Faga'alu, Tafananai, Alega, Faga'itua, Aoa, Masefau, Afono, Vatia, Fagasa, Massacre, Maloata, Poloa, Amanave, Nua-Se'etaga and Leone, Asili, Pala Lagoon, Matuu and around Aunu'u Island. Structurally the most important reefs occur at Pala Lagoon, Faleapoi Pt (south of Leone), Coconut Pt, outer Pago Pago Harbor and Faga'itua Bay on the south coast and Aoa, Masefau, Fagasa and Poloa Bays on the north coast. Much of the north coast lacks structural reefs altogether. Important submerged structural reefs occur as the two offshore banks, Taema and Nafanua (Maragos, 1986; AF and AECOS, 1980).

The broadest fringing reef extends about 1000 m from the shore in the south mid portion of the island, at Nu'u'uli, near Pago Pago Airport. It protects the shallow estuary, Pala Lagoon, separated from it in part by the sandy peninsula, Coconut Pt. Prior to construction work for the airport, the deeper parts of this area had large thickets of staghorn acroporid corals some of which were over 30 m across and 2 m high (Helfrich *et al.*, 1975).

A fifth of the Tutuila reef front is found in Pago Pago Harbor which was extensively studied between 1917 and 1920 by Mayor (1918, 1924 a,b,c and d) and by Bramlette (1926) who described marine bottom samples. Of the reef transects made by Mayor, only those of Aua and Utulei now cross living reef, the others having disappeared as a result of dredging and filling (Dahl and Lamberts, 1977) and sedimentation (Maragos, 1986). A reduction in total numbers of corals, a change in the

relative proportions of different genera and a probable reduction in the average size of individual colonies was recorded. *Acropora* is still the dominant coral but *Psammocora* abundance has been reduced by two thirds. *Pocillopora* has become more abundant replacing *Porites*.

Brief surveys were made at two sites on Tutuila by Dahl *et al.* (1974), prior to the *Acanthaster* infestation of 1978-79. At Leone Bay on the windward south-west side, the reef was surveyed out to 350 m from the shore near Logologo Pt. At that time the reef structure was irregular, somewhat resembling a spur and groove system, with a shallow reef flat and large reef patches in deeper water extending down to 25 m. Coral cover was very variable, sometimes reaching 100%, but with many heliopores. A large flat of white, completely detritus-free, coarse sand occurred at 15 m depth. The water was clear and warm with abundant fish. This area was apparently not affected by the *Acanthaster* outbreak and in 1987 was reported as still having a dense cover of *Acropora hyacinthus* and *A. irregularis*, with other corals also being abundant and fish diversity high (Itano *in litt.*, 29.7.87). On the north side of Tutuila, at Ogegasa Pt, Dahl *et al.* (1974) described a vertical basaltic rock slope from the surface to 3 m, followed by a rocky flat with small scattered corals extending 50 m off shore to 7 m depth. From here a slope with spur and groove formations dropped to 15 m and then a very steep slope dropped down to 30 m, ending in an extensive sand flat. This slope had extremely high coral cover and species diversity.

Sita Bay on the north coast still has good coral growth after the *Acanthaster* infestation (Itano *in litt.*, 29.7.87; Lamberts *in litt.*, 7.2.85). Fagatele Bay, on the south-west, is described in a separate account.

Taema Bank is a drowned barrier reef, similar to Nafanua Bank, about 7 km off the entrance to Pago Pago Harbor. Both banks are the remains of a barrier reef which enclosed a former lagoon extending from the vicinity of the airport to the channel between Tutuila and Aunu'u Island. Water depth varies from 100 m in the lagoon to 6 m over the banks which are cut by passages. The inner slopes of the banks are reported to be heavily silted and mostly devoid of conspicuous marine life but currents keep the seaward slopes free of silt (Maragos, 1986). The seaward slope and crest of the shallower sections of the banks have high coral abundance (Maragos *in litt.*, 10.7.87). The banks and the area between Tutuila and the Manu'a Group are major feeding grounds for birds and have abundant commercial fish (Maragos, 1986).

Reefs around Ofu and Olosega are mentioned briefly above in the table. On Ta'u, the reef between Faga on the central part of the northern coast and Lepula in the north-east is characterized by spur and groove formation extending gradually offshore for 100 m from the outer reef flat margin before sloping to meet basalt pavement areas at ca 20 m depth. The channels are lined with boulders and sand pockets and generally have no coral cover; coral growth on the higher parts of the basalt spurs is generally sparse but increases to ca 30% below 10 m depth. Corals are generally robust, low growing or encrusting forms capable of withstanding high wave energies, the community being dominated by *Acropora humilis*, *Porites lutea*, encrusting *Millepora* and small colonies of massive faviids. Similar coral communities occur along the western half of the north coast, between Utumanua Pt and Avatele Cove, coral cover increasing off Faleasao Village (Itano, 1987). Buckley (1987) gives a brief description of the reef fronting Faleasao Village. Dahl *et al.* (1974) briefly surveyed a reef on the north-west coast, at a leeward but exposed site about 300 m off shore. They described a rocky flat with huge boulders at 20-25 m depth; *Porites* and *Pocillopora* were the most common corals, but diversity, abundance and colony size were low. Itano (1987) surveyed the northern half of the west coast of Ta'u. Here a broad shallow shelf extends out from the fringing reef at Luma to encompass Maafee Rock and has relatively high live coral cover dominated by *Acropora humilis* and other sturdy coral species; the area off Ta'u village had spur-and-groove formations similar to those between Lepula and Faga.

Leatherback Turtles *Dermochelys coriacea*, Olive Ridleys *Lepidochelys olivacea* and Loggerheads *Caretta caretta* have been recorded in American Samoan waters, and both Green Turtles *Chelonia mydas* and Hawksbills *Eretmochelys imbricata* nest, mainly on Rose Atoll with apparently scattered nesting elsewhere, principally in the Manu'a group, although it is possible that *Eretmochelys* also nests sparsely on Tutuila (Amerson *et al.*, 1982; Anon., 1979; Johannes, 1986).

Reef Resources

Tutuila is the most populated island with 90% of the total population of American Samoa (35 000 in 1980); one third lives around Pago Pago Bay and the remainder live mainly in small villages on the limited flat coastal areas. Much of the inner part of the island is still relatively pristine as a result of the inaccessibility of the central ridge system. The Manu'a group have a considerably lower population density and no industry and are therefore little changed, but the population is again concentrated on the coast in several small villages.

The Samoan people have historically relied on reef and lagoon organisms for a substantial part of their diet. Fishing practices were surveyed by the Office of Marine Resources between 1977 and 1980 on Tutuila and the results are described by Wass (1983). Additional information is provided by Hill (1978). Shoreline recreational fishing is traditionally important and about 300 tonnes of fish a year are taken in this way. The Office of Marine and Wildlife Resources is conducting a bottom fish stock assessment in conjuction with the U.S. National Marine Fisheries Service (Itano, 1987). Both eggs and meat of turtles (mainly *Eretmochelys imbricata*) are eaten, and tortoiseshell is used locally for jewellery and decoration. Johannes (1986) estimated that ca 50 turtles a year were taken on Tutuila and Olosega, although in general there was little interest in catching turtles. In the early 1980s the 50 or so inhabitants of Swains Atoll (*see separate account*) apparently relied to some extent on turtle eggs and meat for subsistence (Balazs, 1982). Tourism and reef-related recreational activities are popular in some areas, particularly at Goat Island Pt and Fagatele Bay (*see separate accounts*).

Disturbances and Deficiencies

The reefs of American Samoa have suffered disturbance from a variety of causes, and a 1979 coral reef inventory showed relatively few areas with more than 50% coral cover. The reefs of Tutuila were subjected to a severe infestation of *Acanthaster planci* in 1978-1979 (Beulig *et al.*, 1981; Birkeland and Randall, 1979). Damage, particularly to *Acropora*, was widespread with reefs at the following sites being severely damaged or wiped out: Taema, Nafanua, Faga'itua Bay, Aunu'u, Onenoa, Aoa, Sailele, Masefau, Afono, Vatia, Tafeu Cove, Fagasa, Fagafue, Aasu, Aoloau, Fagamalo, Maloata, Poloa, Vailoatai, Fagatele, Fagalua (Larsen) Bay and the airport. Recolonization has been slow and new colonies are still small. Reefs of the Manu'a islands were unaffected. Surveys in 1987 revealed fair concentrations of *Acanthaster* on Ofu although these are not considered to represent a major infestation (Itano, 1987 and *in litt.*, 29.7.87).

Silt-laden freshwater from torrential rain often overlays the Pago Pago reefs and in 1924 was responsible for considerable coral death (Mayor, 1924a). In 1966 a hurricane caused terrestrial damage but only minor harm to the reefs (Dahl and Lamberts, 1977). Hurricane Tusi, on 17th January 1987, caused considerable terrestrial damage on the Manu'a islands but did not result in large destructive swells and thus caused relatively little direct damage to the reefs. The vegetation has reportedly recovered well in general, with little excess erosion, although two drainage basins on the north coast of Ta'u (Auauli and Avatele) were gutted by excessive runoff, leading to increased sediment load on adjacent reefs with some evidence of coral smothering (Itano, 1987 and *in litt.*, 29.7.87).

In many reef moats on Tutuila, large thickets of *Acropora formosa*, the lower parts of which are often dead and the upper parts killed in a sharply demarcated line, have been found presumably corresponding to extreme low water tide level (Dahl and Lamberts, 1977).

There has been some coral mortality where it is not known whether the cause was natural or man-induced. On Tutuila, extensive coral death occurred in 1973 on the reefs bounded by Coconut Pt, Pago Pago Airport and out to the reef edge. All the corals of the dominant suborder Astrocoeniina, including *Acropora*, *Montipora* and *Pocillopora*, died to a depth of 6 m within an area of at least 8 ha. Fungiids and faviids remained healthy (Lamberts, 1983b). It was postulated that the mortality was caused by some event connected with the erection of a fish trap in the area, such as the addition of poisons to the water, but subsequent experiments proved nothing conclusively. Lamberts (1983b) studied recolonization rates of this reef area. Most of the species that had died out have re-established themselves, but the natural processes are being modified by the dredging of borrow pits, which may also account for increased beach erosion along Coconut Pt. The *Acropora* thickets previously found in these areas will probably never be fully re-established because of changes in substrate (increased sediment) but it is thought that the reef may eventually recover. Similar areas of coral kill were observed in the north shore bay of Masefau and at the edge of moderately deep water in Faga'itua Bay.

Human activities are having an increasing impact on the reef. Coastal areas, particularly Pago Pago Harbor, Pala Lagoon and the more populated inlets, have been fairly extensively degraded through pollution, coral smothering through sedimentation and siltation, fish dynamiting and fish poisoning. Littoral erosion arising from inadequate protection against wave energy has, in the past, removed considerable portions of the narrow coastal platform. Increased agricultural usage of steep slopes and increasing numbers of dredging and construction projects including road construction have resulted in erosion with deposition of terrigenous silt in sheltered bays and on reef areas adjacent to stream outfalls; significant portions of Pago Pago Bay, Faga'itua Bay, Leone Bay and Pala Lagoon have been silted over. Dredging and blasting activities in at least 20 sites around Tutuila and the Manu'a islands have resulted in direct destruction of the reef; dredging in some cases has been to provide access to isolated villages, but has often been carried out to provide a source of road-bed materials (SPREP, 1980; Swerdloff, 1973). There are also problems with the containment of oil spills and the dislocation of coastal outline due to irregular reclamation practices including land filling with rubbish. In recent years there has been a serious problem of littering and rubbish dumping on reefs and in lagoons adjacent to villages.

On Tutuila, the reefs in Pago Pago Bay have suffered particular damage. Between 1942 and 1945, the military dredged several inshore areas for landfill, there was an increase in harbour traffic and shipping converted from coal to oil. Tuna canneries were established on the north shore of the harbour in 1956 and dredging operations were expanded in 1960. By 1973 the tuna canneries and a marine railway in the harbour serviced an ocean-going fleet of over 250 fishing ships and the port facilities are increasingly visited by tour ships and used as a freight transhipment point. High turbidity and siltation caused the death of most corals in the inner half of the bay as described by Dahl and Lamberts (1977). Approximately 95% of the reefs at the back of the bay fronting the villages of Fagatogo and Pago Pago have been filled (Wass, 1983). Only the shallow reefs near the mouth of the bay, with a relatively high rate of water exchange, have remained viable, and a portion of that area has been destroyed by dredging. Organic pollution is also a problem; untreated sewage, polluted streams and untreated wastes from two fish canneries flow into the bay which has changed from a clear-water coral reef regime to a turbid silty area. The problem has been accentuated by the increased use of agricultural fertilizers. Oil streaks often cover the bay, originating from the bilge residue of commercial vessels (primarily the long line fishing fleet), spillage at the fuel dock and leakage from deteriorating underground fuel oil pipe lines (Swerdloff, 1973). Maragos (1972) monitored the growth and survival of transplanted corals at several sites in the harbour and reported greater mortality and least growth at the innermost site, Goat Island Reef (*see separate account*), concluding that poorer water quality inside the bay was the cause. Dahl and Lamberts (1977) reported occasional small oil spills and the drainage of urban and industrial wastes into the harbour. Treatment projects are, however, under way (see below) and Dahl and Lamberts (1977) suggested that Aua Reef is gradually recovering from earlier stress.

Significant damage was caused to reefs adjacent to Pala Lagoon by the extension of the commercial airfield out into the lagoon. An assessment of the expected impact of dredging in Pala Lagoon for a proposed boat basin was carried out by Helfrich *et al.* (1975). Devaney and Suzumoto (1977) carried out a survey of Auasi Harbor in the context of a harbour development project; some reef

damage may have occurred. The reefs at Leone Bay are potentially threatened by a planned boat harbour (Itano *in litt.*, 29.7.87).

The Manu'a Islands, although free from extensive construction and agricultural usage, have suffered some reef destruction. Channels have been dredged at Ofu and Ta'u for small boat harbours and blasting of the Asaga Strait between Ofu and Olosega as part of a bridge and road project has caused some damage to reefs (Maragos, 1986). A large land slide caused by road construction on Olosega had covered some 5 sq. km of reef flat by July 1987 and was expected to cause more damage as construction and associated erosion continued (Buckley *in litt.*, 4.8.87). Small boat harbour dredging and construction on Aunu'u, Ofu, and Fusi (Ta'u) has probably caused some reef damage but also created coral and fish habitat (Itano *in litt.*, 29.7.87).

The recent rapid population growth has put considerable pressure on reef resources and traditional conservation practices have largely been discarded. The dependancy of American Samoans on their marine resources is decreasing, due to the availability of canned and frozen foods, but seafood consumption remains high. Increased mobility has led to a tendency to fish reefs of neighbouring villages. Reef fishermen consider that yields are less than they were previously, but it is believed that while fish varieties have been reduced, there has probably not been a reduction in biomass. Fish poisoning and dynamiting have caused a reduction in the variety of fish stocks. Dynamiting has caused considerable damage to reefs around Tutuila and in the Manu'a group (Swerdloff, 1973; Thomas *in litt.*, 9.7.87); it is less widespread than previously, but in the past three years has been recorded several times in Fagatele and Fagasa Bays and other areas around Tutuila (Thomas *in litt.*, 9.7.87). Giant Clams (*Tridacna* spp) have effectively been fished out in Samoan waters, other than at Rose Atoll (Maragos *in litt.*, 10.7.87). Turtle numbers around Tutuila are said to have declined considerably in the five years up to 1981 (Johannes, 1986).

Legislation and Management

A general overview of conservation activities is given in Eaton (1985) and OTA (1987) provides an analysis of coastal resource development and management.

Ownership of the reefs and their resources was traditionally vested in the chiefs of each village and a complex system of taboos, restricting efforts to certain seasons and locations arose, which served to protect the reefs from over-exploitation. These rights have been largely abandoned but some elements remain. At present, village councils occasionally limit fishing on the reefs fronting the village through temporary bans on fishing or by prohibiting fishermen from other villages (Johannes, in press; Wass, 1983). It is still customary for outsiders to request permission to fish these reefs (Maragos *in litt.*, 10.7.87). Several villages do not allow fishing on Sundays and most prohibit the use of dynamite or bleach (Johannes, in press; Wass, 1983).

Under Federal Public Law 93-435, the American Samoan Government owns all submerged lands from the mean highwater mark out to the limit of the territorial sea. Executive Order 3-80, which established the American

Samoa Coastal Management Program (ASCMP) in 1980, contains 16 policies which govern the use of the coastal zone, including reef protection, marine resource protection, protection of unique areas, improvement of recreational opportunities and control of shoreline development (USDC, 1980). A detailed description of legislation relevant to reefs is given in USDC (1984). Many of the U.S. Federal laws and regulations apply, including the Marine Protection, Research and Sanctuaries Act (1972), the Marine Mammal Protection Act (1972), the Fishery Conservation and Management Act (1976), the Endangered Species Act (1973), the Coastal Zone Management Act (1972), the River and Harbor Act (1899), the Clean Water Act (1977) and the National Environmental Policy Act.

Pollution control is now greatly improved (Maragos, 1986). The Environmental Protection Agency is represented and is active in controlling effluent from the tuna canneries; stricter controls affecting the dumping of nitrogen- and phosphate-rich wastes into Pago Pago Bay have been imposed and will have to be complied with by 1991. A sewerage project is in progress for Pala Lagoon which should improve water quality in the estuary. The Office of Coastal Zone Management actively enforces U.S. Coast Guard oil pollution regulations and funds a project responsible for oil pollution control and cleaning up debris and oil spill in Pago Pago Harbor. It runs an Island Wide Metal Cleanup programme and a Marine Awareness programme for students which stresses the importance of the sea to Samoa, and the damage done by pollution, siltation and other activities. The office is also active in trying to control illegal landfills, mangrove cutting and beach sand mining, although with limited success, and has also contracted teachers to develop and adapt the school science curriculum to make it more relevant to the country (Itano *in litt.*, 29.7.87). Seawalls are being rebuilt along eroded sections of coast (SPREP, 1980). The Office of Marine and Wildlife Resources (OMWR) is responsible for fisheries development and wildlife management in cooperation with U.S. Federal Resource Agencies and American Samoa Agencies (Buckley *in litt.*, 30.9.87).

At present Public Law 16-58 prohibits the use of poison in territorial waters and provides punishment by fines and/or imprisonment; territorial law also prohibits the use of dynamite to harvest fish and other marine resources. Attempts are being made to control the use of dynamite and bleach with a public awareness campaign using newspapers, television and lectures (Itano *in litt.*, 29.7.87).

Executive Order 3-80 (see above) provides for the establishment of Special Management Areas. Two protected areas, Rose Atoll National Wildlife Refuge and Fagatele Bay National Marine Sanctuary, include reefs and are described in separate accounts.

A small Tridacnid clam nursery has been started in Faga'itua Bay on Tutuila by the village of Alofau in cooperation with the Office of Marine and Wildlife Resources, with seed clams obtained from the Micronesian Mariculture Demonstration Center in Palau. The village leaders have closed the immediate reef area around the clams to swimming or fishing. The long term aim is to increase Giant Clam populations on Tutuila's reefs (Anon., 1987b; Buckley *in litt.*, 4.8.87).

Recommendations

A number of reef areas have been recommended for protection.

1. A National Park has been proposed for the northern part of Tutuila from Siufaga Pt east to Craggy Pt, apparently excluding Vatia Bay; restrictions on the use of reefs and marine resources have not been defined in the proposals (Anon., 1988). As of March 1988 the proposals had not been endorsed by the U.S. National Park Service (Harry *in litt.*, 29.2.88). This would include the area of Pola Islet - Pola'uta Ridge, recommended as a protected area for breeding birds and as a marine reserve by Amerson *et al.* (1982) and (Dahl, 1980).

2. Goat Island Reef - Utulei Reef on Tutuila (*see separate account*).

3. Coastal and reef reserves have been recommended at Lepisi Pt and Ogegasa Pt on Tutuila by Dahl (1980).

4. Pago Pago Bay and Pala Lagoon on Tutuila have been identified as areas of particular concern and importance (Maragos, 1986); recommendations for the management of Pala Lagoon are given in Yamasaki *et al.* (1985) and there are plans to rehabilitate both areas.

5. A National Park has been proposed for the area covering much of the southern part of Ta'u, from Fatatele Pt east to Papasao Pt; it would include waters and reefs up to 0.25 mi. (0.4 km) from shore, although restrictions on the use of reefs and marine resources have not been defined in the proposals (Anon., 1988). As of March 1988 the proposals had not been endorsed by the U.S. National Park Service (Harry *in litt.*, 29.2.88).

4. Papaloloa Pt proposed national marine sanctuary on Ofu (*see separate account*).

5. Protection of Nu'utele islet, off Ofu, for its seabird colony has been recommended (Dahl, 1980).

6. Swains Atoll (*see separate account*).

The American Samoan Natural Resources Commission made up of "fono" (the bicameral legislative body of the Territory of American Samoa) members and resource management agency representatives is drawing up a list of endangered species for the Territory (Buckley *in litt.*, 30.9.87; Thomas *in litt.*, 9.7.87). It is hoped that fisheries regulations drafted by the Office of Marine and Wildlife Resources (OMWR) will be implemented in 1988, when OMWR enforcement officers should also be appointed (Itano *in litt.*, 10.7.87). Wass (1983) recommended that village councils should be encouraged to take a more active role in future management schemes and suggested that a management plan for the island of Tutuila as a whole should be formulated. The effects of uncontrolled destruction on the outer edges of the 223 acres (90 ha) of mangrove and wetland areas on the coast need assessment.

It has been suggested that aid for fisheries in the Manu'a group should concentrate on controlling erosion, developing the Fish Aggregating Devices, developing support facilities for "alias" (simple aluminium-hulled catamaran styled fishing boats), and reef enhancement projects (Itano, 1987).

Itano (1987) recommends further surveys on the reefs of Ofu, with particular emphasis on *Acanthaster* and the potential need to implement controlling measures.

References

* = cited but not consulted

*AF and AECOS (Aquatic Farms, Inc. and AECOS, Inc.) (1980). American Samoa Coral Reef Inventory. Part A. Text. Part B. Atlas. Prep. for U.S. Army Corps of Engineers and American Samoa Development Planning Office, Honolulu. 314 pp.

Amerson Jr, A.B., Whistler, W.A. and Schwaner, T.D. (1982). Wildlife and Wildlife Habitat of American Samoa. 1. Environment and Ecology. Fish and Wildlife Service, U.S. Department of the Interior, Washington D.C.

Anon. (1979). Country Statement: American Samoa. Paper presented at Joint SPC-NMFS Workshop on Marine Turtles in the Tropical Pacific Islands. Noumea, New Caledonia, 11-14 December 1979. SPC-NMFS/Turtles/WP.15.

Anon. (1985). Summary record of proceedings of ministerial and technical sessions. *Report of the 3rd South Pacific National Parks and Reserves Conference, Apia* 1. 95 pp.

Anon. (1987a). Memorandum re Rose Atoll, from Fishery Biologist, Office of Marine and Wildlife Resources, American Samoa Govenment to Refuge Manager, Fish and Wildlife Resources, Honolulu, Hawaii.

Anon. (1987b). Micronesian Mariculture Demonstration Center, Palau. *Clamlines* 3: 2-3.

Anon. (1988). National park feasibility study, American Samoa. Draft. Prepared by the National Park Service and American Samoa Government. 139 pp.

Anon. (n.d. a). Potential Marine Sanctuary Site. Manuscript.

*Anon. (n.d. b). Rose Atoll National Wildlife Refuge, American Samoa. Brochure, USDI, FWS.

Balazs, G.H. (1982). Sea turtles and their traditional usage in Tokelau. Unpublished project report prepared for WWF-US and Office of Tokelau Affairs.

Beulig, A., Beach, D. and Martindale, M.Q. (1981). Influence on daily movements of *Acanthaster planci* during a population explosion on American Samoa. *Proc. 4th Int. Coral Reef Symp., Manila* 2: 755 (Abst.).

Birkeland, C. and Randall, R.H. (1979). Report on the *Acanthaster planci* (Alamea) studies on Tutuila, American Samoa. Prep. by Univ. Guam Mar. Lab. for Office of Marine Resources, Government of American Samoa.

Birkeland, C., Randall, R.H., Wass, R.C., Smith, B.D. and Wilkins, S. (in press). Biological resource assessment of the Fagatele Bay National Marine Sanctuary. Report prepared for the Development Planning and Tourism Office, Government of American Samoa and the Sanctuary Program Division, National Oceanic and Administration, U.S. Department of Commerce.

*Bramlette, M.N. (1926). Some marine bottom samples from Pago Pago Harbour, Samoa. *Carnegie Inst. Washington Rept* 23.

*Bryan, E. (1974). Swains Island. Pacific Information Center, B.P. Bishop Museum. Unpub. rept.

Buckley, R. (1986). Internal memorandum on coral transects at Faleasao Village, Ta'u Island (3.11.86). Office of Marine and Wildlife Resources, American Samoa Government.

*Cary, L.R. (1921). Studies of Alcyonaria and of boring through reefs of Samoa. *Carnegie Inst. Washington Yearbook* 19: 193-194.

*Cary, L.R. (1931). Studies on the coral reefs of Tutuila, American Samoa with special reference to the Alcyonaria. *Carnegie Inst. Wash. Publ. 413, Pap. Tortugas Lab.* 27: 53-98.

Chamberlin, R.T. (1921). The geological interpretation of the coral reefs of Tutuila, Samoa. *Carnegie Inst. Washington Yearbook* 19: 194-195.

*Clapp, R.B. (1968). The birds of Swain's Island, south central Pacific. *Notornis* 15(3): 198-206.

*Dahl, A.L. (1970). Ecological report on Tutuila, American Samoa, Washington D.C. 13 pp.

*Dahl, A.L. (1972). Ecology and community structure of some tropical reef algae in Samoa. *Proc. Int. Seaweed Symp.* 7: 36-39.

*Dahl, A.L. (1973). Ecological report on American Samoa. Washington D.C. 18 pp.

Dahl, A.L. (1980). Regional ecosystems survey of the South Pacific Area. *SPC/IUCN Technical Paper* 179. South Pacific Commission, Noumea, New Caledonia.

Dahl, A.L. (1981). Monitoring coral reefs for urban impact. *Bull. Mar. Sci.* 31(3): 544-557.

Dahl, A.L. (1985). Status and conservation of South Pacific coral reefs. *Proc. 5th Int. Coral Reef Cong., Tahiti* 6: 509-513.

Dahl, A.L. and Lamberts, A.E. (1977). Environmental impact on a Samoan coral reef: A resurvey of Mayor's 1917 transect. *Pacific Science* 31(3): 309-319.

Dahl, A.L., Macintyre, I.G. and Antonius, A. (1974). A comparative survey of coral reef research sites. In: Sachet, M.-H. and Dahl, A.L., Comparative Investigations of Tropical Reef Ecosystems: Background for an integrated coral reef program. *Atoll Res. Bull.* 172: 37-77.

*Davis, W.M. (1921). The coral reefs of Tutuila, Samoa. *Science* 53: 559-565.

Devaney, D.M. and Suzumoto, A.Y. (1977). Marine ecology reconnaissance survey: Auasi Harbor project, Auasi, Tutuila Island, American Samoa. U.S. Army Corps of Engineers, Pacific Ocean Division. 124 pp.

Eaton, P. (1985). Land Tenure and Conservation: Protected areas in the South Pacific. *SPREP Topic Review* 17. South Pacific Commission, Noumea, New Caledonia. 103 pp.

*Helfrich, P., Ball, J.L., Bienfang, P., Foster, M., Gallagher, B., Guinther, E., Krasnick, G. and Maragos, J.E. (1975). An assessment of the expected impact of a dredging project for Pala Lagoon, American Samoa. Univ. of Hawaii Sea Grant Technical Report UNIHI-SEAGRANT-TR-76-02. 76 pp.

Hill, H.B. (1978). The use of nearshore marine life as a food source by American Samoans. *Pacific Islands Program, University of Hawaii, Misc. Work Papers* 1. 170 pp.

*Hoffmeister, J.E. (1925). Some corals from American Samoa and the Fiji Islands. *Carnegie Inst. Wash. Publ. 343, Pap. Dep. Mar. Biol.* 22: 1-90.

Itano, D. (1987). Internal memoranda to Director, coral reef assessments on Ofu Reef (10.2.87), Rose Atoll (26.2.87), Swains Island (21.4.87), Ta'u Island Reefs

(20.5.87). Office of Marine and Wildlife Resources, American Samoa Government.

Johannes, R.E. (1986). A review of information on the subsistence use of Green and Hawksbill Sea Turtles on Islands under United States Jurisdiction in the Western Pacific Ocean. National Marine Fisheries Service, Southwest Region/NOAA.

Johannes, R.E. (in press). The role of Marine Resource Tenure Systems (TURFs) in sustainable nearshore marine resource development and management in U.S.-affiliated tropical Pacific islands. In: Smith, B.D. (Ed.), Topic Reviews in Insular Resource Development and Management in the Pacific U.S.-affiliated Islands. *Univ. Guam Mar. Lab. Tech. Rept* 88.

Lamberts, A.E. (1983a). An annotated checklist of the corals of American Samoa. *Atoll Res. Bull.* 264. 15 pp.

Lamberts, A.E. (1983b). Coral kill and recolonization in American Samoa. *National Geographic Society Research Projects* 15: 359-377.

*Maragos, J.E. (1972). A study of the ecology of Hawaiian reef corals. Ph.D. dissertation, University of Hawaii, Honolulu. 292 pp.

Maragos, J.E. (1986). Coastal resource development and management in the U.S. Pacific Islands: 1. Island-by-island analysis. Office of Technology Assessment, U.S. Congress. Draft.

Maragos, J.E. and Elliott, M.E. (1985). Coastal resource inventories in Hawaii, Samoa and Micronesia. *Proc. 5th Int. Coral Reef Cong., Tahiti* 5: 577-582.

*Mayor, A.G. (1918). The growth rate of Samoan coral reefs. *Nat. Acad. Sci. Proc.* 4: 390-393.

*Mayor, A.G. (1924a). Structure and ecology of Samoan reefs. *Carnegie Inst. Wash. Publ. 340, Pap. Dep. Mar. Biol.* 19: 1-25.

*Mayor, A.G. (1924b). Causes which produce stable conditions in the depth of the floors of the Pacific fringing reef-flats. *Carnegie Inst. Wash. Publ. 340, Pap. Dep. Mar. Biol.* 19: 27-36.

*Mayor, A.G. (1924c). Inability of stream-water to dissolve submarine limestones. *Carnegie Inst. Wash. Publ. 340, Pap. Dep. Mar. Biol.* 19: 37-49.

*Mayor, A.G. (1924d). Growth rate of Samoan corals. *Carnegie Inst. Wash. Publ. 340, Pap. Dep. Mar. Biol.* 19: 51-72.

*Mayor, A.G. (1924e). Rose Atoll, American Samoa. Posthumous Memorial Volume, Alfred Goldsborough Mayor. *Carnegie Inst. Wash. Publ. 340, Pap. Dep. Mar. Biol.* 19: 73-79.

NOAA (1983). Announcement of national marine sanctuary program final site evaluation list. *Federal Register* 48(151): 35568-35577.

*OTA (Office of Technology Assessment) (1987). *Integrated Renewable Resource Management for U.S. Insular Areas.* OTA-F-325. U.S. Government Printing Office, Washington D.C. 443 pp.

*Radtke, R. (1985). Population dynamics of the Giant Clam *Tridacna maxima* at Rose Atoll. Hawaii Institute of Biology, Univ. Hawaii.

*Randall, J.E. and Devaney, D.M. (1974). Marine biological survey and resource inventory of selected sites at American Samoa. Prep. for U.S. Army Corps of Engineers, Honolulu District.

*Sachet, M.-H. (1954). A summary of information on Rose Atoll. *Atoll Res. Bull.* 29. 25 pp.

*Setchell, W.A. (1924). American Samoa: Part 1. Vegetation of Tutuila Island. Part 2. Ethnobotany of the Samoans. Part 3. Vegetation of Rose Atoll. *Carnegie Inst. Wash. Publ. 413, Pap. Tortugas Lab.* 20: 1-275.

SPREP (1980). American Samoa. *Country Report* 1. South Pacific Commission, Noumea, New Caledonia.

*Stearns, H.T. (1944). Geology of the Samoan Islands. *Bull. Geol. Soc. Am.* 55: 1279-1332.

Swerdloff, S.N. (1973). The status of marine conservation in American Samoa - 1971. Paper 4. *Proceeding and Papers, Regional Symposium on Conservation of Nature - Reefs and Lagoons.* South Pacific Commission, Noumea, New Caledonia.

Thomas, W. (1985). Fagatele Bay National Marine Sanctuary. *Report of the 3rd South Pacific National Parks and Reserves Conference, Apia* 2: 79-87.

USDC (1980). American Samoa Coastal Management Program and Final Environmental Impact Statement. U.S. Dept of Commerce, NOAA, Office of Coastal Zone Management.

USDC (1984). Final Environmental Impact Statement and Management Plan for the Proposed Fagatele Bay National Marine Sanctuary. Sanctuary Programs Division, U.S. Dept of Commerce, NOAA and Development Planning Office, Pago Pago.

Wass, R.C. (1978). Fagatele Bay reef front and flat: list of species recorded along reef front on September 25, 1978 and along reef flats on February 15, 1978. Office of Marine Resources, American Samoa Government. Unpub.

Wass, R.C. (1983). The shoreline fishery of American Samoa - past and present. In: *Marine and Coastal Processes in the Pacific: Ecological aspects of coastal Zone Management.* Unesco Seminar, Motupore Island Research Centre, University of PNG, 14-17 July 1980, Unesco, Jakarta, Indonesia.

Wass, R.C. (1987). Rose Atoll National Wildlife Refuge Public Use Policy. USDI Fish and Wildlife Service Circular.

Wass, R.C. (n.d. a). *The Fishes of Rose Atoll.* Unpub. rept.

Wass, R.C. (n.d. b). *The Fishes of Rose Atoll.* Supplement 1. Unpub. rept.

Wass, R.C. (n.d. c). *The Tridacna clams of Rose Atoll.* Unpub. rept.

Whistler, W.A. (1983). The flora and vegetation of Swains Island. *Atoll Res. Bull.* 262. 20 pp.

Yamasaki, G., Itano, D. and Davis, R. (1985). A study of and recommendations for the management of the mangrove and lagoon areas of Nu'u'uli and Tafuna, American Samoa. Final rept. 99 pp.

FAGATELE BAY NATIONAL MARINE SANCTUARY

Geographical Location 12 km south-west of Pago Pago Harbour on the southernmost point of Tutuila, including the whole bay up to mean high high water; 14°23'S, 170°46'W.

Area, Depth, Altitude 66 ha; 26-36 m depth to 122 m altitude.

Land Tenure American Samoan Government.

Physical Features Fagatele Bay is formed by a collapsed volcanic crater and is surrounded by steep cliffs and volcanic rocks. Seumalo Ridge rises over 400 ft (122 m) along the western and northern sides, and the eastern side is bounded by Matautuloa Ridge, over 200 ft (61 m) high. Soils on the steep slopes surrounding the bay are silty clay loams. The beaches are composed primarily of calcareous sand with a small amount of volcanic sand. The sand deposits extend off shore for about 20-30 ft (6-9 m) until they merge with the reef platforms which are composed primarily of consolidated limestone and encrusting algae.

The platforms, approximately 200 ft (61 m) or less wide and 2 ft (0.6 m) deep, have bottom reliefs of 1 ft (0.3 m). They fringe the shore of the bay, the widest platform occurring on the eastern side. The reef front drops almost vertically to 5-10 ft (1.5-3.0 m), then gradually slopes seaward to 15-20 ft (4.5-6.0 m). The reef front slope may extend up to 300 ft (91 m) off shore, and contains widely separated pinnacles rising from depths of 15-20 ft (4.5-6.0 m) to within 4-5 ft (1.2-1.5 m) of the surface. The bay bottom reaches a depth of 120 ft (36 m) approximately 1100 ft (335 m) due west of the pocket beach and is covered with rubble.

Wave action (normally from the east) is damped by the encircling reef platform and because the bay faces south-west. Tides are probably similar to Pago Pago Bay where they are diurnal with mean and spring tidal ranges of 2.5 ft (0.76 m) and 3.1 ft (0.95 m) respectively. Water temperatures is 80-82°F (26.7-27.8°C) with little seasonal or diurnal change; salinity ranges from 35.5 ppt to 36.0 ppt. Visibility is normally at least 50 ft (15 m) (USDC, 1984).

Reef Structure and Corals   A fairly well-developed fringing reef flat exists within the protected portion of the bay, the submerged part of which varies in depth from 0.5 to 2 m. There is a sparse covering of corals *Pavona*, *Porites*, *Acropora*, *Pocillopora* and *Millepora*. Along the eastern edge of the bay, 10% of the reef flat at a depth of about 2 ft (0.6 m) is covered by coral and a further 5% has dead coral heads. The most conspicuous corals are *Pocillopora verrucosa*, *Favia* sp., *Galaxea* sp., *Goniastrea* sp., *A. humilis*, *Porites lutea* and the soft coral *Palythoa* sp. Other species have been recorded in Leone Bay, just west of Fagatele Bay, and may be present in the latter.

The reef terrace varies from 2 to 10 m depth with strong currents and surge in some areas. The substrate is basalt in exposed areas and calcium carbonate in sheltered areas. Prior to *Acanthaster* infestation, there was 30-100% coral cover of *Acropora, Porites, Montipora, Pocillopora* and others. The reef front borders the seaward edge of the reef terrace and consists of a portion of the fore-reef, 5-40 m deep, that slopes steeply to deep water. Prior to *Acanthaster* infestation, the upper portions supported the most luxuriant and diverse coral assemblages in the bay. Nearly vertical basalt cliffs and faces extend from the surface to as deep as 80 m along the exposed outer portions of the bay which are characterized by strong currents and surge in the upper portions. Scattered corals are found on these walls including large fan corals at depths below 40 m.

About 172 coral species have been recorded from the area (USDC, 1984).

Noteworthy Fauna and Flora   The steep cliffs have typical coastal and littoral vegetation. There is an abundant avifauna and the fruit bats *Pteropus samoensis* and *P. tonganus* roost in the area (Amerson *et al.*, 1982;

Dahl, 1980; Itano *in litt.*, 29.7.87). The Humpback Whale *Megaptera novaeangliae* is found in the bay and adjacent waters from July to October, the breeding and calving season. Sperm Whales *Physeter catodon* are occasionally sighted off shore. Other cetaceans, including the Pacific Bottlenose *Tursiops truncatus* and Spinner *Stenella* sp. Dolphins also use the bay and adjacent waters (USDC, 1984). Hawksbills *Eretmochelys imbricata* and Green Turtles *Chelonia mydas* are found frequently, and Leatherbacks *Dermochelys coriacea*, Olive Ridleys *Lepidochelys olivacea* and the Loggerheads *Caretta caretta* have been recorded (USDC, 1984).

The bay's configuration provides a protected habitat for an abundant fish fauna. Surveys of fish located on the reef flat and reef front indicate a high diversity and over 80 species of fish have been recorded. 114 species have been recorded from the waters off the south-eastern tip of the bay (Wass, 1978; USDC, 1984).

Scientific Importance and Research   A brief survey of the flora and fauna was carried out in 1979 and more detailed studies of the fish fauna have been made but these have not been published. A further survey to gather baseline population data, by the University of Guam under contract to the National Oceanic and Atmospheric Administration and the Development Planning and Tourism Office, Government of American Samoa, was completed in April 1985 (Birkeland *et al.*, in press), and the reefs will be re-surveyed in 1988 (Thomas *in litt.*, 9.7.87; Itano *in litt.*, 29.7.87). Reference collections of the fish, corals, algae and invertebrates are housed at the Office of Marine and Wildlife Resources, Pago Pago, and at the University of Guam (Itano *in litt.*, 29.7.87). Research on coral recolonization and changes in the composition and structure of inshore fish communities is under way. The pristine nature of the bay provides ideal conditions for the study of coral regeneration following *Acanthaster* predation. Future research requirements are described in USDC (1984), and range from broad surveys to monitoring of *Acanthaster*.

Economic Value and Social Benefits   The steep cliffs surrounding the bay make it relatively inaccessible from land, although the beaches can be reached via a foot trail. The area is located near the village of Leone (1700 inhabitants), and there are three smaller villages in the immediate vicinity. The sanctuary has a number of invertebrates which serve as important subsistence food sources, including sea anemones, lobsters, limpets, clams, octopuses and sea urchins. Subsistence fishing and recreational activities are both important in the area (USDC, 1984).

Disturbance or Deficiencies   In late 1978 and early 1979, hard corals were heavily predated by *Acanthaster*, 90% of the reef being destroyed; recent surveys indicate that coral cover is regenerating (Thomas, 1985).

Human impact has been minimal, and the bay is considered one of the least disturbed areas on Tutuila, although dynamiting has occurred during the last three years (Thomas *in litt.*, 9.7.87). Water quality is considered to be very high as there is no urban or industrial run-off, agricultural activities on the surrounding ridges are limited and there are no permanent streams discharging into the bay. Increased visitor use of the area could affect the reefs, although the regulations should limit this (USDC, 1984).

Legal Protection Fagatele Bay was declared a marine park in October 1982 by the governor of American Samoa. This gave the Department of Parks and Recreation of the American Samoan Government the authority to enforce and promulgate laws to protect the resources to the 10 fathom (18.3 m) line as well as collect any fees. Protection was enhanced by the designation of the area as a National Marine Sanctuary on 17.4.85 under Title 3 of the Marine Protection, Research and Sanctuaries Act, 1972 (Thomas, 1985 and *in litt.*, 9.7.87) which finally came into force in July 1987 (Buckley *in litt.*, 30.9.87). The sanctuary regulations prohibit activities, such as dredging and discharge of pollutants, which would threaten the bay's resources. Traditional uses such as subsistence fishing and recreation are permitted. In the past, local customary rules restricted fishing by outsiders, especially for commercial purposes, and also reinforced government bans on the use of dynamite and chemical poisons (Eaton, 1985).

Management The Sanctuary Programs Division, National Oceanic and Atmospheric Administration (NOAA), of the U.S. Federal Government is responsible for overall administration and programme implementation within the sanctuary. A management plan has been drawn up for the sanctuary (USDC, 1984) and a manager is to be appointed in the 1988 fiscal year (which begins on 1st October 1987), funding not having been available for this so far. The American Samoan Office of Economic Development Planning is the on-site agency which will take the lead in coordinating with the manager for the day-to-day management of the sanctuary. It will coordinate its Coastal Zone Management Program activities with the Department of Parks and Recreation where appropriate (Thomas 1985 and *in litt.*, 9.7.87). The Department of Parks and Recreation is responsible for providing a park ranger to assist the future manager with enforcement; a local ranger has been hired to police the area and is receiving training in Hawaii (Itano *in litt.*, 29.7.87). The area is divided into two zones: an outer zone in which subsistence fishing is permitted, and an inner core zone which is a strict reserve. It is anticipated that most visitors will come by boat from Leone which would give the park authority greater control (Eaton, 1985).

Recommendations Recommendations are provided in the management plan and include a detailed research programme, provision of mooring buoys, improved access, and the development of an interpretive programme and centre. It was originally recommended that Fagalua (Larsen's) Bay be included in the sanctuary, but after extensive review, both the Territorial and Federal governments decided against it (Thomas *in litt.*, 9.7.87).

GOAT ISLAND POINT - UTULEI REEF

Geographical Location Along the western side of the outer portion of Pago Pago Bay, Tutuila, from eastern edge of oil dock at Fagatogo, around Goat Island Point (Rainmaker Hotel), past Utulei Village to Tulutulu (Blunt's) Point; 170°41'W, 14°17'S.

Area, Depth, Altitude 1800 m of shoreline; 0-50 m depth to base of reef slope.

Physical Features The site originally proposed as a marine sanctuary encompasses the intertidal area, reef flat and associated dredged area (10-250 m, wide) and the steeply sloping reef front to a depth of 50 m (Anon., n.d. a).

Reef Structure and Corals The reefs fronting Goat Island Point and Utulei Village are fairly typical of semi-protected reefs in the region and have a fairly diverse fauna. An extensive dredged area along a portion of the shoreline provides habitat for organisms preferring a silty sand bottom (Anon., n.d. a).

Noteworthy Fauna and Flora No information.

Scientific Importance and Research The area was originally studied between 1917 and 1920 by Mayor (1918, 1924a, b, c and d) and subsequently by Maragos (1972) who monitored the growth and survival of transplanted corals here. It is the innermost remaining structural reef (not yet dredged or filled) on the southern side of Pago Pago Harbour (Maragos *in litt.*, 10.7.87).

Economic Value and Social Benefits The area is important for recreational and subsistence fisheries, particularly for villagers from Utulei. Annual catches are estimated at 4524 kg fish and 3271 kg invertebrates (Wass, 1983). It is the most popular area for recreational diving, snorkelling, boating and water sports. The only two public beach parks on Tutuila and the Pago Pago Yacht Club lie within the proposed site, and the major hotel on the island, the Rainmaker, is adjacent. There are plans for additional watersport activities in the area (Anon., n.d. a).

Disturbance or Deficiencies A variety of pollutants including organic and chemical wastes from fish canneries, oil from vessel traffic and fuelling operations, silt from construction projects and increased run-off, and sewage have adversely affected the habitat over the years. A sewage outfall is located at the southern boundary of the site off Utulei (Anon., n.d. a). Species diversity has decreased, hard coral coverage has been reduced, and turbidity has increased. Maragos (1972) found that coral growth and survival in 1970-71 was suboptimal probably due to sedimentation and other water pollution. The fuelling of large vessels and the offloading of petroleum products on the north-west boundary of the proposed site pose a serious threat should a major oil spill occur.

Legal Protection None at present.

Management None.

Recommendations The area was originally considered by the National Oceanic and Atmospheric Administration as a potential marine sanctuary; it was not, however, included on the Final Site Evaluation List and thus remains unprotected (Thomas *in litt.*, 9.7.87). Management measures included in the original proposal included protection of corals, fish and shells on the reefs surrounding the hotel for the benefit of tourists and residents. The residents of Utulei were to be allowed to continue subsistence fishing (Anon., n.d. a).

PAPALOLOA POINT

Geographical Location   Ofu Island; southernmost tip, along the edge of the runway at Ofu airport, eastward to Asagatai Point; 169°40'W, 14°11'S.

Area, Depth, Altitude   0.75 sq. mi. (1.92 sq. km) with 3 miles (4.8 km) of shoreline; depth to 45 m.

Physical Features   Includes shoreline and adjacent fringing reef, 150-180 m wide. Visibility is good in the nearshore depression (see below) but is reduced near the Point (AF and AECOS, 1980).

Reef Structure and Corals   Most of the fringing reef is shallow (0.6-1.0 m at high tide) and consists of rubble and consolidated limestone. The inner reef flat 275 m north-west of Papaloloa Point is relatively barren with low coral cover. A depression lies just off shore here, about 1.5-2.5 m deep, which has numerous large microatolls of *Heliopora coerulea*, 3-5 m in diameter. Other abundant corals are *Porites lutea*, *Millepora* spp., especially *M. tortuosa*, and large patches of *Montipora* spp. Thick branched *Acropora intermedia* grows in thickets in the deeper, more seaward part of the depression. It has been suggested that the microatolls may be the result of regrowth of corals following dredging for fill material for Ofu airport. Small amounts of coral grow on the relatively flat outer reef platform (AF and AECOS, 1980).

Noteworthy Fauna and Flora   The reef fronting the eastern end of the airport has a very diverse fish fauna, over 70 species having been described. Giant Clams *Tridacna* spp. are present, and the Hawksbill Turtle *Eretmochelys imbricata* has been recorded (Itano *in litt.*, 29.7.87). The area is adjacent to Vaoto Marsh.

Scientific Importance and Research   The reefs at Papaloloa Point were surveyed by the Office of Marine and Wildlife Resources in 1986 and 1987 (Itano *in litt.*, 29.7.87). They are an excellent example of a fringing reef community with a diverse and abundant fauna. Itano (*in litt.*, 29.7.87) considers them to be unique in the country for their beauty and coral and fish diversity; Blue Coral *Heliopora coerulea* is present and is more abundant around Ofu than at any of the other sites at which it is known to occur in American Samoa.

Economic Value and Social Benefits   The area is important for recreational and subsistence fishing. Diving, snorkelling and swimming are popular activities and there is considerable potential for increased recreational use. A small hotel (6-8 beds capacity) has been built near Ofu airport which encourages visitors to enjoy the reef through non-destructive activities such as snorkelling and photography; it is mainly used by contract workers on construction projects (Itano *in litt.*, 29.7.87; Thomas *in litt.*, 9.7.87).

Disturbance or Deficiencies   The area was not affected by hurricane damage from Hurricane Tusi, which struck American Samoa on 17th Jan. 1987 (Itano *in litt.*, 29.7.87). The site is fairly remote and is therefore unaffected by pollution or over-exploitation at present.

Legal Protection   None.

Management   None.

Recommendations   Proposed for national marine sanctuary designation and is on the National Marine Site Evaluation List (NOAA, 1983). Protection of the area is considered a high priority in the Action Strategy for Protected Areas in the South Pacific Region (Anon., 1985).

ROSE ATOLL NATIONAL WILDLIFE REFUGE

Geographical Location   241 km east-south-east of Pago Pago Harbour; 14°32'S, 168°08'W.

Area, Depth, Altitude   Lagoon 2 km wide and 20 m deep; 640 ha reef and lagoon; Rose Island 5.18 ha, 3 m alt. and 1.0 km shoreline length; Sand Island 2.59 ha, 5.2 m alt. and 0.5 km shoreline. The exterior boundary of the refuge is the extreme low waterline outside the perimeter reef except at the entrance channel where the boundary is a line extending between the extreme low waterlines on each side of the entrance channel (Wass, 1987).

Land Tenure   U.S. Federal Government ownership.

Physical Features   One of the smallest coral atolls in the world consisting of two low sandy islets, Rose and Sand, on a coralline algal reef enclosing a lagoon. A single channel 6-50 ft (1.8-15 m) deep links the lagoon to the sea. There is no freshwater. Information on the atoll is given in Sachet (1954) and Mayor (1924e).

Reef Structure and Corals   A brief description of some of the reef habitats of the lagoon is given in Wass (n.d. a). 25-50% of the area between the inner edge of the reef flat and the drop-off to the floor of the lagoon is covered by coral blocks, the larger of which are exposed at low tide. The tops of the larger blocks are flattened and encrusted with coralline algae; their sides are profusely covered with a diverse assortment of hard and soft corals, and algae. The remainder of the area consists of rubble flat, encrusted with coralline algae and a few scattered small colonies of branching *Acropora*. The lagoon floor has an undulating sandy bottom at 12-20 m, much of which is covered with algae; there are occasional small colonies of *Acropora*. Jutting up from the floor of the lagoon to its surface are several coral pinnacles with flat tops and very steep sides. The flattened tops are encrusted with coralline algae, while hard and soft corals and algae are found on the often vertical or undercut sides.

The reef front described by Wass (n.d. b) begins at a depth of 4 m and consists of an irregular and often steep slope to a depth of about 50 m. The upper portion may be bisected by ridges and surge channels. In some areas a narrow terrace occurs at 5-20 m, before the bottom plunges steeply to greater depths. The irregular substrate is calcareous and compacted with coralline algae predominating. Corals are abundant and diverse, but table and staghorn *Acropora* are lacking.

The outer reef was surveyed by the Office of Marine and Wildlife Resources and five monitoring sites were chosen in 1986 (Itano, 1987). The results of these surveys are not available at present.

Noteworthy Fauna and Flora   There is a *Pisonia* grove on Rose Island and some coconuts. Vegetation is described by Setchell (1924) and Amerson *et al.* (1982) and is the least diverse of that on any island in American Samoa. There are only seven plant species. Sand Island has no vegetation.

Hawksbills *Eretmochelys imbricata* and Green Turtles *Chelonia mydas* nest on the atoll. About 97% of the total seabird population of American Samoa is resident on the atoll, with about 312 000 birds of 20 species. There are large nesting colonies of seabirds including Greater *Fregata minor* and Lesser *F. ariel* Frigatebirds, Red-footed *Sula sula*, Brown *S. leucogaster* and Blue-faced *S. dactylatra* Boobies, Red-tailed Tropicbirds *Phaeton rubricauda*, White *Gygis alba* and Sooty *Sterna fuscata* Terns, Brown *Anous stolidus* and Black *A. minutus* Noddies, Reef Herons *Egretta gularis* and a variety of shore birds (Amerson *et al.*, 1982).

Over 200 species of fish have been recorded. The lagoon fish fauna is similar to that of the rest of the Samoan Islands although there is a lack of damselfish species and biomass within the lagoon, and relatively few herbivorous fish (Wass, n.d. a). The abundance of carnivorous fish is high, possibly due to the lack of fishing pressure. The fish fauna of the reef front has a low diversity compared with other reef fronts around Tutuila. Surveys in early 1987 indicated little change in the reef fish community over the past six or seven years. The Giant Clam *Tridacna maxima* is present throughout the shallow areas of the lagoon; in 1980 about 10% of the population was dead (Wass, n.d. c).

Scientific Importance and Research   The atoll is considered one of the most isolated and least disturbed in the world. Annual resource surveys are carried out by U.S. Fish and Wildlife Service (USFWS) and American Samoan government personnel; these include surveys of both the outer reef and lagoon as well as of the terrestrial biota of Sand and Rose Islands. Itano (*in litt.*, 29.7.87) suggests that the benthic community of Rose Atoll may be unique in American Samoa because of the overwhelming presence of encrusting coralline algae; soft corals are also abundant whilst hard corals are poorly represented. The *Tridacna maxima* (Giant Clam) population is being studied by USFWS to determine management strategies (Radtke, 1985; Itano *in litt.*, 29.7.87). The atoll has considerable scientific and educational value and has been visited by school teachers from the American Samoan Department of Education (Anon., 1987a).

Economic Value and Social Benefits   The atoll has apparently never been inhabited (Swerdloff, 1973). Prior to its establishment as a wildlife refuge, the Samoans fished Giant Clams on the atoll.

Disturbance or Deficiencies   The islands are virtually undisturbed apart from the presence of a concrete marker, a U.S. Fish and Wildlife Service Refuge sign and a few introduced coconut trees *Cocos nucifera* which are reportedly not thriving. Introduced Polynesian rats *Rattus exulans* feed on bird eggs, and perhaps birds and their young and hatchling turtles; there is however, little direct evidence of this and the rats do not appear to constitute a major threat to the island's biota (Amerson *et al.*, 1982). Hurricane Tusi in 1987 does not appear to have had any adverse impact on the reefs (Itano, 1987).

Legal Protection   Established as a National Wildlife Refuge on 5.7.73, primarily for the protection of turtles and seabirds.

Management   Administered as a strict natural area by the Hawaiian Islands and Pacific Islands National Wildlife Refuge Complex, Honolulu, in cooperation with the American Samoan Government. A permit is required for entrance and all activities within the refuge generally require the issue of a Special Use Permit; these are issued only for activities which are beneficial to fish and wildlife resources and management of the refuge. Fishing is permitted within the refuge but the catch must be consumed on site or released; any Giant Clam fishing requires a special permit. Birds and turtles may not be disturbed or harvested and no animals or plant material may be taken ashore. The number of people allowed to camp overnight at once on the islands is strictly limited (six on Rose Island and two on Sand Island) and no camping on beaches is allowed during the Green Turtle nesting season (Wass, 1987). A brochure has been produced for visitors (Anon., n.d. b).

Recommendations   The island was proposed as an Island for Science under the IBP programme. Wass (1987) outlines policy and recommendations for the use of Rose Atoll to increase environmental awareness and appreciation amongst Samoan teachers. It was recommended by the Government that harvesting of Giant Clams be resumed (Wass, n.d. c). However it has been stressed that the value of Rose Atoll as a refuge and study site for an undisturbed population of *Tridacna* considerably outweighs any benefits which may be derived from a commercial harvest of the clams (Itano *in litt.*, 29.7.87; Maragos *in litt.*, 10.7.87).

SWAINS ATOLL

Geographical Location   270 km north of Samoa; geographically and floristically part of the Tokelau Islands, 160 km to the north-west; 11°03'S, 171°03'W.

Area, Depth, Altitude   Island is 210 ha; max. alt. less than 6 m.

Land Tenure   Privately owned by the Jennings family and a sovereign (flag) possession of the U.S.A.

Physical Features   The island is a ring-shaped atoll with a large, completely enclosed brackish water lagoon in the centre. Rainfall is about 2500 mm/year (Whistler, 1983). The island is described by Bryan (1974).

Reef Structure and Corals   The reefs were surveyed by Itano (1987 and *in litt.*, 29.7.87). Prior to devastating storm damage in February 1987, the reefs were apparently virtually pristine, with 80-100% live coral cover. The lagoon is land-locked and very little sediment reaches the reefs, leading to underwater visibilites exceeding 150 ft (46 m). The community is dominated by *Pocillopora* and *Montipora* with isolated massive colonies of *Porites lutea*. *Stylophora mordax* is fairly common and *Pavona*, *Psammocora* and *Acropora* were noted but not common; an explanate *Porites* was very common on the outer reef slope from 15 to 40 m depth on the north-east coast but was not observed on the west coast. *Millepora* was present and *Heliopora caerulea* was not uncommon on the reef flat. The north-east coast appeared to be the only area unaffected by the storm.

Noteworthy Fauna and Flora   The flora and vegetation of the island are described by Whistler (1983); there is some scrub and forest but the rest of the island is cultivated. Birds are described by Clapp (1968). *Hernandia* forest, uncommon elsewhere in American Samoa, is found on the island. Green Turtles *Chelonia mydas* reportedly used to nest; in 1982 it was stated that they had only been recorded off shore in recent years (Amerson *et al.*, 1982).

Prior to storm damage (and still on the north-east coast), the fish community was dominated by large predatory fishes such as carangids, serranids, lutjanids and large barracuda, and dogtooth tuna (*Gymnosarda unicolor*). Fishing pressure was evidently low with very large and old wrasse (*Cheilinus undulatus*) and giant grouper present, and abundant jack populations. Pomacentrids were low in diversity and abundance with the exception of *Chromis acares* which was common (Itano *in litt.*, 29.7.87).

Scientific Importance and Research   The reefs have been surveyed by the Office of Marine and Wildlife Resources of the American Samoan Government and may more closely resemble those of Tokelau than American Samoa. Prior to storm damage, they were considered among the most beautiful and productive in American Samoa (Itano, 1987).

Economic Value and Social Benefits   Until recently there was an important copra industry on the island, employing several hundred Tokelau Islanders (Amerson *et al.*, 1982). Copra export has now apparently ceased and by 1987 the population had declined to 18-20 Samoans. The storm of early 1987 destroyed all buildings at Taulaga, the only village on the island, and the inhabitants have moved to Tutuila to await the construction of pre-fabricated buildings which will be taken to the island in September 1987 (Itano *in litt.*, 29.7.87). Landing is made on the west side of the island, at Taulaga (Whistler, 1983). In the early 1980s the inhabitants were to some extent reliant on turtle eggs and meat for subsistence (Balazs, 1982).

Disturbance or Deficiencies   The reefs suffered devastating damage as a result of a storm in early 1987. Damage was most severe around the western half of the island with near total destruction of living corals down to 60 ft (18 m) in some areas. At one transect site facing due west, at Taulaga, live coral cover down to 50 ft (15 m) had decreased from 95-100% (recorded in December 1986) to 7-12% (April 1987). Large branching *Pocillopora* colonies had been broken off near their bases, and flat *Montipora* colonies had also been broken or apparently killed by sedimentation. Many of the deeper corals were being smothered by coral rubble generated further up the reef. The north-west facing reef was believed likely to be particularly susceptible to damage as this area is sheltered from the prevailing winds (the south-east trades) and swell, allowing development of more delicate coral growth forms. The zone facing south-east had apparently not been greatly affected. This area had previously not had very high coral cover, the coral fauna being dominated by soft corals and low growing robust forms which could withstand the south-east tradewinds and associated swells. The eastern part of the island was least affected and a transect off the north-east shore showed live coral cover of nearly 100% with virtually no coral breakage. There appeared to have been a concommitant reduction in fish diversity, other than off the east side of the island, with a notable lack of large predatory fishes; many of the smaller fishes were apparently still present but at lower densities. There is concern that the storm damage to the reefs may lead to a reduction in available food for the villagers in the future (Itano, 1987).

The coconut trees on the island are slowly being replaced by littoral forest species, chiefly *Hernandia* and *Pandanus* (Amerson *et al.*, 1982). The Polynesian Rat *Rattus exulans* is common on the island and feral pigs are found in the coconut plantations (Amerson *et al.*, 1982).

Legal Protection   None.

Management   None.

Recommendations   Dahl (1980) recommended that the outer reefs and lagoon should be surveyed; this has now been carried out and recommendations for management should be drawn up.

INTRODUCTION

General Description

Reefs of the east coast of Australia are dominated by the Great Barrier Reef, the world's largest reef system, extending 2000 km along the north-east coast of Queensland and including some 2500 major reefs and almost as many islands. The Great Barrier Reef Region is described in the following accounts. In addition to the barrier reef system itself, the inshore high islands from the southern part of the Region to Princess Charlotte Bay, usually have complex coastlines, parts of which have strong tidal currents and extensive areas of relatively shallow, turbid water, resulting in a wide range of fringing communities adapted to muddy conditions and often with a very high coral species diversity (Hopley et al., 1983; Veron, 1986).

There are also reefs outside this region. Reef building corals (at least six genera) are found as far south as Sydney and Nambucca Heads, 430 km to the north (Wells, 1955a). Reef building corals have also been found at Arrawarra and Woolgoolga headlands, adjacent to Nambucca Heads (Veron, 1973). Reefs around the Solitary Islands, Lord Howe Island and Elizabeth and Middleton Reefs are described in separate accounts. Detailed studies of shallow-water reef-building corals have been made in Moreton Bay (27°S) by Wells (1955a and b) and in Port Phillip Bay (38°S) by Squires (1966). Flinders Reef near Moreton Bay has a dense cover of corals but the reef itself is of sandstone. Coral diversity is very high. The reef is similar to Elizabeth and Middleton Reefs although the area covered is only 10 ha (Veron, 1986).

To the east, the Coral Sea consists of two parts: the northern half is a broad abyssal plain between the northern part of the Great Barrier Reef and Vanuatu and the southern half comprises a number of seamounts, some capped by reefs, cays and atolls, lying between the central and southern Great Barrier Reef and New Caledonia. The reefs in this area are widely separated and have little in common. Most are very rich in *Acropora* and a few other coral genera but none has the diversity of the Great Barrier Reef (Veron, 1986). Flinders Reef, north-east of Townsville (not to be confused with the Moreton Bay reef of the same name), is described by Done (1982), Williams (1982), Sheppard (1985) and Wilkinson (1987). The Lihou and Coringa-Herald reefs are described in separate accounts.

North of the Great Barrier Reef Marine Park, the barrier line of reefs forms deltaic reefs (*see account for* Far Northern Section) and dissected reefs. The latter are found at the northern limit of the barrier line where the deltaic reefs break up into a complex of reef strips running east-west. At the northern extremity they form a series of downward sloping steps which eventually become submerged in mud in the Gulf of Papua. During periods of monsoonal flooding large masses of low salinity water from Papua New Guinean rivers move down to these reefs and can be seen underwater as lenses but appear to have no effect on the reefs.

Extensive reefs are found in the Torres Strait (*see separate account*). Further west, reefs occur in the Cobourg Peninsula Marine National Park and Sanctuary (*see separate account*) and possibly in the Crocodile Islands and Castlereagh Bay Area (Davis, 1985) in the Northern Territory. The coral fauna of Eastern Australia is described by Veron and Pichon (1976, 1980 and 1982), Veron et al. (1977) and Veron and Wallace (1983).

A useful popular account of the reefs is given in Reader's Digest (1984). The Australian Coral Reef Society have produced a coral reef handbook which gives general information (Mather and Bennett, 1984), and the June 1986 issue of *Oceanus* is devoted to the Great Barrier Reef.

Reef Resources

Subsistence use of the reefs of Eastern Australia by Aboriginal communities was important in the past but is now largely restricted to the northernmost reefs, in particular those in the Torres Strait (*see separate account*). The Great Barrier Reef however has major economic importance for commercial fisheries and the tourist industry. Recreational use of the reef has been developed here to a greater extent than anywhere else in the world (*see separate accounts*).

Disturbances and Deficiencies

Much of the Great Barrier Reef Region, as well as many reefs outside this area, are relatively undisturbed by human activities, although some areas are susceptible to damage from intensive recreational use. Many of the reefs however have been damaged by population outbreaks of the Crown-of-Thorns Starfish; the extent to which this may have caused long-term damage is discussed in the following accounts. Fringing reefs may be damaged periodically by freshwater run-off; for example, the 1974 Brisbane floods covered the reefs of Moreton Bay with up to 5 cm of silt.

Legislation and Management

Most of the Great Barrier Reef lies within the Great Barrier Reef Marine Park, which is described in the following accounts. The Australian National Parks and Wildlife Service has general responsibility for marine protected areas in Commonwealth waters outside the Great Barrier Reef Region (Anon., 1986). An inventory of these areas has been produced (Ivanovici, 1984), and the more important reef sites are described in the following accounts.

Recommendations

Recommendations specific to particular locations are given in the following accounts.

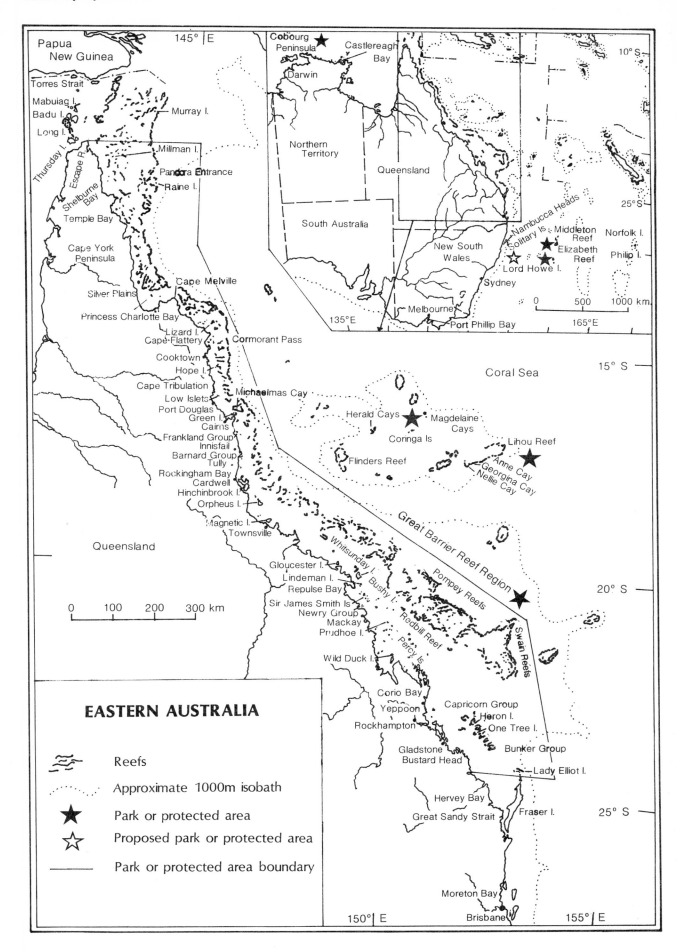

Papua New Guinea

145° E

Cobourg Peninsula
Castlereagh Bay
Darwin

10° S

Torres Strait
Mabuiag I.
Badu I.
Long I.
Thursday I.
Escape R.
Shelburne Bay
Temple Bay
Cape York Peninsula

Murray I.

Millman I.
Pandora Entrance
Raine I.

Silver Plains
Cape Melville
Princess Charlotte Bay
Lizard I.
Cape Flattery
Cooktown
Hope I.
Cape Tribulation
Low Islets
Port Douglas
Green I.
Cairns
Frankland Group
Innisfail
Barnard Group
Tully
Rockingham Bay
Cardwell
Hinchinbrook I.
Orpheus I.
Magnetic I.
Townsville

Cormorant Pass
Michaelmas Cay

Queensland

Northern Territory

Queensland

South Australia

New South Wales

135°E

Melbourne
Port Phillip Bay

Nambucca Heads
Solitary Is
Lord Howe I.
Sydney

Middleton Reef
Elizabeth Reef

Norfolk I.
Philip I.

25° S

0   500   1000 km.

165°E

Coral Sea

15° S

Herald Cays
Coringa Is
Flinders Reef

Magdelaine Cays

Lihou Reef
Anne Cay
Georgina Cay
Nellie Cay

Great Barrier Reef Region

Gloucester I.
Lindeman I.
Repulse Bay
Sir James Smith Is
Newry Group
Mackay
Prudhoe I.

Whitsunday I.
Bushy I.
Pompey Reefs
Redbill Reef
Percy Is

Swain Reefs

20° S

Wild Duck I.

Corio Bay
Yeppoon
Rockhampton

Capricorn Group
Heron I.
One Tree I.
Bunker Group

Gladstone
Bustard Head

Lady Elliot I.

Hervey Bay
Great Sandy Strait

Fraser I.

25° S

0   100   200   300 km

EASTERN AUSTRALIA

Reefs

Approximate 1000m isobath

★ Park or protected area

☆ Proposed park or protected area

Park or protected area boundary

Moreton Bay
Brisbane

150° E

155° E

-16-

References

AIMS (1986a). Annual Report 1985-1986.

AIMS (1986b). Projected Research Activities 1986-1991. First Annual Edition 1986-87.

AIMS (1987). Annual Report 1986-1987.

Allen G.R., Hoese D.F., Paxton J.R., Randall J.E., Russell B.C., Stark W.A. and Talbot F.H. (1976). Annotated checklist of the fishes of Lord Howe Island. *Records Aust. Mus.* 306(15): 365-454.

Allen G.R. and Paxton J.R. (1974). A tropical outpost in the South Pacific. *Aust. Nat. Hist.* 18(2): 50-55.

Anon (1973). New parks and reserves. *Parks and Wildlife* 1(3): 74-78. National Parks and Wildlife Service.

Anon. (1977). Marine Reserves in New South Wales, Australia. In: Gare, N.C., Review of progress in the creation of marine parks and reserves. *Collected Abstracts Papers Int. Conference Marine Parks and Reserves, Tokyo, Japan*, 12-14 May 1975. Pp. 45-48. (Sabiura Marine Park Research Station, Japan, 1977).

Anon. (1981). *Nomination of the Lord Howe Island Group for inclusion in the World Heritage List.* Prepared by New South Wales Government, Australian Heritage Commission, Australian National Parks and Wildlife Service.

Anon (1984). Conservationists miss out in Far Northern Reef zoning. *Queensland Conservation Council Inc. Newsletter* 5(3).

Anon, (1986). Australian National Parks and Wildlife Service Report 1984-85 Canberra.

Baker, J.T. (1977). Management of the Great Barrier Reef Marine Park. *Proc. 3rd Int. Coral Reef Symp. Miami* 2: 597-605.

Baker, J.T., Carter, R.M., Sammarco, P.W. and Stark, K.P. (1983). Proceedings of the Great Barrier Reef Conference, Townsville, Aug 29 - Sept 2, 1983. James Cook University and Australian Institute of Marine Sciences.

Baldwin, C.L. (Ed.) (1987). Fringing Reef Workshop. Science, Industry and Management. *GBRMPA Workshop Series* No. 9.

Bennell, N. (1980). Coastal land development. A shadow over the Great Barrier Reef (Australia). *Tigerpaper* 7(2): 9-12.

Birkeland, C. and Wolanski, E. (in press). Two roles of current patterns in determining the distribution and abundance of *Acanthaster planci. Proc. 2nd Int. Symp. on Indo-Pacific Biology, Guam.*

Bradbury, R.H., Done, T.J., English, S.A., Fisk, D.A., Moran, P.J., Reichelt, R.E., and Williams, D.Mcb. (1985a). The Crown-of-Thorns: Coping with uncertainty. *Search* 16 (3-4): 106-109.

Bradbury, R.H., Hammond, L.S., Moran, P.J. and Reichelt, R.E. (1985b). The stable points, stable cycles and chaos of the Acanthaster phenomenon: A fugue in three voices. *Proc. 5th Int. Coral Reef Cong., Tahiti* 5: 303-308.

Bradbury, R.H., Hammond, L.S., Moran, P.J. and Reichelt, R.E. (1985c). Coral reef communities and the Crown-of-Thorns Starfish: Evidence for qualitatively stable cycles. *J. theor. Biol.* 113: 69-80.

Braley, R.D. (1984). Reproduction in the Giant Clams *Tridacna gigas* and *T. derasa in situ* on the North-Central Great Barrier Reef, Australia and Papua New Guinea. *Coral Reefs* 3(4): 221-227.

Bustard, R. (1972). *Sea Turtles: Natural History and Conservation.* Collins, London.

Bustard, R. (1974). Barrier Reef sea turtle populations. *Proc. 2nd Int. Coral Reef Symp.* 1: 227-234.

Cameron, A.M. and Endean, R. (1981). Renewed population outbreaks of a rare and specialized carnivore (the starfish *Acanthaster planci*) in a complex high-diversity system (the Great Barrier Reef). *Proc. 4th Int. Coral Reef Symp. Manila* 2: 593-596.

Claasen, D. van R. and Kenchington, R. (1987). Managing coral reefs: Operational benefits of remote sensing in marine park planning. *Proc. 10th Canadian Symp. on Remote Sensing*, Edmonton, Alberta, Canada. Pp. 489-496.

Connell, D.W. (1970). Inquiry into the advisability of oil-drilling in the Great Barrier Reef. *Biol. Cons.* 3(1): 60-61.

Craik, G.J.S. (1981a). Recreational fishing on the Great Barrier Reef. *Proc. 4th Int. Coral Reef Symp., Manila* 1: 47-52.

Craik, G.J.S. (1981b). Underwater survey of coral trout *Plectropomus leopardus* (Serranidae) populations in the Capricornia Section of the Great Barrier Reef Marine Park. *Proc. 4th Int. Coral Reef Symp., Manila* 1: 53-58.

Craik, W. (1986). Research on marine mammals sponsored by the Great Barrier Reef Marine Park Authority. *CSIRO Div. Wildl. Rangelands Res. Tech. Mem.* 26: 38-41.

Craik, W. and Dutton, I. (1987). Assessing the effect of sediment discharge on the Cape Tribulation fringing coral reefs. *Coastal Management* 15(3): 213-228.

Davis, S. (1985). Traditional management of the littoral zone among the Yolngu of North Australia. In: Ruddle, K. and Johannes, R.E. (Eds), *The Traditional Knowledge and Management of Coastal Systems in Asia and the Pacific*. Unesco/ROSTEA, Jakarta. Pp. 101-124.

Dinesen, Z.D. (1983). Patterns in the distribution of soft corals across the Central Great Barrier Reef. *Coral Reefs* 1(4): 229-236.

Done, T.J. (1982). Patterns in the distribution of coral communities across the central Great Barrier Reef. *Coral Reefs* 1: 95-107.

Done, T.J. and Fisk, D.A. (1985). Effects of two *Acanthaster* outbreaks on coral community structure - the meaning of devastation. *Proc. 5th Int. Coral Reef Cong., Tahiti* 5: 315-320.

Driml, S. (1987a). *Economic Impacts of Activities on the Great Barrier Reef.* GBRMPA Research Publication.

Driml, S. (1987b). *Great Barrier Reef Tourism: A Review of Visitor Use.* GBRMPA Research Publication.

Dutton, I.M. (Ed.) (1985). Contaminants in the waters of the Great Barrier Reef. Proc. GBRMPA Workshop. Townsville.

Dutton, I.M. (1987). Environmental management of the proposed floating hotel at John Brewer Reef. *Marit. Stud.* 32: 16-21.

Edgecombe, J. (1980). Tangerra - Green Island National Park: Profile of a Barrier Reef tourist island. *Habitat* 8(8): 20-23.

Endean, R. (1973). Population Explosions of *Acanthaster planci* and associated destruction of hermatypic corals in the Indo-West Pacific Region. In: Jones, O.A. and Endean, R. (Eds), *Biology and Geology of Coral Reefs. Vol. 2. Biology 1.* Pp. 389-438.

Endean, R. (1974). *Acanthaster planci* on the Great Barrier Reef. *Proc. 2nd Int. Coral Reef Symp., Brisbane* 1: 185-191.

Endean, R. (1977). *Acanthaster planci* infestations of reefs of the Great Barrier Reef. *Proc. 3rd Int. Coral Reef Symp., Miami*: 563-576.

Endean, R. (1982). Crown-of-thorns Starfish on the Great Barrier Reef. *Endeavor, New Series* 6(1): 10-14.

Endean, R. and Stablum, W. (1975). Population explosions of *Acanthaster planci* and associated destruction of the hard coral cover of reefs of the Great Barrier Reef, Australia. *Env. Cons.* 2: 247-256.

Fisk, D.A. and Done, T.J. (1985). Taxonomic and bathymetric patterns of bleaching in corals, Myrmidon Reef, Queensland. *Proc. 5th Int. Coral Reef Cong., Tahiti.*

Flood, P.G. (1979). Heron Island erosion problems. *Reeflections* 3. Great Barrier Reef Marine Park Authority.

Flood, P.G. and Heatwole H. (1986). Coral cay instability and species turnover of plants at Swain Reefs, southern Great Barrier Reef, Australia. *J. Coastal Res.* 2(4): 479.

Frankel, E. (1975). *Acanthaster* in the past: Evidence from the Great Barrier Reef. *Proc. Crown-of-thorns Starfish Seminar, Brisbane* 6 September 1974. Australian Government Publishing Service, Canberra, Australia. Pp. 159-166.

Frankel, E. (1977). Previous *Acanthaster* aggregations in the Great Barrier Reef. *Proc. 3rd Int. Coral Reef Symp., Miami:* 201-208.

Frith, C.A. (1981). Circulation in a platform reef lagoon, One Tree Reef, southern Great Barrier Reef. *Proc. 4th Int. Coral Reef Symp., Manila* 1: 348-354.

Gare, N.C. (1976). Review of progress in the creation of marine parks and reserves. In: *An International Conference on Marine Parks and Reserves, Tokyo, Japan, 12-14th May 1975. IUCN Publications New Series* 37: 65-71.

Glynn, P.W. (1981). *Acanthaster* population regulation by a shrimp and a worm. *Proc. 4th Int. Coral Reef Symp., Manila* 2: 607-612.

GBRMPA (1980a). Great Barrier Reef Marine Park: Capricornia Section: Zoning Plan. P. 29.

GBRMPA (1980b). Great Barrier Reef Marine Park: Capricornia Section: Understanding the zoning plan. P. 10.

GBRMPA (1981a). Seasonal Closure: Fairfax Islands and Hoskyn Islands Reefs. *Bulletin. Issue No. 3.* November 1981.

GBRMPA (1981b). Nomination of the Great Barrier Reef by the Commonwealth of Australia for inclusion in the World Heritage List.

GBRMPA (1983a). Cairns Section zoning plan and the Cormorant Pass Section zoning plan.

GBRMPA (1983b). *Reeflections* 13.

GBRMPA (1983c). Replenishment areas: North Reef and Boult Reef. *GBRMPA Bulletin* 6.

GBRMPA (1983d). *Australian Marine Research in Progress.* Victorian Institute of Marine Sciences, Melbourne.

GBRMPA (1983e). *Workshop on the Northern Sector of the Great Barrier Reef.* GBRMPA Workshop Series No. 1.

GBRMPA (1985a). AMRIP: GBRR.

GBRMPA (1985b). Far Northern Section Zoning Plan.

GBRMPA (1986). Australian Marine Research in Progress: Great Barrier Reef Region 1985-86. GBRMPA Research Publication.

GBRMPA (1987a). Annual Report 1986-1987.

GBRMPA (1987b). Central Section Zoning Plan.

GBRMPA (1988). Mackay-Capricorn Zoning Plan (In Press).

Hall, D. (1984). Conservation by ecotourism. *New Scientist* March 1: 38-39.

Harriott, V.J. (1985a). Mortality rates of scleractinian corals before and during a mass bleaching event. *Mar. Ecol. Prog. Ser.* 21: 81-88.

Harriott, V.J. (1985b). The potential for a beche-de-mer fishery. *Australian Fisheries* June 1985.

Hegerl, E. (1981). Cod saved - reef lost? *Queensland Conservation Council Newsletter*, October 1981.

Hopley, D., Davies, P.J., Harvey, N. and Isdale, P.J. (1981). The geomorphology of Redbill Reef, Central Great Barrier Reef. *Proc. 4th Int. Coral Reef Symp., Manila* 1: 541-547.

Hopley, D. and Partain, B. (1986). The Structure and Development of Fringing Reefs off the Great Barrier Reef Province, Fringing Reef Workshop, GBRMPA Workshop Series No. 9.

Hopley, D., Slocombe, A.M., Muir, F. and Grant, C. (1983). Nearshore fringing reefs in North Queensland. *Coral Reefs* 1(3): 151-160.

Hundloe, T.H. (1985). Fisheries of the Great Barrier Reef. GBRMPA, Special Publication Series (2).

Hundloe, T.H., Driml, S.M., Shaw, S. and Trigger, J. (1981). Proposed Cairns Section of the Great Barrier Reef Marine Park: Some economic characteristics and multipliers. Report to GBRMPA.

Ivanovici, A.M. (Ed.) (1984). *Inventory of Declared Marine and Estuarine Protected Areas in Australian Waters.* Special Publication 12, Australian National Parks and Wildlife Service. 2 vols.

Kelleher, G.G. (1981). Research needs for coral reef management planning. *Proc. 4th Int. Coral Reef Symp., Manila* 1: 231-236.

Kelleher, G. (1986). The Great Barrier Reef: A World Heritage Site. Educating for the Environment. *Proc. Seminar and Workshops* 6-7 May, 1985, Canberra, ACT.

Kelleher, G.G. and Dutton, I.M. (1985). Environmental effects of offshore tourist development on the Great Barrier Reef. *Proc. Third South Pacific National Parks and Reserves Conference, Apia* 3: 200-207.

Kelleher, G. and Kenchington, R. (1984). Australia's Great Barrier Reef Marine Park: Making development compatible with conservation. In: McNeely, J.A. and Miller, K.R. (1984). *National Parks, Conservation and Development: The Role of Protected Areas in Sustaining Society.* Smithsonian Institution Press, Washington D.C. Pp. 267-273.

Kenchington, R.A. (1978). The Crown-of-Thorns crisis in Australia: A retrospective analysis. *Env. Cons.* 5(1): 11-20.

Kenchington, R. (in press). *Acanthaster planci* and management of the Great Barrier Reef. *Proc. 2nd Int. Symp. on Indo-Pacific Biology, Guam.*

Kenchington, R.A. and Pearson, R. (1981). Crown-of-Thorns Starfish on the Great Barrier Reef: A situation report. *Proc. 4th Int. Coral Reef Symp., Manila* 2: 597-600.

Kettle, B.T. and Lucas, J.S. (in press). Size-related morphological and physiological phenomena in *Acanthaster planci* (L.). *Proc. 2nd Int. Symp. on Indo-Pacific Biology, Guam.*

Kikkawa, J. (1976). The birds of the Great Barrier Reef. In: Jones, O.A. and Endean, R. (Eds), *Biology and Geology of Coral Reefs* 3: 279-342.

Kinsey, D.W. (1972). Preliminary observations on community metabolism and primary productivity of the pseudo-atoll reef at One Tree Island, Great Barrier Reef. *Proc. Symp. Corals and Coral Reefs* 1969: 13-32. (West Biol. Assoc. India).

Kuchler, D.A. (1986). Geomorphological nomenclature. Reef cover and zonation on the Great Barrier Reef. *GBRMPA Tech. Mem.* 7.

Lanyon, J. (1986). Seagrasses of the Great Barrier Reef. *GBRMPA Special Publ. Ser.* 3.

Lavery, H.J. and Grimes, R.J. (1971). *Queensland Agric. Jour.* 97: 106-113.

Leis J.M. (1981) Distribution of fish larvae around Lizard Island, Great Barrier Reef: Coral reef lagoon as refuge? *Proc. 4th Int. Coral Reef Symp., Manila* 2: 471-477.

Limpus, C.J. (1981) The status of Australian sea turtle populations. In: Bjorndal K.A. (Ed.), *Biology and Conservation of Sea Turtles. Proc. World Conference Sea Turtle Cons., Washington D.C., 27-30 November 1979.* Smithsonian Institution Press, Washington. Pp. 297-303.

Limpus, C.J., Miller, J.D., Baker, V. and McLachlan, E. (1983a). The Hawksbill turtle, *Eretmochelys imbricata* (L.) in North-eastern Australia: The Campbell Island Rookery. *Aust. Wildl. Res.* 10: 185-197.

Limpus, C.J., Parmenter, C.J., Baker, V., and Fleay, A. (1983b). The Crab Island Sea Turtle Rookery in the North-eastern Gulf of Carpentaria. *Aust. Wildl. Res.* 10: 173-184.

Lucas, J.S. (1973). Reproductive and larval biology of *Acanthaster planci* in Great Barrier Reef Waters. *Micronesica* 9(2): 197-203.

McGinnity, P. (1981). Whitsunday Area Economic Impact Study. Honours Thesis, Griffith University.

McMichael, D.F. (1974). Growth rate, population size and mantle coloration in the small Giant Clam *Tridacna maxima* (Röding) at One Tree Island, Capricorn Group, Queensland. *Proc. 2nd Int. Coral Reef Symp., Brisbane* 1: 241-254.

Marsh, H.D. (1985). Results of the aerial survey for dugongs conducted in the Cape Bedford/Cape Melville area in November 1984. Report to GBRMPA, April 1985.

Marshall, J.F. and Davies, P.J. (1982). Internal structure and holocene evolution of One Tree Reef, southern Great Barrier Reef. *Coral Reefs* 1(1): 21-28.

Mather, P. and Bennett, I. (Eds) (1984). *A Coral Reef Handbook.* Handbook Series No. 1, 2nd Ed., The Australian Coral Reef Society. 144 pp.

Mayor, (1918). Ecology of the Murray Island coral reefs. *Pap. Dep. Mar. Biol. Carnegie Inst. Washington* 9(213): 1-48.

Moran, P. (1986). The *Acanthaster* phenomenon. *Oceanogr. Mar. Biol. Ann. Rev.* 24: 379-480.

Moran, P., Bradbury, R. and Reichelt, R. (1984). Information on coral recovery and the Crown-of-Thorns starfish in the Central Section of the Great Barrier Reef. Report No. 10. Australian Institute of Marine Science.

Moran, P.J., Bradbury, R.H. and Reichelt, R.E. (1985). Mesoscale studies of the Crown-of-Thorns/coral interaction: A case history from the Great Barrier Reef. *Proc. 5th Int. Coral Reef Cong., Tahiti* 5: 321-326.

Nash, W. and Zell, L.D. (1981). *Acanthaster* on the Great Barrier Reef: Distribution on five transects between 14°S and 18°S. *Proc. 4th Int. Coral Reef Symp., Manila* 2: 601-605.

Nietschmann, B. (1976). Hunting and ecology of dugongs and green turtles, Torres Strait, Australia. *National Geographic Society Research Reports*: 625-651.

Nietschmann, B. (1985). Torres Strait Islander Sea resource management and sea rights. In: Ruddle, K. and Johannes, R.E. (Eds) *The Traditional Knowledge and Management of Coastal Systems in Asia and the Pacific.* Unesco/ROSTEA, Jakarta Pp. 125-154.

Oliver, J. (1985). Recurrent seasonal bleaching and mortality of corals on the Great Barrier Reef. *Proc 5th Int. Coral Reef Cong., Tahiti* 4: 201-206.

Oliver, J. and McGinnity, P. (1985). Commercial coral collecting on the Great Barrier Reef. *Proc. 5th Int. Coral Reef Cong., Tahiti* 5: 563-568.

Parnell, K.E. (1986). Water movement within a fringing reef flat, Orpheus Island, north Queensland, Australia. *Coral Reefs* 5(1): 1-6.

Pearson, R.G. (1974). Recolonisation by hermatypic corals of reefs damaged by *Acanthaster. Proc. 2nd Int. Coral Reef Symp., Brisbane* 2: 207-215.

Pearson, R.G. (1977). Queensland Barrier Reef study: Impact of foreign vessels poaching giant clams. *Australian Fisheries* 36(7): 8-11, 23.

Pickard, G.L. *et al.* (1977). A review of the physical oceanography of the Great Barrier Reef and Western Coral Sea. *Aust. Inst. Mar. Sci. Monog. Ser.* 2: 134.

Pollard, D. and Burchmore, J. (1985). Lord Howe Island Regional Environmental Study. Marine Environment, with a Proposal for an Aquatic Reserve. Lord Howe Island Board.

Pollard, D.A. (1977). A proposed Lord Howe Island Marine Reserve: Protecting the southernmost coral reef in the World. *Collected Abstracts Papers Int. Conference Marine Parks and Reserves, Tokyo, Japan, 12-14 May 1975.* Pp. 71-79. (Sabiura Marine Park Research Station, Japan, 1977).

Pollard, D.A. (1981). Solitary Islands Marine Park Proposal. N.S.W. Fisheries, Sydney.

Randall, J.E. (1976). In: *An International Conference on Marine Parks and Reserves, Tokyo, Japan, 12-14 May 1975. IUCN Publications New Series No. 37*: 63.

Randall, J.E. (1977). Lord Howe Island, a Land and Sea Preserve. *Collected Abstracts and Papers of the International Conference on Marine Parks and Reserves, Tokyo, Japan, 12-14 May 1975.* (Sabiura Marine Park Research Station, Japan, 1977).

Reader's Digest (1984). *Reader's Digest Book of the Great Barrier Reef.* Reader's Digest. Sydney. 384 pp.

Russ, G. (1984a). Distribution and abundance of herbivorous grazing fishes in the central Great Barrier Reef. 1. Levels of variability across the entire continental shelf. *Mar. Ecol. Prog. Ser.* 20: 23-34.

Russ, G. (1984b). Distribution and abundance of herbivorous grazing fishes in the central Great Barrier Reef. 2. Patterns of zonation of mid-shelf and outer shelf reefs. *Mar. Ecol. Prog. Ser.* 20: 35-44.

Russell, B.C. (1983). Checklist of Fishes. Great Barrier Reef Marine Park Capricornia Section. *GBRMPA Special Publication* 1.

Sheppard, C.R.C. (1985). Unoccupied substrate in the central Great Barrier Reef: Role of coral interactions. *Mar. Ecol. Prog. Ser.* 25: 259-268.

Shipway, A.K. (1969). *Emu* 69: 108-109.

Squires, D.F. (1966). Scleractinia. Port Phillip survey 1957-1963. *Mem. natn. Mus. Vict.* 27: 167-174

Stoddart, D.R., Gibbs, P.E. and Hopley, D. (1981). Natural History of Raine Island, Great Barrier Reef. *Atoll Res. Bull.* 254. 44 pp.

Sutherland, L. and Ritchie, A. (1974). Defunct volcanoes and extinct horned turtles. *Aust. Nat. Hist.* 18(2): 44-49.

Talbot, F.H. and Gilbert, A.J. (1981). A comparison of quantitative samples of coral reef fishes latitudinally and longitudinally in the Indo-West Pacific. *Proc. 4th Int. Coral Reef Symp., Manila* 2: 485-490.

Talbot, F.H., Goldman, B., Anderson, G. and Russell, B. (n.d.). The Fishes of the Capricorns and Bunkers. (Unpublished).

Veron, J.E.N. (1973). Southern geographical limits to the distribution of Great Barier Reef hermatypic corals. *Proc. 2nd Int. Symp. Coral Reefs.*

Veron, J.E.N. (1974). Southern geographic limits to the distribution of Great Barrier Reef hermatypic corals. *Proc. 2nd Int. Coral Reef Symp., Brisbane* 1: 465-473.

Veron, J.E.N. (1978a). Deltaic and dissected reefs of the far Northern Region. *Phil Trans R. Soc. Lond. B* 284: 23-37.

Veron, J.E.N. (1978b). Evolution of the far northern barrier reefs. *Phil Trans. R. Soc. Lond. B* 284: 123-127.

Veron, J.E.N. (1981). The species concept in "Scleractinia of Eastern Australia". *Proc. 4th Int. Coral Reef Symp., Manila* 2: 183-186.

Veron, J.E.N. (1986). *Corals of Australia and the Indo-Pacific.* Angus and Robertson.

Veron, J.E.N. and Done, T. (1979). Corals and coral communities of Lord Howe Island. *Aust. J. Mar. Freshwater Res.* 30: 203-236.

Veron, J.E.N. and Hudson R.C.L. (1978). Ribbon reefs of the Northern Region. *Phil. Trans. R. Soc. Lond. B* 284: 3-21.

Veron, J.E.N. and Pichon, M. (1976). Scleractinia of Eastern Australia, Part 1. Families Thamnasteriidae, Astrocoeniidae, Pocilloporidae. *Aust. Inst. Mar. Sci. Monog. Ser.* 1: 86 pp.

Veron, J.E.N. and Pichon, M. (1980). Scleractinia of Eastern Australia. Part 3. Families Agariciidae, Siderastreidae, Fungiidae, Oculinidae, Merulinidae, Mussidae, Pectiinidae, Caryophylliidae, Dendrophylliidae. *Aust. Inst. Mar. Sci. Monog. Ser.* 1: 459 pp.

Veron, J.E.N. and Pichon, M. (1982). Scleractinia of Eastern Australia. Part 4. Family Poritidae. 159 pp.

Veron, J.E.N., Pichon, M. and Wijsman-Best, M. (1977). Scleractinia of East Australia, Part 2. Families Faviidae, Trachyphyllidae. *Aust. Inst. Mar. Sci. Monog. Ser.* 3: 233 pp.

Veron, J.E.N. and Wallace, C.C. (1983). Scleractinia of Eastern Australia. Part 5. Family Acroporidae. 490 pp.

Veron, J.E.N., How, R.A., Done T.J., Zell, L.D., Dodkin, M.J. and O'Farrell, A.F. (1974). Corals of the Solitary Islands, Central New South Wales. *Aus. J. Mar. Freshwater Res.* 25: 193-208.

Vine, P.J. (1970). Field and laboratory observations of the Crown-of-thorns Starfish *Acanthaster planci. Nature* (London) 228: 341.

Walsh, R.J., Harris, C.L., Harvey, J.M., Maxwell, W.G.H., Thompson, J.M. and Tranter, D.J. (1971). *Report of the Committee on the Problem of the Crown-of-thorns Starfish (Acanthaster planci).* Commonwealth Government Printing Office, Canberra, Australia. 45 pp.

Weber, J.N. and Woodhead, P.M.J. (1970). Ecological studies of the coral predator *Acanthaster planci* in the South Pacific. *Mar. Biol.* 6: 12-17.

Wells, J.W. (1955a). A survey of the distribution of reef coral genera in the Great Barrier Reef region. Rep. Great Barrier Reef Committee No. 6 (2).

Wells, J.W. (1955b). Recent and subfossil corals of Moreton Bay, Queensland. *Pap. Dep. Geol. Univ. Qd* 4 (10).

Wilkinson, C.R. (1987). Productivity and abundance of large sponge populations on Flinders Reef flats, Coral Sea. *Coral Reefs* 5(4): 183-188.

Williams. D. McB. (1982). Patterns in the distribution of fish communities across the Central Great Barrier Reef. *Coral Reefs* 1(1): 35-43.

Williams, D. McB. and Hatcher, A.I. (1983). Structure of fish communities on outer slopes of inshore, mid-shelf and outer shelf reefs of the Great Barrier Reef. *Mar. Ecol. Prog. Ser.* 10: 239-250.

Williams, R. (1980). Recreation management of the Great Barrier Reef. Paper presented to 53rd Conference of Royal Australian Institute of Parks and Recreation.

Wilson, J. (1978). Tourist and recreation activity in the northern sector of the Great Barrier Reef. In: Workshop on the Northern Sector of the Great Barrier Reef. Townsville, Queensland, GBRMPA: 148-158.

Wolanski, E. (1981). Aspects of physical oceanography of the Great Barrier Reef lagoon. *Proc. 4th Int. Coral Reef Symp., Manila* 1: 375-381.

Woodland, D.J. and Hooper, J.N.A. (1977). The effect of human trampling on coral reefs. *Biol. Cons.* 11: 1-4.

Zann, L. and Eager, E. (Eds) (1987). The Crown of Thorns Starfish. *Aust. Science Mag.* 3.

Zell L.D. (1981). Constraints for management and interpretive activities on the Great Barrier Reef. *Proc. 4th Int. Coral Reef Symp., Manila* 1: 237-241.

COBOURG PENINSULA MARINE NATIONAL PARK AND SANCTUARY

Geographical Location Northern Territory; 200 km north-east of Darwin, forming the northern shore of Van Diemen Gulf; 11°15'S, 132°15'E.

Area, Depth, Altitude The marine national park covers ca 229 000 ha, of which 29 000 ha is the Sanctuary (Ivanovici, 1984).

Physical Features There are a wide range of benthic habitats including coral reefs (Ivanovici, 1984).

Noteworthy Fauna and Flora The area includes populations of Dugong *Dugong dugon*, Green, Leatherback, Olive Ridley and Hawksbill (*Chelonia mydas, Dermochelys coriacea, Lepidochelys olivacea* and *Eretmochelys imbricata*) Turtles, and Estuarine Crocodiles *Crocodylus porosus* (Ivanovici, 1984; Limpus, 1981).

Scientific Importance and Research No information.

Economic Value and Social Benefits An important area for commercial fishing and prawn trawling. The Peninsula is also used extensively by Aborigines for traditional exploitation or resources. There are some recreational activities (Ivanovici, 1984).

Disturbance or Deficiencies There is some concern about over-exploitation of fishery resources. Crocodile poaching occurs (Ivanovici, 1984).

Legal Protection The waters surrounding the Peninsula were declared a Marine National Park on 1.7.83 under the Territory Parks and Wildlife Conservation Act 1951, Section 12; the boundaries of the Park are shown in Ivanovici (1984). The intertidal area adjoining the Peninsula and adjacent islands, from high water to low water mark, was declared a Sanctuary on 3.9.81 under the Cobourg Peninsula Aboriginal Land and Sanctuary Act 1981 (Ivanovici, 1984).

Management The Conservation Commission of the Northern Territory is responsible for management of the Park and enforcement of Sanctuary regulations. The

Cobourg Peninsula Sanctuary Board is responsible for the management of the Sanctuary (Ivanovici, 1984).

CORINGA-HERALD NATIONAL NATURE RESERVE

Geographical Location   Coral Sea Islands Territory, off the northern coast of Queensland in the central region of the Coral Sea (within 16°-18°S, 149°-151°E), 420 km east of Cairns, 350 km east of the Great Barrier Reef. The reserve includes Coringa Islets, Herald Cays and Magdelaine Cays.

Area, Depth, Altitude   8856 sq. km.

Land Tenure   Commonwealth of Australia.

Physical Features   Islands, reefs, cays and surrounding waters.

Reef Structure and Corals   Corals are described in Veron and Wallace (1983). The *Phyllospongia* community in the lagoon is larger than those found on the Great Barrier Reef but coral and fish diversity is lower (Anon., 1986).

Noteworthy Fauna and Flora   There are extensive seabird colonies, including Red-footed Boobies and Greater Frigatebirds (Anon., 1986), important nesting sites and a diverse marine fauna. Seabirds found nowhere else in the Coral Sea Islands Territory occur in large numbers at Coringa-Herald. It also supports the only stands of *Pisonia grandis*. Cetaceans include Southern Right Whale and Humpback Whale *Megaptera novaeangliae*.

Scientific Importance and Research   Australian National Parks and Wildlife Service (ANPWS) have conducted five biological surveys in the Coral Sea Islands Territory, including the Coringa-Herald Reserve. These have obtained information on breeding and nesting of seabirds and on the cetacean populations. In October 1984 a major resource survey was carried out by the ANPWS in collaboration with the Australian Survey Office (Anon., 1986).

Disturbance or Deficiencies   Foreign fishing vessels take fish and clams in the area. There is a potential threat from pollution from shipwrecks. Rats are adversely affecting the birdlife (Anon., 1986).

Legal Protection   The area was declared as a reserve 3.8.82 under the National Parks and Wildlife Conservation Act 1975, by the Governor-General of the Commonwealth of Australia. The legislation provides complete protection for all marine wildlife.

Management   The Australian National Parks and Wildlife Service manages the reserve. Management plans are being prepared. In 1985, a management patrol was initiated as well as a programme of rat eradication (Anon., 1986).

ELIZABETH AND MIDDLETON REEFS MARINE NATIONAL NATURE RESERVE

Geographical Location   600 km south-east of Brisbane; 95 km north of Lord Howe Island; 29°21'-30°03'S, 158°55'-159°14'E.

Area, Depth, Altitude   Total area ca 105 km in length; each reef covers about 2000 ha.

Land Tenure   State.

Physical Features   Both reefs are open ocean platform reefs. Elizabeth Reef has a deeper, more extensive lagoon. They have broad reef flats exposed at low tide, when the lagoonal waters are well above the level of the surrounding ocean.

Noteworthy Fauna and Flora   An important breeding area for the Black Cod *Epinephelus daemeli*; the last breeding area in Australian waters in which this species has not been overfished. The reefs are also an important feeding and resting area for sea birds and turtles. Three new species of crab were discovered in 1987.

Reef Structure and Corals   Many species of coral, rarely if ever recorded from other reefs are common, but coral diversity is lower than on tropical reefs (Veron, 1986).

Scientific Importance and Research   The reefs are the southernmost open ocean platform reefs in the world. They feature an unusual variety of species as they are located where tropical and temperate currents meet. Visited by researchers from the Queensland Museum in 1987.

Disturbance or Deficiencies   There are few if any signs of environmental stress (Veron, 1986).

Legal Protection   The reefs were declared a National Nature Reserve on 11.12.87 under the National Parks and Wildlife Conservation Act, 1975.

GREAT BARRIER REEF REGION

Geographical Location   Off Queensland coast; the Great Barrier Reef (GBR) Region lies between 10°41'S and 24°30'S from north Cape York Peninsula to south of Gladstone.

Area, Depth, Altitude   348 700 sq. km. in area stretching along 2000 km of the coast, running from the low tide mark to beyond the 200 m bathymetric contour (Kelleher, 1981; Kelleher and Kenchington, 1984). Approximately 2600 individual reefs varying in size from 1 ha to 100 sq. km. The Great Barrier Reef Marine Park (GBRMP) occupies about 98.5% of the Region.

Land Tenure   Title to the sea bed beneath the 3 naut. mi. (5.6 km). Territorial Sea is vested in the Queensland Government, subject to the operation of the Great Barrier Reef Marine Park Act, 1975. There is no title to the seabed seawards of the Territorial Sea. Port areas and areas adjacent to urban or intensive agriculture are administered by the State of Queensland and are generally not included in the Great Barrier Reef Marine Park (amounts to less than 2% of the Region). Public title to islands is vested in the State of Queensland, apart from public lands owned by the Commonwealth. Some land is held by private persons (GBRMPA, 1981b).

Physical Features   South of 15°30'S the reefs are generally 30 km or more off shore. Those that are closer, coastal fringing and nearshore continental islands, have poor water visibility due to terrestrial runoff and sediment (Hopley and Partain, 1986). The coastal lagoon between the main body of the GBR and the mainland has a maximum depth of 145 m but rarely exceeds 60 m (Wolanski, 1981). For physical oceanography see Pickard *et al.* (1977).

Climatic details are given in GBRMPA (1981b). Average summer temperatures range from 30°C in Cairns to 26°C in Gladstone; winter temperatures range from 26°C in Cairns to 15°C in Gladstone. Cyclones may occur in the summer; there are rough seas through autumn and winter with the south-east trade winds, and calm water in the spring (Zell, 1981).

In some areas the tidal range is great, such as the area including Redbill Reef 90 km east of Mackay (20°58'S) where the semi-diurnal tidal range exceeds 5 m (Hopley *et al.*, 1981). The water circulation of the GBR is immensely complicated, governed by properties of the Coral Sea, land runoff, evaporation, the south-east trade winds, forced upwellings due to strong tidal currents in narrow reef passages, and coastal waters including, significantly, mangroves.

Reef Structure and Corals   The GBR is the largest system of coral reefs in the world. A detailed list of reefs and their associated features is given in the Great Barrier Reef Gazetteer. The following accounts give details of the reefs in different Sections of the Region; this account gives a brief overview. In the north the reef is narrow with ribbon reefs on its eastern edge, extensive coastal fringing reefs and patch reefs. In the south it broadens giving a vast wilderness of patch reefs separated by open water or narrow channels. There are 243 coral cays (Kelleher and Kenchington, 1984). The GBR supports over 300 species of hard coral. Approximately 70% of the hermatypic corals of the Indo-Pacific are found along eastern Australia. A full description is given in Veron (1974). Fringing reefs are described by Hopley and Partain (1986).

Noteworthy Fauna and Flora   The GBR supports 1500 species of fish, over 4000 species of molluscs and a variety of endemic species (Kelleher and Kenchington, 1982). 252 species of birds nest and breed on the coral cays (Kelleher, 1981; Kelleher and Kenchington, 1982).

Six turtle species occur in Queensland waters, five on the GBR. Three are closely associated with the Reef: the widespread and abundant Green Turtle *Chelonia mydas* and Loggerhead *Caretta caretta* which have undergone marked fluctuations in the numbers in recent years, cause unknown. The Hawksbill *Eretmochelys imbricata* is widespread but not abundant and breeds predominantly in low densities on inner shelf areas of the northern GBR and islands of the central to eastern Torres Strait. Long Island in the Torres Strait is the major rookery and has several hundred turtles annually. Campbell Island and

Millman Island are important sites (Bustard, 1972; Limpus 1981; Limpus *et al.*, 1983a). Low density nesting of the widespread Flat-back Turtle *Chelonia depressa*, an Australian endemic, occurs as far south as Mon Repos (25°). It is the dominant nesting species between Bustard Head (24°) and Townsville (19°), important concentrations occurring on Peake Island (23°), Wild Duck Island and Avoid Island (both 22°). There are major feeding grounds throughout the GBR (Bustard, 1974; Limpus, 1981; Limpus *et al.*, 1983b). *Lepidochelys olivacea* is poorly known with low density nesting recorded in isolated areas (Limpus, 1981). *Dermochelys coriacea* migrates along the southern areas of the Queensland coast in appreciable numbers but little nesting occurs (Limpus, 1981).

The largest known populations of Dugong *Dugong dugon* occur in Australia. The total number in Queensland is unknown and southern populations may be increasing. Aerial counts are currently being carried out to estimate populations (Marsh, 1985). Large herds (over 100) have been sighted just south of the Capricornia Section in Great Sandy Strait and Hervey Bay, in the central GBR Region from Townsville to Cardwell, and in the northern GBR Region from Cape Flattery to Shelburne Bay.

Several species of cetaceans are found within the Region. The most frequently reported whales are the Humpback *Megaptera novaeangliae*, Minke *Balaenoptera acutorostrata* and Killer Whales *Orcinus orca*. The Humpback Whale is considered to be an endangered species due to whaling earlier this century. Whaling ceased in 1962 and the Humpback population appears to be recovering. Feeding in Antarctic waters during the summer months, Humpbacks migrate north along the east coast of Australia in winter to calve and mate. Sightings are reported throughout the GBR Region as far north as the Cairns area.

Dolphins abound throughout the Region and those species usually seen in coastal areas are the Bottlenose *Tursiops truncatus*, Irrawaddy River *Orcaella brevirostris* and the Indo-Pacific Humpback *Sousa chinensis*. Off shore, the Spinner Dolphin *Stenella longirostris* is also sometimes sighted.

The Saltwater Crocodile, *Crocodylus porosus* is found in mangrove swamps and river estuaries along the coastal fringe of the Region. Past hunting activity and removal of large crocodiles perceived as a threat in areas of frequent human activity has placed crocodile populations in many areas under considerable pressure. Significant populations are now only found in suitable habitat at remote locations mainly towards the northern end of the Region. As they tend to avoid areas of strong wave action, few saltwater crocodiles are seen around offshore coral reefs. There are important populations of Giant Clams (Tridacnidae) which are currently the subject of a long-term research programme (Braley, 1984).

Scientific Importance and Research   With its enormous faunal, floral and geomorphological diversity, the GBR offers unparalleled opportunities for scientific research. Research activities in the Region have accelerated and assumed even greater importance since the formation of the Great Barrier Reef Committee (now the Australian Coral Reef Society) in 1922 and the British GBR Expedition to the Low Isles in 1928-29. The need for such research has become more critical in recent years with the GBR's inclusion on the World Heritage List in 1981 (Kelleher, 1986), concern resulting from Crown-of-thorns Starfish outbreaks and intensifying demands placed on the resource by human use.

While research on the GBR continues to be conducted by scientists from most Australian universities and institutions, concerted efforts are concentrated, for logistical reasons, on field research stations and North Queensland mainland centres. For local and visiting overseas scientists, field stations are operated by the University of Queensland (Heron I.), the University of Sydney (One Tree I.), James Cook University (Orpheus I.) and the Australian Museum (Lizard I.). Details of research activities in the GBR Region are given in Baker *et al.*, (1983), GBRMPA (1983d) and GBRMPA (1986).

Because of their strategic location, the James Cook University and the Australian Institute of Marine Science (AIMS), both in Townsville, have extensive coral reef research programmes that cover the full gambit of scientific disciplines (see AIMS, 1986a,b and 1987). The existence of AIMS and the Great Barrier Reef Marine Park Authority (GBRMPA) has facilitated development of widely-encompassing, multi-disciplinary research programmes with numerous benefits over individual discrete *ad hoc* projects.

The Research and Monitoring Section of the GBRMPA obtains and interprets information to assist in the planning, administration and education functions of managing the Marine Park (GBRMPA, 1987a). To obtain this information the Authority is empowered to commission research or undertake research itself. The GBRMPA Research Programme is split into eleven categories: Analysis of Use (e.g. Driml, 1987a and b); Biology (e.g. Craik, 1986; Lanyon, 1986); Environmental Design (e.g. Dutton, 1987); Information Systems; Management Strategies (e.g. Claasen and Kenchington, 1987; Baldwin, 1987); Marine Geosciences (e.g. Kuchler, 1986); Oceanography; Survey and Bathymetry and Crown-of-thorns Starfish (see Moran, 1986 for review; Zann and Eager, 1987). Most research is carried out by outside agencies under contract with the Authority (Kelleher, 1981; Kelleher and Kenchington, 1984).

Economic Value and Social Benefits   The GBR has great commercial and recreational value. Use of the Reef for a great variety of activities is increasing rapidly. Prawns and demersal and pelagic fish are fished and commercial fisheries yielded a landed value of approximately $436 million annually in 1982. Small quantities of coral are collected (Oliver and McGinnity, 1985). The potential for bêche-de-mer exploitation has been investigated (Harriott, 1985b). From April 1979 to March 1980 there were approximately two million visitor trips to the reefs, islands and adjacent mainland. In 1981/82, tourists spent approximately $110 million on recreational activities within the GBR Region including money on island resorts, charter boats, and activities such as island camping (Driml, 1987a). Further details are given in Driml (1987b), Hundloe (1985), McGinnity (1981), Williams (1980) and Wilson (1978).

Disturbance or Deficiencies   Population outbreaks of the Crown-of-thorns Starfish, *Acanthaster planci* occurred from 1962 to 1974 on approximately 300 reefs in the middle section of the Great Barrier Reef (14°-20°S)

destroying a high proportion of hard coral cover (Endean, 1974; Endean and Stablum, 1975; Kenchington and Pearson, 1981). The resulting disintegration of the coral reef communities and increased algal cover (Walsh *et al.*, 1971) became the subject of intense controversy and international concern. A Committee of Inquiry was set up (Kenchington, 1978; Walsh *et al.*, 1971). Rapid recovery was seen in the 1970s on reefs off Innisfail (e.g. Feather Reef) (Cameron and Endean, 1981; Pearson, 1974). There were new outbreaks in 1978 (Cameron and Endean, 1981; Kenchington and Pearson, 1981; Nash and Zell, 1981). Reefs between 16°30' and 17°30'S were infested in 1980/1981, e.g. Howie Reef, 40 km off Innisfail (Endean, 1982). Survey results suggest that the aggregations may not have been as intense as those in the 1960s (Kenchington and Pearson, 1981; Nash and Zell, 1981). The cause of the outbreaks, either a natural phenomenon or the result of manmade perturbations, has been the subject of numerous publications (Baker, 1977; Birkeland and Wolanski, in press; Bradbury *et al.*, 1985a,b and c; Cameron and Endean, 1981; Done and Fisk, 1985; Endean, 1973, 1974, 1977 and 1982; Frankel, 1975 and 1977; Glynn, 1981; Kenchington in press; Kettle and Lucas in press; Lucas, 1973; Moran *et al.*, 1984 and 1985; Moran, 1986; Randall, 1976; Vine, 1970; Walsh *et al.*, 1971; Weber and Woodhead, 1970; Zann and Eager, 1987). There have been attempts to control the outbreaks, including Queensland Government legislation to protect the Triton *Charonia tritonis* and to ban spearfishing with SCUBA following Endean's report in 1969, though these are not strongly enforced (Endean, 1982; Kenchington, 1978). Other control measures include collecting the starfish by hand or injecting them with compressed air or CuSO4. These methods are only reasonably effective for small areas, and have not been sufficient at Green Island (Kenchington and Pearson, 1981). The Commonwealth Government is funding a major research programme on this issue. Coral bleaching has been observed in several areas (Done and Fisk, 1985; Harriott, 1985a; Oliver, 1985).

Agriculture, predominantly sugar cane, and industry have been developing rapidly, producing waste and water discharge, the introduction of pesticides and fertilizers, soil erosion, dredging, accidental spillage and increased shipping. Unsubstantiated observations by locals indicate increased algal cover, decreased water clarity and changes in fish populations and coral communities, all changes associated with disturbed and polluted coral reefs (Bennell, 1980; Kelleher and Kenchington, 1984). However, a workshop conducted by GBRMPA on contaminants in GBR waters concluded that measured levels of most contaminants (in the groups of heavy metals, organochlorines and hydrocarbons) within reef waters are generally close to the lower limits of detection, although in some adjacent coastal waters (notably harbours), concentrations indicative of low to moderate pollution have been recorded (Dutton, 1985). The environmental effects of offshore tourist development is discussed by Kelleher and Dutton (1985).

The poaching of protected Giant Clams *Tridacna gigas* and *T. derasa* by Taiwanese fishing boats used to be severe; 4600 *T. derasa* were poached from the Swain Reefs during the period 1969-1977 (Pearson, 1977). With regular surveillance and the introduction of a 200 mi. (320 km) Australian Fishing Zone this poaching has almost been eliminated (Kenchington pers. comm.).

Legal Protection   The Great Barrier Reef Marine Park Act was passed in 1975 by the Federal Government establishing the Great Barrier Reef Marine Park Authority (GBRMPA).   This Act provides for the establishment, control, care and development of a marine park in the Great Barrier Reef Region as defined in that Act.   Areas of the Region may be declared as part of the Marine Park.   Operations for the recovery of minerals are prohibited, except for the purpose of research and investigations relevant to the establishment, care and development of the Marine Park, or for scientific research (GBRMPA, 1981b).   GBRMPA recommends areas for declaration as part of the Great Barrier Reef Marine Park, prepares zoning plans for these areas, and arranges and undertakes research and investigation relevant to the marine park (Baker, 1977; GBRMPA, 1981b; Kelleher and Kenchington, 1984).

The Great Barrier Reef Marine Park covers 345 000 sq. km or 98.5% of the Region (GBRMPA, 1987a).   There are also marine reserves on Crown land in the form of 13 Fisheries Habitat Reserves.   In 1981 the Great Barrier Reef was declared a World Heritage Site (GBRMPA, 1981b).   Drilling for petroleum in the Great Barrier Reef Region is prohibited by the Great Barrier Reef Marine Park (Prohibition of Drilling for Petroleum) Regulations.   Oil drilling is discussed by Connell (1970).   Dugong and all sea turtles are totally protected with the exception that indigenous people may take them for their own use (Limpus, 1981).

Management   The Great Barrier Reef Marine Park Act outlines and directs the preparation and application of zoning plans for declared sections of the Marine Park and outlines the matters for which regulations may be made to give effect to the Act.   The Regulations provide for the enforcement of zoning plans and describe the scope of the power of the Authority to permit various activities within the Park.

The Marine Park is managed through a co-operative arrangement between the Federal and Queensland Governments.   There is a Ministerial Council comprising two Ministers from each of the two Governments to co-ordinate the policies of the Governments on Marine Park matters.   The Authority and the Federal Minister to whom the Authority is responsible are assisted by an independent Consultative Committee, made up of representatives of government, industry and community bodies.   Officers of the administrative arm of the Authority are Federal Government employees, whereas day to day management is the responsibility of the Queensland National Parks and Wildlife Service.   Two members of the Authority itself are appointed by the Federal Government and the third member is appointed by the Queensland Government.

The GBRMPA policy is to maximize opportunity for human enjoyment by keeping regulation of activity at the minimum considered necessary to achieve conservation objectives (Kelleher, 1981).   The Great Barrier Reef is divided into sections and these are divided into zones with different protection status.   Zoning plans are a form of broad scale management plan and the planning process leading to the application of a zoning plan to a Park section involves a number of discrete steps which are outlined in the Act.   The major feature of the zoning process is the public participation programme.   Divided into two stages, the programme first invites public input

early in the planning process to help construct a draft zoning plan and later invites comments on the suitability of the draft plan produced.   Public participation assists the Authority in gaining information on the resources and uses of the Park, to identify management issues and to separate conflicting uses.   The programme allows interested members of the public to have a say in the construction of a zoning plan and encourages Marine Park users to take some responsibility for the management of the Park.   Day to day management of such a huge area depends to a large degree on public co-operation.

Zoning plans have now been prepared for all sections of the Marine Park.   These are the Mackay-Capricorn, Central, Cairns and Far Northern Sections.   It is the policy of the Great Barrier Reef Marine Park Authority to review zoning plans after five years of operation to take account of changing circumstances.   The area formerly known as the Capricornia Section has been incorporated into the initial zoning plan for the surrounding Capricorn Section, and these two areas together have been re-named the Mackay-Capricorn Section.   The Plan for the Capricornia area has some significant changes to the previous zoning plan.   The review process for the Cairns Section is due to begin in 1989.

The major categories of zones into which a Park Section is divided are:

- General Use "A" Zone (GUA)
- General Use "B" Zone (GUB)
- Marine National Park "A" Zone (MNPA)
- Marine National Park "B" Zone (MNPB)
- Preservation Zone

Generally, about 75% of a Park section is zone General Use "A", another 20% General Use "B", 4% Marine National Park zones and the remainder Preservation.   It should be borne in mind that the great majority of a Park section is open water and the area covered by reefs is relatively small.   The restriction on activities that may be pursued in each zone is as follows (see table overleaf).

Other specific management conditions may be applied to nominated areas within the zoning plan.   These include Replenishment Areas that provide for the replenishment of natural resources in heavily used areas by restricting for a specified period the activities which cause removal of those resources.   There is also provision for Seasonal Closure Areas, designed to protect such areas as bird nesting sites during egg laying and fledgling rearing.   Special Management Areas do not have to be specified in the zoning plan and are designed to allow for specific management at sites where contingencies arise.   These may only be declared after public representations have been considered on their proposed declaration.

The second major management tool available to the Marine Park Authority apart from zoning plans is the power to issue permits for a broad spectrum of activities that are pursued within the Great Barrier Reef Marine Park.   These activities include tourist facilities and programmes, education programmes, aircraft operations, discharge of wastes, collecting, installation and operation of moorings and traditional hunting and fishing.   The main purposes of the permit system are to encourage

	GUA	GUB	MNPA	MNPB	Pres
Trawling	Yes	No	No	No	No
Line Fishing	Yes	Yes	Limited	No	No
Collecting (comm.)	Permit	Permit	No	No	No
Collecting (rec.)	Limited	Limited	No	No	No
Spearfishing	Yes	Yes	No	No	No
Netting (comm.)	Yes	Yes	No	No	No
Bait Netting	Yes	Yes	Yes	No	No
Research	Permit	Permit	Permit	Permit	Permit
Boating, Diving	Yes	Yes	Yes	Yes	No
Tourist Facilities	Permit	Permit	Permit	Permit	No
Traditional Hunting and Fishing	Permit	Permit	Permit	No	No

responsible behaviour in users, separate potentially conflicting uses, limit the time an activity may be conducted, limit the area in which an activity may be conducted, limit the quantity of resources collected, limit the number of people engaged in an activity and gather data for management. Appropriate conditions under which a permitted activity may be conducted are attached to each permit through a permit assessment procedure developed by the Authority.

The Great Barrier Reef Marine Park Regulations (1985) supersede previous regulations and provide for enforcement of zoning plans in effect in mid-1985. These regulations also cover general matters throughout the Park and in unzoned sections, including:

- Prohibiting the taking of specimens greater than 1200 mm of the Grouper *Promicrops lanceolatus* and the Potato Cod *Epinephelus tukula*;
- Prohibiting littering in the Park;
- Prohibiting the use of spearguns with SCUBA or surface-supplied breathing equipment, and of power heads; and
- Providing for control of prescribed activities including the development of offshore structures and the establishment of mariculture operations.

Regulations are made from time to time to give effect to zoning plans and to address other management issues which arise.

GREAT BARRIER REEF MARINE PARK, CAIRNS AND CORMORANT PASS SECTIONS

Geographical Location The Section extends from just north of Lizard Island past Cooktown, north of Cairns, south to Beaver Reef near Tully, on the Queensland coast.

Cormorant Pass Section lies within the outer boundaries of the Cairns Section, mid-point at approximately 14°40'S, 145°39'E. It is 250 km north of Cairns between No Name Reef and Ribbon Reef No. 10.

Area, Depth, Altitude The Cairns Section covers an area of 35 500 sq. km about 10% of the Region, and the air space above to 3000 ft (912 m). Cormorant Pass Section is 341 ha.

Land Tenure See GBR Region

Physical Features The Cairns Section includes 231 known individual coral reefs, banks, patches and shoals; and 25 islands of continental origin with fringing reefs surrounding 23 of the islands. There are 18 low wooded islands and 19 sand and single cays of reefal origin. Cormorant Pass Section is an area of inter-reefal waters.

Within the outer boundaries of the section are several islands of note, including: Lizard Island; Green Island 27 km north-east of Cairns and Michaelmas Cay 40 km from Cairns. Green Island on the inner part of the GBR is a low (approx. 1 m above H.T.) coral cay of 13 ha, with dense forest vegetation and a coral sand beach on its leeward side. Michaelmas Cay, a major bird rookery, is very small and exposed to the south-east trade winds giving long rolling swells for ten months of the year. In November and December it experiences light northerly winds (Edgecombe, 1980). Lizard Island has a small (1 sq. km) lagoon which links four other adjacent islands and is separated into an inner and an outer lagoon by a narrow channel, which also connects them to the main GBR lagoon (Leis, 1981).

Reef Structure and Corals Within the Cairns Section there are banks, patches and shoals, and reefs including wall reefs (ribbon reefs), plug reefs, patch reefs and coastal and island fringing reefs. Ribbon reefs form a near solid wall of barrier reefs along the edge of the continental shelf from Cooktown (for further details *see account for* Far Northern Section).

Fringing reefs are found along parts of the coast including near Kurrimine in the Tully district, Alexander Reef between Cairns and Port Douglas (with heavy sediment cover), the Cape Tribulation area and on the rocky sections of the coast north of Cooktown.

Noteworthy Fauna and Flora More than 130 bird species, including some 35 seabirds, have been recorded within the Section and 77 are known to breed.

Michaelmas Cay is recognized as one of the most important seabird nesting sites in Queensland (GBRMPA, 1983b) and Green Island has thousands of birds including Common Noddies, Sooty Terns, Crested and Lesser Crested Terns (Edgecombe, 1980).

The Dugong *Dugong dugon* feeds and breeds in the seagrass beds and sheltered bays of the northern part of the Section; Green Turtles *Chelonia mydas*, Loggerheads *Caretta caretta*, Leatherbacks *Dermochelys coriacea* and Hawksbills *Eretmochelys imbricata* are also seen (Ivanovici, 1984). The nearshore open waters around Lizard Island are important nursery grounds for several taxa of fish (Leis, 1981); species richness is high and similar to that at One Tree Island (Talbot and Gilbert, 1981). About 850 species of fish have been recorded from the Section (GBRMPA, 1983b). Giant Clams are relatively common throughout the Section.

The Cormorant Pass Section is an important tourist destination due to its group of Potato Cod *Epinephelus tukula* (Ivanovici, 1984), which are very tame, having been hand fed by divers for at least six years.

Scientific Importance and Research   Lizard Island Research Station was established in 1973 and has a platform for field work on Carter Reef. A smaller field station, established by the Queensland Department of Primary Industries, Northern Fisheries Research Centre, is located on Green Island. Low Isles (off Port Douglas) was the site of the base for the Royal Society Expedition of 1928, which carried out one of the first comprehensive studies on the Reef; it was used as a base for subsequent expeditions in 1958 and 1973.

Economic Value and Social Benefits   The area is adjacent to a number of rapidly growing tourist centres between Tully and Cooktown. Approximately 130 000 tourists visit Green Island each year (GBRMPA, 1983b) which has an underwater observatory in 5 m of water, built in 1953. Tourist facilities and services are also operated on Lizard Island, out of Port Douglas to Agincourt Reef and Low Isles (GBRMPA 1983b; Hundloe *et al.*, 1981), and also to Hastings and Norman Reefs and Michaelmas Cay. The group of Potato Cod at Cormorant Pass is a popular attraction for SCUBA divers (Ivanovici, 1984). Dive tours operate throughout the area. High speed catamarans and semi-submersibles are very common and have the potential for putting up to 2000 people per day on the reefs off Cairns. Recreational fishing yields a demersal fish catch with an estimated value of over 3-4 million Australian dollars, comparable to the most important commercial fishery in the area, the prawn fishery; marlin fishing is also important in this Section (Craik, 1981a). There is some traditional hunting and fishing for turtles and dugong.

Disturbance or Deficiencies   The first reported outbreaks of the Crown-of-Thorns Starfish *Acanthaster planci* in the GBR Region were recorded from Green Island in 1962. This resulted in high coral mortality continuing until 1967 (Endean and Stablum, 1975). By 1979 a good coral cover had been re-established, but renewed outbreaks occurred, again destroying the coral. Queensland Fisheries Service surveys recorded large populations of *Acanthaster* giving an estimate of 1-2 million in 1980 (Nash and Zell, 1981). A survey in 1981 reported 90% mortality (Cameron and Endean, 1981;

Endean, 1982). Limited control measures included injecting with CuSO4 but this was not effective at the scale attempted and is very expensive (estimated cost for killing 1 million starfish exceeds $800 000). Whether massive control programmes should be attempted is hotly debated (Kenchington and Pearson, 1981).

About 70% of the Section is close to population centres, which has led to heavy use of the reefs, particularly off Cairns. By 1976 coral damage and sedimentation caused by boat propellers was evident at Green Island and there was destruction of the reef flat from walking and shell and coral collecting. The lengthening of the jetty has resulted in erosion of the foreshore and a deepening of the channel. Sewage from the hotels and litter pollution is seen (Edgecombe, 1980). There is evidence to suggest that some fish populations, particularly on nearshore reefs, are under heavy fishing pressure. Many outer reefs of the Section exhibit the effects of clam removal by foreign (mainly Taiwanese) fishermen. The long-term effects of this activity are largely unknown. There has been considerable concern about the impact of soil run-off on the fringing reef along the Daintree coast (Craik and Dutton, 1987).

Legal Protection   The Cairns Section was declared in 1981 and the zoning plan came into effect two years later (GBRMPA, 1983a). In operation for five years, the Zoning Plan is now due for review and the re-zoning process has begun. A survey of the Section user groups conducted by the Consultative Committee in 1987 found general acceptance of the Zoning Plan and its objectives.

Cormorant Pass Section was proclaimed on 21.10.82 to protect an areas of 3 sq. km near Lizard Island on account of the group of tame Potato Cod *Epinephelus tukula* (Hegerl, 1981; Ivanovici, 1984). Spearfishing and line fishing other than trolling are forbidden (GBRMPA, 1983b). Both sections are protected under the provisions of the Great Barrier Reef Marine Park Act, 1974, the Cairns Section Zoning Plan and the Great Barrier Reef Marine Park Regulations.

In 1938 the Queensland Government formed Green Island Marine National Park giving full protection to marine organisms but allowing limited recreational fishing (Gare, 1976). Green Island is also designated as a bird sanctuary along with neighbouring Upolo Cay and Michaelmas Cay (Edgecombe, 1980). In 1960 the Park came under the National Parks - Forestry Act. Other National Parks in the area and adjacent to coral reefs include the Barnard Group of islands, the Frankland Islands, the Hope Islands, Cape Tribulation and several islands in the Lizard Island area. A Scientific Research Zone is located adjacent to the Starcke River to protect an area of seagrass and dugong.

Management   See GBR Region.

Recommendations   It is anticipated that many intertidal areas around these cays and islands will be declared Marine Parks under the Queensland Marine Parks Act of 1982, and will be zoned to complement the Great Barrier Reef Marine Park.

GREAT BARRIER REEF MARINE PARK, CENTRAL SECTION

Geographical Location   Extends from the Whitsunday/Lindeman Island Groups to Dunk Island off the city of Innisfail.

Area, Depth, Altitude   Approximately 77 000 sq. km; comprises about 22% of the Great Barrier Reef Region.

Land Tenure   see GBR Region

Physical Features   There are 596 individual coral reefs, patches, shoals and banks including fringing reefs, 13 coral cays and 194 islands of continental origin (193 with fringing reefs). There are mangroves and seagrass beds.

Reef Structure and Corals   Reefs of the Central Section of the Great Barrier Reef Marine Park vary according to their distance from the coastline. The outermost or shelf-edge reefs are most exposed and have the clearest deepest water. The innermost are influenced by the flood waters of rivers and the middle or mid-shelf reefs are of varying shapes and sizes, often with very different combinations of coral communities (Veron, 1986). Soft corals on several reefs in the Section are described by Dinesen (1983). Parnell (1986) provides information on the fringing reef at Orpheus Island.

Noteworthy Fauna and Flora   Noteworthy fauna and flora found in the Central Section are similar to that of the other sections of the Marine Park. Fish communities in the Section have been studied by Russ (1984a and b) and Williams and Hatcher (1983). The Humpback Whale *Megaptera novaeangliae* is seen throughout the Section between July and October. Humpbacks are known to calve in the area as females with new born calves are regularly sighted and an occasional birth is observed, usually in shallow water close to land or a reef.

Large numbers of *Dugong dugon* have been sighted during aerial surveys in several bays along the coast although populations are less significant than those in the Far Northern Section. Dugong are often found in sheltered bays as they provide habitat for seagrasses which are the staple diet for Dugong and the Green Turtle *Chelonia mydas*. Over a dozen species of seagrass have been recorded in bays in and adjacent to the Central Section, including *Halodule uninervis* and *Zostera capricorni*.

Twenty two species of sea birds of eight families have been recorded in the Section and nine of these have been found breeding. These records cover 16 islands and groups of islands of which the more significant are Eshelby Island and the Brook Group.

Scientific Importance and Research   AIMS research has concentrated on Pandora, Rib, Myrmidon, Flinders, Britomart and Davies reefs; the James Cook University of Queensland also carried out research on reefs in this Section (GBRMPA, 1983b). There is a small research station on Orpheus Island in the Palm Island Group, run by James Cook University.

Economic Value and Social Benefits   The small number and ephemeral nature of the cays limits opportunities for island-based activities, and the relatively large distances between reefs and from the mainland, combined with many periods of rough seas, can limit access to the reefs. However, there is a high diversity of marine life supporting recreational activities (particularly in the Whitsunday/Lindeman groups), and commercial activities. Over 200 charter boats and 4000 private boats are based in centres adjacent to the Section and these provide the main access. The main recreational activities are diving, snorkelling, photography, reef fishing and light tackle game fishing. There are three Historic Shipwrecks, the Yongala, the Guthenberg and the Foam. The Yongala is a popular dive location. Reef walking is only possible occasionally as few reefs are exposed at low tide. Several semi-submersible coral viewing vessels are located off Townsville and in the Whitsundays. There are numerous resorts, many of which cater for diving. The major commercial activities are trawling for prawns, bugs and scallops, reef fishing, trolling for mackerel and the provision of tourist services (GBRMPA, 1983b).

Disturbance or Deficiencies   Coral bleaching has been observed on Myrmidon Reef (Fisk and Done, 1985). The advent of large, high-speed catamarans is causing increasing day visitor pressures on the resources of the Section. The use of smaller high-speed charter catamarans for fishing activities is also expected to increase pressure on the area's fish resources. The floating hotel installed and operating at John Brewer Reef since January 1988 is probably the first of several such developments and its impact on the natural resources of the Reef is being closely monitored by the Authority. Environmental protection measures of a high standard have been incorporated into the design of and operational procedures for the Hotel and its environmental impact is not expected to be significant, although there is considerable concern.

Legal Protection   Proclaimed a Section in 15.10.84; protected under the provisions of the Great Barrier Reef Marine Park Act, 1975 and the Great Barrier Reef Marine Park Regulations. A zoning plan has been prepared (GBRMPA, 1987b). The following National Parks occur within the area: Magnetic Island, Whitsunday Islands, Gloucester Island, Repulse Islands, Orpheus Island, Hinchinbrook Island and Channel, Nypa Palms and the Rockinham Bay area off Cardwell (including Dunk Island, the Family Islands and Brook Islands). The following areas, all in the Rockingham Bay area, are Fishery Habitat Reserves: Hinchinbrook Channel, Dallachy Creek, Meunga Creek, Murray River, Wreck Creek and Hull River. The "Yongala" and "Foam" Historic Ship Protected Zones lie within the area.

Management   *See account for* Great Barrier Reef Region.

GREAT BARRIER REEF MARINE PARK, FAR NORTHERN SECTION

Geographical Location   Extends from just north of Lizard Island to the northern tip of Cape York.

Area, Depth, Altitude   83 000 sq. km, covering about 23% of the Great Barrier Reef region.

Land Tenure   (see GBR Region)

Physical Features The Section includes 573 individual coral reefs, 168 coral cays and 58 continental islands (Anon., 1984). The continental shelf drops sharply east of the reef. There are seagrass beds and mangroves along the coast.

Reef Structure and Corals The reefs of the Section consist of an outer barrier of ribbon and plug reefs, vast areas of large submerged shoals and, on the coast, an inner line of reefs. Many of the cays and islands and much of the mainland have fringing reefs (GBRMPA, 1983b). The coral reef system varies from north to south and from the coast seawards. Ribbon reefs form a near solid wall of barrier reefs along the edge of the continental shelf north to Pandora Entrance. The eastern side is fully exposed to the open ocean swell and for ten months of each year is pounded by heavy surf. This outer face plunges into the abyssal depths of the Queensland Trench and has some of the most spectacular underwater scenery in the Region. The water is very clear and corals are found to 50 m depth where they are still diverse and some have even been dredged from 100 m. North of Pandora Entrance the barrier line of reefs develops a pattern that looks like river deltas and are known as deltaic reefs. These form such a barrier to tidal water movements that standing waves 2 m high may be formed (Veron, 1986).

North of Princess Charlotte Bay there is an immense complex of reefs, cays and high islands. These are shallower than those further south and the water is relatively turbid but as there are few rivers, reefs are found right up to the coastline (Veron, 1986). Additional information is given in GBRMPA (1983e).

Noteworthy Fauna and Flora Raine Island is a major seabird rookery in the GBR Region. Raine Island and Pandora Cay form one of the largest breeding grounds for the Green Turtle *Chelonia mydas*. Islands in the north of the area are nesting sites for the Hawksbill Turtle *Eretmochelys imbricata*. The seagrass beds and sheltered bays provide breeding grounds for significant numbers of Dugong *Dugong dugon* (GBRMPA, 1983b).

Scientific Importance and Research Contains some of the most important turtle and seabird nesting sites in the West Pacific and is the least disturbed part of the Great Barrier Reef. Probably the only area where the opportunity exists to protect a near-pristine complete cross section of the Reef, from outer Barrier to inshore reefs (Anon., 1984). However, little research has been carried out (GBRMPA, 1983b), although the natural history of Raine Island is described by Stoddart *et al.*, (1981).

Economic Value and Social Benefits The least used part of the Barrier Reef; some commercial fishing, shipping, recreational and traditional fishing, collecting, cruising, yachting, diving and research (Anon., 1984; GBRMPA, 1983e). The area lies on an important commercial shipping route and contains the historic wreck *Pandora*. There is a long history of aboriginal use of the area; tribal ownership has been established over particular areas of coastline and reef (GBRMPA, 1983b). Additional information is given in GBRMPA (1983e).

Disturbance or Deficiencies Generally slight although the activities mentioned above could have some adverse impacts, particulars bottom trawling.

Legal Protection Proclaimed a Section of the GBR Marine Park 31.8.82 and protected under the provisions of the Great Barrier Reef Marine Park Act, 1975, the Great Barrier Reef Marine Park Regulations and the Far Northern Section Zoning Plan (GBRMPA, 1983b) which came into effect in February 1986. Raine Island is a reserve under the Trusteeship of the Director of the Queensland Department of Community Services. Cape Melville is a National Park on the adjacent mainland and the Pandora Historic Shipwreck Protected Zone (under the Federal Historic Shipwrecks Act 1975) lies within the area. There are three Fishery Habitat Reserves in the area: Princess Charlotte Bay, Silver Plains, Temple Bay and Escape River (Ivanovici, 1984).

Management *See account for* Great Barrier Reef Region.

GREAT BARRIER REEF MARINE PARK, MACKAY-CAPRICORN SECTION

Geographical Location The Mackay-Capricorn Section lies off the central Queensland coast and extends from just south of the Lindeman Island Group to the southern boundary of the Great Barrier Reef Region and from the coastline for a large part of its length up to 280 km offshore; it is adjacent to the coastal centres of Mackay, Yeppoon and Gladstone. This is a new section combining most of the former Southern and Inshore Southern Sections proclaimed in 1983, except for the part now included in the Central Section.

The Capricornia Section, which formerly covered the reefs and waters around the Capricorn and Bunker group of islands, is now part of the Mackay-Capricorn Section.

Area, Depth, Altitude The Mackay-Capricorn Section covers 149 000 sq. km, comprising about 41% of the Great Barrier Reef Region. The area formerly known as the Capricornia Section covers approximately 11 800 sq. km and covers just over 3% of the GBR Region.

Land Tenure see GBR Region.

Physical Features Outside the area formerly covered by the Capricornia Section, the Mackay-Capricorn Section includes 1059 individual coral reefs, shoals and banks, including fringing reefs; there are 58 coral cays and 405 islands of continental origin, as well as Blue Holes. Swain Reefs and Pompey Reefs are dense areas of patch reefs separated by deep narrow channels. Cays at Swain Reefs are described by Flood and Heatwole (1986).

The area formerly known as the Capricornia Section is a subtropical group of reefs including 14 cays (the Capricorn (northern) and Bunker (southern) Groups) lying on the Tropic of Capricorn. They include 21 major reefs each with a sharply defined margin and generally rounded shape. Most have shallow lagoons and many have cays, some with dense vegetation (Veron, 1986). One Tree Island, 100 km east of Rockhampton, lies approximately 18 km from the edge of the continental shelf. It is a lagoonal platform reef with a shallow complex lagoon system. The two larger lagoons cover an area of 12 sq. km and are separated at low tide by a narrow neck of reef and sand. Broad sub-tidal sand sheets behind the eastern and southern reef flats grade into the lagoon which has a low tide depth ranging from 2.5 m (south) to 4 m (east) to 5-10 m (north-west). The island lies on the south-east corner of the sand sheets and measures 4.7 km x 2.7 km (Frith, 1981; Kinsey 1972; Marshall and Davies, 1982). Heron Island, a coral cay, lies 73 km off the mainland on the Tropic of Capricorn. Fairfax Island is a coral cay consisting of two small islands on an egg-shaped reef. Hoskyn Island is a double cay. Lady Musgrove Island is a cay surrounded by an extensive coral reef. Lady Elliot Island, the only other cay in the former Capricornia Section lies somewhat further south and is not considered to belong to either the Bunker or Capricorn Groups. It is described in Hall (1984).

Reef Structure and Corals The former Capricornia Section has 12 shoals, two drying wall reefs, six drying closed ring reefs, 13 drying platform reefs and one platform reef. The reefs are separated by deep water allowing good circulation and usually luxuriant growth; 72% of all Great Barrier Reef coral species occur here (Veron, 1986). The reef at Heron Island extends 5 miles (8 km) to the south, giving a reef flat largely exposed at low tide. One Tree Island has a well developed, unbroken reef crest formed by a narrow coral-algal rim on the north-west and southern sides, and by a higher, broader series of ephemeral shingle banks along the eastern side. The reef crest is of compacted coral rock with small pools and scattered dead coral boulders and is exposed at low tide (McMichael, 1974). Llewellyn Reef is a good example of a raised lagoonal platform (Kelleher and Kenchington, 1984).

Swain Reefs is a vast expanse of reef patches of varying shapes and sizes, some with small cays. The eastern and southern reefs are small and have a rugged, wave-washed appearance as they are exposed to big ocean swells. The western reefs are larger, less exposed and many have lagoons with extensive coral growth. The Pompey Complex includes reefs, channels, sandbars and lagoons. It is surrounded by deep water and broken by deep channels with very strong tidal currents; the reefs make a major barrier to tidal movement resulting in short-lived cascading torrents during tidal changes (Veron, 1986).

Noteworthy Fauna and Flora In the former Capricornia Section 859 species of fish have been recorded (Russell, 1983; Talbot *et al.*, n.d.). The richness of the fauna of Heron Island is well known, particularly the diversity and tameness of the fish at the bommie. Clams and holothurians abound at One Tree Island (McMichael, 1974) and the species richness of the coral reef fishes is high and similar to that seen at Lizard Island (Talbot and Gilbert, 1981).

The Capricorn/Bunker Group includes a third of the islands in the Great Barrier Reef Region known to be important to nesting turtles. Among these, Wreck Island is one of the three largest nesting sites in the world for the Loggerhead Turtle *Caretta caretta* (Bustard, 1974), over 1000 female Loggerheads congregating there in some seasons (Limpus, 1981). Wreck Island is also an important nesting site for the Green Turtle *Chelonia mydas* as is Hoskyn Island. Elsewhere in the Mackay-Capricorn Section, Avoid and Wild Duck Islands

are the second and third largest nesting sites in the GBR Region for the Flatback Turtle *Chelonia depressa*.

Over 30% of the important seabird nesting sites of the whole GBR Region are situated on the Capricornia Islands (GBRMPA, 1983b). Masthead, One Tree, North West and Wilson Islands are the most important; other major sites are Lady Musgrave, Fairfax, Hoskyn, Tryon and Heron (Lavery and Grimes, 1971). The principal breeding ground of the Brown Booby *Sula leucogaster* is on Fairfax Island; other birds on this island include the Wedge-tail Shearwater *Puffinus pacificus*, estimated at 16 600, and Eastern Reef Herons (Reef Herons) Reef Herons *Egretta sacra*, Roseate Terns *Sterna dougallii*, Black-naped Terns *Sterna sumatrana*, Crested Terns *Sterna bergii*, Little Terns *Sterna albifrons*, Lesser Crested Terns *Sterna bengalensis* and 17 000 White-capped Noddy Terns *Anous minutus* (Shipway, 1969). Hoskyn Island supports the Brown Booby, Wedge-tail, Bridled Tern *Sterna anaethetus*, and White-capped Noddy Tern. Many other birds occur including sea eagles and Ospreys *Pandion haliaeetus*. Over 50 species have been found on Heron Island, 30 occurring regularly (Kikkawa 1976; GBRMPA, 1983b).

Scientific Importance and Research The Capricorn and Bunker Groups are important as the most southerly groups of coralline islands in the Indo-Pacific except for Lord Howe Island. The high diversity of coral reef biota and easy accessibility of the reefs make them an important scientific resource. Llewellyn Reef is a fine example of a raised lagoonal platform characteristic of the Capricorn/Bunker Group of Reefs, and is therefore of importance as a reference site (Kelleher and Kenchington, 1984). The vegetation of the coral cays is of intrinsic interest and there have been long-term studies on the dynamics of the flora. The bird life has been extensively researched (Kikkawa (1976). Swain Reefs are the least studied of the GBR Region on account of their remoteness (Veron, 1986).

In 1951 the Great Barrier Reef Committee initiated the Heron Island Research Station, now operated by the University of Queensland. A large number of scientists, both national and international, visit the station (over 4000 in 1979) and there are facilities to provide for up to 50 scientists. The Australian Museum established a Research Station on One Tree Island in 1966, and it has been operated since 1975 by Sydney University. These two stations provide one of the world's significant sources of base-line data on coral reefs. Research is initiated by the GBRMPA for management purposes. For example surveys have been carried out on the coral trout *Plectropomus leopardus*, a popular angling species, to estimate relative population size and density (Craik, 1981b).

Economic Value and Social Benefits Historically the reefs and islands of the area have been subject to commercial exploitation, including guano mining and turtle harvesting. The Capricorn and Bunker Groups are heavily used for recreation. Large numbers of people from the neighbouring coastal towns visit the islands by boat either to camp or on day trips. There is a resort on Heron Island catering for about 180 guests (Kelleher and Kenchington, 1984). Lady Elliott Island supports another tourist resort (Hall, 1984) and camping and reef associated activities have become increasingly popular on several other islands in the area (GBRMPA, 1983b). Line fishing involving the chartering of boats is a popular activity. A total estimate of 390 000 kg of reef fish are caught by recreational fishermen; there is a commercial catch of 130 000 kg (Hundloe *et al.*, 1981). Popular fish include coral trout *Plectropomus* spp., Red Emperor *Lutjanus sebae*, Sweet Lip *Lethrinus chrysostomus* and cod *Epinephelus* spp. (Craik, 1981a).

Disturbance or Deficiencies The area is under increased use owing to its accessibility from growing coastal towns and the high population density of the southern Queensland coast. Heavy use by people camping and fishing has resulted in reef damage from anchors and walking, rubbish, coral and shell collecting and a reduction in fish populations including a decrease in individual size and age. This is particularly noticeable on Masthead, North West and Lady Musgrave Islands.

Human impact is particularly high on Heron Island because of the resort, and this extends to nearby reefs such as Wistari. A drop in live coral cover from 41% to 8% due to reef walking has been estimated in some areas (Woodland and Hooper, 1977). Intense research activity on Heron Island (850 man-weeks in 1976) and One Tree Island may also disturb the reefs. Heron Island is being eroded with a consistent westward movement of sediment within the beach zone. Dredging and blasting of the reef to provide boat access has amplified the erosion, and the walls constructed are insufficient (Flood, 1979).

Many animals have been introduced (chickens, rats, dogs, etc.) which have had a significant impact on the turtle population, and trees have been cut down.

Legal Protection The Capricornia Section was declared part of the GBR Marine Park in 17.10.79 and is protected under the provisions of the Great Barrier Reef Marine Park Act 1975, the Capricornia Section Zoning Plan (GBRMPA, 1980a and b, 1981a) and the Great Barrier Reef Marine Park Regulations. It has been fully zoned and managed since July 1981. The Mackay-Capricorn Section was proclaimed on 15.10.84 and regulations came into operation on 7.11.83; a zoning plan has recently been prepared (GBRMPA, 1988).

Heron Island Reef and Wistari Reef were reserved in 1965 as a Marine Park under Queensland Fisheries legislation (Gare, 1976; Ivanovici, 1984) and fishing restrictions were imposed in 1974 (Craik, 1981b). The islands of Fairfax, Hoskyn, Heron and Lady Musgrave are Queensland owned and were established as National Parks in 1937. Other National Parks within the Section or adjacent to the reefs in the area include Keppel Bay Islands (20 km off the Yeppoon coast), Tryon, North-west and Masthead Islands within the Capricorn Group, Sir James Smith Islands and adjacent Brampton Island, the Newry Group, Bushy Island, Prudhoe Island, and the nearby Beverley Group and Guardfish Cluster, the Percy Group and Wild Duck Island.

There are Fish Habitat Reserves at Repulse Bay, Corio Bay, Bustard, Gott Colosseum, Innes, Rodd's harbour, Eurimbula and Round Hill (the last seven of these in the Gladstone region north of Bustard Head), and Fish Sanctuaries at Hook Island (off Whitsunday Islands), Middle Island (off Yeppoon) and Eurimbula (Ivanovici, 1984).

Management See Great Barrier Reef Region Account

LIHOU REEF NATIONAL NATURE RESERVE

Geographical Location Coral Sea Islands Territory; off the northern coast of Queensland in the central region of the Coral Sea (within 16°-18°S, 151°-153°), 560 km east-north-east of Townsville, 350 km east of the Great Barrier Reef.

Area, Depth, Altitude 8436 sq. km.

Land Tenure Commonwealth of Australia.

Physical Features Sandy cays, vegetated islets and reefs.

Reef Structure and Corals There is a horseshoe-shaped reef system with spectacular topography. Corals are described in Veron and Wallace (1983).

Noteworthy Fauna and Flora 'Several of the cays provide important nesting sites for turtles and seabirds, including Red-footed Boobies and Greater Frigatebirds. Anne and Nellie Cays are major seabird nesting sites, and Georgina Cay has a breeding colony of the threatened Little Tern (Anon., 1986). Southern Right and Humpback Whales may occur.

Scientific Importance and Research Australian National Parks and Wildlife Service (ANPWS) have conducted five biological surveys in the Coral Sea Islands Territory covering the Lihou Reef Reserve. These have obtained information on breeding and nesting of seabirds, and on the cetacean populations. In August 1984, six cays were surveyed by the ANPWS, and in October 1984 a major resource survey was carried out in collaboration with the Australian Survey Office (Anon., 1986).

Disturbance or Deficiencies Foreign fishing vessels take fish and clams in the area. Pollution from shipwrecks is a threat.

Legal Protection The area was declared a reserve 3.8.82 under the National Parks and Wildlife Conservation Act 1975 by the Governor-General of the Commonwealth of Australia. The legislation provides complete protection for all marine wildlife.

Management ANPWS manage the reserve. Management plans are being prepared.

LORD HOWE ISLAND PERMANENT PARK PRESERVE

Geographical Location South Pacific Ocean, 700 km north-east of Sydney, New South Wales; 31°30'-31°50'S, 159°00'-159°17'E. The Group includes Balls Pyramid, the Admiralty Islands and adjacent islets to the south of Lord Howe Island.

Area, Depth, Altitude The islands cover about 15 sq. km of which Lord Howe Island represents 14 sq. km (Anon., 1981). The lagoon at Lord Howe Island is about 5 x 2 km wide at its broadest point and is very shallow, 1-2 m at low tide (Allen and Paxton, 1974; Pollard, 1977; Sutherland and Ritchie, 1974). Outside the lagoon the reef drops to 15-20 m and then gradually slopes to deeper water. The 200 m line is approximately 7-12 km off shore (Allen and Paxton, 1974).

Land Tenure State of New South Wales, private ownership by leasehold subject to a number of conditions (Anon., 1981; Randall, 1977).

Physical Features The islands are the eroded remnants of a volcano situated on the western ridge of the Lord Howe Rise, emerging from depths of 2000 m. The Admiralty Islands comprise seven islets north-east of Lord Howe. Ball's Pyramid is a monolithic spire approximately 20 km south-west of Lord Howe and rises to 550 m (Allen and Paxton, 1974). Lord Howe Island has two mountains, Mount Gower (875 m) and Mount Lidgbird (777 m) in the south (Anon., 1981). The western concave shore borders a shallow lagoon which is protected from the open sea by a barrier reef which has two gaps: Erscott's Passage (3-5 m deep) in the south and North Pass (4-6 m deep) (Allen and Paxton, 1974). Water clarity is good at the reef crest, which is exposed at low tide, but poorer in the lagoon which has correspondingly depauperate coral growth (Pollard, 1977; Anon., 1981). The outside face of the reef experiences strong tidal currents (Pollard, 1977; Randall, 1977). Away from the lagoon the shoreline is steep with rocky cliffs extending to the water's edge and down to depths of 20 m or more. Caves, ledges, fissures and archways are common features. Surface water temperatures range from 17°C in winter to 25°C in summer. Shallow lagoon waters are usually warmer (Allen and Paxton, 1974).

Reef Structure and Corals In former geological time, the reef was more extensive (Veron, 1986). The alga *Lithothamnion* and the coral *Acropora cuneata* are the main reef building organisms (Anon., 1981; Veron and Done, 1979). The range of habitats at Lord Howe Island is illustrated by species such as *Pocillopora damicornis*, *Stylophora pistillata*, and *Acropora* spp. which exhibit shallow water reef crest forms and deep water reef face forms. A total of 57 coral species in 33 genera have been recorded, two of which are undescribed and not known from the Great Barrier Reef. On the whole, species and growth forms are similar to those found further north, e.g. the common *Goniastrea australis*, *Favia speciosa*, *F. abdita*, *F. halicora*, and *Plesiastrea cf. versipora* form spherical colonies of various sizes as seen on the Great Barrier Reef (Veron, 1974). However growth forms and species composition can differ from their tropical counterparts, and there is a range of tropical and temperate species (Anon., 1981). For example, *Scolymia australis*, a non-tropical species occuring on the southern coast of Australia, has its northern limit at Lord Howe Island where it is indistinguishable from small specimens of *S. vitiensis*, a tropical species (Veron, 1981; Veron and Pichon, 1976). *Acanthastrea* sp. and *Seriatopora hystrix* are normally restricted to tropical waters. Coral growth on the outer reef face extends to 20 m with low lying forms (Pollard, 1977). A list of stony corals occuring at Lord Howe Island is given in Veron (1974) and Veron and Done (1979).

Noteworthy Fauna and Flora There is a rich diversity of fish and invertebrates, particularly echinoderms (Pollard, 1977). There are a number of zoogeographical components to the fish fauna: tropical, oceanic and southern temperate (Anon., 1981). Tropical species dominate, the four common groups being the Labridae, Pomacentridae, Gobiidae and Chaetodontidae which also dominate on the Great Barrier Reef (Allen and Paxton, 1974), but the richness is lower here possibly due to the more variable temperature and more temperate facies of the reef slopes (Talbot and Gilbert, 1981). Fishes are listed in Allen *et al.* (1976); 447 species in 107 families have been recorded, 4% of which have only been found in Norfolk Island and Middleton Reef waters (Anon., 1981). *Chaetodon tricinctus* is known only from Lord Howe Island. Others are wide-ranging Indo-Pacific species. Large shoals of mullet, *Myxus elongatus* and Silver Drummer *Kyphosus fuscus*, which can be hand fed, occur at Ned's Beach (Randall, 1977). The commonest shark is *Carcharhinus galapagensis*; Grey Whaler Sharks occur outside the reef. At North Islet in the Admiralty Group there is a high diversity of fish (Allen and Paxton, 1974). The most diverse fauna is found localized in "holes" just inside the reef crest, e.g. Erscott's Hole in the south, Blunt's Hole in the north and Sylph's Hole and Comet's Hole in the centre of the lagoon. Further information is given in Pollard and Burchmore (1985).

At least two species of turtle occur (Pollard, 1977). There are significant populations of seabirds, particularly on the Admiralty Islets (Randall, 1977). 12 species nest on the island group and a further 18 species visit the area. There is an estimate of 100 000 pairs of Sooty Terns, *Sterna fuscata* (one of its most southerly breeding sites). The area is the only known breeding locality for the Providence Petrel *Pterodroma solandri* with an estimate of 96 000 breeding pairs in 1975. There are significant numbers of breeding Flesh-footed Shearwaters, *Puffinus carneipes* and Wedge-tailed Shearwaters *P. pacificus*; an estimate of 4000 Little Shearwaters *P. assimilis* on Roach Island; and other breeding seabirds are Black-winged Petrel *Pterodroma nigripennis*, White-bellied Storm Petrel *Fregatta grallaria*, Masked Booby *Sula dactylatra* (most southerly breeding site), tropicbird *Phaethon rubricauda*, noddy *Anous stolidus* (one of most southerly breeding sites), Grey Ternlet *Procelsterna albivittata*, White Tern *Gygis alba* (Anon., 1981).

Scientific Importance and Research The Lord Howe Islands are considered to be an outstanding example of an island system developed from submarine volcanic activity. The reef is unique in that it is a transition between an algal and coral reef caused by oscillations of hot and cold water around the island, and is one of the southernmost reefs in the world. A number of zoogeographic components are represented by the fauna and the Group is central to a region of significant endemism among fish and other organisms (Anon., 1981; Randall, 1977). A more recent assessment of this area can be found in Pollard and Burchmore (1985).

Economic Value and Social Benefits Tourism is important and about 4000 visitors are flown in each year (Anon., 1981; Randall, 1977). A commercial dive operation has recently been established on the island. Recreational line fishing is popular, and the catch is used for the visitors (Pollard, 1977). The Saddled Rock

Cod *Ephinephelus daemeli*, bluefish, double-header, trevally, large numbers of kingfish, salmon, garfish and sandmullet are caught in the lagoon and on the reef. Wahoo, spanish mackerel, tuna and marlin are caught beyond the reef, and prawns, crab and lobster are also taken (Pollard, 1977).

Disturbance or Deficiencies   Periodic sub-antarctic currents denude the coral reefs, but tropical currents provide recolonizing larvae (Anon., 1981). Human impact is neglible due to local measures preventing overfishing and the absence of any commercial fishing industry (Anon., 1981). There is possibly some pollution from sewage effluent and a rubbish dump at the south end of Lord Howe Island Lagoon (Pollard, 1977). Coral die-off has been observed on some parts of the reef which may be due to seepage of polluted groundwater (Pollard, pers. comm. 1988).

Legal Protection   The land areas are protected by the Lord Howe Island Permanent Park Preserve established in 1981. In 1961, local concern led to a ban on spearfishing, netting and dynamiting in the lagoon, and complete protection for Lionfish *Pterois volitans* and a large wrasse, *Coris aygula*. Since the early sixties the islanders have actively enforced restrictions on fishing and collecting and established an unofficial marine preserve at Ned's Beach (Anon., 1981). In 1967 a marine reserve was established for the main section of the reef (Randall, 1977). The Lord Howe Island Group was declared a World Heritage Site in 1982.

Management   The Lord Howe Island Act 1953 established a Board to oversee the management of the islands resources (Lord Howe Island Board, Lands Department Building, 23 Bridge Street, Sydney 2000). The Board includes an officer of the National Parks and Wildlife Service (P.O. Box N189, Grosvenor Streeet, P.O. Sydney 2000).

Recommendations   Legislation should be provided for the protective measures taken by the islanders and a marine park should be established (Randall, 1977), involving multiple use zonation. A full marine ecological survey should be conducted and possibly a small research station built on the lagoon shore at Lord Howe Island. The ecological implications of constructional work on jettys and slipways and the establishment of moorings must be carefully considered (Pollard, 1977). Some of these recommendations have been carried out in the course of a marine environmental study, following which a proposal for an aquatic reserve was drawn up (Pollard and Burchmore, 1985).

SOLITARY ISLANDS PROPOSED MARINE RESERVE

Geographical Location   2-11 km North-north-east of Coffs Harbour off the northern coast of New South Wales, between 29°55'S and 30°14'S. A group of six islands: Split Solitary, South Solitary (or Lighthouse), South-west Solitary (or Groper), North-west Solitary and North Solitary Islands, and North Rock. The most northern island, North Solitary lies approximately 11 km off shore (Anon., 1977).

Area, Depth, Altitude   The islands stretch for a distance of approximately 46 km. The two highest islands, North and South Solitary are over 40 m high (Anon., 1977).

Physical Features   The Reserve comprises the five rocky islands together with a number of smaller exposed rocks and reefs (Anon., 1977). North-west Solitary, South-west Solitary and Split Solitary are located near the 20 m depth contour between 2 and 5.2 km offshore and are approximately similar in size and appearance, each having steep rugged walls and a protected rocky seafloor on their northern and western sides. North and South Solitary are 7 and 11 km offshore respectively and are situated near the 40 m depth contour. They are much larger than the inshore islands, have a more irregular, complex shape and are exposed to rougher although markedly less turbid seas (Veron *et al.*, 1974).

Reef Structure and Corals   Corals are the dominant sedentary fauna and occur over wide areas and depth ranges and are frequently intermixed with sponges, ascidians, soft corals and algae (Veron *et al.*, 1974): 17-18 genera including 34 species of hermatypic corals have been recorded. At least 11 of the genera are at their southernmost limit. In some localities there is 100% coral cover, in other areas an assemblage of corals, sponges, algae and ascidians occurs (Veron *et al.*, 1974). *Acropora* and *Turbinaria* are dominant (Veron, 1986). Corals cover the rocky seafloor in partly sheltered embayments. Subjective observation suggests that the coral formations around North-west and Split Solitary Islands are essentially similar to those around South-west Solitary Island (Veron *et al.*, 1974). There are four sectors, each characterized by different coral, ascidian, sponge and algal associations. Dominant species form a 100% cover in some places. There is little inter-island variation in coral distribution apart from North Solitary, which has a very different faunal assemblage with large areas dominated by giant sea anemones. Nine of the species of coral found around this island are not found at the others. The lush growth forms and large coral size of several other coral species indicate that environmental conditions are generally more favourable there than at the other islands (Veron *et al.*, 1974).

Noteworthy Fauna and Flora   Kelp beds are extensive, usually intermixed with corals (Veron, 1986). Information is available on fish records. Fourteen species of bird have been recorded. North Solitary Island has a large colony of muttonbirds and the largest colony of crested terns in New South Wales (Anon., 1973).

Scientific Importance and Research   The area represents the southern extreme for several of its coral species and has the most southerly coral communities in Australian waters, although there are no true reefs (Veron, 1986). It is of interest that the coral diversity here is poor compared with that seen at Lord Howe Island which is approximately 135 km south of the Solitary Islands (Pollard, 1977). The area has been studied for some time by the University of New England (Armidale) working from their field research station at Arrawarra (Anon., 1977).

Disturbance or Deficiencies   Rabbit grazing may have disturbed the vegetation on the islands but North Solitary is unique in being free of exotic plants and animals (Anon., 1973).

Legal Protection   North Solitary Island is a Nature Reserve (Anon., 1973).

Recommendations   A Marine Park proposal for the islands has been drawn up by Pollard (1981).

TORRES STRAIT REEFS

Geographical Location   Between Papua New Guinea and Cape York Peninsula; 142°-144°E, 9°-11°S.

Area, Depth, Altitude   Average depth 10-15 m in west to 30-50 m in east (Nietschmann, 1985).

Land Tenure   Much of the area is Queensland State and Australian national territory; 13 outer islands are state reserves, jointly administered by the Department of Aboriginal and Islander Advancement (DAIA) and elected leaders on each island. Part of the area has been in dispute because of Papua New Guinean territorial claims (Nietschmann, 1985).

Physical Features   The Torres Strait area includes volcanic, continental, coral and alluvial islands, with fringing, platform and barrier reefs, and lies at the end of the Sahul shelf. Mabuiag Island is described in Nietschmann (1985). Tides are exceptionally strong, with a 3.5 m tidal range at Mabuiag Island. There are two dominant seasons: *Sagerau tonar*, from May to September, which is the south-east tradewind season with gusty winds, sporadic rain squalls and rough seas; and *Kukiau mutaru*, from December to April, which is the north-west monsoon and rainy season (over 1500 mm rain at Thursday Island), with frequent strong storms and extended periods of calm. In October and November there is a period of calm weather, *Naigai mutaru*, with gradually increasing winds from the north and little rain (Nietschmann, 1985).

Reef Structure and Corals   The western side of the Strait is bordered by huge reefs which are little more than mud banks encircled by coral. Coral growth here is poor and limited in depth because the strong westerly currents cause turbidity. Eastwards, as the water becomes deeper and clearer, reefs and cays are more numerous and richer. Some of the most extensive reefs are the Orman Reef Cluster, between Mabuiag I. and Burk Is and Kuiku Pad, between Mabuiag and Badu Is. The eastern Torres Strait has high islands with extensive reef flats; the eastern edge is the northern extremity of the barrier line and has dissected reefs (Veron, 1986). An early account of the reefs around Murray Island is given in Mayor (1918). Additional information on the reefs is given in Veron (1978a and b), Veron and Hudson (1978) and GBRMPA (1983e).

Noteworthy Fauna and Flora   The Torres Strait has important populations of Green Turtle *Chelonia mydas*, Dugong *Dugong dugon* and a diverse fish and marine invertebrate faune (Nietschmann, 1976 and 1985). There is low density breeding of the Flatback Turtle *Chelonia depressa* in the South-west islands (Limpus, 1981). Occasional nesting of Hawksbill *Eretmochelys imbricata* and Olive Ridley *Lepidochelys olivacea* is seen on Crab Island, off the west coast of the northern tip of the Cape York Peninsula; this island is also a major nesting site for the flatback turtle (Limpus *et al.*, 1983b).

Scientific Importance and Research   One of the most ecologically complex areas of the Sahul Shelf (Nietschmann, 1985).

Economic Value and Social Benefits   The reefs are vitally important to the islanders and Aborigines for subsistence, providing fish, green turtles and dugong in particular, and also crayfish, crabs and a variety of molluscs. There is a human population of about 5500, living on 17 islands, Thursday Island being the governmental and commercial centre. Commercial fishing for trochus, pearl shell and bêche-de-mer also takes place (the Strait was one of the most important pearl shell producing areas at the end of the last century), but crayfish is now the main economic activity, and the industry is developing fast. Crayfish are taken by islanders by free diving and the tails are frozen and sold to buyers on Thursday Island. Commercial crayfishing boats also use the area (Nietschmann, 1985). Additional information is given in GBRMPA (1983e).

Disturbance or Deficiencies   There is considerable concern about over-exploitation of many marine resources. The pearl shell beds, bêche-de-mer, trochus and hawksbill resources were over-exploited at the beginnning of the century. Currently there is heavy pressure on turtles, dugong and crayfish. Subsistence exploitation has declined as increasing numbers of islanders have moved to the mainland, but there is no control over the islanders' commercial fishing activities. Crews of crayfishing boats often take turtles and dugong and Taiwanese trawlers are known to fish illegally in the area (Nietschmann, 1985).

Legal Protection   There is a long history of traditional marine conservation in the area which is documented by Nietschmann (1985). Several of these methods are still practised such as rotation of fishing areas, self-imposed limitations on quantities taken, and traditional rights to reefs, islands and waters. The islanders have rights over "sea territories"; for example, that of the islanders on Mabuiag covers about 640 sq. km, including 190 sq. km of reefs.

Under the 1978 Torres Strait treaty, the islanders and Papua New Guineans bordering the Strait may continue their traditional fishing activities, subsistence fishing is given precedence over commercial fishing, and there is an embargo on seabed mining and oil drilling for ten years. A protected zone, in which commercial fishing is controlled, covers almost all the islands and reefs of the strait. In 1968/69 Queensland state legislation was passed limiting exploitation of dugong and green turtles to islanders and aborigines for subsistence; subsistence hunting by non-indigenous people and all commercial exploitation is prohibited.

There are 74 islands, including 13 inhabited ones, listed as reserves under Queensland legislation. The area also includes Possession Island National Park and Round Island Environmental Park, off Thursday island.

Management   Surveillance is minimal to non-existent, although Taiwanese trawlers are expelled from the area if caught (Nietschmann, 1985).

BELAU

INTRODUCTION

General Description

Currently part of the Trust Territory of the Pacific Islands, the Republic of Belau (Palau) is to become an independent nation in free association with the U.S.A. The archipelago, the westernmost group of the Carolines lying between 6°53'N and 8°12'N, and 134°08'E and 134°44'E, about 741 km east of Mindanao in the Philippines, consists of eight large and 18 small, high volcanic and low limestone islands and about 350 islets, surrounded by a complex of fringing and patch reefs (Dahl et al., 1974; Taylor and Fielding, 1978).

The islands have a wet summer (maximum rain in July) and a drier winter (February - April), with temperatures in the range 70-90°F (21-32°C). Soils are either volcanic or of coralline-limestone and there is little freshwater on the low islands (Johnson, 1972a). General information on the area is given in Office of Ministry of Social Services (1985) and Maragos (1986). The islands have a rich flora with many endemics but are considerably disturbed (Douglas, 1969).

Table of Islands

Helen Reef (see separate account).

Tobi 0.25 sq. mi. (0.65 sq. km); one flat islet; no lagoon.

Merir 0.33 sq. mi. (0.85 sq. km); atoll with one flat islet; lagoon; fringing reef; some of best Green Turtle *Chelonia mydas* nesting beaches.

Pulo Anna (X) 0.33 sq. mi. (0.85 sq. km); small islet; no lagoon.

Sonsorol Is (X) 0.75 sq. mi. (1.9 sq. km); two small islets (Sonsorol and Fana); no lagoon; fringing reef.

Ngeaur (Angaur) (X) 3.25 sq. mi. (8.4 sq. km); raised lime-stone, 200 ft (61 m); reefs described below.

Beliliou (Peleliu) (X) 4.9 sq. mi. (12.7 sq. km); raised limestone island 100 ft (30 m); well-wooded; 2 tidal creeks with mangroves; barrier and fringing reefs; two small islets offshore - Ngedbus (Ngesebus) and Ngebad (Ngabad).

Ngercheu (Ngergoi) (see account for Chelbacheb (Rock Islands) proposed National Park).

Ngerechong (Ngeregong) (see account for Chelbacheb (Rock Islands) proposed National Park).

Ngemlis (see account for Chelbacheb (Rock Islands) proposed National Park).

Chelbacheb (Rock Islands)

- *Ulebsechel (Auluptagel)* (X) 16 sq. mi. (41 sq. km); raised limestone island; thickly wooded;

- *Ngeteklou (Gologugeul)* between Oreor and Ulebsechel;

- Also: Bukrrairong (Kamori), Ngeruktabel (Urukthapel), Tlutkaraguis (Adorius), Butottoribo, Ongael, Ngebedangel (Ngobasangel), Ulong (Aulong), Mecherchar (Eil Malik or Amototi), Bablomekang (Abappaomogan), Ngerukeuid (Ngerukewid or Orokuizu) (*see account for* Chelbacheb (Rock Islands) proposed National Park).

Ngerekebesang (Arakabesan) (X) 2 sq. mi. (5.2 sq. km); volcanic; reefs described by Randall *et al.* (1978).

Ngemelachel (Malakal) (X) marine environment described by Birkeland *et al.* (1976).

Oreor (Koror) (X) 3.6 sq. mi. (9.3 sq. km); volcanic in W., raised limestone in E; birds; western barrier reef drops off steeply; outer reef flats usually submerged at low tide; other parts of barrier reef shallower.

Babeldaob (Babelthuap) (X) 332 sq. km - largest island; volcanic, some limestone to S; 400 ft; mangroves in bays; fringing reef on east side and on part of west coast, barrier reef/small lagoon pattern to south-east.

Ngcheangel (Kayangel) (X) (see separate account).

Ngeruangel (Ngaruangl) 80 x 35 m; atoll with single islet of coral rock; 1 m high; described by Gressitt (1953); reef; abundant terns.

(X) = Inhabited

Orthography follows Motteler (1986), derived from 1983-84 U.S. Geological Survey topographic maps.

Belau is considered to have the richest reefs in the Pacific with the highest species diversity (Faulkner, 1974; Smith, 1977); 300 species of coral have been recorded (Johannes, 1977). The Japanese carried out extensive work on the reefs, concentrating in particular on Iwayama Bay on Oreor (Koror), where they ran the Palao Tropical Research Station from 1935 until World War II (Smith, 1977; Maragos, 1986). Eguchi (1935 and 1938) provided early descriptions of the reefs and corals. In 1968 the IBP (International Biological Programme) recommended that a Tropical Research Institute be set up in the country. The Micronesian Mariculture Demonstration Center was established in the early 1970s on Ngemelachel (Malakal) and carries out extensive work on trochus and giant clams. The Palau Marine Research Institute on Oreor is an internationally governed, non-profit corporation designed to promote research and service. Smith (1977) provides a brief review of other research efforts in the islands.

Dahl *et al.* (1974) described the reefs of Ngeaur (Angaur) at the southern end of Belau beyond the barrier reef. The island lacks major reef development except on the

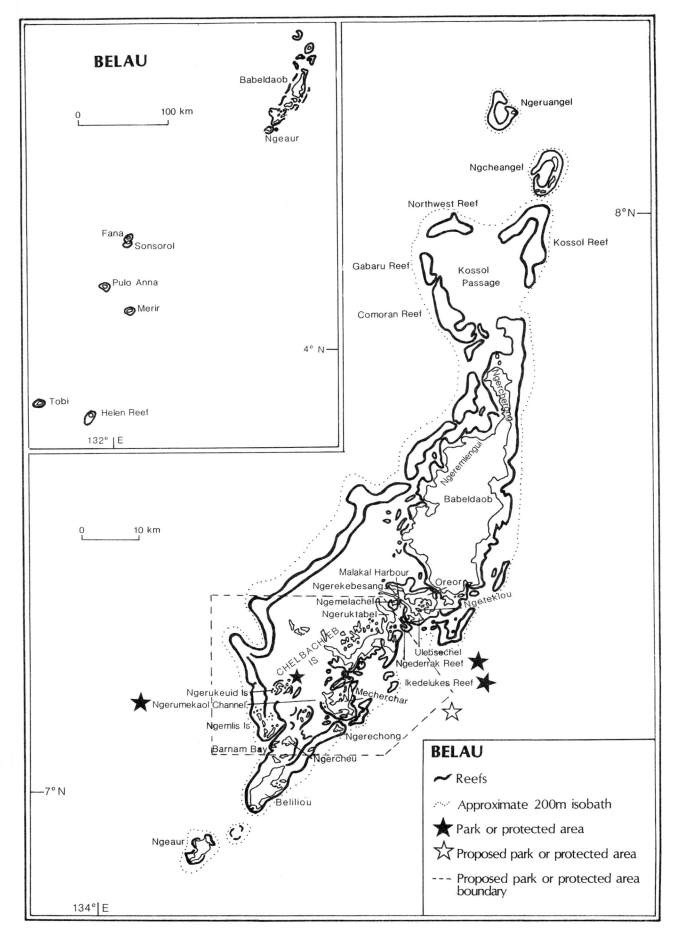

BELAU

Babeldaob

Ngeaur

0   100 km

Fana
Sonsorol

Pulo Anna

Merir

4° N

Tobi   Helen Reef

132° E

Ngeruangel

Ngcheangel

Northwest Reef

8° N

Kossol Reef

Gabaru Reef   Kossol
Passage

Comoran Reef

Ngercheong

Ngeremlengui

Babeldaob

Malakal Harbour
Ngerekebesang   Oreor
Ngemelachel   Ngeteklou
Ngeruktabel

CHELBACHEB
IS   Ulebsechel
Ngederrak Reef

Ikedelukes Reef

Ngerukeuid Is   Mecherchar
Ngerumekaol Channel

Ngemlis Is

Barnam Bay   Ngerechong
Ngercheu

0   10 km

7° N

Beliliou

Ngeaur

134° E

BELAU

〜 Reefs

⋯ Approximate 200m isobath

★ Park or protected area

☆ Proposed park or protected area

--- Proposed park or protected area
boundary

south and west sides, and there is a certain amount of sedimentation. On the west (leeward) side, towards the centre of the reef development, a smooth rocky flat was found with small scattered corals extending 700 m off shore to a depth of 5 m. This was followed by a gentle slope 200 m wide, down to 8 m with larger and more abundant corals. A steeper slope, 100 m wide with lush living coral coverage, extends down to 38 m where it is interrupted by a sand flat 30 m wide. A slope with patches of coral interspersed with sand continues down from 40 m at 1 km off shore. Helen Reef is described separately, but its reefs have never been studied in detail.

The high island cluster of Babeldaob, Oreor and Beliliou (Peleliu), is surrounded by a large and spectacular barrier reef. This is nearly a mile (1.6 km) wide in places, about 451 km long and drops off steeply on its outer edge, enclosing a lagoon of about 1450 sq. km (Johnson, 1972a). There are wide passages through the reef on the east coast including Ngemelachel Harbor. One small atoll (Ngcheangel or Kayangel) is located immediately north of the barrier reef. The northern extremity of the barrier reef is represented by Kossol, Northwest (Ngerael) and Cormoran-Gabaru reefs which are surrounded by relatively deep water. A complex intermingling of barrier and fringing reef structures runs south past Babeldaob to Beliliou and a barrier reef runs north from Beliliou past the Ngerukeuid (Ngerukewid) Islands and to the west of Babeldaob, from which it is separated by a shallow lagoon. Details of fringing reefs around the islands are given in the table above (Smith, 1977).

Reefs around Ngcheangel (Kayangel) Atoll in the north and in the Chelbacheb (Rock Islands) area to the south of Oreor (Koror) are described in separate accounts. Reefs in the south of Ngemelachel (Malakal) Island were described in the course of an environmental study for the site of a secondary treatment sewage plant (Birkeland et al., 1976). A broad fringing reef extends 400 m south of the island itself and is divided into four distinct zones: a sand-and-rubble intertidal zone, a narrow seagrass *Enhalus acoroides* band, a ramose coral *Montipora* zone, and a *Montipora-Lobophora* zone. The coral reef community is very diverse. A total of 48 genera and 163 species of scleractinian corals were observed; it is thought that there may be more species of scleractinian corals near the sewer outfall site on Ngemelachel Island than could be listed for the entire tropical Atlantic to depths of 100 m. Some 66 species of fish (including 25 pomacentrids) were observed on the transects studied.

Reefs on the east coast of Ngerekebesang (Arakabesan) Island were studied in the course of an environmental survey for a resort (Randall et al., 1978). A lagoon fringing reef as wide as 300 m is developed at the north and south margins of the embayment. This reef contains a moderately diverse coral fauna, 40 genera and 117 species. More than 125 fish species were observed, and the diversity of the macroinvertebrates was high. Several species of bivalve gastropods were exceedingly abundant.

Marine habitats are particularly varied with marine lakes (Hamner, 1982), seagrass beds and numerous marine caves which have evolved geologically from marine lakes (Bozanic, 1985). Smith (1977) provides an overview of the major marine and coastal ecosystems in Belau. Seventy-seven species of damselfish have been recorded

(Johannes, 1977). Plankton recorded in Belau are discussed in Yamaguchi (1972) and ecology and distribution of shallow-water crinoids in Meyer and Macurda (1980).

Reef Resources

Fishing activities on Ngcheangel and in the area south of Oreor are described in separate accounts. Fishing methods in use in the 1940s are described in Smith (1947). The exposed reefs in Kossol Passage are little used, the residents of Babeldaob restricting their fishing activities to the fringing reefs. South of Babeldaob commercial fishing predominates, using net, trap and spearfishing. On the western fringing reefs from Beliliou to Ngemlis, there is a fishery for several open water and reef fish. Mollusc fisheries are also important (Smith, 1977). Trochus are harvested and are an important source of income (Heslinga et al., 1984). The Black-lip Pearl Oyster *Pinctada margaritifera* was extensively harvested in the 1930s, with recorded production in 1939 of ca 2500 tons (2540 t), although this probably includes some catches made elsewhere in Micronesia. Experiments in pearl culture using the same species were also carried out at that time (Smith, 1947). A survey of precious coral stocks has been carried out but it is thought that the existing black and gold corals would only support a small industry (Grigg, 1975).

Belau is gaining an increasing revenue from a diver-oriented tourist industry, with the Chelbacheb (Rock Island) area and some of the passages in the leeward barrier reef being popular. The Ngemlis/Barnam Bay area on the south-western portion of the barrier reef has especially luxuriant and spectacular coral formations (Smith, 1977). Warner et al. (1979) carried out a socionomic-economic-ecological impact study of tourism. A large resort has been built on Ngerekebesang (Arakabesan).

Disturbances and Deficiencies

Acanthaster planci infestations have occurred, particularly on Ngeruktabel (Birkeland, 1979; Marsh and Bryan, 1972). Outbreaks before World War II are described in Hayashi (1938).

Most of the human population is concentrated on the three islands immediately to the south of Babeldaob. Oreor is a largely residential and commercial district with hotel and resort facilities. Ngemelachel is largely occupied with maritime and fishing industries, and Ngerekebesang is the District Administration Center and a further residential area. Currently the marine environment is relatively healthy (Maragos, 1986). However, the area is under considerable threat on account of its strategic location which is of interest both for military purposes and for the establishment of a major Pacific port (Smith, 1977; Caulfield, 1986). In 1976, the Save Palau Organisation was formed to oppose the construction of a major port and oil transhipment facility (Port Pacific), and implementation of the proposals was delayed (Falanruw, 1980; NRDC, n.d.). Subsequently, the proposal was abandoned apparently for economic reasons, but new plans have arisen to make Belau the main multiple use port for the western Pacific (Faulkner, 1983).

In 1968, a project to survey and prospect 1500 sq. mi. (3885 sq. km) of land and water in southern Belau with a view to large-scale phosphate mining, was proposed, but this presumably did not go ahead. A community dock harbour was constructed in Beliliou but the area was subjected to an environmental survey first (Aecos, Inc., 1979). Studies were also carried out at Ngerekebesang and an environmental impact assessment prepared prior to the building of the Palau Resort Hotel (Brewer *et al.*, n.d.; Randall *et al.*, 1978). An environmental survey was carried out for the Babeldaob-Oreor Airport (Environmental Consultants Inc., 1978). It is not known whether any of these developments has a deleterious impact on the reefs. Domestic pollution and solid waste disposal may become more serious threats throughout the islands in the future (Maragos, 1986). SPREP (1980) reported that in the 1970s dynamite fishing was causing serious damage to the reefs, but Maragos (1986) found no very clear evidence of this in the 1980s.

Legislation and Management

Under traditional custom, each village cluster or municipality in Belau exercises the right to limit access to adjacent fishing grounds; these rights are still maintained to just beyond the outer reef drop-off and are controlled by the chiefs. Fishermen may be allowed to use their neighbours' waters for local use, if permission is granted. Surplus reefs may also be given to more needy villages. Some grounds are shared; for example the Kossol and Northwest reefs are traditionally jointly fished by the people of Ngcheangel and Ngercherong. Subsidiary family and individual rights are no longer recognised (Johannes, in press).

Under Chapter 2 of the Palau (Belau) Code, there are a variety of regulations relating to reef resources (Anon., 1985). Section 201 provides for the establishment of the Ngerukeuid Islands Wildlife Preserve, Section 203 for the conservation of dugongs, Section 205 for the protection of marine life, and Section 208 for the prohibition of fishing in Ngerumekaol Channel, in the Ngerukeuid Islands, during the grouper spawning season. Under the Trust Territory Code Title 45, Chapter 1, there are provisions for the control of fishing with explosives, poisons and chemicals (Section 1), for limited turtle collecting (Section 2), and for controlling the take of sponges (Section 3) and *Pinctada margaritifera* (Section 4). Chapter 4 covers control of trochus harvesting (Section 51). Under the Republic of Belau Public Law, trochus harvesting can be restricted to certain areas (RPPL No. 1-30), export of clam meat is prohibited (RPPL No.1-9) and the Palau (Belau) Lagoon Monument was established (RPPL No.5-6-5). Under the Koror (Oreor) State Ordinance trochus breeding sanctuaries may be established (Ord. No. 150-69 (48-69)) and shelling may be prohibited (Ord. No. 49-1969 (157-69)).

The trochus management policy includes minimum size limits, restricted seasons, sanctuaries and a moratorium system in which states or villages voluntarily stop collecting shells for one or more years. Moratoriums were introduced in Ngcheangel (1979, 1983), Ngeremlengui (1980) and Oreor and Beliliou (1983). Currently there are two sanctuaries in Koror (Oreor) State (*see account for* Chelbacheb (Rock Islands) proposed National Park) (Heslinga *et al.*, 1984).

Until the Trust Territory arrangement for Belau is terminated, the country will also be subject to several U.S. environmental laws and regulations, including the Clean Water Act and the Endangered Species Act. It is intended that counterpart laws and regulations will be in effect at the time of termination but this has not happened in other areas formerly part of the Trust Territory (the Federated States of Micronesia and Republic of the Marshall Islands) (Maragos *in litt.*, 10.8.87).

Ngerukeuid Islands Wildlife Preserve is described in the account for Chelbacheb (Rock Islands) proposed National Park, and includes Ngerumekaol Channel, which operates as a preserve during the grouper spawning season, and the two trochus sanctuaries mentioned above. There are occasionally moratoriums on the take of trochus in other parts of Belau (*see account for* Ngcheangel). The Micronesian Mariculture Demonstration Center plays a major role in the International Giant Clam Mariculture Project, and is also active in raising trochus.

Recommendations

Various proposals have been made for Belau. In 1979 NRDC (Natural Resources Defense Council) with other organizations filed a petition with NOAA (National Oceanographic and Atmospheric Administration) requesting that a marine sanctuary be designated (NRDC, n.d.). One proposal was that the entire Belau Archipelago, excluding Malakal Harbour and the waters immediately surrounding Oreor should be a marine sanctuary (Tsuda, n.d.). A rather smaller area covering Chelbacheb (Rock Islands) has been proposed as a National Park which would include the Ngerukeuid Islands; these have potential for World Heritage nomination. Part of this area is a World War II battle site and has potential as a National Historical Park (Harry *in litt.*, 29.2.88). Helen Reef was proposed as an Island for Science by the International Biological Programme in the 1960s and is recommended for protection on account of its giant clam populations (Dahl, 1986). Maragos (1986) gives general recommendations for the improved management of marine resources in Belau and stresses that the major requirement is for a coastal zone management programme and a coastal resources inventory. In particular, proposals for the development of a port in this area need very careful consideration. Further recommendations are given in Smith (1977).

References

* = cited but not consulted

*Aecos, Inc. (1979). Baseline environmental surveys and impact assessment relative to dredging of Peleliu community dock harbor and access channel. Prepared for Parsons, Honolulu, Hawaii.
Anon. (1976). Endangered species in Trust Territory and Northern Marianas. *Territorial Register* 1(12): 5.
Anon. (1985). Country review - Republic of Palau. *Report of the 3rd South Pacific National Parks and Reserves Conference, Apia* 3: 162-174.
Anon. (1986). Palau's "Seventy Rock Islands" given more protection. *Environment Newsletter* (SPREP) 5: 1-2.

*Birkeland, C. (1979). Report on the *Acanthaster planci* (*rrusech*) survey of Palau, 18-26 May 1979. *Univ. Guam Mar. Lab. Misc. Rept* 25. 30 pp.

*Birkeland, C., Tsuda, R.T., Randall, R.H., Amesbury, S.S. and Cushing, F. (1976). Limited current and underwater biological surveys of a proposed sewer outfall site on Malakal Island, Palau. *Univ. Guam Mar. Lab. Tech. Rept* 25. 59 pp.

Bozanic, J. (1985). Palau 1985 National Science Foundation cave diving expedition. *National Speleological Society News* (U.S.) October: 311-315.

*Brewer, W.A. and Associates, Tsutsui, A. and Caderas, P. (n.d.). Environmental assessment for Palau Resort Hotel, Ngerakabesang Hamlet, Arakabesan Island, Republic of Palau. Prep. for Pacific Islands Development Corporation.

Brownell, R.L. Jr, Anderson, P.K., Owen, R.P. and Ralls K. (1981). The status of dugongs at Palau, an isolated island group. In: Marsh, H. (Ed.), *The Dugong*. Proceedings of a Seminar/Workshop. James Cook University, Townsville. Pp. 19-42.

Bryan, P.G. and McConnell, D.B. (1976). Status of Giant Clam stocks (Tridacnidae) on Helen Reef, Palau, Western Caroline Islands, April 1975. *Mar. Fish. Rev.* 38: 15-18.

Caulfield, C. (1986). Peace makes waves in the Pacific. *New Scientist* 10 April 1986: 51.

Dahl, A.L. (1980). Regional Ecosystems Survey of the South Pacific Area. *SPC/IUCN Technical Paper* 179. South Pacific Commission, Noumea, New Caledonia.

Dahl, A.L. (1986). *Review of the Protected Areas System in Oceania*. IUCN/UNEP. 239 pp.

Dahl, A.L., Macintyre, I.G. and Antonius, A. (1974). A comparative survey of coral reef research sites. (CITRE). *Atoll Res. Bull.* 172: 37-77.

Douglas, G. (1969). Checklist of Pacific Oceanic Islands. *Micronesica* 5(2): 327-463.

*Eguchi, M. (1935). On the corals and coral reefs in the Palao Islands of the south sea islands. *Sci. Rep. Ser. 2 Geol., Tohoku Imperial Univ.* 16.

*Eguchi, M. (1938). A systematic study of the reef-building corals of the Palao Islands. *Palao Trop. Biol. Stn Stud.* 3: 325-390.

Elliott, H. (1973). Pacific oceanic islands recommended for designation as Islands for Science. *Proceeding and Papers, Regional Symposium on Conservation of Nature - Reefs and Lagoons*. South Pacific Commission, Noumea, New Caledonia. Pp. 287-305.

*Engbring, J. (1983). Avifauna of the south-west islands of Palau. *Atoll Res. Bull.* 267. 22 pp.

*Environmental Consultants Inc. (1978). Environmental survey for the proposed Babelthuap-Koror Airport and Peleliu Island. Interim Report. Prep. for R.M. Parsons, Honolulu, Hawaii. 177 pp.

Falanruw, M.V.C. (1980). Marine environment impact of land-based activities in the Trust Territory of the Pacific Islands. In: *Marine and Coastal Processes in the Pacific: Ecological Aspects of Coastal Zone Management*. Unesco, ROSTEA, Jakarta. Pp. 19-47.

Faulkner, D. (1974). *This Living Reef*. Quadrangle/New York Times Book Co., N.Y. 179 pp.

Faulkner, D. (1983). Belau from above. *Sea Frontiers* 29(1): 33-39.

Gressitt, J.-L. (1952). Description of Kayangel Atoll, Palau Islands. *Atoll Res. Bull.* 14. 6 pp.

*Gressitt, J.-L. (1953). Notes on Ngaruangl and Kayangel Atolls, Palau Islands. *Atoll Res. Bull.* 21. 5 pp.

*Grigg, R.W. (1975). The commercial potential of precious corals in the Western Caroline Islands, Micronesia. *Univ. Hawaii Seagrant Rept* AR-75-03. 15 pp.

Hamner, W.M. (1982). Strange world of Palau's salt lakes. *Nat. Geog.* 161(2): 264-282.

*Hayashi, R. (1938). *Palao Trop. Biol. Stud.* 1. 417 pp.

Heslinga, G.A., Orak, O. and Ngiramengior, M. (1984). Coral reef sanctuaries for Trochus shells. *Mar. Fish. Rev.* 46(4): 73-80.

Hester, F.J. and Jones, E. (1974). A survey of giant clams, Tridacnidae, on Helen Reef, a western Pacific atoll. *Mar. Fish. Rev.* 36: 17-22.

Hirschberger, W. (1980). Tridacnid clam stocks on Helen Reef, Palau, Western Caroline Islands. *Mar. Fish. Rev.* 42(2): 8-15.

*Johannes, R. (1977). The Natural Resources. Will this Mecca bow to the oil merchants? *Oceanic* Symposium (Quoted in NRDC (n.d.)).

Johannes, R.E. (in press). The role of Marine Resource Tenure Systems (TURFs) in sustainable nearshore marine resource development and management in U.S.-affiliated tropical Pacific islands. In: Smith, B.D. (Ed.), Topic Reviews in Insular Resource Development and Management in the Pacific U.S.-affiliated Islands. *Univ. Guam Mar. Lab. Tech. Rept* 88.

Johannes, R. E. (1985). The role of marine resource tenure systems (TURFs) in sustainable nearshore marine resource development and management in U.S.-affiliated tropical Pacific islands. Office of Technology Assessment, U.S. Congress. Draft.

Johnson, S.P. (1972a). Palau: conservation frontier of the Pacific. *National Parks and Conservation Magazine* 46(4): 12-17.

Johnson, S.P. (1972b). Palau: Exploring the Limestone Islands. *National Parks and Conservation Magazine* 46(7): 4-8.

Johnson, S.P. (1972c). Palau and a Seventy Islands Tropical Park. *National Parks and Conservation Magazine* 46(8): 9-13.

Maragos, J.E. (1986). Coastal resource development and management in the U.S. Pacific Islands: 1. Island-by-island analysis. Office of Technology Assessment, U.S. Congress. Draft.

*Marsh, J.A. and Bryan, P.G. (1972). *Acanthaster planci*, Crown-of-thorns Starfish. Resurvey of Palau. *Univ. Guam Mar. Lab. Misc. Rept* 7. 8 pp.

*Meyer, D.L. and Macurda, D.B., Jr (1980). Ecology and distribution of shallow-water crinoids of Palau and Guam. *Micronesica* 16(1): 59-99.

Motteler, L.S. (1986). Pacific Island Names. *B.P. Bishop Mus. Misc. Publ.* 34. 91 pp.

NRDC (Natural Resources Defense Council) (n.d.). Proposal to implement the World Conservation Strategy in the Caroline Islands of Micronesia.

Pritchard, P.C.H. (1981). Marine turtles of Micronesia. In: Bjorndal, K.A. (Ed.), *Biology and Conservation of Sea Turtles*. Smithsonian Institution Press, Washington D.C. Pp. 263-274.

Randall, R.H., Birkeland, C., Amesbury, S.S., Lassuy, D. and Eads, J.R. (1978). Marine survey of a proposed resort site at Arakabesan Island, Palau. *Univ. Guam Mar. Lab. Tech. Rept* 44. 73 pp.

*Read, K.R.H. (1974a). Kayangel Atoll. *Oceans* 7(1): 10-17.

*Read, K.R.H. (1974b). The Rock Islands of Palau. *Oceans* 7(1).

Smith, R.O. (1947). Fishery resources of Micronesia. *Fishery Leaflet* 239. Fish and Wildlife Service, U.S. Dept of the Interior. 46 pp.

Smith, S.V. (1977). Palau environmental study: A

planning document. Contrib. IUCN Mar. Prog. 3.7.70. 102 pp.

Taylor, L.R. and Fielding, A. (1978). A wealth of habitat. *Oceans* 11(1): 30-32.

Tsuda, R.T. (1976). Occurrence of the genus *Sargassum* (Phaeophyta) on two Pacific atolls. *Micronesica* 12(2): 279-282.

*Tsuda, R.T. (1981). Marine benthic algae of Kayangel Atoll, Palau. *Atoll Res. Bull.* 225: 43-48.

Tsuda, R.T. (n.d.). Report to IUCN Coral Reef Group. Unpub. rept. 9 pp.

Tsuda, R.T., Fosberg, F.R. and Sachet, M.-H. (1977). Distribution of seagrasses in Micronesia. *Micronesica* 13(2): 191-198.

*Warner, D.C., Marsh, J.A. and Karolle, B.G. (1979). The potential for tourism and resort development in Palau: A socio-economic-ecological impact study. *Univ. Guam Mar. Lab. Misc. Rept* 25. 82 pp.

*Yamaguchi, M. (1972). Preliminary report on a plankton survey in Palau, December 1971 to January 1972. *Univ. Guam Mar. Lab. Tech. Rept* 5. 14 pp.

CHELBACHEB (ROCK ISLANDS) PROPOSED NATIONAL PARK, INCLUDING NGERUKEUID ISLANDS (SEVENTY ISLANDS) WILDLIFE PRESERVE

Geographical Location Southern Belau, between Oreor and Beliliou, including Ngeruktabel (Urukthapel), Tlutkaraguis, Butottoribo, Ongael, Ulong (Aulong), Ngebedangel (Ngobasangel), Mecherchar (Eil Malik), Bablomekang (Abappaomogan), Ngerukeuid (Ngerukewid or Orukuizu), Ngemlis, Ngerechong, Ngercheu and associated islands; the Ngerukeuid Islands are about 18 mi. (29 km) south-west of Oreor; 7°10'-7°20'N, 134°15-134°30'E.

Area, Depth, Altitude Ngerukeuid Reserve covers 2.6 sq. km; Ngeruktabel covers 7.3 sq. mi. (19 sq. km), max. alt. 680 ft (207 m); Mecherchar covers 3.4 sq. mi. (8.8 sq. km).

Land Tenure Government.

Physical Features The Chelbacheb or Rock Islands (also called the Limestone Islands), to the south of Oreor, are formed from limestone on top of the peaks of submerged volcanoes on an arc-shaped ridge system. They are generally low with impenetrable forest cover (Johnson, 1972b). A popular account is given in Read (1974b). Current patterns in this area are complex (Dahl *et al.*, 1974). The Ngerukeuid Islands are a group of relatively remote raised limestone islands, mostly very small, with markedly undercut shores covering an area of one square mile (2.6 sq. km). The average height is 100 ft (30 m) and they are well wooded (Douglas, 1969). The Ngemlis are 40 km south-west of Oreor and consist of parts of the slightly elevated coral platform; they front on the outer reef margin on the west, a sheltered channel on the south-east and a lagoon to the north. Ngerechong Island is situated on the windward (east) side of the barrier reef, about 32 km south of Oreor.

Brief descriptions of the other islands in the proposed National Park are given in Douglas (1969). Mecherchar and Ngeruktabel are inhabited raised limestone islands, the latter with many offshore islets in the west.

Reef Structure and Corals A number of sites were surveyed in the course of preliminary investigations for the proposed CITRE project (Dahl *et al.*, 1974). In the Ngemlis, surveys were carried out on reefs on the south-east and west. In the south-east there is a shallow 300 m wide reef flat. Corals are scarce inshore but become abundant towards the reef edge at 0.5 m depth, with soft corals predominating. Beyond the edge, there is a vertical drop-off, much of which is overhanging, to 260 m. Alcyonarians are dominant down to 30 m, and scleractinians and gorgonians are both well represented. At around 40 m, the slope begins to project outwards and collects calcareous sediment; corals become scarcer and appear not to grow beyond 60 m. The west side has a broader, 500 m-wide reef flat with stony corals dominant at the seaward edge at 0.5 m depth. There is a steep drop-off to 12 m with good coral coverage, with big buttresses dominated by *Porites* heads and rubble-filled chutes extending down to 20 m. A gentle slope extends down to 40 m with *Pachyseris* dominant, after which the slope becomes increasingly sandy and corals dwindle. The site has high species diversity and a good drop-off but sedimentation limits coral growth at around 50 m depth. At both sites the drop-offs are very near the surface and sponges are almost completely lacking. Ngerechong Island has a poorly developed reef; the outer reef slope on the south-east side of the island consists of a gentle sandy slope, with scattered coral growth, becoming even sandier at depth (Dahl *et al.*, 1974). Mecherchar has a fringing reef on the east coast and Ngeruktabel has a narrow fringing reef (Douglas, 1969).

Noteworthy Fauna and Flora The islands are inhabited by the Micronesian megapode *Megapodius laperouse senex* which nests on Ngeruktabel and in the Ngerukeuid Islands Preserve; the Belau Scops Owl *Otus podarginus* is also present (Johnson, 1972b). Brownell *et al.* (1981) noted that 34 Dugongs *Dugong dugon* were sighted in 1978. Reproductive rates are high but poaching rates are so great that it is suggested that the Belau Dugongs could be exterminated by the end of the century. There is some limestone forest of interest in the Ngerukeuid Reserve (Dahl, 1980). *Pandandus* and some endemic palms are found on the islets (Johnson, 1972a). Giant Clams occur (Johnson, 1972b). Turtles are common (Pritchard, 1981), in particular the Hawksbill *Eretmochelys imbricata* and Green Turtle *Chelonia mydas*. The Saltwater Crocodile *Crocodylus porosus* is found around these islands (Johnson, 1972b).

Scientific Importance and Research The reefs at the Ngemlis are spectacular, although there appears to be a constant flow of sediment (Dahl *et al.*, 1974). Faulkner (1974) provided extensive photographic coverage of the reefs.

Economic Value and Social Benefits The area is becoming important for tourism, having some 25 beaches, and is also of historical and anthropological interest.

Disturbance or Deficiencies There was a major outbreak of *Acanthaster planci* on Ngeruktabel in 1977 (Birkeland, 1979). The inaccessibility and remoteness of most of the islands have ensured that they have remained pristine. However they are coming under increasing pressure with the expansion of the Belau population and the suitability of the area for recreation and fishing (Johnson, 1972b). Many of the islands are important for phosphate; in 1968 they were threatened by a proposed survey for possible large-scale mining of phosphates. It was recommended that the Ngerukeuid Islands Reserve be excluded. Fishing with poisons and explosives became a serious problem in the 1970s. Soil erosion is beginning to lead to siltation in the lagoon which is also being affected by sewage pollution. Dredging of the lagoon for construction materials could cause further problems (Johnson, 1972a). In the late 1970s a spill of coconut oil from a reefed tanker caused localised damage off Mecherchar (Falanruw, 1983). Shell collecting by tourists may deplete mollusc populations (Johnson, 1972a). The eggs of the megapode are considered a delicacy and in the past were collected by local people. Turtles are still caught illegally and there is excessive poaching of Dugong. The Saltwater Crocodile population was heavily exploited in the past (Johnson, 1972b) but lack of firearms has curtailed this.

Legal Protection   The Ngerukeuid Islands were established as a Preserve by District Order in 1956 through the Trust Territory of the Pacific Islands Conservation Law Section 201 of the Trust Territory Code. Fishing, plant collecting and fires are prohibited within the Preserve (Douglas, 1969). Hawksbill and Green Turtles and the Dugong are protected (Anon., 1976). The area also includes Ngerumekaol Channel, 17 km ESE of Oreor and 1.7 km NNE of the Ngerukeuid Islands, which was established under the same legislation as the Ngerukeuid Preserve, as a protected spawning ground for groupers. It is closed each year from April 1st to July 31st (Anon., 1985). In May 1982, Ikedelukes Reef (south-east of Ngeruktabel) and Ngederrak Reef (between Ngeruktabel and Ulebsechel) were created trochus sanctuaries, in place of the seven small trochus sanctuaries originally established in Koror (Oreor) State (Heslinga *et al.*, 1984).

Management   The Preserve is managed by the Conservation Officer of the Bureau of Resources and Development in the Ministry of Natural Resources (Offices of Ministry of Social Services). In 1980 enforcement was reported as variable (Dahl, 1980). The Ngerukeuid Islands were formerly protected by ancient taboos (Douglas, 1969). The trochus sanctuaries are patrolled by local Marine Resources Division personnel and harvesting is permitted within them for a week (Heslinga *et al.*, 1984).

Recommendations   A National Park was proposed at the Technical Meeting of IBP/CT, Nov. 1968, to include rock islets, coasts and lagoons from south of Oreor to north of Beliliou and westwards to the barrier reef. Strictly controlled fishing regulations are recommended to protect the islanders hereditary baitfish rights (Johnson, 1972b). The Ngerukeuid Islands Wildlife Preserve would remain a strict nature reserve within the Park (Johnson, 1972c). An IUCN project is currently being developed to improve management of this area (Anon., 1986). The area, in particular the Ngerukeuid Islands, has potential for World Heritage Nomination. Dahl (1986) suggests protection of additional parts of the marine environment such as the Ngemlis.

HELEN REEF

Geographical Location   South of Belau; 2°43'N, 131°46'E.

Area, Depth, Altitude   Land area 194 ha; reef and lagoon area ca 216 sq. km, with reef extending ca 24 km north-south, 9 km east-west; max. depth in lagoon over 60 m.

Physical Features   An atoll with a submerged reef and lagoon, and a small low island (Helen Island) at its northern end. The reef is about 1200 m wide except in the north where there is an extensive sandy area between the lagoon and the sea. Outside the lagoon, the reef slope drops off steeply. A navigable passage on the western side permits access to the lagoon for vessels of moderate draft. Winds and currents are variable and there is no well-defined windward side. The island is sandy and covered with typical atoll vegetation including coconut trees (Elliott, 1973; Hester and Jones, 1974; Hirschberger, 1980).

Reef Structure and Corals   Coral growth is most luxuriant on the west side of the reef (Hester and Jones, 1974). The reefs are expected to be rich.

Noteworthy Fauna and Flora   At one time the reef supported large populations of giant clams Tridacnidae (Hirschberger, 1980). Helen Reef has some of the best *Chelonia mydas* nesting beaches in the Belau system (Pritchard, 1981). The avifauna is described by Engbring (1983); large numbers of seabirds nest.

Scientific Importance and Research   Although no visit was made, the reef was identified by Dahl *et al.* (1974) as having rich and undisturbed coral growth, but its inaccessibility made it unsuitable for the proposed CITRE research program. In May 1971, the NOAA research vessel, *Townsend Cromwell*, operated by the National Marine Fisheries Service, conducted a survey of the Trust Territory's marine resources in the course of which a brief stop was made. In 1972, a return visit was made to survey clam stocks; resurveys of clams were made in 1975 and 1976 by the Palau (Belau) District Marine Resources Office (Hirschberger, 1980).

Economic Value and Social Benefits   The island is uninhabited although there are some indications of possible former occupation (Elliott, 1973). Helen Reef receives only occasional visits from U.S. Trust Territory outer island support ships and foreign fishing vessels (Hirschberger, 1980).

Disturbance or Deficiencies   The clam populations have been extensively reduced through exploitation by foreign fishing vessels (Bryan and McConnell, 1976; Hester and Jones, 1974; Hirschberger, 1980). The reefs are said to be relatively untouched (Hester and Jones, 1974).

Legal Protection   None.

Management   None.

Recommendations   Proposed as an "Island for Science" by the International Biological Programme in 1968; Dahl (1986) recommends protection against poaching.

NGCHEANGEL (KAYANGEL) ATOLL

Geographical Location   Northernmost land area of Belau Archipelago (no other atoll within 150 miles (241 km)); 20 mi. (32 km) north of Babeldaob; ca 8°10'N, 134°42'E.

Area, Depth, Altitude   2.3 sq. mi. (6 sq. km); atoll has a north-south diameter of less than 3.5 mi. (5.6 km), east-west diameter of 2 mi. (3.2 km).

Physical Features   The atoll consists of an almost complete circle of reef with four islets (Ngajangel, Ngariungs, Ngaralpas, Gorak) on the east side, occupying a little less than half of the perimeter. The islets decrease markedly in size from north to south. The entrance to the lagoon is shallow and not very distinct, located on the west (leeward) side. It consists of a sand-bottomed break in the reef a number of yards wide, 1-2 fathoms (1.8-3.6 m) deep at low tide and dotted with coral heads of varying size (Dahl *et al.*, 1974; Gressitt, 1952). The gaps between the islets are rough and rocky on the seaward side and sandy on the lagoon side and at low tide it is possible to walk beween the two northern and two southern ones. They are described in Gressitt (1952). The atoll has a rather wet climate with an estimated rainfall of perhaps 150 in. (3810 mm) (Gressitt, 1952). A popular account of the atoll is given in Read (1974a).

The lagoon has mainly a sand bottom 4-6 m deep. Most of the south-eastern part is less than 2 fathoms (3.6 m) deep and has large scattered coral heads except in the shallower parts. In the central and northern parts, the water is deeper and the coral heads less visible. In the south-western part the coral heads are large and widely separated and the water a few fathoms deep. In the west the lagoon becomes shallower towards the inlet and the coral heads are closer together (Gressit, 1952). The reef forming the atoll is in general not very wide on the west side. Depth increases rapidly, particularly near the south end. At low tide, the reef is emergent in some areas. On the east, south of the islets, there is for the most part a fairly flat platform exposed at low tide. The sea bottom slopes off at an initial gradient of 10-25°.

Reef Structure and Corals   The patch-reefs and coral heads in the lagoon have high species diversity and coral cover (Dahl *et al.*, 1974). In some shallow areas east of the centre of the lagoon or near the centre of the main island, *Acropora* grows on the bottom (Gressitt, 1952). The leeward (west) outer reef is topped by a 50 m wide flat of calcareous algae, leading to a reef edge with lush coral growth. There is a steep drop-off down to 10 m, also with very high coral coverage and species diversity. This is followed by a gentle slope to 40 m on which *Pachyseris* is dominant. The slope flattens

between 40 and 50 m and becomes sandy beyond 50 m. Temperature decreases markedly with depth and the water becomes increasingly murky, the living reef ending at 50-60 m. The east side is very exposed, the outer reef starting with a broad algal turf-covered flat which drops to a platform about 1000 m wide and 7-10 m deep with little coral growth. There is no drop-off (Dahl *et al.*, 1974).

Noteworthy Fauna and Flora   The islets are almost entirely covered with vegetation, largely coconut palms and crops but also some natural vegetation, as at the north end of Ngajangel. Details in Gressitt (1952). Sea turtles were nesting on Ngaralapas in 1951; Gorak had nests of terns and other seabirds. Megapodes nest on the islets, particularly on Ngariungs; other birds include kingfishers *Halcyon chloris teraokai* and the Morning Bird *Colluricincla tenebrosa*. Marine benthic algae have been described by Tsuda (1981). The alga *Sargassum crassifolium* is known from only two central Pacific atolls, Ngcheangel and Ulittri (Tsuda, 1976), and the seagrass *Thalassodendron ciliatum* is known in Micronesia only from the lagoon at Ngcheangel (Tsuda *et al.*, 1977).

Scientific Importance and Research   The atoll was visited in the early 1950s (Gressitt, 1952 and 1953) and in 1971 during preliminary surveys for the proposed CITRE project (Dahl *et al.*, 1974). The lagoon is considered to be a typical example of Pacific lagoons and has rich coral formations (Dahl *et al.*, 1974).

Economic Value and Social Benefits   In 1980 the human population was 140 (Dahl *in litt.*, 27.10.87). The village is situated on Ngajangel, the largest island (Gressitt, 1952). Harvesting of trochus is an important activity (Heslinga *et al.*, 1984). Fishing on atoll reefs takes place, mainly using V-shaped set nets, spearfishing techniques and set-traps (Smith, 1977). Edible polychaete worms are taken from sand flats around Ngajangel (Read, 1974a).

Disturbance or Deficiencies   A substantial amount of coral in the lagoon had been broken by canoes or fishing operations when the atoll was visited in 1951 (Gressit, 1952). Giant clams were formerly common but have declined through overfishing (Smith, 1977). The reef pass was enlarged with explosives in the 1970s (Dahl *in litt.*, 27.10.87).

Legal Protection   There were moratoriums on trochus harvesting in 1979 and 1983 (Heslinga *et al.*, 1984).

Management   None.

Recommendations   No information.

CHINA

INTRODUCTION

General Description

The coast of mainland China with its numerous offshore islands extends for over 6000 km from about 41°N to about 20°N and is bordered by the Huang Hai (Yellow Sea), Dong Hai (East China Sea), Taiwan Haixia (Taiwan Strait) and Nan Hai (South China Sea). Details of the hydrography of the northern part of the Nan Hai are given in Watts (1971). In winter, warm water from the Kuroshio Current enters the Nan Hai through the Luzon Strait and moves southwards; in summer there is a reversal and the waters enter through the Selat Gasper and Selat Karimata between Borneo and Sumatra in the south and leave through the Taiwan Haixia and Luzon Strait. The Zhujiang (Xi Jiang) (Pearl River) flows into the Nan Hai below Guangzhou (Canton).

The mainland coast has only patchy coral growth because of turbidity, variable salinity and low winter temperatures. Coral communities are restricted to offshore islands such as Namao, Dagan, Shangchuan and Xiachuan, and in Taiya and Taipeng Gulfs (Liang, 1985d). The ecology of reef corals in Tai-A Bay in eastern Guangdong is described by Liang and Zhu (1987).

Coral communities are principally found around the offshore islands and archipelagos in the Nan Hai (Liang, 1985a, b, c and d). The most important area for reefs is probably the Xisha Qundao (Paracel Shoals) (see separate account) which lie to the south of Hainan. Fringing reefs are also found around Hainan (see separate account), and to the south around the atolls of Dongsha and Nansha Qundao (see separate accounts), and Huangyan. The geomorphology of Huangyan, which is an oceanic atoll with raised reefs, rising steeply from the sea bottom at 3500 m, is described by Xu and Zhong (1980); the atoll is triangular with a reef front slope, a reef flat and lagoon which connects with the sea through a navigable channel to the south-east corner of the atoll. Barrett-Smith (1890) and Brook (1892 and 1893) reported corals from Zhenghe Qunjiao Qundao and Zhongsha Qundao (Macclesfield Bank) in the Nan Hai. Zhongsha Qundao lies 183 km south-east of Xisha Qundao and consists of two large atolls: Zhongsha Atoll (53 km in diameter) is a drowned atoll, whereas "Yellow Rock" Atoll (15 km in diameter) is elevated to 1-1.5 m above sea level. Species diversity in Zhongsha Qundao is lower than in Xisha Qundao as most of the reefs lie 7 m below the surface. However, coral distribution is interesting because of the depth at which some species occur (e.g. *Pavona papryracea* and *Rhodarea largrenii* at 70 m) and the isolation of some reef patches (Liang, 1985d).

Zou and Chen (1983) list a total of 179 scleractinians in 45 genera and 14 families in China, but Liang (1985a) lists 325 species in 62 genera. Forty-five species in 27 genera have been described from Huangyan and 45 in 21 genera from the mainland coast of Guangdong and Guangxi Zhuangzu Zizhiqu. There is a tendency towards a decrease in abundance and diversity of pocilloporids and *Fungia* from north to south and a corresponding increase in diversity of *Turbinaria*. Deep water ahermatypic corals from the northern shelf of the Nan Hai are decribed by Zou et al. (1983).

Mangroves are described by Yang (1984). Four species of marine turtle, the Green Turtle *Chelonia mydas*, Loggerhead *Caretta caretta*, Hawksbill *Eretmochelys imbricata* and Leatherback *Dermochelys coriacea* occur in Chinese waters (Huang, 1981).

Reef Resources

There is little information on the economic importance of reefs in China but traditional uses of marine resources are described by Wu (1985b) and the wildlife of the islands of the Nan Hai has been exploited for centuries. Turtles are apparently heavily exploited, especially in the Xisha Qundao (see separate account); originally taken mainly for their meat, and, in the case of *Eretmochelys imbricata*, for tortoiseshell, virtually all parts are now used (Frazier and Frazier, 1985; Huang, 1981).

Disturbances and Deficiencies

Some reefs may suffer natural damage from cyclones (see account for Xisha Qundao). Damage due to human activities on Hainan and in Xisha Qundao is described in separate accounts. Oil pollution does not seem to be a problem at present, but the current programme of oil exploration on the South China continental shelf could result in more frequent spills. Turtle numbers are believed to have declined through over-fishing and habitat loss; Frazier and Frazier (1985) noted that the incidental turtle catch in Fujian and Guangdong could account for a very high percentage of the turtles nesting there. Mangroves are threatened by land reclamation for agriculture, construction of dikes for shrimp and eel culture, and cutting for fuel-wood (Zhu, 1987).

Legislation and Management

The Ministry of Agriculture, Animal Husbandry and Fishery is responsible for the protection of fishing resources and monitoring of pollution of fishing ports. The Institute of Marine Environmental Protection of the National Bureau of Oceanography carries out research on a variety of aspects of marine environmental protection. A multi-disciplinary, comprehensive investigation has been organized to obtain basic information for use in the formulation of the Coastal Zone and Shallow Sea Areas Protection and Management Law, as well as for conducting scientific management of the coastal zone (Wu, 1985b). A coastal management law has reportedly been passed and includes provision for mangrove preservation and management (Zhu, 1987).

Traditional and historical methods of regulating the use of marine resources are described in Wu (1985b). The first Fishery Law was promulgated in 1939 and included regulations to control the use of coastal land, to protect breeding areas of marine biota and to establish fishing grounds. In 1979, regulations were issued for the

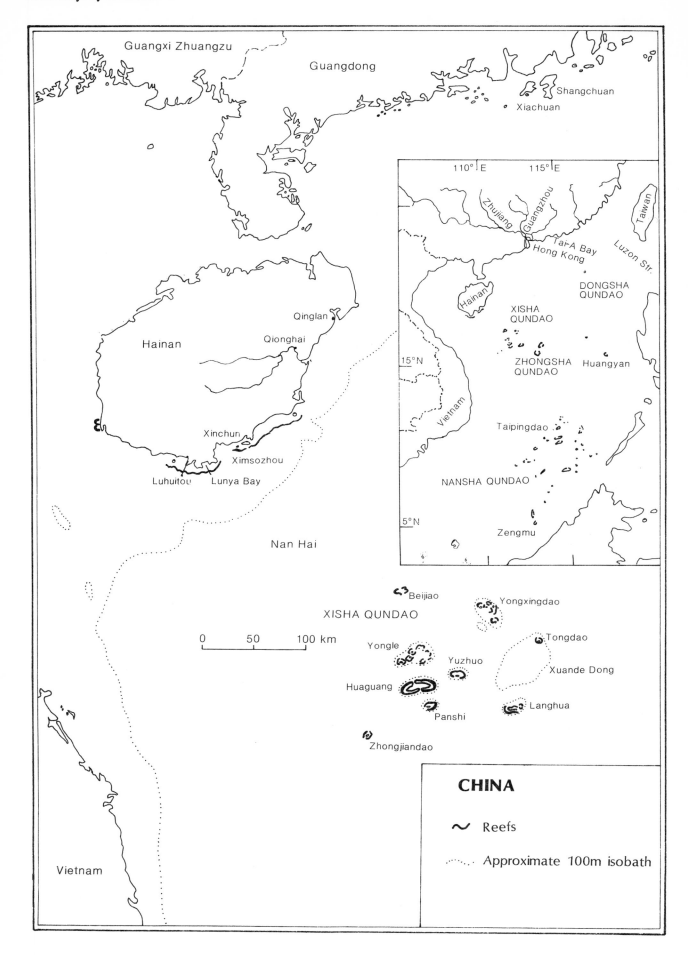

Guangxi Zhuangzu

Guangdong

Shangchuan

Xiachuan

110° E

115° E

Zhujiang

Guangzhou

Tai-A Bay

Hong Kong

Luzon Str.

Taiwan

DONGSHA QUNDAO

Hainan

XISHA QUNDAO

Qinglan

Qionghai

15°N

ZHONGSHA QUNDAO

Huangyan

Hainan

Vietnam

Taipingdao

Xinchun

Ximsozhou

NANSHA QUNDAO

Luhuitou

Lunya Bay

5°N

Zengmu

Nan Hai

Beijiao

Yongxingdao

XISHA QUNDAO

0   50   100 km

Yongle

Tongdao

Yuzhuo

Xuande Dong

Huaguang

Langhua

Panshi

Zhongjiandao

Vietnam

CHINA

~ Reefs

.... Approximate 100m isobath

"Propagation and Protection of Fishery Resources". In 1982 a "Marine Environmental Protection Law" came into force which covers a wide range of issues and is administered by the Environmental Protection Department under the State Council.

The Huidong Nature Reserve in Guangdong protects a sea turtle nesting beach (Wang *in litt.*, 20.11.86); it is not known if there are any corals in the area. Tuntze Nature Reserve at 20°N, 110°20'E on the north coast of Hainan protects 3737 ha of mangroves (*see separate account*). There is also Chinese legislation stipulating reserves in Zhongsha Qundao, Xisha Qundao and Nansha Qundao (Wang *in litt.*, 20.11.86 and *see separate accounts*). It is not known to what extent any of these reserves cover reefs.

Recommendations

Zhu (1987) states that preservation of mangrove areas has become a task of top priority in coastal conservation in China, in view of the threats facing mangroves and their value in increasing coastal productivity, protecting the coastline and decreasing sediment deposition in navigation channels. There also appears to be considerable need to protect and manage reef resources, as evidenced by the recent work on Hainan (*see separate account*).

References

* = cited but not consulted.

Bassett-Smith, P.W. (1890). Report on the corals from Tizard and Macclesfield Banks. *Ann. Mag. Nat. Hist.* 6(6): 353-374, 443-458.

Brook, G. (1892). Preliminary description of new species of *Madrepora* in the collections of the British Museum. Part 2. *Ann. Mag. Nat. Hist.* 6(10): 451-465.

Brook, G. (1893). *Brit. Mus. (Nat. Hist.) Cat. Madre. Corals* 1: 1-212.

Brown, B.E. (Ed.) (1986). Human induced damage to coral reefs. *Unesco Rep. Mar. Sci.* 40. 180 pp.

Chang, K.H., Jan, R.Q. and Hua, C.S. (1981). Inshore fishes at Tai-ping Island (South China Sea). *Bull. Inst. Zool. Acad. Sinica* 20(1): 87-93.

Chen, G.W. (1981). Studies on the dinoflagellates in adjacent waters of the Xisha Islands. *Oceanol. Limnol. Sinica* 12(1).

Chen, K.-H. and Sung, Y.-C. (1977). Biological resources of the Hisha Islands. *China Reconstructs* Feb-March 1977: 31-32.

Frazier, J. and Frazier, S.S. (1985). Preliminary Report: Chinese - American Sea Turtle Survey, June-August 1985. Fujian Teachers University Fuzhou, Fujian and National Program for Advanced Study and Research in China, CSCPRC, National Academy of Sciences, Washington D.C.

Gurjanova, E.F. (1959). The marine zoological expedition to Hainan Island. *Vestinik Academy of Science of the U.S.S.R.* 3.

He, J. and Zhong, J. (1982). On the growth rate of coral reefs in Xisha Islands from archeological evidence. *Nanhai Studia Marina Sinica* 3: 37-43. (In Chinese with English abstract).

Huen-pu, W. (1980). La conservation des espaces naturels en Chine: le point sur la situation actuelle. *Parks* 5(1): 1-10.

Huang, C.-C. (1981). Distribution and population dynamics of the Sea Turtles in China seas. In: Bjorndal, K. (Ed.), *Biology and Conservation of Sea Turtles*. Smithsonian Institution Press, Washington D.C. Pp. 321-322.

*Klebovitsch, V.V. and Wu, B.L. (1963). The Nereidae from Hainan Island. *Studia Marina Sinica* 3: 61-74.

Liang, J.-F. (1985a). Holocene reef corals of China. In: Zeng, Z.X. (Ed.), *Coral Reefs and Geomorphological Essays of South China*. Geography Series No. 15, Institute of Geomorphology, South China Normal University, Guangzhou, China. Pp. 1-54.

Liang, J.-F. (1985b). The distribution of the recent coral reefs of China and its significance in quaternary study. In: Zeng, Z.X. (Ed.), *Coral Reefs and Geomorphological Essays of South China*. Geography Series No. 15, Institute of Geomorphology, South China Normal University, Guangzhou, China. Pp. 55-60.

Liang, J.-F. (1985c). The reef building corals of the recent reef of the south sea islands. In: Zeng, Z.X. (Ed.), *Coral Reefs and Geomorphological Essays of South China*. Geography Series No. 15, Institute of Geomorphology, South China Normal University, Guangzhou, China. P. 61. (Abst.).

Liang, J.-F. (1985d). Ecological regions of the reef corals of China. *J. Coast. Res.* 1(1): 57-70.

Liang, J.-F. and Zhu, J.-X. (1987). Study on the ecological types of the reef corals of Tai-A Bay. *Tropical Geomorphology* 8(1): 1-3.

Melville, D.S. (1984). Seabirds of China and the surrounding seas. In: Croxall, J.P., Evans, P.G.H. and Schreiber, R.W. (Eds), *Status and Conservation of the World's Seabirds*. ICBP Technical Publication 2, Cambridge. Pp. 501-511.

Naymov, D.V., Jingsong, Y. and Mingxian, H. (1960). On the coral reefs of Hainan Island. *Oceanol. Limnol. Sinica* 3(3): 157-176.

*Qi, Z.Y., Ma, X.Y. and Lin, Q.Y. (1984). A preliminary study on the benthic mollusks from Sanya Harbor, Hainan Island. *Nanhai Studia Marina Sinica* 5: 98-108. (In Chinese with English abstract).

Struhsaker, T. (1987). Unpublished report on a visit to China.

*Tseng, C.K. (1983). *Common Seaweeds of China*. Science Press, Beijing, China.

*Wang, C.X. (1981). Studies on the fish fauna of the South China Sea Islands, Guangdong Province, China. *Oceanol. Limnol. Sinica* (suppl.).

Watts, J.C.D. (1971). A general review of the oceanography of the northern sector of the South China Sea. *Hong Kong Fisheries Bulletin* 2: 41-50.

Wu, B.L. (1985a). Coral reefs of Hainan Island in the South China Sea. *Proc. 5th Int. Coral Reef Cong., Tahiti* 2: 412 (Abst.).

Wu, B.L. (1985b). Traditional management of coastal systems in China. In: Ruddle, K. and Johannes, R.E. (Eds), *The Traditional Knowledge and Management of Coastal Systems in Asia and the Pacific*. Unesco/ROSTEA, Jakarta. Pp. 181-189.

*Wu, B.L. and Hutchings, P.A. (1986). Coral Reefs of Hainan Island, South China Sea. *Collected Oceanic Works* 9: 76-80.

Xu, Z. and Zhong, J. (1980). The geomorphological characteristics of Huangyuan Island. *Nanhai Studia Marina Sinica* 1: 11-16. (In Chinese with English abstract).

Yang, H. (1984). The mangroves in China. In: *Man's Impact on Coastal and Estuarine Ecosystems.*

MAB/COMAR Regional Seminar. MAB Co-ordinating Committee of Japan. Pp. 9-14.

Yang, H. (1987). Nature reserves in seacoast of China. *Camp Network Newsletter* August 1987. P. 2.

Yang, R.T., Chiang, Y. and Huang, T. (1975). A report of the expedition to Tung-Sha reefs. *Inst. Oceanogr. Natl Taiwan Univ., Spec. Publ.* 8. 33 pp. (In Chinese).

*****Yang, Z.D. (1978).** A preliminary study on the intertidal ecology of benthic marine algae of Hainan Island. *Studia Marina Sinica* 14: 138-140.

*****Yang, Z.D. (1985).** Seagrasses of the Xisha Islands. *Studia Marina Sinica* 24: 124-131. (In Chinese with English abstract).

Zeng, Z. and Qiu, S. (1987). Evolutional patterns of the coral sand islands in the Xisha area. *Acta Oceanologica Sinica* 6(2): 235-248.

Zheng, S. (1979). The recent Foraminifera of the Xisha Islands. *Studia Marina Sinica* 15: 200-229.

Zhong, J. and Huang, J. (1979). A preliminary analysis of the grain size and composition of the loose sediments in the Xisha Islands, Guangdong Province, China. *Oceanol. Limnol. Sinica* 10(2): 125-135. (In Chinese with English abstract).

Zhu, D. (1987). Mangrove coast of Hainan Island, China. Paper presented at Regional Workshop and International Symposium on the Conservation and Management of Coral Reef and Mangrove Ecosystems, Okinawa, Japan. 5 pp.

Zhu, Y. (1981). Biogenic reefs of the Xisha Islands, Guangdong. *Nanhai Studia Marina Sinica* 2: 34-47. (In Chinese with English abstract).

Zhu, Y. and Zhong, J. (1984). A preliminary study on the dune rock of Shidao Island, Xisha Islands and Hainan Island. *Tropic Oceanology* 3(3): 65-72. (In Chinese with English abstract).

Zhu, Y., Zhong, J., and Nie, B. (1982). A preliminary study on *Alcyonacea spiculis* limestone in Luhuiton, Hainan Island, Guangdong. *Tropic Oceanology* 1(1): 35-41. (In Chinese with English abstract).

Zou, R.-L. (1978). A preliminary analysis of the community structure of the hermatypic corals of the Xisha Islands, Guangdong Province, China. Science Press Academia Sinica. Pp. 125-132. (In Chinese with English abstract).

Zou, R.-L. (1979). Further analysis on th community structure of the hermatypic corals of the Xisha Islands, Guangdong Province, China. *Oceanic Selections* 2(2): 113-129.

Zou, R.-L. (1981). A mathematical model of the hermatypic coral community of the Xisha Islands, Guangdong Province, China. *Proc. 4th Int. Coral Reef Symp., Manila* 2: 329-331.

Zou, R.-L. (1984). Studies on corals from Xisha Islands. 5. The deep-water *Acropora* with a description of a new species. *Tropic Oceanology* 3(2): 52-55.

Zou, R.-L. (1975a). *Studies on the corals of the Xisha Islands, Guangdong Province, China 2. The genus Millepora, with the description of a new species.* Science Press, Academia Sinica. (In Chinese with English abstract).

Zou, R.-L. (1976b). *Studies on the corals of the Xisha Islands, Guangdong Privince, China 3. An illustrated catalogue of scleractinian, hydrocorallien, Heliporina and Tubiporina.* Science Press Academia Sinica. (In Chinese with English abstract).

Zou, R. and Chen, Y. (1983). Preliminary study on the geographical distribution of shallow-water Scleractinia corals from China. *Nanhai Studia Marina Sinica* 4: 89-95. (In Chinese with English abstract).

Zou, R.-L., Ma, J.H. and Sung, S.W. (1966). A preliminary study on the vertical zonation of the coral reef of Hainan Island *Oceanol. Limnol. Sinica* 8: 153-161.

Zou, R., Meng, Z. and Guan, X. (1983). Ecological analyses of ahermatypic corals from the northern shelf of South China Sea. *Tropic Oceanology* 2(3): 1-6. (In Chinese with English abstract).

Zou, R.-L., Song, S.W. and Ma, J.H. (1975). *The Shallow-water Scleractinian Corals of Hainan Island.* Scientific Publisher, Peking. 66 pp.

DONGSHA QUNDAO (TUNG-SHA REEF, PRATAS)

Geographical Location   170 mi. (274 km) south-east of Hong Kong and 240 mi. (386 km) south-west of Kaohsiung City, Taiwan; 20°40'-43'N, 116°42'-44'E.

Area, Depth, Altitude   The Dongsha reefs cover ca 100 sq. km; Dongsha Qundao (Tung-Sha Tao) is 1.7 sq. km; max. alt. 5 m.

Land Tenure   Sovereignty of the area is disputed by China and Taiwan.

Physical Features   The Dongsha reefs are a large submerged atoll with the small island of Dongsha Qundao situated on its west. The island is covered with sandy coral debris and traces of guano and has a lagoon (0.6 sq. km) open on the west side. It is surrounded by a shallow terrace which drops to 40 m or more at the outer edge (Fang *in litt.*, 14.3.84).

The climate is subtropical oceanic, heavily influenced by the monsoon. Average air temperature ranges from 20.2°C to 28.7°C, water temperature from 21°C to 30°C. A thermocline is found at 30-75 m depth in waters outside the atoll. Surface currents flow north-east with a velocity of 0.2-0.5 mi./hr (0.3-0.8 km/hr). Currents at depths of 300 m turn to the north-west with a speed below 0.2 mi./hr (0.3 km/hr). Seasonal winds also produce surface currents (Fang *in litt.*, 14.3.84).

Reef Structure and Corals   The reef structure is typical of atolls (Yang *et al.*, 1975). Over 70 species of reef coral have been recorded (Liang, 1985a and d). In a three day trip, 45 coral species in 17 genera were collected: 60% of the specimens (125 pieces) were *Acropora*, 10% were *Porites* (Yang *et al.*, 1975). The coral fauna appeared to be similar to that in Kenting National Park and Hsiao-liu-chiu in Taiwan (Fang *in litt.*, 14.3.84 and see section on Taiwan).

Noteworthy Fauna and Flora   Twenty-five species of fish in 17 genera (86 specimens) have been collected. Apogonidae, Labridae, and Pomacentridae are dominant families. Twenty-six species of mollusc in 23 genera have been recorded. *Strombus*, *Monetaria* and *Tridacna* are abundant, while *Tutufa bubo*, *Oliva arnata*, *Terebra dimidiata* and *Arca ventricosa* are rare (Yang *et al.*, 1975). The seagrasses *Halophila ovalis* and *Thalassia hemprichii* flourish. Eleven Chlorophycophyta, seven Phaeophycophyta and ten Rhodophyta have also been found (Fang *in litt.*, 14.3.84).

Scientific Importance and Research The weather has been recorded since 1925 due to the presence of the lighthouse. An expedition sponsored by the National Science Council, Taiwan, carried out a survey in March 1975 and studied the hydrology, biology, geology, fishery and chemical composition of the reefs (Fang *in litt.*, 14.3.84).

Economic Value and Social Benefits There is a permanent station established by Taiwan on the island which is an important site for navigation and ocean/weather observations. *Diegenea simplex*, an alga used pharmacologically, is collected in large quantities in this area. The adjacent fishing ground (finfish and shellfish) is visited by fishermen from southern Taiwan and commercial shell and lobster fishing is under way (Fang *in litt.*, 30.4.87).

Disturbance or Deficiencies There is a potential threat from overfishing of shell fish and populations of Giant Clams *Tridacna elongata* are reported to be decreasing (Fang *in litt.*, 14.3.84).

Legal Protection None.

Management Reef resources are not managed.

HAINAN (HUAN CHU-CHIM)

Geographical Location Nan Hai (South China Sea); Hainan Province, separated from the mainland by the 15 mi. (24 km) wide Hainan Strait; 18°09'-20°20'N, 108°36'-112°02'E.

Area, Depth, Altitude 33 556 sq. km; max. alt. over 1000 m.

Physical Features The geology of Hainan is described by Zhu *et al.* (1982) and Zhu and Zhong (1984). The island is generally high in the south-west and centre, lower to the north-east. Average surface water temperature in February is 22°C, in August 29°C. Tides are irregular and mixed, with tidal range usually not more that 3 m (Wu *in litt.*, 4.6.87).

Reef Structure and Corals There have been no major studies of reef ecology but there is a fringing reef on the south and south-east coasts (Wu and Hutchings, 1986; Zou *et al.*, 1966; Zou *et al.*, 1975). Early descriptions of the reefs are given in Gurjanova (1959) and Naymov *et al.* (1960). The reefs are believed to be less than 10 000 years old and are less than 10 m thick. Although well-developed reefs were present along the southern coast in the 1950s, a brief survey in 1984 in the region of Sanya indicated that little of these remained. Brief surveys were made at Simau ehou, Shen Chow, Hsiao Tun Hai and Lunya Bay. The last-named of these, a sheltered bay north-east of Sanya, had the best developed reefs with a high percentage of living coral, although extensive areas of dead coral were also present (Wu and Hutchings, 1986). There are now only scattered coral colonies but a substantial number of genera have been found (Wu, 1985a). Liang (1985a and d) lists some 166 species of scleractinian corals in 43 genera from Hainan and associated islands; Zou *et al.* (1966) list 110 species and subspecies in 34 genera. Twenty-six species have been recorded in the cooler waters off the northern part of Hainan. The most diverse areas are at the southern tip and include Luhuitou (87 species recorded), Ximsozhou (77 species) and Xinchun lagoon (53 species) (Liang, 1985 a and d). However Brown (1986) reports that only 20 species are now found.

Noteworthy Fauna and Flora There are 10 species of mangrove, covering ca 40 sq. km (Zhu, 1987), and 9 species of sea grass. Marine algae, numbering ca 500 species, are described by Yang (1978). Around 700 species of molluscs have been recorded, including *Trochus niloticus*, *Pinctada maxima* and *P. martensis* (Qi *et al.*, 1984) and *Tridacna* spp. (Wu and Hutchings, 1986). Over 1000 species of fish, 100 species of echinoderms and 120 species of polychaetes are known to occur (Klebovitsch and Wu, 1963; Wu *in litt.*, 4.6.87).

Scientific Importance and Research Research has been carried out for many years on Hainan by the South China Sea Institute of Oceanography which has a marine research station at Sanya in the south (Zhu *in litt.*, 3.11.87). The reefs were surveyed in the 1950s by Gurjanova (1959) and Naymov *et al.* (1960).

Economic Value and Social Benefits Turtles are exploited; in 1985, 97 tonnes were landed in Qionghai county (Wang *in litt.*, 20.11.86). Collection of coral for the building trade (see below) can provide an income 10-30 times the average income of a farmer (Struhsaker, 1987).

Disturbance or Deficiencies The reefs appear to have suffered significant damage from siltation and overfishing using fine meshed nets. There is extensive algal cover in some areas and high densities of the urchin *Diadema* sp. Dredging probably accounts for the siltation. Fish are more abundant in naval restricted areas where there is lower fishing pressure. Overfishing is particularly evident at Shen Chow and Haia Tun Hai in the south (Wu, 1985a; Wu and Hutchings, 1986). The mining of corals for the construction industry, dredging and the collection of live corals for the aquarium trade represent additional threats (Brown, 1986; Wu and Hutchings, 1986). Struhsaker (1987) describes the collecting of coral from a reef near the town of Qinglan on the north-east coast. A considerable number of people collected both live and dead coral, prising it free with a heavy metal bit at the end of a 2 m pole. A single person could collect 1000-2500 kg of coral each week and it was thought that in total very large amounts were being taken from the reef. The great majority was used for building houses, although some was sold to tourists. Mangroves are threatened by land reclamation for agriculture, construction of dikes for shrimp and eel culture, and cutting for fuel-wood (Zhu, 1987).

Legal Protection According to Yang (1987), Tuntze Nature Reserve at 20°N, 110°20'E on the north coast of Hainan protects 3737 ha of mangroves; it is not known if reefs are included. Fishing has been prohibited for several years in Lunya Bay on the south coast (Wu and Hutchings, 1986).

Management None.

Recommendations It is suggested that at least part of the south coast reef area should be closed to allow recovery of the reef; dredging in the vicinity of the reefs should be banned; size restrictions should be introduced for commercial fish species; and a public awareness programme should be introduced as a high priority. Future work in the area should carefully document the distribution and composition of the reefs (Wu, 1985a; Wu and Hutchings, 1986).

Economic Value and Social Benefits Projects on the exploitation of new fishing grounds and commercial shell and lobster fisheries are now under way in the area (Fang *in litt.*, 30.4.87).

Disturbance or Deficiencies Not known.

Legal Protection There is reportedly Chinese legislation stipulating a reserve or reserves in the area (Wang *in litt.*, 20.11.86), but details are lacking.

Management None known.

NANSHA QUNDAO (NAN-SHA REEFS, SPRATLY ISLANDS)

Geographical Location The main island is Taiping dao (or Itu Aba Island), 10°23'N, 114°22'E, and is located north-west of Tizard banks and reefs; the southernmost, at 4°N, is Zengmu (Tsan-Mou Reef); 4°-11°30'N, 109°30'-117°50'E.

Area, Depth, Altitude Taiping dao is 1289 x 366 m; 3.8 m above sea-level.

Land Tenure Sovereignty over the reefs is disputed by a number of countries in the area, including China and Taiwan.

Physical Features The Nansha Qundao are a series of 104 emergent reefs and countless submerged reefs and include fringing reefs and atolls. The climate is oceanic tropical. Air temperature ranges from 26.1°C to 28.8°C. Average water temperature is 28.1°C. The north-east monsoon blows from October to March, the south-west monsoon from May to October. The current flows south-west during the former and east or north during the latter (Fang *in litt.*, 14.3.84).

Reef Structure and Corals Taiping dao is surrounded by a reef terrace and has sandy shoals, reef flats, reef barriers and reef fronts. Coral communities are flourishing and diverse, and there is high coral cover, with *Acropora* the dominant species (Fang *in litt.*, 14.3.84). Bassett-Smith (1890) first described corals from Tizard Bank. Over 100 species of coral have now been recorded at Tizard Atoll. There are two distinct coral zones, an upper terrace to 18 m depth and a lower terrace, from 37 to 56 m depth, with different coral composition. On the intervening slope, from 18 to 37 m, only one species was found. Some species have been recorded at considerable depth, for example *Leptoseris striata* at 65 m, *Montipora* sp. at 81 m and *Favia* sp. at 83 m. Eighteen species of reef coral have been recorded on coral knolls in the lagoon (Liang, 1985a and d).

Noteworthy Fauna and Flora A total of 111 species of fish in 26 families have been identified (Chang *et al.*, 1981). Giant clams *Tridacna* spp. have been seen. A flourishing eel grass bed was found on a sandy shoal (Fang *in litt.*, 14.3.84).

Scientific Importance and Research Research expeditions, which concentrated on fish, were undertaken to Taiping Dao from the Fishery Research Institution of Taiwan and the Institute of Zoology, Academia Sinica, Taiwan, in 1976 and 1981 respectively (Fang *in litt.*, 14.3.84).

XISHA QUNDAO (PARACEL SHOALS)

Geographical Location Nan Hai (South China Sea) south-east of continental shelf and Hainan; in the region 15°30'-17°00'N, 111°-113°E.

Area, Depth, Altitude There are 31 islets. The largest is Yongxingdao (1.8 sq. km); three others, Tongdao, Zhongjiandao and Taipingdao, are over 1.5 sq. km and the remaining 25 islets are all smaller than 0.5 sq km. Depth of the lagoon in the Yongle complex atoll is 43 m, in the Xuande complex atoll 67 m, and in the Xuande Dong (East Xuande) complex atoll 58 m. Max. alt. (15 m) is reached by Shidao islet; 10 islets reach ca 10 m and the remainder are under 7 m.

Land Tenure Sovereignty of the area is disputed by China and Vietnam.

Physical Features The islands and reefs have an atoll-like structure: complex atolls, simple atolls and table reefs have been described. The complex atolls are Yongle, Xuande and Xuande Dong; the simple atolls include Beijiao, Yuzhuo, Huaguang, Panshi and Langhua; the table reefs include Jinyin and Zhonjian (Wu *in litt.*, 4.7.87). Their geology is described by Zhu (1981) and Zhu and Zhong (1984), their geomorphology and origin by Zeng and Qiu (1987), and sediments by Zhong and Huang (1979). Average temperature on the islands is 26°C, with a minimum of 15.3°C. Annual rainfall averages 1400 mm (range 1300-2000 mm), and there are clearly marked wet and dry seasons. During the wet summer months winds are generally weak and from the south-west whilst during the drier winter months winds are strong and from the north-east or north-west; typhoons are frequent from June to January, with a peak in November. Annual mean surface water temperature is 26.8°C. Mean surface salinity is 33.7 ppt (range 33.1-34.2 ppt) and tides are irregular diurnal with average tidal range of 0.6-1.5 m. Currents are mainly from easterly directions (south-east to north-east) (Wu *in litt.*, 4.7.87).

Reef Structure and Corals The community structure and species diversity of these reefs is described by Zou (1978, 1979 and 1981). On the north-east reef flats there is luxuriant hermatypic coral growth and diverse populations of other reef organisms. In contrast, the south-west reef flats have a sparse fauna. There is no algal ridge and the coral fauna on both the north-east and south-west seaward slopes is poor. More detailed studies were carried out at Zhongjiandao (Zhong Jian), a wooded island in the south-west, and Zhao Shu, a sandy island in the north-east of the group. Transects across the reefs of these islands are described in Zou (1979) and a mathematical model of the coral communities is given in Zou (1981). Coral growth rates are described by He and Zhong (1982). The most recent studies give 123 (Liang, 1985a) or 127 (Zou and Chen, 1983) species and subspecies of scleractinian corals in 37 or 38 genera respectively; Zou (1975b) listed 113 in 33 genera. *Millepora* is particularly abundant and Zou (1975a) describes a new species, *M. xishaensis*. Deep water acroporids from Yongle Qundao are described by Zou (1984).

Noteworthy Fauna and Flora The islands are covered with low bushes, *Pisonia grandis, Guettarda speciosa* and *Scaevola sericea*. Many have major seabird colonies (Huen-pu, 1980; Melville, 1984), including Red-footed Boobies *Sula sula*, Brown Boobies *S. leucogaster* and Greater Frigatebirds *Fregata minor*. The Green Turtle *Chelonia mydas* nests in the islands, apparently in significant numbers, but this population is subject to exploitation, largely by fishermen from Hainan (Frazier and Frazier, 1985; Huang, 1981; Wang *in litt.*, 20.11.86). Over 500 species of molluscs are present, including *Haliotis, Pinctada maxima, Turbo* and Giant Clams *Hippopus* and *Tridacna*. More than 1000 species of crustacea have been recorded, as well as ca 100 genera of Foraminifera (Zheng, 1979), 160 species of polychaete, 150 species of echinoderms (Wu *in litt.*, 4.7.87), and 552 fish species (Wang, 1981). Three species of sea grass are present (Yang, 1985) and over 200 species of seaweed, the main species being the same as those found off Hainan (Tseng, 1983; Wu *in litt.*, 4.7.87). Dinoflagellates are described in Chen (1981).

Scientific Importance and Research The reefs have been extensively studied, including an ecological survey from 1973 to 1976 by the South China Sea Institute of Oceanology, Academia Sinica (Zou, 1979).

Economic Value and Social Benefits The biological resources of the islands are described by Chen and Sung (1977). At present only Yongxingdao and Shindao are inhabited (Wu *in litt.*, 4.7.87).

Disturbance or Deficiencies There were considerable changes in morphology of some of the islands as a result of a typhoon which passed between Yongle Qundao and Zhongjiandao in August 1975 (Zou, 1979). Damage to the reefs due to human activities is less serious than on Hainan but the escalating coral trade and overfishing are beginning to cause problems (Brown, 1986). Guano from the booby colonies is used as fertilizer and its extraction caused considerable devastation earlier in the century, leading to the disappearance of the Red-footed Booby. Extraction is now regulated and the Brown Booby is protected (Chen and Sung, 1977). Rats have been introduced and the cats that have been subsequently introduced to control them have also affected the bird populations (Melville, 1985).

Legal Protection There is reportedly Chinese legislation stipulating a reserve in the islands, but details are lacking (Wang *in litt.*, 20.11.86).

Recommendations Given the scientific interest of the Xisha reefs, it would appear that this should be considered an area of conservation concern.

COOK ISLANDS

INTRODUCTION

General Description

The fifteen Cook Islands are scattered in the area 8°-23°S, 156°-167°W and are self-governing but linked to New Zealand. There is considerable information on the oceanography of the region, gathered in the course of survey work by the New Zealand Oceanographic Institute to identify manganese deposits in the south-west Pacific (Summerhayes, 1967; Landmesser *et al.*, 1976).

The Southern Cook Islands consist of a linear series of volcanic and uplifted islands (with both limestone and volcanic facies) of different ages (Turner and Jarrard, 1982), extending for ca 650 km from Mauke to Palmerston, and the two more isolated islands of Rarotonga and Mangaia, all rising from depths of 4500-5000 ft (1372-1524 m). The islands are described in Stoddart (1975c) and their geology in Wood and Hay (1970). Rarotonga is the largest and is an isolated volcanic island, now deeply dissected by erosion. The other high islands in the Southern Group have mixed volcanic and limestone rocks, some being elevated to form makatea (fossil reef); their geology is discussed in Spencer *et al.* (1987). The Southern Group is influenced by winds from the north-east, east and south-east. Rainfall ranges from 1500 mm to 2800 mm (Anon., 1985). Hurricanes and storms occur mainly from January to March and approach from the north-east. The mean sea surface temperature varies from 27.3°C in January to 25.5°C in June. Tides are semi-diurnal with a small amplitude. On Rarotonga, spring tidal range is 0.85 m and neap range 0.33 m (Henry, 1977; Stoddart, 1975c).

The area between Aitutaki and Penrhyn (Tongareva) is known as the South Penrhyn Basin and reaches depths of over 4500 m. The Northern Group are low atolls, the western ones lying on the Manihiki Plateau, and are sparsely settled and little changed, with only 25 sq. km of land area in total.

Table of Islands

Southern Group

Mangaia (X) 27.3 sq. mi. (71 sq. km); low, volcanic, 554 ft (169 m); uplifted island surrounded by makatea; described by Davis (1928); geology and surrounding fringing reef described by Marshall (1927) and Stoddart *et al.* (1985a and b); narrow living reef flat; outer reef slope steepens precipitously but levels off to wide terrace at 30-40 m depth (Paulay, in press); 28 coral genera known (Paulay, 1985); fish described by Clerk (1981).

Rarotonga (X) 25.8 sq. mi. (67 sq. km); high, volcanic, 650 m; described by Doran (1961); geology described by Marshall (1908 and 1930) and Wood (1967); further information in Stoddart (1975c); reefs described below; those of Ngatangiia Harbour described in separate account; 41 coral genera known (Paulay, 1985).

Mauke (Parry) (X) 7.1 sq. mi. (18.4 sq. km); low, volcanic, 100 ft (30 m), raised island surrounded by makatea; completely surrounded by 50-100 m wide reef flat, much intertidal or supratidal but some submerged; reef flat dominated by hard reef rock pavement and therefore very depauperate; outer reef slopes gently to 8-15 m depth, then steeply to at least 80 m, generally without shelves (Paulay, in press); 32 coral genera known (Paulay, 1985).

Mitiaro 8.6 sq. mi. (22.3 sq. km); low volcanic; raised island surrounded by makatea; geology described by Wood and Hay (1970); fringing reef with narrow living reef flat; 23 coral genera known (Paulay, 1985).

Atiu (X) 10.9 sq. mi. (28.2 sq. km); volcanic plateau 300 ft (91 m); raised island surrounded by makatea; geology described by Marshall (1908 and 1930) and Wood and Hay (1970); fringing reef with narrow living reef flat; long stretch of coastline along north side lacks reef flat (Paulay, in press); 19 coral genera known (Paulay, 1985).

Takutea 0.5 sq. mi. (1.3 sq. km); low lying, elongated, sand cay (Wood and Hay, 1970). No anchorages; surrounded by reef; turtle nesting site (Brandon, 1977).

Manuae (Hervey) 8.5 sq. mi. (22 sq. km); atoll with twin, flat, coral, sand islets: Manuae and Au-o-to; shallow closed lagoon (Wood and Hay, 1970); surrounded by continuous reef with single narrow boat passage; lagoonal sedimentation described by Summerhayes (1971); turtle nesting site (Brandon, 1977); first island of the Cook Islands to be discovered by Captain Cook.

Aitutaki (X) (*see separate account*).

Palmerston (X) land area 1 sq. mi. (2.6 sq. km); atoll with elongated lagoon and 8 islets (Wood and Hay, 1970); underwater morphology described by Irwin (1985); surrounded by reef; fish described by Grange and Singleton (1985); major nesting site for Green Turtle *Chelonia mydas* (Balazs, 1981); additional information in Helm and Percival (1973).

Northern Group

Suwarrow Atoll (Suvarov) (*see separate account*).

Penrhyn (Tongareva, Mangarongaro) (X) 3.8 sq. mi. (9.8 sq. km); largest atoll with many islets; described by Doran (1961) and Wood and Hay (1970); surrounded by reef; Green Turtle *Chelonia mydas* and Hawksbill *Eretmochelys imbricata* nest (Balazs, 1981); surveyed in 1985 by Ministry of Marine Resources (Sims *in litt.*, 11.12.87).

Manihiki (X) (*see separate account*).

Rakahanga (X) 1.55 sq. mi. (4 sq. km); atoll; small lagoon which has closed in last 25 years; 2 main islets to north and south, with 7 smaller islets between; surrounded by reef; *Chelonia mydas* nest (Balazs, 1981).

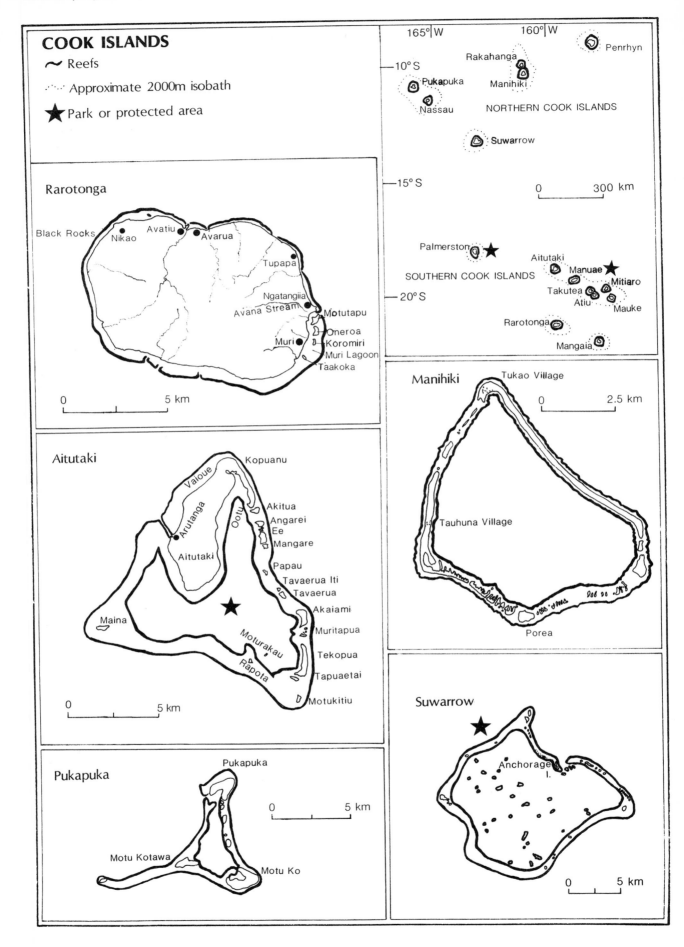

COOK ISLANDS

~ Reefs

..... Approximate 2000m isobath

★ Park or protected area

Rarotonga

Black Rocks
Nikao
Avatiu
Avarua
Tupapa
Ngatangiia
Avana Stream
Motutapu
Oneroa
Koromiri
Muri
Muri Lagoon
Taakoka

0   5 km

Aitutaki

Valoue
Kopuanu
Arutanga
Ootu
Akitua
Angarei
Ee
Mangare
Aitutaki
Papau
Tavaerua Iti
Tavaerua
Akaiami
Maina
Muritapua
Moturakau
Tekopua
Rapota
Tapuaetai
Motukitiu

0   5 km

Pukapuka

Pukapuka
Motu Kotawa
Motu Ko

0   5 km

165° W   160° W

Penrhyn
10° S
Rakahanga
Pukapuka
Manihiki
Nassau
NORTHERN COOK ISLANDS
Suwarrow
15° S
0   300 km

Palmerston ★
Aitutaki
Manuae ★
SOUTHERN COOK ISLANDS
Mitiaro
Takutea
20° S
Atiu
Mauke
Rarotonga
Mangaia

Manihiki

Tukao Village
0   2.5 km
Tauhuna Village
Porea

Suwarrow

★
Anchorage I.
0   5 km

Nassau (X) 0.45 sq. mi. (1.2 sq. km); islet without lagoon; oval, flat sand cay with few dunes (Wood and Hay, 1970); narrow reef flat; owned and used by Pukapuka people.

Pukapuka (Danger, Pakapuka) (X) (*see separate account*).

(X) = Inhabited

Reefs of the Cook Islands have been studied by both the Eclipse Expedition of 1965 (McNight, 1972) and the 1969 Cook Bicentenary Expedition to the South-west Pacific (Gibbs *et al.*, 1971); reefs of Aitutaki (*see separate account*) and Rarotonga received particular attention. Fringing and lagoon reefs are common. A barrier reef is found at Aitutaki, and atoll reefs at Manuae and Palmerston (Dahl, 1980a). Reefs at Manihiki, Suwarrow and Pukapuka are described in separate accounts.

Crossland (1928) and Dana (1898) gave early descriptions of the fringing reefs around Rarotonga and more recent work has been carried out by Stoddart (1972 and 1975c), Dahl (1980b) (who compared these reefs with those of Aitutaki), Gauss (1982) and Lewis *et al.* (1980), the last in the course of a survey to assess the potential for offshore sand and gravel aggregate. A brief description is given in Paulay (in press). The fringing reef increases in size from the south-east corner of the island counterclockwise, such that the widest and narrowest reefs are adjacent at the south-east corner. The reefs are about 400 m, and up to 800 m, wide along the south coast and 100-200 m wide along the west and north coasts; on the windward east coast, the reefs are 50-100 m wide with large intertidal portions and large amounts of accumulated rubble. The reef edge is continuous except for steep-sided narrow inlets at Avatiu, Avarua, Ngatangiia and several places along the south coast. The reef flats are covered with sand, and are rarely covered by more than 1.5 m of water even at high tide (Stoddart, 1972). Growing corals are common in many areas and the holothurian *Holothuria atra* is locally abundant, mostly on the inner southern reefs of Ngatangiia (Paulay *in litt.*, 10.7.87). There is an algal ridge on the reef which is thought to be higher in the south than in the north (Marshall, 1930). Work during the 1967 Cook Bicentenary Expedition was concentrated mainly at Ngatangiia Harbour (*see separate account*).

Fifty-seven scleractinian corals were recorded in 1971. The most abundant genera collected at Aitutaki and Ngatangiia (Rarotonga) in the course of the Cook Bicentenary Expedition were *Porites, Pocillopora, Montipora, Acropora, Favia, Favites, Hydnophora* and *Lobophyllia. Cyphastrea, Turbinaria, Galaxea* were common and *Fungia* and *Herpolitha* were rare (Gibbs *et al.*, 1971; Stoddart and Pillai, 1973). A more recent list of genera for Rarotonga, Aitutaki, Atiu, Mauke and Mitiaro is given in Paulay (1985), 41 genera (35 hermatypic) having been recorded from Rarotonga.

A history of the reef-related research work carried out in the islands is given in Stoddart (1975c) and additional references are given in Cowan (1980), Thompson (1983) and Paulay (in press). Mangrove vegetation and seagrasses are absent throughout the Cook Islands.

Marine biology is discussed in Randall (1978) and a checklist of the marine fauna, including 14 species of fish, is given in Gibbs *et al.* (1975). Paulay (in press) describes the heterodont bivalve fauna in detail, echinoderms are described in Devaney (1973 and 1974), Marsh (1974) and McKnight (1972), polychaetes in Gibbs (1972) and alpheid shrimps in Banner and Banner (1967). Shrimps were also collected by Operation Raleigh in 1986 (Richmond, 1986).

Reef Resources

Reefs, reef flats and lagoons play a leading role in protein supply in the Cook Islands (Hambuechen, 1973b). Islanders on the atolls of the Northern Group may be as much as 90% dependent on protein from the sea, much of which is from the reefs; those in the Southern Group may be 60% dependent. On Palmerston, fish are the main protein source and almost all reef fish are eaten, most of them speared on the reef crest at low tide, although gill nets may also be put across small reef passages (Grange and Singleton, 1985). Reef fishing production is currently static with sporadic shipments made to Rarotonga (Sims, 1985 and *in litt.*, 11.12.87). Lawson and Kearney (1982) provide an assessment of skipjack and baitfish resources. Turtles, but not their eggs, are regularly eaten on some of the islands (Palmerston, Pukapuka, Manihiki and possibly Penrhyn). Turtle meat is said to be not readily acceptable on Rarotonga and turtles on Penrhyn are taken principally for their shell (Balazs, 1981; Sims *in litt.* to B. Groombridge, 28.8.86).

Trochus *Trochus niloticus* was first introduced to the Cook Islands in 1957; its abundance, distribution and exploitation are described by Sims (1985), greatest exploitation taking place on Aitutaki (*see separate account*). It has also become established on Palmerston and Manuae, an introduction programme having been started in 1981 when the value of the trochus industry was first appreciated. However, probably only the atoll islands have reefs suitable for this species. Attempts to introduce the Goldlip Pearl Oyster *Pinctada maxima* on Rakahanga and Pukapuka are under way, although this has proved difficult in the past (Paulay *in litt.*, 10.7.87; Sims *in litt.*, 11.12.87).

The tourist industry is increasing, particularly on Rarotonga and Aitutaki (*see separate account*). Tourists snorkel throughout the Rarotongan lagoon and there is one Rarotongan SCUBA-dive operator who usually operates off Nikao, where a wreck has recently been sunk in 90 ft (27 m) of water, and off Tupapa (Sims *in litt.*, 11.12.87).

Disturbances and Deficiencies

Hurricanes periodically cause damage but this has not been studied; unspecified damage has been recorded on Pukapuka, on motus on the south and east sides of Penrhyn and on the north coasts of most Southern Group islands. Reefs around Rarotonga and Aitutaki were surveyed in 1971 for *Acanthaster planci* and numbers were found to be high. Although there were coral patches at about 12 m depth on the southern terrace of Rarotonga, elsewhere 50-90% of the coral was dead (Devaney and Randall, 1973). In 1978, no starfish were found, but the reef-front terrace around Rarotonga was generally floored with dead, algal-covered coral with small patches of coral/algal sand between the dead coral heads. Only a few massive *Porites* heads were found to be alive, and it was suggested that this poor coral cover was also associated with the *Acanthaster* outbreak (Lewis *et al.*, 1980). There have also been some reports of damage from *Acanthaster* on Manuae, Manihiki and Penrhyn (Hambuechen, 1973a) although these conflict with local reports (Sims *in litt.*, 11.12.87).

Most problems due to human activities have arisen on Rarotonga, the most densely populated and technologically developed island, where airport and hotel construction, port improvement, pollution and soil run-off have contributed to coastal degradation. The reefs are considered to be in an advanced stage of degradation (SPREP, 1980; Dahl, 1980b). An island near Avatiu is much altered by reclamation of the reef flat (Stoddart, 1972). Chemical run-off may be a significant problem (Hambuechen, 1973a and b), and problems created by the increased use of pesticides are described by Hambuechen (1973b). Fish poisoning and dynamiting are believed to be relatively unimportant (Paulay *in litt.*, 10.7.87) although they have occurred (Dahl, 1980b; Hambuechen, 1973a).

Reefs elsewhere are said to be in good condition. However, soil erosion as a result of pineapple cultivation is occurring on Mangaia and Atiu. The lagoons of Aitutaki and Manuae are reported to be steadily silting up (Summerhayes, 1971). Phosphate mining may be investigated on Rakahanga and Manihiki (Eldredge *in litt.*, 18.2.85) and could cause damage. Pearl oyster stocks on Penrhyn have declined from previous levels (Sims *in litt.*, 11.12.87).

Legislation and Management

Traditional legislation is still important on the Cook Islands and operates with increasing effectiveness in the outer islands. The *Ra'ui* concept enables restricted access to be enforced on land and lagoon areas. This effectively inhibited tree cutting to the seaward side of the road on Rarotonga for several years (SPREP, 1980; Anon., 1985), but has scarcely been used in the Southern Group in the past thirty years (Paulay *in litt.*, 10.07.87).

General conservation information is given in SPREP (1980), an early description is given in Gare (n.d.), and information on protected areas and land tenure is given in Anon. (1985). The Territorial Sea and Exclusive Economic Zone Act (1979) includes provision for regulations to be promulgated for protection and preservation of the marine environment. Legislation prohibiting the use of poisons and dynamite is reported to exist (Hambuechen, 1973a). The Trochus Act 1975 regulates the harvest of trochus by licenses within reserves (Anon., 1985). Trochus regulations are enforced by the Ministry of Marine Resources, through the *ad hoc* appointment of Trochus Inspectors. A fully effective system is still being developed (Sims *in litt.*, 11.12.87). Legislation controlling the exploitation of trochus on Aitutaki is described in a separate account. The Island Council of Palmerston reportedly forbad the use of spearguns in 1977 in response to an apparent decline in turtle numbers (Balazs, 1981); however turtles are now apparently freely hunted there (Sims *in litt.*, 11,12,87).

The Conservation Act 1975, devised to conserve nature and natural resources and to protect historic sites and the environment, has not been used to any extent and there are no regulations deriving from it. It provides for any land, lagoon, reef, and island to be declared a National Park, World Park or Historic Site. In 1987, a new Conservation Act was passed which includes provision for marine parks (Dahl *in litt.*, 27.10.87).

Suwarrow Atoll National Park is described in a separate account. On Rarotonga there is a foreshore reserve between the main road and the beach from Avatiu Harbour to the airport (SPREP, 1980). The original purpose of this is not known; it consists only of a narrow "nature strip" between the road and beach and is becoming built up in parts (Sims *in litt.*, 11.12.87). Takutea was designated a bird sanctuary in 1905 but not yet legislated as such although there are plans to do so (Anon., 1985). There are three fishing reserves protected under the Trochus Act (1975) where diving for trochus without a licence is prohibited: Aitutaki, Palmerston and Manuae lagoons.

Recommendations

A Coastal Zone Management Programme (SOPACOAST) is currently being developed for the South Pacific in collaboration with South Pacific regional organisations and the Commonwealth Science Council. Pukapuka has been selected as a study site to develop a management plan for low islands (*see separate account*). Information and experience gained from this project will be used in the development of management programmes in other parts of the Cook Islands. Furthermore, the environment and resources of the Northern Group will be documented, an archipelagic resource conservation strategy will be developed for the area and training for its effective development and implementation will be carried out (Anon., 1986).

Dahl (1980a) recommended the creation of a number of protected areas, but these have not been followed-up: on Rarotonga, the Black Rocks (Tuoro) area and the islets and reefs of Ngatangiia Harbour have been recommended as reserves; the lagoon and eastern motus of Aitutaki have been recommended as a reserve; Manuae was proposed as an international maritime park but this is no longer considered a priority. A more detailed proposal for a coastal protected area system on Rarotonga is given in Dahl (1980b), including the establishment of a marine fisheries management reserve for all the reefs, lagoons and adjacent coastal waters. Additional work is required to implement the national park at Suwarrow.

References

* = cited but not consulted

*Agassiz, A. (1903). The coral reefs of the tropical Pacific. The Cook Islands. *Mem. Mus. Comp. Zool. Harvard* 28: 168-170.

Andrews, G.J. (1987). Marine ecological survey of Pukapuka Atoll. Final Report. Commonwealth Science Council/Australian Institute of Marine Science.

Anon. (1985). Country Review - Cook Islands. *Report of the 3rd South Pacific National Parks and Reserves*

Conference, Apia 3: 60-75.

Anon. (1986). Environmental Planning Programme - Coastal Zone Management of Tropical Islands. South Pacific Coastal Zone Management Programme (SOPACOAST). *CSC Technical Publ. Ser.* 204. Commonwealth Science Council, London. 28 pp.

Balazs, G.H. (1981). Status of Sea Turtles in the Central Pacific Ocean In: Bjorndal, K.A. (Ed.), *Biology and Conservation of Sea Turtles*. Smithsonian, Institution Press, Washington D.C. Pp. 243-252.

*Banner, A.M. and Banner, D.M. (1967). Contributions to the alpheid shrimp fauna of the Pacific Ocean. 2. Collections from the Cook and Society Islands. *Occ. Papers B.P. Bishop Mus.* 23: 253-286.

*Beaglehole, E. and Beaglehole, P. (1938). Ethnology of Pukapuka. *B.P. Bishop Mus. Bull.* 150. 419 pp.

Brandon, D.J. (1977). Turtle Farming project in the Cook Islands. *Proc. SPC 9th Tech. Meeting on Fisheries*. Working Paper 21: 1-12.

Bullivant, J.S. (1962). Direct observation of spawning in the blacklip pearl shell oyster (*Pinctada margaritifera*) and the thorny oyster (*Spondylus sp.*). *Nature* 193(4816): 700-701.

Bullivant, J.S. (1974a). Crabs from Manihiki. *N.Z. Oceanographic Institute Memoirs* 31: 41-44.

Bullivant, J.S. (1974b). Lagoon and reef morphology of Manihiki Atoll. *N.Z. Oceanographic Institute Memoirs* 31: 17-21.

Bullivant, J.S. (1974c). Manihiki Atoll survey 1960: General account and station list. *N.Z. Oceanographic Institute Memoirs* 31: 5-16.

Bullivant, J.S. and McCann, C. (Eds) (1974a). Contributions to the natural history of Manihiki Atoll, Cook Islands. *N.Z. Oceanographic Institute Memoirs* 31: 1-63.

Bullivant, J.S. and McCann, C. (1974b). Fishes from Manihiki Atoll. *N.Z. Oceanographic Institute Memoirs* 31: 49-60.

*Clerk, C.C. (1981). The animal world of the Mangaian. Ph.D Diss., Univ. College London. 551 pp.

Cowan, G. (1980). Bibliography of research on the Cook Islands. New Zealand MAB Report 4. N.Z. National Commission for Unesco. N.Z. Soil Bureau, Lower Hutt.

*Crossland, C. (1928). Coral reefs of Tahiti, Moorea and Rarotonga. *J. Linn. Soc., London* 36: 577-620.

Dahl, A.L. (1980a). Regional ecosystems survey of the South Pacific Area. *SPC/IUCN Technical Paper* 179. South Pacific Commission, Noumea, New Caledonia.

Dahl, A.L. (1980b). Report on marine surveys of Rarotonga and Aitutaki (November 1976). South Pacific Commission, Noumea, New Caledonia.

*Dana, J.D. (1898). *Corals and Coral Islands*. Dodd, Mead and Co., New York.

*Davis, W.M. (1928). The coral reef problem. *Am. Geogr. Soc. Spec. Publ.* 9: 1-596.

*Devaney, D.M. (1973). Zoogeography and faunal composition of South-eastern Polynesian asterozoan echinoderms. In: Fraser, R. (Comp.). *Oceanography of the South Pacific, 1972* N.Z. National Commission for Unesco, Wellington, New Zealand. Pp. 357-366.

*Devaney, D.M. (1974). Shallow water asterozoans of southeastern Polynesia 2. Ophiuroidea. *Micronesica* 10: 105-204.

*Devaney, D.M. and Randall, J.E. (1973). Investigations of *Acanthaster planci* in south-eastern Polynesia during 1970-1971. *Atoll Res. Bull.* 169. 35 pp.

*Doran, E. (1961). Cook Islands landscape. *Atoll Res. Bull.* 85: 51-53.

Douglas, G. (1969). Checklist of Pacific Oceanic Islands. *Micronesica* 5(2): 327-463.

Fosberg, F.R. (1975). Vascular plants of Aitutaki. *Atoll Res. Bull.* 190: 73-84.

*Gare, N.C. (n.d.). The Cook Islands - a conservation report. Mimeo. Australian National Parks and Wildlife Service, Canberra, Australia. 25 pp.

*Gauss, G.A. (1982). Sea bed studies in nearshore areas of the Cook Islands. *South Pacific Mar. Geol. Notes, CCOP-SOPAC, ESCAP* 2(9): 131-154.

*Gibbs, P.E. (1972). Polychaete annelids from the Cook Islands. *J. Zool., London* 168: 199-220.

Gibbs, P.E. (1975). Survey of the macrofauna inhabiting lagoon deposits on Aitukaki. *Atoll Res. Bull.* 190: 123-31.

Gibbs, P.E., Stoddart, D.R., Vevers, H.G. (1971). Coral reefs and associated communities in the Cook Islands. *Bull. Roy. Soc. New Zealand* 8: 91-105.

Gibbs, P.E., Vevers, H.G. and Stoddart, D.R. (1975). Marine fauna of the Cook Islands: Checklist of species collected during the Cook Bicentenary Expedition in 1969. *Atoll Res. Bull.* 190: 133-148.

Gilmour, A.E. (1974). Tidal measurements at Manihiki Atoll. *N.Z. Oceanographic Institute Memoirs* 31: 29-30.

Grange, K.R. and Singleton, R.J. (1985). *A Guide to the Reef Fishes of Palmerston and Suwarrow Atolls, Cook Islands*. NZOI Oceanographic Field Report 21, DSIR, New Zealand. 24 pp.

Hambuechen, W.H. (1973a). Cook Islands. Paper 5. *Proceedings and Papers, Regional Symposium on Conservation of Nature - Reefs and Lagoons, 1971*. South Pacific Commission, Noumea, New Caledonia.

Hambuechen, W.H. (1973b). Pesticides in the Cook Islands. Paper 16. *Proceedings and Papers, Regional Symposium on Conservation of Nature - Reefs and Lagoons, 1971*. South Pacific Commission, Noumea, New Caledonia.

*Helm, A.S. and Percival, W.H. (1973). *Sisters in the Sun - the story of Suwarrow and Palmerston Atolls*. Robert Hale and Co., London/Whitcomb and Tombs Ltd, New Zealand.

Henry, T.A. (1977). Situation Report - Cook Islands. In: *Collected Abstracts and Papers of the International Conference on Marine Parks and Reserves, Tokyo May 1975*. Sabiura Marine Park Research Station, Kushimoto, Japan.

*Irwin, J. (1985). *The Underwater Morphology at Palmerston and Suwarrow Atolls, Cook Islands*. Oceanographic Institute Field Report, DSIR, New Zealand.

Khristoforova, N.K. and Bogdanova, N.N. (1981). Environmental conditions and heavy metal content of marine organisms from atolls of the Pacific Ocean. *Proc. 4th Int. Coral Reef Symp., Manila* 1: 161-162.

Landmesser, C.W., Kroenke, L.W., Glasby, G.P., Sawtell, G.H., Kingan, S., Utanga, E., Utanga, A. and Cowan, G. (1976). Manganese nodules from the South Penrhyn Basin, Southwest Pacific. *South Pacif. Mar. Geol. Notes* 1(3): 17-40.

Lawson, T.A. and Kearney, R.E. (1982). An assessment of the skipjack and baitfish resources of the Cook Islands. *Skipjack Survey and Assessment Programme Final Country Report* 2. South Pacific Commission, Noumea, New Caledonia.

Lewis, K.B., Utanga, A.T., Hill, P.J. and Kingan, S.G. (1980). The origin of channel-fill sands and gravels on an algal-dominated reef terrace, Rarotonga, Cook Islands. *South Pacif. Geol. Notes* 2(1): 1-23.

*Marsh, L.M. (1974). Shallow water asterozoans of

Southeastern Polynesia 1. Asteroidea. *Micronesica* 10: 65-104.

*Marshall, P. (1908). Geology of Rarotonga and Atiu. *Trans. N.Z. Institute* 4: 98-100.

*Marshall, P. (1927). Geology of Mangaia. *B.P. Bishop Mus. Bull.* 36: 1-48.

*Marshall, P. (1930). Geology of Rarotonga and Atiu. *B.P. Bishop Mus. Bull.* 12: 1-75.

McCann, C. (1974a). Mollusca from Manihiki Atoll. *N.Z. Oceanographic Institute Memoirs* 31: 35-40.

McCann, C. (1974b). Scleractinian corals from Manihiki Atoll. *N.Z. Oceanographic Institute Memoirs* 31: 31-32.

*McKnight, D.G. (1972). Echinoderms collected by the Cook Islands Eclipse Expedition 1965. *N.Z. Oceanographic Institute Records* 1: 36-45.

McKnight, D.G. (1974). Echinoderms from Manihiki Lagoon. *N.Z. Oceanographic Institute Memoirs* 31: 45-47.

Miller, J.M. (1980). Marine Science: A survey of the literature. In: *Bibliography of Research on the Cook Islands*. New Zealand MAB Report 4. N.Z. National Commission for Unesco, N.Z. Soil Bureau, Lower Hutt.

*Neale, T. (1966). *An Island to Oneself*. Collins.

Paulay, G. (1985). The biogeography of the Cook Island's coral fauna. *Proc. 5th Int. Coral Reef Congr., Tahiti* 4: 89-94.

Paulay, G. (in press). Biology of Cook Islands' bivalves. Part 1: heterodont families. *Atoll Res. Bull.*

*Randall, J.E. (1978). Marine biological and archaeological expedition to Southeast Oceania. *Nat. Geogr. Soc. Res. Rep. 1969 Projects*: 473-495.

Richmond, M. (1986). Shrimp distribution analysis report. Preliminary field report, Operation Raleigh. 5 pp.

Ridgeway, N.M. (1974). Hydrology of Manihiki lagoon. *N.Z. Oceanographic Institute Memoirs* 31: 23-28.

Scoffin, T.P., Stoddart, D.R., Tudhope, A.W. and Woodroffe, C.D. (1985a). Exposed limestone of Suwarrow Atoll. *Proc. 5th Int. Coral Reef Congr., Tahiti* 3: 137-140.

Scoffin, T.P., Stoddart, D.R., Tudhope, A.W. and Woodroffe, C.D. (1985b). Rhodoliths and coralliths of Muri Lagoon, Rarotonga, Cook Islands. *Coral Reefs* 4(2): 71-80.

Sims, N.A. (1985). The ecology, abundance and exploitation of *Trochus niloticus* L. in the Cook Islands. *Proc. 5th Int. Coral Reef Congr., Tahiti* 5: 539-544.

*Sims, N.A. (1986). Report on the attempted introduction of Green Mussels (*Perna viridis*) into the Cook Islands. Internal rept, Ministry of Marine Resources, Cook Islands.

*Sims, N.A. and Charpy, L. (in prep.). Hydrology, productivity and bivalve culture potential in Aitutaki and Rarotonga, Cook Islands.

*Spencer, T., Stoddart, D.R. and Woodroffe, C.D. (1987). Island uplift and lithospheric flexure: observations and cautions from the South Pacific. *Z. Geomorph. Suppl.* 63: 87-102.

SPREP (1980). Cook Islands. *Country Report* 3. South Pacific Commission, Noumea, New Caledonia.

*Stoddart, D.R. (1969). Geomorphology of the Solomon Islands coral reefs. *Phil. Trans. Roy. Soc.* B 255: 355-82.

Stoddart, D.R. (1972). Reef islands of Rarotonga. *Atoll Res. Bull.* 160: 1-7.

Stoddart, D.R. (1975a). Almost-atoll of Aitukaki: Geomorphology of reefs and islands. *Atoll Res. Bull.* 190: 31-57.

Stoddart, D.R. (1975b). Reef islands of Aitukaki. *Atoll Res. Bull.* 190: 59-72.

Stoddart, D.R. (1975c). Scientific studies in the Southern Cook Islands: Background and bibliography. *Atoll Res. Bull.* 190: 1-30.

Stoddart, D.R. (1975d). Vegetation and floristics of the Aitutaki motus. *Atoll Res. Bull.* 190: 87-116.

Stoddart, D.R. (1975e). Mainland vegetation of Aitutaki. *Atoll Res. Bull.* 190: 117-122.

*Stoddart, D.R. and Pillai, C.S.G. (1973). Coral reefs and reef corals in the Cook Islands, South Pacific. In: Fraser, R. (Comp.), *Oceanography of the South Pacific, 1972*. N.Z. National Commission for Unesco, Wellington, New Zealand. Pp. 475-488.

Stoddart, D.R., Scoffin, T.P., Spencer, T., Harmon, R.S. and Scott, M. (1985a). Sea level change and karst morphology, Mangaia. *Proc. 5th Int. Coral Reef Congr., Tahiti* 3: 201.

*Stoddart, D.R., Spencer, T. and Scoffin, T.P. (1985b). Reef growth and karst erosion on Mangaia, Cook Islands: A reinterpretation. *Z. Geomorph. Suppl.* 57: 121-140.

Summerhayes, C.P. (1967). Bathymetry and topographic lineation in the Cook Islands. *N.Z. J. Geol. Geophysics* 10(6): 1382-1399.

Summerhayes, C.P. (1971). Lagoonal sedimentation at Aitukaki and Manuae in the Cook Islands: A reconnaissance survey. *N.Z. J. Geol. Geophysics* 14: 351-363.

Thompson, R.M. (1983). Bibliography of geology and geophysics of the Cook Islands. In: Jouannic, C. and Thompson, R.M. (Eds), Bibliography of Geology and Geophysics of the South-western Pacific. *UN, ESCAP, CCOP/SOPAC Tech. Bull.* 5: 17-27.

Tudhope, A.W., Scoffin, T.P., Stoddart, D.R. and Woodroffe, C.D. (1985). Sediments of Suwarrow Atoll. *Proc. 5th Int. Coral Reef Congr., Tahiti* 6: 611-616.

*Turner, D.L. and Jarrard, R.D. (1982). K-Ar dating of the Cook-Austral chain: A test of the hot-spot hypothesis. *J. Volcanol. Geothermal Res.* 12: 187-220.

*Wood, B.L. (1967). Geology of the Cook Islands. *N.Z. J. Geol. Geophysics* 10: 1429-1445.

*Wood, B.L. and Hay, R.F. (1970). Geology of the Cook Islands. *N.Z. Geological Survey Bull.* 82: 1-103.

AITUTAKI

Geographical Location North-west of Manuae, 225 km north of Rarotonga; 18°51'S, 159°48'W.

Area, Depth, Altitude 106 sq. km, of which about half is the lagoon (66 sq. km lagoon proper and 10 sq. km reef flat); max. alt. 119 m; max. depth of lagoon 10.5 m. Aitutaki Island is 16.8 sq. km; other islands have a total area of 2.2 sq. km.

Physical Features Aitutaki is a roughly triangular volcanic cone, forming an almost-atoll, rising from depths of more than 4000 m and is the northernmost of the group of similar cones running south to Mauke. Geomorphology of the reefs and lagoons is described by Stoddart (1975a). The large lagoon is generally shallow; over three quarters of it is less than 4.5 m deep and maximum depth is 10.5 m (Stoddart, 1975c). Lagoonal sedimentation is described by Summerhayes (1971) and Gibbs (1975). Tidal range is small, 0.12 m at neaps and 0.49 m at springs. Annual rainfall is 1830-1984 mm, with a maximum in December to February and a pronounced dry season between June and September (Stoddart, 1975c). Mean annual temperature is 25.6°C.

Aitutaki is the main island and lies on the north-west rim. Twelve detrital reef islands or motus (Akitua, Angarei, Ee, Mangere, Papau, Tavaerua Iti, Tavaerua, Akaiami, Muritapua, Tekopua, Tapuaetai and Motukitiu) along the eastern reef and Maina to the west are described by Stoddart (1975b). Akaiami and Tekopua are the largest. There is a sand cay north of Tapuaetai. Moturakau and Rapota are volcanic islands on the south of the atoll.

The reef is a barrier reef forming a triangle, the sides of which are 13-15 km long; the total length of the peripheral reef is 45 km. The reefs are mostly 600-1000 m wide, with a maximum of 1700 m. The prevailing trade winds cause the outer edge to be steep along the south and east, and of intermediate slope on the west. There is a deep pass at Arutanga in the north-west navigable by small vessels. Water exchange between the lagoon and ocean occurs over the peripheral reef flats (Miller, 1980).

Reef Structure and Corals The reefs have been studied by Stoddart and Pillai (1973) and the lagoon by Gibbs (1975); brief descriptions are given by Dahl (1980b) and Paulay (in press). Stoddart and Pillai studied representative transects across the eastern windward reef and the western reefs. A total of 28 coral genera have been described (Paulay, 1985).

The eastern reef forms a continuous rock flat, 0.6-1 km wide, which is higher opposite the islands and lower between them (probably because of the scouring which can occur there and the rubble build-up in front of the islands). Sedimentation fans are found between the motus (Summerhayes, 1971). The reefs are narrower adjacent to the islands, so that the lagoon edge of the reef is lobate unlike the seaward edge which is straight. Most of the reef is covered at low water but it is not an area of active reef construction. Corals grow in the rock-floored moats near the seaward margin and in the deeper, more sheltered areas between the islands. A typical transect as at Kopuanu, Ootu, shows the following (Stoddart, 1975a):

a) An algal rim, deeply dissected by surge channels and coated with crustose *Porolithon*, a turf of filamentous algae, and larger algae (e.g. *Sargassum* and *Turbinaria*); corals are limited to encrustations, mainly of *Acropora*, and *Pocillopora*; *Millepora* occurs on the seaward side of the rim and in small colonies on the sides of the surge channels; b) a zone of green algae on the inner part of the algal rim; c) a zone of encrusting pink algae and *Turbinaria* on the inner slope of the algal rim; d) a moat 75-85 m wide with scattered patches of coral including large microatolls of *Porites lutea*; and e) a narrow zone of sand passing to the island beach.

The western reefs are 0.8-1.7 km wide and in the north abut against the main volcanic island. There are some well-defined surge channels but the reef edge is irregular. The depth of the flat increases from the land to the sea, where there is a narrow constructional barrier of corals rising from depths of 3-4 m. In the south the reef is formed by coalescing coral colonies generally rising from 3-4 m depth with winding intersecting channels in

between. A transect taken at Vaioue in the north, where the flat is 900 m wide, shows the following:

a) an irregularly indented reef edge, without an algal ridge, formed by coalescing corals and rubble forming an open-work structure superficially bounded with encrusting pink algae; *Acropora*, soft corals and *Turbinaria* algae are conspicuous; b) an outer flat, mainly of ramose *Acropora* with winding channels up to 2 m deep between; c) the main reef flat, 500 m wide, with a sandy surface with coalescing and anastomosing linear coral patches which are dead on the upper surfaces but rimmed with *Millepora* and living corals on the sides; and d) an inner zone, 50-100 m wide, consisting mainly of dead corals but with living microatolls of *P. lutea*, 1-2 m in diameter, covered with bushy *Turbinaria* algae on the upper surface. Reef blocks are common near the edge of the reef and are no doubt of storm origin.

The southern reefs have no algal rim, are protected from both prevailing trade winds and hurricanes, and there are few storm blocks. The reef flat consists of a sand flat, with corals forming an open framework interrupted by pools. The pools become fewer and the reef patches more continuous near the reef edge and there is much delicate coral growth.

There are many small reef patches in the south-west and south of the lagoon, but fewer in the north-east and extreme south-east. Large surface reefs are not common except in the south-west where there are irregular, flat-topped, steep sided ridges with abundant coral growth on their sides, extending transversely across the lagoon from the main island to Maina.

Noteworthy Fauna and Flora *Tridacna maxima* is very common on all reefs (Miller, 1980). The holothurian *Holothuria atra* is common on reef flats. There are abundant Trochus *Trochus niloticus* stocks. The lagoon fauna and terrestrial vegetation of the reef islands is described by Gibbs *et al.* (1971). Additional information on the vegetation and flora of the atoll is given in Fosberg (1975) and Stoddart (1975d and e).

Scientific Importance and Research The atoll was first described by Agassiz (1903) and was subsequently visited in the course of the 1967 Cook Bicentenary Expedition. Currently studies are under way on trochus stocks by the Ministry of Marine Resources (Sims, 1985). It is also the site of introductions of *Tridacna derasa* from Belau and *Eucheuma* culture growth trials. The Green Snail *Turbo marmoratus* has been introduced to the north-west outer reef slope (Sims *in litt.*, 6.5.87). Studies of lagoon water hydrology and productivity in relation to prospects for commercial bivalve culture have been undertaken by Sims and Charpy (in prep.).

Economic Value and Social Benefits *Tridacna maxima* forms the basis of a major food-gathering and commercial industry and there is some commercial exploitation of *Caulerpa* sp. and *Turbo setosus* (Sims *in litt.*, 6.5.87). The crab *Scylla serrata* and banded stomatopod *Lysiosquilla maculata* are trapped by fishermen in the muddier parts of the lagoon (Miller, 1980). Trochus was introduced in 1957 from Fiji and rapidly became well established. Its economic potential was largely ignored at first and commercial harvesting was not undertaken until 1981 (Sims, 1985). Since then considerable quantities have been taken for export during regulated and specified harvest seasons (see below). There is an airstrip (Douglas, 1969). Tourists are taken on snorkelling trips on the lagoon (Sims *in litt.*, 11.12.87).

Disturbance or Deficiencies   In the early 1970s, 80-90% of the corals on the southern seaward slope of the western reef were found to be dead in the course of an *Acanthaster planci* survey by the *Westward* expedition from Honolulu (Devaney and Randall, 1973). At several points around the lagoon margins, particularly near Maina and along the east rim, the coral patches were being buried by marginal sediments and many of the patches close to the main island, particularly in the turbid water of the north-east arm, were dead and covered with *Lopophylla* communities. This was thought to be possibly due to *Acanthaster* predation, although only small numbers of starfish were seen. Stoddart (1969) suggested that such damage might have been due to hurricanes. *Acanthaster* are now present in sporadic localized concentrations (Sims *in litt.*, 6.5.87).

The discharge into the lagoon of the chemical dip used in banana packaging is potentially serious (SPREP, 1980). The lagoon is said to be silting up (Summerhayes, 1971). The central lagoon was the site of extensive dynamiting during the early 1950s for a flying-boat landing strip. Arutanga passage was blasted by the U.S. Airforce during World War II. Since then there has been periodic blasting and dredging at Arutanga, Papau and for canoe passages (Sims *in litt.*, 6.5.87). In 1981 it was reported by locals that populations of the small Green Snail *Turbo setosus* had declined, perhaps as a result of the proliferation of the introduced trochus (Sims, 1985). Such declines, if indeed they occurred, appear to have halted (Sims *in litt.*, 6.5.87).

Legal Protection   Regulations for trochus harvesting have been arbitrarily declared at the beginning of each season since 1981. In 1984, there was a quota of 20 tonnes, no time limit for the season, an upper size limit of 11 cm and a reserve was declared over 3 km of windward reef (Sims, 1985). Currently there is a lower size limit of 8 cm. Harvests in 1985 and 1987 were regulated by a time limit of three days and one day respectively. There is a 2 km reserve area centred on Akaiami within which all collection is banned (Sims *in litt.*, 6.5.87).

Recommendations   Aitutaki lagoon and eastern motus, including the adjacent reef, has been recommended for reserve status (Dahl, 1980a), and general recommendations for improved management of lagoon and coastal resources are given in Dahl (1980b). It has been recommended that quota levels for trochus should be set at 30% of the estimated standing stock and that monitoring of stocks should continue (Sims, 1985).

MANIHIKI

Geographical Location   Northern Cook group; 10°25'S, 161°02'W.

Area, Depth, Altitude   2.0 sq. mi. (5.2 sq. km); lagoon about 5 km. diam., max. depth 72 m.

Physical Features   Manihiki is a pear-shaped atoll which faces the prevailing winds. There are two large islets, Tukao and Tauhunu, to the north-east and west and many smaller islets in the south, the largest of which is Porea. Lagoonal islets are found at different stages of development, miniature reef flats having developed around some of them. Tides are described by Gilmour (1974) and hydrology by Ridgeway (1974). A general description of the atoll is given in Bullivant (1974c) and Bullivant and McCann (1974a). Mean annual rainfall is 2482 mm.

Reef Structure and Corals   The lagoon and reefs are described by Bullivant (1974b and c) and corals by McCann (1974b). The outer reef is typical with buttresses and surge channels and deep water close in. The greatest development of coral within the lagoon is at the south-east perimeter, where the waves wash over the reef. Pinnacles of living and dead coral are abundant in the lagoon, some breaking the surface as sandbanks and islets. *Acropora* sp. and *Fungia* sp. were especially abundant on the slopes of the miniature fringing reefs surrounding the islets.

Noteworthy Fauna and Flora   Molluscs are described by McCann (1974a), echinoderms by McNight (1974), crabs by Bullivant (1974a) and fish by Bullivant and McCann (1974b). Green *Chelonia mydas* and Hawksbill *Eretmochelys imbricata* turtles nest (Balazs, 1981). Blacklip Pearl Oysters *Pinctada margaritifera* are abundant in the lagoon (Bullivant, 1962).

Scientific Importance and Research   The atoll was visited by an expedition from the New Zealand Oceanographic Institute (NZOI) in 1960 in order to determine the hydrological characteristics of the lagoon so that data of use in the pearl shell fisheries investigations might be available (Bullivant, 1974c). Stock assessment and growth studies of *P. margaritifera* are currently under way (Sims *in litt.*, 11.12.87). Breeding of this species and the thorny oyster are described by Bullivant (1962).

Economic Value and Social Benefits   Good quality pearl shell is found on Manihiki and the Blacklip Pearl Oyster is exploited commercially. Commercial pearl-farms are now in operation (Sims *in litt.*, 11.12.87).

Disturbance or Deficiencies   Reefs are reported to have been damaged by *Acanthaster planci* (Hambuechen, 1973a) although local people assert that this is not true (Sims *in litt.*, 11.12.87). There were fears that phosphate mining investigations would disrupt the lagoon (Eldredge *in litt.*, 18.2.85), but these were not allowed because of local fears that drilling would disturb the pearl oyster stocks (Sims *in litt.*, 11.12.87).

Legal Protection   None.

Management   The Ministry of Marine Resources and the Island Council are responsible for the management of the pearl-shell fishery (Sims *in litt.*, 11.12.87).

Recommendations   There were no specific recommendations from the NZOI study.

NGATANGIIA HARBOUR AND MURI LAGOON

Geographical Location South-eastern side of Rarotonga; 21°15'S, 159°44'W.

Physical Features The harbour and lagoon are sheltered by a chain of islets, mainly of rubble and sand except for Taakoka which is volcanic. Motutapu (600 x 360 m), the largest and northernmost island, is a simple cay with a rubble beach on the seaward side, a wide intertidal expanse of *Uca* dominated sand and silt on the leeward side and a makatea overlooking the deeper water of the harbour entrance. Oneroa (500 x 200 m), the second largest island, south of Motutapu, is similar but lacks the makatea. Koromiri (320 x 120 m) is the smallest of the cays and is also similar. Taakoka, to the south, is not a cay but is set well back from the reef edge and consists of a low basalt hill. On each side of the harbour, there is a pronounced coastal indentation where Avana Stream, the largest stream on Rarotonga, and Turangi Stream reach the coast, creating a deep gap in the reef north of Motutapu. The harbour entrance is bordered with 3 m high undercut cliffs. Inside the harbour is the muddy delta of Avana Stream, which filled in much of the harbour after forest clearance early in the century (Dahl *in litt.*, 27.10.87). Outside the harbour, a gently troughed channel is strewn with coral heads for 150 m offshore. Beyond this a double 20 m deep channel system is floored with sand to the edge of the reef-front terrace. There are beaches of fine sand at Muri. The islands are described in more detail by Stoddart (1972) and additional information is given in Lewis *et al.* (1980).

Reef Structure and Corals The reef lies about 500 m from the coast. There is a poorly developed algal ridge (Gibbs *et al.*, 1971). The lagoonal reefs had almost no living coral in 1976, although the state of the coral skeletons suggested that this was a recent phenomenon (Dahl, 1980b). Rhodoliths and coralliths in Muri Lagoon are described by Scoffin *et al.* (1985b).

Noteworthy Fauna and Flora The reef flats at Ngatangiia Harbour have particularly large numbers of the holothurian *Holothuria atra* (up to 10 per sq. m). Vegetation of the islands is described by Stoddart (1972).

Scientific Importance and Research Ngatangiia Harbour is the site of unsuccessful growth trials of the green mussel *Perna viridis* in 1985 (Sims, 1986). Studies of estuarine and lagoon hydrology and productivity in relation to prospects for commercial bivalve culture have been carried out (Charpy and Sims, in prep.).

Economic Value and Social Benefits Artisanal and subsistence fishing are important activities in the area. Fish traps are still used. There is extensive use of the area by tourists and water sports operators (Sims *in litt.*, 6.5.87). The area has great potential for tourism and recreation, provided high density facilities are avoided (Dahl, 1980b).

Disturbance or Deficiencies The lagoon was considered considerably degraded in the 1970s (Dahl, 1980b). There has been extensive dredging in the harbour, north of Avana Stream mouth. There was some dynamiting in the lagoon in 1985 near the dinghy sailing club (Sims *in litt.*, 6.5.87).

Legal Protection None.

Management None.

Recommendations The islets of Motutapu, Oneroa, Koromiri and Taakoko, with the adjacent reefs were recommended as reserves by Dahl (1980a), and it is recommended that no buildings should be permitted on these motus. Comprehensive planning should be instituted for the whole area (Dahl, 1980b).

PUKAPUKA (DANGER ISLAND, PAKAPUKA)

Geographical Location 67 km north-west of Nassau, about 460 km south-west of Manihiki and Rakahanga and more than 1300 km from Rarotonga; 10°53'S, 165°49'W.

Area, Depth, Altitude Total land area about 700 ha; max. alt. 4 m; lagoon about 8 km x 3-5 km and 10-50 m deep.

Physical Features The atoll is triangular with three islets at the reef apexes and several sand banks (Andrews, 1987; Wood and Hay, 1970). The northern islet, Pukapuka, is horseshoe shaped and has a maximum width of 1600 m. The two southern islets (Motu Ko and Motu Kotawa) consist of unconsolidated coral debris with coarse slabby boulders toward the seaward margin, finer coral fragments in the interior and fine sand with a thin layer of carbonate cement in places along the lagoon shore. The atoll is surrounded by a closed reef, apart from an artificial passage at the north-western edge, and has a shallow lagoon. The western slope of the atoll descends into the Samoan Basin which reaches depths of more than 6000 m. The seaward reef flat and occasional beach are bordered at many places by low ledges or small ramparts of partly cemented boulders and coral debris carried by storm waves from the outer slope and ridge over the reef flat. Many coral boulders lie on the reef flat. The lagoon is strewn with heads and patches of coral that rise from the flat sandy bottom to just below the surface or are actually exposed at low tide. A description of the reef-lagoon complex is given in Andrews (1987).

There is a relatively constant temperature (27-29°C) throughout the year. Annual rainfall in the region averages 2811 mm with a high of 3940 mm and a low of 1812 mm. Cyclones generally do not affect Pukapuka although there are sporadic storm surges which cause considerable damage to the atoll and islands (Anon., 1986; Andrews, 1987).

Reef Structure and Corals A description of the reefs is given in Andrews (1987). The outer reef has high coral cover (60%+) dominated by *Montipora aequituberculata*, *Pocillopora eydouxi* and *Pavona minuta* with significant areas of *Acropora* in shallower water and massive *Porites* at depth. An area of the outer reef on the northern extreme of Motu Ko was significantly different from the relatively consistent outer reef complex, probably related to the presence of a wrecked fishing vessel in 1981. The reef flats are comprised of algal mats and sand, generally covered by a maximum of one metre of water at high tide. Live corals are scarce with a few small *Porites* colonies on the flat and occasional stunted *Pocillopora* and *Acropora* species on a well-defined algal ridge. In the lagoon, the area of high coral cover is restricted mostly to within the 3 m contour. Small faviids dominate the sandy substrate with some large *Porites* species. Larger patch reefs display considerable diversity with areas of *Montipora monasteriata* being significant. An area consisting mainly of monospecific pinnacles of *Astreopora listeri* is located between Te alo Pukupuku and Motu Ko.

Noteworthy Fauna and Flora   The Green *Chelonia mydas* and Hawksbill *Eretmochelys imbricata* Turtle nest on the atoll (Balazs, 1981). There is a large and diverse population of fish.   The Coconut Crab *Birgus latro* occurs.   *Trochus niloticus* (which did not previously occur) and *Pinctada margaritifera* have been introduced to establish commercial stocks (Andrews, 1987).

Scientific Importance and Research   The metal content of tridacnids has been studied (Khristoforova and Bogdanova, 1981).   Pukapuka has been selected as pilot project for a low island within a regional project to improve coastal zone management in the South Pacific. In 1986 a survey was conducted by the Australian Institute of Marine Sciences, with the support of the Commonwealth Science Council and the Cook Islands Government, under the South Pacific Coastal Zone Management Project (SOPACOAST). Maps have been prepared of the reef resources and hydrological parameters, and benchmarks are being established as a basis for measuring ecological changes (Andrews, 1987).

Economic Value and Social Benefits   The entire population of about 790 lives in three villages on the lagoon side of the northern islet and relies predominantly on the reefs for food. Fishing practices are traditionally based but have taken advantage of modern technology; modern dinghies and outboard motors are preferred when available.   The main fishing techniques are described in Andrews (1987).

Disturbance or Deficiencies   Periodic storm surges may cause damage. There is potential for increased pressure on reef resources (Anon., 1986).   Stocks of *Pinctada margaritifera* declined dramatically in the 1950s, probably owing to overfishing. There was a decrease in abundance and individual size of some fish species until spearfishing was prohibited and stocks of clams, particularly *Tridacna maxima*, have declined.   There is some evidence of eutrophication (from sewage) and an increase of turbidity in the lagoon (Andrews, 1987).

Legal Protection   Spearfishing has been banned in the lagoon by the Island Council since 1985 to protect the small, easily speared *Epinephelus* and *Cephalopholis* which are highly valued as food (Andrews, 1987; Sims *in litt.*, 11.12.87).

Management   The uninhabited part of the northern islet, and the two southern islets are divided amongst the three villages as food reserves ('motu's') and each village controls the reef and fishing grounds near its motu.   Each reserve has a small settlement of houses occupied only during visits for food supplies and copra breaking (Andrews, 1987; Beaglehole and Beaglehole, 1938). Fishing on the outer reefs is open to all islanders (Anon., 1986).   Conservation practices are regularly reviewed by the traditional governing body, the Island Council, which includes two representatives from each village.   The Council enforces such issues as the spearfishing "tapu" within the lagoon. The villages govern the opening and

closing of their motus to visitors and the taking of coconut crabs and seabirds; such closures, which also apply to adjacent areas within the lagoon, may last for up to six years.

Recommendations   The SOPACOAST project, having defined and described the resources of the atoll, will aim to better apply existing knowledge of the area to coastal zone management and assist in improving community capability to administer, manage and monitor coastal resources and the environment. Training and educational activities related to the project theme will be arranged (Anon., 1986).   Andrews (1987) gives recommendations for further research, such as investigations of the clam stocks and reasons for their decline, and monitoring of the lagoon resources. Conservation measures for turtles and clams should probably be introduced.

SUWARROW (SUVAROV) ATOLL NATIONAL PARK

Geographical Location   950 km NNW of Rarotonga on the south-west rim of the Manihiki Plateau; 13°14'S, 163°05'W.

Area, Depth, Altitude   40.5 ha land; max. alt. of islands 2-3 m; lagoon area 133 sq. km (10 km diameter); max. depth of lagoon 80 m.

Land Tenure   Crown land.

Physical Features   The atoll has 22 vegetated islets situated on the almost continuous rim, 0.5-1.0 km wide, and a diamond-shaped reef. The lagoon has active water exchange with the sea through a wide, 10 m deep, pass near Anchorage Island in the north-east.   Throughout most of the lagoon patch reefs give the seabed a very irregular morphology.   Tidal range at springs is about 1 m.   Trade winds blow from the south-east but most storms and cyclones approach from the north-west (Douglas, 1969; Scoffin *et al.*, 1985a; Tudhope *et al.*, 1985). Sediments are described by Tudhope *et al.* (1985) and other geological aspects by Scoffin *et al.* (1985a). General information on the atoll is given in Helm and Percival (1973) and Neale (1966).

Reef Structure and Corals   There is a well developed algal ridge and a broad reef flat 100-800 m wide surrounding the atoll (Scoffin *et al.*, 1985a).   Several patch reefs, more than 100 m in diameter, reach up to sea level in the central lagoon while small patch reefs are particularly abundant around the western margin (Tudhope *et al.*, 1985).

Noteworthy Fauna and Flora   Some wet atoll forest remains (Douglas, 1969; Wood and Hay, 1970).   The atoll is a turtle nesting site (Brandon, 1977) and of major importance for nesting seabirds (Anon., 1985).   Coconut Crabs *Birgus latro* occur on the islands and sparse clam *Tridacna maxima* beds in the lagoon and on the reef (Anon., 1985).   Reef fish are described by Grange and Singleton (1985).

Scientific Importance and Research   The heavy metal content of *Tridacna* and *Caulerpa* in the lagoon has been studied by Khristoforova and Bogdanova (1981). A joint

New Zealand DSIR/Royal Society of London Northern Cook Islands cruise visited the atoll in 1981. Black lip Pearl Oyster *Pinctada margaritifera* stocks are being surveyed and growth studies are underway by the Ministry of Marine Resources. Trochus *Trochus niloticus* introductions were successfully carried out in 1985 and 1987 (Sims *in litt.*, 6.5.87). There was an unsuccessful attempt to introduce the Goldlip Pearl Oyster *Pinctada maxima* (Paulay *in litt.*, 10.7.87).

Economic Value and Social Benefits   Pearl fishing was once an important activity.

Disturbance or Deficiencies   There are illegal visits in small boats as the atoll is a popular stopover for yachts (Anon., 1985; SPREP, 1980). There has been some hurricane damage, particularly on the islets (Douglas, 1969). Coconut crabs have been reported to attack nesting seabirds. Pearl oyster stocks have significantly declined (Sims *in litt.*, 11.12.87)

Legal Protection   Declared a National Park 29.6.78 under Section 11(1) of the Conservation Act 1975. Fishing is permitted for immediate use but not for commercial purposes. Licenses may be issued for mother-of-pearl collection, and for making of copra, culling of Coconut Crabs and controlling of rats on Anchorage Island (Anon., 1985).

Management   The National Park is under the joint control of the Conservation Service of Internal Affairs and Marine Resources (Fisheries) but there is no active management. There is reportedly a caretaker and his family (SPREP, 1980; Anon., 1985), although other sources indicate that, at least between 1982 and 1986, this was not the case (Paulay *in litt.*, 10.7.87). The area is zoned for pearl culture and commercial pelagic fishing (Anon., 1985).

Recommendations   The atoll was recommended as an Island for Science (Douglas, 1969) but this programme was never followed through. A warden has been appointed (Sims *in litt.*, 11.12.87). Base line studies must be completed so that a comprehensive management plan can be developed and implemented (Anon., 1985); this is considered a priority in the Action Strategy for Protected Areas in the South Pacific Region, drawn up at the Third South Pacific National Parks and Reserves Conference in 1985.

FEDERATED STATES OF MICRONESIA

INTRODUCTION

General Description

The Federated States of Micronesia (F.S.M.) consists of the states of Yap, Truk, Ponape and Kosrae and became a freely associated independent nation in close association with the U.S.A. in 1986, having previously been part of the Trust Territory of the Pacific Islands. The F.S.M. includes most of the Caroline Islands, running west-north-west from Kosrae to Yap, and consists of volcanic and metamorphic islands and atolls. "Almost-atolls" are also present, making the group one of the more typical line island chains. The development and origin of the Caroline Islands is described by Scott and Rotondo (1983); general information is given in SPREP (1980). Belau (Palau), the westernmost group of the Caroline Islands, is described in a separate section.

Table of Islands

Yap State

Yap Islands Proper (Waqab)   4 high volcanic and metamorphic islands separated by ditches or narrow channels through mangroves:

- *Yap*   (X) 21.68 sq. mi. (56 sq. km); elongated volcanic and metamorphic, 155 ft (47.2 m) in south, 579 ft (176 m) centre and north; reefs described below; echinoderms described by Grosenbaugh (1981), seagrasses by Kock and Tsuda (1978) and marine benthic algae by Tsuda and Belk (1972); over 650 fish species recorded, including some genera and families not recorded elsewhere in the F.S.M. (Gawel *in litt.*, 25.6.87);

- *Gagil Tamil (Gagil-Tomil)*   (X) 11.13 sq. mi. (28.8 sq. km); volcanic; very indented coastline fringed with mangrove; separated from Yap by Tagareng Canal; broad fringing reef to east and south;

- *Maap (Map)* (X) 4.1 sq. mi. (10.6 sq. km); volcanic 200 ft (61 m) with mangroves along north-west and south coasts; broad fringing reef to east;

- *Rumung*   (X) 1.66 sq. mi. (4.3 sq. km); volcanic; fringing reef to north-west and north-east.

Outer Yap Islands

Ulithi   (X) 1.8 sq. mi. (4.7 sq. km); largest atoll in Carolines, with 40 islets scattered in four main groups; most reefs unsurveyed but probably undisturbed; survey of site at Falalop Island carried out by Tsuda *et al.* (1978); sea turtle nesting beaches; wrecks in lagoon dating from World War II.

Ngulu (X) 0.165 sq. mi. (0.43 sq. km); atoll with 8 islets; reef; sea turtle nesting beaches.

Fais   (X) 1.1 sq. mi. (2.8 sq. km); raised atoll, 60 ft (18 m); phosphate mining (Gawel *in litt.*, 25.6.87).

Sorol   (X) 0.36 sq. mi. (0.9 sq. km); atoll with 10-11 islets; enclosed reef; periodically inhabited (Gawel *in litt.*, 25.6.87).

Eauripik   (X) 0.09 sq. mi. (0.23 sq. km); atoll with six small islets all covered with coconuts except Edarepe I. which is awash at high tide.

Woleai   (X) 1.75 sq. mi. (4.5 sq. km); large atoll with double form and 21 islets; reefs well developed; totally undisturbed (Dahl *et al.*, 1974).

Ifalik (Ifaluk)   (X) 0.57 sq. mi. (1.5 sq. km); atoll with three islets; atoll described by Tracey *et al.* (1961) and Bates and Abbott (1958); check-list of marine algae (Abbott, 1961).

Faraulep (Fechaulep)   (X) 0.16 sq. mi. (0.4 sq. km); atoll with three low small islets.

Gaferut   0.043 sq. mi. (0.11 sq. km); island 100 ft (30 m) (Niering, 1961); fringing reef; no lagoon; Red-footed Booby *Sula sula*, frigatebirds, Coconut Crab *Birgus latro*; Green Turtle *Chelonia mydas* nesting.

Olimarao   0.085 sq. mi. (0.22 sq. km); atoll with 2 islets on reef surrounded by lagoon.

Elato   (X) 0.2 sq. mi. (0.5 sq. km); atoll with 2 groups of islets (Elato and Toas); mangroves; turtle nesting.

Lamotrek   (X) 0.38 sq. mi. (0.98 sq. km); atoll with 3 wooded islets.

West Fayu (Pigailoe)   0.24 sq. mi. (0.62 sq. km); coral islet; extensive reef with lagoon; sea turtle nesting beaches.

Satawal   (X) 0.5 sq. mi. (1.3 sq. km); small coral island, 15 ft (4.6 m); no lagoon; plants described by Fosberg (1969); few reef fish resources but tuna fishing important (Gillett, in press)..

Pikelot   0.04 sq. mi. (0.1 sq. km); small islet, fringed by extensive reef; no lagoon; turtle nesting; visited for turtle fishing.

Truk State

Truk Lagoon (Chuk)   almost-atoll (described in main text), with following principal islands:

- *Moen (Wono)*   (X) 7.25 sq. mi. (18.8 sq. km); high volcanic 1215 ft (370 m); mangroves along N and SE coasts; reefs described below;

- *Dublon (Tonowas)*   (X) 3.4 sq. mi. (8.8 sq. km); high volcanic, 1145 ft (349 m); some mangroves along N, NE and NW coasts; fringing reef; reefs,

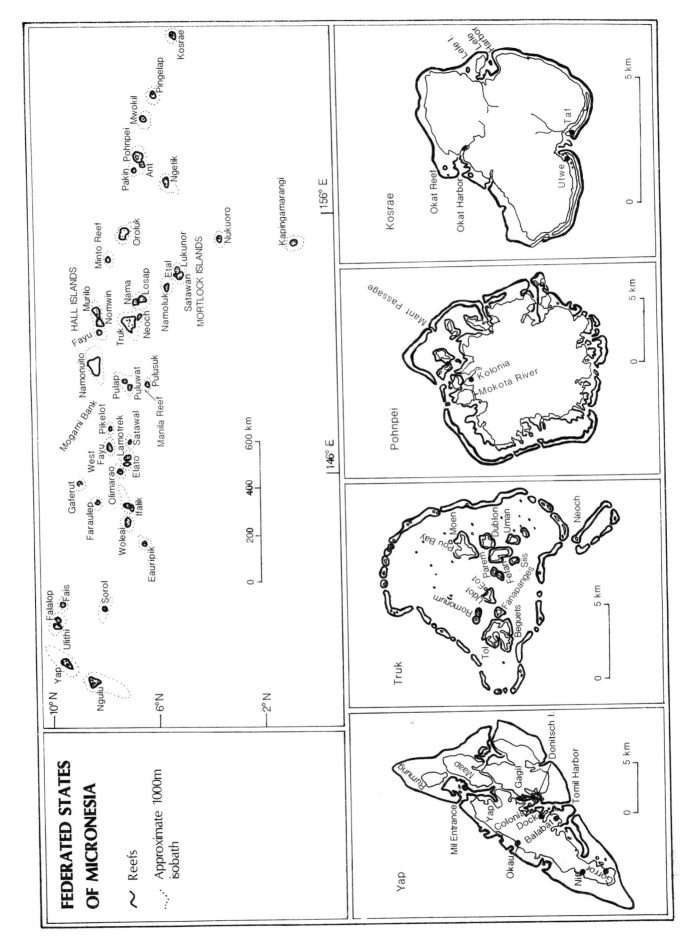

FEDERATED STATES OF MICRONESIA

Reefs

Approximate 1000m isobath

Kosrae

Pohnpei

Truk

Yap

mangroves and seagrasses mapped, and corals, fishes and algae listed in OPS, TTPI (1978a);

- *Etten* (X) small volcanic island, mostly man-made by coral fill on reef creating airstrip in World War II; fringing reef; reefs, mangroves and seagrasses mapped, and corals, fishes and algae listed in OPS, TTPI (1978a);

- *Uman* (X) 1.8 sq. mi. (4.7 sq. km); high volcanic, 800+ ft (244 m); small patches of mangroves to N and S; fringing reef; reefs, mangroves and seagrasses mapped, and corals, fishes and algae listed in OPS, TTPI (1978b);

- *Fefan (Fefen)* (X) 5.1 sq. mi. (13.2 sq. km); high volcanic, 978 ft (298 m); very small mangrove areas on W coast; fringing reef; reefs, mangroves and seagrasses mapped, and corals, fishes and algae listed in OPS, TTPI (1978c);

- *Siis (Tsis)* (X) 0.25 sq. mi. (0.65 sq. km); small island, 300 ft (91 m); extensive fringing reef on east coast;

- *Parem (Param)* (X) 200+ ft (61 m); small volcanic island; mangroves on S. coast; fringing reef;

- *Eot* (X) 0.19 sq. mi. (0.5 sq. km); 200+ ft (61+ m); linked to Udot by fringing reef; reefs, mangroves and seagrasses mapped, and corals, fishes and algae listed in OPS, TTPI (1978d);

- *Udot* (X) 1.9 sq. mi. (4.9 sq. km); moderately high volcanic, 400-500 ft (122-152 m); fringing reef; reefs, mangroves and seagrasses mapped, and corals, fishes and algae listed in OPS, TTPI (1978d);

- Romonum (Ulalu) (X) 0.29 sq. mi. (0.75 sq. km); low island; fringing reef;

- *Fanapanges (Fala-Beguets)* (X) 0.6 sq. mi. (1.6 sq. km); 200 ft (61 m); extensive fringing reef;

- *Tol* (X) 13.2 sq. mi. (34.2 sq. km); high volcanic; 1440 ft (439 m); consists of four virtually connected islands (Tol, Wonei, Pata, Polle); deeply indented coastline, mangrove-lined; fringing reef particularly to west;

- *Pis* (X) small island on northern part of barrier reef.

Neoch (Kuop) (X) 0.19 sq. mi. (0.49 sq. km); atoll near Truk Lagoon with 4 islets; marine benthic algae described (Tsuda, 1972).

Nama (X) 0.3 sq. mi. (0.8 sq. km); coral islet; no lagoon; fringing reef.

Losap (X) 0.4 sq. mi. (1 sq. km); atoll with 8 islets; semi-circular reef.

Westerns

Namonuito (X) 1.7 sq. mi. (4.4 sq. km); large atoll with 10 islets; fringing reefs.

Pulusuk (X) 1 sq. mi. (2.6 sq. km); atoll with low coral island; fringing reef.

Puluwat (X) 1.3 sq. mi. (3.4 sq. km); atoll with 2 large islets, 3 smaller (Niering, 1961); reef.

Pulap (X) 0.33 sq. mi. (0.85 sq. km); atoll with 3 wooded islets.

Manila Reef

Hall Islands

Nomwin (X) 0.7 sq. mi. (1.8 sq. km); atoll with 9 islets, circular reef.

Murilo (X) 0.5 sq. mi. (1.3 sq. km); atoll with chain of 5+ islets on north of reef.

Fayu 0.15 sq. mi. (0.39 sq. km); small low-lying coral island; no lagoon; central depression which collects water; surrounded by fringing reef; many seabirds.

Mortlock Islands (Nomoi)

Satawan (X) 1.75 sq. mi. (4.5 sq. km); large atoll with numerous islets and 11 main ones.

Lukunor (X))/atoll with 6+ islets on oval lagoon.

Etal (X) 0.7 sq. mi. (1.8 sq. km); small atoll with 13 islets enclosing small lagoon.

Namoluk (X) 0.83 sq. mi. (2.2 sq. km); atoll with 5 islets and closed lagoon (7.5 sq. km), one of deepest in the Pacific (Marshall, 1975); reefs 1.5 km on each side, encircle lagoon; reefs undescribed; marine molluscs listed by Marshall (1975); Green Turtles and Hawksbills *Eretmochelys imbricata* found in lagoon and nest on Amwes islet; Coconut Crab *Birgus latro* present; many seabirds (Marshall, 1975).

Pohnpei (Ponape) State

Pohnpei (Ponape) (*see separate account*).

Ant (X) 0.7 sq. mi. (1.8 sq. km); atoll near Pohnpei with several small islets; periodically inhabited; excellent reef (described below); privately owned; Green Turtles may nest.

Pakin (X) 0.42 sq. mi. (1.1 sq. km); atoll with 5+ islets, no channel into lagoon; periodically inhabited; northern reef surveyed; good reefs (described below); Green Turtle recorded.

Ngetik (Sapwuahfik) (X) 0.66 sq. mi. (1.7 sq. km); atoll with 3 islands and several small islets.

Oroluk (X) 0.2 sq. mi. (0.5 sq. km); atoll with ring of very small, wooded islets; Polynesia; good reefs; Green Turtle nesting.

Minto Reef

Nukuoro   (X) 0.66 sq. mi. (1.7 sq. km); atoll with 40+ islets; circular reef; inaccessible.

Kapingamarangi   (X) 0.5 sq. mi. (1.3 sq. km); atoll with 33 islets on circular reef; geology described by McKee (1958); geography by Wiens (1956) and Niering (1956); heavily populated, Polynesian; turtles once important, now scarce.

Mwokil (Mokil)   (X) 0.5 sq. mi. (1.3 sq. km); atoll with 3 main islets, small lagoon small, rectangular reef.

Pingelap   (X) 0.66 sq. mi. (1.7 sq. km); atoll with 2 islands and 1 islet; 20 ft (6 m); small lagoon; square reef; a few mangroves on south; flora described in St John, 1948; general descripiton in Murphy (1949).

Kosrae State

Kosrae (Kusaie)   (X) 42.33 sq. mi. (100 sq. km); high volcanic 2061 ft (628 m) island with chain of mtns to S and isolated peak to N, 1943 ft (592 m); deeply dissected; narrow coast plain with some mangroves particularly S and NW, best examples in Lele Harbour to E; also four smaller volcanic islands (most important Lele (Leluh), population centre of Kosrae); reefs described below; sparse nesting of Green Turtles and Hawksbills.

(X) = Inhabited

Yap State includes the Yap Islands proper, a number of outlying atolls and low coral islets and the raised coral island of Fais. Most of the literature concerns the Yap Islands proper (Maragos *in litt.*, 10.8.87). In 1987, the Universities of Hawaii and Guam and the U.S. Army Corps of Engineers were conducting an extensive inventory, at over 60 stations, of these as part of the Yap Islands coastal resources inventory and atlas project (Maragos *in litt.*, 10.8.87). The Yap Islands proper consist of a group of metamorphic and old volcanic islands of moderate slope and lobed shorelines which are surrounded by broad fringing reefs and lie within a triangular shaped reef complex with limited lagoon waters. Road causeways now connect all the high island complex except Rumung. The islands have interesting reef faunas, mangroves, endemic plants, and some evergreen forest but are very disturbed (Douglas, 1969).

Yap has excellent broad fringing reefs especially on the north-west with some lagoon development inside the reef. Tsuda (1978) mapped the general current patterns and the biological characteristics of the lagoon. Corals have been described by Neudecker (1978) and Sugiyama (1942); over 200 species have now been recorded (Maragos *in litt.*, 10.8.87). Reef sites at Balabat and Pelak on the east coast are described by Tsuda *et al.* (1978), and the reef around Donitsch Island in Tomil Harbour is described by Amesbury *et al.* (1976). Dahl *et al.* (1974) surveyed the reefs at four sites on the west coast. The lagoon off Gorror, on the south-west coast, was shallow with a sandy bottom with turtle grass and occasional coral patches of high diversity (ca 50 species). The lagoon off Nif was deep. The surrounding reef walls were covered with large corals down to a sandy bottom at 10 m. The reef was emergent at low tide, the top covered with algal

turf and occasional small colonies of *Favites*. The leeward outer reef off Okau began with a rocky fore-reef flat 1500 m wide, cut by long surge channels 2 m deep and 1-2 m wide. Coral coverage increased to the reef edge at a depth of 7 m with only a few species less than in Belau. The slope then dropped steeply to 25 m with good coral growth, beyond which a more gentle slope was dominated by *Pachyseris*. Coral growth ended at around 40 m. At Mil Entrance, on the west between Yap and Rumung islands, the reef flat on the north side of the channel extended down to 2 m with good coral coverage. There was a drop-off with good coral growth down to 10 m followed by a gentle slope with turtle grass. In general reef quality was high but reefs were subject to terrestrial influences, a result of Yap being a high island.

Truk Lagoon is an "almost-atoll" (Scott and Rotondo, 1983), consisting of a deeply eroded, volcanic submergence having left embayed basaltic islands and stacks in a lagoon surrounded by a shallow barrier reef with a few coral islands and cays. The lagoon, nearly 300 ft (91 m) deep in places (Maragos *in litt.*, 10.8.87), is considered to be at an intermediate stage of evolution between a true barrier reef and a true atoll reef. The 19 volcanic islands within the lagoon are steep-sided with little coastal plain, and most are fringed by narrow coral reef flats (Maragos, 1986). Most have little remaining natural vegetation except on the summits of the highest islands (Douglas, 1969) and some have mangrove stands. The World War II Japanese wrecks in the lagoon have provided a substrate for abundant growth of reef organisms. Tsuda *et al.* (1977a) carried out preliminary observations on the algae, corals and fish inhabiting the sunken ferry *Fujikawa Maru*. Other reefs have been described in the course of environmental impact studies (see below), including those on the east coast of Tol (in the western part of Truk Lagoon) and fringing reef on the west, south-east and north-east sides of Moen (Clayshulte *et al.*, 1978). A Moen coastal resource inventory was completed recently (Cheney *et al.*, 1982). Around 100 species of holothurians have been recorded in Truk Lagoon (Beardsley, 1971). Virtually no scientific studies have been carried out in outer Truk State (Maragos *in litt.*, 10.8.87).

Pohnpei State consists of a high island complex and some outer atolls. The island of Pohnpei is described in a separate account. Dahl *et al.* (1974) give brief details for some of the reefs around Ant and Pakin. Both atolls have excellent reef development. On the leeward side of Ant, beyond the northern tip, the reef crest is 100 m wide with a fore-reef flat 200 m wide extending down to the reef edge at 10 m depth. Coral growth begins at 2 m depth, 30 m off shore, increasing to 90% coverage at 50 m off shore and 100% at 100 m. From the reef edge there is a steep slope down to 30 m depth with coral cover of over 100% on account of the dominant overlapping table *Acropora*. The slope continues down to 50 m with interspersed sand cover increasing to 50%, although coral development continues much deeper. Pakin has a steep drop-off and coral development to beyond the normal SCUBA range. The centre of the northern reef was surveyed. An 80 m wide reef crest merges with a sloping fore-reef 50 m wide and 7 m deep at the edge. The first 20 m is bare of coral cover, but coral density increases to the edge with head-shaped *Porites* dominant. From the edge an almost vertical drop-off descends, with good coral cover of large specimens and a considerable amount of algae. Sand patches begin at 50 m but at 60 m, coral cover is still 30% and continues to the limit of visibility. Surveys at both atolls have been carried out recently but the results have not yet been published (Birkeland *in litt.*, 10.11.87).

Kosrae State includes the fourth largest island in the Carolines. In the geological past, there was probably a shallow lagoon or moat and reef system around the island (either a barrier reef or very wide fringing reef (Maragos *in litt.*, 10.8.87)), but the lagoon is now nearly filled in to sea level with sandy loams, swamp and forests. Raised sand and rubble beaches mark the position of the outer edge of the reef. Beyond this, the reef forms a flat extending to the breaker zone. Good reef growing conditions have permitted the full development of a typical fringing reef (Scott and Rotondo, 1983). Over 50 sites, covering all reef areas around Kosrae, were surveyed in 1986 as part of a coastal resources inventory and atlas project sponsored by the U.S. Army Corps of Engineers (Maragos *in litt.*, 10.8.87). The reef flat at Okat is described by Eldredge *et al.* (1979). Wide fringing reefs are found particularly around the northern half of the island and at embayments on the south at Taf and Utwe (Maragos, 1986); other large embayments occur at Lele (Leluh) in the east and Okat in the north (Maragos *in litt.*, 10.8.87). The most valuable concentrations of marine resources in the State are at Okat (Maragos, 1986) and Utwe (Maragos *in litt.*, 10.8.87).

Very little is known about the reefs of other islands in the F.S.M. but Oroluk Atoll reefs and Minto Reefs are considered particularly important (Gawel *in litt.*, 29.1.87).

Green Turtles *Chelonia mydas* and Hawksbills *Eretmochelys imbricata* are the principal species present, with Olive Ridleys *Lepidochelys olivacea* and Leatherbacks *Dermochelys coriacea* recorded on rare occasions. Nesting, mainly of Green Turtles, is largely confined to the more distant coral atolls and islands, although both Green Turtles and Hawksbills are found year-round in the lagoons of the high islands (McCoy, 1981; Pritchard, 1981). Seagrass distribution in Micronesia is described in Tsuda *et al.* (1977b).

Reef Resources

Fisheries, agriculture and tourism are the three major industries (SPREP, 1980). Smith (1947) described fishing techniques in use in Micronesia, including the Caroline Islands, in the 1940s. Fish and shellfish were the most important source of protein, with inshore and reef fisheries being of much greater importance than offshore fisheries. The most important finfish groups were angelfish, barracuda, crevalle, goatfish, parrot fish, squirrel fish, surgeon fish and wrasse. Among invertebrates the more important were *Anadara* cockles, conchs, crabs, especially *Scylla serrata*, octopus, *Spondylus*, spiny lobster, sea anemone, sea urchin, clams, including giant clams *Tridacna*, several species of trochus, and turbos. Reef fishing is still important throughout the States. In Yap State, a subsistence lifestyle dominates, fishing is important and tourism is still a relatively minor industry. In Truk, the fringing reefs around the islands are important for day to day resources, most fishing being carried out by women hand collecting on the flats. Pou Bay is the most popular fishing spot on Moen (Maragos, 1986). Truk has been a source of trepang, or beche-de-mer; in 1941, 520 tonnes wet weight of sea cucumbers were harvested from Truk Lagoon and nearly 32 000 pounds (14.5 tonnes) of trepang were exported from Truk to Japan (Beardsley, 1971; Smith, 1947); since then there has been no fishery although large numbers of holothurians are still present (Beardsley, 1971). Truk Lagoon is famous for its World War II Japanese wrecks which are a major attraction to SCUBA divers (Rosenberg, 1981) and account for a large proportion of the tourist industry. On Kosrae, the main subsistence fishery is at Okat; there is considerable potential for tourism and fishery development on this island (Maragos, 1986), and several hotels have been built or are being planned (Maragos *in litt.*, 10.8.87). Subsistence harvest of turtles in the outer atolls and low islands of the F.S.M. is discussed in McCoy (1981).

Disturbances and Deficiencies

Acanthaster planci infestations have occurred in a number of areas (Antonius, 1971a and b; Cheny, 1973; Marsh and Tsuda, 1973; Tsuda, 1971). In May 1971, twelve islands (Pulap, Puluwat, Pulusuk, Satawal, Lamotrek, Elato, Olimarao, Eauripik, Woleai, Ifalik, Faraulep and Mogami Bank) were visited very briefly to investigate this problem (Tsuda, 1971). *Acanthaster* has also been recorded on Yap, Ant, Pohnpei, Mwokil, Nukuoro, Kapingamarangi, Pingelap (Antonius, 1971a and b), and Kosrae (Maragos *in litt.*, 10.8.87).

Most of the F.S.M. are inhabited, with the high and large islands in each state serving as urban and government centres. Many of the islands are highly disturbed (Douglas, 1969) and may come increasingly under pressure with the withdrawal of U.S. assistance (NRDC, n.d.) and the possible development of military activities. Fisheries and tourism may have increasing impacts on the reefs (Maragos, 1986). A number of environmental impact studies have been carried out around sewage outfalls. Pollution of this kind was considered to be a minor problem where the sewage undergoes secondary treatment before release (Tsuda, n.d.), but Maragos (1986 and *in litt.*, 10.8.87) considers it among the major problems in these islands, especially if treatment is not properly maintained and if outfalls discharge into sluggish lagoon waters. Moreover, many islanders do not use the new sewer systems due to cost (Maragos *in litt.*, 10.8.87).

Reefs on Yap are subject to influence from terrestrial run-off, and coastal dredging is of major concern; this group of islands was also seriously affected by World War II (Maragos, 1986). Sewage pollution has been a problem but efforts are being made to improve this (Maragos, 1986). Environmental impact studies carried out in Yap include those at Donitsch sewer outfall (Amesbury *et al.*, 1976), at Balabat and Pelak on the east coast of Yap (Tsuda *et al.*, 1978), at the Colonia Dock site and for the siting of proposed dredging projects in Yap Lagoon (Amesbury *et al.*, 1977b; Strong *et al.*, 1982; Tsuda, 1978). In Outer Yap an environmental impact study has been carried out at a site on Falalop Island, Ulithi Atoll (Tsuda *et al.*, 1978). Oil spills and pesticides have had localised impact on Yap (Falanruw, 1980). A comparison by Falanruw (1980) of the results of reef surveys by Sugiyama (1942) and Neudecker (1978) shows that there has been a considerable reduction of the coral zone in Yap Harbour.

Truk is the most populous and crowded district and the reefs are considered to be seriously disturbed (Dahl *et al.*, 1974) and were considerably damaged during World War II; recovery of bombed corals and those affected by dynamite fishing appears to have been fairly minor. Other problems in Truk include dredge and fill activities, the construction of a road causeway across Pou Bay, and road building using reef materials (Maragos, 1986). Environmental impact studies have been carried out at an airport runway expansion site on Moen, built along the north-west coast using material from the reef flat at Pou Bay (Amesbury, 1981; Devaney *et al.*, 1975; Amesbury *et al.*, 1978, 1979, 1980 and 1982); considerable damage was caused as the screens to limit the turbidity caused by dredging and filling were not properly applied (Maragos, 1986). Studies were also carried out on Moen at a sewage outfall at Point Gabert (Tsuda *et al.*, 1975), and on the north-east fringing reef to monitor sediment load and reef fish following dredging (Zolan *et al.*, 1981; Amesbury, 1981). Various other environmental impact studies have been carried out in Truk, including one for a proposed fishery complex site on Tol (Clayshulte *et al.*, 1978) and one for the Truk Tuna Fishery Complex on Dublon (Amesbury *et al.*, 1977a).

In Pohnpei State there is concern that airports being built on reefs on the outer islands and atolls may have a deleterious effect on the reefs (Maragos, 1986). On Pakin there were plans in the 1970s to open a channel into the lagoon and build a pier (Dahl *et al.*, 1974). Problems on Pohnpei Island are described in a separate account.

On Kosrae, reefs at Okat have been seriously damaged and new road development projects are now threatening others. A marine environmental survey of Okat was carried out in the course of development of a port and runway (Eldredge *et al.*, 1979), but unnecessary sedimentation occurred through dredging and filling and the reefs have been damaged; a decline in fishery yields in this area has been reported (Maragos *in litt.*, 10.8.87). Some remedial work has been accomplished but these developments will have long term consequences (Maragos, 1986). The construction of a circumferential road at Utwe has caused significant damage as "borrow" sites have been opened in reef areas for fill and aggregate (Maragos, 1986 and *in litt.*, 10.8.87). As yet, other types of pollution and waste disposal are still a minor problem in this State and a sewage treatment plant project has been initiated (Maragos, 1986).

Coastal resources are heavily exploited for subsistence purposes. Dynamiting the reefs for fish seems to be a widespread problem (SPREP, 1980), particularly in Truk State where reefs around Moen and other islands have been seriously damaged (Maragos, 1986). Dynamiting and fishing with bleach is also reported from Yap (Johannes, in press). In Ulithi there was a major ciguatera problem in the 1960s (Randall and Jones, 1968), and fish poisoning problems are described in Gawel (1984). Giant Clams have been heavily over-exploited. *Tridacna gigas* is now extinct and *Hippopus hippopus* is very rare although fossil shells in recent sediments suggest that these species used to be abundant (Anon., 1987). Turtles in the F.S.M. are coming under increasing hunting pressure; for example those on Oroluk in Pohnpei are considered threatened by the recent settlement of Kapingamarangi islanders who have reported declining catches since they started taking turtles in the late 1960s (Gawel *in litt.*, 25.6.87; McCoy, 1981).

Legislation and Management

Traditional customs play an important role in reef management in some areas. In Yap, traditional fishing rights are the most complicated of all the islands. Village boundaries extend through the lagoon to the open sea and outsiders are strictly prohibited. Fishing rights in reef flat areas are controlled by particular families but are overseen by the village as a whole. Fishing areas are sometimes put off limits until fish populations have increased. Although some customs have died out, most are maintained and the villagers themselves may prohibit fishing with dynamite and bleach (Falanruw, 1984). On Ulithi, the chiefs of the eight districts control marine resources and the reefs, lagoons and islands are divided among the clans. On Satawal, the men are free to fish anywhere but the women may only fish within the reef. The "Chief of the Sea" may sometimes prohibit the use of spears in certain areas. Fishing activities on more distant reefs are more strictly regulated; for example, fishing on Wenimong Reef is restricted to special occassions. On Woleai, the lagoon is a controlled area divided among the chiefs of the atoll and the large coral heads within the lagoon are also owned. In Truk State, the reef flats surrounding the islands in the lagoon are generally owned by the adjacent land owners and lineages and access by others is restricted. This helps to control exploitation on "owned" reefs although it may increase exploitation on "open" reefs (Johannes, in press; Maragos, 1986). On Namonuito, in the Western Islands of Truk, there are traditional rights and tenure on at least one (Ulul) of the five islands. On Etal, in the Mortlocks, the reef slopes and flats are divided into small named tracts owned by clans who have exclusive fishing rights; taboos may be placed on some reefs for conservation purposes. On Lukunor, rights have relapsed, although in the 1940s unsuccessful attempts were made to reintroduce them. On Kosrae, traditional rights have died out (Johannes, in press).

There are now few legislative environmental controls in the F.S.M. Under the Trust Territory of the Pacific Islands, a large number of Trust Territory and U.S. environmental laws and permit programmes applied, as described in Dahl (1980), in particular the National Environmental Policy Act (the U.S. law prescribing environmental impact statements or EISs). With the termination of the Trust Territory, these provisions now only apply to U.S. actions within the F.S.M. (now very few in number) and do not apply to private individual or F.S.M. state actions. Counterpart environmental laws and regulations were intended to be introduced for use by the F.S.M. goverment, but these had not materialized by July 1987 (Maragos *in litt.*, 10.8.87).

The ships in Truk Lagoon have been designated a historical monument (Rosenburg, 1981) and have potential for development as a National Historical Park (Harry *in litt.*, 29.2.88). An inventory of reef resources has been compiled for Moen (Cheney *et al.*, 1982; Maragos and Elliott, 1985) and an atlas and report of reef resources have been prepared for Pohnpei (*see separate account*). Coastal resource inventories and atlases for Kosrae and Yap Islands proper are being sponsored by the U.S. Army Corps of Engineers, and the Universities of Guam and Hawaii are preparing coastal resource management plans for these islands. The Government of Truk State has requested a similar inventory for Truk Lagoon and the Governor of Yap State has requested funds to continue the inventory and atlas programme in the outer Yap Islands and atolls (Maragos *in litt.*, 10.8.87).

Recommendations

There is a serious need for environmental legislation to replace that which applied under the Trust Territory of the Pacific Islands (Maragos *in litt.*, 10.8.87). Some degree of protection has been recommended for Gaferut (Dahl, 1980). Turtle reserves have been recommended for Elato, Pikelot, West Fayu and Oroluk. East Fayu was proposed as an "Island for Science" by the International Biological Programme as it is largely undisturbed. Ulithi is one of the very few relatively undisturbed atolls directly accessible by plane; Woleai is considered totally undisturbed (Dahl *et al.*, 1974). Recommendations for the management of Pohnpei reefs are given in a separate account. A giant clam mariculture project is being considered for Kosrae and nursery trials are to be carried out on shallow subtidal and intertidal fringing reef flats (Anon., 1987).

Dahl (1980 and 1986) stresses that there is an urgent need to inventory the biomes of the F.S.M. in view of the richness of the area and the likelihood of great development pressure in the near future. The National Resources Defence Council, with support from IUCN/WWF, has developed a conservation plan of action for consideration by the governments of the F.S.M. (NRDC, n.d.). Truk State is considered to be particularly urgently in need of attention, and the coastal resource inventory carried out for Moen should be expanded to other islands and atolls. Funds have been requested from the government of the F.S.M. and the U.S. National Congress for additional inventories of Truk Lagoon and outer Yap State (Maragos *in litt.*, 10.8.87).

References

* = cited but not consulted

*Abbott, I.A. (1961). A checklist of marine algae for Ifaluk Atoll, Caroline Islands. *Atoll Res. Bull.* 77. 5 pp.

Amesbury, S. (1981). Effects of turbidity on shallow-water reef fish assemblages in Truk, Eastern Caroline Islands. *Proc. 4th Int. Coral Reef Symp., Manila* 1: 155-159.

*Amesbury, S., Clayshulte, R.N., Determan, T.A., Hedlund, S.E. and Eads, J.R. (1978). Environmental monitoring study of airport runway expansion site Moen, Truk, Eastern Caroline Islands. Part 1. Baseline Study. *Univ. Guam Mar. Lab. Tech. Rept* 81. 60 pp.

*Amesbury, S., Clayshulte, R.N., Grimm, G.R. and Rosario, G.P. (1982). Biological monitoring study of airport runway expansion site Moen, Truk. Eastern Caroline Islands. *Univ. Guam Mar. Lab. Tech. Rept.* 81. 60 pp.

*Amesbury, S.S., Colgan, M., Braley, R. and Bowden, A. (1980). Environmental monitoring study of airport runway expansion site, Moen Truk, eastern Caroline Islands. Part B. Monitoring Study Report 2: 1980 survey. *Univ. Guam Mar. Lab. Misc. Rept* 30. 59 pp.

*Amesbury, S.S., Lassuy, D.R., Chernin, M.I. and Smith, B.D. (1979). Environmental monitoring study of airport runway expansion site, Moen Truk, eastern Caroline Islands. Part B. Monitoring Study Report 1: 1979 survey. *Univ. Guam Mar. Lab. Misc. Rept* 27. 45 pp.

*Amesbury, S.S., Marsh, J.A., Randall, R.H. and Stojkovich, J.O. (1977a). Limited current and underwater biological survey of proposed Truk Tuna Fishery Complex, Dublon Island, Truk. *Univ. Guam Mar. Lab. Tech. Rept* 34. 49 pp.

Amesbury, S.S., Tsuda, R.T., Randall, R.H., Birkeland, C.E. and Cushing, F. (1976). Limited current and underwater biological survey of the Donitsch sewer outfall site, Yap, western Caroline Islands. *Univ. Guam Mar. Lab. Tech. Rept* 24. 49 pp.

*Amesbury, S.S., Tsuda, R.T., Randall, R.H. and Birkeland, C.E. (1977b). Marine biological survey of the proposed dock site at Colonia, Yap. *Univ. Guam Mar. Lab. Tech. Rept* 35. 22 pp.

Anon. (1987). Micronesian Mariculture Demonstration Centre, Palau, *Clamlines* 3: 2-3.

Antonius, A. (1971a). Das Acanthaster-Problem im Pazifik (Echinodermata). *Inst. Revue ges. Hydrobiol.* 56(2): 283-319.

Antonius, A. (1971b). Die Riff-Seestern-Expedition. *Umschau in Wissenschaft und Technik, Fankfurt/Main* 3: 95-96.

*Baker, R.H. (1951). The avifauna of Micronesia, its origin, evolution and distribution. *Univ. Kansas Publ. Mus. Nat. Hist.* 3(1): 1-359.

*Bascom, W.R. (1965). Ponape: A Pacific economy in transition. *Anth. Rec.* 22: 1-156.

Bates, M. and Abbott, D.P. (1958). *Coral Island. Portrait of an Atoll.* Charles Scribner's Sons, N.Y. 254 pp.

Beardsley, A.J. (1971). "Beche-de-mer" fishery for Truk? *Commercial Fisheries Review* 33(7-8): 64-66.

*Birkeland, C. (Ed) (1980). Marine biological survey of northern Ponape Lagoon. *Univ. Guam Mar. Lab. Tech. Rept* 62. 102 pp.

*Cheney, D.P. (1973). An analysis of the *Acanthaster* control programs in Guam and the Trust Territory of the Pacific Islands. *Micronesica* 9(2): 171-180.

Cheney, D.P. (1974). Spawning and aggregation of *Acanthaster planci* in Micronesia. *Proc. 2nd Int. Coral Reef Symp., Brisbane* 1: 591-594.

*Cheney, D.P., Ives, J.H. and Rocheleau, R. (1982). Inventory of the coastal resources and reefs of Moen Island, Truk Atoll. U.S. Army Corps of Engineers, Honolulu. 106 pp.

Clayshulte, R.N., Marsh, J.A., Randall, R.H., Stojkovich, J.O. and Molina, M.E. (1978). Limited current and underwater biological survey at the proposed fishery complex site on Tol Island, Truk. *Univ. Guam Mar. Lab. Tech. Rept* 50. 117 pp.

Dahl, A.L. (1980). Regional ecosystems survey of the South Pacific Area. *SPC/IUCN Technical Paper* 179. South Pacific Commission, Noumea, New Caledonia.

Dahl, A.L. (1986). *Review of the Protected Areas System in Oceania.* IUCN/UNEP. 239 pp.

Dahl, A.L., Macintyre, I.G. and Antonius, A. (1974). A comparative survey of coral reef research sites (CITRE). *Atoll Res. Bull.* 172: 37-77.

*Devaney, D.M., Losey, G.S. and Maragos, J.E. (1975). A marine biological survey of proposed construction sites for the Truk runway. R.M. Parsons, Co. 69 pp.

Douglas, G. (1969). Checklist of Pacific Oceanic Islands. *Micronesica* 5(2): 327-463.

Eldredge, L.G., Best, B.R., Chernin, M.I., Kropp, R.K., Myers, R.F. and Smalley, T.L. (1979). Marine environmental survey of Okat, Kosrae. *Univ. Guam Mar. Lab. Tech. Rept* 63. 101 pp.

Falanruw, M.V.C. (1980). Marine environment impact of land-based activities in the Trust Territory of the Pacific Islands. In: *Marine and Coastal Processes in the Pacific: Ecological Aspects of Coastal Zone Management.* Unesco, ROSTEA, Jakarta. Pp. 19-47.

Falaruw, M.V.C. (1984). People pressure and management of limited resources on Yap. In: McNeely, J.A. and Miller, K.R. (Eds), *National Parks, Conservation and Development: The Role of Protected Areas in Sustaining Society.* Smithsonian Institution Press, Washington D.C. Pp. 348-354.

*Fosberg, F.R. (1969). Plants of Satawal Island, Caroline Islands. *Atoll Res. Bull.* 132.

Gawel, M. (1984). Fish poisoning related to human impacts on coral reefs in the Federated States of Micronesia. In: Man's Impact on Coastal and Estuarine Ecosystems. *Proc. MAB/COMAR Regional Seminar, Tokyo.* Pp. 43-45.

Gillett, R. (in press). Traditional tuna fishing: A study at Satawal, central Caroline Islands. *Bull. Anthropol. B.P. Bishop Mus.*

*Glassman, S.F. (1952). The flora of Ponape. *B.P. Bishop Mus. Bull.* 209. 151 pp.

*Grosenbaugh, D.A. (1981). Qualitative assessment of the asteroids, echinoids and holothurians in Yap Lagoon. *Atoll Res. Bull.* 255: 49-54.

Holthus, P.F. (1985). A reef resource conservation and management plan for Ponape Island. *Proc. 5th Int. Coral Reef Cong., Tahiti* 4: 231-236.

Holthus, P.F. (1987). Pohnpei coastal resources: Proposed management plan. Draft Report. South Pacific Regional Environment Programme, South Pacific Commission, Noumea, New Caledonia. 80 pp.

*Johannes, R.E. (1978). Improving Ponape's reef and lagoon fishery. Mar. Resource Div., Ponape. Unpub. rept. 28 pp.

Johannes, R.E. (in press). The role of Marine Resource Tenure Systems (TURFs) in sustainable nearshore marine resource development and management in U.S.-affiliated tropical Pacific islands. In: Smith, B.D. (Ed.), Topic Reviews in Insular Resource Development and Management in the Pacific U.S.-affiliated islands. *Univ. Guam Mar. Lab. Tech. Rept* 88.

*Kock, R.L. and Tsuda, R.T. (1978). Seagrass assemblages of Yap, Micronesia. *Aquatic Bot.* 5: 245-249.

*Manoa Mapworks (1985). Pohnpei Coastal Resource Atlas. Prepared for the U.S. Army Corps of Engineers, Pacific Ocean Division under contract no. DACW-83-84-M-0577. Honolulu, Hawaii. 78 pp.

McCoy, M.A. (1981). Subsistence hunting of turtles in the western Pacific: The Caroline Islands. In: Bjorndal, K.A. (Ed.), *Biology and Conservation of Sea Turtles.* Smithsonian Institution Press, Washington D.C. Pp. 275-280.

*McKee, E.D. (1958). Geology of Kapingamarangi Atoll, Caroline Islands. *Bull. Geol. Soc. Amer.* 69: 241-278.

Maragos, J.E. (1986). Coastal resource development and management in the U.S. Pacific Islands: 1. Island-by-island analysis. Office of Technology Assessment, U.S. Congress. Draft.

Maragos, J.E. and Elliott, M.E. (1985). Coastal resource inventories in Hawaii, Samoa and Micronesia. *Proc. 5th Int. Coral Reef Cong., Tahiti* 5: 577-582.

*Marsh, J.A. and Tsuda, R.T. (1973). Population levels of *Acanthaster planci* in the Mariana and Caroline Islands, 1969-1972. *Atoll Res. Bull.* 170: 1-16.

*Marshall, M. (1975). The natural history of Namoluk Atoll, Eastern Caroline Islands. *Atoll Res. Bull.* 189.

*Murphy, R.E. (1949). "High" and "low" islands in the eastern Carolines. *Geog. Rev.* 39: 425-439.

Neudecker, S. (1978). Qualitative assessment of coral species composition of reef communities in Yap Lagoon. In: Tsuda, R.T. (Ed.), Marine Biological Survey of Yap Lagoon. *Univ. Guam Mar. Lab. Tech. Rept* 45: 43-80.

*Niering, W.A. (1956). Bioecology of Kapingamarangi Atoll, Caroline Islands: Terrestrial aspects. *Atoll Res. Bull.* 49. 32 pp.

*Niering, W.A. (1961). Observations on Puluwat and Gaferut, Caroline Islands. *Atoll Res. Bull.* 76: 1-10.

NRDC (Natural Resources Defense Council) (n.d.). Proposal to implement the World Conservation Strategy in the Caroline Islands of Micronesia.

*OPS, TTPI (Office of Planning and Statistics, Trust Territory of the Pacific Islands) (1978a). *Tonowas/Etten Land Use Guide.*

*OPS, TTPI (Office of Planning and Statistics, Trust Territory of the Pacific Islands) (1978b). *Uman Land Use Guide.*

*OPS, TTPI (Office of Planning and Statistics, Trust Territory of the Pacific Islands) (1978c). *Fefan Land Use Guide.*

*OPS, TTPI (Office of Planning and Statistics, Trust Territory of the Pacific Islands) (1978d). *Udot/Eot Land Use Guide.*

Pritchard, P.C.H. (1981). Marine turtles of Micronesia. In: Bjorndal, K.A. (Ed.), *Biology and Conservation of Sea Turtles.* Smithsonian Institution Press, Washington D.C. Pp. 263-274.

*Randall, J.E. and Jones, R.S. (1968). Report on ciguatera at Ulithi Atoll, Caroline Islands. *Univ. Guam Mar. Lab. Misc. Rept* 1. 4 pp.

Rosenberg, P.A. (1981). *Shipwrecks of Truk.* Privately printed. 102 pp.

*St John, H. (1948). Report on the flora of Pingelap Atoll, Caroline Islands, Micronesia, and observations on the native inhabitants: Pacific Plant Studies 7. *Pac. Sci.* 2: 96-113.

Scott, G.A.J. and Rotondo, G.M. (1983). A model for the development of types of atoll and volcanic islands on the Pacific lithospheric plate. *Atoll Res. Bull.* 260. 33 pp.

Smith, R.O. (1947). Fishery resources of Micronesia. *Fishery leaflet* 239. Fish and Wildlife Service, U.S. Dept of the Interior. 46 pp.

SPREP (1980). Trust Territory of the Pacific Islands. *Country Report* 14. South Pacific Commission, Noumea, New Caledonia.

*Strong, R.D., Randall, R.H., Smalley, T.L., Bumoon, B. and Bowoo, O. (1982). Environmental assessment for proposed dredging operations in Yap Lagoon. *Univ. Guam Mar. Lab. Tech. Rept* 78. 88 pp.

*Sudo, K.I. (1984). Social organization and types of sea tenure in Micronesia. *Senri. Ethno. Stu.* 17: 203-230.

*Sugiyama, T. (1942). Reef-building corals of Yap Island and its fringing reefs. *Mem. Palaeont. Inst. Fac. Sci. Tohoku Imp. Univ.* 39: 7-26.

*Tracey, J.I., Abbott, D.P. and Arnow, T. (1961). Natural History of Ifaluk Atoll: Physical environment. *B.P. Bishop Mus. Bull.* 222: 1-75.

*Tsuda, R.T. (compiler) (1971). Status of *Acanthaster planci* and coral reefs in the Mariana and Caroline Islands, June 1970-May 1971. *Univ. Guam Mar. Lab. Tech. Rept* 2. 127 pp.

*Tsuda, R.T. (1972). Marine benthic algae from Truk and Kuop, Caroline Islands. *Atoll Res. Bull.* 155: 1-10.

Tsuda, R.T. (Ed.) (1978). Marine biological survey of Yap Lagoon. *Univ. Guam Mar. Lab. Tech. Rept* 45. 162 pp.

Tsuda, R.T. (n.d.). Report to IUCN Coral Reef Group. Unpub. 9 pp.

Tsuda, R.T., Amesbury, S.S. and Moras, S.C. (1977a). Preliminary observations on the algae, corals and fishes inhabiting the sunken ferry *Fujikawa Maru* in Truk Lagoon. *Atoll Res. Bull.* 212: 1-6.

Tsuda, R.T., Amesbury, S.S., Moras, S.C. and Beeman, P.P. (1975). Limited current and underwater biological survey at the Point Gabert wastewater outfall on Moen, Truk. *Univ. Guam Mar. Lab. Tech. Rept* 20. 39 pp.

Tsuda, R.T. and Belk, M.S. (1972). Additional records of marine benthic algae from Yap, western Caroline Islands. *Atoll Res. Bull.* 156: 1-5.

Tsuda, R.T., Chernin, M.I., Stojkovich, J.O., Lassuy, D.R. and Smith, B.D. (1978). Current and underater biological survey of selected sewer outfall sites in the Yap Central Islands and on Falalop Island, Ulithi Atoll, Yap Outer Islands. *Univ. Guam Mar. Lab. Tech. Rept* 46. 101 pp.

Tsuda, R.T., Fosberg, F.R. and Sachet, M.-H. (1977b). Distribution of seagrasses in Micronesia. *Micronesica* 13(2): 191-198.

Tsuda, R.T., Randall, R.H. and Chase, J.A. (1974). Limited current and biological study in the Tuanmokot Channel, Ponape. *Univ. Guam Mar. Lab. Tech. Rept* 15. 58 pp.

USACE (U.S. Army Corps of Engineers, Pacific Ocean Division) (1986). Pohnpei Coral Reef Inventory. Fort Shafter, Hawaii.

Wiens, H.J. (1956). The geography of Kapingamarangi Atoll in the eastern Carolines. *Atoll Res. Bull.* 48. 86 pp.

Zolan, W.J. and Clayshulte, R.N. (1981). Influence of dredging discharge on water quality, Truk Lagoon. *Proc. 4th Int. Coral Reef Symp., Manila* 1: 213. (Abst.).

POHNPEI (PONAPE)

Geographical Location   Pohnpei State is in the eastern Caroline Islands, between Truk and Kosrae; nearest islands to Pohnpei are Ant and Pakin; ca 6°45'N, 157°15'E.

Area, Depth, Altitude   334 sq. km; diameter about 20 km; max. alt. 2595 ft (791 m).

Physical Features   Pohnpei is a roughly pentagonal high basalt volcanic island with a number of high peaks from which radiate sharp ridges, steep cliffs and narrow valleys which often reach to the deeply indented shore. There are only limited coastal lowlands and plains. Rainfall is almost 5000 mm a year on the windward coast and an estimated 10 000 mm a year in the interior, resulting in numerous streams which discharge into nearshore waters. Reefs and the coast are described in USACE (1986). The island is surrounded by a complex reef and lagoon system. The highly convoluted fringing reef, 20-2000 m wide, supports a nearly continuous belt of mangrove. A deep, 1-2 mile (1.6-3.2 km) wide lagoon encircles most of the island except in the south-east where there is a continuous fringing reef platform. The lagoon has many linear, patch and pinnacle reefs and a

few basalt islands with fringing reefs. An extensive barrier reef, dissected by over 15 deep passes, encloses the lagoon and supports a few sand islets (Baker, 1951; Glassman, 1952; Holthus, 1985; Maragos, 1986). Physical features of the northern part of the lagoon are described by Birkeland (1980).

Reef Structure and Corals   A coral reef inventory and atlas covering the whole island have been prepared (Manoa Mapworks, 1985; Maragos, 1986; USACE, 1986). Prior to this, the barrier reef had been studied at Mant Passage, the north-east entrance through the windward side of the reef. Species diversity was low and the water turbid (Dahl *et al.*, 1974). Corals and reefs of the northern lagoon are described in Birkeland (1980). An environmental study was carried out in the region of Tuanmokot Channel (Tsuda *et al.*, 1974).

Noteworthy Fauna and Flora   The flora is described in Glassman (1952).

Scientific Importance and Research   In 1984 the reef and lagoon resources were inventoried semi-quantitatively during a two month survey by the U.S. Army Corps of Engineers and the University of Hawaii in order to compile a coral reef inventory and atlas (see above). The reefs are being analysed as part of the proposed coral reef management plan (Maragos, 1986; Holthus, 1985 and 1987). Bibliographies of coastal Pohnpei are provided in Birkeland (1980) and USACE (1986).

Economic Value and Social Benefits   Fish are an important subsistence resource (Johannes, 1978) and there is good potential for further development (Maragos, 1986). Trochus shell is harvested and a small artisanal fishery provides the local market with fresh reef fish (Holthus, 1985).

Disturbance or Deficiencies   *Acanthaster* outbreaks were studied in the early 1970s (Cheney, 1974). Most of the reefs are in a relatively healthy state but there are an increasing number of threats as most of the population of Pohnpei State is based on this island at Kolonia, and the island is the capital of the F.S.M. (Holthus, 1985). There has been recent reef destruction from dredge and fill activities on shallow flats, especially near the port and airport area of Kolonia. Sediment from upland construction has also caused reef degradation and reef "borrow" sites for aggregate for building of the circumferential road have caused problems. Domestic pollution and disposal of solid waste are becoming of increasing concern. Increased fishing pressure is becoming apparent in the lagoon. Dynamite fishing has been reported but is not said to be serious (Maragos, 1986).

Legal Protection   There are no protected areas, although *Trochus* reserves have been designated to control overharvesting (Holthus, 1985).

Management   There are at present very few environmental controls in effect, those that previously operated having largely ceased to apply with the termination of the Trust Territory of the Pacific Islands in October 1986 (Maragos *in litt.*, 10.8.87). The traditional use of marine resources, sea tenure and their demise have been briefly documented (Bascom, 1965; Johannes, in press; Sudo, 1984). The five municipalities used to have exclusive fishing rights in adjacent waters but these traditional customs have now died out.

Recommendations   Following preliminary recommendations made by Holthus (1985), a detailed reef and lagoon resource management plan has been prepared (Holthus, 1987) which could serve as a model for other high islands.   Using the Pohpei Coastal Resource Atlas (USACE, 1986) as a map base, a zoning plan for the mangrove reef and lagoon areas has been developed.   The zones are: 1) development/general use (with specific sand mining and reef flat dredging sites); 2) sustainable use; 3) seasonal preserves; 4) species preserves; 5) marine parks. Management measures set out include: permits; water and environmental quality monitoring; fisheries conservation including the establishment of Giant Clam preserves and monitoring of depleted species; facilities siting; emergency planning; land use controls; and hazardous substances management.   A team of various Pohnpei State government agencies is reviewing the plan, and organising its distribution and public review. It will then be printed by the University of Hawaii Sea Grant (Holthus *in litt.*, 27.10.87).   Appropriate measures to control excessive soil erosion and reef sedimentation are suggested for coastal areas adjacent to extensive land clearing and road construction (Holthus, 1985).   Earlier recommendations for improvement of the reef fishery are given in Johannes (1978).

FIJI

INTRODUCTION

General Description

Fiji comprises about 844 islands and islets (of which about 106 are inhabited) scattered in the area 15°-23°S, 177°-178°W. The country is divided into 14 provinces and Rotuma, and the islands may be divided into several distinct groups: Rotuma; Vanua Levu and associated islands, including Taveuni and the Ringgold Isles; the Lau Group; the Lomaiviti Group; the Yasawas; Viti Levu and associated islands; and Kadavu and associated islands. The main archipelago comprises a total land area of 18 333 sq. km, of which 87% is accounted for by Viti Levu (10 386 sq. km) and Vanua Levu (5534 sq. km); other large islands are Taveuni, Kadavu and Gau. Apart from Kadavu and islands in the Koro Sea, the islands consist almost entirely of volcanic and plutonic rocks of various ages which have been subjected to degeneration and soil formation under typical tropical conditions of intense weathering. The Fiji Plateau consists of two submerged platforms, the Viti Levu and Vanua Levu platforms, and is surrounded by deep water, except at the Kermadec Ridge which links it with North Island, New Zealand. Depths of 2000-3000 m are found within the Lau Group in the east but are generally less than 2000 m around Viti Levu and Vanua Levu. The 1830 m (1000 fathom) isobath is situated less than 18 km off Viti Levu in the south-west.

The surface current flows south-westerly through the group. Water temperatures are always above 20°C, with a summer ocean maximum of about 30°C and a mean annual variation of about 6°C. Tidal range is very small, neap tides having a mean range of 0.9 m and springs of 1.30 m (Ryland, 1981). Tides are semi-diurnal with the lower low-water springs falling during the night in summer but during the day in winter (Ryland, 1979).

The climate is tropical with high humidity and temperatures may rise to 35°C, but these are modified by the south-east trade winds from May to November. The summer is hot and wet with several tropical cyclones while the winter months are drier and cooler. Mean monthly temperatures range from 23°C in July and August to 27°C in January. Annual rainfall is unevenly distributed owing to the rain shadow caused by the mountains and high plateaux (1200+ m) of the larger islands (mean 3000 mm on the east coast, 1650 mm on the west coast) (Guinea, 1981).

In the following table, reefs taken from Ordinance Survey maps are referenced O.S. Considerable information has been provided by Rodda (*in litt.*, 14.9.87). However, the tables are not comprehensive for all reefs and islands, particularly in the more complex systems.

Table of Islands

ROTUMA GROUP

Rotuma is the northernmost island group and lies ca 400 km north of Vanua Levu. The group is volcanic in origin (the most recent eruptions are post human

settlement) and consists of Rotuma and nine associated islets, five of which are situated on the reef surrounding Rotuma (Woodhall, 1985). The climate is more equatorial than the rest of Fiji (warmer and wetter). There are no mangroves (Dunlap and Singh, 1980). For most of the year a current flows west-north-west at ca 1 kph (Smith *in litt.*, 5.10.87).

Rotuma (X) 44 sq. km; volcanic; a few swampy areas, one perennial stream and freshwater seeps on some beaches; east and west parts connected by possibly recent, narrow, sandy vegetated isthmus; well-developed fringing reef; reef flat generally narrow, up to 400 m wide at Noa'tau and in a few other areas (Smith *in litt.*, 5.10.87); extensive submerged sand and coral bank extends ca 8 km from Malhaha towards north-west of Rotuma, and submarine bank (Whale Bank), ca 5 x 1.5 km with mean depth 30-33 m, lies to the west; 5 small islets on reef around coast: Hauatiu, Haua-Meamea, Solnohu, Solkope, Afgaha.

Uea 77 ha; cone to 860 ft (262 m), cliff-bound; no reef.

Hatana 4 ha; 60 ft (18.3 m); volcanic; booby *Sula* sp. nesting; small islet, Hofhavunglola, on same reef platform with nesting Noddy Terns *Anous* sp. (Smith *in litt.*, 5.10.87).

Hofliua less than 1 sq. km; 190 ft (58 m); cliff-bound; volcanic; no reef.

VANUA LEVU AND RINGGOLD GROUPS

Vanua Levu is the second largest island in Fiji; reefs along the south and west coasts include Cakau Levu, the Vanua Levu Barrier Reef and Rainbow Reef. Taveuni, the third largest island and part of the Vanua Levu Group, is separated from Vanua Levu by Somosomo Strait (9 km wide at its narrowest). There are a number of offshore islands and a complex offshore reef system, including the Great Sea Reef. Descriptions of geology and reefs are given in Woodhall (1985b). The Ringgold Isles and associated reefs extend east and north-east of Vanua Levu and are sometimes considered part of the Lau Group, although they are separated from them by the Nanuku Passage and are not geologically part of them.

Vanua Levu 5534 sq km; high volcanic, 1032 m; some limestone along south-east coast; arid plains; mountain chain creates rainshadow, with dry north-west side (mean rainfall 160 cm p.a.) and wet south side (420 cm rain p.a.); average temperature 25°C; river system is less developed than Viti Levu; only one river over 25 km long (Dreketi); three rivers (Wailevu, Labasa and Qawa) merge to form the Labasa delta on the central northern coast; mangroves found near all river mouths (Dunlap and Singh, 1980); reefs along coast of Somosomo Str. to south-east (Davis, 1920); several areas surveyed for Giant Clams (Lewis *et al.*, 1985).

Namenalala (Namena) 43.2 ha; hilly and forested; 105 m rounded summit; volcanic; on Namena barrier reef; also fringing reef; turtles; important seabird nesting colony (Clunie, in press); fishing; resort.

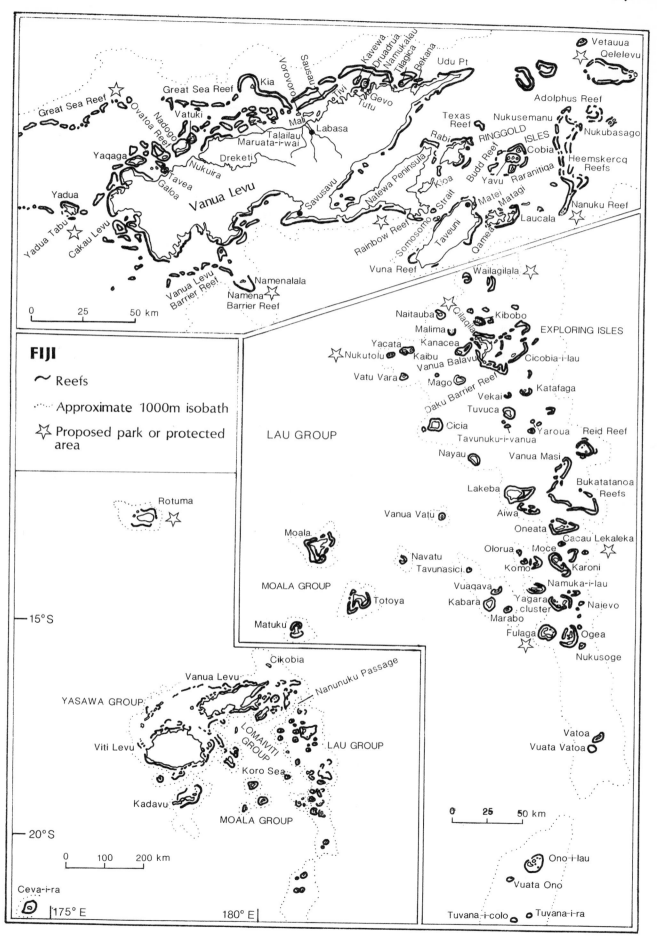

Yadua   5.26 sq. mi. (13.6 sq. km); volcanic; 195 m; fringing reefs (O.S.).

Yadua Tabu   0.7 sq km; volcanic; 100 m; within Yadua fringing reef (O.S.); endemic iguana *Brachylophus vitiensis* (Singh, 1985).

Yaqaga   3.73 sq. mi. (9.7 sq. km); 887 ft (270 m); volcanic; rocky; fringing reef (O.S.).

Galoa   190 acres (77 ha); 90 m, volcanic; fringing reef (O.S.).

Tavea   39 acres (15.8 ha); 79 m, volcanic; surrounded by mangroves; fringing reefs (O.S.).

Nukuira   mangrove-covered cay surrounded by circular Ovatoa Reef (6-10 km diameter) within Great Sea Reef.

Nadogo (Nadogu)   mangrove-covered cay within circular reef (6-10 km diameter) (O.S.).

Vatuki   1.6 ha; mangrove-covered sand and beach-rock cay within circular reef (ca 3 km diameter) (O.S.).

Macuata-i-wai   1.25 sq. mi. (3.2 sq. km); 500 ft (152 m); volcanic; stony; fringing reef; Mavuva (18 ha) lies off north tip within fringing reef (O.S.).

Talailau   mangrove-covered cay within circular Cakau Talailau Reef (3-7 km diameter).

Kia   400 acres (162 ha); 780 ft (238 m); volcanic, very rugged; surrounded by Great Sea Reef; no fringing reefs (O.S.).

Vorovoro   35.2 ha; 292 ft (89 m); no reefs (O.S.).

Mali   (X) 6.7 sq. km; 562 ft (171 m); conical peak; volcanic; fringing reef mainly along north coast (O.S.).

Sausau   6 acres (2.4 ha); volcanic; situated on 25 km reef parallel to Great Sea Reef (O.S.); sea snakes.

Tivi   343 ft (105 m); volcanic; fringing reef (O.S.).

Tutu   2.8 sq. km; 634 ft (193 m); volcanic; joined to mainland by mangroves; fringing reef along ca 1/2 of coast (O.S.).

Kavewa   (X) 210 acres (85 ha); volcanic; within circular reef (O.S.).

Gevo   (X) 190 acres (77 ha), 463 ft (141 m); volcanic; fringing reef (O.S.).

Druadrua   (X) 1.5 sq. mi. (3.9 sq. km); 438 ft (134 m); volcanic; flat-topped; rocky; fringing reef (O.S.).

Namukalau   (X) 50.2 ha; volcanic; fringing reef on north side; mangroves.

Tilagica   on Great Sea Reef (O.S.); surrounded by mangroves.

Bekana   12.2 ha; 6 m; volcanic; within circular reef (2-4 km diameter) (O.S.).

Rabi   (X) 26.56 sq. mi. (68.8 sq. km); volcanic; 1529 ft (466 m); steep slopes; reef encircled (Davis, 1920); settled by former Banaban islanders; Florida Reefs off south-east coast; Texas Reef ca 15 km north-west (O.S.).

Kioa   (X) 18.6 sq. km; 1 km from Vanua Levu; volcanic, 305 m; sheer south and south-east face to eroded volcanic cone; fringing reef mainly on east and south coasts (O.S.).

Taveuni   (X) 168 sq. mi. (435 sq. km); volcanic backbone of dormant cones; 4072 ft (1241 m); crater lake (Davis, 1920); drier north-west side; south-east coast has ca 700 cm annual rainfall (highest in Fiji); steep sides and high rainfall contribute to scarcity of reefs - absent from much of east and west coasts of southern part of island, elsewhere largely fringing reefs (Ryland, 1979); Vuna Reef at south-western extremity encloses a lagoon which is a surviving part of more extensive structure (Davis, 1920); Korolevu with associated reef complex lies off north-west coast; Viubani lies within reef off northern tip (O.S.); described by Brookfield (1978a).

Qamea   (X) 13 sq. mi. (33.7 sq. km); volcanic; 996 ft (304 m); deeply dissected coastline; mangrove-filled islets; fringing reef round much of coast; Qamea and Laucala both part-encircled by same barrier reef (Davis, 1920); small resort.

Laucala   (X) 4.7 sq. mi. (12.2 sq. km); volcanic, 440 ft (134 m); private; Motualevu Reef ca 5 km north-east.

Matagi   (X) 232 acres (94 ha); volcanic, ca 100 m; sited on Qamea barrier reef; privately owned; small resort.

Cikobia   (X) 5.78 sq. mi. (15 sq. km); narrow limestone ridge with 2 volcanic areas; west very rugged; 630 ft (192 m); east lower and sandy, 430-450 ft (131-137 m); fringing reef; seabird colony (Clunie, in press).

Ringgold Isles

Heemskercq Reefs   large elongate barrier reef system with deep lagoon.

Nukusemanu   reefs lying to north, with sand cay (2 ha) with some beach rock; seabird colony (Clunie, in press).

Nanuku Reef Islets   three small reef islets of sand and beach-rock to south; bank of sandy coral; 150 sq. yds (123 sq. m) of reef; seabirds; turtles (Bustard, 1970).

Budd Reef Islets   almost-atoll, 12 mi. (19.3 km) diameter; narrow barrier reef circling 16 x 10 km lagoon, 50-80 m deep (Clunie, in press; Davis, 1920); include:

- *Yanuca*   (X) 102 ha; 137 m; volcanic;

- *Yavu*   62 ha; 112.8 m; volcanic;

- *Maqewa*   26 ha; volcanic; rocky, covered with low scrubby bush; 280 ft (85 m); joined to Yanuca by reef;

- *Beka*   2.8 ha; volcanic; 110 ft (33.5 m);

- *Cobia*   70 ha; little-eroded volcanic remains of crater rim with wet, steep slopes, 580 ft (177 m);

almost circular but breached over 300 m distance; tides flow over reef into sandy bottomed lagoon (800 m broad, 44 m deep); rich marine life; mangroves fringe coast; reef slopes steeply into deep water; 4 small sandy beaches; forest seriously damaged by goats, serious erosion; reef unaffected at present but threatened (Clunie, in press);

- *Raranitiqa and Tovuka*   44 ha and 1.5 ha volcanic double islet; Rarantiqa low and flat; Tovuka small and rugged (45 m) with bad soil erosion from introduced goats (Clunie, in press).

Adolphus Reef   circular lagoon awash at high tide.

Nukubasaga   20 ha; atoll; 3 m; rockbound; beach-rock and sand; 3 m; major seabird colony, turtles.

Nukubalati (Nukupureti)   3.2 ha; atoll; 3 m; sand cay with beach rock; connected to Nukubasaga by reef; seabirds, turtles.

Qelelevu (Naqelelevu)   (X) 147 ha; atoll, 2 islets; 30-40 ft (9-12 m); continuous barrier reef to west of islets; islets described by Clunie (in press); reefs described and mapped in Phipps and Preobrazhensky (1977); seabird nesting colony; Coconut Crab *Birgus latro* present (Clunie, in press); adjoined by limestone islets Taulalia/Tauraria (13.5 ha).

Vetauua   38.3 ha; 2 m; sand and beach-rock with 2 small limestone stacks ca 3 m high on reef flat; fringing reef described and mapped in Phipps and Preobrazhensky (1977); major seabird nesting colony (Clunie, in press).

LAU ISLANDS

The Lau Archipelago is the most easterly part of Fiji and consists of about 40 small islands and over 250 islets and cays covering an area of ca 440 sq. km in 113 900 sq. km of ocean, between 16°30'-20°S and 178-180°W. Their geology is described by Ladd and Hoffmeister (1945). The northern islands are mainly high and of composite volcanic and limestone, the southern ones mainly low and limestone only but there are some islands of solely volcanic formation (Rodda *in litt.*, 14.10.87). The largest is Lakeba (*see separate account*). Although many of the islands are only a few kilometres apart, depths between them frequently reach 200-300 m, and sometimes over 1500 m; bathymetric studies indicate the sea bed to be an irregular volcanic terrain (Phipps and Preobrazhensky, 1977). Mean annual temperature is 26°C; mean rainfall is 200 cm with most parts of the region experiencing a moderate dry season of three months. Most islands lack freshwater (Dunlap and Singh, 1980). Geology and reef systems of this group are described and mapped in Woodhall (1984a, b and c; 1985a and c).

Wailagilala   (X) true atoll surrounded by very deep water; lagoon; 2 islets of sand and beach-rock, variable in size (30 ha in 1950s); barrier reef; reef system covers 24 sq. km, described and mapped by Phipps and Preobrazhensky (1977); fore-reef drops vertically to 200-300 m; lighthouse.

Kibobo   land area 16.6 ha in reef system of 25 sq. km; 3 islets in middle of lagoon; largest is limestone, 58 m;

second is volcanic, 37 m; both on one reef in lagoon; third volcanic with fringing reef joining it to main barrier reef; turtles; fishing.

Naitauba   (X) 8.8 sq. km in reef system of 26 sq. km; circular volcanic and limestone; 610 ft (186 m); private; fringing reefs described by Brodie and Brodie (1987).

Malima   3 volcanic islets on one reef in lagoon; land area 33 ha; largest 39.6 m, others lower; reef system 17 sq. km.

Exploring Isles   reef system 860 sq. km; includes:

Vanua Balavu   (X) 20.54 sq. mi. (53.2 sq. km); limestone/ volcanic, 930 ft (283 m) cliffs; Qilaqila (Bay of Islands) lies south of north-western tip and encloses numerous islets; Daku barrier reef runs from Namalata Pt for 8.5 mi. (13.6 km) to Muamua Passage; surveyed for Giant Clams, as well as Qilaqila Reefs (Lewis *et al.*, 1985); larger barrier reef complex to east, enclosing:

- *Sovu Islets*   3 steep-sided limestone masses; total area 16.6 ha; 230 ft (70 m); fringing reef; seabirds;

- *Avea (Yavea)*   (X) 2.16 sq. km; volcanic and limestone; 600 ft (183 m); fringing reef joining island to Vanua Balavu;

- *Namalata (Malata)*   (X) 2.1 sq. km; high backed limestone ridge with some volcanic rock; 420 ft (128 m);

- *Susui*   (X) 3.3 sq. km; end of elevated reef; 430 ft (131 m); limestone in west, volcanic rocks and sandflat in east;

- *Munia*   (X) 4.5 sq. km; high ridge, 950 ft (290 m); steep scarps; volcanic, very fertile; fringing reef joined to main barrier reef; privately owned;

- *Cikobia-i-lau*   (X) 2.98 sq. km; volcanic and limestone; 550 ft (168 m); barrier reef surveyed for Giant Clams (Lewis *et al.*, 1985); fringing reef separated from barrier reef.

Kanacea   (X) 5 sq. mi. (13 sq. km); reef system 53 sq. km; volcanic with 7 peaks; 850 ft (259 m) summits and higher slopes; fringing reef with lagoon to the east; private.

Kaibu   1.5 sq. km; mainly limestone with low plain underlain by volcanic rock; 150 ft (388 m); rugged; on same reef system (20 sq. km) as Yacata.

Yacata   (X) 8.3 sq. km; limestone and volcanic with terraces; 256 m; fringing reef encloses Kaibu as well; many limestone islets on north coast.

Nukutolu Islets   Koronidre (limestone) and Korosiga (sand and beach rock) on one elongated reef; Nasaqa (sand and beach rock) on separate reef to east; lighthouse; visited from Yacata for coconuts and turtles; seabird rookery.

Vatu Vara   3.84 sq. km in 8.5 sq. km reef system; high limestone; 314 m; flat-topped; cliff-bound; fringing reef; privately owned.

Mago   (X) 22 sq. km in 36.5 sq. km reef system; limestone overlain and underlain by volcanics; 204 m; central basin; fringing reef; privately owned.

Katafaga   80 ha in 17 sq. km reef system; volcanic plus limestone; 180 ft (55 m); twin summits; reefs described and mapped by Phipps and Preobrazhensky (1977); privately owned; visited during turtle season.

Vekai   1 ha in reef system of 8 sq. km; low limestone islet on inner edge of reef; turtle fishing; seabirds.

Tuvuca   (X) 5 sq. mi. (13 sq. km) in 21 sq. km reef system; high limestone, undercut cliffs along much of coast and inland; central basin with several depressions; 800 ft (244 m); barrier reef with narrow lagoon; fringing reef around north; reefs described and mapped by Phipps and Preobrazhensky (1977); geology described by Rodda (1981).

Cicia   (X) 13 sq. mi. (34 sq. km) in reef system of 49 sq. km; mainly volcanic; 540 ft (165 m) with rim of limestone cliffs; fringing reef, described and mapped by Phipps and Preobrazhensky (1977).

Tavunuku-i-vanua   55 ha patch reef with 4 ha sand cay.

Yaroua   80 ha patch reef with 12 ha sand cay.

Reid Reef   almost circular central lagoon; reef system 89 sq. km; 3 limestone islets each with fringing reef; Late-i-Viti (18 m), Late-i-Tonga (15 m) and Booby Rock (4 m); surveyed by Fisheries Dept Giant Clam project (Lewis *et al.*, 1985).

Nayau   (X) 7 sq. mi. (18 sq. km) in reef system of 26 sq. km; limestone with some areas of volcanic rock exposed beneath; 179 m; hills around central basin, undercut sea cliffs; fringing reef.

Bukatatanoa Reefs   barrier reef complex awash at high tide; second largest reef system in Lau covering ca 750 sq. km (over 200 km of reef); main reef system has limestone islet, Vanua Masi (15 ha, 24.4 m); also limestone Bacon I. on barrier reef; surveyed by Fisheries Dept Giant Clam project (Lewis *et al.*, 1985).

Lakeba   (X) (*see separate account*).

Aiwa   121 ha land area in 59 sq. km reef system; 2 narrow limestone islets, ca 1.6 km long; alt. 200 ft (61 m); low bluffs and sheer cliffs; surveyed by Unesco/UNFPA project; reefs, molluscs and fish described by Salvat *et al.* (1976); no true fringing reefs; submerged 12 km barrier reef on southern side; fishing.

Oneata   (X) 4 sq. km in 95 sq. km reef system; low limestone ridge, 160 ft (49 m) with some volcanic rocks in extensive lagoon; swampy central depression; mushroom rocks; joined by reef to Loa; used as U.S. observation post.

Cacau Lekaleka   barrier reef with enclosed lagoon.

Moce   (X) 4 sq. mi. (10.4 sq. km) in 55 sq. km reef system; volcanic cone 590 ft (180 m); wet; fringing reef.

Karoni   40.4 ha in same reef system as Moce; limestone, 120 ft (37 m); on reef attached to main barrier reef.

Komo   (X) 1.53 sq. km in 34 sq. km reef system; narrow volcanic ridge, 270 ft (82 m), within lagoon; 8 ha limestone rocky islet (Komo Driki) off shore on separate reef to west.

Olorua   16.2 ha in 6 sq. km reef system; small steep island; remains of volcanic cone, 250 ft (76 m).

Vanua Vatu   (X) 4.1 sq. km on reef system of 12.5 sq. km; circular limestone, 102 m; 1 small village, known for "sacred" red prawns.

Tavunasici   40 ha in 4 sq. km reef system; limestone island ca 800 m diameter with low cliffs, coral-sand beaches, beach rock; 61 m; encircled by barrier reef with lagoon 400-800 m across; surveyed by Unesco/UNFPA project; reefs described by Salvat *et al.* (1976).

Vuaqava   8.1 sq. km in 17 sq. km reef system; limestone atoll; 300-350 ft (91-107 m) sea cliffs, central basin with tidal saltwater; fringing reef; lake used as turtle pen by Kabara islanders.

Kabara   (X) 12 sq. mi. (31 sq. km) in 44.5 sq. km reef system; basin-shaped; limestone; 300-350 ft (91-107 m) cliffs and 470 ft (143 m) volcanic hills; central basin 100 ft (91 m) pitted and karstic; reefs described by Galzin *et al.* (1979) and Salvat *et al.* (1976); no fringing reef; continuous barrier reef, emergent at low tide on east, surrounds lagoon ca 500 m wide, 2-3 m deep; lagoon filling with sand and outer reef dying (98% of barrier reef dead) presumed to be result of recent uplift; surveyed by Unesco/UNFPA project.

Marabo   (X) 1.29 sq. km in 3.5 sq. km reef system; oval limestone island, 49 m; steep undercut cliffs; no landing places.

Yagasa Cluster   4 limestone islets in 90 sq. km reef system; under-cut cliffs; barrier reef surrounds islets; surveyed by Unesco/UNFPA project; reefs, molluscs, fish described by Salvat *et al.* (1976):

-   *Yagasa-levu*   2.6 sq. km, 119 m; rising in terraces, inaccessible;

-   *Navutu-i-loma*   limestone, many mushroom islets;

-   *Navutu-i-ra*   270 ft (82 m); low cliffs and mushroom islets; attached by fringing reef to barrier reef;

-   *Yavuca*   very rugged, inaccessible; attached by fringing reef to same barrier reef as Navutu-i-ra.

Naievo (Nayabo)   25 ha limestone islet with sand flat, surrounded by reefs covering 1.4 sq. km; 3 m.

Namuka-i-lau   (X) 5 sq. mi. (13 sq. km) in 50 sq. km reef system; narrow limestone ridge around oval island; 260 ft (79 m); fairly well developed barrier reef around most of coast, except north; surveyed by Unesco/UNFPA project; reefs, molluscs, fish described by Salvat *et al.* (1976).

Fulaga   (X) 18.5 sq. km on reef system of 55 sq. km; crescent of limestone surrounding lagoon; 79 m; mushroom islets; marine cave at Qaranitoa; fringing reef; seagrass beds; land crab breeding area; bays and islets very attractive.

Ogea 90 sq. km reef system with ca 200 islands and islets including:

- *Ogea Driki* 5.23 sq. km; limestone; steep cliffs; 300 ft (91 m); landing place choked with mangroves; seabirds;

- *Ogea Levu and Yanuya* (X) 13.3 sq. km; limestone; 270 ft (82 m); fringed with mushroom islets.

Teteika ca 1.5 sq. km patch reef; limestone with thin veneer of algae and coral (Ladd and Hoffmeister, 1945).

Nukusoge ca 5 sq. km; atoll reef with sand bank normally above high tide level (Ladd and Hoffmeister, 1945).

Vatoa (X) 4.45 sq. km in 20.5 sq. km reef system; limestone; 209 ft (64 m); caves; barrier reef.

Vuata Vatoa circular reef 4 x 3 mi. (6.4 x 4.8 km) with deep lagoon; awash at high tide; moderate Giant Clam populations; surveyed by Fisheries Dept Giant Clam project (Lewis *et al.*, 1985).

Ono-i-lau (X) ca 7.9 sq. km total land area in 80 sq. km reef system; 6 main islands; 3 are volcanic remains of breached crater; one sand islet (Udui); 2 are clusters of limestone islets and stacks; 113 m; over 100 islands in total, within barrier reef.

Vuata Ono ring reef ca 3 x 2 mi. (4.8 x 3.2 km); awash at high tide; Giant Clams very abundant; surveyed by Fisheries Dept Giant Clam project (Lewis *et al.*, 1985).

Tuvana-i-ra 70 ha in 4.5 sq. km reef system; small sand cay with small areas of limestone.

Tuvana-i-colo 38 ha in 2.25 sq. km reef system; atoll, small sand and limestone cay; barrier reefs.

Navatu shallowly submerged reef 3 x 2 mi. (4.8 x 3.2 km) with central lagoon; may be submerged atoll; surveyed by Unesco/UNFPA project; reefs described by Salvat *et al.* (1976); 95% dead on inner part but good corals on outer slope; coral fauna poor; clams surveyed by Fisheries Dept Giant Clam project (Lewis *et al.*, 1985).

Moala Group

Volcanic with reefs (sometimes included in Koro Sea islands).

Matuku (X) 11 sq. mi. (28.5 sq. km); volcanic, old crater rim breached by sea; steep slopes to 1262 ft (385 m) (Davis, 1920); good harbour on west; reef-bound.

Totoya (X) 11 sq. mi. (28.5 sq. km); volcanic eroded crater rim up to 361 m, almost completely closed; central lagoon; barrier reef 2 mi. (3.2 km) off shore; reef encircled (Davis, 1920).

Moala (X) 23.98 sq. mi. (62.1 sq. km); volcanic, 8 peaks; 1535 ft (468 m); rugged with high relief; mangroves; fringing and barrier reef (Davis, 1920).

LOMAIVITI GROUP/KORO SEA ISLANDS

The Lomaiviti group consists of about 15 islands in the Koro Sea, east of Viti Levu, covering a land area of ca 415 sq. km. Most of the islands are of volcanic origin and all the larger islands are high islands. The region has a weak dry season, with mean rainfall of 300 cm p.a. and mean annual temperature of 20°C. The group has very complex reef systems particularly at Wakaya, Makogai, Ovalau, Nairai and Cakau Momo (Dunlap and Singh, 1980).

Gau (X) 140 sq. km; volcanic backbone ridge; 2345 ft (715 m); highland to north; 54% dense forest cover; birds described by Watling (1985); mangroves on north coast; leeward barrier reef (Ryland, 1979); reef encircled (Davis, 1920).

Yaciwa ca 2 ha; limestone cay; emerged reef; lighthouse; seabird roost (Watling, 1985).

Mabulica Reef shallow lagoon.

Nairai (X) 9.4 sq. mi. (24.3 sq..km); volcanic; 1104 ft (336 m) ridge; extensive barrier reefs (Davis, 1920); reef irregularly shaped with elongations to north, south and west; surveyed for Giant Clams (Lewis *et al.*, 1985); Sucuni Levu and Lailai both volcanic islets, part of old, eroded Nairai volcano.

Cakau Momo circular, non-emergent atoll, 1.5 mi. (2.4 km) diameter; reefs; surveyed for Giant Clams (Lewis *et al.* (1985).

Koro (X) 104 sq. km; massive volcano with often steep seaward faces and central plateau; 522 m; fringing reefs.

Batiki (X) 3.6 sq. mi. (9.3 sq. km); volcanic, 600 ft (183 m); mangroves; surrounded by reefs.

Wakaya (X) (*see separate account*).

Makogai (X) 8.4 sq. km; volcanic, 4 summits, 850 ft (259 m) (Davis, 1920); mangroves; linked to Wakaya by barrier reef - unusual figure of 8 shape; fringing reef; surveyed for Giant Clams (Lewis *et al.*, 1985) and has Giant Clam Project marine station (Anon., 1986b); forest badly destroyed (Clunie, in press).

Makodroga 81 ha; high volcanic, 138 m; fringing reef; visited for fish and firewood; turtle nesting; forest badly destroyed (Clunie, in press).

Ovalau (X) 103 sq. km; volcanic, central basin surrounded by peaks, 2053 ft (626 m) (Davis, 1920); reef encircled; reef system to west is maze of fringe and patch reefs; Ladotagane (114 ft (34.7 m), Yanuca Levu (76 ha) (X) and Yanuca Lailai (30 ha) lie off south coast, all are volcanic, hilly and steep.

Moturiki (X) 10.9 sq. km; volcanic; 436 ft (133 m); mangroves; fringing reef.

Cagalai 8.8 ha; sand and beach rock cay south-west of Moturiki; platform reefs (Ryland, 1981).

Leleuvia 10 ha; sand and beach rock cay; fringing and platform reefs (Ryland, 1981).

Naigani (X) 470 acres (190 ha); volcanic with one area of limestone, rounded, 186 m; mangroves; fringing reefs; surveyed for Giant Glams (Lewis *et al.*, 1985); has resort hotel.

VITI LEVU AND ASSOCIATED ISLANDS

Viti Levu is the largest island in Fiji, covering over half the total land area; it is the only high continental island in Fiji.

Viti Levu (X) 4011 sq. mi. (10 388 sq. km); largely volcanic but also sedimentary rocks, 4341 ft (1323 m); central plateau running approx. north/south causing rainshadow to leeward north-west; prevailing winds are south-east trades, causing a wet windward south-east and drier rain-shadow north-west; annual rainfall exceeds 3000 mm over much of the south-east; high rainfall drains mainly through system of four rivers which merge as the Rewa, the outflow of which greatly influences marine ecosystems in the south-east; other important rivers are the Navua, Sigatoka and Ba (Ryland, 1981); mangroves along coast; general desciption of reefs given below; *see separate accounts* for descriptions of Suva Barrier Reef, and fringing reefs on Coral Coast and in Nadi Waters.

Vatialailai 71 ha.

Macuata 40.4 ha; 400 ft (122 m); volcanic.

Nanuyakato 4.9 ha; 240 ft (73 m); volcanic; dry.

Tovu (X) 41 ha; 62 m; volcanic; hilly; part privately owned; Tovulailai (7.8 ha), volcanic, within same reef system.

Malake (X) 1.75 sq. mi. (4.5 sq. km); volcanic; hilly, 775 ft (236 m); mangroves.

Nananu-i-ra 2.69 sq. km; volcanic; resorts on south coast.

Nananu-i-cake 3.0 sq. km; volcanic.

Qoma 16.2 ha; 120 ft (36.6 m); volcanic.

Nukulevu 1.3 ha; sandy islets; 30 ft (9.1 m).

Omini 1.6 ha; calcareous sandstone; fringed with mangroves.

Qata 2.1 ha; calcareous sandstone; fringed with mangroves.

Tawainave 3.2 ha; calcareous sandstone; fringed with mangroves.

Vatulami calcareous sandstone; fringed with mangroves; east of Tawainave.

Telau 3.2 ha; 100 ft (30 m); calcareous sandstone.

Viwa (X) 60 ha; 160 ft (48.7 m); calcareous sandstone.

Bau (X) 8.6 ha; 80 ft (24.4 m); calcareous sandstone; sheltered by barrier reef.

Toberua (X) 1.6 ha; sand; 2-3 m; resort hotel.

Mabualau 2.1 ha; limestone; 3-5 m; seabird rookery; sea snakes (Guinea, 1980).

Vuo (*see account for* Suva Barrier Reef).

Draunibota (X) (*see account for* Suva Barrier Reef).

Serua

Beqa (X) 14 sq. mi. (36.3 sq. km); 439 m, volcanic; indented coast, radially dissected; mangrove; lagoon 390 sq. km with 5 islands and patch reefs; barrier reef undescribed; fringing reef; surveyed for Giant Clams (Lewis *et al.*, 1985); islands in lagoon are:

- *Nanuku (Storm I.)* 0.8 ha; beach rock and sand cay; on barrier reef;

- *Moturiki* 18 ha; volcanic, 320 ft (97.5 m); land crabs;

- *Yanuca* (X) 154 ha; 450 ft (137 m); volcanic;

- *Bird I. (Cakaunibuli)* 4 ft (1.2 m); sand; on Cakau Nisici reef in Beqa Lagoon; seabirds;

- *Ugaga* 3.2 ha; volcanic with fringing reef, south-west of Beqa.

Vatulele (X) 12.19 sq. mi. (31.6 sq. km); limestone with area of volcanic rocks in east; 110 ft (33.5 m); vertical cliffs on west limestone, slope and caves on east; fringing reef on west, extending into barrier reef on north and east; red prawns in pool-cave; three nearby islets:

- *Vatu Levu* 30 ft (9.1 m); limestone; rugged, lies in lagoon of Vatulele Reef;

- *Vatu Lailai* 10 ft (3 m); limestone; lies in lagoon of Vatulele Reef;

- *Vatu Savu* 20 ft (6 m); limestone; rugged.

YASAWA GROUP

The Yasawa group consists of an arc of over 65 islands and islets covering ca 150 sq. km extending west and north-west of Viti Levu from the Nadi Bay region. They are predominantly high volcanic islands (although some of the small islands in the southern Mamanuca sub-group are low islands), and the group has the appearance of a drowned mountain range. There are many fine sandy beaches. The region has a marked dry season running from June to October; mean annual temperature is 25°C, mean rainfall 150 cm p.a. The islands are particularly vulnerable to tropical storms which occur yearly. There are many small fringing and patch reefs but no major reef systems, apart from Ethel Reef, a 20 km long barrier reef 10-20 km to the west.

Mamanuca Group (*see separate account*).

Kuata 1.46 sq. km; 570 ft (174 m); volcanic.

Wayasewa 2.5 sq. mi. (6.5 sq. km); 1160 ft (354 m); volcanic; joined to Waya by sand spit exposed at low tide.

Waya (X) 8.5 sq. mi. (22 sq. km); 1874 ft (571 m); volcanic; rugged, steep west coast in south.

Viwa (X) 200 acres (81 ha); ca 18 m; limestone; emerged coral-algal reefs; bank of coral debris.

Narara 49.7 ha; mainly sandstone.

Naukacuvu 48.5 ha; mainly sandstone; between Narara and Nanuya Balavu.

Nanuya Balavu 80 ha; mainly sandstone.

Drawaqa 67.6 ha; mainly sandstone.

Naviti (X) 34 sq. km; 1272 ft (388 m), volcanic; rugged; mangroves in 2 main bays; many small islets on margin of tidal sand flat along north part of east coast.

Yaqeta (X) 2.82 sq. mi. (7.3 sq. km); 220 m; volcanic; mangrove swamps.

Matacawa Levu (X) 9.5 sq. km; 980 ft (299 m); windward coast with mangroves; volcanic; extensive tidal sand flat in bay to windward.

Tavewa 400 acres (162 ha); 180 m; volcanic; private plantation; minor mangroves in SW.

Nacula (X) 22 sq. km; 270 m; volcanic; mangroves on windward side.

Sawa-i-lau 79.2 ha; limestone peak with sheer rock face; 210 m; lower area of limestone and volcanic rock; marine cave.

Yasawa (X) 32 sq. km; 233 m; volcanic; mangroves.

Yalewa Kalou 27.4 ha; 180 m; slightly eroded crescentic volcanic cone with some sheer rock faces on outer coast.

KADAVU AND ASSOCIATED ISLANDS

Kadavu and its smaller associated islands lie to the south of Viti Levu; they are high volcanic islands and, like the Moala group in the Koro Sea, do not arise from the Fiji Platform. Solo lies in the centre of a lagoon bounded by North Astrolabe Reef, ca 10 km north of the northernmost part of Great Astrolabe Reef. The other islands are either immediately off Kadavu or lie within Great Astrolabe Reef which extends ca 35 km northwards off the north-eastern edge of Kadavu. Most of the islands are described in the account for Great and North Astrolabe Reefs.

Kadavu (X) 408 sq. km; elongated, volcanic, to over 600 m; irregular coastline (Davis, 1920); mangroves; fringing reef, absent at Buke Levu where coastline is cliffs; geology described by Woodrow (1980).

Solo (X) (*see account for* Great and North Astrolabe Reefs).

Great Astrolabe Reef (*see separate account*).

Tawadromu 18.6 ha; volcanic; 70 ft (21 m).

Yadatavaya 50.3 ha; volcanic; 65 ft (19.8 m).

Galoa (X) 3.34 sq. km; volcanic; 114 m; small resort.

Matanuku 570 ft (174 m); linked by sandy flat; mangroves.

Nagigia (Denham) 10 ft (3 m) coral rock; honey combed; emerged reef.

Ceva-i-ra on Conway Reef, most isolated reef in Fiji; 21°46'S, 174°31'E; small sand islet with young coconut palms, probably inundated during very high swells.

(X) = Inhabited

––––––––––––––––

Ryland (1979 and 1981) provides an overview description of the reefs of Fiji and early descriptions are given by Agassiz (1899), Gardiner (1898), Hoffmeister (1925) and Davis (1920). A pilot project was carried out between 1974 and 1976 in the Lau Group, involving surveys of Lakeba, Aiwa, Kabara, Tavunasici, and Namuka-i-lau, the Yagasa cluster and the Navatu Reef, as part of the Unesco/UNFPA project on Population and Environment in the eastern islands of Fiji as a contribution to the MAB (Man and the Biosphere) Program (Anon., 1983; Brookfield, 1978a, b and 1979; MAB, 1977; Salvat *et al.*, 1976 and 1977). Reefs around seven islands in the northern Lau Group and Ringgold Isles were studied by Phipps and Preobrazhensky (1977). The Institute of Marine Resources at the University of the South Pacific in Suva is studying the Laucala Bay region (*see account for* Suva Barrier Reef), and has studied some of the reefs along the Coral Coast of Viti Levu, and reefs around some of the Nadi Waters cays (*see separate accounts*). It also has a research field station on Dravuni in Great Astrolabe Reef off Kadavu (*see separate account*). Reef-related research currently in progress includes: interactions between coral reefs, sea grass beds and mangroves (as part of a project funded by SPREP); monitoring of *Acanthaster*; and ecology and taxonomy of opisthobranchs (Brodie *in litt.*, 10.11.87). Reefs on Malololailai have been the studied by the University of Toronto (*see account for* Nadi Waters and Mamanuca Group).

The following account is based largely on Ryland (1979 and 1981). Reefs are found associated with all the island groups; many of the reef systems are extensive and complex and include barrier, fringing and platform reefs. There are only two shelf atolls: Wailagilala, in the Lau Group, and Qelelevu, east of Vanua Levu. Two types of barrier reefs are found. Oceanic ribbon reefs include the Great Sea Reef, Beqa Barrier Reef, Great Astrolabe Reef and some of the Lau Group barrier reefs; these enclose lagoons or sea areas of normal salinity and their entire character is oceanic, but they are poorly known. The Great Sea Reef, extending for over 200 km, is one of the world's major barrier reefs. It delimits the continental shelf as a westwards-widening wedge from Udu Point, the north-eastern tip of Vanua Levu. Approaching the northernmost islands of the Yasawa chain, it deepens and swings south and then east, around Yalew Kalou, and continues as the Pascoe Reefs which reach the surface intermittently. Between the reef west of Yalewa Kalou and the Yasawa Group is a channel more than 550 m deep, Round Island Passage, which gives access to the enclosed Bligh Water between Vanua

Levu and Viti Levu. To the west of the Yasawa and Mamanuca Groups is a string of reefs, some rather deep but probably still living, enclosing an extensive area of shelf waters, scattered with platform reefs; the boundary reefs west of the Yasawa Group are called the Ethel Reefs, whilst in the south, approaching Viti Levu and rising to the surface again, is the Malolo Barrier Reef. Parts of the shelf margin in the Viwa area may not have much of a reef build-up. The second type of barrier reef includes reefs which may be exposed, with well developed spurs and grooves, as off Suva (*see separate account*), or more sheltered, as off Ba. Such reefs are separated from the mainland by a relatively narrow and shallow lagoon channel of neritic nature, which is generally turbid with surface water of low salinity extending out to or even over the reef.

Fringing reefs are found from the southern end of the Mamanuca Group almost to Beqa, south of Viti Levu, where the 100 fathom (183 m) isobath is adjacent to the coast. The shelf is present but rather narrow south of Bligh Water and around eastern Viti Levu, and fragmented offshore reefs are found here, separated from the mainland only by narrow, shallow lagoon channels. The smaller islands to the south and east of Viti Levu and Vanua Levu have barrier reefs or exposed fringing reefs depending on the closeness to shore of the 100 fathom isobath. Often fringing reefs line one coast but the reef swings off shore as a barrier on the other. Reefs are largely absent off the southern coast of Taveuni where the coast plunges very steeply into the sea.

Platform reefs are restricted to shelf waters and are common inside the Great Sea Reef, including the Yasawa-Mamanuca arc, and within the Mabualau-Ovalau series of barrier reefs off eastern Viti Levu. Their configuration may be influenced by neighbouring reefs, especially where the system is complex, as at Cagalai and Leleuvia to the south-west of Ovalau, or off the Rewa delta. The platform reefs in Nadi waters and towards the Mamanuca Group are more typical (*see separate account*).

The reefs around Viti Levu are best known and are described in Ryland (1979 and 1981). Corals at the mouth of the Rewa River, where the greatest disruption of the fringing reef occurs, have been described by Cooper (1966) and Squires (1962). Islands at the mouth of the river split the flow up into several channels: the Navuloa mouth, the Nasoata mouth and the Nukulau mouth. The Navuloa mouth runs into Bau Water, forming a lagoon-like area with luxuriant corals to the east, which joins Bligh Water to the north of Viti Levu (Cooper, 1966). Its effect on the Nukulau reefs is described in the account for Suva Barrier Reef.

Fringing reefs within the shelter of the barrier reef are different in character from those of the exposed south-west coast. Good examples are seen at Tailevu Point (north-east Viti Levu) and Vuda Point, near Lautoka. Much of a fringing reef within the shelter of a barrier reef consists of a well consolidated flat of barren appearance. Where the reef flat is overlain with sand, as at Vuda, zoanthids may cover large areas but the most typical animals are brittle stars *Ophiocoma scolopendrina* and active, snapping squillid shrimps. Near the reef edge are various zoanthids *Palythoa*, soft corals *Xenia, Sinularia, Sarcophyton*, Organ-pipe Coral *Tubipora* sp., Fire Coral *Millepora tenera* and *M. platyphylla*, and small faviids, *Pocillopora* and *Acropora*. Tailevu Point

Reef is notable for the abundance of *Fungia*. There is no raised crest but the edge of the reef becomes increasingly honey-combed with channels and caves. Coral growth is rich but not particularly diverse. The reef edge is sinuous and drops almost vertically for 2-4 m to a sandy bed. Water tends to be murky. The Tailevu Reef has many gorgonians just below low water mark of spring tides and Pacific Harbour is characterized by an abundance of stinging hydroids *Macrorhynchia* (= *Lytocarpus*) *philippinus*. No detailed analysis of the seaward slope fauna has yet been made (Boschma, 1950; Ryland, 1979). Within the shelter of the offshore reefs, wide grey beaches of gentle slope are deposited. These are backed by mangals dominated by *Bruguiera gymnorhiza* and fringed with *Rhizophora stylosa*. Where the degree of wave exposure is higher, the beaches are whiter and steeper, as at Pacific Harbour. Westward of Beqa and its barrier reef, deep water comes close to shore and the area is known as the Coral Coast (*see separate account*). Other marine ecosystems are described briefly by Bajpai (1979), in particular the Sigatoka sand dunes on the south coast of Viti Levu; these are also described by Kirkpatrick and Hassall (1981).

Cernohorsky (1965), Chapman (1971) and Fowler (1959) describe the mitres (molluscs), marine algae and fishes of Fiji respectively. Parkinson (1982) lists 106 species of bivalve in 25 families recorded in Fiji waters and also lists shells found at dive sites along southern Viti Levu (*see accounts for* Suva Harbour and Coral Coast) and in the Yasawa Islands. Although Fiji lies to the south and east of the main centre of marine mollusc diversity in the Papua New Guinea - Indonesia - Philippines region, it still supports a wide range of species, including many popular with shell collectors such as cones (*Conus* spp.), cowries (*Cypraea* spp.), mitres (*Mitra* spp.), and augers (*Terebra* spp.), as well as olives (*Oliva* spp.), ceriths (*Cerithium* spp.), strombs (*Strombus* spp.) and murexes (*Murex* spp.). Fiji has long been known as a source of the Golden Cowry *Cypraea aurantia* and *Cypraea valentina*, the most valuable shell found in Fijian waters. Fifteen species of zoanthids have been collected around Viti Levu (Muirhead and Ryland, 1981 and 1985) and Muzik and Wainwright (1977) describe five species of sea fan from Fiji. Dilly and Ryland (1985) describe *Rhabdopleura* from Suva and Great Astrolabe Reefs. Kott (1981) describes ascidians of reef flats in Fiji and Ryland *et al.* (1984) describe didemnid ascidians in Viti Levu.

Dunlap and Singh (1980) list endemic marine fish, such as *Atherina ovalana* and *Engyprosopon fijiensis* known only from the Lomaiviti waters. Sea snakes on the islands Mubualau, Namuka and Sausau have been described by Guinea (1980). Green Turtles *Chelonia mydas* and Hawksbills *Eretmochelys imbricata* may be widespread although not common, but little information is available (Bustard, 1970; Hirth, 1971). The waters of the Lomaiviti group are breeding and calving grounds for Humpback *Megaptera novaeangliae* and Sperm Whales *Physeter macrocephalus* (Dunlap and Singh, 1980). Mangrove resources are described by Baines (1979 and 1980a) and Lal (1983); the hybrid mangrove *Selala* is of interest (Tomlinson, 1978). Seabird colony distribution is described in Garnett (1984); colonies on Namenalala and the Ringgold Isles are discussed by Clunie (in press), and seabirds (including the recently rediscovered Fiji Petrel *Pseudobulweria macgillivrayi*) of Gau by Watling (1985).

Reef Resources

Early publications on the Fijian fisheries include Deane (1910), Hornell (1940), Levy (1940/44) and Toganivalu (1914). Subsistence fishing still predominates. In 1984, subsistence harvest was estimated at ca 14 800 tonnes, with commercial fisheries at ca 5800 tonnes and industrial fish production (essentially tuna) at 6000 tonnes (De Backer, 1985). Subsistence fisheries traditionally take place in inshore (mainly reef) areas. Government loans may be granted for fishery enterprises. Fishery resources are briefly reviewed by Baines (1982b) and Kearney (1982) provides an assessment of skipjack and baitfish resources.

Trochus, mother-of-pearl, bêche-de-mer and giant clams are important commercial resources (Baines, 1982b; Burrows, 1940/44; Dunlap and Singh, 1980; King and Stone, 1979; Lewis *et al.*, 1984 and 1985; South Pacific Commission, 1980; Turbet, 1940/44). Almost all trochus and mother-of-pearl is exported; in 1983 export of the former amounted to 343 tonnes, of the latter 23 tonnes, making Fiji one of the largest producers in the South Pacific region for both these. A factory has reportedly been set up to produce button blanks from trochus. In 1983, 33 tonnes of bêche-de-mer were exported and ca 11 tonnes marketed locally; species consumed locally (mainly that known as "dairo") differ from those exported ("sucuwalu", "loaloa"). Giant Clams are an important subsistence and local commercial item in many areas; it is estimated that ca 17 tonnes passed through municipal markets and other outlets in 1983 (Anon., 1984). A Giant Clam fishery project has been underway since 1984 with the Fisheries Division of the Ministry of Primary Industries (Lewis *et al.*, 1984 and 1985). Local and export markets for lobsters (of which five species occur in Fiji) are reportedly strong, although commercial exploitation has apparently not been encouraged; an estimated 29 tonnes was marketed in Fiji in 1983, mostly through hotels and restaurants (Anon., 1984). Turtle meat is in heavy demand, for consumption both by native Fijians and in tourist restaurants; eggs are also collected and in 1970 there was reported to be a major shell carving industry, using both Hawksbill and Green Turtle shells (Bustard, 1970; Hirth, 1971). Black coral is the basis of a small local cottage industry, selling to tourists; Fiji's black coral resources may be considerable but at present are only lightly exploited (Anon., 1984). The Seaking Trading company runs an ornamental coral export trade, operating mostly in Bau waters (Brodie *in litt.*, 10.11.87). In the 1970s a single exporter of aquarium fish was operating (Fiji Biomarine Ltd.) but had become inactive by 1984. In 1978, 24 000 live fish were exported (Dunlap and Singh, 1980). An aquarium fish company, operating from Pacific Harbour on Viti Levu, has however been active since 1985 (Brodie *in litt.*, 10.11.87). Parkinson (1982) provides a commercial catalogue of specimen mollusc shells found in the country. Currently, shell collecting by villagers is generally of the larger, more spectacular species for sale to tourists, while the smaller valuable species sought by specialist shell-collectors are largely neglected. Marine algal resources are described in Prakash (1987).

Since 1982 tourism has been the largest source of foreign exchange, in gross terms, although in net terms it still comes second to sugar (De Backer, 1985). In 1985 there were an estimated 250 000 tourists. Much of the tourism is concentrated on Viti Levu, along the Coral Coast, at Suva, Deuba, Nadi, at a new major development at Pacific Harbour, west of the Navua River, and in the Yasawa Group. Many of the small islands are being developed as resorts such as Toberua, north-east of Suva, and Nananu-i-ra adjacent to Storm Island (north Viti Levu) (*see also account for* Nadi Waters and Mamanuca Group). There are diving centres at Suva, Sigatoka, Pacific Harbour and Lautoka on Viti Levu, at Matei on Taveuni, and on Mana and Tai ("Beachcomber") in the Mamanucas, catering mainly for expatriates and tourists (Brodie *in litt.*, 10.11.87; Dunlap and Singh, 1980). The lagoon at Beqa is rated as one of the world's top diving areas, Frigate, Sulphur and Cutter Passages being particularly popular sites. Fish Patch, off Suva, is another popular diving locality on account of the high fish densities and interesting corals found there (Evans, n.d.). Tourism on Vanua Levu is still at a low level and is concentrated around Savusavu and eastwards. There is a diving centre at Mua Beach and cruise ships occasionally visit Savusavu. There is some tourism on the west coast of Taveuni. The Kadavu Group has very little tourism although there is a small resort on Galoa (Brodie *in litt.*, 26.11.87). Other areas, particularly the outer islands, such as the Lau Group, with diverse reef resources, have virtually no tourism at present (Dunlap and Singh, 1980). A dive operation (Marine Pacific) using a large boat is to begin dive tours in the Lau Group in 1988 (Brodie *in litt.*, 10.11.87).

Disturbances and Deficiencies

Hurricanes cause periodic damage to the Fijian reefs. Following one in 1965, reefs and large numbers of fish were killed in the mouth of the Navuloa, in the southern part of Bau water, on the east coast of Viti Levu, probably due to freshwater inundation (Cooper, 1966). Although the Rewa delta was flooded, the barrier reef on either side of Nukulau was not affected. In 1975 Hurricane Val damaged reefs in the Lau group but Phipps and Preobrazhensky (1977) found evidence of rapid coral regeneration. Cyclone Fay in 1978 caused extensive damage to reefs at the northern end of Taveuni (Dunlap and Singh, 1980). Cyclone Wally severely damaged the fringing reef at Deuba near Pacific Harbour in 1980 (Ryland *et al.*, 1984), as well as other reefs in this area and there were extensive mud slides and siltation (Guinea *in litt.*, 6.3.85). Reefs on Malololailai suffered some visible damage in a 1983 hurricane (*see account for* Nadi Waters and Mamanuca Group). Two hurricanes (Eric and Nigel) hit Fiji within 48 hours of each other in January 1985 and a further one (Hina) in March 1985. Terrestrial damage was severe, particularly in western Viti Levu and there was some damage to reefs in the Mamanuca Group (*see separate account*). Large scale flooding following exceptional rainfall in early 1986 caused massive damage to reef corals in the Suva area (*see account for* Suva Barrier Reef).

Other causes of natural damage include earthquakes such as the earthquake of November 1979 which caused damage on Taveuni and the Natewa Peninsula, and tsunamis, such as that resulting from the Suva earthquake of September 1953 (Everingham, 1987). In the 1960s, *Acanthaster* infestation was considered a serious enough problem to warrant permanent monitoring (Owens, 1971; Robinson, 1971) and studies were carried out at Natadola and Wakaya Island (*see separate accounts*). However by 1978/80 infestation was not serious (Ryland *in litt.*, 25.8.87). More recent work has been carried out on *Acanthaster* on the Suva Barrier Reef (*see separate account*).

The rugged interior of the main islands confines extensive agricultural production and urban development to a fertile but vulnerable coastal fringe of flatter land. Major environmental problems in coastal waters (SPREP, 1980; Lal, 1984) include oil pollution (especially in Suva and Lautoka); sand dredging affecting turtle grass beds and fauna (*see account for* Suva Barrier Reef); overfishing (e.g. turtles); insensitive placement of groynes, jetties or reef blasting interfering with water flow and sand deposition; depletion of mangrove areas through urban growth and drainage schemes; depletion of coral communities through tourist resort sewage outfalls and, to some extent, collection of live organisms. Perhaps most important is the threat of sedimentation from logging activities and coastal development. For example, in 1982 during construction of Queen's Road (the Suva-Nadi highway) along the south coast of Viti Levu, soil was washed from one site into Serua Bay, smothering mangroves and affecting adjacent coral reefs (Lal, 1984). Problems caused by the rapid development of the tourist industry along the Coral Coast and at Nadi are discussed in separate accounts and by Baines (1977b and 1982a). Siltation resulting from construction of the Pacific Harbour complex of luxury villas caused the death of corals in the small fringing reef at the mouth of Qaraniqio Creek (Bajpai, 1979) and construction of groynes and artificial harbours has increased sedimentation loads in a number of areas along the Coral Coast (Lal, 1984). The reef around Cobia, one of the Budd Islets, in the Ringgold Isles, is potentially threatened by soil erosion and runoff, a result of overgrazing by introduced goats (Clunie *in litt.*, 16.4.85).

Coral sand is an essential raw material for cement production in Fiji. Penn (1981 and 1982) found that coral sand extracted from seagrass reef flats was causing destruction of seagrass beds within the dredge pit areas (*see account for* Suva Barrier Reef). This could ultimately have an impact on reefs as well.

The disposal of tailings from copper mining operations at Namosi in the offshore area south of Viti Levu could pose a threat but environmental impact studies will be carried out before such activities commence. It was envisaged that tailings should be disposed of in deep water to minimise effects on the reef (Rodda *in litt.*, 14.9.87). Petroleum exploration is being carried out extensively in Fijian waters in the Bligh Water area, the Great Sea Reef area, the Yasawas area and the Lomaiviti Group area. Efforts are being made to minimize effects of drilling and oil spills in the course of this activity (Dunlap and Singh, 1980; Richmond, 1979). Clunie (*in litt.*, 16.4.85) and Lal (1984) point out the potential for increased pollution problems near large towns and industrial centres and from sewage and waste disposal near tourist resorts especially in the Mamanuca Group and along the Coral Coast.

Over-fishing is generally not a problem in the rural areas except near population centres (Lal, 1984). Depletion of shellfish in localized areas and possible over-collection of aquarium fish are growing problems. Bait fishing is intensive in some lagoons (Clunie *in litt.*, 16.4.85) but this is unlikely to have any long term effect. There is some collection of corals and shells for tourists but this does

not appear to be intensive. Turtles (taken by setting nets or by divers with spears) are being depleted. Dynamite fishing has been reported (Lal, 1984) as well as fishing with poisons such as derris (Ryland *in litt.*, 25.8.87). Ciguatera poisoning is a problem in Bau Water (Cooper, 1966).

Giant Clams have been seriously depleted. *T. gigas* and *Hippopus hippopus* are now very rare or locally extinct; for example *T. gigas* was fished at Wailagilala fifteen years ago but is no longer found there. *Tridacna maxima*, *T. derasa* and *T. squamosa* are found in decreasing order of abundance. The Bulia and Dravuni reef systems near Kadavu and those at Vanua Balavu (Lau Group) are almost devoid of Giant Clams due to artisanal fishing. Bukatatanoa and Reid reefs (east of Lakeba, Lau) have low densities of clams and evidence of foreign poaching. Vuata Vatoa and Navatu reefs (south Lau) have moderate clam populations but these are not suitable for exploitation. Vuata Ono (south Lau) is very isolated and has a high density of clams in a limited area. The area around Suva is devoid of clams, possibly as much to do with water turbidity as over-exploitation. In general, isolated reefs harbour a larger population of clams than those with associated islands (Adams *in litt.*, 7.2.85; Lewis *et al.*, 1984 and 1985).

Legislation and Management

Extended family groups called "mataqalis" have traditional fishing rights over reefs in Fijian waters from mean high water mark to the outer edge of the associated fringing or barrier reef; the boundaries to these areas are generally seaward extensions of boundaries on land, although rights over the marine areas are rights of use rather than ownership. Fishing rights boundaries have been recorded for all the islands. In the Lau Group, one "clan" or "ngonendau" in each grouping of villages consisted of fishing specialists, the head of which was in charge and had the knowledge and power to declare restrictions on the times and areas of fishing (Baines, 1982a and b; Eaton, 1985; Kunatuba, 1983; Siwatibau, 1984). Under British tidal law, however, all land below MHWM and extending outwards to the ocean edge of the outer reef is legally defined as the property of the nation. Nevertheless, in fulfilment of a pledge by Governor Sir Arthur Gordon (1881) "all" reefs and shellfish beds have been assigned by the Native Lands and Fisheries Commission to members of the indigenous Fijian race for purposes of subsistence fishing and harvesting. They may be licensed to fish commercially and have the right to permit or refuse applications for commercial fishing by others. The Minister of State and Lands and Mineral Resources has been empowered to waive these rights for development of special projects by the Government (Dunlap and Singh, 1980), but for any development which can be shown to have adverse effects on fisheries or fishing rights, compensation has to be paid to the native fishing rights owners after an environmental impact assessment has been carried out (Lal, 1984). In 1978, the Fijian Fisheries Commission was appointed to draft terms of reference for a study of fishing rights but this had not been completed by 1985.

There is at present an inadequate legislative infrastructure for the conservation of critical marine habitats and environmental controls are dispersed amongst several different acts and regulations. Aspects of legislation and management relating to the coastal zone are briefly described in SPREP (1980).

The Environmental Management Committee (EMC), which has representatives of most government departments, the National Trust for Fiji and the University of the South Pacific, advises the government on environmental matters, and reviews development projects and environmental impact assessments of projects (Brodie *in litt.*, 26.11.87). Mining is the only activity for which an environmental impact assessment is necessary by law (Lal, 1984) and provision for environmental protection in the Mining Act is reasonable, but enforcement and monitoring are generally weak (Brodie *in litt.*, 10.11.87). Environmental impact assessments are also required for some hotel constructions and are sometimes passed to the EMC for evaluation (Brodie *in litt.*, 11.10.87).

Leases for coral sand mining are granted by the Director of Mineral Development for only short periods at a time to assist in natural rehabilitation of seagrasses and reefs. Rehabilitation efforts in the form of artificial reefs and seagrass transplants are under way. Management plans include the estimation of potential extractable sand resources, prediction of changes in local lagoonal hydrography and careful design of dredging operations (Lal, 1984).

The Town Planning Act 1946 (Chap. 109) provides for the preparation of Town Planning Schemes including the conservation of natural beauties of the area including lakes, banks or rivers, foreshore or harbours, and other parts of the sea, hill slopes, summits and valleys. Act No. 3 1974 Harbour Act (Chap. 160) enables heavy penalities to be imposed for the pollution of harbour and coastal waters but this has never been used. The Public Health Act (Chap. 91) controls sewage discharge.

The 1978 Fisheries Act (Chap. 158) includes regulations prohibiting the use of dynamite and other explosives and poisons (including derris, rotenone, plant extracts and chemicals) for fishing. It provides minimum size limits for harvestable *Scylla serrata*, turtles, Trochus, and mother-of-pearl, completely protects *Charonia tritonis* and *Cassis cornuta* and prohibits the taking of eggs and killing of turtles during the 4-month nesting season. There are at present no legal controls on clam fishing, except for the licensing of fishermen and vessels. In 1984 a memorandum from the Minister for Primary Industries setting out a broad policy for the exploitation of living sedentary reef resources was approved by cabinet. It contains preliminary guidelines for the controlled exploitation of Giant Clam (vasua), lobster (urau), bêche-de-mer (dri), trochus (sici), mother-of-pearl (civa), corals (including Black Coral), collectors shells, aquarium fish and seaweeds. For clams, these include the restriction of harvesting to Fijian nationals with the written approval of the owners of the customary fishing rights, minimum size limits, monitoring of catches, licensing of exports and refusal of licences unless all of the clam is utilized, and a fair price to be paid to the local supplier. There are similar regulations for lobster stocks but those for the other resources are generally less stringent as there is believed to be less danger of over-exploitation (Anon., 1984).

The National Trust for Fiji Ordinance 1970 (Chap. 265) provides for the permanent preservation of lands (including reefs) for the benefit of the nation; the protection and augmentation of such lands and their surroundings and to preserve their natural aspect and features; to protect animal and plant life; and to provide for the access to and enjoyment by the public of such lands; and for the development of parks and reserves. The Forestry Ordinance (1953) provides for establishment of Nature Reserves within Reserved Forest Areas (Dahl, 1980).

There are no marine reserves at present. Dunlap and Singh (1980) list a number of recreational areas which have been established but few of these include reefs. There are a few protected islands but the surrounding reefs are unprotected. On Viti Levu, the islands of Draunibota (Cave), Labiko (Snake) and Vuo (Admiralty) in the Bay of Islands are protected. Yadua Tabu off the west coast of Vanua Levu is managed by the National Trust for Fiji in cooperation with the landowners as a sanctuary for the endangered Fiji Crested Iguana *Brachylophus vitiensis* (Singh, 1985). A mangrove management plan has been completed and is being acted on (Brodie *in litt.*, 10.11.87).

Recommendations

One of the principal aims of the Government's Development Plan Eight for the period 1980-85 was: "to establish a system of regional and national parks in Fiji which will serve a dual purpose; first the provision of managed outdoor recreation facilities for the local population and, second, the conservation of areas of outstanding beauty or special scientific (botanical or marine) interest". It was stressed that National Parks and Reserves needed to be established in a manner which avoided conflict with development schemes. To minimize costs, it was recommended that where possible protected areas should be established by leasing Crown Lands, and negotiating for rights with native land owners who were willing to consider compensation in the form of economic benefits (such as employment and income) rather than outright cash payments. The initial programme was to be concentrated in Viti Levu, the region where the natural environment was considered to be under greatest pressure and where the need for provision of outdoor recreation was greatest; priority was also to be given to sites near existing urban areas. The particular need to preserve mangroves, through the creation of coastal marine reserves and the incorporation of mangrove areas into coastal parks where necessary, was stressed.

Much of this policy was contained in a plan for a National Parks and Reserves System for Fiji, submitted to the government by the National Trust for Fiji in 1980 (Dunlap and Singh, 1980). The report made a number of further recommendations for the management of the country's resources including marine ones, reviewed the functions of government agencies, statutory authorities and other institutions most concerned with planning and use of Fiji's natural resources and proposed methods for dealing with communally owned lands. The plan proposed a network of protected areas and a suggested timetable for setting them up. The following have marine components:

* see separate account.

T = terrestrial, M = marine, P = park, R = reserve.

Viti Levu, Lomaiviti, Kadavu and Yasawas

- *Natadola Bay: 1980; TP for public recreation.
- *Makuluva and reefs: 1981-85; MR (Suva Barrier Reef account).
- *Suva Barrier Reef and cays: 1981-85; MP and MR.
  Bird I., Beqa Lagoon: seabird rookery.
  Vatu-i-ra: seabird rookery.
  Mubualau: seabird rookery.
  Beqa Lagoon: 1986; MP and MR.
- *Nadi Bay reefs: 1981-85; MP and MR.
- *Coral Coast reefs: 1981-85; MP and MR.
- Makogai and reefs: 1981-85; TP and TR.
- *Wakaya and reefs: 1986; historical and MP.
- Cakau Momo (Horseshoe Reef): 1986; MP.
- *North Astrolabe Reef: 1981-85; MR.
- *Great Astrolabe reef: 1986; MP and MR.
- *Yabu: seabird rookery (Gt and North Astrolabe Reefs account).
- Taqa Rocks: seabird rookery.
- *Mamanuca Group: 1986; MR (Nadi Bay and Mamanuca Group account).
- *Malamala and reefs: 1986; TP, TR, MP, MR, (Nadi Bay and Mamanuca Group account).
- White Rock: seabird rookery.

Vanua Levu and Taveuni

- Namenalala and reef: 1986; TR and MR; also seabird rookery.
- Nanuku islets: 1981-85; turtle nesting beach reserve.
- Great Sea Reef: 1986; MP and MR.
- Qelelevu Atoll: 1986; TR and MR.
- Rainbow Reef: 1986; MP and MR.
- Rotuma: 1986; TP and MP.

Lau Group

- Wailagilala: 1986; TR and MR.
- Qilaqila (Bay of Islands): 1981-85; TP and MP.
- Fulaga Bay of Islands: 1986; MP and PR; seagrass.
- Sovu islands: seabird rookery.
- Cakau Lekaleka Barrier Reef: 1986; MP and MR.
- Nukutolu islets and reefs: 1981-85; MR.

There is also a critical need for a beach park in the Deuba area on Viti Levu for use by people in the Suva-Nausori region (Dunlap and Singh, 1980). The plan contained a detailed draft management plan for the Natadola Bay area, which was proposed as Fiji's first national park (*see separate account*). This has not been pursued, and the Garrick Memorial Reserve, an inland area of rain forest in Viti Levu is now being developed as the first national park. In 1985 there were plans that Tai and Luvuka and their surrounding reefs in Nadi Bay should become the first marine park (*see separate account*).

However the National Trust for Fiji, the parastatal body chiefly charged with the setting up of reserves (excepting forest reserves), is hampered by insufficient funds and manpower: up to 1985, at least, it received 45 000 Fiji Dollars per annum (its intended budget is 90 000 Fiji Dollars per annum) (Anon., 1985a). Legislation for the establishment of protected areas is to be contained in a National Parks and Reserves Act, as proposed by the National Trust for Fiji, but this has not yet been drafted. A new Town and Country Planning Act, with improved provision for environmental protection, has also been

proposed or drafted, but details are lacking. A proposal has also been drawn up for the development of a National Conservation Strategy for Fiji (Prescott-Allen, 1986), which is currently being revised (Fernando *in litt.*, 2.11.87).

Dahl (1980) recommended some additional reef areas for protection:

- Leleuvia: sand cay, fringing reef.
- Ra coast: fringing reef.
- Mana: fringing reef.
- Yasawa-i-rara: fringing reef.
- Makodroga: fringing reef.
- Vuaqava: marine lake.
- Yasawa: marine caves.
- Vatulele: red prawn pool-cave.
- Koro: red turtle pool-cave.
- Balolo Pt, Ovalau: balolo rise area.
- Moturiki: land crab breeding area.
- Rewa delta: for seagrasses and *Rhizophora*.

A number of other proposed protected marine areas have also been reported. Several informally protected areas for the tagging of clams will be set up in the near future (e.g. Nairai, Koro Sea), but these will have no official status (Adams *in litt.*, 7.2.85). In 1985 it was reported that the possibility was being investigated of declaring the reefs around Yadua Tabu off western Vanua Levu a reserve, with the cooperation of the traditional landowners (Singh pers. comm., 1985). Clunie (in press) recommends that the rich marine life of the lagoon at Cobia in the Budd Reef Islets (Ringgold Islands) should be surveyed, and that action should be taken to preserve this island. Recommendations are also made for the protection of Vetauua, Nukubasaga and Nukubalati as seabird reserves.

The biology of the three sea snakes, *Laticauda colubrina*, *L. laticaudata* and *Hydrophis melanocephalus* requires close investigation to ensure that the proposed marine reserves are effective to protect them (Guinea, 1980). All three species occur in Laucala Bay and Bau waters.

A mangrove management plan has been completed (Brodie *in litt.*, 10.11.87); earlier recommendations for mangrove management are given in Baines (1979). The importance of mitigating effects of coastal tourist development, both in minimising environmental impacts and ensuring that local people can still retain access to foreshore and inshore areas is discussed in Baines (1982a).

In general, adequate legislative provision for the conservation of critical marine habitats is still lacking. Procedures have reportedly been prepared for the establishment of marine parks and reserves which take into account traditional fishing rights. Compensation (for loss of fishing revenues) may be an important consideration in the conservation of marine species (Anon., 1985a). Projects of high priority for further research include coral reefs, turtles, seabird colonies, fish and shellfish resources (SPREP, 1980). New approaches to the management of subsistence fisheries are discussed in Baines (1982b).

Parkinson (1982) has made recommendations for the development of a small scale specimen shell industry. A small group from Central Division should be trained in the collection and marketing of shells, to act as a nucleus for training other groups. A vessel should be obtained and fitted out to be used for extension work on shells and related products in the islands. A draft extension booklet entitled "Collecting shells as a business" has been prepared, which includes basic conservation measures, such as only collecting perfect adult specimens, not destroying coral and replacing in their original position any boulders which have been moved.

Additional general recommendations for management of the marine habitat are given in Anon. (1985a). The efforts of the Fisheries Division to protect endangered or seriously depleted species (e.g. turtles) should be re-enforced, through the preservation of critical habitat and development of measures to protect these species during that part of their life cycle when the habitat is being used. Inshore marine areas near population and tourist centres in need of protection or which provide recreation opportunities should be identified. The latter should be chosen only where there is reasonable assurance that pollution or other external conditions over which a park authority has little, if any control, will not adversely affect marine life in the future. Emphasis should be given to the protection of marine habitats containing species such as aquarium fish, soft corals and sponges, etc., and to promoting research on these species by organisations such as the Institute of Marine Resources, University of the South Pacific, to assist the Fisheries Division with their management. Reviews of the Unesco/UNFPA project in the Lau Group are provided in Anon. (1983), Brookfield (1983) and Bayliss-Smith (1983). Planning recommendations of the project were geared mainly to improving economic development. Account was taken of the vulnerability of the islands to cyclone damage, and diversification of production was proposed as a means to reduce this vulnerability. The best management methods were found within the indigenous system. It is not known if any recommendations were followed up.

References

* = cited but not consulted

Agassiz, A. (1899). The islands and coral reefs of Fiji. *Bull. Mus. Comp. Zool. Harv.* 33: 1-167.
Anon. (n.d.). Day trip to Dick's Place. Brochure.
Anon. (1983). Programme on Man and the Biosphere (MAB). Expert consultations on Project 7: Ecology and rational use of island ecosystems. Final report. *MAB Report Series* 47. 39 pp.
Anon. (1984). Broad policy exploitation of living sedentary reef resources - preliminary consideration. Draft Cabinet Memorandum by the Minister for Primary Resources.
Anon. (1985a). Country report - Fiji. *Report of the 3rd South Pacific National Parks and Reserves Conference, Apia* 3: 76-92.
Anon. (1985b). Summary record of proceedings of ministerial and technical sessions. *Report of the 3rd South Pacific National Parks and Reserves Conference, Apia* 1. 95 pp.
Anon. (1986a). Fiji coral communities. *Earthwatch News* 7(4): 21.
Anon. (1986b). Fisheries Division, Fiji. *Clamlines* 1: 10-11.
*Baines, G.B.K. (1977a). Draft outline of a study of the impacts of current and proposed coral sand dredging in the Suva area. Unpub. ms. 4 pp.
*Baines G.B.K. (1977b). The environmental demands of tourism in coastal Fiji. In: Winslow, J.H. (Ed.), *The Melanesian Environment*. Australian National University Press, Canberra. Pp. 448-457.
Baines, G.B.K. (1979). Mangroves for National Development: A report on the mangrove resources of Fiji. Prep. for Government of Fiji. Griffith University, Australia.
*Baines, G.B.K. (1980a). Mangrove resource management in a Pacific Island nation: Fiji. Proc. Asian Symp. on Mangrove Environment, Research and Management, Kuala Lumpur, Malaysia.
Baines, G.B.K. (1980b). Coastal tourism development in Oceania. In: *Marine and Coastal Processes in the Pacific: Ecological Aspects of Coastal Zone Management*. Unesco, ROSTEA, Jakarta. Pp. 191-207.
Baines, G.B.K. (1982a). South Pacific Island Tourism: Environmental costs and benefits of the Fijian example. In: Rajotte, F. (Ed.), *The impact of tourism development in the Pacific: Papers and proceedings of a Pacific-wide conference held by satellite*. Environmental and Resource Studies Programme, Trent University, Ontario.
Baines, G.B.K. (1982b). Pacific Islands: Development of coastal marine resources of selected islands. In: Soysa, C., Chia, L.S. and Coulter, W.L. (Eds), *Man, Land and Sea: Coastal Resource Use and Management in Asia and the Pacific*. The Agricultural Development Council, Bangkok, Thailand. Pp. 189-198.
Bajpai, I. (1979). The Coastal Ecosystem - Man's Impact. Keynote Address. *Proc. 2nd South Pacific Conference on National Parks and Reserves, Sydney* 1: 140-147.
Bayliss-Smith, T. (1983). Modernisation and island ecosystems: Insights from the Unesco/UNFPA eastern Fiji project. Annex 4. Programme on Man and the Biosphere (MAB). Expert consultations on Project 7: Ecology and rational use of island ecosystems. Final report. *MAB Report Series* 47: 29-34.
Birkeland, C. (Ed.) (1987). Comparison between Atlantic and Pacific tropical marine coastal ecosystems: Community structure, ecological processes, and productivity. *Unesco Rep. Mar. Sci.* 46. 262 pp.
*Boschma, H. (1950). Notes on the coral reefs near Suva in the Fiji Islands. *Proc. K. ned. Akad. Wet.* (C)53(3): 1-6.
*Brodie, J.E. and Brodie (1987). The state of Naitauba Island fringing reef. *Inst. Nat. Res. Tech. Rept* 87/4. Univ. S. Pacific, Suva. 17 pp.
Brodie, J.E. and Ryan, P.A. (1987). Environmental monitoring of the Treasure Island marine sewage outfall. *Environmental Studies Rept* 36. Institute of Natural Resources, Univ. S. Pacific, Suva. 23 pp.
*Brookfield, H.C. (Ed.) (1977). Koro in the 1970's: Prosperity through diversity? Island repts. no. 2, Unesco/UNFPA Population and Environment Project in Eastern Islands of Fiji, sponsored by Man and the Biosphere (MAB) Programme, Project 7: Ecology and Rational Use of Island Ecosystems. Australian National Univ., Development Studies Centre, Canberra, Australia.
*Brookfield, H.C. (Ed.) (1978a). Taveuni: Land, population and production. Island repts. no. 3, Unesco/UNFPA Population and Environment Project in Eastern Islands of Fiji, sponsored by Man and the Biosphere (MAB) Programme, Project 7: Ecology and Rational Use of Island Ecosystems. Australian National Univ., Development Studies Centre, Canberra, Australia.
*Brookfield, H.C. (Ed.) (1978b). The small islands and the reefs. Island repts. no. 4, Unesco/UNFPA Population and Environment Project in Eastern Islands of Fiji, sponsored by Man and the Biosphere (MAB)

Programme, Project 7: Ecology and Rational Use of Island Ecosystems. Australian National Univ., Development Studies Centre, Canberra, Australia.

*Brookfield, H.C. (Ed.) (1979).** Lakeba: Environmental change, population dynamics and resource use. Island repts. no. 5, Unesco/UNFPA Population and Environment Project in Eastern Islands of Fiji, sponsored by Man and the Biosphere (MAB) Programme, Project 7: Ecology and Rational Use of Island Ecosystems. Australian National Univ., Development Studies Centre, Canberra, Australia.

Brookfield, H.C. (1983). The Unseco/UNFPA project on population and environment in the eastern islands of Fiji. Annex 3. Programme on Man and the Biosphere (MAB). Expert consultations on Project 7: Ecology and rational use of island ecosystems. Final report. *MAB Report Series* 47: 27-28.

***Burrows, W. (1940/1944).** Notes on molluscs used as a food by the Fijians. *Proc. Fiji Society* 2: 12-14.

Bustard, H.R. (1970). Turtles and an iguana in Fiji. *Oryx* 10: 317-322.

***Cernohorsky, W.O. (1965).** The Mitridae of Fiji. *Veliger* 8(2): 70-159.

***Chapman (1971).** The marine algae of Fiji. *Revue algologique*: 164-170.

Clunie, F. (in press). Seabird breeding colonies on Namena Island and in the Ringgold Isles of north-eastern Fiji. *ICBP Study Report*.

Cooper, M.J. (1966). Destruction of marine fauna and flora in Fiji caused by the hurricane of February 1965. *Pac. Sci.* 20: 137-141.

Dahl, A.L. (1980). Regional ecosystems survey of the South Pacific Area. *SPC/IUCN Technical Paper* 179. South Pacific Commission, Noumea, New Caledonia. 99 pp.

Davis, W.M. (1920). The islands and coral reefs of Fiji. *Geo. Journ.* 55(1): 34-45, (3): 200-220, (5): 377-388.

***Deane, W. (1910).** Fijian fishing and its superstitions. *Trans. Fijian Society*: 57-61.

De Backer, R. (1985). Unbending after the economic storm. *The Courier* 92: 34-40.

Dilly, P.N. and Ryland, J.S. (1985). An intertidal *Rhabdopleura* (Hemichordata, Pterobranchia) from Fiji. *J. Zool., Lond.(A)* 205: 611-623.

Dunlap, R.C. and Singh, B.B. (1980). A National Parks and Reserves System for Fiji. A report to the National Trust of Fiji.

Eaton, P. (1985). Land Tenure and Conservation: Protected areas in the South Pacific. *SPREP Topic Review* 17. South Pacific Commission, Noumea, New Caledonia.

Evans, L. (n.d.). Fiji from a fish's eyes. (Unreferenced popular magazine).

Everingham, I.B. (1987). Tsunamis in Fiji. *Fiji Mineral Resources Department Report* 62. 19 pp.

***Fairbridge, R.W. and Stewart, H.B. (1960).** Alexa Bank, a drowned atoll on the Melanesian border plateau. *Deep Sea Res.* 7: 100-116.

***Fowler, H.W. (1959).** *Fishes of Fiji*. Government Printer, Suva. 670 pp.

Galzin, R., Ricard, M., Richard, G., Salvat, B. and Toffart, J.L. (1979). Le complexe récifal de Kabara (Lau Group, Fiji). Geomorphologie, Biologie et Socio-Ecologie. *Ann. de l'Inst. Océanographique, Paris* 55(2): 113-134.

***Gardiner, J.S. (1898).** The coral reefs of Funafuti, Rotuma and Fiji together with some notes on the structure and formation of coral reefs in general. *Proc. Phil. Soc. Camb.* 9(8): 417-503.

Garnett, M.C. (1984). Conservation of seabirds in the South Pacific Region: A review. In: Croxall, J.P., Evans, P.G.H. and Schreiber, R.W. (Eds), *Status and Conservation of the World's Seabirds*. ICBP Technical Publication 2. Pp. 547-558.

***Gawel, M. and Seeto, J. (1982).** *Limited marine investigations at the Fijian Hotel, Yanuca Island*. Inst. Mar. Res., Univ. S. Pacific.

Gibbons, J.R.H. (1981). The biogeography of *Brachylophus* (Iguanidae) including the description of a new species, *B. vitiensis* from Fiji. *J. Herpet.* 15(3): 255-273.

Gibbons, J. and Penn, N. (19). Dimensions in diving: An account from Fiji. *Skindiving* 13(2): 56-62.

Guinea, M.L. (1981). The sea snakes of Fiji. *Proc. 4th Int. Coral Reef Symp., Manila* 2: 581-585.

Hirth, H.F. (1971). South Pacific Islands - Marine Turtle Resources. Report to Fisheries Development Agency Project. FAO. Report F1: SF/SOP/REG/102/2. 34 pp.

***Hoffmeister, J.E. (1925).** Some corals from American Samoa and Fiji Islands. *Pap. Dep. Mar. Biol. Carnegie Inst. Wash. (Publ. 340)* 22: 1-90.

***Holmes, R. (1980).** A preliminary evaluation of the geology, engineering environment and coral sand resources associated with the fringing and barrier reefs adjacent to Suva. *Fiji Mineral Resources Department Rept* 20. 22 pp.

***Hornell, J. (1940).** *Report on the Fisheries of Fiji*. Government Printer, Suva. 87 pp.

Kearney, R.E. (1982). An assessment of the skipjack and baitfish resources of Fiji. *Skipjack Survey and Assessment Programme Final Country Report* 1. South Pacific Commission, Noumea, New Caledonia.

King, M.G. and Stone, R.M. (1979). Commercial fisheries in Western Pacific islands. *Proc. Seminar Workshop on utilization and management of inshore marine ecosystems of the Tropical Pacific Islands, November 24-30, 1979*. Univ. S. Pacific, Suva.

Kirkpatrick, J.B. and Hassal, D.C. (1981). Vegetation of Sigatoka sand dunes, Fiji. *N.Z. J. Bot.* 19: 285-287.

Kobluk, D.R. and Lysenko, M.A. (1987). Impact of two sequential Pacific hurricanes on sub-rubble cryptic corals: The possible role of cryptic organisms in maintenance of coral reef communities. *J. Paleont.* 61(4): 663-675.

Kott, P. (1981). The ascidians of the reef flats of Fiji. *Proc. Linn. Soc. New South Wales* 105: 147-212.

Krishna, R. (1979). Coral sand dredging. *Proc. Seminar Workshop on utilization and management of inshore marine ecosystems of the Tropical Pacific Islands, November 24-30, 1979*. Univ. S. Pacific, Suva. Pp. 50-51.

Kunatuba, P. (1983). Traditional knowledge of the marine environment in Fiji. 2. Traditional sea tenure and conservation in Fiji. *Institute of Marine Resources Technical Paper*. Univ. S. Pacific, Suva.

***Ladd, H.S. and Hoffmeister, J.E. (1945).** Geology of Lau, Fiji. *B.P. Bishop Mus. Bull.* 171: 1-399.

Lal, P.N. (Ed.) (1983). Mangrove Resource Management. Proc. of an Interdepartmental Workshop held 24.2.83 at MAF Headquarters Conference Room, Suva. Technical Report No. 5. June 1983. Fisheries Division, MAF, Fiji.

Lal, P.N. (1984). Environmental implications of coastal development in Fiji. *Ambio* 13(5-6): 316-321.

***Levy, N. (1940/1944).** The turtle and the tortoiseshell industry. *Proc. Fiji Society* 2: 183-185.

Lewis, A.D., Adams, T.J.H. and Ledua, E. (1984). Giant Clam Project Progress Report. Fisheries Division, Ministry of Primary Industry, Fiji.

Lewis, A.D., Adams, T.J.H. and Ledua, E. (1985). Giant Clam Project Progress Report. Fisheries Division, Ministry of Primary Industry, Fiji.

*MAB (1977). Population, resource and development in the eastern islands of Fiji: Information for decision making. General Report Vol. 1, Unesco/UNFPA Population and Environment Project in the Eastern Islands of Fiji.

*Mayer, A.G. (1924). Structure and ecology of Samoan reefs. *Pub. Carnegie Inst.* 340: 1-25.

Morton, J. and Raj, U. (1980a). *The shore ecology of Suva and South Viti Levu. Book 1. Introduction to zoning and reef structure - soft shores.* Univ. S. Pacific, Suva.

Morton, J. and Raj, U. (1980b). *The shore ecology of Suva and South Viti Levu. Book 2. Suva Barrier Reef and its communities.* Univ. S. Pacific, Suva. 152 pp.

Muirhead, A. and Ryland, J.S. (1981). Zoanthidea of south and west Viti Levu. (Abstract) *Future Trends in Reef Research.* Ann. meeting of International Society for Reef Studies.

Muirhead, A. and Ryland, J.S. (1985). A review of the genus *Isaurus* Gray, 1828 (Zoanthidea), including new records from Fiji. *J. Nat. Hist.* 19: 323-335.

*Muzik, K. and Wainwright, (1977). Morphology and habitat of five Fijian sea fans. *Bull. Mar. Sci.* 27(2): 308-337.

National Trust for Fiji (1978). National Parks and Related Reserves - Fiji Islands. Unpub. Situation Rept.

*Owens, D. (1971). *Acanthaster planci* starfish in Fiji: Survey of incidence and biological studies. *Fiji Agric. J.* 33: 15-23.

Parkinson, B.J. (1982). *The specimen shell resources of Fiji.* Report prepared for the South Pacific Commission and the Government of Fiji. South Pacific Commission, Noumea, New Caledonia.

Penn, N. (1981). The environmental consequences and management of coral sand dredging in the Suva region, Fiji. PhD Thesis, University of Wales.

*Penn, N. (1982). The environmental consequences and management of coral sand dredging in the Suva Region, Fiji. Fieldwork Rept, Institute of Marine Resources, Univ. S. Pacific, Suva, Fiji.

Phipps, C.V.G. and Preobrazhensky, B.V. (1977). Morphology, development and general coral distribution of some reefs of the Lau Islands, Fiji. *Memoires du Bureau de Recherches Geologiques et Minières* 89: 440-455.

Prakash, J. (1987). The marine algae resources of Fiji. In: Furtado, J.I. and Wereko-Brobby, C.Y. (Eds), *Tropical Marine Algal Resources of the Asia-Pacific Region: A status report.* Commonwealth Science Council Technical Publication 181: 23-28.

Prescott-Allen, R. (1986). Sustaining Fiji's Development. Draft Project Proposal for a National Conservation Strategy for Fiji. Ministry of Housing and Urban Affairs, Government of Fiji and CDC/IUCN, Gland. 31 pp.

*Raj, U., Southwick, G. and Stone, R. (1981). Komave Reef Platform: A preliminary biology study. Institute of Marine Resources, Univ. S. Pacific, Suva.

Richmond, R.N. (1979). Mineral/petroleum development in the South Pacific. *Proc. Seminar Workshop on utilization and management of inshore marine ecosystems of the Tropical Pacific Islands, November 24-30, 1979.* Univ. S. Pacific, Suva. Pp. 45-49.

Robinson, D.E. (1971). Observations on Fijian coral reefs and the Crown-of-Thorns Starfish. *J. Roy. Soc. N.Z.* 1: 99-112.

Rodda. P. (1981). The phosphate deposits and geology of Tuvutha. *Fiji Mineral Resources Department Economic Investigation* 3. 43 pp.

Ryland, J.S. (1979). Introduction to the coral reefs of Fiji. *Proc. Seminar Workshop on Utilization and Management of Inshore Marine Ecosystems of the Tropical Pacific Islands, November 24-30, 1979.* Univ. S. Pacific, Suva. Pp. 13-22.

Ryland, J.S. (1980). "Cascade" feature of South Viti Levu fringing reefs, Fiji. Paper given at "Reefs-Past and Present", meeting of Int. Soc. for Reef Studies, Cambridge.

Ryland, J.S. (1981). Reefs of south-west Viti Levu and their tourism potential. *Proc. 4th Int. Coral Reef Symp.,* Manila 1: 293-298.

Ryland, J.S., Wrigley, R.A. and Muirhead, A. (1984). Ecology and colonial dynamics of some Pacific reef-flat Didemnidae (Ascidiacea). *Zool. J. Linnean Soc.* 80: 261-282.

*Salvat, B., Ricard, M., Richard, G., Galzin, R. and Toffart, J.L. (1976). The Ecology of the Reef-Lagoon complex of some islands in the Lau Group. Unesco/UNFPA Population and Environment Project in the Eastern Islands of Fiji. Project Working Paper No. 2. 23 pp.

Salvat, B., Richard, G., Toffart, J.L., Ricard, M. and Galzin, R. (1977). Reef lagoon complex of Lakeba Island (Lau group - Fiji): Geomorphology, biotic associations and socio-ecology. *Proc. 3rd Int. Coral Reef Symp.,* Miami 2: 297-303.

Singh, B. (1985). Owner involvement in the establshment of parks. *Report of the 3rd South Pacific National Parks and Reserves Conference, Apia* 2: 269-270.

Siwatibau, S. (1984). Traditional environmental practices in the South Pacific: A case study of Fiji. *Ambio* 13(5-6): 365-368.

SPREP (1980). Fiji. *Country Report* 4. South Pacific Commission, Noumea, New Caledonia.

Squires, D.F. (1962). Corals at the mouth of the Rewa River, Viti Levu, Fiji. *Nature* 195(4839): 361-362.

*Toganivalu, D. (1914). Fishing. *Trans. Fijian Society.* 5 pp.

*Tomlinson, P.B. (1978). Rhizophora in Australasia: Some clarification of taxonomy and distribution. *J. Arnold Arboretum* 59: 156-169.

*Turbet, C.R. (1940/1944). Beche-de-mer; trepang. *Proc. Fiji Soc.* 2: 147-154.

Watling, D. (1985). Notes on the birds of Gau Island, Fiji. *Bull. Brit. Orn. Club* 105(3): 96-102.

Woodhall, D. (1984a). *Geology of Vanua Balavu, Namalata, Susui, Munia, Cikobia-i-Lau, Sovu and Avea.* Mineral Resources Department, Suva.

Woodhall, D (1984b). *Geology of Nukutolu, Vatuvara, Yacata and Kaibu, Naitauba, Kibobo, Malima, Wailagi Lala, Katafaga, Vekai, Cicia, Mago, Kanacea, Tuvuca, Tavunuku-i-vanua, and Yaroua.* Mineral Resources Department, Suva.

Woodhall, D. (1984c). *Geology of Vanua Vatu, Nayau, Lakeba, Reid Reef, Moce and Karoni, Aiwa, Oneata, Komo, Olorua and Bukatatanoa Reef.* Mineral Resources Department, Suva.

Woodhall, D. (1985a). *Geology of Namuka, Yagasa, Fulaga, Kabara, Tavu-na-sici, Marabo, Vuaqava, Vatoa, Naievo, Tuvana-i-colo, Tuvana-i-ra, Ono-i-Lau and Ogea.* Mineral Resources Department, Suva.

Woodhall, D. (1985b). *Geology of Taveuni, Qamea, Laucala, Cikobia and nearby islands.* Mineral Resources Department, Suva.

Woodhall, D. (1985c). Geology of the Lau Ridge. In: Scholl, D.W. and Vallier, T.W. (Eds), *Geology and offshore resources of Pacific island arcs - Tonga region.* Circum-Pacific Council for Energy and Mineral Resources Earth Science Series 2. Circum-Pacific Council for Energy and Mineral Resources, Houston, Texas. Pp. 351-378.

Woodhall, D. (1987). Geology of Rotuma. *Fiji Mineral Resources Dept Bull.* 8. 40 pp.

Woodrow, P.J. (1980). Geology of Kandavu. *Fiji Mineral Resources Dept Bull.* 7. 31 pp.

Zann, L., Brodie, J., Berryman, C. and Naqasima, M. (in press). Recruitment, ecology, growth and behaviour of juvenile *Acanthaster planci* L. Proceedings of a Symposium on Recent Findings in *Acanthaster* Biology and Implications for Reef Management, Guam, 1986. *Bull. Mar. Sci.*

CORAL COAST

Geographical Location   South coast of Viti Levu, west of Serua (18°16'S, 177°55'E) to, but not including, Natadola (18°08'S, 177°18'E).

Area, Depth, Altitude   About 60 km of coastline.

Physical Features   The 100 fathom (183 m) isobath lies approximately 1 km offshore; soundings of 550-1280 m are indicated within 5.5 km of the shore. The well-developed fringing reef extends almost unbroken for 63 km with a seaward extension of 500-1000 m. The only major gap coincides with the mouth of the Sigatoka River where terrigenous sands have built up into high dunes. At intervals, corresponding to the creeks which descend from the hills of the Southern Coastal Range, the continuity of the reef is broken by passages 100-300 m across. These provide shelter for a wealth of corals and other sessile forms at and below low water mark of spring tides. While the moat may become very hot by day (35°C+), the passages preserve the cooler ocean temperature. Landwards in these passages, water from the moats flows swiftly out through the channels paralleling the shore or cascades over a miniature cliff (Ryland, 1979). The reefs are backed by carbonate sand beaches (Ryland, 1981). Other coastal habitats are described by Morton and Raj (1980).

Reef Structure and Corals   The reefs at Cuvu (Thuvu) and Korolevu have been described by Morton and Raj (1980). The reef just east of Cuvu has a moat, extremely rich in corals forming micro-atolls of *Porites, Seriatopora, Pavona* and *Euphyllia*. Seaward of this is the reef platform, dominated by *Porites* and dissected by irregular channels with strong currents. Beyond this there is an emergent reef crest with a gentle seaward slope on which heavy surf breaks. *Tubipora* is abundant, with *Favia, Favites* and *Acropora* becoming more common seawards. A shallow stretch of standing water of variable width, the summit moat, runs the whole length of the reef immediately behind the crest and is filled with brown algae. The seaward slope has three zones; a *Sargassum* zone; a rich zone of diverse corals and algae, and the seaward edge which has a strong overlay of *Lithophyllum*, with *Porolithon onkodes* becoming dominant seawards. *Pocillopora eydouxi* is the main

coral. The surge channels on the reef front are dominated by *P. verrucosa* and *P. eydouxi* with some *Acropora*.

At Namatakula, the fringing reef is up to 800 m wide and differs from the Cuvu-Naevuevu Reef in that there is no well developed moat with micro-atolls and no clear cut seaward *Sargassum* zone. A *Montipora* zone is found between the lower beach and a wide, shallow moat where other corals begin to appear. This extends out about 800 m, to the summit slope which has strong currents and rich coral cover, with abundant *Montipora*, faviids and *Acropora*. The seaward slope also has diverse corals, with *Platygyra, Goniastrea, Millepora* and *Acropora*. The seaward edge of the reef is covered with *Acropora* and *Pocillopora*, and there is abundant *Palythoa*.

The reef 200 m west of Korolevu Beach Hotel is similar to the Cuvu-Naevuevu Reef. Its distinctive feature is in the middle region where the moat is in many places roofed over by large tables of *Porites, Millepora*, faviids, *Symphyllia* and *Galaxea*. These canopies at low water level overhang living micro-atolls below, which include a variety of species.

The reef at Malevu, 10 km east of Sigatoka, has been studied in some detail by Ryland (1981) and is also briefly described by Morton and Raj (1980). It is bounded by the outflows of Bulu Creek to the east and Korotogo Creek to the west. To seaward it is fully exposed to the south-easterly weather: a substantial swell and rough sea appear normal, particularly during the cooler months. Several zones have been identified. The seaward slope or wave break zone (Zone 1) is intermittently exposed and consists of a hard *Lithophyllum/Porolithon* crust supporting a sward dominated by *Amphiroa* spp. and *Caulerpa peltata*. There is little coral apart from small mounds of *Pocillopora verrucosa*. It rises gently to the reef crest zone (Zone 2) which has a permanent belt of low growing *Sargassum cristaefolium* and no corals. It dips to the reef flat (Zone 3) at a level closely approximating MHWN. About 500 m from the beach, the flat becomes broken by tunnels and crevices which become deeper and more numerous. These are characterized by a variety of Milleporidae, Alcyonacea and Zoanthidea and a range of less abundant corals. In Zone 4, the crevices break into numbers of rounded, flat-topped pedestals separated by gulleys, which become more isolated, some of them forming large micro-atolls, particularly of *Porites lutea* and *Diploastrea heliopora*. In Zone 5, in the central part of the reef flat lagoon nearer the beach, the micro-atolls are associated with elongated drifts of rubble and large clumps of *P. andrewsi* and *Pavona divaricata* in particular and change to irregular lines of massive and branching corals, separated by sandy drainage channels. Within 50-500 m of the beach, coral cover between the channels becomes discontinuous. Approaching the beach there is generally a distinct gutter containing tall loose clumps of *Sargassum* sp. On most days during the low tide period, waves continue to break over the reef crest, keeping the lagoon full. This water cascades over the flanking ridge on which a community of endemic didemnid ascidians, *Diplosoma multipapillatum* is found (Kott, 1981). The important features of the reef are the clear zonation of corals, zoanthids, algae and ascidians and the high diversity of scleractinians encountered in the reef flat lagoon. The reef differs in detail and species composition from those described at Cuvu and Korolevu, but is similarly rich and flourishing (Ryland, 1981).

Morton and Raj (1980) describe Komave Reef, between Korolevu Bay to the west and Naicobocobo river mouth to the east. It is about 4.5 km long and 700 m wide. The surge zone, 50-100 m wide, is very exposed and the seaward slope is broken by numerous surge channels. Corals are diverse and are dominated by *Pocillopora*, *Favia*, *Platygyra* and *Montipora*. The surge platform is consolidated with *Lithophyllum moluccense* and *Porolithon onkodes*. From 30 to 60 ft (9-18 m) depth, secondary platforms with luxuriant coral cover, dominated by *Porites* and faviids, are found before the drop-off. The reef summit is dominated by algae, and to the landward edge *Palythoa* and the ascidian *Diplosoma similis* (Rylands *in litt.*, 25.8.87) become dominant. The reef flat is almost devoid of corals, although in pools *Porites andrewsi* is dominant. The reef moat with its micro-atolls has moderate currents which permit growth of a diverse coral community, characterized by large table colonies of *Porites*, *Favia*, *Symphyllia* and *Galaxea*, which often shelter other corals. The shallow, landward lagoon has been artificially deepened and has no corals of significance.

Noteworthy Fauna and Flora Ryland *et al.* (1984) studied didemnid ascidians at Malevu. Parkinson (1982) lists 85 species of specimen-shell mollusc (including 22 *Terebra* species) in 11 families recorded in a series of dives in 1981 along the coast from Sigatoka east to Navua.

Scientific Importance and Research The Coral Coast has the longest chain of fringing reefs in Fiji, all of which are considered rich and flourishing (Morton and Raj, 1980; Ryland, 1981). The U.S.P. Institute of Marine Resources has worked on some of them, such as Cuvu-Naevuevu (Dunlap and Singh, 1980). Surveys have been undertaken of the foreshore area of the Hyatt Regency Hotel development (Raj *et al.*, 1981) and of the Fijian Hotel reef near Sigatoka (Gawel and Seeto, 1982).

Economic Value and Social Benefits The Coral Coast is developing as a prime tourist area on account of the attractive reefs and beaches and the high insolation and moderate rainfall. Hotels are concentrated mainly in the vicinity of Korolevu, Malevu-Korotogo and Cuvu but new unit-accomodation resorts opened near Namada and Tagaqe during 1980. The shallow lagoon-like nature of the reef-flat permits snorkelling without causing damage to beds of intertidal coral. There is a diving centre at Sigatoka (Dunlap and Singh, 1980).

The reefs are also important as sources of fish and shellfish to the coastal villages (Ryland, 1981).

Disturbance or Deficiencies In the late 1970s, *Acanthaster* were present at Malevu but were not causing a problem (Robinson, 1971). Morton and Raj (1980) described *Acanthaster* as being common on Komave Reef. Cyclone Wally in 1980 had a major impact on the reefs in this area (Ryland *et al.*, 1984).

In general planning regulations are sound and relatively few tourists venture on or over the reef (Ryland, 1981). However, coastal development is intense in some areas. For example, the Hyatt Regency Hotel is situated by the Komave Reef; an artificial island has been built out to the west of the hotel, with spits to the east and west,

apparently to trap coral sand for the beach. Immediately in front of the beach a swimming hole has been excavated (Morton and Raj, 1980). Additional damage to reefs in the 1970s as a result of tourism is briefly described by Baines (1980).

Legal Protection None.

Recommendations Dunlap and Singh (1980) recommend the Coral Coast Reefs for marine park and reserve status. Pollution sources should be investigated to determine the practicality of protecting the fringing reefs. Those most suitable should be protected. Cuvu Beach is recommended as a terrestrial reserve.

GREAT AND NORTH ASTROLABE REEFS

Geographical Location North-east of Kadavu, south of Viti Levu; Solo is an island on North Astrolabe Reef at 18°40'S, 178°30'E; Great Astrolabe includes a number of islands and extends about 35 km north of the north-eastern coast of Kadavu at ca 18°45'S, 178°30'E.

Area, Depth, Altitude Solo (600 sq. m, max. alt. 10 ft (3 m). Great Astrolabe Reef includes Yaukuvelevu (62 ha, 400 ft (122 m)), Yaukuvelailai (18 ha, 210 ft (64 m)), Yabu (6 ha, 170 ft (52 m)), Dravuni (200 acres (81 ha)), Bulia (425 acres (172 ha), 460 ft (140 m)), Ono (11.7 sq. mi. (30 sq. km), 344 m), Vanuakula (17.3 ha, 250 ft (76 m)), Yanuyanu-i-sau (28 m) and Yanuyanu-i-loma (13.5 ha, 140 ft (42.7 m)), Namara (19 ha, 230 ft (71 m)), Qasibale (60 ft (18.3 m)), Vuro (21 ha, 270 ft (82 m)) and Vurolailai (90 ft (27 m)).

Physical Features North Astrolabe Reef is a circular atoll barrier reef with a rocky islet, Solo, the eroded peak of a volcano, in the middle of the open lagoon. Great Astrolabe Reef is also a barrier reef (classified by Ryland (1981) as an oceanic ribbon reef), extending north-east and west of Kadavu to encompass a number of small islands (Dunlap and Singh, 1980). All the islands are volcanic. Davis (1920) briefly described Ono which has steep cliffs on the west coast.

Reef Structure and Corals Morton and Raj (1980a) describe the fringing reef at Yaukuve which has a basalt volcanic shore. The sublittoral fringe has rich and diverse corals including several species of *Acropora*, *Porites lutea*, *Pocillopora damicornis* and faviids. On the vertical face beyond this, there are broad foliose sheets of *Montipora foliosa*. The first part of the reef platform is overlaid with silty sand and has few living corals. At the edge, where the surge breaks at lowest spring tides, there are good *A. reticulata* tables. A popular account of the reef at Solo Lighthouse is given by Gibbons and Penn (n.d.).

Noteworthy Fauna and Flora Yabu is a seabird rookery.

Scientific Importance and Research The Institute of Marine Resources, University of the South Pacific, has a research field station on Dravuni with accommodation for 15 visiting scientists. An unpublished fish checklist and some comments on fisheries have been prepared following a fish collecting trip organized by the

University of the South Pacific and the Ontario Museum, Canada (Brodie *in litt.*, 10.11.87).

Economic Value and Social Benefits   Private yachts sometimes visit the islands; groups of SCUBA divers occasionally explore the reefs and tour operators arrange charter cruises to the reefs (Dunlap and Singh, 1980). Solo has a lighthouse.

Disturbance or Deficiencies   Giant Clams have almost disappeared from the Bulia and Dravuni reef systems due to artisanal fishing (Lewis *et al.*, 1984; Adams *in litt.*, 7.2.85)

Legal Protection   None as yet.

Management   Traditional management and customs are still practiced on the Dravuni reefs (Birkeland, 1987).

Recommendations   North Astrolabe Reef was recommended for development as a marine reserve between 1981 and 1985; Great Astrolabe Reef was recommended for development as a marine park and reserve in 1986.   The Yabu seabird rookery is recommended for additional protection (Dunlap and Singh, 1980). None of these recommendations appears to have been acted upon.

LAKEBA

Geographical Location   Central Lau Islands; 178°48'W, 18°15'S.

Area, Depth, Altitude   Emergent island 55.9 sq. km; total lagoon area 82 sq. km; lagoon around island 0-10 m deep, 36 sq. km; Great Lagoon 46 sq. km; mangrove 5 sq. km; max. alt. 219 m.

Physical Features   Lakeba is the largest island of the Lau Group, and is almost circular and mainly volcanic but with the remains of a limestone rim. Tidal range is less than 1 m. East to south-east trade winds predominate. The south and west coast is bordered by limestone cliffs reaching 250 ft (76 m) in height.   Sea surface temperatures range from 27.9 to 29.3°C; salinity is 33-34 ppt. The island is surrounded by a continuous barrier reef demarcating a lagoon, 500-700 m wide, with a single large opening on the south-west and sandy beaches. The north, south and west sides of the island are bordered by a shallow lagoon, except in front of Tubou where depths are great enough for ships to anchor. Off the eastern end of the island there is a very deep and extensive lagoon called the Great Lagoon which has not been studied. The eastern shore of the lagoon is bordered by mangroves. Two sink holes, Waci Waci and Yadrana, of karstic origin are located in the south-east and north of the lagoon. These are subcircular with an area of about 1 sq. km and are areas of strong turbidity (Salvat, 1976 and 1977).

Reef Structure and Corals   On the west coast, corals cover less than 50% of the lagoon bottom. Outside the barrier reef flat, the reef is well developed, terminating seaward in a dense algal crest of *Caulerpa, Halimeda* and *Turbinaria*. On the north coast, seaweed beds and algae are found nearshore and the frontal zone of the barrier reef is covered with corals. On the south coast corals are abundant on the barrier reef near the pass but not on the channel bottom.

Sixty-five species of scleractinian corals and three species of *Millepora* have been recorded. Salvat *et al.* (1976 and 1977) describe the zonation on the barrier reef. The outer slope is dominated by encrusting *Acropora*, providing 20-70% coverage and 100% on some vertical walls. The frontal zone has only 5-50% living coral cover, with *Acropora* dominating. The outer reef flat has the greatest diversity of corals. *Porites* usually makes up only 1% coral cover. Species diversity around the sinkholes is low, although corals occur around Waci Waci.

Noteworthy Fauna and Flora   Along the eastern shore of the island is a 9 km stretch of mangroves, between Nukunuku and Waitabu villages.   It consists of a monospecific stand of *Rhizophora* and reaches a maximum width of 1500 m. Salvat *et al.* (1977) recorded 218 species of mollusc and 145 fish species.

Scientific Importance and Research   The island was surveyed as part of a Unesco/UNFPA project on population and environment in the eastern islands of Fiji under the MAB Program (Salvat *et al.*, 1976 and 1977).

Economic Value and Social Benefits   The population of Lakeba numbers about 2100. A small amount of fishing is carried out but there are only about 10 motor boats. *Trochus niloticus* and mother-of-pearl shells are exported to Suva, and the Giant Clam *Tridacna derasa* and other molluscs are much sought after for food (Salvat *et al.*, 1977).

Disturbance or Deficiencies   The reefs and marine environments are relatively undisturbed as the lagoon and reefs are less important food sources than the island and the local people do not dive below 10 m when collecting marine organisms. Most fishing is done by women, within the limits of the lagoon and reefs. However, Giant Clams (Tridacnidae) have been overexploited and *Trochus* exploitation should not be increased. A road was built round the island in 1969 but coral sand mining, and mangrove cutting have so far had very little impact (Salvat *et al.*, 1977).

Legal Protection   None.

Management   Fishing is regulated by traditional laws.

Recommendations   Not mentioned for protection in the Fiji National Trust Plan (Dunlap and Singh, 1980). General recommendations from the Unesco/UNFPA project are given in Anon. (1983), Brookfield (1983) and Bayliss-Smith (1983).   It is not known if any recommendations were followed up.

NADI WATERS AND MAMANUCA GROUP

Geographical Location   Western Viti Levu, within the approximate area 17°20'-17°55'S, 177°00'-177°25'E, including the Mamanuca Group (the southern part of the Yasawa Group), and several scattered islands between these and Viti Levu, in the Nadi Bay region. Local inhabitants use the term Mamanuca to refer to the southern part (Mamanuca-i-cake) of the Mamanuca Group as it is shown on maps, referring to the northern part of the group, or Mamanuca-i-ra, as the Narokorokoyawa Group.

Area, Depth, Altitude The Mamanuca-i-ra (Narokorokoyawa) Group comprises Eori (16.2 ha, 64 m), Navadra (30.3 ha, 128 m), Vanua Levu (50.5 ha, 350 ft (107 m)), Vanualailai (ca 4 ha, joined to Vanua Levu by sand spit), Naniukalele, Kadomo (30.3 ha, 330 ft (100 m)), Yavuriba (0.8 ha, 100 ft (30 m)). The Mamanuca Rocks lie between Mamanuca-i-ra and Mamanuca-i-cake. The Mamanuca-i-cake (Mamanuca) Group comprises Monu (73 ha, 730 ft (222 m)), Yanuya (300 acres (122 ha), 340 ft (104 m)), Monuriki (40 ha, 590 ft (180 m)), Matamanoa (14.2 ha, 230 ft (70 m)), Nautanivono (240 ft (73 m)), Tavua (470 acres (190 ha), 570 ft (174 m)), Tokoriki (99 ha, 310 ft (95 m)). Islands within the Malolo Group include Malolo (3.73 sq. mi. (9.7 sq. km), 750 ft (229 m)) which is joined to Malololailai (Malolosewa or Plantation I.) (500 acres (235 ha), 230 ft (70 m)), Wadigi (3 islets, 60 ft (18.3 m)), Qalito ("Castaway") (56.7 ha, 390 ft (119 m)), Mociu ("Honeymoon") (0.8 ha, 180 ft (55 m)), Mana (132 ha, 69.5 m), Vomo (109 ha, 380 ft (116 m)) and Vomo Lailai (6 ha, 200 ft (61 m)). Islands lying within Nadi Waters include Namotu (6 ha in the 1950s but now greatly reduced after hurricanes), Tavarua, (10 ha, flat-topped), Tivoa, Luvuka ("Treasure") (10 ha), Tai ("Beachcomber") (2 ha), Navini (2.1 ha), Malamala (4.9 ha), Kadavu ("Bounty") (16.2 ha) and Yakuilau.

Land Tenure A private company (Islands in the Sun (Fiji) Ltd.) owns 99-year leases for Kadavu, Luvuka, Tai and Vomo. Malololailai is privately owned (Kobluk *in litt.*, 23.11.84).

Physical Features An area of insular shelf, protected from the prevailing winds, with high islands and sand cays, most with fringing and some with barrier reefs, and numerous small patch reefs. Islands of the Mamanuca-i-ra and Mamanuca-i-cake Groups are almost wholly volcanic, apart from some limestone on Eori. The islands in Nadi Waters are all of sand and beach-rock. Being sheltered, the beaches of Nadi Bay itself are not of white sand and are often backed by mangals. There is little detailed published information on the islands or reefs.

There is an extensive well-protected lagoon on the west side of Malololailai that is shared with Malolo. The lagoon centre is deep and is fringed by a typical lagoon patch reef network. It is open to the north and south, the southern opening being protected from the sea by an east-west trending small barrier reef that lies between the island and the large Malolo Barrier Reef about 6 km to the south. Information on the geology of the island is available (Kobluk *in litt.*, 23.11.84). Malolo has a fine 3 km sand beach and is hilly.

Navini is a small oval patch reef about 20 km off shore, almost due west from Nadi Airport, or WSW from Vuda Point, which has been described by Ryland (1981). It is 700 m in longest dimension and supports a cay 300 m in length. The seabed slopes away from the reef to surrounding depths of 37 m. The sea is rough on the windward side under normal trade wind conditions.

Reef Structure and Corals Most of the islands in the Mamanuca group are surrounded by fringing reefs. On Malololailai, the best reef growth is on the south-east and east where there is a broad intertidal reef platform that extends across sandy intertidal flats, to a sandy beach along the shore. There is a good shallow water coral community and a high diversity of molluscs (with ca 300 species recorded) and coral biota (Anon., 1986; Kobluk *in litt.*, 23.11.84). Further details of the reef here are given in Kobluk and Lysenko (1987).

Ryland (1981) describes corals on Navini patch reef. To leeward the reef flat is about 50 m wide and shelves very gently to the region of drop-off. Luxuriant coral growth here breaks surface at low water springs. The approach to the edge is marked by the abundance of *Acropora* tables, often 1 m or more across, thickets of *Acropora* spp. and *Pocillopora damicornis*, micro-atolls of *Diploastrea heliopora*, and large colonies of *Sarcophyton trocheliophorum*. The fore-reef descends steeply but irregularly for 5-10 m with coral heads, bare cliffs, or slabs of *Diploastrea*, but without large growths of *Acropora hyacinthus* or similar forms. There is no groove and buttress fore-reef or coralline reef crest.

Noteworthy Fauna and Flora The tree vegetation on Navini cay includes *Hernandia peltata*, *Terminalia littoralis* and *Calophyllum inophyllum* mixed with coconuts and *Acacia simplicifolia* scrub. Some bushes of the rare *Suriana maritima* are present on the lee side (Ryland, 1981).

Scientific Importance and Research The islands have been mapped by the Mineral Resources Department of Fiji and a geological map is in preparation (Rodda *in litt.*, 14.9.87). Reefs on Malololailali have been the subject of study over at least three seasons by the University of Toronto in collaboration with Earthwatch (Anon., 1986a).

Economic Value and Social Benefits This is the most important tourist location in Fiji. There are several large hotels on the mainland in the vicinity of Nadi Airport, but much of the attraction of the area lies in day trips to or short stays on the islands and cays. The latter, despite their lack of ground water and minimal land area, are in great demand as sites for resorts. Most of those between Nadi and the Mamanuca Group have now been, or are being developed, often to a high density of units; these include Navini, Mana, Malolo, Matamanoa; Malololailai, Qalita, Tai, Kadavu, Luvuka, Tavarua and Nanuyalevu, with plans (in 1985) for the development of Vomo. The reefs surrounding these islands are major attractions, with snorkelling and coral viewing from boats being popular activities (Dunlap and Singh, 1980; Ryland, 1981). The lagoon on the west side of Malololailai is used as a yacht club anchorage; the island is often visited by tourists, a landing strip permitting regular light plane flights from Nadi Airport (Kobluk *in litt.*, 23.11.84). There are diving centres on Mana and Molololailai (Anon., n.d.). Cruise ships (Blue Lagoon and Seafarer) call at various sites in the Mamanuca Group and further north in the Yasawas to buy shells and marine curios. Fishing is an important activity on many of the islands.

Disturbance or Deficiencies There was some hurricane damage to the Malololailai reefs in 1983 (Kobluk *in litt.*, 23.11.84). The same area was hit by two cyclones in 1985, 57 days apart, which caused a reduction of about 50% in the number of boulders sheltering cryptic coral on the

Molololailai reefs. However, there was little change in the composition of the remaining cryptic coral fauna (Kobluk and Lysenko, 1987). Minor damage to reefs was also noted around Luvuka, Tai and Qalito (Brodie *in litt.*, 10.11.87; Wells pers. obs., 1985).

Developments on the low islands are recent and the reefs are unspoilt but their small size renders both the cay and the coral vulnerable to damage and exploitation (Ryland, 1981). The high density of tourist traffic year round is considered to have had some impact on the reefs at Malololailai (Kobluk *in litt.*, 23.11.84). In the south and south-west of the lagoon mollusc populations have reportedly been depleted by overcollecting. The main reef on the east and south-east has suffered less damage as it is not easily accessible. On Luvuka, there had been some evidence of algal overgrowth caused by sewage pollution on the reefs, but steps are being taken to control this (Brodie pers. comm., 1985).

Legal Protection None.

Management The "vanua" of Vuda on the mainland south of Lautoka have the fishing rights on Tai and Luvuka. Since the development of the resorts, subsistence fishing has improved and tourist activity expanded, with holders of traditional fishing rights becoming involved in resort management and boat hire. A new sewage disposal system has been installed on Luvuka, to carry sewage from the resort to deep water beyond the reef. A study of the effects of the new system concluded that there was no immediate threat either to public health or to the island fringing reefs from the new system (Brodie and Ryan, 1987).

Recommendations The reef around Tai was recommended for reserve status by Dahl (1980). Dunlap and Singh (1980) recommended both the reefs of Nadi Bay and of the Mamanuca Islands for marine park and reserve status, with Nadi Bay accorded higher priority. It was noted that the Mamanuca group of islands and reefs should be given higher priority if the Nadi Bay proposal did not go through. It was recommended that potential pollution sources be investigated to determine the practicality of protecting the reefs and if suitable, designation and classification should be carried out. Designation was originally timetabled for 1985. Dunlap and Singh (1980) recommended that Malamala amd the surrounding reefs be given park and reserve status (terrestrial and marine) in 1986.

The development of a marine national park around Tai and Luvuka as a pilot project, with the help of the fishing right owners, is now considered a high priority (Anon., 1985b) and is considered a priority in the Action Strategy for Protected Areas in the South Pacific Region, drawn up at the Third South Pacific National Parks and Reserves Conference in 1985. The company which owns the leases on Tai and Luvuka is enthusiastic and in 1985 negotiations were under way with the fishing rights owners. The reserve is to extend 500 m from high water mark around both islands (although it is unclear if this also included all water between the two islands) (Rowles pers. comm., July, 1985). No progress appeared to have been made in this by late 1987 (Brodie *in litt.*, 10.11.87). The proprietor of the Musket Cove Resort on Malololailai is also concerned that the development of tourism should not damage the reefs (Kobluk *in litt.*,

23.11.84). Long-term monitoring of the sewage disposal system for Luvuka is recommended (Brodie and Ryan, 1987).

NATADOLA BAY PROPOSED NATIONAL PARK

Geographical Location South-west Viti Levu, west of Sigatoka; 18°08'S, 177°18'E.

Land Tenure The area close to the beach and around the lake and mangroves is Crown Land. Sixteen mataqalis own native land, and there is some freehold land owned by Fiji Macambo Holdings.

Physical Features Natadola Bay is surrounded by rocks of the Wainimala Group, a deformed sequence of volcanic and sedimentary rocks overlain by siltstone and limestone, and forms a wide bay bordered with white sand and surrounded by low hills. Navo, an island on the eastern border of the Bay, is composed of limestone and has undercut cliffs. A branch of the Tuva River flows into the area via a brackish lake with no direct connection to the sea. There is a barrier reef 1.5-2 miles (2.4-3.2 sq. km) off shore with a central gap, a fringing reef on the western side of the bay and an apron extension of the fringing reef at the eastern side of the bay. Between the two fringing reefs, there is a gently sloping sand bottom. The lagoon within the barrier reef is shallow enough in many parts to wade across at low tide. Rainfall averages 1850 mm and there is a marked dry season (Robinson, 1971; Singh, 1980).

Reef Structure and Corals The barrier reef has not been studied; Robinson (1971) gives a brief description of the lagoon. Patch reefs are scattered across the sandy lagoon bottom and at extreme low tide are covered by 6-12 in. (15-30 cm) of water. The flat tops of the patch reefs are covered in many places by dead zoantharian coral or living alcyonarians, predominantly *Sinularia*. The sides of the patch reefs have a rich growth of stony and soft corals. A transect across the lagoon revealed a zone, corresponding to the path of the main tidal current, where soft coral flourishes and living hard coral is decreased.

Noteworthy Fauna and Flora The terrestrial vegetation of the area is described by Singh (1980). There is a dense mangrove forest in the lake area. Shore birds are the predominant wildlife. Singh (1980) provides a preliminary list of marine fish.

Scientific Importance and Research An initial investigation was carried out during the preparation of the National Parks and Reserves Plan for Fiji. This was followed by other investigations by the National Trust of Fiji.

Economic Value and Social Benefits The area has considerable archaeological interest. At present there are no visitor facilities, but recreational activities such as swimming, snorkelling, horse racing and camping are popular. Fishing and shellfish collecting is prevalent around the barrier reef. The area is of value to visitors from the nearby hotels and tourist centres (Singh, 1980).

Disturbance or Deficiencies   The area may be vulnerable due to its small size and the proximity of agricultural land which lies as close as 10 m to the beach in some places. Runoff of chemical fertilizers could cause problems. Shell collecting was extensive in the 1960s (Robinson, 1971).   Robinson (1971) discusses the possibility of the reef having been damaged by *Acanthaster* between 1965 and 1969, resulting in the appearance of a zone of soft coral in the lagoon and the almost complete dominance of the reef crest by the soft coral *Sinularia* where once there had been an abundance of ahermatypic corals.   Sedimentation in Natadola Harbour had increased while turbulence was reduced, and a number of *Acanthaster* were observed.   The land behind Natadola Lagoon had been intensely cultivated by the 1960s, probably resulting in increased runoff, fertilizer enrichment and sedimentation, and it is not clear whether these activities or the *Acanthaster* had caused coral damage. Baines (1982) noted that speculators had paid relatively high prices for farm land backing Natadola Bay with the prospect of developing the area into a large-scale tourist resort; as of 1985 no such development had begun (Wells and Jenkins, pers. obs., 1985).

Legal Protection   None.

Management   The native fishing rights are held by the Vanua of Nasoqo and the Vanua of Nasoni of Nadroga.

Recommendations   In Dunlap and Singh (1980) the area was considered to be the most suitable for the establishment of Fiji's first National Park. A plan for its development and management was drawn up (Singh, 1980). A system of zonation was proposed to include Natural Areas, Development Areas and Historic Areas; traditional usage of the area would still be permitted.   It was intended to install interpretive facilities and undertake survey and research work, and to monitor carefully the increased usage of the area once it was gazetted. Initially only the terrestrial part of the park was to be developed, primarily for the purpose of recreation. As of 1985 no progress had been made in the further development of this area.

SUVA BARRIER REEF AND NUKULAU AND MAKULUVA CAYS

Geographical Location   South-east of Suva Harbour (18°08'S, 178°25'E), south coast of Viti Levu.

Land Tenure   Nukulau is government-owned.

Physical Features   Suva Barrier Reef is part of the south-eastern reef chain of Viti Levu and encloses Suva Peninsula and Laucala Bay. It forms a crescent protecting Suva Harbour to the south-east and skirting the tip of the peninsula at a distance of about one km. The eastern part encloses Laucala Bay and is broken midway by Nukubuco Passage, adjacent to which lies the islet of Nukubuco. At its eastern extremity, cut off from the main reef by the Nukulau Passage, are the sand cays of Nukulau and Makuluva, each with its own fringing reef (Morton and Raj, 1980b). Makuluva, the outermost cay has a wide fringing reef, extending south-east of the islet independent of the barrier reef. Most of the Rewa flows

into Laucala Bay, near Nukulau. The shores of the Bay are described by Morton and Raj (1980a).   Further information is given in Ryland *et al.* (1984).

Reef Structure and Corals   Morton and Raj (1980b) describe a transect across the barrier reef at Laucala Bay. Five zones are identified. Starting on the landward side, there is a seagrass bed of *Syringodium* and *Halophila*. A shallower zone of soft corals and *Porites* follows.   In addition to several *Porites* species, *Alveopora* and *Seriatopora* are also found. The principal soft coral species are *Sarcophyton* and *Sinularia*. The next part of the reef flat is dominated by echinoderms.   Seaward of this is a long stretch of loose consolidated rubble intermingled with patches of coarse sand, up to 700 m wide. Towards the summit of the reef crest, there are dead coral heads, especially *Acropora* tables and pieces of rubble, the largest being blocks up to a metre across. These provide habitats for diverse invertebrate communities described by Morton and Raj (1980b). The seaward slope of the reef is exposed and subjected to heavy surf. The upper part is dominated by *Acropora*, *Montipora viridis*, *Pocillopora eydouxi* and *Palythoa*, with numerous dead rubble patches between the living coral. In deeper water *Acropora*, *Pocillopora verrucosa*, *Millepora* and *Montipora viridis* predominate.

The west, or channel section of the barrier reef, just outside Suva Harbour, has a sheltered, shallow lagoon area with scattered corals in the seaward part, followed by an *Amphiroa-Turbinaria-Sargassum* flat with limited coral growth on the seaward edge. Further out is a zoanthid zone dominated by *Palythoa*, but with *Acropora* and *Montipora* becoming increasingly abundant. Seaward of this is the reef crest which is less exposed than on the Laucala Bay side. *Acropora* dominates, with *Montipora*, faviids, *Tubipora musica* and *Pavona decussata*.   Surge channels seaward of the crest are covered with *Millepora tenera* and *Pocillopora verrucosa*, with a wide diversity of other corals.

The lagoon between the main reef and the Suva Peninsula forms a channel with a fairly strong tidal current, and there is a scatter of patch reefs here, in about 5 m of water.   Elaborate growths of branched corals reach up to a metre from the bottom, *Acropora* predominating. Faviid head corals are lacking but there is abundant *Echinopora lamellosa* and *Hydrophora exesa*.   *Merulina*, *Pavona*, *Psammocora*, *Porites*, *Seriatopora* and *Stylophora* are also found.

Makuluva fringing reef has a particularly well developed rubble flat, described in detail by Morton and Raj (1980b), loosely consolidated and with a fragile canopy of *Amphiroa* and *Lithophyllum* calcareous algae. At 30-40 m from the seaward edge, the reef crest is covered mainly with *Amphiroa* and green algae, with low-profile *Acropora* heads, *Porites*, *Favia* and *Goniastrea*. The seaward slope and surge zone have deep surge channels. The surge zone is dominated by zoanthids, *Palythoa* and *Xenia*, and Organpipe Coral *Tubipora musica* is very common.   Of the scleractinians, *Favia* and *Favites* may cover 10-20% of the surface, and *Porites andrewsi*, *Acropora* sp. and *Montipora viridis* are common.   *Millepora pocillopora* and *Distichopora violacea* are also found.

On Nukulau, corals are rare, absent or dead off the landward shore but increase in diversity towards the east and south, presumably due to the decrease in sediment and increase in salinity (Squires, 1962). Corals are rare or absent in Laucala Bay.

Noteworthy Fauna and Flora   Morton and Raj (1980a and b) give details of the marine fauna found in the area. Parkinson (1982) lists 30 mollusc species (including 12 *Conus* spp) in 7 families recorded in survey dives in 1981 in the Suva Harbour region.   There are some Tridacnids present on Suva reef, particularly *T. maxima* (Brodie *in litt.*, 10.11.87).   The seasnake *Laticauda colubrina* is found on Nukulau (Guinea, 1980). Ryland *et al.* (1984) studied didemnid ascidians on Nukubuco Reef. There is a Mangrove Park on the island of Laucala, and the Rewa River is fringed with mangroves in its lower reaches.

Scientific Importance and Research   The reefs in this area are used extensively for teaching purposes, as the University of the South Pacific campus is situated on the shore of Laucala Bay.

Economic Value and Social Benefits   Tourism is becoming increasingly important in this area, and most diving in Fiji takes place around Suva. Day trips to Nukulau are a popular attraction, and the reefs can be viewed there with glass panelled boats.

Disturbance or Deficiencies   In 1984 there was an outbreak of *Acanthaster planci* on the barrier reef, the population dynamics of which had been studied since the juvenile stage (Zann *et al.*, in press). However, the extent of coral damage is not yet known.

Suva is the main international port and population centre of Fiji and there has been considerable death of stony corals in this area since the beginning of the century. Mayer (1924) mentions that the reefs had deteriorated by the 1920s and attributed this to silting following the removal of vegetation from the watershed of the Rewa River. Silting at the mouth of the Rewa has continued, increased from time to time by heavy flooding.   For example in January 1965, there was extensive coral death, followed by mangrove establishment on Nukulau after severe flooding in 1964 (Robinson, 1971). The 1965 hurricane did not appear to damage the reefs (Cooper, 1966).   Cyclone Wally in April 1980 had a noticeable impact on the reefs, causing severe flooding and discolouration of the water (Ryland *et al.*, 1984) and there was massive damage to reef corals in the Suva area following large scale flooding caused by exceptional rainfall in early 1986 (Brodie *in litt.*, 10.11.87).

From 1960 to 1979, around 123 million tons (dry weight) of coral sand were extracted from seven different sites in Laucala Bay; in 1979 extraction from seagrass beds was proceeding at the rate of 2000 tons a week (Bajpai, 1979).   Coral sand is also dredged from the sand banks near Nukuboco. The full impact of this activity is not known but it is potentially damaging to adjacent reefs and has already caused extensive mortality of turtle grass *Syringodium isoetifolius* (Baines, 1977a; Holmes, 1980; Krishna, 1979; Lal, 1984; Penn, 1981 and 1982). Nukubuco was used for target practice in World War II. Makuluva has changed extensively over the years, partly by destruction and partly by movement. Large concrete water tanks, originally on the island are now on the reef, the sand of the island having moved away to the

south-east (Rodda *in litt.*, 14.9.87)

Legal Protection   Vuo (Admiralty Island) (1.0 ha, calcareous sandstone), Draunibota (Cave Island) (1.9 ha, calcareous sandstone and limestone) and Labiko (Snake Island) are three small reserves in the Bay of Islands which are popular for recreational purposes (Dahl, 1980; Eaton, 1985). These are under the administration of the Ministry of Forests but there are plans to transfer them to the National Trust (Dunlap and Singh, 1980).   The adjacent reefs are not protected.

Management   None.

Recommendations   Suva Barrier Reef has been recommended for marine park and reserve status (Dunlap and Singh, 1980). Investigations should be made to determine whether the reefs are likely to be affected by pollution; if not, protected areas should be designated and classified and applications for foreshore leases made. Makuluva and its reefs are recommended as a marine reserve.   Both developments were originally intended to take place over the period 1981-1985.   Dahl (1980) recommends protection of the Rewa River delta and mangroves.   Guinea (1980) suggested that the seasnake *Laticauda colubrina* required protection which could be achieved by the creation of a reserve on Nukulau.

WAKAYA

Geographical Location   In the Lomaiviti Group; 17°39'S, 179°01'E.

Area, Depth, Altitude   3 sq. mi. (7.8 sq. km).

Land Tenure   Privately owned.

Physical Features   The island is volcanic, with high bluffs reaching 600' (183 m) on the west (Davis, 1920).

Reef Structure and Corals   Robinson (1971) surveyed three areas on the northern reefs of Wakaya in the course of a study of *Acanthaster* infestation, but gives few details of reef structure.

Noteworthy Fauna and Flora   Red Deer from New Caledonia have been introduced to the island. The island is covered largely by coconut plantations.

Scientific Importance and Research   The reef is considered to be of particular interest (Dunlap and Singh, 1980).

Economic Value and Social Benefits   Trochus is collected for food from the northern reefs, and the shell used to be collected for a button factory at Levuka until this closed in 1970 (Robinson, 1971).

Disturbance or Deficiencies   Robinson (1971) described high densities of *Acanthaster* on parts of the Wakaya reefs in the 1960s. The change in reef structure from stony corals to predominantly soft corals was attributed to *Acanthaster* predation since the island is comparatively isolated and has had no history of agricultural chemicals in recent years.

Legal Protection None yet.

Management None.

Recommendations Wakaya and its reefs are recommended for terrestrial and marine park status. The potential for developing a historical and marine park in co-operation with freehold landowners and the Fiji Museum should be investigated (Dunlap and Singh, 1980).

FRENCH POLYNESIA

INTRODUCTION

General Description

French Polynesia extends over ca 2 500 000 sq. km of ocean from 134°28'W (Temoe) to 154°40'W (Manuae or Scilly) and from 7°50'S (Motu One) to 27°36'S (Rapa). Emergent land totals about 4000 sq. km and there is about 7000 sq. km of lagoon. The islands are situated in a general NW-SE orientation, their age decreasing from north-west to south-east, and they form five distinctive archipelagos: Society, Tuamotu, Gambier, Marquesas and Tubuai or Austral Islands. There are around 130 islands, of which 84 are atolls; most of the remainder are high volcanic islands, many being very mountainous with inaccessible interiors. Salvat (1985) provides a classification of them based on geomorphologial and economic characteristics. Clipperton Island lies much further to the east and is under the authority of the French Government; it is described in a separate account in this section.

Annual rainfall affects the mean temperature through a warm rainy season from November to April and a relatively cool and dry season from May to October. The eastern trade winds predominate from October to March. From April to June, there are long calm periods broken by occasional cyclones, which generally arrive from the north-east and north-west. Within this general pattern there are significant differences between the archipelagoes (Teissier, 1969; SPREP, 1980). Cyclones have been rare in the past, averaging one per century to the north of the Marquesas, one to three per century from the Marquesas to the region north of the Tuamotu group, four to eight per century from the Tuamotu group to the Gambiers and one every two or three years in the Austral areas (Gabrié and Salvat, 1985). 1982/3 was exceptional in that five cyclones occurred (Nano, Orama, Reva, Veena and William), probably related to the abnormal El Niño of that period, and were accompanied by abnormally low sea levels (Rougerie and Wauthy, 1985; Harrison and Cane, 1984). Gabrié and Salvat (1985) provide a summary of the hydrology of the region. Tides are semi-diurnal, with an amplitude rarely exceeding 40 cm. Seawater temperature decreases southward and eastward to Rapa where the minimum temperature suitable for coral growth is found. Summer temperatures are 26-30°C and winter temperatures are 20-22°C.

Table of Islands

SOCIETY ARCHIPELAGO

The Society archipelago covers 720 sq. km and consists of nine high, volcanic islands and five atolls, divided into the Windward (Tahiti, Moorea, Maiao, Mehetia and Tetiaroa) and the Leeward Islands (Huahine, Raiatea-Tahaa, Bora-Bora, Maupiti, Tupai, Maupihaa (Mopelia), Manuae and Motu One) for the purposes of administration. They provide a classic example of a complete sequence of island forms, from active volcanoes to small atolls, moving northwards from Mehetia. Their

origins and geology are discussed by Pirazzoli (1985a) who summarises earlier studies, and provides a brief description of the vegetation and flora.

There are numerous publications on the reefs of Tahiti and Moorea, including early descriptions by Crossland (1927b, 1928a and b, 1939). Chevalier (1971 and 1977) describes reef origin in these areas.

Mehetia (Mahetia, Meetia) 2 sq. km; high volcanic, still active; 433 m; youngest island in Societies situated close to "hot spot" (Talandier, 1983); no true fringing reefs but coral colonies found on submarine volcanic slopes, particularly *Pocillopora*; research expedition in 1983 (Anon., 1983); privately owned; introduced pigs and goats.

Tahiti (X) 1042 sq. km; high volcanic; 2241 m and 1323 m; marine environment extensively studied; reef origins described by Chevalier (1971); discontinuous fringing reefs with a chain of barrier reefs enclosing a lagoon in some places, frequently interrupted (Chevalier, 1973); early description by Crossland (1928a, b and 1939); primary production in lagoons of Faaa and Vairao described (Ricard, 1974 and 1976a); Baie de Port Phaëton and other urban areas studied 1981 (see main text); barrier reef slope on north-west surveyed by Salvat *et al.* (1985) between 70 and 1100 m depth using submersible as part of OTEC power plant project: 0-100 m, slope covered with living coral reef; 100- 200 m a subvertical cliff of dead coral; 200- 250 m encrusting corals on a volcanic substrate; 250-500 m basaltic outcrops and sediment channels; 500-1100 m a gentler slope with sedimentary deposits and coral debris; species are being identified from photographs; avifauna described in Thibault (1975a).

Tetiaroa (X) 1288 ha; atoll with 13 islets around enclosed lagoon; 6 islets have seabird rookeries, including Tahuna Rahi and Tahuna Iti; privately owned with hotel; avifauna described in Thibault (1976).

Moorea (X) 136 sq. km; high volcanic; 1207 m; 25 km north-west of Tahiti; triangular in shape with ca 61 km of coastline and two distinct bays (Opunohu and Cook) on north coast; barrier reef, intersected by twelve passes corresponding to the principal valleys and submerged at high tide, circles the island enclosing shallow lagoon 500-1500 m wide and generally 0.5-3 m deep; deeper channels in lagoon run parallel to the coast, opening into the passes; four coral islets (Fareone, Tiahura, Irioa and Ahi); marine environment extensively studied; Galzin and Pointier (1985) provide detailed description of the island including the lagoon at Vaipahu, Paroa, Paevaeva and Afareaitu; corals described by Chevalier and Kühlmann (1984), lagoonal zooplankton by Lefevre (1985), planktonic productivity by Sournia and Ricard (1975b), foraminiferal distribution by Venec-Peyre (1985); less sedimentation than on Tahiti; (*see separate accounts for* Tiahura and Temae).

Maiao (Tubuai-Manu) (X) 9.5 sq. km; almost atoll with volcanic ridge; 154 m; 7 islets on barrier (Tapuaemanu) reef; ridge flanked by coral flats and barrier reef; avifauna described in Thibault (1973a).

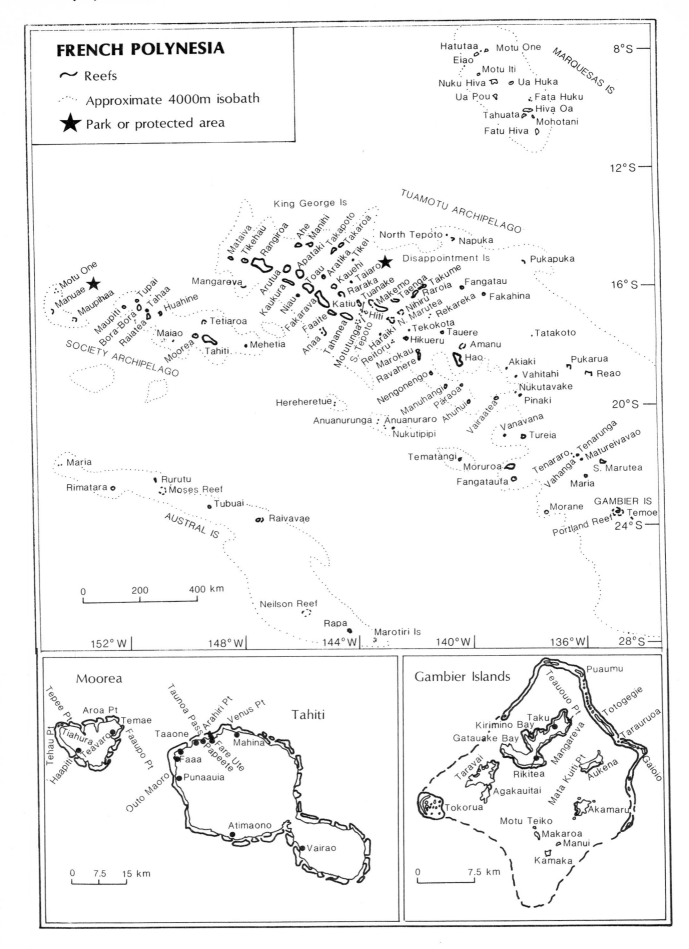

FRENCH POLYNESIA

~ Reefs

⋯ Approximate 4000m isobath

★ Park or protected area

8°S

Hatutaa • Motu One
Eiao ⬡
• Motu Iti
Nuku Hiva ⬡ ○ Ua Huka
Ua Pou ○
• Fata Huku
Hiva Oa
Tahuata ◦ Mohotani
Fatu Hiva ◊

MARQUESAS IS

12°S

King George Is

TUAMOTU ARCHIPELAGO

Mataiva
Tikehau
Rangiroa
Ahe Manihi
Arutua Apataki Takapoto
Kaukura Toau Aratika Takaroa
Niau Tikei
Fakarava Kauehi Taiaro
Faaite Raraka
Katiu Tuanake
Anaa Hiti
Tahanea Makemo
Motutunga S. Tepoto
Reitoru Haraiki N. Marutea
Marokau Rekareka
Ravahere Tekokota
Nengonengo Hikueru Amanu
Manuhangi Hao
Paraoa
Ahunui
Nukutipipi
Anuanurunga Anuanuraro

North Tepoto • Napuka

Disappointment Is ★

• Pukapuka

Taenga Takume
Nihiru Raroia
Tauere

Fangatau
Fakahina

Akiaki
Vahitahi
Nukutavake
Pinaki

Pukarua
Reao ⌐

Vairaatea
Vanavana

Motu One
Manuae ★
Maupihaa
Maupiti
Bora-Bora Tupai
Raiatea Tahaa
Huahine
Tetiaroa
Maiao
Moorea
Tahiti Mehetia

SOCIETY ARCHIPELAGO

Mangareva

16°S

Hereheretue

20°S

Tenararo Tenarunga
Vahanga Matureivavao
S. Marutea

Tematangi
Moruroa
Fangataufa

Maria

GAMBIER IS
Morane Temoe
Portland Reef
24°S

Maria
Rimatara
Rurutu
Moses Reef
Tubuai
Raivavae

AUSTRAL IS

0   200   400 km

Neilson Reef

Rapa
Marotiri Is

152°W   148°W   144°W   140°W   136°W   28°S

Moorea

Tepee Pt
Tehau Pt
Aroa Pt
Tiahura Temae
Haapiti Pt Teavaro
Faaupo Pt
Outo Maoro

Taunoa Pass Arahiri Pt
Taaone
Fare Ute
Faaa Papeete
Punaauia

Venus Pt
Mahina

Tahiti

Atimaono

Vairao

0   7.5   15 km

Gambier Islands

Teauouo Pt
Puaumu
Taku
Kirimino Bay
Gatauake Bay
Taravai
Rikitea
Agakauitai
Tokorua
Motu Teiko
Makaroa
Manui
Kamaka

Mangareva
Mata Kuiti Pt
Aukena
Akamaru

Totogegie
Tararuruoa
Gaioio

0   7.5 km

Huahine Nui and Huahine Iti (X) 73 sq. km; twin islands; volcanic; 435 m and 669 m; described by Pirazzoli (1985a); surrounded by narrow barrier reef with five passes; barrier reef raised at north to form cultivated terrace; fringing reef around island largely dead; major interdisciplinary research study currently underway (Gabrié *in litt.*, 16.11.87); marine survey carried out by Randall and Amesbury (1987); avifauna described in Thibault (1973b).

Raiatea and Tahaa (X) 194 sq. km and 88 sq. km; high volcanic; 1017 m and 590 m; barrier reef encircling both islands; lagoon contin ous apart from two short sections on west and penetrates deeply into bays, becoming larger north of Tahaa; avifauna described in Thibault (1973b).

Bora-Bora (Pora-Pora) (X) 30 sq. km; high volcanic; 727 m; deep large lagoon with one pass; well developed reef islands on wide barrier reef; reef and lagoon described by Guilcher *et al.* (1969); Baie de Nunue described by Gabrié and Porcher (1985a); Pte Matira described by Porcher (1984); considered one of the most beautiful islands in the world; described by Pirazzoli (1985a).

Tupai (Motu Iti) (X) 8 sq. mi. (21 sq. km); only atoll in Society Islands; barrier reef with two motu; narrow shallow closed lagoon with numerous coral patches and abundant phytoplankton; described in Pirazzoli (1985a); scientific studies carried out in 1983 (Delesalle, 1984); avifauna described in Thibault (1973c); partly private.

Maupiti (X) 13.5 sq. km; almost-atoll with small residual volcanic island (380 m) separated from a wide barrier reef flat by shallow, partially reticulated lagoon; barrier reef with well developed islands; central island flanked by well-developed fringing reef; considerable freshwater inflow; abundant reef life and lagoon fish fauna; described in Pirazzoli (1985a).

Maupihaa (Mopelia, Mopihaa) (X) 1 sq. mi. (2.6 sq. km); atoll with many islets; reef with narrow pass; coralline algal ridge; coral plays very small role in reef formation; described by Guilcher *et al.* (1966 and 1969); water movements described by Berthois *et al.* (1963).

Manuae (Scilly, Fanuaura) (X) (*see separate account*).

Motu One (Bellingshausen) (X) closed atoll with 4 islets; triangular reef.

TUAMOTU GROUP

The Tuamotu group consists of 76 islands which, apart from Makatea, are low atolls, either closed or with one or more passes or "hoa". They cap the tops of cones which rise steeply, not from the ocean floor, which is 4000-4500 m deep in this region, but from a huge ridge which forms wide shelves ranging in depth from 1500 to 3000 m deep. Geomorphological and geochronological evidence suggests that they are much older than the other groups of islands (Montaggioni, 1985). They lie within the hurricane belt. There are some mangroves along lagoon shores. The lagoons have high benthic populations and biomass of invertebrates, *Tridacna maxima* generally being the most important with the sea cucumber *Halodeima atra* (Salvat, 1971a and b). The hoa are described by Chevalier (1972).

Early descriptions of the reefs included Ranson (1958). The exterior reefs of the atolls are very exposed and have an algal ridge, followed by a spur and groove formation and a small drop-off which receives the impact from the waves. The outer slope drops from 5 to 40 m and has massive corals, with *Acropora* and *Millepora*. There is little or no coral on the reef flat. The inner slopes on the lagoonside are mainly sandy with a few corals. Patch reefs may be present in the lagoon. The reefs of Rangiroa, Takapoto, Tikehau, Taiaro, Mataiva and Moruroa (Mururoa) are described in separate accounts.

Morane atoll; enclosed lagoon, 3 islets.

Maria atoll; enclosed lagoon.

S. Marutea atoll; 11 mi. (18 km) long with islets; 1 pass into lagoon.

Actaeon Group:

- *Matureivavao (Maturei Vavao)* atoll; low with enclosed lagoon; benthos of lagoon studied (Renaud-Morant *et al.*, 1970);

- *Tenarunga* atoll; low, enclosed lagoon;

- *Vahanga* atoll; low, enclosed lagoon;

- *Tenararo (Tenaroa)* atoll; low, enclosed lagoon.

Fangataufa oblong atoll; lagoon 40-42 m deep (10 km); 1 pass into lagoon; reefs consist of coralline edge, reef-flat, inshore belt; hydrozoan described (Redier, 1967); molluscs studied on four reefs (Salvat, 1970b and 1972); terrestrial fauna described in Thibault (1987).

Moruroa (Mururoa) (*see separate account*).

Tematangi atoll; low with enclosed lagoon.

Tureia (X) atoll; low, with enclosed lagoon.

Vanavana atoll; narrow strip of land enclosing lagoon.

Duke of Gloucester Group:

- *Nukutipipi* atoll; 4 km; enclosed lagoon; badly damaged by hurricanes; privately owned; proposed for future studies (Salvat, 1987);

- *Anuanurunga* atoll; 4 islets on reef;

- *Anuanuraro* (X?) 5 km enclosed lagoon; privately owned.

Hereheretue (X) atoll; enclosed lagoon; molluscs studied on 2 reefs (Richard, 1970).

Reao (X) narrow atoll and enclosed lagoon (22 x 4 km); molluscs studied on 5 reefs (Richard, 1970); benthic fauna of lagoon described - densities of *T. maxima* 50-70 per sq. m over 370 ha of lagoon coast (Salvat, 1971b); *Porites mordax* and *Acropora formosa* very abundant in lagoon but coral diversity low (Chevalier, 1981).

Pukarua (Pukaruha) (X) atoll; enclosed lagoon (13 x 3 km).

Tatakoto   (X) atoll; low with enclosed lagoon (15 x 6 km).

Pinaki   (X) atoll; 3 islets to north-west of reef.

Nukutavake   (X) coral island with no lagoon (5 x 2 km).

Vairaatea   (X) 2 islands; barrier reef.

Vahitahi   (X) long atoll with enclosed lagoon.

Akiaki   small round island with enclosed lagoon.

Ahunui   enclosed lagoon.

Paraoa   enclosed lagoon; turtles.

Manuhangi   enclosed lagoon.

Nengonengo   nearly circular atoll; pearl-rich lagoon.

Hao   (X) atoll (56 x 15 km); 1 pass into lagoon; considerable research work (Salvat, 1982); growth study of *Turbo* by Villiers and Sire (1985).

Amanu   (X) atoll.

Ravahere   atoll; enclosed lagoon.

Marokau   (X)

Reitoru   atoll; enclosed lagoon.

Haraiki

Hikueru   (X) atoll (12 km); no passes into lagoon.

Tekokota   atoll; enclosed lagoon.

Tauere

Rekareka (Rekareta)   atoll; enclosed lagoon; no freshwater.

N. Marutea   atoll; submerged barrier reef.

Nihiru   (X) circular atoll with enclosed circular lagoon.

Pukapuka   (X?) atoll; enclosed, very shallow (less than 5 m) lagoon.

Fakahina (Fangahina)   atoll.

Fangatau (Angatau)   (X) atoll; enclosed lagoon (7 x 4 km).

Disappointment Group (Pukarua):

-   *Napuka*   (X) irregularly shaped atoll with closed lagoon and narrow reef; described by Crossland (1927a); study of fishing carried out since 1981 (Conte, 1985); proposed for future research (Salvat, 1987);

-   *N. Tepoto*   (X) no lagoon but central depression, 1 mi. (1.6 km) diameter; proposed for future research (Salvat, 1987).

Takume   (X) atoll (23 x 7 km); 2 passes into lagoon.

Raroia   (X) 334 ha (9 sq. mi. (23.3 sq. km)); oval atoll with many islets around lagoon, 6 ft (1.8 m) deep; geology described by Newell (1956); general information in Doty and Morrison (1954) and Newell (1954b); reefs and sedimentary processes described (Newell, 1954a); molluscs described by Morrison (1954) and fish by Harry (1953).

Taenga   (X) southern reef awash.

Makemo   (X) 2 passes into lagoon.

Katiu   (X) atoll (24 x 13 km); low; 2 passes into lagoon.

Hiti   atoll; enclosed lagoon.

Tuanake   small boat entrance only to lagoon.

S. Tepoto (Eliza)   atoll; small boat entrance only to lagoon.

Motutunga   atoll.

Tahanea   atoll; 3 passes into lagoon.

Anaa   (X) atoll; 11 islets, with enclosed lagoon (30 x 10 km); ca 600 million individuals of *Cardium fragum* estimated in lagoon (Richard, 1982a); proposed for future research (Salvat, 1987).

Faaite   (X) atoll.

Fakarava (Fakareva)   (X) rectangular atoll with islets confined to east of lagoon (56 x 24 km); 2 passes into lagoon.

Raraka   (X) circular atoll.

Taiaro   (X) (*see account for* W.A. Robinson Integral Reserve and Biosphere Reserve).

Kauehi (Kaueki)   (X) circular atoll.

Aratika   triangular atoll (37 x 24 km) with 2 passes into lagoon.

Toau (Toua)   untouched by ciguatera poisoning.

Niau   (X) atoll; 5 m; elliptical with completely enclosed lagoon; fringing reef.

Kaukura   (X) atoll; 2 narrow passes into lagoon (47 x 13 km).

Apataki   (X) atoll (30 x 24 km) with 3 passes into lagoon; described by Blanchet (1978).

Arutua   (X) circular atoll (28 km); 1 pass into lagoon.

King George Islands:

-   *Tikei*   (X) small low coral island (3 m); fringing reef;

-   *Takapoto*   (X) (*see separate account*);

-   *Takaroa*   (X) atoll (28 x 8 km); 1 pass into lagoon;

- *Manihi* (X) atoll; shoaly lagoon; coral diversity greatest near pass; *Leptoseris* and *Pachyseris* found only here (Chevalier, 1981);

- *Ahe* (X?) atoll (24 x 9 km); 1 pass through lagoon.

Makatea (X) 28 sq. km (7 x 4.5 km); only raised atoll in French Polynesia, 113 m; terraced with central hollow; partly surrounded by fringing reef extending out ca 100 m from base of vertical cliffs which flank almost all coast; described by Doumenge (1963); coral cover low; coral diverse on outer slopes only; geology and reefs described by Montaggioni (1985), Montaggioni *et al.* (1985a and b); research project in 1982 (Montaggioni *et al.*, 1983; Salvat, 1982); avifauna described in Thibault (1972) and Thibault *et al.* (in press).

Rangiroa (X) (*see separate account*).

Tikehau (X) (*see separate account*).

Mataiva (Matahiva) (*see separate account*).

GAMBIER ISLANDS

The Gambier Islands are often considered part of the Tuamotu archipelago, and are the easternmost part of French Polynesia, lying between 23° and 23°18'S, and 134°51' and 135°07'W (550 km from Pitcairn but some 1600 km from Tahiti). The islands and reefs cover an area of 35 km from north to south and 30 km from east to west. Mangareva (the largest), Taravai, Aukena and Akamaru are the four principal islands and are of particular interest in that all four volcanic islands are surrounded by a single barrier reef. They are very rugged and highly dissected, with large bays dissecting the coasts and narrow insular shelves on which fringing reefs are found. Brousse *et al.* (1974) describe the submarine platforms and terraces, the climate and the geomorphology of fringing and barrier reefs. The dominant winds are the north-easterly trade winds. Cyclones are rarer than in the Tuamotu group, although one in 1906 caused considerable damage (Teissier, 1969). Tides are semi-diurnal with a greater range than on the Tuamotu atolls, reaching 80 cm at Rikitea on Mangareva. The main current in the area is the South Equatorial, flowing north south or north-east/south/west. Currents in the passes may be greater than 2 knots. Water temperatures average about 26°C in March and 22°C in August. In all seasons there is a thermocline at 200 m or more depth. There is a considerable swell on the exposed eastern side of the island group and in the south (Brousse *et al.*, 1974). Further general information is given in Brousse (1974), Chevre (1974) and Moniod (1974).

The barrier reef which surrounds the Mangareva group is continuous over a distance of 90 km. The crest is emergent or very shallow over 42% of this distance in the north-west, north, east and the south-west corner. Elsewhere it is submerged to depths of 15 m, and there are no deep passes. Unlike most of the Tuamotu atolls, there are few motus (about 25) on the reef (Brousse *et al.*, 1974). The south is the most open, so that the southern sides of the islands tend to be very exposed, particularly Kamaka and the islets of Manui and Makaroa. The western sides of the islands are the most sheltered. Sedimentation near the coast, particularly in the bays, is considerable. The large bays of Gatavake, Kirimiro and Taku are turbid and there is little calcareous algal or coral growth. Pirazzoli (1985b) has mapped the northern part of the lagoon using satellite imagery.

Chevalier (1974) describes the distribution of corals on the barrier reef, fringing reefs, pinnacles and the lagoon bottom. Brousse *et al.* (1974) describe the outer slope of the reef at a number of sites, concentrating on the north, east and north-west where there is a full reef system. The slope is gentle to 25-30 m and then drops increasingly steeply, with submarine platforms occurring in some places at depths of 10-12 m and 18-22 m. A spur and groove system has been observed to depths of about 20 m, but is not as diverse and rich as on other atolls in the Pacific, possibly due to the gentle slope of the outer slope. The barrier reef is 200-800 m in width, and is widest in the north-east between Puaumu and Gaioio, and particularly wide (1100 m) at Tokorua. Coral fauna is poorer than that of Tahiti. On the external slope of the barrier reef, corals are rare and disappear completely below 30-40 m, perhaps due to upwellings although this has not been proven (Chevalier, 1974 and 1981). The coral fauna of the fringing reef is variable and in general rather poor; that of the lagoon is rich due to the numerous pinnacles. The Gambier Lagoon is unusual for the numerous soft coral communities found there; it is suggested that some environmental change (cyclones, temperature drops) caused replacement of the scleractinian fauna by more resistant alcyonarians (Chevalier, 1981; Tixier-Durivault, 1969). Octocorals are described by Tixier-Durivault (1974).

Temoe (Timoe) 6.9 sq. km; low coral atoll (1.8 m) with lagoon (23 m max. depth) enclosed by reef 100 yds (91 m) wide; lagoon mapped by Pirazzoli (1985b).

Mangareva (X) 5 sq. mi. (13 sq. km); high volcanic (445 m), fringing reef on exposed south coast, 150 m wide, sometimes discontinuous; well developed reef on east south of Rikitea, near Pointe Teonekura and to north-east of village between Pointe Kureru and Pointe Teauouo; in Rikitea Bay reefs are largely covered with sand; wide (100-500 m) reefs are found on the gentle slope, in very shallow water at low tide, on the north and west sides, mainly around headlands (Brousse *et al.*, 1974); molluscs described by Richard (1974b).

Aukena 0.5 sq. mi. (13 sq. km); volcanic (198 m), reefs best developed (unusually) on more exposed south-east coast; fringing reef extends from point at Teanakoporo to point at Mata Kuiti, about 500 m wide with some rich coral and algal growth; on north-west small discontinuous reefs found (Brousse *et al.*, 1974).

Akamaru (X) 0.7 sq. mi. (1.8 sq. km); volcanic (246 m), few reefs due to exposure, always non-emergent.

S. Lagoon Islets:

- *Manui* volcanic,; reefs virtually absent;

- *Kamaka* volcanic, 176 m; few reefs;

- *Makaroa* volcanic; reefs virtually absent; visited for fishing;

- *Tarauruoa* volcanic; reefs virtually absent;

- *Totegegie* volcanic; reefs virtually absent;

- *Motu-Teiko* volcanic; reefs virtually absent.

Taravai (X) 2.2 sq. mi. (5.7 sq. km); volcanic (256 m); south coast exposed with only narrow reefs, sometimes discontinuous; east coast also exposed, reefs well developed only to north of village of Taravai; a reef also occurs in the shallow channel between this island and Agakauitai; discontinuous fringing reefs on west coast (Brousse *et al.*, 1974).

Agakauitai (X) volcanic, 479 ft (146 m); reefs same distribution as those around Taravai (Brousse *et al.*, 1974).

AUSTRAL ISLANDS

The Austral Islands lie between 144° and 154°W and comprise seven islands lying on a NW-SE axis crossing the Tropic of Capricorn, to the south and south-east of the rest of French Polynesia. The two main islands are Rapa and Raivavae, which make interesting comparisons, the former having a temperate climate and the latter a tropical climate.

The coral fauna of the Australs is considered to be particularly interesting (Pichon, 1985a). Despite slightly cooler waters, some genera, such as *Galaxea, Goniastrea* and *Turbinaria*, which are absent from more typically tropical islands are found there. This may be due to larval transport from more faunally diverse areas than French Polynesia, such as the Cook Islands to the west.

Marotiri Is (Bass Is) 0.1 sq. mi. (0.26 sq. km); 9 volcanic rock pinnacles (105 m) without vegetation, including Marotiri Nui, Marotiri Iti and Vairiavai; corals described by Faure (1985 and 1986).

Rapa (X) (see *separate account*).

Raivavae (Raevavae) (X) 16 sq. km; 9 km long; high volcanic (437 m); fringing reefs and almost continuous barrier reef and reef islets; molluscs of lagoon and barrier reef studied (Salvat, 1971c and 1973a); lagoon fauna poorer than that of Gambiers, despite similar latitude; fauna of outer reefs similar; avifauna described in Thibault (1974c).

Tubuai (X) 48 sq. km; high volcanic (422 m); surrounded by barrier reef with 7 islets; geomorphology described by Brousse *et al.* (1980); rich coral fauna cf. other Austral Is. (77 species, including ssp. not found elsewhere in Polynesia) (Chevalier, 1980 and 1981); algae described by Denizot (1980), fish by Plessis (1980) and avifauna by Thibault (1974c).

Rurutu (X) 29 sq. km; high volcanic (389 m), some elevated reef limestone makatea; geology described by Bardintzeff *et al.* (1985).

Rimatara (X) low (95 m), volcanic and makatea; fringing reef.

Maria (Hull) 0.5 sq. mi. (1.3 sq. km); 4 islets, dense atoll forest triangular reef, shallow lagoon.

MARQUESAS

The Marquesas are the most northerly islands in French Polynesia and are situated in the west-flowing South Equatorial Current, some 5500 km west of the Galapagos between 7°50' and 10°35'S and 138°20' and 140°30'W. They are at a greater distance from a continent than any other island in the world (Brousse *et al.*, 1978). They form two groups with a total land area of 1274 sq. km: one in the north-west with three principal islands (Ua Huka, Nuku Hiva and Ua Pou) and several islets (Motu Iti, Eiao, Hatutaa) and banks (Hinakura, Motu One, Clark and Lawson); and a group 111 km to the south-east with four main islands (Fatu Hiva, Hiva Oa, Tahuata, Mohotani) and several islets, rocks and banks (Thomasset, Fatu Huku and Dumont d'Urville). The islands are volcanic and their geology is described by Brousse *et al.* (1978). The coasts are generally heavily embayed with high cliffs, and are largely rocky. Sandy beaches are rare except at the mouth of bays, apart from a large beach on the east coast of Eiao. Once forested, the islands are largely denuded except at altitudes over 900 m (Halle, 1978). Depths between the islands, apart from the channel between Hiva Oa and Tahuata, average about 2000 m.

The climate is described by Cauchard and Inchauspe (1978). Air temperature varies very little and is usually above 25°C; annual rainfall is between 700 and 1400 mm, most falling in June, unlike neighbouring archipelagos where most rainfall falls at the beginning of the year. However it is very irregular and there are often long periods of virtual drought. The west coasts of Nuku Hiva, Ua Huka and Hiva Oa, the north-west of Ua Pou and the north of Tahuata receive little rain and in general have a desert appearance. Cyclones and tropical storms are extremely rare unlike the rest of French Polynesia. Easterly trade winds predominate. In general there is little seasonal variation although usually the first six months of the year are warm and wet compared with the second half of the year. A brief history of the islands is given by Brousse *et al.* (1978).

The Marquesas are the only volcanic islands in French Polynesia not to be surrounded by uninterrupted fringing or barrier reefs. Brousse *et al.* (1978) and Chevalier (1978) and Glynn and Wellington (1983) discuss the poor development of reefs. Crossland (1927a) provides an early description of reefs on Fatu Hiva and Tahuata in the southern group and Nuku Hiva in the northern group. There are small reefs, of fringing or patch morphology, mostly confined to shallow depths, usually 10 m or less, in bays or along protected shores (Brousse *et al.*, 1978; Chevalier, 1978). Corals and calcareous algae are also found in dispersed patches around most of the coast, including a scattering on the external slope and submarine cliffs. Two types of reef are distinguished: those that are "embryonic" and poorly developed (possibly comparable to the apron reefs of other authors) as at Taiohae (Nuku Hiva), Hane (Ua Huka) and Hanamate, Punahe and Taaoa (Hiva Oa); and true reefs of less than one km in length, some of which (Baie du Controleur and Hakatea on Nuku Hiva) are rudimentary and others which are well developed (Anaho on Nuku Hiva, Hanaiapa on Hiva Oa; and reefs on Tahuata, some of which (Motopu, Hana Hevane and Hana te Toi) have reached maturity.

Chevalier (1978) suggests that these reefs are very young, the reef framework being only 3-5 m thick. The coral fauna is poor, 26 species having been identified. The most common genera are *Millepora, Pocillopora* and *Porites*, only three or four species being abundant everywhere, and the genus *Acropora* and the families Mussidae and Faviidae are unknown (Chevalier, 1978 and 1981). One or a few corals (e.g. *Porites lobata, Pocillopora spp.*) are often the predominant structural elements and crustose algae may be important (Chevalier, 1978; Crossland, 1927a). Algal crests are absent although other calcareous algae communities have been found. Phytoplankton productivity is high (Sournia, 1976).

Motu One (Ilot de Sable)   cay situated on a volcanic plug, with no vegetation; consists of sand and formation regularly changes; fringing reef (Brousse *et al.*, 1978; Chevalier, 1978; Crossland, 1927a); the only coral island in the Marquesas; to the east of the islet is a large stand of *Porolithon* and calcareous algae considered unique in French Polynesia; *Chelonia mydas* nesting area.

Hatutaa (Hatutu)   1813 ha (8 x 3 km); high volcanic; 420 m; seabirds.

Eiao   52 sq. km (13 x 7 km); high volcanic; 577 m.; seabirds (Thibault *in litt.*, 17.2.88).

Motu Iti   3 low barren dry islets.

Nuku Hiva   (X) 330 sq. km (32 x 20 km); high volcanic, 1208 m; fringing reef in some bays; those at Baie de Taiohae, Crique des Tai-Oa, Baie de Controleur, Baie d'Anaho, Baie de Hatiheu, and Baie de Haaopu described by Brousse *et al.* (1978) and Chevalier (1978); red tide recorded (Sournia and Plessis, 1974).

Ua Huka   (X) 77 sq. km (15 x 8 km); high volcanic, 854 m; reefs in Baie de Hane, Baie de Vaipaee and Baie Hatuana described by Brousse *et al.* (1978) and Chevalier (1978); islet (Motupapa) with seabirds (Thibault *in litt.*, 17.2.88).

Ua Pou   (X) 105 sq. km. (15 x 10 km); high, volcanic, 1252 m; described by Brousse (1978); reefs in bays of Hakahetau, Hakanahi, Hakahau, Hohoi, Hakatao and Hakamaii described by Brousse *et al.* (1978) and Chevalier (1978); flat-topped islet (Motuoa) with seabirds (Thibault *in litt.*, 17.2.88).

Fatu Huku (Fatu Hutu)   1.3 sq. km (4.5 x 0.8 km); dry rocky islet; 359 m; seabird rookery

Hiva Oa   (X) 320 sq. km (35 x 13 km); high, volcanic; 1190 m; fringing reef in Baie Taaoa, Baies de Punahe, Hanamate, Puamau, Hanaiapa described by Brousse *et al.* (1978) and Chevalier (1978).

Tahuata   (X) 50 sq. km (15 x 9 km); high, volcanic; 1050 m; described in detail by Brousse (1978); fringing reef in Baie de Motopu, Hana Hevane, Hana de Toi and Baies de Vaitahu and Hapatoni described by Brousse *et al.* (1978), Crossland (1927) and Chevalier (1978); reefs most abundant at Hana Hevane and Motopu.

Mohotani (Motane, Mohotane)   15 sq. km (8 x 2 km); 520 m; dry; described by Brousse (1978) and Sachet *et al.* (1975); seabirds (Thibault *in litt.*, 17.2.88).

Thomasset Reef (Ariane Rock)   isolated rocky islet.

Fatu Hiva   (X) 80 sq. km (15 x 7 km); high, volcanic; 960 m; described by Brousse (1978); very few corals; reefs at Pointe Tataaihoa and Baie d'Omoa described by Brousse *et al.* (1978) and Chevalier (1978).

(X) = Inhabited

The geomorphology and geology of the reefs of French Polynesia have been described by Chevalier (1973). Gabrié and Salvat (1985) provide a summary of general reef information. The main reef formations are found around the high islands and atolls although there are several oceanic banks of variable forms (e.g. Ebrill Reef in the Gambiers and Moses Reef in the Australs). Pichon (1985a) describes organic production in some of the reefs. Some mapping of the reefs using satellite imagery has been attempted (Pirazzoli, 1985b) in the Gambiers, the Tuamotu group and of the submerged atoll Portland Reef, south of the Gambiers.

Lying at the easternmost extremity of the Indo-Pacific Province, French Polynesia is at the limit of the axis of decreasing species richness and has a comparatively poor coral fauna. This is accentuated by the prevailing currents and winds which hinder the dispersal of larvae from the Western Pacific, the comparatively low water temperature compared with the Western Pacific, and the remoteness of the islands from continental masses. Many Western Pacific genera are not found (e.g. *Symphyllia, Oulophyllia, Seriatopora, Goniopora* and the families Merulinidae and Euphyllidae) and only 18 species seem to be endemic to the region. Other characteristics typical of a marginal area are a high species diversity within some genera (e.g. *Psammocora, Pocillopora, Leptoseris,* and to a lesser extent *Montipora*) and a comparative abundance of some taxa which are uncommon or absent in the central Indo-west Pacific such as *Sandalolitha* and *Porites irregularis*.

Chevalier (1981) discusses the origin of the coral fauna. Scleractinia are described by Charaber (1979) and Chevalier (1979a and 1981) and listed in Pichon (1985b). 168 species in 51 genera have been identified, including a few ahermatypic forms such as *Culicia* and *Tubastraea* but not including deep-water ahermatypic corals. Although the Acroporidae show the highest species diversity, the Poritidae and Agaricidae are dominant in biomass, particularly around high islands, and the Pocilloporidae and some Faviidae are abundant in the atolls.

In general, the coral fauna of the high volcanic islands surrounded by fringing or barrier reefs is richer than that of low atolls of banks. Many species are found only on the reefs adjacent to volcanic islands. *Psammocora, Synaraea, Pachyseris, Pavona* and closely related genera are more abundant on fringing or barrier reefs than on atolls. Generally the fauna of the outer slope and reef rim varies only slightly from one atoll to another but the coral fauna of the lagoons may be very different depending on depth and degree of exchange with the open ocean. Open atolls have a richer fauna than closed atolls, and in the former, greatest coral diversity is to be found near the passes. In closed or semi-closed lagoons, the coral fauna may be much impoverished and dominated by *Porites* and *Acropora*. The Tuamotu and Society Islands have a moderately high coral diversity, with large reefs and a variety of reef types. These two groups of islands are of particular interest as they were the subject of Darwin's early studies of reef morphology and evolution (Glynn and Wellington, 1983).

Richard (1985b) summarises the history of coral reef research in French Polynesia. Following an agreement between the Muséum National d'Histoire Naturelle in Paris and the Centre d'Expérimentation Nucléaires in 1965, numerous missions were undertaken to study the marine life of the area within the context of an interdisciplinary study of island ecosystems (Salvat, 1976a). The Antenne du Muséum National d'Histoire Naturelle et de l'Ecole Pratique des Hautes Etudes (Antenne Muséum-EPHE) was established in 1971 on Moorea. The main objectives of its research programme are to study the wealth and productivity of reef ecosystems (Galzin and Pointier, 1985; Salvat, 1982). A bibliography of the main references on the reefs of French Polynesia is given in Gabrié and Salvat (1985). Around 70 islands have been visited by scientists from the Antenne Muséum-EPHE and publications have appeared on 30. Current reef research activities are described by Salvat (1987) and include studies using satellite imagery aimed at reef management.

The Tuamotu islands are best known scientifically (30 out of 76 islands visited), particularly Takapoto, Rangiroa, Reao, Taiaro, Moruroa and Fangataufa, which are now the object of a number of more detailed and long term studies. Facilities for research work on each island in the Tuamotu group are summarized in Salvat (1982). Moorea, Tahiti and Manuae in the Society Islands are also well known. Studies have been carried out in the Gambiers under the auspices of the Comité Scientifique du Service Mixte de Contrôle Biologique in the context of a programme to look at all islands within a certain distance of the nuclear testing being carried out on Moruroa (Brousse *et al.*, 1974). Missions were undertaken to the Marquesas in 1972 and 1973 also under the auspices of the Direction des Centres d'Expérimentations Nucléaires (Brousse *et al.*, 1978). There is a NASA Satellite Laser Ranging Station on Huahine in the Society Islands. Extensive work is also carried out by ORSTOM (Institut français de Recherche Scientifique pour le Développement en Coopération) which is based on Tahiti. This is largely fisheries oriented. Monnet *et al.* (1986) provide a bibliography of ORSTOM publications. Additional work on fisheries, particularly the pearl fishery, is carried out by EVAAM (Etablissement pour la Valorisation des Activités Aquacoles et Maritimes). In 1985 the University of California at Berkeley established the South Pacific Biological Research Station on Moorea in Cook Bay which will be carrying out work in collaboration with the Antenne Muséum-EPHE.

Richard (1985a) provides a list of all marine invertebrate, fish and algae species recorded in French Polynesia and a bibliography of references related to marine fauna and flora. The marine molluscs have been particularly well studied and publications include Richard (1974a and b), Salvat and Rives (1975) and Salvat (1967). A malacological expedition to the Australs took place in 1968 (Salvat, 1973a). Richard (1974b) and Salvat (1970a) describe the molluscs of the Gambiers. The marine

mollusc fauna of the Marquesas is described in Lavondes *et al.* (1973) and Richard and Salvat (1973). *Cypraea obvelata* is endemic to French Polynesia. *Tridacna maxima* is the only member of the family Tridacnidae to occur in French Polynesia (Richard, 1977); it is found in all the archipelagos except the Marquesas but finds conditions for its optimal development in the lagoons of the Tuamotu atolls. Although not exploited to any great extent at present, its high productivity has considerable potential.

Randall (1973 and 1985) describes the fish, 800 species in 90 families having been recorded. A total of 246 fish species have been recorded in the Gambiers (Fourmanoir *et al.*, 1974). The fish fauna of the Marquesas is described by Plessis and Mauge (1978); it seems to resemble that of the rest of Polynesia but is slightly less diverse.

Marine algae are listed by Payri and Meinesz (1985). Holyoak and Thibault (1984), Thibault and Thibault (1975), Thibault and Rives (1975) and Thibault (in press) describe the birds of French Polynesia and Garnett (1984) gives a brief overview of seabird distribution. The seabird fauna of the Marquesas is described by Ehrhardt (1978). Birds of the Gambiers are described by Lacan and Mougin (1974) and of the Society archipelago by Thibault (1974a).

Reef Resources

Fishing is an important activity, and a general review of fisheries is given by Ugolini *et al.* (1982). Improved marketing and management of the fishing industry are discussed by Grand (1985). The catch is generally consumed locally although exports are starting with the growth of Papeete as a port and the development of domestic flights. Traditional subsistence fishing is still carried out on many atolls and islands such as Napuka, in the Tuamotu group (Conte, 1985 and 1988), and the Marquesas (Baines, 1982). In the Tuamotu group, numerous studies have been carried out on lagoonal productivity and potential for aquaculture and exploitation of fish stocks (Plessis, 1969; Salvat, 1971a; Grand, 1983). Results of deep fishing trips on the outer slopes of some of the atolls are discussed in Manac'h and Carsin (1985). Fishing methods in the Gambiers are described by Fourmanoir *et al.* (1974). In the Society Islands, crabs from Maiao lagoon are sent to the Papeete market. Lagoon fisheries on Raiatea and Tahaa are described by James (1980). Gillett and Kearney (1983) provide an assessment of skipjack and baitfish resources.

Pearl oysters were intensively collected for their mother-of-pearl from the early 19th century to about 1950 when production started to fall due to overexploitation. Exploitation peaked at 1400 tons although maximum sustainable yield is probably about 1000 tons. Both the mother-of-pearl and pearl industries are now thriving as a result of the development of culture techniques (Coeroli, 1983; Real-Testud and Richard, 1984). Exports of the highly valued black pearls from the Black Lip Oyster *Pinctada margaritifera* have increased since 1972 due to increased culture operations following a decline in natural stocks in the 1970s. The pearl industry has had the highest export value of any product in Polynesia since 1983. Of the 152 fishing co-operatives in French Polynesia, 91 are concerned with mother-of-pearl and pearl culture. These are found on 18 atolls in the Tuamotus (Ahe, Amanu, Apataki, Aratika, Arutua, Faaite, Hao, Hikueru, Katiu, Kaukura, Makemo, Manihi, Marokau, Raraka, Taenga, Takapoto, Takaroa and Takume) and in the Gambiers.

Hatchery production has been found to be unsuccessful and natural stocks are used, either adults collected by divers or spats which are collected and reared. Growth of pearl farming has created a demand for living oysters which are able to withstand the grafting operation. Experiments on spat collection have been carried out largely in the Tuamotu and Gambier Islands, early experiments having been carried out in 1963 at Bora-Bora and in 1968 in Manihi and Takapoto. In 1976 a programme was established with the help of the Territorial Fishery Service and pilot stations were established on Takapoto, Hikueru and Rikitea. Since 1981 EVAAM and ORSTOM have been carrying out studies to improve the quality of production (Coeroli and Mizuno, 1985; Cabral *et al.*, 1985; Intes and Coeroli, 1985). Studies are being carried out on Takapoto, Manuae, Gambiers, Hikueru and Manihi (atolls listed in descending order of abundance of stocks). Culture experiments have been carried out on Tetiaroa. *Trochus niloticus* was introduced to Tahiti in the late 1950s and commercial exploitation began in 1971, when populations were estimated at 2500 tonnes. The industry is described by Yen (1985).

The tourist industry is expanding rapidly and tourists have increased in number from 82 822 in 1975 to 160 000 in 1986 (Anon., 1987; Gabrié *in litt.*, 16.11.87). This growth is facilitated by the improved domestic flight service and the construction of hotels on many islands. Reef-related activities are of growing importance. Resorts are found in the Society Islands on Tetiaroa, Tahiti, Moorea, where there are numerous hotels on the coast and reef-related recreational activities are catered for (*see accounts for* Temae and Tiahura), Huahine, Raiatea and Bora-Bora; in the Tuamotu group on Manihi and Rangiroa; in the Australs on Tubuai and Rurutu; and in the Marquesas on Nuku Hiva (Gabrié *in litt.*, 16.11.87; Thibault *in litt.*, 17.2.88). On Nuku Hiva there are two small hotels and one large one; however at present tourism is at a very low level because of the difficulty in obtaining flights to the island (Thibault *in litt.*, 17.2.88). As of January 1987, Tahiti and Moorea together accounted for 81% of total hotel capacity (Gabrié *in litt.*, 16.11.87).

Disturbances and Deficiencies

French Polynesia periodically suffers severe hurricane damage, as has been recorded at Hao (1903), Hikueru, Kaukura, Marokau and the Gambiers (1906) (Teissier, 1969). However, it is not known to what extent the reefs were damaged in these early events. Three of the five cyclones which affected French Polynesia in 1982-1983 passed close to the Society Islands and caused significant reef damage, as well as in other areas (Dupon, 1986). Furthermore, from mid-March to the end of May 1983, the mean sea level dropped by as much as 20-25 cm below normal in the Society Islands (Rougerie and Wauthy, 1983) causing extensive death to corals, algae and reef biota close to the surface. For example, large areas of reef died at Moorea. This prolonged period of low sea-levels and cyclonic disturbances was probably

related to the abnormal El Niño which occurred at that time (Glynn, 1984; Pirazzoli, 1985a). There is some evidence of recovery on the outer reef slopes (Salvat *in litt.*, 1986). The impact of these events on seabird colonies is not known, although Sooty Terns *Sterna fuscata* in the Marquesas deserted their nesting colonies for several months during 1983 (Thibault, *in litt.*, 17.2.88). There have been some outbreaks of *Acanthaster planci*, for example in Moorea (*see account for* Tiahura).

The population of French Polynesia has increased rapidly since the 1920s, reaching about 166 700 in 1984. Bellwood (1978) gives information on the early Polynesians, and Bayliss-Smith (1974) describes population growth on the atolls. Over 70% of the population is concentrated on the leeward islands, particularly Tahiti and then Raiatea, placing the reefs and marine environments of this region under greatest pressure. The other high volcanic islands are less at risk, due to their lower populations, but atoll environments are very vulnerable. Industrial activity is still slight but the reefs are being affected by dredging, coastline alteration, filling, and discharge of sewage (Porcher and Dupuy, 1985). The Gambier Islands are considered to be 98% devastated by man. The establishment of the Centre d'Experimentation du Pacifique in 1963 caused major changes on the islands; many new buildings were erected and an airport was constructed on the reef at Totegegie. A brief history of the people and culture of these islands is given by O'Reilly (1974) and Egron (1974).

74% of the population of Tahiti live between Mahina and Punaauia which includes a large stretch of lagoon, bays, rivers and major industrial and urban concentrations. A year long study has revealed widespread pollution (Fraizier *et al.*, 1985). The area between Faaa Airport and Venus Pt, which includes the harbour and its industrial zone, is particularly at risk, as well as North Punama, which is threatened by the increasing urban zone to the south of Tahiti. The harbour is being extended to include new storage tanks, an electric power plant operating on fuel oil, and various other facilities which could have an adverse impact on the environment. The concrete covering of the reef between the passes of Papeete and Taaone will increase the confinement of the lagoon waters and accelerate pollution which is already reported to be severe in the harbour, from ship traffic and hydrocarbon spillage, the discharge of urban sewage, and sedimentation caused by accelerating erosion and excavation of coral marl. There is clear evidence of reef degradation (Deneuebourg, 1971; Ricard *et al.*, 1981; Salvat, 1982; Poli *et al.*, 1984). The fringing reef at Taunoa, from Fare Ute to Arahiri Pt to the east of the pass at Taaone has been the subject of environmental impact studies (Fraizier, 1980; De Nardi *et al.*, 1983). The reef has coral cover of up to 65-85% at 3-4 m depth, but in deep water, there is noticeable siltation with much dead coral and living coral cover drops to 10%. The central part of the lagoon has no corals and much sediment. On the barrier reef, near the construction works, cover is as low as 20-50%, although it reaches 80% in more pristine environments. Environmental impact studies on the reefs and lagoon waters of the urban zone include Poli *et al.* (1983), Porcher *et al.* (1986) and Porcher (1986), and on the Papeete Port include Ricard *et al.* (1981 and 1986). Studies have also been carried out on the lagoon at Punaauia (Delesalle *et al.*, 1982) in preparation for a channel project, and at Outo

Maoro (Porcher *et al.*, 1985) in the north-west and Atimaono (Porcher *et al.*, 1987) in the south. IFREMER (Institut français de recherche pour l'exploitation de la mer) has been studying the site of a shore based OTEC pilot plant since 1982 and has carried out surveys of the reefs (Equipe de projet E.T.M., 1985b) and temperatures and currents (Equipe de projet E.T.M., 1985a) to assess potential environmental effects.

A major problem in the Society Islands is the use of coral sand from the lagoon as a source of road building materials. A number of studies to investigate the damage this causes have been carried out (Vaugelas, 1979; Gabrié *et al.*, 1985; Masson and Simon, 1985; Porcher *et al.*, 1985; Porcher and Gabrié, 1987).

There are about 12 dredging sites in the lagoon at Moorea; Galzin and Pointier (1985) describe those at Teavaro, Haapiti and Tiahura. Although currently illegal on Moorea, dredging is still taking place on the north-west coast (Gabrié *in litt.*, 16.11.87). Reefs around Moorea are also suffering from sedimentation resulting from agriculture on steep slopes and from tourist developments, particularly in the north-west; reefs from Pointe Tiahura through to the Club Mediterranée development are said to be badly damaged (*see separate account*). A number of environmental impact studies have been carried out on Moorea including Anon. (1977), Porcher and Bouilloud (1984) and Porcher and Gabrié (1987) and a study is under way to investigate the impact of human activities on the reef (Gabrié *in litt.*, 16.11.87).

On Maupiti coral extraction is a problem, and there is also pollution from agricultural runoff into lagoon, coral rock from the lagoon is used for road building and for ground levelling for tourist developments. It is thought that insecticides used in water melon cultivation may cause pollution (Plessis, 1973b). On Huahine, the fringing reef is largely dead and is becoming silted up and there is reef damage from coral and sand dredging for construction of the coastal road (Pirazzoli, 1985a). Environmental impact studies have been carried out on Bora-Bora at the sites of hotel developments (Porcher, 1984 and 1985a; Gabrié, 1986).

In the Tuamotu group, Makatea was intensively mined for phosphate from 1917 to 1966; it was once the most populated island in the Tuamotu group but is now inhabited by only 30 copra workers. There has been some reef damage in Manihi (Hicks pers. comm. 1985). In many of the Tuamotu atolls reef damage has been caused by military activities involving nuclear test sites, particularly on Moruroa (*see separate account*), Anaa, Fangataufa and Hao. Primary forest in the Marquesas has been destroyed on many islands by introduced herbivores (Brousse *et al.*, 1978; Halle, 1978).

Gabrié (*in litt.*, 16.11.87), writing specifically about Moorea, but by extension applying to other parts of French Polynesia, notes that the following recreational activities may have impacts on reefs: walking on the reef; collecting shells; diving and snorkelling; and a variety of motorised vessels used by tourists. These activities in general only take place in limited areas of reef, immediately in front of hotels, but the impact they have, and the relative importance of the different activities, is difficult to assess as no studies have been carried out. Tourists may also have disruptive effects on seabird colonies if they are allowed uncontrolled access, as at

Tahuna Rahi and Tahuna Iti on Tetiaroa (Thibault *in litt.*, 17.2.88).

Overfishing has been reported in a number of areas including Hikueru, Takume and Takaroa (SPREP, 1980). Ciguatera is a widespread problem and major research on this is carried out at the Louis Malarde Institute on Tahiti (Bagnis *et al.*, 1985). Incidences of red tides in the Marquesas are described by Taxit (1978) and Bagnis and Denizot (1978) discuss the problems associated with widespread ciguatera.

Legislation and Management

Traditionally, fishing rights in the Society Islands were monopolized by the upper classes who were both leaders and fishermen and the lower classes were rarely permitted to fish (Baines, 1982) but this has now ceased. Baines (1982) briefly mentions traditional fishing practises in the Marquesas where fishermen form a distinct class, accorded an inferior position in the social scale.

No overall environmental policy exists, although there is legislation for a certain number of ad hoc problems (SPREP, 1980). The Territory has full power over environmental matters. The law of 26 December 1964, modified on 16 May 1973 and 2 January 1979, concerning marine pollution by oil, is enforced, but the International Convention for the Prevention of Pollution of the Sea by Oil has not yet been enforced.

Legislation covers research in the marine environment, fisheries and aquaculture (SPREP, 1980). Arrêté no. 208/AA/Pêche of 29.1.69 prohibits the use of SCUBA equipment for spearfishing. Arrêté no. 2125 of 2.10.81 prohibits the use of SCUBA for all fishing or collection of all marine animals. Arrêté no. 150/SG of 18.2.46 prohibits the use of certain toxic substances for fishing. Arrêté no. 1942/AA of 9.7.70 controls fishing in the lagoon of Faaura Rahi on Huahine.

The Fisheries Department controls the marine environment, particularly pearl exploitation. Exploitation quotas are set every year for each lagoon and studies concerning restocking of over-exploited lagoons have been made with the assistance of CNEXO. Three fishing sectors for trochus have been defined on the Tahiti and Moorea reefs, one of which is exploited each year. Size limits and a fishery quota have been defined on the basis of work carried out by EVAAM. Nevertheless there was a marked decline in populations in 1984 and current efforts are directed at reversing this trend (Yen, 1985).

A number of land use planning studies, the responsibility of the Service de l'Aménagement, have been carried out, particularly on Tahiti (Porcher and Dupuy, 1985). Environmental impact assessments (detailed above) are the responsibility of the Délégation à l'Environnement. There have been some voluntary efforts on Moorea to remove *Acanthaster planci* from the reefs (Hicks *in litt.*, 1986). Coral dredging has been prohibited around the island, but enforcement is poor (Gabrié *in litt.*, 16.11.87).

Forestry Regulations contain provisions for the establishment of Strict Nature Reserves (Réserves Intégrales) but there is no provision for their supervision.

There are no other provisions for protected areas. The following areas have been established:

1. W.A. Robinson Réserve Intégrale and Biosphere Reserve (Tuamotu group) (*see separate account*).
2. Manuae (Scilly) Atoll Reserve (Society Islands) (*see separate account*).
3. The islets of Mohotani, Eiao, Hatutaa and Motu One (Marquesas); declared reserves under arrête no. 2559 of 28 July 1971 (Decker, 1973); it would seem that the lagoons around these islets are also protected.

Recommendations

Dahl (1986) stresses the importance of setting up a representative series of marine reserves throughout the country. Areas recommended for protection in Dahl (1980) are as follows:

Society Islands

- Tetiaroa: seabird sanctuary on Tahuna Iti; reserve with 400 m protective belts
- Moorea: representative selection of reef and lagoon habitats; Tiahura and Temae have been suggested (*see separate accounts*)
- Raiatea: reserve for complete estuary - lagoon - reef sequence in one of least devastated bays e.g. Faatema
- Tupai: reserve for seabird rookery, internal lagoons and barrier reef
- Maupihaa: reserve for seabird rookeries, turtle nesting areas and a selection of marine biomes
- Motu One: reserve for seabird rookeries, turtle nesting areas and a selection of marine biomes

Tuamotu group

- Anuanuraro: reserve for lagoon
- Anuanurunga: reserve for lagoon
- Apataki: reserve for seabirds and turtles
- Hereheretue: reserve for lagoon
- Kauehi: reserve for seabirds and turtles
- Matureivavao: reserve for birds and vegetation
- Napuka: reserve for seabirds and turtles
- Nukutipipi: reserve for lagoon
- Pukapuka: reserve for seabirds, turtles and lagoon
- Tekokota: reserve for seabirds and turtles
- Toau: reserve for atoll

Gambiers

- Manui: reserve for seabirds
- Motu-Teiko: reserve for seabirds

Marquesas

- Ua Huka: reserve for seabirds (Motupapa); proposed by Salvat (1974)
- Ua Pou: reserve for seabirds (Motuoa); proposed by Salvat (1974)

The Service de l'Amenagement has set up a project to preserve part of the coral reef ecosystem in the north-west part of Tahiti, including the barrier and channel reef (Porcher and Dupuy, 1985). The Délégation à l'Environnement is working with SPREP to develop reef and lagoon management plans for Tahiti and Bora-Bora (Holthus *in litt.*, 4.3.88). In Huahine, a major interdisciplinary project is under way aimed at developing a management plan for the island. This involves institutions and organizations in French Polynesia, France and other countries including the Universities of Guam and California (Gabrié *in litt.*, 16.11.87), and has included a marine survey of proposed protected areas (Holthus *in litt.*, 4.3.88). Studies are under way to develop improved management of the reefs of Moorea (Gabrié *in litt.*, 16.11.87). Salvat (1973b) stresses the importance of the Tuamotu archipelago and the need for protection of some of the atolls. Salvat (1974) has described conservation measures for the Marquesas; recommending the protection of Motupapa, an islet of Ua Huka, and Motuoa, an islet of Ua Pou, for seabirds. Motu Paio, on Rangiroa, has also been recommended as a seabird reserve (Thibault *in litt.*, 17.2.88). Clipperton is being considered for protection by the French Government (Bourrouilh-Le Jan *in litt.*, 28.9.87).

The exploitation of marine and terrestrial aggregates in Tahiti, Huahine, Raiatea, Tahaa, Bora-Bora and Maupiti is described by Porcher and Gabrié (1985 and 1987) in the context of two studies requested by the Service de l'Equipement de Polynésie française in 1984 and 1985. The results will include an inventory of dredging activities and recommendations for a limited number of dredging sites and the restoration of damaged areas. Additional studies will attempt to identify alternative materials (Gabrié *et al.*, 1985, Porcher et Gabrié, 1985 and 1987).

References

* cited but not consulted.

Aissaoui, D.M. and Purser, B.H. (1985). Reef diagenesis: Cementation at Mururoa Atoll. *Proc. 5th Int. Coral Reef Cong., Tahiti* 3: 257-262.
***Allison, E.C. (1959).** Distribution of *Conus* on Clipperton I. *Veliger* 1: 32-34.
***Anon. (1975).** Lutte contre la pollution du lagon de Tahiti. Rapport BCEOM, French Polynesia. 38 pp.
***Anon. (1977).** Moorea - mission Benthyplan (Benthos, Hydrologie, Plancton). Rapport Antenne Muséum-EPHE. 150 pp.
***Anon. (1979a).** Scilly - atoll de l'archipel de la Société Polynésie française. Compte rendu préliminaire d'une expedition scientifique interdisciplinaire et interorganismes. *Bulletin 1, Muséum/E.P.H.E.*
*Anon. (1979b).** Exploitation of coral sand in lagoons - French Polynesia and French West Indies. IUCN/WWF Project 1442.
Anon. (1983). Biologie marine et terrestre: Mehetia (juillet 1983). *Ofai* 5: 19-21.
Anon. (collective work) (1985). Contribution à l'étude de l'atoll de Tikehau. *ORSTOM Tahiti, Not. Doc. Océan.* 24. 138 pp.
***Anon. (1987).** Bilan et perspectives du tourisme en Polynésie française. Avril 1987. Service du Tourisme, Ministère du Tourisme de Polynésie française.
Anon. (n.d.a). Taiaro Reserve, W.A. Robinson Sanctuary Booklet.
Anon. (n.d.b). Situation Report. French Polynesia.
Atkinson, H. *et al.* **(1984).** Report of a New Zealand, Australian and Papua New Guinea Scientific Mission to Mururoa Atoll. Mimeo rept., New Zealand Ministry of Foreign Affairs, Wellington.

Bagnis, R. (1976). Faune ichtyologique du lagon de Taiaro. *Cah. Pacif.* 19: 283-286.

Bagnis, R. and Bennett, J. (1979). Distribution des Dinoflagellés potentiellement ciguatérigènes et contribution à l'inventaire de la faune ichtyologique de Scilly. *Bull. Ant. Tahiti Mus. Nat. Hist. Nat. et E.P.H.E.* 1: 49-51.

Bagnis, R., Bennett, J., Barsinas, M., Chebret, M., Jacquet, G., Lechat, I., Chanteau, S., Chungue, E., Drollet, J.H., Legrand, A.M., Mitermite, Y., Pompon, A., Rongeras, S. and Tetaria, C. (1985). Criteria of ciguateric risk evaluation in humans. *Proc. 5th Int. Coral Reef Cong., Tahiti* 4: 475-482.

Bagnis, R. and Denizot, M. (1978). La ciguatera aux Iles Marquises: aspects humains et biomarins. *Cah. Pacif.* 21: 293-314.

Bagnis, R., Galzin, R. and Bennett, J. (1979). Poissons de Takapoto. *J. Soc. Océan.* 35(62): 69-74.

Baines, G.B.K. (1982). Pacific Islands: Development of coastal marine resources of selected islands. In: Soysa, C., Chia, L.S. and Coulter, W.L. (Eds), *Man, Land and Sea: Coastal Resource Use and Development in Asia and the Pacific*. The Agricultural Development Council, Bangkok, Thailand. Pp. 189-198.

Bakus, G.J. (1975). Marine zonation and ecology of Cocos Island, off Central America. *Atoll Res. Bull.* 179. 8 pp.

Bardintzeff, J-M., Brousse, R. and Gachon, A. (1985). Conditions of building of coral reefs on a volcano: Mururoa in Tuamotu and Rurutu in Australes (French Polynesia). *Proc. 5th Int. Coral Reef Cong., Tahiti* 6: 401-405.

Bayliss-Smith, T.P. (1974). Constraints on population growth; the case of the Polynesian outlier atolls. *Human Ecology* 2: 259-295.

Bellwood, P. (1978). *The Polynesians: Prehistory of an island people*. Thames and Hudson, London. 180 pp.

*Berthois, L. *et al.* (1963). Le renouvellement des eaux du lagon dans l'atoll de Maupihaa-Mopelia (Iles de la Société). *C.R. Acad. Sci. Paris* 257: 3992-3995.

*Blanchet, G. (1978). L'atoll d'Apataki et la SCEP. *ORSTOM Tahiti Not. Doc. Sci. Humaines*: 1-51.

Blanchet, G. (1985). Socio-economic study of small-scale fishing in the atoll of Tikehau. *Proc. 5th Int. Coral Reef Cong., Tahiti* 5: 583-587.

*Bouchon, C. (1983). Les peuplements de scléractiniaires de l'atoll de Takapoto (Polynésie française). *J. Soc. Océan.* 39(77): 35-42.

Bouchon, C. (1985). Quantitative study of scleractinian coral communities of Tiahura Reef (Moorea Island, French Polynesia). *Proc. 5th Int. Coral Reef Cong., Tahiti* 6: 279-284.

*Bouchon-Navaro, Y. (1983). Distribution quantitative des principaux poissons herbivores (Acanthuridae et Scaridae) de l'atoll de Takapoto (Polynésie française). *J. Soc. Océan.* 39(77): 43-54.

Bouchon-Navaro, Y., Bouchon, C. and Harmelin-Vivien, M.L. (1985). Impact of coral degradation on a chaetodontid fish population (Moorea, French Polynesia). *Proc. 5th Int. Coral Reef Cong., Tahiti* 5: 427-432.

*Bourrouilh-Le Jan, F.G. (1971a). Existence et conditions d'une dolomitisation précoce à partir d'aragonite dans les eaux isolées de l'atoll de Clipperton (Pacifique Oriental) et sa liaison avec la synthèse d'hyrocarbures polycycliques dans ces mêmes eaux. *8th International Sedimentological Congress, Heidelberg, 1971. Add. Abstr.* P. 2.

*Bourrouilh-Le Jan, F.G. (1971b). Phosphates, sols bauxitiques et karsts dolomitiques du Centre et SW Pacifique. Comparaisons sédimentologiques et géochimiques. Colloque International. *Doc. BRGM* 24: 113-128.

Bourrouilh-Le Jan, F.G. (1983). Modern phosphate sedimentation in the enclosed lagoon of Clipperton Island. ENE Pacific. Programme International de Corrélations Géologiques 156 - Phosphorites. *6ème Réunion Internationale en salle et sur le terrain.* Abst.

Bourrouilh-Le Jan, F.G., Albouy, Y. and Benderitter, Y. (1985). A tool for a better knowledge of an atoll: The magnetic field at Clipperton Island (E-NE Pacific); geophysical and geological results. *Proc. 5th Int. Coral Reef Cong., Tahiti* 6: 407-412.

Bourrouilh-Le Jan, F. and Tallandier, J. (1985). Sédimentation et fracturation de haute énergie en milieu récifal: tsunamis, ouragans et cyclones et leurs effets sur la sédimentologie et la géomorphologie d'un atoll, motu et hoa, à Rangiroa, Tuamotu, SE Pacifique. *Marine Geology* 67: 263-333.

Brousse, R. (1974). Géologie et pétrologie des Iles Gambier. *Cah. Pacif.* 18(1): 159-244.

Brousse, R. (1978). Elements d'analyses de quelques îles des Marquises, Fatu Hiva, Tahuata, Motane, Ua Pou. *Cah. Pacif.* 21: 107-144.

*Brousse, R., Chevalier, J.P., Denizot, M., Richer de Forges, B., and Salvat, B. (1980). Etude géomorphologique de l'île de Tubuai (Australes). *Cah. Indo-Pacif.* 2(3): 1-54.

Brousse, R., Chevalier, J.P., Denizot, M. and Salvat, B. (1974). Etude géomorphologique des îles Gambier. *Cah. Pacif.* 18(1): 9-119.

Brousse, R., Chevalier, J.P., Denizot, M. and Salvat, B. (1978). Etude géomorphologique des Iles Marquises. *Cah. Pacif.* 21: 9-74.

Brousse, R. and Gelugne, P. (1986). Géologie et pétrologie de l'île de Rapa. In: DIRCEN (1986): 9-61.

Buigues, D. (1985). Principal facies and their distribution at Mururoa Atoll. *Proc. 5th Int. Coral Reef Cong., Tahiti* 3: 249-255.

Burlot, R., Faissole, F., Humbert, L., Leblanc, P., Pelissier Hermitte, G., Pouchan, P., and Giron, A. (1985). Relation between fresh and saline water in carbonate reef systems of Temae (Moorea). *Proc. 5th Int. Coral Reef Cong., Tahiti* 6: 1-5.

Cabral, P., Coeroli, M. and Mizuno, K. (1985). Preliminary data on the spat collection of mother-of-pearl (*Pinctada margaritifera*, Bivalve, Mollusc) in French Polynesia. *Proc. 5th Int. Coral Reef Cong., Tahiti* 5: 177-182.

Caire, J.F., Raymond, A. and Bagnis, R. (1985). Ciguatera: A study of the setting up and the evolution of a *Gambierdiscus toxicus* population on an artificial substrate introduced to an atoll lagoon, with following associated ambiance factors. *Proc. 5th Int. Coral Reef Cong., Tahiti* 4: 429-435.

Carsin, J.-L. (1985). The natural eutrophication of the waters of Clipperton lagoon: Equipments, methods, results, discussions. *Proc. 5th Int. Coral Reef Cong., Tahiti* 3: 359-364.

Cauchard, G. and Inchauspe, J. (1978). Climatologie de l'archipel des Marquises. *Cah. Pacif.* 21: 75-105.

*Charaber, J.P. (1979). La faune corallienne (Scléractiniaires et Hydrocoralliares) de la Polynésie française. *Cah. Indo-Pacif.* 1(2): 129-151.

Charpy, L. (1985). Distribution and composition of particulate organic matter in the lagoon of Tikehau. *Proc. 5th Int. Coral Reef Cong., Tahiti* 3: 353-358.

*Chevalier, J.P. (1971). Origine des formations récifales de l'Ile de Tahiti. *Bull. Soc. Et. Océan* 15(2): 53-58.

Chevalier, J.P. (1972). Observations sur les chenaux incomplets appelés Hoa dans les atolls des Tuamotu. *Proc. Symp. Corals and Coral Reefs, India*: 477-488.

*Chevalier, J.P. (1973). Geomorphology and geology of coral reefs in French Polynesia. In: *Geology and Biology of Coral Reefs* 1. Academic Press, New York. Pp. 113-141.

Chevalier, J.P. (1974). Aperçu sur les Scléractiniaires des Iles Gambier. *Cah. Pacif.* 18(2): 615-627.

Chevalier, J.P. (1976). Madréporaires actuels et fossiles du lagon de Taiaro. *Cah. Pacif.* 19: 253-264.

*Chevalier, J.P. (1977). Origin of the reef formations of Moorea Island (Society Archipelago). *Proc. 3rd Int. Coral Reef Symp., Miami* 2: 283-288.

Chevalier, J.P. (1978). Les coraux des Iles Marquises. *Cah. Pacif.* 21: 243-283.

*Chevalier, J-P. (1979a). La faune corallienne (Scléractiniaires et Hydrocoralliaires) de la Polynésie française. *Cah. Indo-Pacif.* 1(2): 129-151.

*Chevalier, J-P. (1979b). Scilly atoll de l'archipel de la Société, Polynésie française. Géomorphologie et coraux. *Bull. Ant. Tahiti Mus. Nat. Hist. Nat. et E.P.H.E.* 1: 31-33.

*Chevalier, J.P. (1980). La faune corallienne de l'île Tubuai (Australes). *Cah. Indo-Pacif.* 2: 55-68.

Chevalier, J.P. (1981). Reef scleractinia of French Polynesia. *Proc. 4th Int. Coral Reef Symp., Manila* 2: 177-182.

Chevalier, J.P. and Denizot, M. (1979). Les organismes constructeur de l'atoll de Takapoto. *J. Soc. Océan.* 35(62): 31-34.

Chevalier, J.P., Denizot, M., Mougin, J.L., Plessis, Y. and Salvat, B. (1968). Etude géomorphologique et bionomique de l'atoll de Mururoa (Tuamotu). *Cah. Pacif.* 12: 1-144.

Chevalier, J.P., Denizot, M., Ricard, M., Salvat, B., Sournia, A. and Vasseur, P. (1979). Géomorphologie de l'atoll de Takapoto. *J. Soc. Océan.* 35(62): 9-18.

*Chevalier, J.P. and Kühlmann, D.H.H. (1983). Les Scléractiniaires de Moorea, Ile de la Société (Polynésie française). *J. Soc. Océan.* 9(77): 55-75.

Chevalier, J.P. and Richard, G. (1976). Les récifs extérieurs de l'atoll de Taiaro: bionomie et évaluations quantitatives. *Cah. Pacif.* 19: 203-226.

Chevalier, J.P. and Salvat, B. (1976). Etude géomorphologique de l'atoll fermé de Taiaro. *Cah. Pacif.* 19: 169-201.

Chevre, H. (1974). Aperçu sur la météorologie des Iles Gambier. *Cah. Pacif.* 18(1): 143-158.

Coeroli, M. (1983). Développement de la production nacrière et perlière en Polynésie française. *La Pêche Maritime* 1268: 629-631.

Coeroli, M. and Mizuno, K. (1985). Study of different factors having an influence upon the pearl production of the black lip pearl oyster. *Proc. 5th Int. Coral Reef Cong., Tahiti* 5: 551-556.

Conte, E. (1985). Ethno-archaeological study of the uses of Napuka's marine environment. *Proc. 5th Int. Coral Reef Cong., Tahiti* 5: 589-593.

Conte, E. (1988). L'exploitation traditionelle des ressources marines à Napuka (Tuamotu-Polynésie française). Thèse, Université de Paris 1, Pantheon-Sorbonne.

Crossland, C. (1927a). Marine ecology and coral formations in the Panama region, the Galapagos and Marquesas Islands and the atoll of Napuka. The

expedition to the South Pacific of the S.Y. *St George*. *Trans. Roy. Soc. Edinburgh* 55(2): 531-554.

*Crossland, C. (1927b). Barrier reefs of Tahiti and Moorea. *Nature* 119: 618-619.

*Crossland, C. (1928a). Notes on the ecology of the reef builders of Tahiti. *Proc. Zool. Soc. Lond.* 717-735.

*Crossland, C. (1928b). Coral reefs of Tahiti, Moorea, and Rarotonga. *J. Linn. Soc. London, Zool.* 36(248): 577-620.

*Crossland, C. (1939). Further notes on the Tahitian barrier reef and lagoons. *J. Linn. Soc.* 40: 459-474.

Dahl, A.L. (1980). Regional ecosystems survey of the South Pacific Area. *SPC/IUCN Technical Paper* 179. South Pacific Commission, Noumea, New Caledonia.

Dahl, A.L. (1986). *Review of the Protected Areas System in Oceania.* IUCN/UNEP. 239 pp.

Dana, T.F. (1975). Development of contemporary eastern Pacific coral reefs. *Mar. Biol.* 33: 355-374.

Danielsson, B. (1984). Under a cloud of secrecy: the French nuclear tests in the Southeastern Pacific. *Ambio* 13(5-6): 336-341.

Decker, B.G. (1973). Unique dry-island biota under official protection in northwestern Marquesas Islands (Iles Marquises). *Biol. Cons.* 5(1): 66-67.

Defarge, C., Trichet, J. and Siu, P. (1985). First data on the biogeochemistry of kopara deposits from Rangiroa Atoll. *Proc. 5th Int. Coral Reef Cong., Tahiti* 3: 365-370.

Delesalle, B. (1982). Un atoll et ses problèmes: Mataiva et ses phosphates. *Oceanis* 8(4): 329-337.

Delesalle, B. (1984). Atoll de Tupuai, Archipel de la Société. *Ofai* 6: 29-30.

Delesalle, B. (1985a). Phytoplankton abundance and diversity of Mataiva Atoll. *Proc. 5th Int. Coral Reef Cong., Tahiti* 2: 104 (Abst.).

Delesalle, B. (1985b). Mataiva Atoll, Tuamotu archipelago. *Proc. 5th Int. Coral Reef Cong., Tahiti* 1: 269-322.

Delesalle, B. *et al.* (1981). Mataiva. Etude de l'environnement lagonaire et récifal de l'atoll de Mataiva (Polynésie française). Ecole Pratique des Hautes Etudes, Antenne de Tahiti. 139 pp.

*Delesalle, B., Gabrié, C., Montaggioni, L., Monteforte, M., Naim, O., Odinetz, O., Payri, C., Poli, G. and Richard, G. (1982). Punaauia. Le Lagon de Punaauia: étude de l'environnement lagonaire géomorphologie, plancton et benthos, du secteur concerné par le projet du chenal. Rapport RA 7-1982, Muséum EPHE, Antenne du Tahiti.

*De Nardi, J.L., Raymond, A. and Ricard, M. (1983). Etude des conséquences pour le lagon de Taunoa des travaux d'extension du port de Papeete. Etude descriptive du site actuel. Rap. CEA-R 5221: 108 pp.

Deneuebourg, G. (1969). Les forages de Mururoa. *Cah. Pacif.* 13: 47-58.

*Deneuebourg, G. (1971). Etude écologique du port de Papeete, Tahiti. *Cah. Pacif.* 15: 75-82.

*Denizot, M. (1979). Scilly, atoll de l'archipel de la Société, Polynésie française. Algues de Scilly. *Bull. Ant. Tahiti Mus. Nat. Hist. Nat. et E.P.H.E.* 1: 55-58.

*Denizot, M. (1980). La végétation algale de Tubuai. *Cah. Indo-Pacif.* 2(4): 241-254.

DIRCEN (Direction des Centres d'Expérimentations Nucléaires) (1986). *Rapa*. DIRCEN/SMCB (Service Mixte de Contrôle Biologique). 237 pp.

*Doty, M. and Morrison, J.P.E. (1954). Interrelationships of the organisms on Raroia, aside from man. *Atoll Res. Bull.* 35. 61 pp.

*Doumenge, F. (1963). L'île de Makatea et ses problèmes (Polynésie française). *Cah. Pacif.* 5: 41-68.

Dufour, H. (1985). Contribution to the study of the Hoa Vaimati Lagoon (Rangiroa Atoll, North Tuamotu). *Proc. 5th Int. Coral Reef Cong., Tahiti* 6: 7-12.

*Dufour, H., Raymond, A. and Siu, P. (1984). Contribution à l'étude des lagunes de l'atoll de Rangiroa (Tuamotu nord - Polynésie française). Rapport CEA. 148 pp.

Dupon, J.F. (1986). Atolls and the cyclone hazard: a case study of the Tuamotu islands. Environmental Case Study 3. SPREP, South Pacific Commission, Noumea, New Caledonia.

Egron, R.P.D. (1974). Mangaréviens - peuple de l'archipel des Gambier. *Cah. Pacif.* 18(1): 131-142.

Ehrhardt, J.P. (1978). L'avifaune des Marquises. *Cah. Pacif.* 21: 389-407.

Elliot, H.F.I. (1973). Past, present and future conservation status of Pacific Islands. In: Costin, A.B. and Groves, R.H. (Eds). *Nature Conservation in the Pacific.* IUCN Pubs N.S. 25. Pp. 209-227.

England, K.W. (1971). Actinaria from Mururoa atoll (Tuamotu, Polynesia): Hormatiidae: *Calliactis polypus* Sagartiidae *Verrillactis* n. gen. *paguri. Cah. Pacif.* 15: 23-40.

Equipe de projet E.T.M. (1985a). Temperature and currents on the outer reef slope of Tahiti Island (French Polynesia): A description of their variability. *Proc. 5th Int. Coral Reef Cong., Tahiti* 2: 11 (Abst.).

Equipe de projet E.T.M. (1985b). A survey of the outer reef slope in the vicinity of the proposed site for an OTEC power plant at Tahiti (French Polynesia). *Proc. 5th Int. Coral Reef Cong., Tahiti* 2: 12 (Abst.).

Faure, G. (1985). Reef scleractinian corals of Rapa and Marotiri, French Polynesia (Austral Islands). *Proc. 5th Int. Coral Reef Cong., Tahiti* 6: 267-272.

Faure, G. (1986). Faune corallienne des îles Rapa et Marotiri, Polynésie française (îles Australes). In: DIRCEN (1986): 175-186.

Faure, G. and Laboute, P. (1984). Formations récifales de l'atoll de Tikehau (Tuamotu, Polynésie française, Océan Pacifique). 1. Définition des unités récifales et distribution des principaux peuplements de Scléractiniaires. *ORSTOM Notes Doc.* 22: 108-136.

Fontes, J.C., Kulbicki, G. and Letolle, R. (1969). Les sondages de l'atoll de Mururoa: aperçu géochimique et isotopique de la série corbonatée. *Cah. Pacif.* 13: 69-74.

Fourmanoir, P., Griessinger, J.M. and Plessis, Y. (1974). Faune ichtyologique des Gambiers. *Cah. Pacif.* 18(2): 543-559.

*Fraizier, A. (1980). Etude des conséquences des travaux de dragage du chenal de Taunoa sur le milieu lagonaire. Note CEA-N-2168. 32 pp.

Fraizier, A., Franck, D., Benente, P., Jouen, R. and Debiard, J.P. (1985). Observations on the various forms of pollution of the Tahiti lagoon. *Proc. 5th Int. Coral Reef Cong., Tahiti* 6: 445-451.

*Gabrié, C. (1986). Projet d'implantation de l'hotel Tangaroa en baie de Nune - Bora Bora: étude d'environnement. Caractéristiques écologiques du site récifal concerné par le projet: comparaison avec l'ensemble de la baie. Rapport Antenne Muséum-EPHE RA 16. 41 pp.

Gabrié, C. and Salvat, B. (1985). Generalités sur les îles de la Polynésie française et leurs récifs coralliens. *Proc 5th Int. Coral Reef Cong., Tahiti* 1: 1-15.

Gabrié, C., Porcher, M. and Masson, A. (1985). Dredging in French Polynesian coral reefs: towards a general policy of resources exploitation and sites management. *Proc. 5th Int. Coral Reef Cong., Tahiti* 4: 271-277.

Gachon, A. and Buigues, D. (1985). Volcanic erosion and reef growth phases. *Proc. 5th Int. Coral Reef Cong., Tahiti* 3: 185-191.

Galzin, R. (1976). Biomasse ichtyologique dans les écosystèmes récifaux. Etude préliminaire de la dynamique d'une population de *Pomacentrus nigricans* du lagon de Moorea (Société, Polynésie française). *Rev. Trav. Inst. Pêch. Mar.* 40(314): 575-578.

Galzin, R. (1977). Ichthyological biomass and diversities indices in a lagoon complex of the island of Moorea (Tiahura transect) in French Polynesia. *Proc. 3rd Int. Coral Reef Symp., Miami.*

*Galzin, R. (1979). Scilly, atoll de l'archipel de la Sociétç, Polynésie française. Contribution à l'inventaire de la faune ichtyologique de Scilly. *Bull. Ant. Tahiti Mus. Nat. Hist. Nat. E.P.H.E.* 1: 52-54.

Galzin, R. (1983). Annual variation of coral reef fish community (Moorea - French Polynesia). *Biologie et Geologie des Récifs Coralliens.* Colloque Annuel, International Society for Reef Studies, Nice, 8-9 December 1983. Abstract.

Galzin, R. (1985). Spatial and temporal community structure of coral reef fishes in French Polynesia. *Proc. 5th Int. Coral Reef Cong., Tahiti* 5: 451-455.

Galzin, R. and Pointier, J.P. (1985). Moorea Island, Society archipelago. *Proc. 5th Int. Coral Reef Cong., Tahiti* 1: 73-102.

Garnett, M.C. (1984). Conservation of seabirds in the South Pacific Region: A review. In: Croxall, J.P., Evans, P.G.H. and Schreiber, R.W. (Eds), *Status and Conservation of the World's Seabirds.* ICBP Technical Publication 2. Pp. 547-558.

Gillett, R.D. and Kearney, R.E. (1983). An assessment of the skipjack and baitfish resources of French Polynesia. *Skipjack Survey and Assessment Programme Final Country Report* 7. South Pacific Commission, Noumea, New Caledonia.

Glynn, P.W. (1984). Widespread coral mortality and the 1982-83 El Nino warming event. *Env. Cons.* 11(2): 133-146.

Glynn, P.W. and Wellington, G.M. (1983). *Corals and Coral Reefs of the Galapagos Islands.* University of California Press, Berkeley/Los Angeles/London.

*Grand, S. (1983). Exploitation des poissons récifo-lagonaires aux Tuamotu. Document ORERO 4. 25 pp.

Grand, S. (1985). The importance of the reef-lagoon fishery in French Polynesia. *Proc. 5th Int. Coral Reef Cong., Tahiti* 5: 495-500.

*Grand, S., Siu, P. and Yen, S. (1983). La pêche à Rangiroa. Document ORERO 6. 26 pp.

Gros, R., Jarrige, F. and Fraizier, A. (1980). Hydrologie de la zone nord-ouest du lagon de Rangiroa. Rapport CEA-R-5028. 11 pp.

Gueredrat, J.A. and Rougerie, F. (1978). Etude physicochimique et planctonique du lagon de l'atoll de Takapoto. Rapport ORSTOM Noumea. 39 pp.

*Guilcher, A., Denizot, M. and Berthois, L. (1966). Sur la constitution de la crête externe de l'atoll de Maupelia on Maupihaa (îles de la Société) et de quelques autres récifs voisins. *Cah. Océanogr.* 18(10): 851-856.

*Guilcher, A., Berthois, L., Doumenge, F., Michel, A., Saint-Requier, A. and Arnold, R. (1969). Les recifs et lagons coralliens de Mopelia et de Bora-Bora (îles de la Société) et quelques autres récifs et lagons de comparaison (Tahiti, Scilly, Tuamotu occidentales). *Mémoires ORSTOM* 38. 107 pp.

Hallé, F. (1978). Arbres et forêts des îles Marquises. *Cah. Pacif.* 21: 315-358.

Harmelin-Vivien, M. (1985). Tikehau atoll, Tuamotu archipelago. *Proc. 5th Int. Coral Reef Cong., Tahiti* 1: 211-268.

Harmelin-Vivien, M.L. and Laboute, P. (1983). Preliminary data on underwater effects of cyclones on the outer reef slopes of Tikehau Isalnd (Tuamotu, French Polynesia) and its fish fauna. (Abst.) *Biologie et Ecologie des Recifs Coralliens*, Colloque Annuel, International Society for Reef Studies. Nice, 8-9 December 1983.

Harrison, D.E. and Cane, M.A. (1984). Changes in the Pacific during the 1982-1983 event. *Oceanus* 27(2): 21-28.

Harry, R.R. (1953). Ichthyological field data of Raroia Atoll, Tuamotu Archipelago. *Atoll Res. Bull.* 18. 190 pp.

Hertlein, L.G. and Emerson, W.K. (1953). Mollusks from Clipperton Island (eastern Pacific) with the description of a new species of gastropod. *Trans. San Diego Soc. Nat. Hist.* 11(13): 345-364.

Hertlein, L.G. and Emerson, W.K. (1957). Additional notes on the invertebrate fauna of Clipperton I. *Am. Mus. Novit.* 1859: 1-9.

Holyoak, D.T. and Thibault, J.-C. (1984). Contribution à l'étude des oiseaux de Polynésie orientale. *Mem. Mus. Natn. Hist. Nat. Ser.A, Zool.* 127: 1-209.

*Intes, A. (1984). Présentation générale de l'atoll. In: L'atoll de Tikehau (archipel des Tuamotu, Polynésie française). Premiers résultats. *ORSTOM Tahiti Not. Doc. Océan.* 22: 4-12.

Intes, A. and Coeroli, M. (1985). Evolution and condition of natural stocks of pearl oysters (*Pinctada margaritifera* Linné) in French Polynesia. *Proc. 5th Int. Coral Reef Cong., Tahiti* 5: 545-550.

*James, P. (1980). Rapport de la mission d'étude de la pêche lagonaire à Raiatea et Tahaa (îles sous le vent). *ORSTOM Tahiti, Not. Doc. Océan.* 25 pp.

*Jaubert, J. (1977a). Light, metabolism and growth forms of the hermatypic scleractinian coral *Synarea convexa* (Verrill) in the lagoon of Moorea (French Polynesia). *Proc. 3rd Int. Coral Reef Symp., Miami* 1: 483-488.

Jaubert, J. (1977b). Light, metabolism and the distribution of *Tridacna maxima* in a south Pacific atoll: Takapoto (French Polynesia). *Proc. 3rd Int. Coral Reef Symp., Miami* 1: 489-494.

*Jaubert, J., Thomassin, A. and Vasseur, P. (1976). Morphologie et étude bionomique préliminarie de la pente externe du récife de Tiahura, île de Moorea (Polynésie française). *Cah. Pacif.* 19: 199-323.

Kühlmann, D.H.H. and Chevalier, J.-P. (1986). Les coraux (Scléractiniaires et Hydrocoralliaires) de l'atoll de Takapoto, îles Tuamotu: aspects écologiques. *Mar. Ecol.* 7(1): 75-104.

Labeyrie, J., Lalou, C. and Delibrias, G. (1969). Etude des transgressions marines sur l'atoll de Mururoa par la datation des différentes niveaux de corail. *Cah. Pacif.* 13: 59-68.

Laboute, P. (1985). Evaluation of damage done by the cyclones of 1982-1983 to the outer slopes of the Tikehau and Takapoto Atolls (Tuamotu Archipelago). *Proc. 5th Int. Coral Reef Cong., Tahiti* 3: 323-329.

*Lacan, F. and Mougin, J.L. (1974). Les oiseaux des Iles Gambier et de quelques atolls orientaux de l'Archipel des Tuamotu (Océan Pacifique). *L' Oiseau et R.F.O.* 44(3): 191-280.

*Lavondes, H., Richard, G. and Salvat, B. (1973). Noms vernaculaires et usages traditionnels de quelques coquillages des Marquises. *J. Soc. Océan.* 29(39): 121-137.

Lebeau, A. (1985). Breeding evaluation trials in the Green Turtle *Chelonia mydas* (Linné) on Scilly Atoll (leeward Islands, French Polynesia) during the breeding seasons 1982-1983 and 1983-1984. *Proc. 5th Int. Coral Reef Cong., Tahiti* 5: 487-493.

*Lefevre, M. (1984). Répartition de la biomasse zooplanctonique autour de l'île de Moorea (Polynésie française). *J. Rech. Océan.* 9(1): 20-22.

Lefevre, M. (1985). Spatial variability of zooplankton populations in the lagoons of a high island. *Proc. 5th Int. Coral Reef Cong., Tahiti* 6: 39-45.

Manac'h, F. and Carsin, J.L. (1985). Deep fishing on the outer slope of atolls. *Proc. 5th Int. Coral Reef Cong., Tahiti* 5: 469-474.

*Marquet, G. (1987). Etude de l'impact des rejets de porcheries sur la faune d'eau douce de Tahiti (Polynésie française). Rapport Ministère Santé et Environnement Polynésie française. 43 pp.

*Masson, M. and Simon, J.P. (1985). Schéma général d'exploitation des granulats et protection de l'environnement de l'île de Tahiti. Rapport, CETE, Mediterranée.

Michel, A. (1969). Plancton du lagon et des abords extérieurs de l'atoll de Mururoa. *Cah. Pacif.* 13: 81-132.

Michel, A., Colin, C., Desrosieres, R. and Oudot, C. (1971). Observation sur l'hydrologie et le plancton des abords et de la zone des passes de l'atoll de Rangiroa (Archipel des Tuamotus, Océan Pacifique Central). *Cah. ORSTOM, sér. Océanogr.* 9(3): 375-402.

Moniod, F. (1974). Etude hydrologique des Iles Gambier. *Cah. Pacif.* 18(1): 291-235.

Monnet, C., Sodter, F. and Vigneron, E. (1986). *Répertoire Bibliographique de l'ORSTOM en Polynésie française 1955-1986.* Centre ORSTOM de Tahiti, Papeete. 130 pp.

Montaggioni, L.F. (1985). Makatea Island, Tuamotu Archipelago. *Proc. 5th Int. Coral Reef Cong., Tahiti* 1: 103-158.

Montaggioni, L.F., Richard, G., Bourrouilh, F., Gabrié, G., Humbert, L., Monteforte, M., Naim, O., Payri, C. and Salvat, B. (1983). Aspects of the geology and marine biology of Makatea, an uplifted atoll, Tuamotu Archipelago, French Polynesia. *Biologie et Geologie des Recifs Coralliens.* Colloque Annuel, Int. Soc. for Reef Studies. (Abst.).

Montaggioni, L.F., Richard, G., Bourrouilh-Le Jan, F., Gabrié, C., Humbert, L., Monteforte, M., Naim, O., Payri, C. and Salvat, B. (1985a). Geology and marine biology of Makatea, an uplifted atoll, Tuamotu Archipelago, Central Pacific Ocean. *J. Coastal Res.* 1(2): 165-172.

Montaggioni, L.F., Richard, G., Gabrié, C., Monteforte, M., Naim, O., Payri, C. and Salvat, B. (1985b). Les récifs coralliens de l'île de Makatea, archipel des Tuamotus, Pacifique Central: geomorphologie et repartition des peuplements. *Ann. Inst. Oceanogr.* 61(1): 1-25.

Morize, E. (1985). Study of a small-scale fishery in the atoll of Tikehau (Tuamotu Archipelago, French Polynesia). *Proc. 5th Int. Coral Reef Cong., Tahiti* 5: 501-506.

*Morrison, J.P.E. (1954). Animal ecology of Raroia atoll, Tuamotu. Part 1. Ecological notes on the molluscs and other animals of Raroia. *Atoll Res. Bull.* 34. 18 pp.

*Naim, O (1980). Bilan quantitatif de la petite faune associée aux algues du lagon de Tiahura, île de Moorea (Polynésie française). *C.R. Acad. Sci.* 291: 549-551.

Naim, O. (1981). Effect of coral sand extractions on the small mobile fauna associated with the algae of a fringing reef (Moorea-French Polynesia). *Proc. 4th Int. Coral Reef. Symp., Manila* 1: 123-127.

Newell, N.D. (1954a). Reefs and sedimentary processes of Raroia. *Atoll Res. Bull.* 36: 1-35.

Newell, N.D. (1954b). Expedition to Raroia, Tuamotus. *Atoll Res. Bull.* 31: 1-21.

Newell, N.D. (1956). Geological reconnaissance of Raroia (Kon Tiki) atoll. Tuamotu Archipelago. *Bull. Am. Mus. Hist. Nat.* 109(3): 311-372.

*Niaussat, P.-M. (1978). *Le lagon et l'atoll de Clipperton.* Travaux et Mémoires de l'Académie des Sciences d'Outre-Mer, Paris. 189 pp.

O'Reilly, R.P.P (1974). Aperçu sur l'histoire de l'archipel des Gambier. *Cah. Pacif.* 18(1): 121-130.

Ottino, P. (1972). *Rangiroa, parenté, étendue, résidence et terres dans un atoll polynésien.* Cujas ed., Paid. 530 pp.

Payri, C. (1983). Features of the algal community of Tiahura Lagoon (Moorea Island - French Polynesia). *Biologie et Geologie des Récifs Coralliens,* Colloque Annuel, International Society for Reef Studies, Nice, 8-9 December 1983. (Abst.).

Payri, C. (1987). Variabilité spatiale et temporelle de la communauté des macrophytes des récifs coralliens de Moorea (Polynésie française). Thèse, USTL, Montpellier.

Payri, C.E. and Meinesz, A. (1985). Bibliography of marine benthic algae in French Polynesia. *Proc. 5th Int. Coral Reef Cong., Tahiti* 1: 498-518.

*Peyrot-Clausade, M. (1976). Polychètes de la cryptofaune du récif de Tiahura, Moorea. *Cah. Pacif.* 19: 325-336.

*Peyrot-Clausade, M. (1977). Décapodes Brachyoures et Anomoures (à l'exception des Paguridae) de la cryptofaune de Tiahura, Moorea. *Cah. Pacif.* 20: 211-222.

Peyrot-Clausade, M. (1985). Motile cryptofauna modifications related to coral degradations on Tiahura coral reef flat. *Proc. 5th Int. Coral Reef Cong., Tahiti* 6: 459-464.

Pichon, M. (1985a). Organic production and calcification in some coral reefs of Polynesia. *Proc. 5th Int. Coral Reef Cong., Tahiti* 6: 173-177.

Pichon, M. (1985b). Scleractinia. *Proc. 5th Int. Coral Reef Cong., Tahiti* 1: 399-403.

*Pickering, C. (1876). *The geographical distribution of animals and plants. Part 2. Plants in their wild state.* Salem, Mass. Pp. 227-228.

Pirazzoli, P.A. (1982). Télédétéction en milieu récifal. Utilisation d'une image LANDSAT pour évaluer la bathymetrie dans l'atoll de Rangiroa (Polynésie française). *Oceanis* 8(4): 297-308.

Pirazzoli, P.A. (1985a). Leeward islands (Maupiti, Tupai, Bora Bora, Huahine), Society Archipelago. *Proc. 5th Int. Coral Reef Cong., Tahiti* 1: 17-72.

Pirazzoli, P.A. (1985b). Bathymetric mapping of coral reefs and atolls from satellite. *Proc. 5th Int. Coral Reef Cong., Tahiti* 6: 539-544.

Pirazzoli, P.A. and Montaggioni, L.F. (1984). Variations récentes du niveau de l'ocean et du bilan hydrologique dans l'atoll de Takapoto (Polynésie française). *C.R. Acad. Sc. Paris* 299(2)7: 321-326.

Plessis, Y. (1969). Les atolls des Tuamotu en tant qu'écosystème marin. *Bull. Mus. Nat. Hist.* 2ème ser. 40(6): 1232-1236.

Plessis, Y. (1970). Note préliminaire sur la faune de Rangiroa (Polynésie). *Bull. Mus. Nat. Hist. Nat.* 2ème ser. 11(5): 1306-1319.

Plessis, Y. (1972a). Ichtyologie corallienne: écologie et exploitation. *Proc. Symp. Corals and Coral Reefs* 1969. Mar. biol. ass. India. Pp. 457-468.

*Plessis, Y. (1972b). L'analyse ichthyologique quantitative du récif frangeant de Tiahura, Ile de Moorea, Polynésie, premiers résultats. *C.R. Soc. Biogeogr.* 427: 26-28.

*Plessis, Y. (1973a). Etude preliminaire de la faune ichtyologique de Moorea. *Cah. Pacif.* 17: 289-298.

*Plessis, Y. (1973b). Maupiti (Polynésie française). Quelques problèmes sur le peuplement ichtyologique et la protection de la nature. *C.R. Soc. Biogeogr.* 438: 44-48.

Plessis, Y. (1980). Etude ichtyologique de Tubuai, archipel des Australes (Polynésie). *Cah. Indo-Pacif.* 2(3): 255-269.

Plessis, Y. (1986). Etude ichtyologique de Rapa. In: DIRCEN (1986): 215-230.

Plessis, Y.B. and Mauge, L.A. (1978). Ichtyologie des îles Marquises. *Cah. Pacif.* 21: 215-235.

*Poli, G., Delesalle, B., Gabrié, C., Montaggioni, L., Monteforte, M., Naim, O., Payri, C., Richard, G. and Trondle, J. (1983). Tahiti, lagon zone urbaine: étude de l'environnement lagunaire du secteur urbain, evolution des pollutions et des degradations. Rapport definitif, Convention Territoire de la Polynésie française/Naturalia et Biologia/Muséum EPHE, Antenne de Tahiti, RA 10: 1-110.

Poli, G. and Richard, G. (1973). Rangiroa, étude des peuplements du lagon et du récif de la partie sud-est de l'atoll (aout 1972). *Rapport Diff. Restr. Antenne Muséum EPHE*: 1-39.

Poli, G. and Salvat, B. (1976). Etude bionomique d'un lagon d'atoll totalement fermé: Taiaro. *Cah. Pacif.* 19: 227-251.

Pomier, M. (1969). Mururoa, état de la cocoteraie en mars 1966. *Cah. Pacif.* 13: 75-79.

*Porcher, M. (1983). Etude du schéma de secteur de Outu Naoro (Tahiti). Aménagement littoral. Rapport CETE-Méditerranée. 60 pp.

*Porcher, M. (1984). Projet hôtelier à Bora Bora: étude prèliminaire relative aux effets du projet sur le milieu récifal. Rapport CETE Mediterranée.

*Porcher, M. (1985a). Projet Hôtelier. Baie de Nunue - Ile de Bora Bora (Polynésie française). Rapport CETE-Mediterranée.

*Porcher, M. (1985b). Hôtel Sofitel à Moorea: chantier de terrassements sous-marins; mise en place de techniques visant à limiter les effets des travaux sur le milieu récifal. Rapport CETE Mediterranée.

*Porcher, M. (1986). Etude d'un point de rejet en mer des eaux usées urbaines à la sortie de la passe de Taopuna. Rapport CETE Mediterranée.

Porcher, M. and Bouilloud, J.P. (1984). Projet hôtelier Sofitel Moorea (Polynésie française). Etude d'impact. Rapport CETE Mediterranée. 131 pp.

*Porcher, M., Bouilloud, J.P. and Dubois, S., (1986). Etude de deux sites de rejet en mer des eaux usées urbaines entre le passe de Taopuna et la Pointe Venus. Rapport CETE Mediterranée.

Porcher, M. and Dupuy, M. (1985). Environment and coastal land use planning in coral reef areas, French Polynesia. *Proc. 5th Int. Coral Reef Cong., Tahiti* 6: 557-562.

Porcher, M. and Gabrié, C. (1985). Effects of extraction of coral materials on the coral reefs of Tahiti. *Proc. 5th Int. Coral Reef Cong., Tahiti* 2: 304 (Abst.).

*Porcher, M. and Gabrié, C. (1987). Schéma général d'exploitation de granulats à Moorea et dans les îles sous le vent. Rapport CETE/Antenne Muséum-EPHE. 120 pp.

*Porcher, M., Gabrié, C. and Bouilloud, J.P. (1985). Schéma d'exploitation des granulats pour l'île de Tahiti (Polynésie française). Les extractions en milieu corallien.

Rapport CETE Mediterranée/Antenne Muséum-EPHE. 150 pp.

*Porcher, M., Gabrié, C., Payri, C. and Salvat, B. (1987). Etude du plan d'aménagement du domaine d'Atimaono: étude d'environnement en milieu récifal - analyse de l'état initial. Rapport CETE/Antenne Muséum-EPHE. 80 pp.

Pouchan, P., Burlot, R., Humbert, L. and Pelissier Hermitte, G. (1985). Water resources and reservoir characteristics of coral islands. *Proc. 5th Int. Coral Reef Cong., Tahiti* 2: 307 (Abst.).

Purser, B.H. and Aissaoui, D.M. (1985). Reef diagenesis: Dolomitisation and dedolomitisation at Mururoa atoll, French Polynesia. *Proc. 5th Int. Coral Reef Cong., Tahiti* 3: 263-269.

Quigier, J.P. (1969). Quelques données sur la répartition des poisson des récifs coralliens. *Cah. Pacif.* 13: 181-185.

Randall, J.E. (1973). Tahitian fish names and a preliminary checklist of the fishes of the Society Islands. *Occ. Pap. B.P. Bishop Mus.* 24(11): 167-214.

Randall, J.E. (1985). Fishes. *Proc. 5th Int. Coral Reef Cong., Tahiti* 1: 462-481.

Randall, J.E. and Amesbury, S. (1987). A marine survey on Huahine Nui, French Polynesia. University of Guam. 71 pp.

Ranson, G. (1958). Coreaux et récifs coralliens. Observations sur les îles coralliennes de l'archipel des Tuamotu (Océanie française). *Cah. Pacif.* 1: 15-36.

*Real-Testud, A.M. and Richard, G. (1984). Quelques données sur les huitres perlières et la perliculture. *Xenophora* 21: 7-17.

Redier, L. (1967). Un nouvel hydraire *Cordylophora solangiae* n.s. (atoll de Fangataufa - Tuamotu. *Cah. Pacif.* 11: 117-128.

*Renaud-Mornant, J. (1977). Rapport micro-meio benthics. *Halodeima atra* (Holothuridae) dans un lagon Polynésian (Tiahura, Moorea, îles de la Société). *Cah. Pacif.* 20: 1-6.

Renaud-Mornant, J., Salvat, B. and Bossy, C. (1970). Macrobenthos and meiobenthos from the closed lagoon of a polynesian atoll: Maturei Vavao (Tuamotu). *Biotropica* 3(1): 36-55.

*Renon, J.P. (1977). Zooplancton du lagon de l'atoll de Takapoto (Polynésie française). *Ann. Inst. Océan.* 53(2): 217-236.

Ricard, M. (1974). Etude taxonomique des Diatomées marines du lagon de Vairao (Tahiti). 1. Le genre *Mastogloia. Rev. Algol.* 11(1-2): 161-177.

*Ricard, M. (1976a). Production primaire planctonique de trois lagons de l'Archipel de la Société (Polynésie française). *Cah. Pacif.* 19: 383-395.

*Ricard, M. (1976b). Premier inventaire des diatomées marines du lagon de Tiahura (île de Moorea, Polynésie française). *Rev. Algol.* 11(3-4): 343-355.

*Ricard, M. (1980). Diminution de la production primaire du lagon de Tiahura (île de Moorea, Polynésie française) sous l'influence de la pollution liée a l'exploitation de sables coralliens. *Cah. Indo-Pacif.* 2(1): 73-90.

Ricard, M. (1981). Some effects of dredging on the primary production of the Tiahura lagoon in Moorea (Society Islands, French Polynesia). *Proc. 4th Int. Coral Reef Symp., Manila* 1: 431-436.

Ricard, M. (1983). Primary productivity of a high island lagoon: Functioning of Tiahura Lagoon, Moorea Island (French Polynesia). *Biologie et Geologie des Recifs Coralliens.* Colloque Annuel, International Society for Reef Studies, Nice, 8-9 December 1983. Abstract.

Ricard, M. (1985). Rangiroa atoll, Tuamotu archipelago. *Proc. 5th Int. Coral Reef Cong., Tahiti* 1: 159-210.

*Ricard, M., Badie, C., Renon, J.P., Simeon, C. and Sournia, A. (1978). Données sur l'hydrologie, la production primaire et le zooplancton du lagon de l'atoll fermé de Takapoto (archipel des Tuamotu, Polynésie française). Rapp. CEA Saclay. 89 pp.

Ricard, M. and Delesalle, B. (1981). Phytoplankton and primary production of the Scilly lagoon waters. *Proc. 4th Int. Coral Reef Symp., Manila* 1: 425-429.

*Ricard, M. and Delesalle, B. (1983). Hydrological and phytoplanctonological features of land-locked Taiaro atoll (Tuamotu archipelago, French Polynesia). *15th Pacif. Sci. Cong., Dunedin*: 197.

*Ricard, M., Delesalle, B., Denizot, M., Montaggioni, L., Renon, J.P., and Vergonzanne, G. (1981). Etude des organismes vivants, plancton et benthos su secteur lagonaire et recifal de Taunoa concerne par le projet d'extension du port de Papeete. Etude descriptive du site actuel, mai 1981. RA3-1981, Muséum/EPHE.

*Ricard, M., Gabrié, C., Harmelin-Vivien, M., Payri, C. and Richard, G. (1986). Pollution du Port de Papeete: aspects des divers peuplements biologiques nectoniques et benthiques. 32 pp.

*Ricard, M., Gros, R. and Delesalle, B. (1979a). Scilly, atoll de l'archipel de la Société, Polynésie française. Hydrologie et phytoplancton. *Bull. Ant. Tahiti Mus. Nat. Hist. Nat. et EPHE* 1: 39-40.

*Ricard, M., Gueredrat, J.A., Magnier, Y., Renon, J.P., Rochette, J.P., Rougerie, F. and Sournia, A. (1979b). Le plancton du lagon de Takapoto. *J. Soc. Océan.* 35(62): 47-67.

*Ricard, M., Richard, G., Salvat, B. and Toffart, J.L. (1977). Coral reef and lagoon research in French Polynesia. *Rev. Algol.* N.S. 1: 1-44.

Richard, G. (1970). Etude sur les mollusques récifaux des atolls de Reao et de Hereheretue (Tuamotu-Polynésie). Bionomie et Evaluations quantitatives. Diplome EPHE Paris: 1-109.

*Richard, G. (1973a). Abondance et dominance des mollusques dans un écosystème corallien (Moorea, Polynésie française). *Bull. Must. Nat. Hist. Nat.* 2ème ser. 163(19): 309-313.

*Richard, G. (1973b). Etudes des peuplements du complexe lagunaire de Tiahura-Moorea, Polynésie française. *Bull. Soc. Et. Océan.* 15(11-12): 309-324.

Richard, G. (1974a). *Adusta (Cribraria) bernardi* sp. n. (Mesogastropoda, Cypraeidae) des îles de la Société et les porcelaines de la Polynésie française. *Bull. Soc. Et. Océan.* 16(1): 377-383.

Richard, G. (1974b). Bionomies des mollusques littoraux des baies envasées de l'Ile de Mangareva, Archipel des Gambiers, Polynésie française. *Cah. Pacif.* 18(2): 605-614.

Richard, G. (1976). Transport de matériaux et évolution récente de la faune malacologique lagunaire de Taiaro (Tuamotu - Polynésie française). *Cah. Pacif.* 19: 265-282.

Richard, G. (1977). Quantitative balance and production of *Tridacna maxima* in the Takapoto Lagoon (French Polynesia). *Proc. 3rd Int. Coral Reef Symp. Miami* 1: 599-605.

*Richard, G. (1978). Abondance et croissance de *Arca ventricosa* dans le lagon de Takapoto. *Haliotis* 9(1): 7-10.

*Richard, G. (1979). Scilly, atoll de l'archipel de la Société: étude des mollusques récifaux et lagunaires peu profondes. *Bull. Ant. Tahiti Mus. Nat. Hist. Nat. et E.P.H.E.* 1: 39-40.

*Richard, G. (1982a). Bilan quantitatif et premières données de production de *Cardium fragum* dans le lagon de Anaa. *Malacologia* 22 (1-2): 347-352.

Richard, G. (1982b). Mollusques lagunaires et récifaux de Polynésie française: inventaire faunistique, bionomie, bilan quantitatif, croissance, production. Thèse de Doctorat d'Etat, Paris. 313 pp.

Richard, G. (1983). Growth and production of *Chama iostoma* in Takapoto Atoll Lagoon (Tuamotu-French Polynesia). (Abst.). *Biologie et Ecologie des Recifs Coralliens* Colloque Annuel, International Society for Reef Studies, Nice, 8-9 December 1983.

*Richard, G. (1984). Rapa, rapport malacologique. Ministère de la Défense, Centre d'expérimentations du Pacifique, Service Mixte de Contrôle Biologique, Ecole Pratique des Hautes Etudes, Antenne de Tahiti. Pp. 1-17.

Richard, G. (1985a). The malacological fauna of Rapa (Austral Islands) examined in the context of French Polynesia and the Indo-Pacific: Special ecological and geographical features. *Proc. 5th Int. Coral Reef Cong., Tahiti* 2: 324. (Abst.).

Richard, G. (1985b). Faune et flore: premier abrégé des organismes marins de Polynésie française. *Proc. 5th Int. Coral Reef Cong., Tahiti* 1: 379-518.

Richard, G. (1986). La faune malacologique de Rapa: originalités écologiques et biogéographiques. In: DIRCEN (1986): 187-202.

Richard, G. (1987). Evolution de l'extension des mortalités massives de mollusques autres que la nacre (Bénitiers, Spondyles, Arches...) à Takapoto. Rapport Antenne Muséum-EPHE. RA 17-1987.

Richard, G. (in press). Croissance et production de *Chama iostoma* dans le lagon de Takapoto (Tuamotu-Polynésie française). *Atoll Res. Bull.* 10 pp.

*Richard, G. and Salvat, B. (1972). Ecologie quantitative des mollusques du lagon de Tiahura, île de Moorea, Polynésie française. *C.R. Acad. Sci. Paris* 275, D. 1547-1550.

*Richard, G. and Salvat, B. (1973). *Conus (Dendroconus) gauguini* sp. n. (Neogastropoda, Conidae) des Iles Marquises (Polynésie française). *Cah. Pacif.* 17: 25-29.

*Richard, G., Salvat, B. and Millous, O. (1979). Mollusques et faune benthique du lagon de Takapoto. *J. Soc. Océan.* 62(35): 59-68.

*Robinson, M. (1972). *Return to the Sea*. John de Graff Inc. Tuckhoe, New York.

*Rossfelder, A. (1976). *Clipperton, l'île tragique*. Albin Michel, Paris. 277 pp.

Rougerie, F. and Gros, R. (1980). Les courants dans la passe d'Avatoru, atoll du Rangiroa (archipel des Tuamotu). *ORSTOM Papeete, Not. Doc. Océan.* 80(17): 8 pp.

*Rougerie, F., Gros, R. and Bernadec, M. (1980). Le lagon de Mururoa. Esquisse. *ORSTOM Not. Doc. Océan.* 80(16). 34 pp.

*Rougerie, F. and Wauthy, B. (1983). Anomalies de l'hydroclimat et cyclogenèse en Polynésie en 1982 and 1983. *Met-Mar.* 121: 27-40.

Ruzie, G. and Gachon, A. (1985). Contribution of geophysical techniques in carbonate studies of the atolls - application to the investigation of the structures of Mururoa Atoll. *Proc. 5th Int. Coral Reef Cong., Tahiti* 6: 381-388.

Sachet, M.-H. (1958). Histoire de l'Ile Clipperton. *Cah. Pacif.* 2: 3-32.

*Sachet, M.-H. (1962a). Geography and land ecology of Clipperton Island. *Atoll Res. Bull.* 86. 115 pp.

*Sachet, M.-H. (1962b). Monographie physique et biologique de l'île de Clipperton. *Ann. Inst. Océanogr. Paris* 40(1): 1-107.

*Sachet, M.-H. (1966). Mission aux îles Marquises. *Cah. Pacif.* 9: 11-13.

Sachet, M.-H. (1969). List of vascular plants of Rangiroa. *Atoll Res. Bull.* 125: 33-44.

*Sachet, M.-H. (1983). Végétation et flore terrestre de l'atoll de Scilly (Fenua Uva). *J. Soc. Océan.* 39(77): 29-34.

*Sachet, M.-H., Schafer, P.A. and Thibault, J.C. (1975). Mohotani: une île protegée aux Marquises. *Bull. Soc. Et. Océan.* 16(6): 557-568.

Salvat, B. (1967). Importance de la faune malacologique dans les atolls Polynésiens. *Cah. Pacif.* 11: 7-49.

*Salvat, B. (1970a). Les mollusques des récifs d'îlots du récif barrière des Iles Gambier (Polynésie). Bionomie et densité du peuplement. *Bull. Mus. Nat. Hist. Nat.* 42(3): 525-542.

Salvat, B. (1970b). Etudes quantitatives (comptages et biomasses) sur les mollusques récifaux de l'atoll de Fangataufa (Tuamotu-Polynésie). *Cah. Pacif.* 14: 1-57.

Salvat, B. (1971a). Les lagons d'atolls polynésiens. Richesse actuelle - possibilités d'exploitation perspectives d'aquaculture. Colloque International sur l'exploitation des Océans. Bordeaux.

Salvat, B. (1971b). La faune benthique du lagon de l'atoll de Reao. *Cah. Pacif.* 16: 30-109.

*Salvat, B. (1971c). Mollusques lagunaires et récifaux de l'île de Raevavae (Australes, Polynésie). *Malac. Rev.* 4: 1-15.

Salvat, B. (1972). Distribution des mollusques sur les récifs extérieurs de l'atoll de Fangataufa (Tuamotu, Polynésie). Radiales quantitatives. Biomasses. *Proc. Symp. Corals and Coral Reefs, India* : 373-378.

Salvat, B. (1973a). Mollusques des îles Tubuai (Australes, Polynésie). Comparaison avec les îles de la Société et des Tuamotus. *Malacologia* 14(1-2): 429-430.

Salvat, B. (1973b). Plaidoyer pour la protection d'atolls de Polynésie. *Proceeding and Papers, Regional Symposium on Conservation of Nature - Reefs and Lagoons*. South Pacific Commission, Noumea, New Caledonia. Pp. 19-21.

Salvat, B. (1974). Mesures en faveur de la protection de la nature aux Iles Marquises. Unpub. rept.

*Salvat, B. (1976a). Un programme interdisciplinaire sur les écosytèmes insulaires en Polynésie française. *Cah. Pacif.* 19: 397-406.

Salvat, B. (1976b). Compte-rendu préliminaire d'une mission scientifique dans l'atoll de Taiaro. Réserve intégrale W.A. Robinson. *Cah. Pacif.* 19: 18-23.

*Salvat, B. (1979a). Le lagon et les peuplements de l'atoll de Scilly. *Bull. Soc. Et. Océan.* 15(11-12): 309-324.

*Salvat, B. (1979b). Scilly, atoll de l'archipel de la Société, Polynésie française. Le lagon et ses peuplements. *Bull. Ant. Tahiti Mus. Nat. Hist. Nat. et E.P.H.E.* 1: 34-38.

Salvat, B. (1981). Geomorphology and marine ecology of the Takapotu Atoll (Tuamotu Archipelago). *Proc. 4th Int. Coral Reef Symp. Manila* 1: 503-509.

Salvat, B. (Ed.) (1982). *Ofai*. Bulletin de Liaison 1, Muséum National d'Histoire Naturelle, Ecole Pratique des Hautes Etudes, Antenne de Tahiti.

*Salvat, B. (1983). La faune benthique du lagon de l'atoll de Scilly, archipel de la Société. *J. Soc. Océan.* 39(77): 3-15.

Salvat, B. (1985). An integrated (geomorphological and economical) classification of French Polynesian atolls. *Proc. 5th Int. Coral Reef Cong., Tahiti* 2: 337 (Abst.).

Salvat, B. (Ed.) (1987). *Ofai* 7. 88 pp.

Salvat, B., Chevalier, J.P., Richard, G., Poli, G. and Bagnis, R. (1977). Geomorphology and biology of Taiaro atoll, Tuamotu archipelago. *Proc. 3rd Int. Coral Reef Symp., Miami* 1: 289-296.

*Salvat, B. and Erhardt, J.P. (1970). Mollusques de l'île Clipperton. *Bull. Mus. Nat. Hist. Nat. (Paris).* Ser. 2. 42(1): 223-231.

Salvat, B. and Renaud-Mornant, J. (1969). Etude écologique du macrobenthos et du meiobenthos d'un fond sableux du lagon de Mururoa (Tuamotu-Polynésie française). *Cah. Pacif.* 13: 159-179.

Salvat, B. and Richard, G. (1985). Takapoto atoll, Tuamotu archipelago. *Proc. 5th Int. Coral Reef Cong., Tahiti* 1: 323-378.

*Salvat, B. and Rives, C. (1975). *Coquillages de Polynésie*. Ed. Pacif. Papeete. 391 pp.

*Salvat, B. and Salvat, F. (1972). Geographic distribution of *Pinna rugosa* Sowerby, 1835 (Mollusca: Bivalvia) and its occurrence on Clipperton Island. *The Veliger* 15(1): 43-44.

Salvat, B., Sibuet, M. and Laubier, L. (1985). Benthic megafauna observed from the submersible "Cyana" on the fore-reef slope of Tahiti (French Polynesia) between 70 and 100 m. *Proc. 5th Int. Coral Reef Cong., Tahiti* 2: 338 (Abstract).

*Salvat, B., Vergonzanne, G., Galzin, R., Richard, G., Chevalier, J.P., Ricard, M. and Renaud-Mornant, J. (1979). Consequences ecologiques des activites d'une zone d'extraction de sable corallien dans le lagon de Moorea (Ile de la Societe, Polynésie française). *Cah. Indo-Pacif.* 1(1): 83-126.

Salvat, B. and Venec-Peyre, M.T. (1981). The living foraminifera in the Scilly Atoll lagoon (Society Islands). *Proc. 4th Int. Coral Reef Symp., Manila* 2: 767-774.

*Sournia, A. (1976). Abondance du phytoplancton et absence de récifs coralliens sur les côtes des îles Marquises. *C.R. Acad. Sci.* 282. Ser. D: 553-555.

Sournia, A. and Plessis, Y. (1974). A red-water diatom *Aulacodiscus kittonii* var. *africanus* in Marquesas Islands, Pacific ocean. *Bot. Mar.* 17: 124.

Sournia, A. and Ricard, M. (1975a). Phytoplankton and primary productivity in Takapoto atoll, Tuamotu Islands. *Micronesica* 11(2): 159-166.

*Sournia, A. and Ricard, M. (1975b). Production primaire planctonique dans deux lagons de Polynésie française (île de Moorea et atoll de Takapoto). *C.R. Acad. Sci. Paris. sér D.* 280(6): 741-743.

*Sournia, A., and Ricard, M. (1976). Phytoplankton and its contribution to primary productivity in two coral reef areas of French Polynesia. *J. Exp. Mar. Biol. Ecol.* 21: 129-140.

SPREP (1980). French Polynesia. *Country Report 5*. South Pacific Commission, Noumea, New Caledonia.

*Squires, D.F. (1959). Results of the Puritan American Museum of Natural History Expedition to Western Mexico. *Bull. Am. Mus. Nat. Hist.* 118(7): 367-432.

Stoddart, D.R. (1969). Reconnaissance geomorphology of Rangiroa atoll, Tuamotu archipelago. *Atoll Res. Bull.* 125: 1-31.

*Talandier, J. (1983). Crise volcanismique de Mehetia en 1981 et ses implications. C.E.A., Lab. Geophys., Tahiti. 27 pp.

Taxit, R. (1978). Les phenomènes d'eaux rouges aux Iles Marquises. *Cah. Pacif.* 21: 285-292.

Teissier, R. (1969). Les cyclones en Polynésie française. *Bull. Soc. Et. Océan.* 14(5-6): 1-48.

Tercinier, G. (1969). Note de synthèse sur les sols du Motu Faucon (étude pédologique d'une portion représentative de l'atoll de Mururoa). *Cah. Pacif.* 13: 17-46.

*Thibault, J.-C. (1972). Observations ornithologiques à Makatea. Rapport MNHN/EPHE. 4 pp.

*Thibault, J.-C. (1973a). Rapport ornithologique sur l'île de Maiao. Rapport MNHN/EPHE. 11 pp.

*Thibault, J.-C. (1973b). Compte-rendu ornithologiques d'un séjour dans les îles Sous-le-vent (Huahine, Raiatea et Tahaa). Rapport MNHN/EPHE. 8 pp.

*Thibault, J.-C. (1973c). Rapport ornithologique sur l'atoll Tupai. Rapport MNHN/EPHE. 11 pp.

*Thibault, J.-C. (1974a). *Le peuplement avien des îles de la Société*. Thesis submitted in fulfillment of Diplôme de l'Ecole Pratique des Hautes Etudes (section Sciences Naturelles). 134 pp.

*Thibault, J.-C. (1974b). Rapport préliminaire sur l'avifaune de Takapoto. Rapport ORSTOM/MNHN. 8 pp.

*Thibault, J.-C. (1974c). Notes de terrain sur l'avifaune de deux îles de l'archipel des Australes: Tubuai et Raivavae. Rapport ORSTOM/EPHE. 4 pp.

*Thibault, J.-C. (1974d). Rapport préliminaire sur les oiseaux de Rapa. Rapport ORSTOM/MNHN. 3 pp.

*Thibault, J.-C. (1975a). *Oiseaux de Tahiti*. Editions du Pacifique, Tahiti. 112 pp.

*Thibault, J.-C. (1975b). Notes de terrain sur les oiseaux de Rapa. Rapport ORSTOM/MNHN. 11 pp.

*Thibault, J.-C. (1976). L'avifaune de Tetiaroa (archipel de la Société, Polynésie française). *L'Oiseau et RFO* 46: 29-45.

*Thibault, J.-C. (1987). Les vertébrés terrestres de Fangataufa (Tuamotu, Pacifique sud): évolution des peuplements, 1965-87. Rept prepared for SMCB, Direction des Centres d'Expérimentation Nucléaires. 46 pp.

*Thibault, J.-C. (in press). Les oiseaux de Polynésie. In: *Atlas de la Polynésie française*. Ed. ORSTOM.

*Thibault, J.-C. *et al.* (in press). Recent changes in the avifauna of Makatea (Tuamotu, South Pacific Ocean). *Atoll Res. Bull.*

*Thibault, J.C. and Rives, C. (1975). *Oiseaux de Tahiti*. Ed. Pacif., Papeete. 111 pp.

*Thibault, B. and Thibault, J.C. (1975). Liste des oiseaux de Polynésie orientale (nouvelles aquisitions faunistiques). *L'Oiseau et RFO* 45(1): 89-92.

Tixier-Durivault, A. (1969). Les Alcyonidae des Tuamotu (Mururoa) et des Gambier. *Cah. Pacif.* 13: 133-157.

*Tixier-Durivault, A. (1974). Les octocalliaires des Gambier. *Cah. Pacif.* 18(2): 629-630.

Trichet, J. (1969). Quelques aspects de la sédimentation calcaire sur les parties émergées de l'atoll de Mururoa. *Cah. Pacif.* 13: 1-14.

*Ugolini, B., Robert, R. and Grand, S. (1982). La pêche en Polynésie française. Document ORERO 2. 13 pp.

*de Vaugelas, J. (1979). Effets d'une extraction de sable corallien sur les zones sédimentaires voisines, changements qualitatifs et quantitatifs de la matière organique vivante et détritique. In: Exploitation of coral sand in lagoons - French Polynesia and French West Indies. Report IUCN/WWF 1442. Annex 9.

de Vaugelas, J. (1985). On the presence of the mud-shrimp *Callichirus armatus* in the sediments of Mataiva lagoon. *Proc. 5th Int. Coral Reef Cong., Tahiti* 1: 314-316.

*de Vaugelas, J. Delesalle, B. and Monier, C. (1986). Aspects of the biology of *Callichirus armatus* (A. Milne-Edwards, 1870) (Decapoda, Thalassinidea) from French Polynesia. *Crustaceana* 50(2): 204-216.

Venec-Peyre, M.T. (1985). The study of living foraminiferan distribution in the lagoon of the high volcanic island of Moorea (French Polynesia). *Proc. 5th Int. Coral Reef Cong., Tahiti* 5: 227-232.

*Vergonzanne, G. (1979). Scilly, atoll de l'archipel de la Société, Polynésie française. Compte rendu préliminaire des observations scientifiques sur la population de tortues vertes (*Chelonia mydas*) de Scilly. *Bull. Ant. Tahiti Mus. Nat. Hist. Nat. et E.P.H.E.* 1: 59-61.

Villiers, L. and Sire, J.-Y. (1985). Growth and determination of individual age of *Turbo setosus* (Prosobranchia Turbinidae), Hao Atoll (Tuamotu, French Polynesia). *Proc. 5th Int. Coral Reef Cong., Tahiti* 5: 165-170.

Yen, S. (1985). The exploitation of Troca (*Trochus niloticus* L.) in French Polynesia. *Proc. 5th Int. Coral Reef Cong., Tahiti* 5: 557-561.

RAPA

Geographical Location   South Austral archipelago; 500 km from Raivavae, 1300 km from Tahiti; 27°35'S, 144°20'W.

Area, Depth, Altitude   40 sq. km (c. 9 x 7 km); 650 m.

Physical Features   A high volcanic island surrounded by a large and gently sloping platform, 0-55 m deep and 2-4.5 km wide. The coastline has 19 bays with deep channels, separated by high cliffs, including a large deep bay, Ahurei Bay, to the east which represents a drowned caldera. The geology of the island is described by Brousse and Gelugne (1986). Sea temperatures are low since the island is outside the tropical zone, with a maximum of 23°C in summer and 20°C in winter. Annual rainfall is about 2-3 m, unevenly distributed throughout the year, often accompanying storms. Winds are frequently strong (DIRCEN, 1986; Faure, 1985).

Reef Structure and Corals   The temperature is not optimal for coral growth and there are no true reefs or lagoon. However coral communities are found on the slope surrounding the island and in the bays and channels. Corals are particularly abundant in Ahurei Bay where monospecific stands of *Acropora formosa* are found. However, there is no *Porites* or *Pachyseris* and algal cover is very high, reaching 70% in some areas (Faure, 1985). Corals are also found in the bays of Hiri and Akao, as well as algal pavement and a reef flat formed of isolated coral colonies with brown algae (Brousse and Gelugne, 1986). Chevalier (1981) described 13 coral genera and Faure (1985) described 61 species in 31 genera; additional information on the corals is given in Faure (1986).

Noteworthy Fauna and Flora   The marine mollusc fauna is considered poor compared with that of Raevavae (Salvat, 1973). 250 species have been recorded (Richard, 1984, 1985a and 1986), and endemicity is about 10%, similar to the isolated Marquesas to the north. A number of species are common to Easter Island. The mollusc fauna is considered of particular interest on account of the dominance of herbivorous species due to the high algal cover; carnivorous species are poorly represented. The fish fauna is described by Plessis (1986). There are important sea bird colonies on the south and west coasts and on Motu Tanturau. The terrestrial wildlife shows high levels of endemism. The avifauna is described in Thibault (1974d and 1975b) and terrestrial vegetation in DIRCEN (1986).

Scientific Importance and Research   A scientific mission visited the island in 1984 under the auspices of the Direction des Centres d'Expérimentation Nucléaires (DIRCEN, 1986).

Economic Value and Social Benefits   Fishing is an important subsistence activity (Plessis, 1986).

Disturbance or Deficiencies   The terrestrial vegetation has been damaged, principally by goats, and by cows in the north-east corner of the island, but also by bush fires and by man (DIRCEN, 1986). Erosion now seems to be contributing to the sedimentation which appears to limit coral growth, in addition to recent volcanic events and cold water temperatures (Brousse and Gelugne, 1986).

Legal Protection   None.

Management   None.

Recommendations   The island is of major scientific interest for its terrestrial wildlife and high levels of endemism, much of which is considered under threat from habitat clearance. There would appear to be urgent need to investigate the conservation status of this island ecosystem, and it is recommended that the entire island system should be protected (Salvat *in litt.*, 1988).

MANUAE (SCILLY OR FENUAURA ATOLL) RESERVE

Geographical Location   In the leeward islands (extreme west) of the Society Islands, 550 km west of Tahiti; 16°30'S, 154°40'W.

Area, Depth, Altitude   200 ha; lagoon elongated north-south, 11 x 9 km; lagoon av. depth 50 m, max. 70 m, deeper on the west than east.

Physical Features   Manuae is a circular enclosed atoll. The eastern part of the atoll consists of sandy, wooded islets separated by shallow channels. The western part is a barrier reef, skirted on the lagoon side by a low sand dune 10 km long. The geomorphology is described by Chevalier (1979b). The lagoon is only accessible to light boats via the pass at the northern point of the island. To the west and south there is water exchange between the lagoon and sea at high tide but during calm periods, water enters only via the few operative hoa. In strong winds and tides, the flow varies according to wind direction and is described by Ricard and Delesalle (1981). Salinity in the lagoon ranges from 35.45 to 35.63 ppt, and there is slight stratification with a lower salinity at the surface. Surface temperature in the shallow margin area is about 28°C; in the centre it is about 27°C. The hydrology of the lagoon is described by Ricard *et al.* (1979a).

Reef Structure and Corals   The western and southern parts of the reef have been divided into the following zones (Chevalier, 1979b): well developed algal ridge; reef flat with corals and algae; eroded fossilized algal crest; immersed eroded organic flagstone, emerged coastal strip and sandy inner slope. The lagoon has many coral patches and pinnacles, none of which reach more than 20 m below the surface. There is heavy sedimentation and many of the patches are covered with silt (Salvat, 1979a and b). Fourteen genera of scleractinia have been recorded; *Porites* (at the surface), *Montipora* and *Psammocora* dominate (Chevalier, 1979b; Salvat and Venec-Peyre, 1981).

Noteworthy Fauna and Flora   There are coconut groves on the eastern part of the atoll and the vegetation is described by Sachet (1983). The Green Turtle *Chelonia mydas* nests on the atoll (Vergonzanne, 1979). The fish fauna is described by Bagnis and Bennett (1979) and Galzin (1979). *Tridacna maxima* and *Pinctada margaritifera* are abundant in the lagoon (Salvat and Venec-Peyre, 1981). The marine molluscs are described by Richard (1979), the foraminifera by Salvat and Venec-Peyre (1981), algae by Denizot (1979) and phytoplankton by Ricard and Delesalle (1981). Benthic fauna of the lagoon is described by Salvat (1983).

Scientific Importance and Research   Visited by an expedition from the Muséum National d'Histoire Naturelle, Paris, in 1979 (Anon., 1979a). Since then there have been several missions by EVAAM to study *Pinctada* stocks and by CNEXO to study turtles (Lebeau, 1985). The atoll is of interest because it is intermediate between an open and a closed atoll.

Economic Value and Social Benefits   The atoll is inhabited.

Disturbance or Deficiencies   There is a potential risk of overexploitation of marine resources and there is regular poaching of turtles by the inhabitants of Maupiti (Thibault *in litt.*, 17.2.88). Potential problems from ciguatera are discussed by Bagnis and Bennett (1979). Enforcement of protection is considered a problem, as at present the Territory of French Polynesia lacks the resources to undertake this adequately (Thibault *in litt.*, 17.2.88).

Legal Protection   Designated as a reserve in 1971 under Arrêté No. 2559/DOM of 28.7.71 (Anon., n.d. b). Alteration of the natural features or ecology of the lagoon is prohibited. Exploitation of fauna and flora within the lagoon, except for that required for staff on the island, is prohibited. A permit from the head of the Territoire en Conseil de Gouvernement is required for scientific research.

Management   None, although there has been one arrest for poaching which went before the court in Papeete.

Recommendations   It has been recommended that the reserve should be extended to include some of the terrestrial areas as turtle and bird sanctuaries (Dahl, 1980). Efforts must be made to convince the local authorities of the interest of the reserve and its resources, and to establish effective protection with a warden, although the difficulty of such an enterprise is recognised (Salvat *in litt.*, 1988).

TEMAE (RECIF D'ILOT)

Geographical Location   North-east coast of Moorea between Aroa and Faaupo Points; 149°46'W, 17°29'S.

Area, Depth, Altitude   4.5 km of barrier reef.

Physical Features   There is no lagoon in this area and the barrier reef is connected directly to the coast. The beach is exposed to strong waves. Pools carpeted with blue green algae are found along the edge of the rubble zone which has a low species diversity. Zones of conglomerate and flagstone are interspersed with the pools. The broken reef flat, in 20 cm of water, consists of joined coral patches. There is considerable turbidity on the outer slope. Nine zones (see next paragraph for deeper zones) have been defined at Aroa Point but towards the south-east some of these become less important. The reef flat is atrophied and the conglomerate of the upper zone is replaced by beach rock. Towards Faaupo Point the reef flat widens again, the conglomerate reappears and the outer edge resembles that at Aroa Point (Galzin and Pointier, 1985). The area is described in greater detail in Anon. (1977).

Inland of the beach is a saline lake, the hydrological aspects of which are described by Burlot *et al.* (1985) and Pouchan *et al.* (1985).

Reef Structure and Corals   The reef flat is colonised in some areas with *Acropora*, *Millepora* and *Porites*. The reef front is dominated by algae (*Turbinaria ornata*, *Sargassum* sp. and *Porolithon onkodes*) and corals (*Pocillopora*, *Acropora*, *Cladiella*). The ridge consists of 95% coralline algae. Other species are *Chlorodesmis* and *Caulerpa*. The outer slope is a platform furrowed by narrow channels (1 m wide and 1.8 m deep), with basins 7-10 m deep and 20-30 m in diameter (Galzin and Pointier, 1985). The reefs in this area resemble outer slope reefs of atolls, unlike the reef at Tiahura, also on Moorea (Anon., 1977).

Noteworthy Fauna and Flora   Some of the fish and mollusc fauna, particularly *Nerita plicata*, are described in Galzin and Pointier (1985). Algae are described by Payri (1987) and molluscs by Richard (1982b).

Scientific Importance and Research   This is the only site on Moorea where the barrier reef joins the coast. The reef was studied in the course of Mission Benthyplan to Moorea in May 1977 (Anon., 1977).

Economic Value and Social Benefits   The bay and reef to the east of Temae, adjacent to the hotel, are popular for recreation, snorkelling and water sports. The reef at Temae is occasionally visited by tourists (Gabrié *in litt.*, 16.11.87).

Disturbance or Deficiencies   The area was exposed to the hurricanes of 1982 and 1983 which caused considerable damage. A large hotel development is scheduled for the area, although tourism is currently declining (Salvat *in litt.*, 1988).

Legal Protection   None.

Management   None.

Recommendations Considered by some to be worthy of protection as one of the few remaining healthy reefs on Moorea. There are plans to create a "lagoonarium" as part of the hotel development, in order to protect part of the reef.

TIAHURA

Geographical Location North-west Moorea, between Tepee and Tehau Points; 17°29'S, 149°52'W.

Area, Depth, Altitude 1750 m of coastline; lagoon 840 m wide and 5-12 m deep.

Physical Features The area includes a fringing reef, channel, barrier reef and ridge and a small islet, motu Irioa, covered with bushy vegetation. The pattern of water circulation is complex. At flood tide oceanic waters flow into the lagoon by the Taotoi Pass and over the barrier reef; at ebb tide, the lagoon waters flow out through the pass. In the lagoon, the greatest current is in the channel but there is a small but important circulation of water on the fringing reef. Tides are semi-diurnal with amplitudes rarely exceeding 0.3 m. The volume of water in the lagoon is comparatively small resulting in wide fluctuations in several parameters depending on the intensity of oceanic exchange. Tidal currents may be strong (currents range from 0.5 to 2 knots) and there may be a disappearance or temporary lapse in the tidal cycle. The area is sheltered from the south-east trade winds but there is normally a fairly large swell on the north-west reefs (Galzin and Pointier, 1985).

Reef Structure and Corals From the beach, a sediment zone, consisting largely of decomposed microatolls descends to a fragmented reef flat which precedes a zone of coral patches some of which have aggregated to form vast coral heads. The fringing reef, 250 m wide at the transect area (where most research has been carried out), is shallow (less than 1.5 m deep). The channel, 9 m deep and 100 m wide at the transect, has coral formations on the shoreward side and a gentle sandy slope on the seaward side. The barrier reef (490 m wide) is shallow (no more than 2.5 m deep at high tide) on the landward side with a zone of fine coral sand. Seaward of this there are sparse coral patches which gradually become denser towards a zone of scattered pebbles and coral heads. There is a mixed reef ridge on slightly elevated flagstone on the outer edge (Galzin and Pointier, 1985).

Jaubert *et al.* (1976) describe the outer slope of the Tiahura Reef. This has an unusual morphology, with three zones between sea-level and 60 m depth. From 0-15 m depth, there are coral buttresses and gullies; this upper reef is divided into a furrowed and low-inclined upper platform (0-7 m depth) colonized by scattered coral colonies, and a spur and groove system ending at 15 m depth in a sediment basin lying parallel to the reef front. From 15-27 m depth, the spurs and grooves of the lower reef slope have dense corals on the upper parts of the spurs, but more scattered corals deeper. From 27-47 m depth, coral sand is found with an enteropneusta and cephalochordate community with sand eels *Taenioconger*. There is a sharp break in the slope at 65 m depth (Galzin and Pointier, 1985).

Reef productivity is described by Ricard (1983). Jaubert (1977a) has described growth forms of the coral *Synarea convexa*. 24 scleractinians have been collected, poritids being most common. Bouchon (1985) described scleractinian coral communities before the outbreak of *Acanthaster planci* and found that species richness and coral coverage increased seawards. The fringing reef flat was characterized by a poor coral community, dominated by *Porites* and *Montipora*. A richer community dominated by *Acropora* and *Pocillopora* was found on the barrier reef flat. The outer reef had three zones: an *Acropora* and *Pocillopora* zone from 0-10 m depth, a *Porites* and *Montastrea* zone from 15 to 20 m depth and a massive *Porites* zone at 30 m depth. Since the *Acanthaster* outbreak there has been a major decrease in coral coverage.

Noteworthy Fauna and Flora The fauna and flora of the reef and lagoon have been intensively studied and are briefly reviewed in Galzin and Pointier (1985). Fish of the lagoon are described by Galzin (1976, 1977, 1983 and 1985) and Plessis (1972b and 1973a). 280 species have been recorded. Fish biomass is greater on the fringing reef by the channel than on the reef flat or fringing reef proper. A less diverse fauna is found near the shore. The pomacentrid *Stegastes nigricans* is the predominant species on fringing reef patches near the channel and the acanthurid *Ctenochaetus striatus* dominates on the reef flat. *Chaetodon* abundance is noticeable as on all Polynesian reefs. Decapod crustaceans and polychaetes are described by Peyrot-Clausade (1976 and 1977) and de Vaugelas *et al.* (1986). The holothurian *Halodeima atra*, the most abundant echinoderm in the lagoon, is described by Renaud-Mornant (1977); molluscs are described by Richard (1973a and b) and Richard and Salvat (1972); 150 species have been collected from the transect. The algal community of the lagoon is described by Payri (1983).

Scientific Importance and Research The area has been studied since 1971 when the Antenne du Muséum et Ecole Pratique des Hautes Etudes was established and it is the most studied reef on Moorea (Ricard *et al.*, 1977). A transect was established along which most of the studies have been carried out. Many of the studies have been directed at determining the productivity of the reef and a preliminary production balance has been determined which is summarised in Galzin and Pointier (1985). Naim (1980) studied the small mobile macrofauna in the algae in the lagoon. Phytoplankton was studied between 1974 and 1978 (Ricard, 1976a and b; Sournia and Ricard, 1975b and 1976) and zooplankton by Lefevre (1984). Between 1971 and 1981, some 178 publications were produced on a variety of subjects.

Economic Value and Social Benefits There are six hotels on the shore facing the reef. The most recently constructed, the Hotel Sofitel, was completed in 1987; all the others are over 10 years old, although the Club Méditerranée at the western end of the area was enlarged in 1985 (Gabrié *in litt.*, 16.11.87; Wells pers. obs., 1985). The whole area, particularly around the motu, is used by tourists from the hotels, especially from the Club Méditerranée (Gabrié *in litt.*, 16.11.87).

Disturbance or Deficiencies Tiahura reef was seriously damaged by an outbreak of *Acanthaster planci* in 1981 which caused a major decline in coral coverage particularly of branching forms such as *Pocillopora* and *Acropora*. In some areas, coverage was reduced by 50%, the worst affected area being the outer reef slope. The loss of scleractinians affected the motile cryptofauna (Peyrot-Clausade, 1985) and fish; the density of chaetodontids which are coral feeders declined by 47% (Bouchon, 1985; Bouchon-Navaro *et al.*, 1985). The reefs may also have suffered coral mortality as a result of the abnormal El Niño of 1983 (Glynn, 1984).

Extraction of coral sand (for building materials) from the lagoon in two parallel dredging channels has seriously damaged the area. Coral sand extraction took place in 1962 and 1975, the latter occurrence causing the almost complete isolation of an 18 ha zone of fringing reef between the dredging dike and Teepe Point. The increased turbidity had a noticeable impact on primary production (Ricard, 1980 and 1981) and on the ichthyological and benthic communities (Anon., 1979; Naim, 1981; Salvat *et al.*, 1979). There has been an increase in water temperature in this area, suppression of water circulation, a large drop in salinity, a marked increase in turbidity and noticeable sedimentation (de Vaugelas, 1979). Corals were reduced by 95% and diversity of other taxonomic groups is also lower; macrophytic algal production increased by 5%. The dredged zone is considered irretrievably destroyed, a study in 1984 indicating that the fringing reef is still in a poor condition. Porcher (1985) and Porcher and Bouilloud (1984) carried out an environmental impact study during the construction of the Hotel Sofitel. A channel was dredged through the reef to allow boats to reach the shore at the hotel, but attempts to manage the dredging operations were partially successful (Gabrié *in litt.*, 16.11.87 and see below). No dredging has been carried out on the Tiahura reef itself since the completion of the hotel. However an illegal dredging operation, which had been running for at least a year, was in operation immediately west of the area in July 1987 and had disturbed the adjacent fringing reefs. No studies have been carried out on the effects of recreational use on the reefs. There are not known to be any plans at present for further hotel development along this stretch of coast (Gabrié *in litt.*, 16.11.87).

Legal Protection   Coral dredging is now banned on Moorea (Gabrié *in litt.*, 16.11.87).

Management   An experimental attempt, which was largely successful (Gabrié *in litt.*, 16.11.87), was made to control the effects of the most recent dredging operation by enclosing the site with a woven curtain which prevented the dispersal of fine particles (Galzin and Pointier, 1985). However other recommended actions (Porcher and Bouilloud, 1984), in particular that of vacuuming away the accumulated sediment before removal of the curtain, were not carried out (Gabrié *in litt.*, 16.11.87).

Recommendations   Given the scientific importance of this site and the long history of research on the transect, it would seem of the utmost importance that attempts are made to stop further deterioration and to ensure that any future developments have as small an impact as possible. Preliminary proposals have been drawn up by the Antenne Muséum-EPHE for the restoration of areas degraded by dredging (Gabrié *in litt.*, 16.11.87) and for monitoring the area on a ten-yearly basis (Salvat, 1987).

MATAIVA

Geographical Location   Westernmost atoll in Tuamotu group, 300 km north of Tahiti; 14°55'S, 148°36'W.

Area, Depth, Altitude   10 x 5.5 km; lagoon depth av. 8 m, max. 30 m.

Physical Features   The atoll has an unusual morphology and geology with a wide (200-1500 m), almost continuous, elliptical atoll rim which permits very little oceanic exchange, and a reticulated lagoon divided into about 70 shallow basins by a network of slightly submerged partitions. A detailed description of this, including the geology and hydrology of the atoll, is given in Delesalle (1985b). There are several hoa to the south and a single pass in the north. The north and east coastline forms a single islet.

The climate is similar to that of the other Tuamotu atolls. There is a dry, cool season from April to September (24-27°C) and a hot, wet season from October to March (28-30°C). Rainfall is rarely more than 25 mm a year. Storms and cyclones are rare although there were three in 1983. Easterly trade winds predominate. Water temperatures in the lagoon do not vary much with a maximum of 29-31°C in the rainy season and a minimum of 25-27°C in the dry season. The lagoon waters have a more or less pronounced milky appearance due to the suspension of fine, silty calcareous particles and have high and variable concentrations of dissolved nutrients, especially nitrates and silicates (Delesalle, 1985b).

Reef Structure and Corals   Coral diversity in the lagoon is very low, only 28 species having been recorded. Highest diversity is found near the pass and hoa. Coral cover is also very low, diminishing from 30% near the hoa to 0% in the eastern part of the atoll. Colonies are mainly located on the partition summits around the edges of the pools. *Porites lobata*, forming microatolls on these shoals, and *Acropora tortuosa* on the pool sides, are the commonest species. *Montipora aequituberculata* and *Leptastrea purpurea* are widely distributed but have very small colonies. The outer slope is more diverse that the lagoon but corals are often small. The eroded reef flat has a few corals near the reef front, *Acropora* and *Pocillopora* being most common (Delesalle, 1985b).

Noteworthy Fauna and Flora   The algae and phyto-plankton are described by Salvat (1982). The phytoplankton is quantitatively dominated by phytoflagellates and qualitatively by diatoms (Delesalle, 1985a). Marine fauna recorded from the atoll include over 220 molluscs, 115 fish species and about 100 crustaceans (Delesalle, 1985b), among which the thalassinid mud-shrimp *Callichirus armatus* is responsible for extensive reworking of the lagoon soft bottom sediments (de Vaugelas, 1985; de Vaugelas *et al.*, 1986). The Coconut Crab *Birgus latro* is rare (Delesalle, 1985b).

Scientific Importance and Research   Considerable work has been carried out because of the potential for phosphate mining on the atoll. Missions to Mataiva were undertaken in 1981 (Delesalle *et al.*, 1981), 1982 and 1983 (Delesalle, 1985b).

Economic Value and Social Benefits   In 1983, the population numbered just over 180, but there is much exchange with Rangiroa. Fishing and copra are the traditional means of support. The sea provides about 50% of the protein requirements, including fish, molluscs, crustaceans and turtles. Fishing techniques include spears, lines and fish parks located in the pass and near the hoa (Delesalle, 1985b).

Disturbance or Deficiencies   There are many dead colonies in the lagoon caused by the high turbidity, particularly in the eastern part.   On the north side however *Porites* is abundant even though the turbidity is still high.   There was some cyclone damage to the outer reefs in 1983. Much of the dead coral may have been caused by events such as occurred in 1978 and 1980, when extreme low tides combined with strong sunlight to cause very high temperatures (32°C) for as long as ten days.   In 1981, coral cover was as low as 10% even near the hoa but had increased by 1983 (Delesalle, 1985b).

A major threat would be the initiation of phosphate mining. Phosphate deposits cover about 5 sq. km and represent some 10 million tons of available ore. Exploitation of this resource would take 10-15 years and necessitate the employment of about 200 people and could have a major impact on the natural environment (Delesalle, 1982 and 1985b). The inhabitants of the atoll are reportedly opposed to the mining (Thibault *in litt.*, 17.2.88) and the plans are currently dormant (Salvat *in litt.*, 1988).

Legal Protection   None.

Management   None.

Recommendations   A turtle and seabird reserve was proposed by Dahl (1980) for the atoll but both birds and turtles are now scarce. Should the proposed phosphate mining plans go ahead, a rigorous environmental impact assessment should be carried out and measures enforced to ensure that these activities have minimum impact on the natural resources.

MORUROA

Geographical Location   South-east Tuamotu group; 21°46'-21°54'S, 138°47'-139°03'W.

Area, Depth, Altitude   28 x 10 km; atoll rim is 65 km long; max. alt. 6.5 m; lagoon 26 x 9 km, max. depth 52 m.

Physical Features   The atoll, oriented WSW-ENE, has one pass into the lagoon in the north-west and 300 islets. The rim is wider in the south (550-700 m) than in the north (400 m). On the west, there are few emergent islands and much of the rim is submerged at low tide; in the east and south there are numerous motu and large portions of the rim are permanently emerged. Islets vary in length from 20 m to several kilometres and in width from 30 to 900 m. They are described by Chevalier *et al.* (1968). Soils are described by Tercinier (1969). The geology and geomorphology of the atoll have been extensively studied, such research being facilitated by the fact that the atoll has been drilled to its volcanic basement in many places (Chevalier *et al.*, 1968; Deneuebourg, 1969; Bardintzeff *et al.*, 1985; Buigues,

1985; Fontes *et al.*, 1969; Gachon and Buigues, 1985; Aissaoui and Purser, 1985; Labeyrie *et al.*, 1969; Ruzie and Gachon, 1985).   Dolomitisation is described by Purser and Aissaoui (1985). Surrounding waters reach depths of over 3000 m, even between the atoll and its nearest neighbour Fangataufa at 50 km distance.

The climate is described in Chevalier *et al.* (1968). It is cooler than much of the Tuamotu archipelago as the atoll is situated well to the south-east. Humidity is high, most rain falling between November and January. Easterly trade winds predominate, although during the winter, westerly winds may prevail. Cyclones are periodically experienced. Tides are semi-diurnal with an amplitude of 40-70 cm.

Hoa vary in width from a few metres to 150 m and in depth from 50 cm at low tide at the outer end to 1-4 m on the lagoon side. About 28% connect directly with the sea; the remainder do so only during high seas and storms. Chevalier *et al.* (1968) distinguish five different types of hoa: those that are occasionally functional and which have a single opening into the lagoon and end in a cul-de-sac on the ocean side; functional hoa which connect with the sea at high tide and may or may not at low tide; non-functional hoa which are entirely isolated from the lagoon and the sea; and two other varieties of the latter. Of the 288 hoa, 79 are functional and are most abundant on the south, their density increasing from east to west on both sides of the atoll.

There is a single pass in the north-west, about 4500 m wide and with a maximum depth of 8-9 m, and the islet, Giroflée, situated in the south-west (Chevalier *et al.*, 1968). The pass is unusually large compared with other atolls in the Tuamotu group and the rapid exchange of water is responsible for the particular characteristics of the lagoon fauna and flora, especially the relatively rich coral fauna. Sedimentation in the lagoon is described by Trichet (1969).   Further information is provided in Rougerie *et al.* (1980).

Reef Structure and Corals   The exterior reef can be divided into three zones: reef flat, algal crest and outer slope. The reef flat in front of the motu has two zones. The internal zone, which ranges from a few metres wide in the north to 60 m wide in the south, may be submerged or emergent at high tide and is irregular with numerous erosion channels and occasional "negro heads".   The external zone is submerged at low tide and, like the inner zone, is formed of reefal conglomerate; living corals are found towards the seaward edge of the zone and are most abundant on the sheltered north and west sides of the atoll. The reef flat between the motu, particularly in the north and west, forms a platform with a gentle slope lagoonwards. The leeward reef flat has more abundant coral fauna than the windward reef flat, coverage reaching 50-60% in some areas (Chevalier *et al.*, 1968).

The algal crest varies in form according to its location on the atoll. It is largest on the windward reefs in the south, 30-50 m wide and emergent by 80 cm at low tide. The exterior part is formed of living calcareous algae, predominantly *Porolithon* and *Chevaliericrusta*; the internal part is largely dead and eroded. There are numerous spurs and grooves which extend down the outer slope to depths of 20 m, the spurs with a width of 3-10 m, the grooves usually remaining narrow. The spurs are covered with calcareous algae on the upper parts and corals in deeper water. At the eastern extremity of the atoll, which is even more exposed, the algal crest is similar but the spurs are broken by ledges. On the more sheltered eastern coast, the algal crest is less well developed, 10-20 m wide with 20-30 cm exposed at low tide. On the western leeward side, the atoll is even less developed, 10-20 m wide and emergent at low tide by 10-20 cm.

The outer slope can be studied by diving only at certain points on the atoll (Camelia, Brigitte, Canon, Grue, Aline, Becasse and Hortensia). The slope extends normally for 40-50 m to depths of 15-25 m. The profile of the slope in other areas has been estimated by soundings and is illustrated in Chevalier *et al.* (1968). Below the algal zone which extends to about 10 m on the windward side of the atoll, the outer slope is dominated by *Acropora* and *Pocillopora*. On the leeward side, the outer slope has a richer coral fauna and a clear zonation can be distinguished. From 1 to 4 m depth, there is still a rich algal cover with a coral coverage of 20-30%, dominated by *Pocillopora brevicornis*, which is richest on the spurs. From 4 to 8 m depth, coral coverage reaches 40, the main species being *P. brevicornis*, *Acropora corymbosa*, *A. danai*, and *Millepora platyphylla*. From 8 to 15 m depth, corals are less abundant although species composition is similar apart from *P. brevicornis* which is less abundant. From 15 to 25 m depth, coral coverage is 30% at maximum and the principal species are *A. humilis*, *A. corymbosa* and *A. syringodes*. Below 25 m depth, corals become progressively rarer and disappear at about 60 m. Between 25 and 30 m depth the main species is *Montipora caliculata*. The hoa have a poor coral diversity but scattered madreporarian corals may occur.

The ecology and benthos of the lagoon is described by Salvat and Renaud-Mornant (1969). Chevalier *et al.* (1968) describe the lagoon reefs, including the fringing reefs of the motu and submerged reef flats, as well as patch reefs around the edge of the lagoon. The latter are particularly numerous in the south-west of the lagoon and become more scattered towards the centre. Maximum diameter is about 120 m. The motu fringing reefs may have coral coverage as high as 80%, the main species being *A. brevicornis*, *A. danai* and *A. corymbosa* (Chevalier *et al.*, 1968). On the patch reefs *Acropora* spp. generally dominate the surface of the reef, *A. danai* and *Millepora platyphylla* occur to 2 m depth and *A. macrophthalma* dominates to 15 m depth.

Eighty species of coral in 26 genera have been recorded from the atoll. Tixier-Durivault (1969) described the soft corals.

Noteworthy Fauna and Flora   In the 1960s the motu were largely vegetated, coconuts palms being dominant (Pomier, 1969). The vegetation and terrestrial fauna are described in Chevalier *et al.* (1968) and Pomier (1969). The bird fauna is considered comparatively poor with less than ten breeding species but there are large numbers, mainly terns and noddies which had colonies on the north-western part of the atoll. Seabirds include Murphy's *Pterodroma ultima* and Kermadec Petrels *P. neglecta*, the Greater Crested Tern *Sterna bergii*, the White Tern *Gygis alba*, and the Brown *Anous stolidus* and Black Noddies *A. minutus*. A number of other migrating species have been recorded. The only mammals are four species of rodent, all introduced by man. The marine fauna is described in Chevalier *et al.* (1968). England (1971) described the actiniarian fauna. Fish distribution is described by Quigier (1969). The plankton is described by Michel (1969).

Scientific Importance and Research   Following the establishment of a Centre d'Experimentations Nucleaires (C.E.P.) on the atoll, a series of detailed studies was carried out prior to nuclear testing on all biological aspects to provide a baseline from which changes due to radioactive events could be monitored. The research was carried out by a number of scientific institutions under the direction of the Service Mixte de Controle Biologique, established in 1964 under the Direction des Centres d'Experimentations Nucleaires. Much of the work was published in a special issue of *Cahiers Pacifique* in 1968. Subsequent missions have studied a variety of aspects although this has not all been published, such as the study visit made by a team comprising Australian, New Zealand, and Papua New Guinea scientists (Atkinson *et al.*, 1984).

Economic Value and Social Benefits   The island was uninhabited following devastation by cyclones in 1904-1906, but was sporadically visited for the collection of copra. A military base with some 3000 employees is situated on the north-eastern end of the atoll.

Disturbance or Deficiencies   Cyclones damaged the atoll in 1980, 1981 and 1983. The establishment of the C.E.P. on the atoll caused considerable alteration to the terrestrial parts of the atoll (Chevalier *et al.*, 1968). Atomic testing commenced in 1966 (Danielsson, 1984). Atmospheric testing ceased in 1974 and was replaced by underground testing, requiring the drilling of bore holes to 800-1000 m depth on the atoll rim, mainly along the southern side. Since 1980 underground testing has taken place in holes drilled in the lagoon. There have been reports of leakage of radioactive substances, physical damage to the structure of the atoll, tidal waves generated by explosions and poorly managed dumping of radioactive material on the northern part of the atoll which is now considered highly contaminated. There has been considerable controversy as to whether the heavy incidence of ciguatera in the region is related to the testing; evidence suggests that radiation itself is not a causal factor but that other military activities carried out in the area such as dredging and rubbish dumping may be a contributory factor (Caire *et al.*, 1985).

Hermit crabs *Coenobita perlata* declined rapidly after the establishment of the C.E.P. as they are used as fishing bait. The lobsters *Parribacus antharticus* and *Panulirus penicillatus* have become increasingly rare.

Legal Protection   None.

Management   None.

RANGIROA

Geographical Location   200 km north-east of Tahiti; 15°00'S, 147°40'W.

Area, Depth, Altitude   1640 sq km (84 x 33 km with a periphery of 225 km); average lagoon depth 20 m, max. 35 m, width 4-30 km.

Physical Features   The atoll rim is widest in the east and narrowest in the south (300-600 m); further details are given in Ricard (1985). There are about 240 islands or motu, mainly on the north-west part of the atoll; the remainder consists of sandy areas more or less submerged. The lagoon has scattered coral pinnacles, sand banks, islets and shallows and a silty bottom (Dufour *et al.*, 1984) and silty narrow beaches in the north, characterised by the presence of marine phanerograms which are rarely found in Polynesia, which form lawns covering dozens of sq. km. There are two main passes (Hotua Uva and Hivia) on the north side, which are 450 and 550 m wide and 14 and 35 m deep respectively. Each has a patch reef situated at the entrance to the lagoon. There are about 150 hoa, some of which are permanently functional. Others are open to the lagoon but blocked from the ocean. Others, particularly in the south, are open to the ocean but blocked to the lagoon and some hoa are completely isolated. There are numerous small fossil reef pinnacles (feo) on the south side, described by Ricard (1985).

Easterly trade winds predominate, but north or south-west gales may occur between November and May which is the cyclone season. Rainfall is between 1200 and 1500 mm a year and is highest between November and January. Hydrological details are given in Ricard (1985), Rougerie and Gros (1980) and Gros *et al.* (1980); the lagoon waters are similar to the ocean due to exchange through the passes and its large size; in some ways the lagoon operates more like an inland sea. Sea level changes are described in Ricard (1985). Stoddart (1969) describes the geomorphology of the atoll. A general description is given by Ottino (1972).

Reef Structure and Corals   The surf zone of the reef flat is most exposed on the northern rim and consists of an algal ridge 10-20 m wide, mainly of *Porolithon craspedium*, *P. onkoides* and some *Lithophyllum*, which forms spurs and grooves. The submerged reef flat is a calcareous platform with small colonies of *Porites* and *Acropora* scattered over its surface. Landward of this is flat rocky calcareous floor and a zone of beachrock. The reef flat is similar in the south with a well developed algal crest but this is followed by a "feo" zone which is dissected by deep channels. Corals are less abundant in the south, only *Acropora* being regularly observed although hydrocorals may be abundant. Coral diversity is greatest at Hao.

Coral diversity and abundance are low in the lagoon, although corals are abundant where the slope drops off, *Acropora* providing 95% cover, and *Porites* 3%, with a scattering of *Pocillopora* and *Montipora*. In the south of the lagoon the sand is less silty and coral patches are abundant in shallow water but become less so with depth. The massive genera *Favia*, *Platygyra* and *Montipora* are most abundant with *Pocillopora* and *Acropora*. Coral patches or pinnacles are scattered throughout in about 20-25 m of water. The sheltered side of a pinnacle (generally the north-west side) has a 100-150 m wide platform of coral rock at 0.5-3 m depth, covered with sand and scattered with coral patches usually of *Acropora*. The exposed side has a narrow sandy shelf with *Acropora* and encrusting algae. The slope below the platform has abundant *Millepora platyphilla* and scattered corals of a variety of species. The lagoon bottom around the pinnacles is sandy with scattered *Acropora*.

Permanently functional hoa have a high coral diversity. Those that are blocked from the ocean have a sparse covering of *Acropora* and *Montipora*. Those that are open to the ocean but blocked from the lagoon have a rich coral community, particularly near the outer reef flat. In completely isolated hoa *Porites* is usually the only coral present. The reefs around Motu Paio are described in Ricard (1985).

Noteworthy Fauna and Flora   A total of 58 species of fish were described by Plessis (1970 and 1972a) but research since then has brought the total to about 600 for the lagoon alone. The lagoon has particularly rich algal and faunal communities and about 95 species of mollusc have been described. Plankton, benthic fauna and flora, particularly molluscs, and fish are described by Michel *et al.* (1971), Poli and Richard (1973) and Ricard (1985). The permanently functioning hoa have a particularly diverse fauna. Sachet (1969) has described the terrestrial flora of the atoll. Motu Paio (Bird Island), 5 mi. (8 km) from the village of Tiputa, is the nesting site for large numbers of birds, mainly noddies and terns.

Scientific Importance and Research   Rangiroa is the second largest atoll in the world and the largest in the Tuamotu group. Pirazzoli (1982 and 1985b) has investigated mapping the lagoon by satellite imagery. The atoll was visited in the 1960s, the results of the expedition being summarised by Stoddart (1969). The atoll rim is covered with "kopara" deposits, cyanobacterial deposits which accumulate in shallow ponds just above the tide mark (Defarge *et al.*, 1985; Ricard, 1985).

The Commissariat à l'Energie Atomique and the Oil and Oleaginous Research Institute both have laboratories on the atoll. The EVAAM station carries out research on lagoon and coastal fishing, and pearl culture. With the introduction of small scale tuna fishing, interest has grown in the possibility of farming baitfish. Hoa and kopara ponds are being studied for this purpose, and the experimental work carried out so far is described briefly in Ricard (1985). A hatchery for *Pinctada margaritifera* has been established and there are plans to expand the research carried out on the pearl oyster.

Economic Value and Social Benefits   Rangiroa is the most populated atoll (about 700 people in 1985) in the Tuamotu archipelago. There are two villages, Avatoru and Tiputa on the north side, each to the east of the two navigable passes. Fishing and fish culture (*Chanos chanos*), coconuts and pearl culture are the main sources of income. Fishing used to be the most important activity but today it provides employment for only about 20-30 fishermen; about half the catch is for local consumption, the remainder being exported to Tahiti (Grand *et al.*, 1983). Fish parks located in the passes are used to catch migrant fish for export to Tahiti. A variety of fishing methods are described in Ricard (1985). The lagoon at Hoa Vaimati is separated from the main lagoon and provides a suitable environment for breeding baitfish *Chanos chanos* (Dufour, 1985). Kopara ponds in the north west at Pavete are also used for this purpose (Ricard, 1985). A pilot station has been built here and nearly 8 t of baitfish have been produced since 1984.

Tourism is becoming increasingly important and there are four hotels, an airstrip and watersport facilities.

Disturbance or Deficiencies   The reefs suffered serious damage from the 1983 cyclones which also altered the formation of the beaches. The north-west and west of the atoll suffers most serious damage during such events (Bourrouilh-Le Jan and Tallandier, 1985).

Legal Protection   None.

Recommendations   Motu Paio has been recommended as a reserve for birds and its owner is willing for this to happen (Thibault *in litt.*, 17.2.88). In the early 1970s the atoll was suggested for National Park status (Elliot, 1973).

TAKAPOTO

Geographical Location   One of the northern Tuamotu atolls, 600 km from Tahiti and 10 km south-west of its sister atoll Takaroa; 14°35'S, 145°13'W.

Area, Depth, Altitude   74 sq. km (17 x 5 km, with a periphery of 44.6 km), of which 23 sq. km is emerged rim and 51 sq. km is lagoon; lagoon depth av. 23 m, max. 60 m.

Physical Features   Oriented SW-NE. The lagoon, with a largely sandy bottom, is more or less closed but some oceanic exchange takes place through "hoa" or reef flat spillways which are about 10 m wide and less than 10 cm deep and permit water exchange depending on the tides, winds and relative levels of ocean and lagoon. There are 25 hoa on the south coast, one of which, at Orapa, provides some degree of water exchange each day; the hoa on the north-west side of the atoll are non-functional except for one which functions during very strong swells (Salvat and Richard, 1985). Lagoon salinity reaches 39.8 ppt. Geomorphology and marine ecology of the lagoon area are described by Chevalier *et al.* (1979), Chevalier and Denizot (1979) and Salvat (1981); physicochemistry and plankton by Gueredrat and Rougerie (1978); and hydrology and primary production by Ricard *et al.* (1978) and Sournia and Ricard (1975a and b). Information on the lagoon is summarised in Salvat and Richard (1985). The climate is typical of the Tuamotu group with a dry cool season from May to October and a hot and rainy one from November to April. Annual rainfall is 1.2-2.5 m and easterly trade winds predominate except during bad weather when winds are from the west. Tidal amplitude is about 40 cm and low tides usually occur in the early morning and late evening.

Reef Structure and Corals   A total of 58 scleractinian genera and five hydrocorals have been recorded from the atoll (Bouchon, 1983; Kühlmann and Chevalier, 1986). The outer reef flat and algal crest have poor coral coverage. On the outer slope coral coverage is usually over 60%; the upper zone is dominated by *Pocillopora verrucosa*, *Acropora danai* and *Pavona minuta*; the middle zone by *Porites*; and the lower zone by *Leptoseris hawaiensis* and *Pachyseris speciosa*. The outer slope near the village is described by Salvat and Richard (1985) and sites near Okukina in the west, Fakatopatere in the

south-west and Orapa in the south-east are described by Kühlmann and Chevalier (1986).

Coral communities are poorer in the lagoon due to sedimentation. There are over 400 coral patches and pinnacles from 10 to 100 m in diameter, the majority of which are very small and are situated in the south-west of the lagoon. Coral coverage of the pinnacles is usually under 1% except for *Porites lobata* and *Millepora platiphylla* and the pinnacles are characterised by the predominance of molluscs. *Acropora* and *Pocillopora* are found on pinnacles near spillways but are not found elsewhere in the lagoon (Chevalier, 1981; Chevalier and Denizot, 1979; Kühlmann and Chevalier, 1986). The deepest parts of the lagoon are dominated by Thamnasteridae. Salvat and Richard (1985) give a detailed description of the Tararo, Orapa and Vairua coral patches.

Noteworthy Fauna and Flora   Fish are described by Bagnis *et al.* (1979) and Bouchon-Navaro (1983) and about 300 species have been recorded; the lagoonal fauna is considered comparatively rich. Takapoto lagoon is of particular interest for the predominance of four species of mollusc, *Tridacna maxima*, *Pinctada margaritifera*, *Arca ventricosa* and *Chama iostoma*, on the coral pinnacles; their biomass is estimated at 1100 tons and about 70 million individuals. In order to study the potential for farming these species, there has been intensive research within the MAB programme on a variety of parameters, summarised in Salvat and Richard (1985). Publications include Jaubert (1977b) and Richard (1977, 1978, 1983, 1987 and in press) and Richard *et al.* (1979). Although these species constitute a large standing crop, research has shown that they have a very slow growth rate and low productivity. The holothurian *Halodeima atra* is very abundant in the lagoon. Plankton are described by Gueredrat and Rougerie (1978), Renon (1977) and Ricard *et al.* (1979b).

The terrestrial fauna and flora are reviewed by Salvat and Richard (1985). Sachet (1985) provides a species list of the flora. Eight km north of the village near Vairua, tall groves of *Pisonia grandis* provide roosting and nesting sites for noddies *Anous*, boobies *Sula* and terns *Gygis*. Seaward of this there is poor open scrub vegetation which may resemble the original atoll vegetation. The bird fauna is poor and there is no true seabird colony (Salvat and Richard, 1985; Thibault, 1974b).

Scientific Importance and Research   One of the best known Pacific atolls, Takapoto was selected for a MAB/Unesco research programme and long term studies have been carried out by the Antenne de Tahiti. Some 40 papers have been published concerning the atoll and are listed by Salvat and Richard (1985) who provide a review of current knowledge. Species diversity is low but abundance is high. The Centre de Recherche Gouvernemental pour la Culture des Perles was set up by the Service de la Pêche in 1970 and carries out research on pearl culture. Pirazzoli (1985b) has investigated mapping the lagoon using satellite imagery and Pirazzoli and Montaggioni (1984) describe recent variations in sea level on the atoll.

Economic Value and Social Benefits   Takapoto is considered to have considerable potential for economic development due to the expansion of the pearl culture industry in 1970 and the construction of a small airstrip in

1973. The population has varied considerably and reached about 400 in 1985. The entire population now lives in the village of Fakatopatere, situated at the south of the atoll. Pearl culture, fishing and copra are the main sources of income. Fish is an important part of the diet, the principal species caught in stone fish traps situated in the more functional hoa being Red Snappers *Lutjanus* and *Lethrinus* spp., parrotfish *Scarus* and sea bass Serranidae. The black lip oyster *Pinctada margaritifera* was first exploited at the beginning of the 19th century, when mother-of-pearl was exported to Valparaiso and Sydney. From 1946 to 1965 Takapoto production, which made up 8% of the total Tuamotu-Gambier production, varied from 1 to 149 tons. Overcollection and the development of pearl culture on the island since 1969 has meant that mother-of-pearl production ceased after 1965. By 1984 there were 13 pearl farms of which six were co-operatives, six private and one state-owned. About 7000 pearls are produced on Takapoto annually with a value of US$ 282 000. Shellcraft provides an additional source of income (Salvat and Richard, 1985).

Disturbance or Deficiencies Takapoto was seriously damaged during the period of cyclones from December 1982 to April 1983, Cyclone Orama-Nima which passed closest causing particularly severe damage to the village (Salvat and Richard, 1985). 50-100% destruction was recorded over 40% of the outer slope of the reef on the east side although the rest remained largely intact. Scleractinian species started to show signs of recovery after two years (Laboute, 1985). Ciguatera poisoning is apparently absent from the lagoon but toxic fish have been caught on the outer reef near the wharf; there was an outbreak of ciguatera between 1960 and 1970.

Legal Protection None.

Management None.

Recommendations Regular monitoring of the reefs and marine resources should be carried out and appropriate management techniques implemented as they become necessary.

TIKEHAU

Geographical Location North-west Tuamotu archipelago, 300 km north of Tahiti, between Mataiva and Rangiroa; 15°00'S, 148°10'W.

Area, Depth, Altitude 25 x 25 km; widest diameter 28 km; reef rim about 78 km long and varying in width from 300 m in the north-west to 1300 m in the south-east; lagoon av. depth 20 m, max. 45 m.

Physical Features A circular open atoll with one pass, 300 m wide and 3.7 m deep, on the west and more than 150 hoa or channels, over 100 of which are situated in the south-east. It is separated from Mataiva by a distance of 37 km and depths of less than 1000 m. The motu are usually 150-300 m wide except the one in the south-west which is 900 m at its widest point. There is a single pass, Tuheiava on the west, about 400 m wide. A seven mile long channel flows from the pass through the lagoon to the village. There are numerous pinnacles in the lagoon (Harmelin-Vivien, 1985).

The climate is typical of the Tuamotu group; there is a hot season from November to April with occasional heavy rainfall and a cooler wet season from May to October. Easterly trade winds predominate. Hurricanes occur in the hot season, usually blowing in the opposite direction to the trade winds. Tides are semi diurnal with a range of less than 1 m. The surface temperature of the lagoon waters is lowest (about 26°C) in July and August and highest (above 30°C) in February and March, and temperatures fluctuate more than that of the ocean. The salinity of the lagoon is variable, depending on the amount of water exchange taking place with the ocean; however inflow of sea water largely seems to counterbalance any salinity increases due to evaporation. Heavy rainstorms may reduce salinity as observed following the hurricane of 1983. The geological history of the atoll and the hydrology of the lagoon are described in Harmelin-Vivien (1985). Additional general information is given in Intes (1984) and Anon. (1985).

Reef Structure and Corals Faure amd Laboute (1984) divide the reefs into the outer slope, the reef flat and the lagoon reef. The outer slope consists of three zones: the fore reef zone from 0-10 m is formed of a spur and groove zone and a fore reef platform; the outer terrace (10-25 m) resembles an old spur and groove system and is an irregularly eroded surface; and the deep slope which starts below 25 m.

The reef flat can be subdivided into an outer reef flat which consists of an algal ridge and the submerged flat, and an inner reef flat.

The lagoon has an inner reef slope which drops to about 5 m and is generally covered with sediments; between 6 m and 12 m the edge of the lagoon is covered with sedimentary deposits with abundant coral patches. The soft bottom of the lagoon consists of fine coralline mud, extensively reworked by Thalassinid mud-shrimps. Coral pinnacles are found scattered throughout the lagoon and are most abundant to the south-west; these usually measure 50-100 m in diameter although some may reach 200 m. The largest ones support small motu or islets and are surrounded by extensive reef flats.

Scientific Importance and Research Tikehau provides a model of a mid-sized open atoll and was selected by ORSTOM for the development of a multidisciplinary research program in collaboration with a number of other French laboratories. This progamme was initiated in 1982 and preliminary results were published in 1984 (Harmelin-Vivien, 1985). Particulate organic matter in the lagoon has been studied by Charpy (1985).

Economic Value and Social Benefits Since the 1970s, after a long period of regression, the population rose slightly to reach 266 in 1983. There is an airstrip and the atoll is also serviced by schooner. The economy is based on copra production, which suffered heavy losses in the 1983 cyclones, and fishing. Fishing has increased in importance as a result of the damage suffered by the copra plantations in the hurricanes, but this may be a temporary phenomenon until the plantations have been restored. An elaborate fish "park" system of traps staked in shallow waters near passes is used and is described by

Blanchet (1985) and Morize (1985). The catch is exported to Papeete.

Disturbance or Deficiencies   The reefs suffered extensive damage from three cyclones between December 1982 and April 1983 which caused 50-100% coral destruction (Harmelin-Vivien and Laboute, 1983). Cyclone Veena passed particularly close, causing destruction of 80% of the outer reef communities, only the north-west sector being spared. There was reported to be some recovery after two years (Laboute, 1985).

Following a rapid rise in catch rates in the early 1970s, there has been a sharp decline in the artisanal fishery yields (Blanchet, 1985).

Legal Protection   None.

Management   None.

W.A. ROBINSON INTEGRAL RESERVE AND BIOSPHERE RESERVE (TAIARO ATOLL)

Geographical Location   In the Tuamotu Archipelago, 540 km ENE of Tahiti, and 230 km from Raroia Atoll; 15°42'S, 144°34'W.

Area, Depth, Altitude   2000 ha; max. alt. 5 m; lagoon max. depth 29 m.

Land Tenure   Private property; owned by W.A. Robinson.

Physical Features   An almost circular closed atoll, 5 km in diameter, which encloses a lagoon with water saltier (reaching 43 ppt) than that of the sea. The emergent belt of the atoll measures 700 m from the sea to the lagoon at its widest part, and has a circumference of 12 km. The lagoon is fairly uniformly deep (20-25 m) with a maximum depth of 29 m. The emergent rim is continuous but there are 18 closed channels (hoa) blocked by boulders probably deposited by a tidal wave between 1878 and 1906. The lagoon level is currently gradually becoming lower, the water saltier and its fauna scarcer (Salvat *et al.*, 1977). The lagoon is permanently isolated except possibly during severe storms (Chevalier, 1976). Geomorphology is described by Chevalier and Salvat (1976) and hydrology and phytoplankton by Ricard and Delesalle (1983). Further information is given in Salvat (1976b).

Reef Structure and Corals   The outer reefs, about 50 m wide (Salvat *et al.*, 1977) have a well developed algal crest on the windward side (south and south-east). On the sheltered western side, the reefs have a much richer and more diverse coral and mollusc fauna (Chevalier and Richard, 1976). Below 20 m, little living fauna is found (Poli and Salvat, 1976).

The lagoonal area has been studied in most detail. The centre of the lagoon is devoid of coral pinnacles (Chevalier and Salvat, 1976), and only one coral species, *Porites lobata*, is found. In the past the lagoon coral fauna was much richer, several fossil species having been recorded (Chevalier, 1976). A study at eight stations along the margin of the lagoon during the 1972 expedition revealed three types of habitat: sandy detritus, coral pavement and algae. The bottom of the lagoon was studied at five stations and was found to be homogenous in its morphology and fauna with abundant *Codakia divergens* (Bagnis, 1976; Poli and Salvat, 1976; Salvat, 1976b; Salvat *et al.*, 1977) on sand and many small dead coral patches covered with the valves of *Chama* and *Crassostraea* (Salvat *et al.*, 1977).

Noteworthy Fauna and Flora   The atoll is unusual in that it is covered, not with coconut palms, but mainly with open bush vegetation. Pickering (1876) lists the varied vegetation found in 1839. *Birgus latro*, the Coconut Crab, occurs on the atoll. The lagoon fauna is well known and includes 104 mollusc species (Poli and Salvat, 1976; Richard, 1976) (of which *Tridacna maxima, Pinctada maculata, Codakia divergens* and *Gafrarium pectinatum* are particularly abundant) three sponges, two echinoderms and 55 fish (Bagnis, 1976; Salvat *et al.* 1977).

Scientific Importance and Research   Botanical and malacological collections were made in the 1830s (U.S. Exploring Expedition, C. Wilkes). In 1970, a study on competition between the mosquitos *Aedes polynesiensis*, a vector of human filariasis, and *A. albopictus*, an introduced species, was initiated by the Pacific Section of the National Institute of Allergy and Infectious Diseases at Honolulu, Hawaii. This study began before the atoll became a reserve. The introduced species does not seem to have become established and further introductions will not be permitted. In 1972 a study mission from the Muséum National d'Histoire Naturelle (Paris) and the Ecole Practique des Hautes Etudes carried out an inventory of the lagoon and reefs, and studies of the ecology, geomorphology and hydrology of the reef and lagoon (Salvat, 1976b). Scientists are encouraged to apply for permission to work there (Anon., n.d. a).

Economic Value and Social Benefits   The history of the island from the time of the first Polynesians is told in Robinson (1972). There is currently no fishing or tourism.

Disturbance or Deficiencies   Coconuts were introduced into the atoll. Once a centre of the Polynesian Kingdom, 200 years ago, the atoll is now only inhabited by a family of caretakers.

Legal Protection   Designated as a permanent reserve for science on 1st August 1972 under decree no. 2456/AA. In February 1973, the area was extended to include the entire atoll, with a protective zone of 1 km surrounding it. Measures have been taken by the Administrative Committee, in collaboration with the Governor of French Polynesia and the owner of Taiaro, to protect the fauna and flora of the atoll, reef, lagoon and surrounding sea. The atoll was declared a Biosphere Reserve in January 1977.

Management   A permit is required to visit the atoll. The reserve is administered by the Management Committee Secretariat, Délégue de la Commission des Monuments Naturels et Sites, BP 866, Tahiti, French Polynesia. A caretaker and his family warden the atoll.

Recommendations   The atoll should be re-surveyed to assess any changes that may have taken place since it was last studied in 1972, and to reaffirm its importance as a protected area (Salvat *in litt.*, 1988).

CLIPPERTON ISLAND

Geographical Location   1300 km from the Pacifc coast of Mexico, 2800 km from Hawaii and 5200 km from Tahiti; 10°18'N, 109°13'W (the only atoll lying east of Ducie).

Area, Depth, Altitude   6 sq. km; average altitude 5 m; volcanic plug 29 m.

Land Tenure   Under the authority of the French Government.

Physical Features   An egg-shaped near atoll with enclosed lagoon oriented NW-SE. The atoll rim consists of a narrow band of rock normally 100-200 m wide, but reaching 400 m in the west and narrowing to 45 m in the north-east where waves may spread into the lagoon. At the south-east end is a small 29 m high volcanic plug covered with lichen and guano. Soils are described by Bourrouilh-Le Jan (1971b). Lagoonal waters are described by Bourrouilh-Le Jan (1971a and 1980), Bourrouilh-Le Jan *et al.* (1985), Carsin (1985) and Niaussat (1978); there is a strong halocline and the waters are highly eutrophic and almost fresh at the surface. Sedimentation is described by Bourrouilh-Le Jan (1983). The magnetic field has been studied by Bourrouilh-Le Jan *et al.* (1985). North-easterly trade winds predominate but are replaced occasionally in the summer by tropical storms and sometimes cyclones from the south-east. Even during moderate winds, the atoll is subject to heavy oceanic swell.

Reef Structure and Corals   The atoll is surrounded by a reef flat exposed at low tide, a reef front cut by channels, a submarine terrace (12-18 m deep) and a precipitous outer slope. Coral, including *Pocillopora verrucosa, P. meandrina* and *Pavona gigantea*, and algal growth is found on the reef front and submarine terrace. The lower edge of the submarine terrace and outer slope are completely covered by corals (Bakus, 1975; Sachet, 1962a and b). The lagoonal fauna is poor but algae and higher plants are abundant (Carsin, 1985; Niaussat, 1978).

Noteworthy Fauna and Flora   The fauna and flora consists of an unusual assemblage including both Panamic and Indo-Pacific forms. The invertebrate fauna of the lagoon is described by Allison (1959) and Hertlein and Emerson (1953 and 1957), Salvat and Erhardt (1970) and Salvat and Salvat (1972). There are seven main species of seabird (two noddies, two terns, two boobies and frigatebirds), a lizard *Emoia cyanura* and the crab *Geocarcinus planatus*. The island is largely covered with scrub vegetation and a few coconut palms (Sachet,

1958). Additional information is given in Niaussat (1978) and Squires (1959 and 1962).

Scientific Importance and Research   The only coral atoll in the East Pacific, Clipperton is an exception to the normal lack of extensive coral reef development west of Ducie, having rich coral growth (Dana, 1975). Sachet (1958) listed scientific expeditions to the atoll up to that date; the most recent ones include the Scripps Institute of Oceanography Acapulco Trench Expedition in 1954, 1956 and 1958, the Bougainville Missions of the French National Navy from 1966 to 1969, and the Cousteau Expedition in March 1980. The closed lagoon is of particular interest. It has permanently deoxygenated water, is a model for modern formation and sedimentation of phosphate and carbonate diagenesis, and apatite has been discovered in the intertidal modern deposits along the shore (Bourrouilh-Le Jan, 1983; Bourrouilh-Le Jan *in litt.*, 28.9.87; Bourrouilh-Le Jan *et al.*, 1985; Carsin, 1985; Niaussat, 1978). The fauna and flora are of biogeographical interest on account of their Indo-Pacific and American relationships.

Economic Value and Social Benefits   The history of the island is described in Rossfelder (1976) and Sachet (1958). Phosphate was worked from 1898 to 1917 but the island is now uninhabited although it is visited by U.S. tuna fishermen (Bourrouilh-Le Jan *in litt.*, 28.9.87). Access is difficult on account of heavy oceanic swell.

Disturbance or Deficiencies   Coconuts were planted in 1897. Introduced pigs were destroyed in 1958 to prevent a decline in bird populations. Meteorological stations have occasionally been set up and the island would be suitable for a satellite observation post; a permanent installation would have a major effect on the atoll and its lagoon although it is argued that this could be controlled (Bourrouilh-Le Jan *in litt.*, 28.9.87). There is also a proposal to open up the lagoon in order to build a port (Thibault *in litt.*, 17.2.88).

Legal Protection   None.

Management   None at present.

Recommendations   The island is to be legally protected by the French Government, but this will take some time to implement. Visits by tuna fishermen should be prohibited (Bourrouilh-Le Jan *in litt.*, 28.9.87). Sachet (1958) and Elliot (1973) pointed out the need to preserve Clipperton as a natural laboratory for future scientific studies. The island may be compared with Henderson (see Pitcairn Group) as one of the least altered island systems in the Pacific.

GUAM

INTRODUCTION

General Description

Guam is an unincorporated territory of the United States and has a land area of about 541 sq. km. It is the largest and southernmost of the Mariana Islands and is composed of raised limestone and old deeply weathered volcanoes, with a maximum altitude of 1151 ft (400 m). Vegetation of Guam is described by Fosberg (1960) and Eldredge (1983a) provides a bibliography of the Mariana Islands, including Guam.

Extensive work has been carried out on the marine environment by the University of Guam Marine Laboratory, which is located on the east coast at Mangilao, and a comprehensive bibliography of publications is available (Marine Laboratory, University of Guam, 1986). References to oceanographic literature are given in Eldredge and Kropp (1981). Tracey et al. (1964) provide an early description of the reefs of the island. Randall and Holloman (1974) provide an overview of geology, hydrology and physiography of the coastal regions, and Easton et al. (1978) describe the shoreline and reefs. The geology is described in greater detail by Emery (1962). Randall and Eldredge (1976) mapped the basic coastal outline, fringing reefs, mangroves and seagrass beds in the course of a survey as part of the Coastal Zone Management Program which began in 1976. Stojkovich (1977) carried out a survey and species inventory of twelve representative pristine marine communities on Guam as part of this programme, the majority of which are described in the following accounts. Marsh (1974) studied reef flat productivity, and Randall (1978 and 1985) conducted transects on reef flat platforms at five locations (see below and separate accounts). Water circulation on reef flats at Pago Bay and Tumon Bay are described by Marsh et al. (1981). The corals of Guam are described by Randall and Myers ((1983); the soft coral Asterospicularia randalli may be endemic (Gawel, 1976).

Most of the reefs are fringing, although there are two barrier reef lagoons, Apra Harbor which is now extensively modified, and Cocos Lagoon in the south-west. Apra Harbor, on the west coast, is the only deep water port. It is bounded to the south by the Orote Peninsula and to the north by Luminao Reef and Cabras Island (see separate accounts for Luminao Reef and Orote Ecological Reserve Area). Four species of reef building corals, Pavona frondifera, Pectinia lactuca, Leptoseris gardineri and Montipora sp., are found in Apra Harbor, near Piti Channel and Sasa Bay, and nowhere else on Guam (Marine Laboratory, University of Guam, 1977). A detailed study of Piti Channel and Piti Bay was carried out from 1972 to 1977 over the period of construction of the Cabras Power Plant (Marsh et al., 1977). Coral growth rates in Apra Harbor and at Cabras Power Plant were studied by Neudecker (1981). Luminao Reef and reefs within the War in the Pacific National Historical Park are described in separate accounts.

Reefs in Agana Bay, to the north of Apra Harbor are described by Randall (1978), Randall and Eldredge (1974) and Tracey et al. (1964). The reef at Tumon Bay, to the north, was described by Tracey et al. (1964) and Randall (1973c) prior to Acanthaster infestation (see below) and subsequently by Randall (1978). Further information on this area is given in Jones and Randall (1972) and Randall and Jones (1973). Tanguisson Reef, further north, was extensively studied during the construction of the Tanguisson Power Plant (Jones et al., 1976; Colgan, 1981). Reefs along the north-west coast are described in the accounts for Haputo Ecological Reserve Area and Pati Point Natural Area.

The north-east coast from Pati Point to Pago Bay is characterized by gentle to steep cliffs and sea-level cut benches of varying widths with no fringing reef, although corals are found in deeper waters (Stojkovich, 1977). Ylig Bay reefs are described by Randall (1978). Reefs of the Guam Territorial Seashore Park in the south-west, including Cocos barrier reef and lagoon, are described in a separate account. An emergent holocene reef at Aga Point, to the east of the Guam Territorial Seashore Park, is described by Siegrist and Randall (1985).

The fish of Guam are described by Amesbury (1978), Amesbury and Myers (1982), Jones and Larson (1974) and Kami (1975); fish grazing effects are described by Birkeland et al. (1985). Holothurians are described by Rowe and Doty (1977), crinoids by Meyer and Macurda (1980), asteroids by Yamaguchi (1975b), and gastropod molluscs by Roth (1976). Colgan (1985) describes the distribution of the vermetid mollusc Dendropoma maxima. The ecology of seagrasses on reefs in Guam is discussed by Tsuda (1972b). Benthic algae are described by Tsuda (1972c), brown algae by Tsuda (1977), and coralline algae by Gordon et al. (1976). Mangroves are described by Moore et al. (1977), Marine Laboratory, University of Guam (1977) and Wilder (1976). The most extensive and mature mangrove communities are found in Apra Harbor, although this area has undergone extensive modification through dredging, landfill and construction. Descriptions and recommendations for their preservation are given in Stojkovich (1977), and discussion of the recovery of the area in Bultitude and Strong (1985).

Reef Resources

In 1980 Guam had a population of more than 106 000 people. Amesbury (1982), Amesbury et al. (1986), Eldredge (1983b) and Kami and Ikehara (1976) describe the fisheries. Several species of gastropod and bivalve are taken for food. A survey of edible reef molluscs was undertaken to assess their conservation status (Stojkovich and Smith, 1978). The extent of coral, shell and algal harvesting is detailed by Hedlund (1977). In the 1970s, several species of reef corals were collected commercially, as well as a small quantity of shells and black coral. Larger quantities of shells were being taken privately. Collecting areas for Acropora included Nimitz Channel and Taleyfac Bay on the west coast south of Orote Point, and, for black coral, near Umatac Bay. Shells are taken in a variety of places; Tumon and Agana Bays were good areas in the 1960s. Edible algae is collected from Pago Bay and Sella Bay. Tumon Bay is an

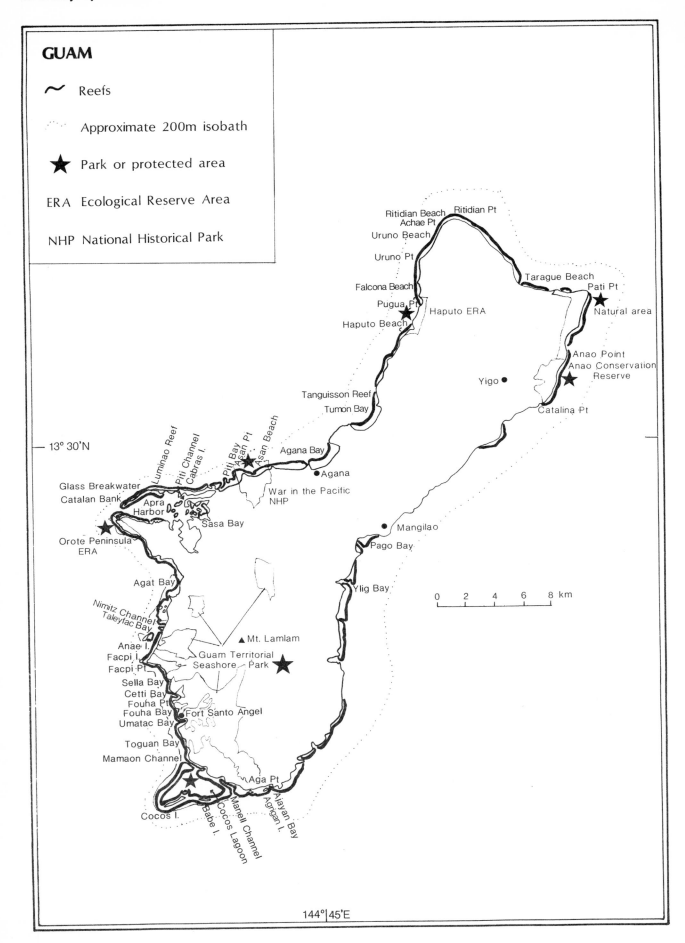

GUAM

~ Reefs

⋯ Approximate 200m isobath

★ Park or protected area

ERA Ecological Reserve Area

NHP National Historical Park

Ritidian Beach
Ritidian Pt
Achae Pt
Uruno Beach
Uruno Pt
Tarague Beach
Pati Pt
Falcona Beach
Natural area
Pugua Pt
Haputo ERA
Haputo Beach
Anao Point
Anao Conservation Reserve
Yigo
Catalina Pt
Tanguisson Reef
Tumon Bay

— 13° 30'N

Luminao Reef
Piti Channel
Cabras I.
Piti Bay
Asan Pt
Asan Beach
Agana Bay
Glass Breakwater
Agana
Catalan Bank
Apra Harbor
War in the Pacific NHP
Sasa Bay
Mangilao
Pago Bay
Orote Peninsula ERA

Agat Bay
Ylig Bay

Nimitz Channel
Taleyfac Bay
0 2 4 6 8 km

Anae I.
Facpi I.
Facpi Pt
▲ Mt. Lamlam
Guam Territorial Seashore Park
Sella Bay
Cetti Bay
Fouha Pt
Fouha Bay
Fort Santo Angel
Umatac Bay
Toguan Bay
Mamaon Channel
Aga Pt
Alayan Bay
Agrigan I.
Manell Channel
Cocos Lagoon
Cocos I.
Babe I.

144° 45'E

important recreational area with most of the hotel developments located along its shoreline. SCUBA diving is popular on a number of reefs. A general description of marine and coastal exploitation is given in Jennison-Nolan (1979).

Disturbances and Deficiencies

The impact of Typhoon Pamela on the reefs is described by Ogg and Koslow (1978) and Randall and Eldredge (1977). Little damage occurred along the reef-flat platforms and reef margins. Yamaguchi (1975) described mass mortalites of reef animals due to sea-level fluctuations. There was a major outbreak of *Acanthaster planci* from 1968 to 1970 (Cheney, 1973 and 1974; Marsh and Tsuda, 1973; Tsuda, 1971). For example, over 90% of the corals on Tanguisson Reef died (Randall, 1973a) but the reef recovered over a period of eleven years (Colgan, 1981; Jones *et al.*, 1976; Randall, 1973b). In 1979, scattered aggregations of up to 200 individuals were observed on the reefs (Birkeland, 1982).

The effects of development by the U.S. military, private industry and more recently tourism and home building have led to pressures from accelerated use and landscape alteration. Shallow, non-contiguous dredging (less than 2 ft (0.6 m)) is about to begin in Tumon Bay to enhance recreation there, and the reefs will be affected (Eldredge *in litt.*, 1987). Storm drainage collected from the hotel parking lots and grounds in Tumon Bay is discharged into the supralittoral reef flat zone producing sediment deposits. The formation of such sediment deposits is described by Clayshulte (1981). Reefs in Agana Bay are already affected and there is sedimentation in Ylig Bay (Randall, 1978). Siltation studies have been carried out at Fouha Bay and Ylig Bay (Randall and Birkeland, 1978). Terrestrial inputs of nitrogen and phosphorus on fringing reefs are described in Marsh (1977). Apra Harbor has also come under increasing stress from industrial activities (*see also account for* Luminao Reef). Studies on the effects of untreated sewage (Jones and Randall, 1971) and secondarily treated sewage (Tsuda and Grosenbaugh, 1977) indicated that little destruction of reefs occurred when the sewage undergoes secondary treatment.

Extensive studies have been carried out on thermal pollution at the Tanguisson Power Plant on the north-west coast by Jones and Randall (1973), Jones *et al.* (1976) and Neudecker (1976, 1977a, b and 1981). The plant became operational in 1971 and thermally enriched discharge waters subsequently destroyed a reef flat and margin area of 20 000 sq. m. Release of plant effluent resulted in an elevation of water temperature on the adjacent reef flat and reef margin and wave action exposed even the deeper parts of the reef margin to temperatures above ambient as well as other potentially detrimental effluent parameters such as chlorine and heavy metals. Introduction of the effluent was shown to be responsible for some destruction of reef margin corals but effluent is stratified beyond the surf zone and no longer threatens benthic organisms (Jones *et al.*, 1976). There has been little recovery of the destroyed community. The Cabras Power Plant also releases heated effluent into the Apra Harbor Lagoon affecting an adjacent tidal flat. However, studies carried out by Marsh and Doty (1975 and 1976), Marsh and Gordon (1972, 1973 and 1974), Marsh *et al.* (1977) and

Marsh *et al.* (1985) suggested that greatest damage occurred during the construction phase due to poorly managed dredging activities, when a coral community was destroyed in West Piti Bay. There had been little regeneration by 1976. Coral transplantation experiments were attempted (Birkeland *et al.*, 1979).

Differences in the population structures and abundances of economicaly important reef fishes were observed between accessible, heavily fished reef flats and inaccessible, lightly fished reef flats at six study sites (Katnik, 1981). Shell populations on the reefs were found to have noticeably declined in the 1970s (Hedlund, 1977).

Legislation and Management

The United States Federal Endangered Species Act and the Coastal Zone Management Act apply, and coastal resources are subject to planning controls and regulations. The Guam Coastal Management Program and Final Environmental Impact Statement was published in 1979; the entire island is considered to be a coastal zone. Environmental problems, management and legislation are discussed in SPREP (1980).

Fishing regulations prohibit poisons and explosives and limit exploitation of live coral, coconut crabs, *Trochus* and spiny lobsters (Dahl, 1980; Hedlund, 1977). The Guam Water Quality Management Plan was mandated under Section 208 of the 1977 U.S. Clean Water Act. Guam Water Quality Standards designate certain coastal waters as conservation areas in which no discharge of pollutants is allowed. Pollution controls are described in Anon. (1973) and SPREP (1980). Hazardous waste material is controlled by the Guam Environmental Protection Agency (Zucker, 1984).

In 1975, the Guam Territorial Park System, administered by the Department of Parks and Recreation, Government of Guam, came into existence and includes three protected areas created under Law 12-209, two of which include reefs (Anon., 1985 and *see separate accounts*):

1. The Guam Territorial Seashore Park
2. Anao Conservation Area

The Federal Government has established four protected areas, all of which include or are adjacent to reefs:

1. The "War in the Pacific National Historical Park"
2. Haputo Ecological Reserve Area
3. Orote Peninsula Ecological Reserve Area
4. Pati Point Natural Area

The Ecological Reserve Areas and Natural Area were established as physical or biological units in which current natural conditions are maintained; all are described in separate accounts.

Recommendations

Major challenges for the future, as identified by Maragos (1986), include stemming of soil erosion and development of more effective programmes for the disposal of solid and hazardous wastes, including unexploded ordnance. The probable application on Guam of a new Defense Environmental Restoration Account, administered by the

U.S. Army Corps of Engineers, should help in the latter effort. Increased legislative controls for waste management and small and deep draft navigation, including adequate environmental controls, and close scrutiny and control of future resort and military construction on the island, will be important in the conservation of coastal resources.

Hedlund (1977) makes recommendations for the improved management of coral, shell and algae harvesting. These include protection of three coral species, *Euphyllia* spp., *Plerogyra sinuosa* and *Tubastrea aurea* and two molluscs, *Cassis cornuta* and *Charonia tritonis*, public education programmes and a variety of other measures specifically for corals. Plucer-Rosario and Randall (1985) discuss the potential for transplantation of three rare coral species from the commercial harbour to areas unaffected by pollution. The development of aquaculture on Guam is discussed in Fitzgerald and Nelson (1979), and of fisheries in Amesbury (1982) and Amesbury *et al.* (1986).

References

* = cited but not consulted

Amesbury, S.S. (1978). Studies on the biology of the reef fishes of Guam. Part 1. Distribution of fishes on the reef flats of Guam. Part 2. Distribution of eggs and larvae of fishes at selected sites on Guam. *Univ. Guam Mar. Lab. Tech. Rept* 49. 65 pp.

Amesbury, S.S. (1982). Fishery development in Guam. Unique problems, unique opportunities. In: *Third Expert Conference for Economic Development in Asia and the Pacific*. Asian-Pacific Development Center, Tokyo.

Amesbury, S.S. and Myers, R.F. (1982). *Guide to the Coastal Resources of Guam. Vol. 1. The Fishes.* University of Guam Press. 141 pp.

Amesbury, S.S., Cushing, F.A. and Sakamoto, R.K. (1986). *Guide to the coastal resources of Guam. Vol. 3. Fishing on Guam.* University of Guam Press.

Anon. (1973). Pollution Control in Guam. Paper presented on behalf of Guam Administration. *Proceeding and Papers, Regional Symposium on Conservation of Nature - Reefs and Lagoons*. South Pacific Commission, Noumea, New Caledonia.

Anon. (1979). Guam Territorial Seashore Park Master Plan. Dept of Parks and Recreation, Agana.

Anon. (1985). Country review - Guam. *Report of the 3rd South Pacific National Parks and Reserves Conference, Apia* 3: 93-114.

Birkeland, C. (1982). Terrestrial runoff as a cause of outbreaks of *Acanthaster planci* (Echinodermata: Asteroidea). *Mar. Biol.* 69: 175-185.

Birkeland, C., Nelson, S.G., Wilkins, S. and Gates, P. (1985). Effects of grazing by herbivorous fishes on coral reef community metabolism. *Proc. 5th Int. Coral Reef Cong., Tahiti* 4: 47-51.

Birkeland, C., Randall, R.H. and Grimm, G. (1979). Three methods of coral transplantation for the purpose of reestablishing a coral community in the thermal effluent area at the Tanguisson Power Plant. *Univ. Guam Mar. Lab. Tech. Rept* 60. 24 pp.

Birkeland, C., Rowley, D. and Randall, R.H. (1981). Coral recruitment patterns at Guam. *Proc. 4th Int. Coral Reef Symp., Manila* 2: 339-344.

Bryan, P.G. (1973). Growth rate, toxicity, and distribution of the encrusting sponge *Terpios* sp.

(Hadromerida; Seberitidae) in Guam, Mariana Islands. *Micronesica* 9(2): 237-242.

Bultitude, D.R. and Strong, R.D. (1985). Mangrove swamps: Nature's nursery. *Glimpses of Micronesia* 25(1): 52-56, 65.

Cheney, D.P. (1973). An analysis of the *Acanthaster* control programs in Guam and the Trust Territory of the Pacific Islands. *Micronesica* 9(2): 171-180.

Cheney, D.P. (1974). Spawning and aggregation of *Acanthaster planci* in Micronesia. *Proc. 2nd Int. Coral Reef Symp., Brisbane* 1: 591-594.

Chernin, M.I., Lassuy, R., Dickinson, "E". and Shepard, J.W. (1977). Marine reconnaissance survey of proposed sites for a small boat harbor in Agat Bay, Guam. *Univ. Guam Mar. Lab. Tech. Rept* 39. 54 pp.

Chesher, R.H. (1969a). Destruction of Pacific corals by the sea star *Acanthaster planci*. *Science* 165: 280-283.

Chesher, R.H. (1969b). Divers wage war on the killer star. *Skin Diver Mag.* 18(3): 34-35, 84-85.

Clayshulte, R.N. (1981). Formation of small marine sediment deltas on a Guam leeward reef flat by storm drain run-off. *Proc. 4th Int. Coral Reef Symp., Manila* 1: 467-473.

Colgan, M.W. (1981). Succession and recovery of a coral reef after predation by *Acanthaster planci* (L.). *Proc. 4th Int. Coral Reef Symp., Manila* 2: 333-338.

Colgan, M.W. (1985). Growth rate reduction and modification of a coral colony by a vermetid mollusc *Dendropoma maxima*. *Proc. 5th Int. Coral Reef Cong., Tahiti* 6: 205-210.

Dahl, A.L. (1980). Regional ecosystems survey of the South Pacific Area. *SPC/IUCN Technical Paper* 179. South Pacific Commission, Noumea, New Caledonia.

Dickinson, R.E. and Tsuda, R.T. (1975). A candidate marine environmental impact survey for the potential development of the Uruno Point reef area on Guam, Mariana Islands. *Univ. Guam Mar. Lab. Tech. Rept* 19. 50 pp.

Easton, W.H., Ku, T.L. and Randall, R.H. (1978). Recent reefs and shore lines of Guam. *Micronesica* 14(1): 1-11.

Eldredge, L.G. (1979). Marine biological resources within the Guam Seashore Study Area and the War in the Pacific National Historical Park. *Univ. Guam Mar. Lab. Tech. Rept* 57. 75 pp.

Eldredge, L.G. (1983a). Mariana's active arc: A bibliography. *Univ. Guam Mar. Lab. Tech. Rept* 82. 19 pp.

Eldredge, L.G. (1983b). Summary of environmental and fishing information on Guam and the Commonwealth of the Northern Marianas: Historical background, description of the islands, and review of climate, oceanography, and submarine topography around Guam and the Northern Mariana Islands. NOAA-TM-NMFS-SWFC-40. 181 pp.

Eldredge, L.G., Dickinson, R. and Moras, S. (Eds) (1977). Marine survey of Agat Bay. *Univ. Guam Mar. Lab. Tech. Rept* 31. 251 pp.

Eldredge, L.G. and Kropp, R.K. (1981). Selected bibliography of the physical, chemical, and biological oceanographic literature for the waters surrounding Guam. *Univ. Guam Mar. Lab. Tech. Rept* 73. 22 pp.

Emery, K.O. (1962). Marine Geology of Guam. *U.S. Geol. Surv. Prof. Paper* 403-B: B1-B76.

Fitzgerald, W.J., Jr and Nelson, S.G. (1979). Development of aquaculture in an island community (Guam, Mariana Islands). *Proc. World Mariculture Soc.* 10: 39-50.

Fosberg, F.R. (1960). The vegetation of Micronesia. 1. General description, the vegetation of the Marianas

Islands, and a detailed consideration of the vegetation of Guam. *Bull. Am. Mus. Nat. Hist.* 119(1): 1-75.

*Gawel, M.J. (1976). *Asterospicularia randalli*: a new species of Asterospiculariidae (Octocorallia: Alcyonacea) from Guam. *Micronesica* 12(2): 303-307.

*Gordon, G.D., Masaki, T. and Akioka, H. (1976). Floristic and distributional account of the common crustose coralline algae on Guam. *Micronesica* 12(2): 247-277.

Hedlund, S.E. (1977). The extent of coral, shell and algal harvesting in Guam waters. *Univ. Guam Mar. Lab. Tech. Rept* 37. 34 pp.

*Jennison-Nolan, J. (1979). Guam: Changing patterns of coastal and marine exploitation. *Univ. Guam Mar. Lab. Tech. Rept* 59. 62 pp.

*Jones, R.S. and Larson, H.K. (1974). A key to the families of fishes as recorded from Guam. *Univ. Guam Mar. Lab. Tech. Rept* 10. 48 pp.

*Jones, R.S. and Randall, R.H. (1971). An annual cycle study of biological, chemical and oceanographic phenomena associated with the Agana ocean outfall. *Univ. Guam Mar. Lab. Tech. Rept* 1. 67 pp.

*Jones, R.S. and Randall, R.H. (1973). A study of biological impact caused by natural and man-induced changes on a tropical reef. *Univ. Guam Mar. Lab. Tech. Rept* 7. 184 pp.

*Jones, R.S., Randall, R.H. and Strong, R.D. (1974). An investigation of the biological and oceanographic suitability of Toguan Bay, Guam as a potential site for an ocean outfall. *Univ. Guam Mar. Lab. Tech. Rept* 11. 97 pp.

*Jones, R.S., Randall, R.H. and Wilder, M.J. (1976). Biological impact caused by changes on a tropical reef. *Univ. Guam Mar. Lab Tech. Rept* 28. 209 pp.

*Kami, H.T. (1975). Check-list of Guam fishes, Supplement 2. *Micronesica* 11(1): 115-121.

*Kami, H.T., and Ikehara, I.I. (1976). Notes on the annual juvenile siganid harvest in Guam. *Micronesica* 12(2): 319-322.

Katnik, S. (1981). The effects of fishing pressures on some economically important fishes on Guam's reef flats. *Proc. 4th Int. Coral Reef Symp., Manila* 1: 111. (Abstract).

*Kock, R.L. (1982). Patterns in the abundance variations of reef fishes near an artificial reef at Guam. *Environ. Biol. Fishes* 7(2): 121-136.

Maragos, J.E. (1986). Coastal resource development and management in the U.S. Pacific Islands: 1. Island-by-island analysis. Prepared for the Office of Technology Assessment, U.S. Congress. 59 pp.

*Marine Laboratory, University of Guam (1977). Marine environmental baseline report, Commercial Port, Apra Harbour, Guam. *Univ. Guam Mar. Lab. Tech. Rept* 34. 96 pp.

Marine Laboratory, University of Guam (1986). Listing of contributions, technical reports, environmental survey reports, miscellaneous reports and M.S. Theses, April 1986. 35 pp.

*Marsh, J.A., Jr (1974). Preliminary observations on the productivity of a Guam reef flat community. *Proc. 2nd Int. Coral Reef Symp., Brisbane* 1: 139-145.

*Marsh, J.A., Jr (1977). Terrestrial inputs of nitrogen and phosphorus on fringing reefs of Guam. *Proc. 3rd Int. Coral Reef Symp., Miami* 1: 331-336.

*Marsh, J.A. and Doty, J.E. (1975). Power plants and the marine environment: Additional observations in Piti Bay and Piti Channel, Guam. *Univ. Guam Mar. Lab. Tech. Rept* 21. 44 pp.

*Marsh, J.A. and Doty, J.E. (1976). The influence of power plant operations on the marine environment in Piti

Channel, Guam: 1975-1976 observations. *Univ. Guam Mar. Lab. Rept* 26. 57 pp.

*Marsh, J.A. and Gordon, G.D. (1972). A marine environmental survey of Piti Bay and Piti Channel, Guam. *Univ. Guam Mar. Lab., Envir. Surv. Rept* 3. 28 pp.

*Marsh, J.A. and Gordon, G.D. (1973). A thermal study of Piti Channel, Guam and adjacent areas, and the influence of power plant operations on the marine environment. *Univ. Guam Mar. Lab. Tech. Rept* 6. 51 pp.

*Marsh, J.A. and Gordon, G.D. (1974). Marine environmental effects of dredging and power plant construction in Piti Bay and Piti Channel, Guam. *Univ. Guam Mar. Lab. Tech. Rept* 8. 93 pp.

*Marsh, J.A. and Tsuda, R.T. (1973). Population levels of *Acanthaster planci* in the Mariana and Caroline Islands, 1969-1972. *Atoll Res. Bull.* 170: 1-16.

Marsh, J.A., Chernin, M.I. and Doty, J.E. (1977). Power plants and the marine environment in Piti Bay and Piti Channel, Guam; 1976-1977 observations and general summary. *Univ. Guam Mar. Lab. Tech. Rept* 38. 93 pp.

Marsh, J.A., Pendleton, D.E., Wilkins, S. de C. and Hillman-Kitalong, A. (1985). Effect on selected reef organisms of a potential seawater SO_2 scrubber system at a power plant on Guam. *Proc. 5th Int. Coral Reef Cong., Tahiti* 4: 177-182.

Marsh, J.A., Ross, R.M. and Zolan, W.J. (1981). Water circulation on two Guam reef flats. *Proc. 4th Int. Coral Reef Symp., Manila* 1: 355-360.

*Meyer, D.L. and Macurda, D.B., Jr (1980). Ecology and distribution of the shallow-water crinoids of Palau and Guam. *Micronesica* 16(1): 59-99.

*Moore, P., Raulerson, L., Chernin, M. and McMakin, P. (1977). Inventory and mapping of wetland vegetation in Guam, Tinian and Saipan, Mariana Islands. Univ. Guam Biosci. 253 pp.

NOAA (1983). Announcement of National Marine Sanctuary Program Final Site Evaluation List. *Federal Register* 48(151): 35568-35577.

*National Park Service (1983). General Management Plan, War in the Pacific National Historical Park. Dept Interior.

*Neudecker, S. (1976). Effects of thermal effluent on the coral reef community at Tanguisson. *Univ. of Guam Mar. Lab. Tech. Rept* 30. 55 pp.

*Neudecker, S. (1977a). Transplant experiments to test the effect of fish grazing on coral distribution. *Proc. 3rd Int. Coral Reef Symp., Miami* 1: 317-323.

Neudecker, S. (1977b). Development and environmental quality of coral reef communities near the Tanguisson Power Plant. *Univ. Guam Mar. Lab. Tech. Rept* 41. 68 pp.

Neudecker, S. (1981). Growth and survival of scleractinian corals exposed to thermal effluents at Guam. *Proc. 4th Int. Coral Reef Symp., Manila* 1: 173-180.

*Ogg, J.G. and Koslow, J.A. (1978). The impact of Typhoon Pamela (1976) on Guam's coral reefs and beaches. *Pac. Sci.* 32(2): 105-118.

Plucer-Rosario, G. (1983). Effect of substrate and light on growth and distribution of *Terpios*, an encrusting sponge which kills corals. M.S. Thesis, University of Guam. 39 pp.

Plucer-Rosario, G. and Randall, R.H. (1985). Preservation of rare coral species by transplantation and examination of their recruitment and growth. *Proc. 5th Int. Coral Reef Cong., Tahiti* 2: 299 (Abstract).

Pritchard, P.C.H. (1981). Marine turtles of Micronesia. In: Bjorndal, K.A. (Ed.), *Biology and Conservation of Sea Turtles*. Smithsonian Institution Press, Washington D.C. Pp. 263-274.

*Randall, R.H. (1973a). Distribution of corals after *Acanthaster planci* infestation at Tanguisson Point, Guam. *Micronesica* 9(2): 213-222.

*Randall, R.H. (1973b). Coral reef recovery following extensive damage by the "crown-of-thorns" starfish *Acanthaster planci* (L.). In: Tokioka, T. and Nishimura, S. (Eds), *Proc. 2nd Int. Symp. on Cnidaria Publ. Seto Mar. Biol. Lab.* 20: 462-489.

*Randall, R.H. (1973c). Reef physiography and distribution of corals at Tumon Bay, Guam, before Crown-of-thorns starfish *Acanthaster planci* (L.) predation. *Micronesica* 9(1): 119-158.

Randall, R.H. (Ed.) (1978). Guam's reefs and beaches. Part 2. Transect Studies. *Univ. Guam Mar. Lab. Tech. Rept* 48. 90 pp.

*Randall, R.H. (1979). Geologic features within the Guam seashore study area. *Univ. Guam Mar. Lab. Tech. Rept* 55. 53 pp.

Randall, R.H. (1982). Corals. In: Randall, R.H. and Eldredge, L.G. (Eds), Assessment of the shoalwater environments in the vicinity of the proposed OTEC development at Cabras Island, Guam. *Univ. Guam Mar. Lab. Tech. Rept.* 79: 63-106.

Randall, R.H. (1985). Habitat geomorphology and community structure of corals in the Mariana Islands. *Proc. 5th Int. Coral Reef Cong, Tahiti* 6: 261-266.

*Randall, R.H. and Birkeland, C. (1978). Guam's reefs and beaches. Part 2. Sedimentation studies at Fouha Bay and Ylig Bay. *Univ. Guam Mar. Lab. Tech. Rept* 47. 77 pp.

*Randall, R.H. and Eldredge, L.G. (1974). A marine survey for the Sleepy Lagoon Marina. *Univ. Guam Mar. Lab. Env. Surv. Rept* 14. 42 pp.

*Randall, R.H. and Eldredge, L.G. (1976). *Atlas of Reefs and Beaches of Guam.* Coastal Zone Mgmt Sect., Bur. Planning, Guam. 191 pp.

*Randall, R.H. and Eldredge, L.G. (1977). Effects of Typhoon Pamela on the coral reefs of Guam. *Proc. 3rd Int. Symp. Coral Reefs, Miami* 2: 525-531.

Randall, R.H. and Eldredge, L.G. (Eds) (1982). Assessment of the shoalwater environments in the vicinity of the proposed OTEC development at Cabras Island, Guam. *Univ. Guam Mar. Tech. Rept.* 79. 208 pp.

*Randall, R.H. and Holloman, J. (1974). Coastal survey of Guam. *Univ. Guam Mar. Lab. Tech. Rept* 14. 404 pp.

*Randall, R.H. and Jones, R.S. (1973). A marine survey for the proposed Hilton Hotel dredging project. *Univ. Guam Mar. Lab. Env. Surv. Rept* 7. 30 pp.

*Randall, R.H. and Myers, R.F. (1983). *Guide to the coastal resources of Guam: Vol. 2. The corals.* Univ. of Guam Press. 128 pp.

*Randall, R.H. and Sherwood, T.S. (Eds) (1982). Resurvey of Cocos Lagoon, Guam, Territory of Guam. *Univ. Guam Mar. Lab. Tech. Rept* 80. 104 pp.

Randall, R.H., Tsuda, R.T., Jones, R.S., Gawel, M.J. and Rechebei, R. (1975). Marine biological survey of the Cocos barrier reefs and enclosed lagoon. *Univ. Guam Mar. Lab. Tech. Rept* 17. 159 pp.

Raulerson, L. (1979). *Terrestrial and freshwater organisms within, and limnology and hydrology of, the Guam Seashore Study Area and the War in the Pacific National Historical Park.* Dept. of Biology, Univ. of Guam. 93 pp.

*Roth, A. (1976). Preliminary checklist of the gastropods of Guam. *Univ. Guam Mar. Lab. Tech. Rept* 27. 99 pp.

*Rowe, F.W.E. and Doty, J.E. (1977). The shallow-water holothurians of Guam. *Micronesica* 13(2): 217-250.

SPREP (1980). Guam. *Country Report* 6. South Pacific Commission, Noumea, New Caledonia.

Siegrist, H.G., Jr. and Randall, R.H. (1985). Community structure and petrography of an emergent holocene reef limestone on Guam. *Proc. 5th Int. Coral Reef Cong., Tahiti* 6: 563-568.

Stojkovich, J.O. (1977). Survey and species inventory of representative pristine marine communities on Guam. *Univ. Guam Mar. Lab. Tech. Rept* 40. 183 pp.

*Stojkovich, J.O. and Smith, B.D. (1978). Survey of edible marine shellfish and sea urchins on the reefs of Guam. Aquatic and Wildlife Resources, Dept. of Agriculture, Guam. Technical Report 2: 65 pp.

*Tracey, J.I., Schlanger, S.O., Stark, J.T., Doan, D.B. and May, H.D. (1964). General geology of Guam. *U.S. Geol. Surv. Prof. Pap.* 403-A: 1-104.

*Tsuda, R.T. (Compiler) (1971). Status of *Acanthaster planci* and coral reefs in the Mariana and Caroline Islands, June 1970-May 1971. *Univ. Guam Mar. Lab. Tech. Rept* 2. 127 pp.

*Tsuda, R.T. (1972a). Proceedings of the University of Guam - Trust Territory *Acanthaster planci* (crown-of-thorns starfish) workshop. *Univ. Guam Mar. Lab. Tech. Rept* 3. 36 pp.

*Tsuda, R.T. (1972b). Morphological, zonational, and seasonal studies on two species of *Sargassum* on the reefs of Guam. *Proc. Seventh Int. Seaweed Symp., Sapporo, Japan*: 40-44.

*Tsuda, R.T. (1972c). Marine benthic algae on Guam. 1. Phaeophyta. *Micronesica* 8(1-2): 63-86.

*Tsuda, R.T. (1977). Zonational patterns of the Phaeophyta (brown algae) on Guam's fringing reefs. *Proc. 3rd Int. Coral Reef Symp. Miami* 1: 371-375.

*Tsuda, R.T. and Grosenbaugh, D.A. (1977). Agat sewage treatment plant: Impact of secondary treated effluent on Guam coastal waters. *Univ. Guam WRCC Tech. Rept* 3. 39 pp.

U.S. Army Corps of Engineers (1983). Planning considerations for use and development of Cocos Lagoon and Merizo shore, Guam. U.S. Army Corps of Engineers. 37 pp.

*U.S. Navy (1985). Draft Management Plan for the Orote Peninsula Ecological Reserve Area. Pearl Harbor, Hawaii.

*Wilder, M.J. (1976). Estuarine and mangrove shorelines. In: Randall, R.H. and Eldredge, L.G., *Atlas of Reefs and Beaches of Guam.* Coastal Zone Mgmt Sect., Bur. Planning, Guam. Pp. 157-191.

*Yamaguchi, M. (1975a). Sea-level fluctuations and mass mortalities of reef animals in Guam, Mariana Islands. *Micronesica* 11(2): 227-243.

*Yagamuchi, M. (1975b). Coral-reef asteroids of Guam. *Biotropica* 7(1): 12-23.

Zucker, W.H. (1984). Hazardous waste management on Guam: a case study. *Ambio* 13(5-6): 334-335.

ANAO CONSERVATION RESERVE

Geographical Location North-west coast in the municipality of Yigo, from Anao Point to Catalina Point, adjacent to Pati Point Natural Area; 130°32'N, 144°56'E.

Area, Depth, Altitude Sea level to 161 m; 263 ha.

Land Tenure Government of Guam.

Physical Features   A rugged limestone cliff borders the coral reef fringed coastline. Inland, the topography rises quickly to the limestone cliffs (Anon., 1985).

Reef Structure and Corals   No information.

Noteworthy Fauna and Flora   Terrestrial flora and fauna are described in Anon. (1985).

Scientific Importance and Research   Not known at present.

Economic Value and Social Benefits   Accessible by footpath.

Disturbance or Deficiencies   There is some illegal fishing.

Legal Protection   Established in 1953 by Public Law 12-209.

Management   Hunting, shelling, crabbing, fishing and outdoor recreation activities are permitted. The area is managed jointly by the Department of Parks and Recreation and the Department of Agriculture to make it accessible to the public while preserving the natural features.

GUAM TERRITORIAL SEASHORE PARK

Geographical Location   The Park covers an extensive but irregularly patterned area in the south-west of Guam, with park areas interspersed by non-park private and Federal lands. It includes a contiguous stretch of coastline from Anae Island and patch reef, 1 km off shore south of Nimitz Beach Park in the west, south to include twenty-two acres (8.9 ha) of Cocos Island and all of the Lagoon and east to include a portion of Ajayan Bay, just north of Manell Channel. The area includes Cetti Bay on the south-west coast between Sella and Fouha Bays. The park lies within the area 13°13'-13°25'N, 144°38'-144°44'E.

Area, Depth, Altitude   The Park covers 3596 ha of land and 2539 ha of water; Cocos Barrier reefs and lagoon cover 10 sq. km (NOAA, 1983); the Park covers marine habitat to the 60 ft (18.3 m) depth contour; max. alt. 396 m at Mt Lamlam.

Land Tenure   The entire park is either owned with title or claimed by the Government of Guam; some areas of the park are claimed by individuals and others are leased to private individuals by the Government for agriculture and livestock grazing (Anon., 1985). Cocos Lagoon and one third of Cocos Island are owned by the Government of Guam; the remaining two thirds of Cocos Island (not within the park) is privately owned.

Physical Features   The south-west part of Guam is largely volcanic and consists of a series of hills of around 400 m in altitude extending along the west of the island. Much of the coastline has low limestone cliffs with scattered beaches. Geological features are described by Randall (1979).

Anae Island and patch reef form a mini-barrier reef system which protects the inside submarine terrace from large swells and strong currents. Anae Island is the only one of the eight islets on the south-west coast which is not associated with a fringing reef. The western and northern sides of the island and patch reef slope steeply to a 30 m terrace while the eastern and southern sides consist of a gently sloping terrace 3-8 m deep. In these protected waters, large coral mounds, pinnacles and ridge, 6-8 m high, are separated by sandy floored channels and holes (Stojkovich, 1977). Randall and Holloman (1974) describe some features of the area.

Randall and Holloman (1974) describe some features of Cetti Bay, which is surrounded by steep slopes and sandy beaches. The shoreline consists of rocky volcanic headlands with steep shorelines, bordered by low-lying narrow limestone terraces (NOAA, 1983). Silt content is high and visibility low in the inner bay, but visibility is good in other areas.

Cocos Barrier Reef and Lagoon comprise a triangular barrier reef, lagoon and associated islands. The area has been divided into three biotopes. The terrestrial biotope includes Cocos Island and a small sand islet at its eastern end, Babe Island, with the landward border along Cocos Lagoon. This consists of a narrow fringing reef, an intertidal zone dotted with mangrove patches, and seagrasses. A second biotope consists of the deep Mamaon and Manell Channels and a third includes the lagoon, barrier reef-flat platform and fringing reef-flat platforms. The barrier reefs are nearly 5 km long on the north-west side, 5-6 km long on the south. On the north-east side of the lagoon there is a 4 km long stretch of coast consisting of steep mountainous land and alluvial coastal lowland (Stojkovich, 1977; NOAA, 1983). A more detailed description is given in Randall *et al.* (1975).

Randall and Holloman (1974) provide a physiographic description of Ajayan Bay which is summarized by Stojkovich (1977). The fringing reef platform bordering most of the south-east shoreline is completely cut by the Ajayan River, forming a small estuary embayment with moderate alluvial silt deposition at the river mouth. A small islet, Agrigan Island, is located on the south-west reef flat. The channel is characterized by progressively steeper fringing reef walls seaward to approximately 18 m in depth. The floor of the channel grades from a silt-mud zone to sand approximately midway out. Water visibility improves seaward. The reef flats are wide and largely covered by seagrass beds.

Reef Structure and Corals   Much of the surface of the patch reef adjacent to Anae Island is exposed at low tide and is largely devoid of live corals. Along the inside patch reef edge there are small colonies of *Acropora*, *Leptastrea* and *Porites*, with *Goniastrea retiformis* in scattered patches. Diversity and colony size increase towards the terrace at 4-9 m. Huge *A. palifera* and hemispherical *Porites* colonies dominate. The algal community is moderately diverse (Stojkovich, 1977).

Stojkovich (1977) gives a brief description of the reefs of Cetti Bay. The reef flat is continuous around the bay with the exception of two breaks occurring at river mouths. The platforms (15-20 m wide) have no moat or algal ridge and are exposed at low tide. They are largely devoid of corals but have a rich algal community, abundant holothurians and many large sea anemones. The reef margin, face and terrace are fairly uniform around the bay apart from a volcanic area on the north side which is cut by irregular cracks and fissures. The margin face typically extends down for 3-4 m and then slopes to a tilted terrace zone 4-10 m deep beyond which the sand floor of the bay begins. The algal and coral communities of the margin, face and terrace are very rich. Massive columns and mounds of *Porites* characterize the terrace and large colonies of *Montipora* and *Acropora* are common. A large bed, 6 m in diameter, of the soft corals *Sinularia* and *Lobophyton* is found to the south.

The reefs at Fouha Bay are described by Randall (1978).

The reefs of Cocos Barrier Reef and Lagoon have been described by Randall *et al.* (1975), Randall and Sherwood (1982) and Stojkovich (1977). Coral cover on the barrier reef-flat platform is variably dense and diverse based on differing degrees of reef-flat exposure. In general, there is an increase in coral cover and diversity from the seaward side to the lagoon side and a total of 59 species have been recorded. The shallow lagoon terrace extends lagoonward to the 3 m contour, varying in width from 200-1000 m. The boundary along the near shore shelf is demarcated by the seagrass *Enhalus acoroides*. Extensive regions of the terrace floor are covered by *Acropora formosa* and thickets ranging in diameter from a few to many metres create a varied range of habitats. In general, coral growth is more dominant on the southern terrace. Towards the eastern end of the lagoon, the *Acropora* thickets become increasingly large with zones of mixed corals between patches. Coral diversity is highest here. The western part of this area is devoid of corals for the most part; 79 species have been recorded in the rest of this area.

The sand floor of the central portion of the lagoon is interrupted by mounds, knolls and knobs, rich in coral, algae, associated invertebrates and fish. The under surface of overhanging mushroom-shaped knolls have *Leptoseris*, *Pavona*, *Plerogyra* and *Porites*, normally found in much deeper habitats. A total of 102 species of coral was recorded. Soft corals are particularly abundant, especially *Sinularia polydactyla*.

The channel margins of Mamaon and Manell barrier reef channels are highly variable with respect to coral density, diversity and physiographic character. In general lagoonward sides of the channels were more highly developed with diversity highest at the channel mouths. Several species of *Porites* dominate, and coral diversity is highest with 104 species recorded. The steep channel slopes and submarine cliffs range in depth from 3-30 m. Turbid water and high sedimentation rates inhibit a rich coral growth but isolated patches occur especially near the channel mouths. *Pavona*, *Acropora* and *Porites* are abundant. In the more cavernous parts of the channels, low light intensities have permitted a variety of deep water species to develop. The channel floors are largely devoid of corals apart from a few small *Porites* colonies. The barrier reefs are nearly 5 km long on the north-west side, 5-6 km long on the south. On the north-east side of the lagoon there is a 4 km long stretch of coast consisting of steep mountainous land and alluvial coastal lowland (NOAA, 1983).

In Ajayan Bay, the east and west side channel walls are considerably different. The east side slopes gradually to the channel floor, while the west side drops almost vertically. Abundance and diversity of algae on the channel walls increases seaward. Coral development is considerably more diverse on the west side and also becomes richer seaward. The reef flat on the west is largely depauperate due to frequent exposure during low tides, but there are a few scattered corals in water-filled crevices and holes (Stojkovich, 1977).

Noteworthy Fauna and Flora Terrestrial and freshwater fauna and flora of the area are described in Raulerson (1979) and briefly in Anon. (1985). Marine fauna and flora checklists for several areas within the Park are given in Stojkovich (1977). At Cocos Barrier Reef and Lagoon, the marine flora is diverse, 91 species having been described. Highest diversity was found on the barrier and patch reefs. The lagoon supports an extremely rich ichthyofauna, with large numbers of juvenile reef fish. The Hawksbill *Eretmochelys imbricata* is often present and the Dugong *Dugong dugon* was reported in 1974 (Randall *et al.*, 1975). The Coconut Crab *Birgus latro* occurs on Cocos Island (as elsewhere), as well as a variety of reptiles. The island has a number of nesting seabirds including White Terns *Gygis alba*. A diverse and abundant fish community exists in the channel in Ajayan Bay, including numerous stonefish *Synanceia verrucosa*. Associated wetland and terrestrial communities are described by Moore *et al.* (1977), the seagrass beds on the reef flats in this bay being some of the most extensive on Guam. The fish community at Anae is very diverse and there are abundant fish and marine algae at Cetti Bay. Turtles used to nest on the beach at the latter (Dahl, 1980), and Green *Chelonia mydas* and Hawksbill *Eretmochelys imbricata* Turtles are reported still to occur (NOAA, 1983).

Scientific Importance and Research There has been extensive marine research in several areas including Nimitz Channel and Taleyfac Bay (Chernin *et al.*, 1977), Fouha Bay (Randall and Birkeland, 1978), Toguan Bay (Jones *et al.*, 1974) and also Sella and Umatac Bays. Shellfish and sea urchin studies were conducted in 1978 and the Division of Aquatic Wildlife Resources carries out periodic offshore aerial surveillance programmes. Two artificial reefs have been constructed in Cocos Lagoon to investigate improved lagoon management techniques (Kock, 1982). The Cocos Lagoon area is unique in the diversity of features found there including barrier reefs, fringing reefs, patch reefs, barrier reef channels, mangroves, seagrass beds and estuaries, and is considered an important scientific study site. The submarine terrace between the patch reef and adjacent fringing reef on Anae supports one of the richest and most diverse coral communities found in Guam's coastal waters (Stojkovich, 1977). Studies of the hydrology, limnology and terrestrial and freshwater biology of the area are described in Raulerson (1979).

Economic Value and Social Benefits At present there are three boat launching sites, the main site being Merizo Pier Park. The Cocos Lagoon area is considered suitable for limited recreational development, and is already popular for boating, snorkelling and diving. Cocos Island, two-thirds of which is privately owned and outside the Park boundaries, is a major day-time resort, with nearly 300 000 visitors in 1986. There is also a resort with 75 cottages and other facilities (Eldredge *in litt.*, 1987). A government park, Dano Park, is being developed at the western end of the island (the portion within the Seashore Park) for picnicking and camping (Lotz *in litt.*, 21.9.87). Ajayan Bay is readily accessible and is a popular place for fishermen, skin divers and picnickers. Anae Island and patch reef are readily accessible by small boat and is popular for SCUBA diving and underwater photography. Cetti Bay is extremely isolated and can only be reached by boat or a long walk but is scenically very attractive and is popular with divers, boaters and fishermen (Stojkovich, 1977).

Disturbance or Deficiencies The sponge *Terpios* was overgrowing corals at Anae Island at a rapid rate in the early 1970s and its growth rate was monitored (Bryan, 1973; Plucer-Rosario, 1983). The increasing popularity of the Merizo coast and of Cocos Island, as a full-time fishing and tourist operation, could be a threat. Randall *et al.* (1975) considered that any physical disruption of habitats within the lagoon or immediately adjacent areas could have serious effects on the fish population in particular. Shell populations have been depleted by collectors, especially popular species such as *Cassis cornuta* (Hedlund, 1977). There is some illegal fishing with dynamite and bleach (Anon., 1985). However, Randall and Sherwood (1982) found that little change had occurred in Cocos Lagoon between 1975 and 1982 although there had been a substantial increase in tourism in the area. The freighter M.V. *Toros Bay* ran aground north-west of Cocos Island on 21st December 1986 and caused damage by scouring two patches of reef totalling ca 1650 sq. m within the park. Damage was assessed by the Guam Environmental Protection Agency (Stillberger *per* Birkeland *in litt.*, 10.11.87). Other problems in the park include poaching, illegal wildfires causing erosion, illegal dumping of waste, including vehicles, and agricultural encroachment (Anon., 1985).

Legal Protection The Guam Territorial Seashore Park was established 12 December 1978 under Executive Order No. 78-42, and is designed to protect the wildlife, marine life and other oceanic resources and natural environment of south-west Guam. Some portions have been protected since 1953. Hunting, shelling, fishing, ranching, boating and outdoor recreation activities are permitted. The Guam Environmental Protection Agency water quality rating for most of this area is "A", recreational, but for Anae Island and Patch Reef and for Cetti Bay is "AA", conservation. There are three Natural Landmarks within the Park: Fouha Point, Mt Lamlam and Facpi Point. Public Law 95-625 of 10 November 1978 called for revision and updating the National Park Service study of the Guam National Seashore, in order to make recommendations for the protection of the natural and historic resources of the area, as well as providing visitor access and interpretive services. A Natural Landmark Survey and the study by Raulerson (1979) were in response to this.

Management The Master Plan was adopted in 1979 and encourages multiple use (Anon., 1979). The U.S. Army Corps of Engineers (1983) has reviewed planning and the management of the entire area. The Department of Parks and Recreation is responsible for co-ordination, planning, facility maintenance, outdoor recreation, historic preservation and scenic resources. The Department of Agriculture is responsible for leases and land registration. The Guam Environmental Protection

Agency is responsible for water and air pollution and solid waste. All agencies have active programs in the Park in their area of responsibility.

Recommendations Cocos Barrier Reef and Lagoon was proposed as a marine reserve under the Coastal Zone Management Act (Dahl, 1980) and listed on the Sanctuary Evaluation List by NOAA (1983). The latter includes within the proposed sanctuary site the barrier reefs, lagoon, three islets (Cocos, Babe and a third sandy island) and the coastal region lying between the mouth of Mamaon and Manell Channels. Additionally, Cocos Lagoon has been proposed as a U.S. National Park Service National Natural Landmark (Eldredge *in litt.*, 1987). Their 2B rating indicates that the site "appears to be nationally significant" and the "site is in some danger". An area from Facpi Point to Fort Santo Angel, on the northern side of Umatac Bay, has also been proposed as a National Marine Sanctuary (NOAA, 1983), and includes Sella, Cetti and Fouha Bays. The area includes the offshore waters to depths of 18.3 m and covers 5 sq. km. As of mid-1987 the sanctuary proposals were not under active consideration (Lotz *in litt.*, 21.9.87).

Stojkovich (1977) gives a list of recommendations for the Cocos Lagoon area which include: fishing, coral harvesting and shell collecting within the proposed sanctuary to be prohibited except by special permit; the GEPA water quality classification to be changed from "A" recreation to "AA" conservation; the establishment of an upper limit on the number and type of point source discharges into Mamaon and Manell Channels; recreational activities to be retained but strictly controlled; the establishment of an upper limit on the number of transport boats and persons using the area at any given time, the establishment of the entire Cocos area as a marine underwater park with trails and basic information on the geology, physiography and biota; the placement of artificial reefs and fish traps for scientific and maricultural purposes should be allowed with the issuance of a special permit; strict litter laws to be implemented especially for waste cans.

Stojkovich (1977) recommended that Ajayan Bay, Anae Island and patch reef, and Cetti Bay be established as natural sanctuaries in which no coral harvesting be allowed; that fishing be allowed only by special permit; that swimming, snorkelling and SCUBA diving activities be retained; that special care be taken to preserve the seagrass beds and that the adjacent wetlands be included in any preservation plan. Mooring buoys should be established and underwater trails developed.

Some of these recommendations have been taken care of in the course of implementation of the park. There are plans for a full range of interpretive facilities and under the General Development Plan there are moves to promote recreational use in accordance with the Land Classification Plan. Overnight accomodation will be restricted to camp grounds. There are plans for intertidal reef flat nature trails.

HAPUTO ECOLOGICAL RESERVE AREA (ERA)

Geographical Location North-west coast; Double Reef (including Pugua Patch Reef) lies between Falcona Beach and Pugua Pt; Haputo ERA extends to the outer edge of the reef line and beyond this to include Double Reef; 130°35'N, 144°50'W.

Area, Depth, Altitude   102 ha of which 73 ha are land and 29 ha are sea; 1 m depth to 122 m altitude.

Land Tenure   The surrounding land is part of the U.S. Naval Communications Finegayan Military Reservation and the Federal Aviation Administration.

Physical Features   The shoreline has rocky limestone cliffs with two sand beach coves. Inland the topography rises rapidly to the top of the limestone cliffs (Anon., 1985). Haputo Beach lies in a small embayment, bordered by cut benches and steep rocky slopes on both the north and south sides, and is described by Randall and Holloman (1974). Double Reef consists of a narrow fringing reef and adjacent patch reef (Pugua) about 350 m off shore. Pugua Patch Reef is about 300 m wide with a wave-washed upper surface which is occasionally exposed during low spring tides. An adjacent developmental reef front and submarine terrace zone extend both north-west and shoreward to the fringing reef platform. A sandy channel floor lies south of the patch reef. Many holes and coral ridges, with relief of 6-15 m, are located shoreward on the submarine terrace (Randall and Holloman, 1974; Stojkovich, 1977). Freshwater springs are abundant on the inner reef flat.

Reef Structure and Corals   Stojkovich (1977) describes four zones at Double Reef and Pugua Patch Reef: an inner reef flat; an outer reef flat and margin face; a submarine terrace with massive coral ridges and sand channels; and the patch reef.

The inner reef flat is poorly developed and largely exposed at low tide. The outer reef flat and margin face are riddled with small channels, indentations and holes in a honeycombed structure. Coral coverage is high and very diverse with much *Pocillopora, Acropora* and encrusting corals. Algal cover is also rich and diverse. The submarine terrace has massive coral ridges up to 10 m high and 100-120 m long, extending to within 3 m of the surface. These are mainly covered by small *Pocillopora* and *Acropora* colonies with a variety of other genera interspersed. In several places, massive *Porites* and *Acropora* mounds and pillars add to the relief. Directly opposite the patch reef, two extensive thickets of *A. formosa* extend from the reef margin to the terrace. In one area, a wreck has created an artificial reef. The patch reef consists of reef-rock pavement with local patches of sand, rubble and scattered coral/algal communities. The extreme western side is almost continually wave-washed. *Tubastrea aurea*, a species rare in Guam, has been reported from Double Reef.

Stojkovich (1977) recognized three zones on Haputo Beach fringing reef: inner reef flat; outer reef flat margin and channels; and submarine terrace. The inner reef flat consists of a smooth reef-rock pavement with sand cover, intermittently exposed at low tide, with little community development apart from a few algae and occasional small corals. The outer reef flat and margin forms a honeycomb matrix; there is a well-developed spur and groove system with some of the grooves 3-4 m deep. In several places, the grooves are closed at the top by calcareous algae and corals forming enclosed overhangs. Coral coverage is outstanding and diversity remarkably

high for north-western Guam, with no one genus dominant although *Pocillopora* and *Acropora* are very evident. Algae cover is rich, turf algae dominating. The submarine terrace begins at a depth of 6 m and slopes gradually to a plateau at 15-20 m, and is not particularly rich. Occasional large *Porites* colonies are found with small colonies of many other genera.

Noteworthy Fauna and Flora   The honeycomb structure of the outer reef flat at Double Reef and Haputo Beach has led to a diverse and very rich fish fauna, 108 species having been recorded at the former and twenty-one at the latter. Large fish are found near the submarine terrace and manta rays have been recorded. Marine fauna and flora species lists for the area are given in Stojkovich (1977). Terrestrial fauna and flora are briefly described in Anon. (1985).

Scientific Importance and Research   The relatively small size and well defined boundaries of Double Reef and Pugua Patch Reef make the area an ideal monitoring site for tropical reef habitats. The calcareous red algal community is one of the finest around the island (Stojkovich, 1977).

Economic Value and Social Benefits   There is a small track to Haputo Beach. Calm seas and good anchorage make water access possible throughout most of the year. The range of topography and attractiveness make Double Reef a favourite recreational spot for SCUBA diving, fishing and photography (Stojkovich, 1977).

Disturbance or Deficiencies   In 1968, there was a population outbreak of *Acanthaster planci* on Double Reef and divers removed large numbers of starfish from a 90 000 sq. m area that year (Chesher, 1969b). *Acanthaster* has also had a visible effect on Haputo Beach fringing reef but this was less apparent on the submarine terrace (Stojkovich, 1977). Disturbances continue to be reported. The sponge *Terpios* was found to be abundant on Double Reef in the early 1980s, encrusting more than 3% of the area (Plucer-Rosario, 1983). There is some illegal fishing with dynamite and bleach (Anon., 1985).

Legal Protection   Haputo ERA was established 15.3.84 by the Chief of Naval Operations, U.S. Navy. Hunting, shelling, fishing and outdoor recreation activities are permitted (Anon., 1985). The Guam Environmental Protection Agency water quality rating for this area is "AA", conservation.

Management   A management plan was approved on 21st January 1986 (Lotz *in litt.*, 21.9.87).

Recommendations   Stojkovich (1977) recommended that Double Reef and Pugua Patch Reef should be created a marine sanctuary with no coral harvesting, fishing or other such activity, although swimming, snorkelling and SCUBA diving should be permitted. Littering should be prohibited. It was also recommended that Haputo Beach fringing reef be established as a marine sanctuary in which no coral harvesting, fishing or other such activity be allowed, but swimming, snorkelling and SCUBA diving should be permitted. Littering should be prohibited. Mooring buoys should be established. With the establishment of the Ecological Reserve Area, most of these recommendations have been implemented, although mooring buoys might be needed.

LUMINAO BARRIER REEF

Geographical Location   The barrier reef extends west from Cabras Island, on the west coast, and is continuous with the submerged Catalan Bank, serving as the foundation for Glass Breakwater; located at ca 13°28'N, 144°39'E.

Area, Depth, Altitude   The barrier reef is 2.5 km long and ranges in width from 240 m at the eastern end to 620 m at the western end (Randall, 1982).

Land Tenure   Lies within the boundaries of Apra Harbor Naval Reservation.

Physical Features   Randall and Eldredge (1982) and Randall and Holloman (1974) describe the area. The barrier reef forms a wide, shallow platform which can be divided into four physiographic zones: inner and outer moats which retain water at low tide; outer limestone pavement which is exposed at low spring tides, and a wave-washed reef margin. The fore-reef slopes to a depths of 5-7 m, an indeterminate submarine terrace dips to 15-20 m and the outer seaward slope drops steeply (Randall and Eldredge, 1982).

Reef Structure and Corals   Randall and Eldredge (1982) and Stojkovich (1977) describe the western end of the seaward side of the barrier reef. A diverse coral community, consisting of 160 species within 45 genera, is found on the shallow reef platform and upper fore-reef slope. The western end of the Luminao Barrier Reef has the greatest species richness (134 species) and surface coverage (30.7%) of any similar reef on Guam. Thirty species of *Acropora* are known, *A. aspera* and *A. formosa* being the most abundant; 22 species of *Montipora*, 11 species of *Pavona*, and eight *Psammocora* are also recorded. The family Fungiidae is not well represented. Randall (1982) describes the reefs and corals.

Noteworthy Fauna and Flora   There is a wide variety of fish. Marine fauna and flora species lists are given in Randall and Eldredge (1982) and Stojkovich (1977).

Scientific Importance and Research   Coral recruitment patterns were studied on the reef by Birkeland *et al.* (1981).

Economic Value and Social Benefits   A diverse coral community, calm waters and easy access from the breakwater has made this a popular area for snorkellers and underwater photographers.

Disturbance or Deficiencies   The lagoon side of Luminao Reef has been largely altered by dredging, filling and construction and is not considered within the recommended marine sanctuary area. However, the seaward side of the reef could also be affected in the long term by industrial activities in the vicinity of Apra Harbor. Oil spills have occurred in the area; for example, in November 1982, about 16 800 gallons of deballast oil floated west along the breakwater.

Legal Protection   The Guam Environmental Protection Agency water quality rating for this area is "A", recreational. A total ban on fishing was implemented on the Luminao Barrier Reef (Guam P.L. 16-114) for one year ending 1 October 1983. There was no attempt to extend this because of lack of enforcement and logistical problems.

Management   None.

Recommendations   Stojkovich (1977) recommended that the area be established as a natural sanctuary in which no coral harvesting, net fishing or other such activity be permitted. A series of underwater trails should be developed.

OROTE PENINSULA ECOLOGICAL RESERVE AREA

Geographical Location   South coast of Orote Peninsula, west central Guam; 13°26'N, 144°38'E.

Area, Depth, Altitude   66 ha; 12 ha land and 54 ha water; 120 ft (36.6 m) depth to 61 m alt.

Land Tenure   United States Navy; part of the U.S. Naval Apra Harbor Reservation.

Physical Features   The limestones drop vertically to the first submarine terrace at 15-20 m depth, which is 20-45 m wide. A second submarine cliff drops to a second terrace at about 80 m. The submerged terraces have interesting tunnel and cave features. A large hole about 7 m in diameter known as the Blue Hole opens on the first terrace at about 18 m depth, and descends vertically to about 80 m with a window at 35 m (Stojkovich, 1977). Throughout the summer months the north-west swell is small, but there are strong currents (Stojkovich, 1977).

Reef Structure and Corals   There is an undisturbed coral and algal community on the submerged cliff face, but no true coral reefs. Coral cover however is minimal. The upper slope and most of the terrace floor consists of scattered small *Pocillopora* colonies. In more protected habitats around fallen blocks and larger rubble, a much richer coral community is evident. *Acropora* and *Porites* are common along with patches of soft corals and crinoids. Diversity is moderate with no single genus dominating. The coral *Pachyclavularia violacea* was observed in semi-protected areas and had not been recorded elsewhere on Guam in 1977. Deep water and cryptic corals are moderately abundant in the Blue Hole (Stojkovich, 1977).

Noteworthy Fauna and Flora   The Marianas fruit bat *Pteropus m. mariannus* is found in the area (Anon., 1985). Species lists for marine algae and fauna recorded in the area are given in Stojkovich (1977).

Scientific Importance and Research   Surveyed briefly by Stojkovich (1977).

Economic Value and Social Benefits   The area is popular with fishermen and SCUBA divers but is inaccessible by land (Stojkovich, 1977).

Disturbance or Deficiencies   The area is virtually pristine (Stojkovich, 1977).

Legal Protection   Established by the Chief of Naval Operations, U.S. Navy on 13.3.1984. Fishing, shelling and outdoor recreational activities are permitted (Anon., 1985). The Guam Environmental Protection Agency water quality classification for this area is "A", recreational.

Management   A management plan for the reserve was approved on 3.3.86 (Lotz *in litt.*, 21.9.87); the draft of the plan was prepared by the U.S. Navy (1985).

Recommendations   Included in the management plan.

PATI POINT NATURAL AREA AND NORTH-WEST COASTLINE

Geographical Location   An area on the extreme north-western coast, extending from Falcona Beach north to Ritidian Point; Tarague Beach lies between Ritidian (13°39'N, 144°51'E) and Pati Points (13°35'N, 144°56'E).

Area, Depth, Altitude   Sea level to 138 m; Pati Point Natural Area covers 112 ha.

Land Tenure   Uruno Beach is privately owned; the remaining area, including Pati Point Natural Area, lies within the Andersen Air Force Base Military Reservation.

Physical Features   Randall and Holloman (1974) describe the geology, hydrology and physiography of the area. The north coast consists of rugged limestone cliffs bordering intermittent long stretches of beach with a wide reef flat platform and convex algal ridge. There is heavy surf and strong rip currents especially near the larger surge channels and near Ritidian.

Reef Structure and Corals   Stojkovich (1977) describes three broad zones in the Uruno-Ritidian area: reef flat and associated moat; reef margin and face with cuestal algal ridge; and submarine terrace and slope.

The sandy inner reef flat is sparsely inhabited apart from the alga *Caulerpa antoensis*. The moat (1-1.5 m deep) extends out to the reef margin and algal ridge. Coral development becomes progressively more luxurious towards the back of the algal ridge and small, low colonies give way to larger thickets and colonies of *Acropora* and *Porites*. *Goniastrea retiformis*, *Pocillopora damicornis*, *Psammocora contigua* and *Millepora* spp. are also abundant. Towards Uruno, high relief colonies were more common. The moat and outer reef platform have a rich composition of turf and fleshy algae.

The algal ridge here is a good example of a cuestal type of margin development. The ridge is typically above the high water line and continuously wave-washed, with a thick orange mat of the alga *Gelidiella acerosa*; *Porolithon onkodes* is also abundant and small *Pocillopora* colonies are found in a few scattered crevices.

The terrace, 5-10 m wide, drops abruptly to about 15 m. Scoured surge channels up to 8 m wide cut through the reef platform in several places. Coral cover is moderate to sparse here due to *Acanthaster planci* predation. Recolonization was evident in the 1970s.

Encrusting *Montipora*, *Favia*, *Platygyra*, *Porites* and *Pocillopora* were common and algal cover was rich with the larger forms predominant.

Stojkovich (1977) also studied an area near East Tarague Beach, near Scout Beach, with a well developed convex algal ridge and reef flat platform, typical of northern Guam. The reef flat platform consists of a poorly defined inner zone on which numerous scattered remnant patches of limestone occur; a middle zone consisting mostly of a thin veneer of sand covering an irregular limestone platform with columnar limestone projections protruding through in many places; and a margin with a well defined convex algal ridge and massive spur and groove system cut in places by large surge channels. The coral community on the latter is predominately low relief due to heavy wave action but very dense *Pocillopora*, *Acropora*, *Montipora* and *Millepora* are well represented.

Noteworthy Fauna and Flora   Chaetodontid, pomacentrid, acanthurid and balistid fish are abundant in the moat area of the reef. Fish are also abundant in the holes, overhangs and crevices of the terrace and slope zones. Larger game fish and grey sharks have been recorded and the Hawksbill Turtle *Eretmochelys imbricata* has been seen in the deeper slope waters and used to nest on Ritidian Beach (Pritchard, 1981). Species lists for marine fauna and flora are given in Stojkovich (1977). A brief description of the terrestrial fauna and flora of Pati Point is given in Anon. (1985).

Scientific Importance and Research   The longest stretch of pristine beach on Guam, with a diverse and attractive reef (Stojkovich, 1977). The Uruno Reef area has been surveyed by Dickinson and Tsuda (1975) and Stojkovich (1977) studied a number of sites in the area.

Economic Value and Social Benefits   Heavy swells and strong currents preclude the development of this area for recreation without major modifications which are not recommended (Dickinson and Tsuda, 1975). There is a single military access road at Tarague Beach and the area is largely restricted to use by military dependents. Access to Pati Point Natural Area is not permitted.

Disturbance or Deficiencies   *Acanthaster planci* predation was heavy in some areas in the late sixties (Chesher, 1969a). Ritidian Beach has been used for the disposal of munitions (Dahl, 1980). Dickinson and Tsuda (1975) investigated the biological and environmental impact of resort development at Uruno.

Legal Protection   Pati Point Natural Area was established by the U.S. Air Force in 1973. The Guam Environmental Protection Agency water quality classification for this area is "AA", conservation.

Management   The area lies mainly within Andersen Air Force Base military reservation and is largely inaccessible to the public. Access to private property is limited.

Recommendations   Stojkovich (1977) recommended that this entire coastal sector be made a natural marine sanctuary in which no coral harvesting, net fishing or other such activity be permitted. The Tarague Beach area could be established as a natural sanctuary in which no coral harvesting, net fishing or other such activity be permitted, but swimming, snorkelling and SCUBA diving activities be retained.

WAR IN THE PACIFIC NATIONAL HISTORICAL PARK

Geographical Location   West central Guam; the site is split in two by the Apra Harbour Naval Station; the park lies within the area 13°22'-13°28'N, 144°38'-144°44'E.

Area, Depth, Altitude   374 ha land, 405 ha water; 65 m depth to 313 m alt.

Land Tenure   Includes land owned by both the Federal and Guam Governments and privately.   Of the water area, 327 ha is owned by the Government of Guam, the remaining 78 ha by the Federal Government.

Physical Features   The Park consists of seven physically separate units: Asan Beach Unit, Asan Inland Unit, Fonte Plateau Unit, Piti Unit, Mt Chachao/Mt Tenjo Unit, Agat Unit and Mt Alifan Unit.   These include sand beaches, offshore reefs, rugged hills and mountain tops (Anon., 1985).

Reef Structure and Corals   Two units, Asan Beach and Agat, contain coral reefs (Anon., 1985).   Those at Agat were described in Eldredge *et al.* (1977) and Randall (1978).   The Piti Unit is adjacent to reefs.   The marine resources of this area are described by Eldredge (1979); some 140 species of coral have been identified in the Agat area.

Noteworthy Fauna and Flora   Terrestrial and freshwater fauna and flora are described by Raulerson (1979), and briefly in Anon. (1985).   A total of 75 species of gastropod, 18 species of bivalves, 45 species of echinoderms and 26 fish species have been recorded (Eldredge, 1979).

Scientific Importance and Research   Studies of the terrestrial and freshwater biology, hydrology and limnology of the area are detailed in Raulerson (1979), and of marine biological resources in Eldredge (1979).

Economic Value and Social Benefits   The primary purpose of the park is to preserve the historic features of World War II, including many Japanese defensive fortifications; the designated areas encompass the major assault beaches and the major beach-head perimeter established in late June 1944; much of the heaviest fighting during the American Invasion of Guam in World War II occurred in these areas (Raulerson, 1979).

Disturbance or Deficiencies   There are some problems from fishing with bleach and dynamite (Anon., 1985). Sections of the shoreline have been altered by sea walls and sewage outfalls.   Agat Bay is moderately heavily developed (Randall, 1978), and in 1987 a boat launch site and marina were under construction there (Eldredge *in litt.*, 1987).   Other problems include illegal dumping, grassland fires and poaching (Anon., 1985). Most of the terrestrial vegetation has been modified by man, although in some areas "benign neglect" over the past 35 years has allowed natural vegetation to recover (Raulerson, 1979).

Legal Protection   Established by the U.S. National Park Service in 1978 under Federal Public Law 95-348 as a "multiple-use management area".   The legislation authorizes the inclusion of the reef areas of Agat, Piti and Asan.   It was established to conserve and interpret outstanding natural, scenic and historic values.   Shelling, fishing, boating and outdoor recreation activities are permitted.

Management   A management plan has been approved and implemented (National Park Service, 1983).

HAWAII

INTRODUCTION

General Description

The Hawaiian Archipelago is the longest and most isolated chain of tropical islands in the world, stretching about 2300 km from Hawaii in the south-east (19°35'N, 155°30'W) to Kure Atoll in the north-west (28°25'N, 178°20'W). Considerable geophysical evidence suggests that the entire chain formed as a result of tectonic motion of the Pacific plate over a relatively stationary hotspot in the mantle of the earth (Wilson, 1963; Morgan, 1972; Jackson *et al.*, 1980; Grigg, 1985a). Hawaii, the youngest island (the oldest rocks are 0.8 million years), is situated over the hotspot and contains the most active volcano in the world, Kilauea. Steady crustal movement is gradually transporting each island mass north-westward, eventually breaking connection with the hotspot. As a result, moving north-west of Hawaii, the islands become progressively older and gradually decrease in size due to subsidence and erosion. All of the islands north-west of Gardner Pinnacles, the last island in the chain with subaerial basalts, are either atolls, coral islands or reefs and shoals of limestone construction. Midway Atoll, the second to last island in the chain, has been dated at 27.7±0.6 million years (Dalrymple *et al.*, 1977). Beyond Kure Atoll, the last island, the chain continues underwater as a series of drowned atolls (Grigg, 1982) or seamounts which extend all the way to Kamchatka (Jackson *et al.*, 1980).

The geography of the Hawaiian Islands is varied and complex. The islands at the south-eastern end are the tops of some of the largest mountains in the world (measured from their bases); Mauna Loa is almost 32 000 ft (9756 m) above the sea floor, making it the highest on earth. Some of the older high islands are deeply eroded with spectactular river valleys such as Waipio Valley on Hawaii, Iao Valley on Maui and Halawa Valley on Molokai. All of these are heavily vegetated and contain highly evolved faunas with many endemic birds and insects. A rise in sea level of approximately 130 m during the Holocene has created many bays and estuaries in the drowned heads of some of the river valley systems, particularly on older islands in the chain. The islands north-west of the high islands are much simpler geographically because they are low rocky islets or coral islands at sea level. All of the stages described by Charles Darwin in his theory regarding the evolution of atolls are represented in the Hawaiian Islands (Scott and Rotondo, 1983). Coastal geology is described in Moberly (1963) and submarine topography in Coulbourn *et al.* (1974).

The climate of Hawaii is mainly tropical but approaches subtropical at the north-western extreme. The atmospheric climate is characterized by mild temperatures all year, moderate humidity, persistent north-easterly tradewinds, highly varied rainfall patterns within short distances and infrequency of severe storms (Armstrong, 1973). In the summer from May to October when the sun is almost overhead, the weather is warmer and drier and the trades more persistent. "Winter" months from October to April are cooler, wetter and characterized by more variable winds. Storm fronts from the south, north and west (Kona conditions) are most frequent during the winter. In general, Hawaii's climate reflects the interplay between four factors; latitude, the surrounding ocean, Hawaii's location relative to storm tracks and the Pacific anticyclone and terrain. Rainfall over the open sea averages about 700 mm/yr but over land it ranges between 200 and 10000 mm/yr, due mainly to orographic effects, and is highest at an elevation of about 750 m. Temperature at sea level varies seasonally between about 23 and 27°C. Air temperature decreases about 1.5°C per 300 m of altitude.

In the ocean, surface temperature differences are seasonal; during the summer there are almost no differences across the chain, whereas in the winter differences average about 3°C but may be as large as 6°C. Salinity is relatively constant everywhere (about 35 ppt) except in nearshore areas subject to land run-off effects. Other oceanographic factors which vary seasonally or latitudinally across the chain and which affect the ecology of coral reefs specifically are considered in more detail below.

From the time of the discovery voyages of Captain James Cook, the Hawaiian Islands have been appreciated for their rich natural history (Kay, 1972). The first studies of Hawaiian coral reefs were those of Dana (1853) who gave a lengthy description. The next major studies were by Vaughan (1907 and 1910), with other early work by MacCaughey (1918) and Edmondson (1924, 1928, 1929 and 1933). Extensive work on the marine environment is carried out at the Hawaii Institute of Marine Biology in Kaneohe Bay, Oahu (*see separate account*). Recent research is described in Bardach (1974), Grigg (1981a) and Grigg and Dollar (1980).

In spite of their low diversity, coral reefs in Hawaii are fairly well developed. Their community structure, succession and development throughout the archipelago have been reviewed by Grigg (1983). Reefs are best developed on leeward (southern and south-western) coasts or in bays sheltered from wave action. These include many sites along the Kona Coast and Kealakekua Bay on Hawaii, Molokini "lagoon", the southern coast of west Maui, the north coast of Lanai, the south-east coast of Molokai, Hanauma Bay and some reefs near Barbers Point on Oahu, and the lagoons of the north-western Hawaiian Islands, including Midway and Kure (*see separate accounts*). Coral growth rates are highest at the south-eastern end of the chain where water temperature and sunlight is optimal. However, islands in this portion of the chain are relatively young (1-2 million years) and fringing reefs with only a weakly developed spur and groove system are found. Moving north-west the first barrier reef encountered is off Kaneohe Bay; this part of Oahu is about 2.5 million years old, although the barrier reef itself is probably only Holocene in origin (i.e. 10 000 years or less). French Frigate Shoals (25°N) is about the youngest atoll (about 12 million years old) in the archipelago. Towards the end of the chain, coral growth rates gradually diminish to a point where rates of limestone accretion due to corals and other carbonate producers fail to keep up with rates of erosion and subsidence. This threshold has been called the Darwin Point (Grigg, 1981b and 1982), and marks the latitude in the archipelago where islands drown. Because of plate

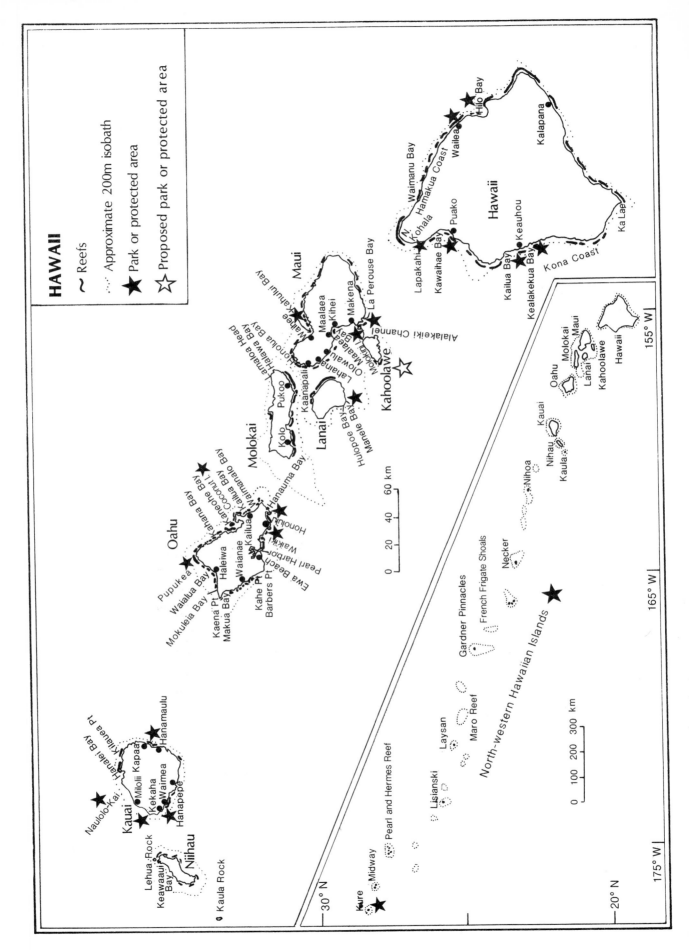

HAWAII

~ Reefs

...... Approximate 200m isobath

★ Park or protected area

☆ Proposed park or protected area

motion to the north-west over the past 70 million years, a number of former drowned Hawaiian islands or guyots continue northward in the form of a submerged chain known today as the Emperor Seamounts.

Coral species composition in the Hawaiian Archipelago is remarkably uniform. With the exception of three species of *Acropora* in the centre of the archipelago, most species occur throughout the archipelago. Given the rather large gradients in temperature and light across the chain, this suggests that most species of coral in the Hawaiian Archipelago are generalized species. Initial taxonomic work was carried out by Vaughan (1907 and 1910) who described 86 species of Hawaiian reef and ahermatypic corals. Maragos (1977), considering only the hermatypic species, revised and updated Vaughan's list and concluded that only 40 were reef-building species. Two more species of acroporid corals were added to the list by Grigg *et al.* (1981) and more recently two additional species and genera have been reported. However some of the earlier species records may be junior synonyms of other species (Maragos and Jokiel, 1986) and the total Hawaiian list may not exceed 40 species. Endemism in the coral fauna is considered to be about 20% (Jokiel, 1987). Relative to the Indo-West Pacific where over 200 species of corals commonly occur on a single coral reef, this fauna must be considered depauperate. Isolation of the Hawaiian Archipelago is generally considered to be the cause of this low diversity (Dana, 1971; Grigg, 1983; Jokiel, 1987; Maragos, 1977; Maragos and Jokiel, 1986).

The effects of isolation on the diversity and endemism of the marine invertebrate and reef fish faunas are discussed in Kay and Palumbi (1987) and Hourigan and Reese (1987) respectively. Both groups show low diversity relative to elsewhere in the Pacific, but high endemism (30%). The reef and shore fauna of Hawaii is described in Devaney and Eldredge (1977) and the ecology and distribution of crustose Corallinaceae in Littler (1973a and b). Powers (1970) provides a numerical taxonomic study of Hawaiian reef corals. Distribution of seabirds in the islands is described in Harrison *et al.* (1984).

Individual data sheets have been prepared for all the major islands in the chain as well as the North-western Hawaiian Islands (NWHI), Midway and Kure Atolls. Individual data sheets have also been prepared for special areas of interest: Molokini Island, Hanauma Bay, Waikiki- Diamond Head and Kaneohe Bay, the last three all on Oahu.

For convenience, accounts for the U.S. dependencies Howland, Jarvis, Baker and Johnston (all reserves), and Palmyra and Wake Atolls are included in this section. Kingman Reef, also a U.S. dependency, is a triangular reef with a deep lagoon and one tiny flat coral islet, often only exposed at low tide. It is under the jurisdiction of the U.S. Navy (Maragos, 1986). It is reported to have a rich marine fauna and was proposed for listing as an "Island for Science" (Douglas, 1969).

Reef Resources

Subsistence and commercial fishing, and many aquatic activities such as snorkelling, SCUBA diving, shell collecting, underwater photography, swimming, surfing, windsurfing, sailing, whale watching and over-water transportation are carried out in Hawaii. Many species belonging to various taxa including algae, black coral, lobsters, crabs, octopus, and reef fish are harvested commercially (Grigg, 1965 and 1976).

Disturbances and Deficiencies

As latitude increases within the chain, temperature and light gradually decline and account for reductions in coral growth rates (Grigg, 1981b). Also, because the north-western islands are less protected from ocean swell, disturbance from storms to these islands is of greater limiting significance than in the high islands. Effects of storm stress on coral community structure are discussed in Dollar (1981). In general, natural disturbances related to storms account for more damage to coral reefs in Hawaii than those caused by anthropogenic sources.

The environmental conditions for reef growth in Hawaii range from optimal to suboptimal, to absolutely limiting. On the major high islands (Hawaii to Kauai), land run-off and sedimentation have limited reef development in many areas. Urbanization and associated anthropogenic pollutants from sewer outfalls, sugar mill discharges, power plant effluents, dredging activities and non-point source run-off, pose serious but generally highly localized impacts. Estuarine pollution is discussed in Cox and Gordon (1970) and non-point pollution in State Department of Health (1978). In general the "health" of coral reefs in Hawaii can be said to be good and sedimentation levels are declining owing to declining urban development, ranching, and plantation-style agriculture. However, overfishing is a persistent problem on the high islands.

impacts on the U.S. dependency islands are described in the following accounts.

Legislation and Management

Prior to western contact, the right to harvest seafood in local coastal waters was controlled by the chiefs. Since then these rights have largely been bought by the government but a few areas of coastal waters (konohikis) remain under nominal private control (Johannes, 1984).

Efforts to regulate and control human impact on the marine environment are gradually improving environmental conditions for the growth of corals and other shallow water reef resources. Management of coral reef resources in the State is the responsibility of the Division of Aquatic Resources, Department of Land and Natural Resources. Regulations concerning resource use apply to all waters of the state except for special areas which are designated Marine Life Conservation Districts, Marine Fisheries Management Areas and Natural Area Reserves which have specific use restrictions and guidelines (*see separate accounts*). The State Department of Health manages coastal water quality and the State Department of Economic Development manages Hawaii's Coastal Zone Management Program, both of which can have positive influence on coral reefs. An ecosystem approach to water quality is discussed in Technical Committee on Water Quality Standards (1977). The U.S. Army Corps of Engineers regulates dredging, filling and construction in coastal waters with its permit programme. The U.S. Fish and Wildlife

Service (Federal Government) shares management responsibility with the State in the North-western Hawaiian Islands (NWHI) for the Hawaiian Islands National Wildlife Refuge. The Western Pacific Regional Fisheries Management Council is responsible for the management of fishery resources in the NWHI (primarily outside 3 miles (4.8 km) offshore).

In 1977 a five-year research programme was started by the State of Hawaii in cooperation with the U.S. Fish and Wildlife Service, the National Marine Fisheries Service and the Sea Grant College at the University of Hawaii. Its aim was to assess the fishery resources of the Hawaiian Archipelago and to develop management plans for preserving unique wildlife and conserving commercially important species. At least 26 research projects were undertaken, touching on all trophic levels of a variety of reef and oceanic ecosystems (Grigg, 1981a).

Coastal resource inventory studies were initiated in 1978. Texts and atlases have been completed for the islands of Oahu, Maui, the west coast of Hawaii, and Kauai and are in preparation for Molokai (Maragos and Elliott, 1985). The following protected areas include reefs:

Hawaii
- Kealakekua Bay Marine Life Conservation District (MLCD)
- Lapakahi State Historical Park MLCD
- Wailea Bay MLCD
- Hilo Bay Marine Fishery Management Area (MFMA)
- Puako Bay and Reef MFMA
- Kailua Bay MFMA

Lanai
- Hulopoe Bay-Palawai and Manele Bay-Kamao MLCD
- Manele Boat Harbour MFMA

Maui
- Honolua and Mokuleia Bays MLCD
- Kahului Harbor MFMA (three separate areas)
- Cape Kinau, Ahihi and La Perouse Bays State Natural Area
- Molokini Islet MLCD

Oahu
- Hanauma MLCD
- Pupukea Beach Park MLCD
- Waikiki-Diamond Head MFMA
- Coconut Island Marine Life Refuge

Kauai
- Waimea Bay and Recreational Pier MFMA
- Hanamaulu Bay and Ahukini Recreational Pier MFMA
- Milolii Reef State Park
- Nualolo-Kai Reef State Park

North-western Hawaiian Islands
- Hawaiian Islands National Wildlife Refuge and MFMA
- Kure MFMA

The following U.S. dependencies are also protected (*see separate accounts*):

- Baker Island National Wildlife Refuge
- Howland Island National Wildlife Refuge
- Jarvis Island National Wildlife Refuge
- Johnston Island National Wildlife Refuge

Recommendations

The adequacy of state and federal regulations concerning the management of marine resources in Hawaii can be said to be generally good, although, in some areas, particularly Oahu, enforcement of the regulations has been ineffective due to lack of sufficient enforcement staff and of effort on the part of many enforcement officers. Many reefs have been seriously overfished. Perhaps the most important problems facing the state is the need for further research documenting the effects of overfishing specific to area, type of gear used and effort applied, so that overfishing can be controlled. A need exists for stronger enforcement of existing regulations as well as more restrictive legislation, particularly regarding gill nets and spear fishing. It is particularly important to separate the effects of fishing from environmental degradation caused by various sources of pollution and natural disturbance. There are some areas which may require additional protection such as Kahoolawe Island. Johannes (1984) suggests that the remaining areas of coastal water under private control might present a novel opportunity for creating additional marine protected areas with few legal or management problems.

References

AECOS, Inc. (1979a). *Hawaii Coral Reef Inventory.* Island of Oahu. Prep. for U.S. Army Corps of Engineers, Pacific Ocean Division. 552 pp.
AECOS, Inc. (1979b). *Maui Island Coral Reef Inventory.* Prep. for U.S. Army Corps of Engineers, Pacific Ocean Division. 303 pp.
AECOS, Inc. (1979c). Post construction water quality, benthic habitat and epifaunal survey for reef runway, Honolulu International Airport. Part B: Benthic Biology. Prep. for Parsons, Hawaii. AECOS 142B. 69 pp.
AECOS, Inc. (1981a). *Oahu Coastal Zone Atlas*: Representing the Hawaii coral reef inventory, Island of Oahu. Prep. for U.S. Army Corps of Engineers, Pacific Ocean Division.
AECOS, Inc. (1981b). *Maui Coastal Zone Atlas*: Representing the Hawaii coral reef inventory, Island of Maui. Prep. for U.S. Army Corps of Engineers, Pacific Ocean Division.
AECOS, Inc. (1982). *Kauai Island Coastal Resource Inventory.* Prep. for U.S. Army Corps of Engineers, Pacific Ocean Division.
Agegian, C.R. and Abbott, I.A. (1985). Deep water macroalgal communities: A comparison between Penguin Bank, Hawaii, and Johnston Atoll. *Proc. 5th Int. Coral Reef Cong., Tahiti* 5: 47-50.
Amerson, A.B. Jr (1971). The natural history of French Frigate Shoals, Northwestern Hawaiian Islands. *Atoll Res. Bull.* 150. 383 pp.
Amerson, A.B. and Shelton, P.C. (1976). The natural history of Johnston Atoll, Central Pacific Ocean. *Atoll Res. Bull.* 192. 385 pp.
Amerson, A.B. Jr, Clapp, R.B. and Wirtz, W.O. II (1974). The natural history of Pearl and Hermes Reef, Northwestern Hawaiian Islands. *Atoll Res. Bull.* 174. 306 pp.

Anderson, B.D. (1978). Coral community structure at Hanauma Bay, Oahu, Hawaii. A model for coral reef management. Ph.D. dissertation. Heed Univ., Florida. 360 pp.

Anon. (n.d.). *Marshall Islands Guidebook.* Republic of the Marshall Islands, Majuro, MI 96960. 110 pp.

Apfelbaum, S.I., Ludwig, J.P. and Ludwig, C.E. (1983). Ecological problems associated with disruption of dune vegetation dynamics by *Casuarina equisetifolia* L. at Sand I., Midway Atoll. *Atoll Res. Bull.* 261. 19 pp.

Armstrong, R.W. (Ed.) (1973). *Atlas of Hawaii.* University Press of Hawaii. 222 pp.

Ashmore, S.A. (1973). The geomorphology at Johnston Atoll. Technical Report TR-237, Naval Oceanogarphic Office, Washington D.C. 25 pp.

Atkinson, M.J. and Grigg, R.W. (1984). Model of a coral reef ecosystem. 2. Gross and net benthic primary production at French Frigate Shoals, Hawaii. *Coral Reefs* 3(1): 13-22.

Banner, A.H. (1968). A freshwater kill on the coral reefs in Hawaii. *Hawaii Inst. Mar. Biol. Tech. Rept* 15.

Banner, A.H. (1974). Kaneohe Bay, Hawaii: urban pollution and a coral reef ecosystem. *Proc. 2nd Int. Coral Reef Symp.* 2: 685-702.

Banner, A.H. and Bailey, J.H. (1970). The effects of urban pollution upon a coral reef system, a preliminary report. *Hawaii Inst. Mar. Biol. Tech. Rept* 25. 66 pp.

Bardach, J.E. (1974). CORMAR: Coral Reef Management and Research at the University of Hawaii. *Env. Cons.* 1(3): 233-234.

Bathen, K.H. (1968). A descriptive study of the physical oceanography of Kaneohe Bay, Oahu, Hawaii. *Hawaii Inst. Mar. Biol. Tech. Rept* 14. 353 pp.

Bathen, K.H. (1978). *Circulation Atlas of Oahu, Hawaii.* UNIHI-SEAGRANT-MR-78-05. 94 pp.

Branham, J.M., Reed, S.A., Bailey, J.H. and Caperon, J. (1971). Coral-eating sea stars *Acanthaster* in Hawaii. *Science* 172: 1155-1157.

Brock, J.H. (1976). Benthic marine communities of shoreline structures in Kaneohe Bay, Oahu. Prep. for U.S. Army Corps of Engineers, Pacific Ocean Division. 161 pp.

Brock, J.H. and Brock, R.E. (1974). *The marine fauna of the coast off northern Kona, Hawaii.* UNIHI-SEAGRANT-AR-74-02. 30 pp.

Brock, V.E., Jones, R.S. and Helfrich, P. (1965). An ecological reconnaissance of Johnston Island and the effects of dredging. Ann. Rep. to U.S. Atomic Energy Commission. *Hawaii Inst. Mar. Biol. Tech. Rept* 5. 25 pp.

Brock, V.E., Van Heukelem, W. and Helfrich, P. (1966). An ecological reconnaissance of Johnston Island and the effects of dredging. *Hawaii Inst. Mar. Biol. Tech. Rept* 11. 56 pp.

Bryan, E.H. (1959). Notes on the geology and natural history of Wake Island. *Atoll Res. Bull.* 66. 22 pp.

Bugglen, R.G. and Tsuda, R.T. (1966). A preliminary marine algal flora from selected habitats on Johnston Atoll. *Hawaii Inst. Mar. Biol. Tech. Rept* 9: 1-29.

Bugglen, R.G. and Tsuda, R.T. (1969). A record of benthic algae for Johnston Atoll. *Atoll Res. Bull.* 120: 1-20.

Carlquist, S. (1980). *Hawaii, A Natural History.* S.B. Printers, Inc., Honolulu, Hawaii. 468 pp.

Chapman, G.A. (1979). Honolulu International Airport Reef Runway Post-construction Environmental Impact Report: Vol. 1. (Executive Summary), Vol. 2. (Technical Report). Prep. by Parsons, Hawaii, Inc. for the State of Hawaii Dept of Transportation, Air Transportation Facilities Division, Honolulu. 27 pp. (Vol. 1) and ca 400 pp. (Vol. 2).

Chave, K.E. and Smith, S.V. (1974). A reef ecosystem under stress. *Env. Cons.* 1(1): 41-42.

Clapp, R.B. (1972). The natural history of Gardner Pinnacles, Northwestern Hawaiian Islands. *Atoll Res. Bull.* 163. 29 pp.

Clapp, R.B. and Kridler, E. (1977). The natural history of Necker Island, Northwestern Hawaiian Islands. *Atoll Res. Bull.* 206. 102 pp.

Clapp, R.B. and Wirtz, W.O., II (1975). The natural history of Lisianski Island, Northwestern Hawaiian Islands. *Atoll Res. Bull.* 186. 196 pp.

Clapp, R.B., Kridler, E. and Fleet, R.R. (1977). The natural history of Nihoa Island, Northwestern Hawaiian Islands. *Atoll Res. Bull.* 207. 147 pp.

Clark, J.R.K. (1977). *The Beaches of Oahu.* Univ. Press of Hawaii, Honolulu. 193 pp.

Clutter, R.I. (1969). Plankton ecology. In: Estuarine pollution in the State of Hawaii. Part 2: Kaneohe Bay study. *Univ. Hawaii, Water Resources Research Center Tech. Rept* 31, Section A: 1-18.

Coles, S.L. (1984). Colonization of Hawaiian reef corals on new and denuded substrate in the vicinity of a Hawaiian power station. *Coral Reefs* 3(3): 123-130.

Coles, S.L. (1985). The effects of elevated temperature on reef coral planula settlement as related to power station entrainment. *Proc. 5th Int. Coral Reef Cong., Tahiti* 4: 171-176.

Conant, S., Christensen, C.C., Conant, P., Gagne, W.C. and Lee Goff, M. (1984). The unique terrestrial biota of the Northwestern Hawaiian Islands. In: Grigg, R.W. and Tanoue, K.Y. (Eds), *Proceedings of the Second Symposium on Resource Investigations in the Northwestern Hawaiian Islands.* UNIHI-SEAGRANT-MR-84-01. Pp. 77-94.

Coulbourn, W.T., Campbell, J.F. and Moberly, R. (1974). Hawaiian submarine terraces, canyons, and Quaternary history evaluated by seismic reflection profiles. *Mar. Geol.* 17: 215-234.

Cox, D.C. and Gordon, L.C. (1970). Estuarine pollution in the State of Hawaii. Vol. 1: Statewide study. *Univ. Hawaii Water Resources Research Center Tech. Rept* 31.

Dalrymple, G.B., Clague, D.A. and Lanphere, M.A. (1977). Revised age for Midway volcano, Hawaiian volcanic chain. *Earth Planet. Sci. Lett.* 37: 107-116.

Dana, J.D. (1853). *Corals and Coral Islands.* London.

Dana, T.F. (1971). On the corals of the world's most northern atoll (Kure: Hawaiian Archipelago). *Pac. Sci.* 25(1): 80-87.

Dawson, E.Y. (1959). Changes in Palmyra Atoll and its vegetation through the activities of man, 1913-1958. *Pac. Not.* 1(2).

Dawson, E.Y., Aleem, A.A. and Halstead, B.W. (1955). Marine algae from Palmyra Island with special reference to the feeding habits and toxicology of reef fishes. *A. Hancock Found. Pub. Occ. Pap.* 17: 1-39.

Department of Planning and Economic Development, State of Hawaii (1974). *Hawaii, the Natural Environment.* 36 pp.

Devaney, D.M., Kelly, M., Lee, P.J. and Motteler, L.S. (1976). Kaneohe: A history of change (1778-1950). Prep. for U.S. Army Corps of Engineers, Pacific Ocean Division. B.P. Bishop Mus. 271 pp.

Devaney, D.M. and Eldredge, L.G. (Eds) (1977). Reef and shore fauna of Hawaii. *B.P. Bishop Mus. Spec. Pub.* 64(1). 278 pp.

Dollar, S.J. (1975). Zonation of reef corals off the Kona Coast of Hawaii. M.S. Thesis. Department of Oceanography, Univ. Hawaii. 181 pp.

Dollar, S.J. (1981). Storm stress and coral community structure in Hawaii. *Proc. 4th Int. Coral Reef Symp., Manila* 1: 214. (Abst.).

Dollar, S.J. (1982). Wave stress and coral community structure in Hawaii. *Coral Reefs* 1(2): 71-81.

Doty, M. (1971). The productivity of benthic frondose algae at Waikiki Beach, 1967-68. *Univ. Hawaii Bot. Sci. Paper* 22: 1-119.

Douglas, G. (1969). Check List of Pacific Oceanic Islands. *Micronesica* 5(2): 327-463.

Dunlap, E. (1987). Poor collecting at Maui, Hawaii. *Hawaiian Shell News* 35(5) N.S. 329: 5.

Easton, W.H. and Olson, E.A. (1976). Radiocarbon profile of Hanauma reef, Oahu, Hawaii. *Geol. Soc. Amer. Bull.* 87: 711-719.

Edmondson, C.H. (1924). Notes on the rates of growth of coral in Hawaii. *Proc. 2nd Pan-Pacific Science Congress (Auldia)* 2: 1553-1555.

Edmondson, C.H. (1928). The ecology of an Hawaiian coral reef. *B.P. Bishop Mus. Bull.* 45: 1-64.

Edmondson, C.H. (1929). Growth of Hawaiian corals. *B.P. Bishop Mus. Bull.* 58. 38 pp.

Edmondson, C.H. (1933). Reef and shore fauna of Hawaii. *B.P. Bishop Mus. Spec. Pub.* 22. 295 pp.

Ely, C.A. and Clapp, R.B. (1973). The natural history of Laysan Island, Northwestern Hawaiian Islands. *Atoll Res. Bull.* 171. 361 pp.

Emery, K.O. (1956). Marine geology of Johnston Island and its surrounding shallows, Central Pacific Ocean. *Geol. Soc. Am. Bull.* 67(11): 1505-1519.

Environmental Consultants, Inc. (1974a). Honolua Bay study: Geological, physical and biological surveys. Prep. for Belt, Collins and Associates. Rept. ECI-106. 62 pp.

Environmental Consultants, Inc. (1974b). Marine biology of the Makena-Ahihi region (Maui). Prep. for Neighbor Islands Consultants, Rept. ECI-112. 50 pp.

Environmental Consultants, Inc. (1975). Observations of the marine environment at Coconut Island, Kanehoe Bay, Oahu. Prep. for Group Architects Collaborative, Inc., Honolulu. ECI-122. 23 pp.

Environmental Consultants, Inc. (1977). A reconnaissance survey of nearshore environments at Kihei, Maui. Prep. for U.S. Army Corps of Engineers, Pacific Ocean Division Rept. ECI-140. 70 pp.

Evans, E.C., III (Ed.) (1974). Pearl Harbor biological survey - final report. Naval Undersea Center, Hawaii Laboratory. NUC TN-1128. 780 pp.

Fan, P.F. and Burnett, W.C. (1969). Sedimentation. In: Estuarine pollution in the State of Hawaii. Part 2: Kaneohe Bay study. *Univ. Hawaii Water Resources Research Center Tech. Rept* 31, Section A: 33-48.

Fosberg, F.R. (1959). Vegetation and flora of Wake Island. *Atoll Res. Bull.* 67. 20 pp.

Fowler, H.W. and Ball, S.C. (1925). Fishes of Hawaii, Johnston Island and Wake Island. *B.P. Bishop Mus. Bull.* 26. 31 pp.

Galtsoff, P.S. (1933). Pearl and Hermes Reef, Hawaii: Hydrological and biological observations. *B.P. Bishop Mus. Bull.* 107. 49 pp.

Glynn, P.W. and Wellington, G.M. (1983). *Corals and Coral Reefs of the Galapagos Islands.* University of California Press, Berkeley/Los Angeles/London.

Gooding, R.M. (1971). Oil pollution on Wake Island from the tanker *R.C. Stoner*. *Spec. Scient. Rept U.S. FWS (Fish.)* 636: 1-10.

Gordon, J.A. and Helfrich, P. (1970). An annotated bibliography of Kaneohe Bay. *Hawaii Inst. Mar. Biol. Tech. Rept* 20. 260 pp.

Gordon, M.S. and Kelly, H.M. (1962). Primary productivity of an Hawaiian coral reef: A critique of flow respirometry in turbulent waters. *Ecology* 43(3): 473-480.

Gosline, W.A. (1965). The inshore fish fauna of Johnston Island, a cental Pacific atoll. *Pac. Sci.* 9: 442-480.

Grace, M.I. (Ed.) (1974). *Marine Atlas of Hawaii.* Bays and Harbors. UNIHI-SEAGRANT-MR-74-01. 239 pp.

Grigg, R.W. (1965). Ecological studies of black coral in Hawaii. *Pac. Sci.* 19: 244-260.

Grigg, R.W. (1975). The effects of sewage effluent on benthic marine ecosystems off Sand Island, Oahu. Abstract in Proc. 13th Pac. Science Congress, Vancouver, Canada.

Grigg, R.W. (1976). Fishery management of precious and stony corals in Hawaii. UNIHI-SEAGRANT-TR-77-03. 48 pp.

Grigg, R.W. (1981a). Coral reef resource management: A five-year research program in the Hawaiian Archipelago. *Proc. 4th Int. Coral Reef Symp., Manila* 1: 243-246.

Grigg, R.W. (1981b). Reef development and community structure of Hawaiian Coral Reefs. (Abstract). *Bull. Mar. Sci.* 31: 809.

Grigg, R.W. (1982). Darwin Point, a threshold for atoll formation. *Coral Reefs* 1: 29-34.

Grigg, R.W. (1983). Community structure, succession and development of coral reefs in Hawaii. *Mar. Ecol. Prog. Series* 11: 1-14.

Grigg, R.W. (1985a). The evolutionary development and paleoecology of coral reefs in Hawaii. *Proc. 5th Int. Coral Reef Cong., Tahiti* 2: 157.

Grigg, R.W. (1985b). Hilo-Hamakua Coast sugar mill ocean discharges: Before and after E.P.A. compliance. UNIHI-SEAGRANT-TR- 85-01. 25 pp.

Grigg, R.W. and Dollar, S.J. (1980). The status of reef studies in the Hawaiian Archipelago. In: Grigg, R.W. and Maragos, J.E. (Eds), *Proc. Symp. Status of Resource Investigations in the Northwestern Hawaiian Islands* UNIHI-SEAGRANT-MR-80-04. Pp. 100-120.

Grigg, R.W. and Dollar, S.J. (1981). Impact of a kaolin clay spill on a coral reef in Hawaii. *Mar. Biol.* 65: 269-276.

Grigg, R.W. and Maragos, J.E. (1974). Recolonization of hermatypic corals on submerged lava flows in Hawaii. *Ecology* 55: 387-395.

Grigg, R.W. and Pfund, R.T. (Eds) (1980). *Proceedings of the Symposium on the Status of Resource Investigations in the Northwestern Hawaiian Islands.* UNIHI-SEAGRANT-MR-80-04. 333 pp.

Grigg, R.W., Polovina, J.J. and Atkinson, M.J. (1984). Model of a coral reef ecosystem. 3. Resource limitation, community regulation fisheries yield and resource management. *Coral Reefs* 3: 23-27.

Grigg, R.W. and Tanoue, K.Y. (Eds) (1984). *Proceedings of the Second Symposium on Resource Investigations in the Northwestern Hawaiian Islands.* UNIHI-SEAGRANT-MR-84-01. Vols 1 and 2. 491 pp. and 353 pp.

Grigg, R.W., Wells, J. and Wallace, C. (1981). *Acropora* in Hawaii. Part I. History of the scientific record, systematics and ecology. *Pac. Sci.* 35: 1-13.

Gunderson, K.R. and Stroup, E.D. (1967). Bacterial pollution of Kaneohe Bay, Oahu. *Univ. Hawaii Water Resources Research Center Tech. Rept* 12.

Harrison, C.S., Naughton, M.B. and Fefer, S.I. ((1984). The status and conservation of seabirds in the Hawaiian Archipelago and Johnston Atoll. In: Croxall, J.P., Evans, P.G.H. and Schreiber, R.W. (Eds), *Status and Conservation of the World's Seabirds.* ICBP Technical Publication 2. Pp. 513-526.

Hirota, J., Taguchi, S., Shuman, R. and Jahn, A. (1980). Distribution of plankton stocks, productivity and potential fishery yield in Hawaiian waters. In: Grigg, R.W. and Pfund, R. (Eds), *Proceedings of the Symposium on the Status of Resource Investigations in the Northwestern Hawaiian Islands.* UNIHI-SEAGRANT-MR-80-04. Pp. 191-203.

Hobson, E.S. (1974). Feeding relationships of teleostean fishes on coral reefs in Kona, Hawaii. *Fish. Bull. U.S.* 72: 915-1031.

Hourigan, T.F. and Reese, E.S. (1987). Mid-ocean isolation and the evolution of Hawaiian reef fishes. *Trends Ecol. Evol.* 2(7): 187-191.

Jackson, E., Koisumi, I., Dalrymple, G., Clague, D., Kirkpatrick, R. and Greene, H. (1980). Introduction and summary of results from DSDP Leg 55, the Hawaiian-Emperor hot-spot experiment. In: Shambach, J. (Ed.), *Init. Rep. of the Deep Sea Drilling Project 55.* U.S. Govt Printing Office, Washington D.C. Pp. 5-31.

Johannes, R.E. (1978). Stony Coral Harvesting on Oahu. Unpub. rept. 27 pp.

Johannes, R.E. (1984). Traditional conservation methods and protected marine areas in Oceania. In: McNeely, J.A. and Miller, K.R. (Eds), *National Parks, Conservation and Development: The Role of Protected Areas in Sustaining Society.* Smithsonian Institution Press, Washington D.C. Pp. 344-347.

Jokiel, P.L. (1987). Ecology, biogeography and evolution of corals in Hawaii. *Trends Ecol. Evol.* 2(7): 179-182.

Jokiel, P.L. and Coles, S.L. (1974). Effects of heated effluent on hermatypic corals at Kahe Point, Oahu. *Pac. Sci.* 28: 1-18.

Kay, E.A. (Ed.) (1972). *A Natural History of the Hawaiian Islands.* Selected readings. Univ. Press of Hawaii, Honolulu. 653 pp.

Kay, E.A. and Palumbi, S.R. (1987). Endemism and evolution in Hawaiian marine invertebrates. *Trends Ecol. Evol.* 2(7): 183-186.

Keating, B.H. (1985). Submersible observations on the flanks of Johnston Island (Central Pacific Ocean). *Proc. 5th Int. Coral Reef Cong., Tahiti* 6: 413-418.

Kimmerer, W.J. and Durbin, W.W. Jr (1975). The potential for additional marine conservation districts on Oahu and Hawaii. Univ. of Hawaii Sea Grant Tech. Rept. TR-76-03. 108 pp.

Kinsey, D.W. (1983). Johnston Atoll - a modern analog of early Holocene reefs? Paper given at 15th Pacific Science Congress, Dunedin, New Zealand.

Kinzie, R.A., III. (1972). A survey of the shallow water biota of Maalaea Bay, Maui, Hawaii. Prep. for Environmental Systems Dept, Westinghouse Electric Corp. 50 pp.

Kohn, A.J. and Helfrich, P. (1957). Primary organic production of a Hawaiian coral reef. *Limn. and Ocean.* 2(3): 241-251.

Krassick, G.J. (1973). Temporal and spatial variations in phytoplankton productivity and related factors in the surface waters of Kaneohe Bay, Oahu, Hawaii. M.S. Thesis, Univ. Hawaii. 91 pp.

Krauss, N.L.H. (1969). Wake Island (W. Pacific) Bibliography. Unpub. ms. 13 pp.

Ladd, H.S., Tracey, J.I. and Grant Cross, M. (1967). Drilling on Midway Atoll, Hawaii. *Science* 156(3778): 1088-1094.

Littler, M.M. (1973a). The population and community structure of Hawaiian fringing reef crustose Corallinaceae (Rhodophyta, Cryptonemiales). *J. Exp. Mar. Biol. Ecol.* 11(2): 103-120.

Littler, M.M. (1973b). The distribution, abundance and communities of deep-water Hawaiian crustose Corallinaceae (Rhodophyta, Cryptonemiales) *Pac. Sci.* 27(3): 281-289.

Lobel, P.S. (1984). Ecological investigation to assess the impact of proposed deep ocean disposal of brine waste off Johnston Atoll. Report to U.S. Army Corps of Engineers, Pacific Ocean Division by Center for Earth and Planetary Physics, Harvard Univ., Cambridge, Mass.

Lobel, P.S. (1985). Oceanographic investigation to assess the impact of proposed deep ocean disposal of brine waste off Johnston Atoll. Report to U.S. Army Corps of Engineers, Pacific Ocean Division by Center for Earth and Planetary Physics, Harvard Univ., Cambridge, Mass.

Losey, G.S. (1976). Kaneohe data evaluation study: reef fishes. In: *Kaneohe Bay Water Resources Data Inventory.* Appendices 1 and 2. Sunn, Low, Tom and Hara, Inc. Prep. for U.S. Army Corps of Engineers, Pacific Ocean Division. Pp. 352-363.

MacDonald, G.A., Davis, D.A. and Cox, D.C. (1960). Geology and groundwater resources of the island of Kauai. *Hawaii Division of Hydrography, Bulletin* 13.

MacCaughey, V. (1918). A survey of the Hawaiian coral reefs. *American Naturalist* 52: 409-438.

Manoa Mapworks (1983). *Kauai Coastline Resource Atlas.* Prep. for U.S. Army Engineer Division, Pacific Ocean. 279 pp.

Manoa Mapworks (1984). *Molokai Coastal Resource Atlas.* Prep. for U.S. Army Engineer Division, Pacific Ocean. 357 pp.

Maragos, J.E. (1972). A study of the ecology of Hawaiian reef corals. Thesis, Univ. of Hawaii, Honolulu, Hawaii.

Maragos, J.E. (1974a). Coral transplantation: A method to create, preserve and manage coral reefs. University of Hawaii Sea Grant Program Advisory Report UNIHI-SEAGRANT-AR-74-03. 30 pp.

Maragos, J.E. (1974b). Reef corals of Fanning Island. *Pac. Sci.* 28(3): 247-256.

Maragos, J.E. (1975). The status of available information on reef coral populations in Kaneohe Bay. Prep. for the Kaneohe Bay Urban Water Resources study. In: *Kaneohe Bay Water Resources Data Evaluation.* Appendices 1 and 2. Sunn, Low, Tom and Hara, Inc. Prep. for U.S. Army Corps of Engineers, Pacific Ocean Division. Pp. 372-422.

Maragos, J.E. (1977). Order Scleractinia, stony corals. In: *Reef and Shore Fauna of Hawaii* Section 2: Protozoa through Ctenophora. *B.P. Bishop Mus. Spec. Pub.* 64(1): 158-241.

Maragos, J.E. (1979a). Preliminary environmental assessment of the construction and operation of a temporary storage facility for foreign nuclear spent fuel at Wake Island. Unpub. ms. 42 pp.

Maragos, J.E. (1979b). Preliminary environmental assessment, Palmyra Atoll. Unpub. ms. 55 pp.

Maragos, J.E. (1986). Coastal resource development and management in the U.S. Pacific Islands: 1. Island-by-island analysis. Office of Technology Assessment, U.S. Congress. Draft.

Maragos, J.E. and Elliott, M.E. (1985). Coastal resource inventories in Hawaii, Samoa and Micronesia. *Proc. 5th Int. Coral Reef Cong., Tahiti* 5: 577-582.

Maragos, J.E., Evans, C. and Holthus, P. (1985). Reef corals in Kaneohe Bay six years before and after termination of sewage discharges (Oahu, Hawaiian Archipelago). *Proc. 5th Int. Coral Reef Cong., Tahiti* 4: 189-194.

Maragos, J.E. and Jokiel, P.L. (1986). Reef corals of Johnston Atoll: One of the world's most isolated reefs. *Coral Reefs* 4: 141-150.

Maragos, J.E., Roach, J., Bowers, R.L., Hemmes, D.E., Self, R.F.L., Macneil, J.D., Ells, K., Omeara, P., Vansant, J., Sato, A., Jones, J.P. and Kam, D.T.O. (1977). Environmental surveys before, during and after offshore marine sand mining operations at Keauhou Bay, Hawaii. *University of Hawaii Sea Grant College Program Working Paper* 28. 65 pp.

McCain, J.C., Coles, S.L. and Peck, J. (1975). The marine biological impact of the Honolulu Power Plant. UNIHI-SEAGRANT-TR-76-01. 50 pp.

Moberly, R.J., Jr. (1963). Coastal geology of Hawaii. Final Report, State Dept. Planning and Economic Development. *Hawaii Institute Geophysics, Rept* 41. 216 pp.

Mogi, H., Planning and Research, Inc. (1977). Makena-La Perouse State Park, Maui, Hawaii. Prep. for State of Hawaii, Div. of State Parks. 59 pp.

Morgan, W. (1972). Deep mantle convection plumes and plate motions. *Am. Assoc. Petrol. Geol. Bull.* 56: 203-213.

Nolan, R.S. and Cheney, D.P. (1981). *West Hawaii Coral Reef Inventory.* Prep. for U.S. Army Corps of Engineers, Pacific Ocean Division. 455 pp.

Parrish, J.D., Callahan, M.W. and Norris, J.E. (1985). Fish trophic relationships that structure reef communities. *Proc. 5th Int. Coral Reef Cong., Tahiti* 4: 73-78.

Pfeffer, P.A. and Tribble, G.W. (1985). Hurricane effects on an aquarium fish fishery in the Hawaiian Islands. *Proc. 5th Int. Coral Reef Cong., Tahiti* 3: 331-336.

Pollock, J.B. (1928). Fringing and fossil reefs of Oahu. *B.P. Bishop Mus. Bull.* 55. 56 pp.

Polovina, J.J. (1984). Model of a coral reef ecosystem. 1. The ECOPATH model and its application to French Frigate Shoals. *Coral Reefs* 3(1): 1-11.

Powers, D.A. (1970). A numerical taxonomic study of Hawaiian Reef Corals. *Pac. Sci.* 24(2): 180-186.

Quan, E.L. (1969). Some aspects of pollution in Kaneohe Bay, Oahu and its effects on selected organisms. M.S. Thesis. Univ. Hawaii.

Ralston, S., Gooding, R.M. and Ludwig, G.M. (1986). An ecological survey and comparison of bottom fish resource assessment (submersible versus handline fishing) at Johnston Atoll. *Fish. Bull.* 84(1): 141-155.

Randall, J.E., Lobel, P.S. and Chave, E.H. (1985). Annotated checklist of the fishes of Johnston Island. *Pac. Sci.* 39: 24-80.

Rastetter, E.B. and Cooke, W.J. (1979). Response of marine fouling communities to sewage abatement in Kaneohe Bay, Oahu, Hawaii. *Mar. Biol.* 53: 271-280.

Reed, S.A., Kay, E.A. and Russo, A.R. (1977). Survey of benthic coral reef ecosystems, fish populations, and micromollusks in the vicinity of the Waianae sewage ocean outfall, Oahu, Hawaii. Summer 1975. *Univ. Hawaii Water Resources Research Center Tech. Rept* 104. 34 pp.

Richmond, T. de A. and Muller-Dubois, D. (1972). Coastline ecosystems on Oahu, Hawaii. *Vegetatio* 25(5-6): 367-400.

Roy, K.J. (1970). Change in bathymetric configuration, Kaneohe Bay, Oahu, 1882-1969. *Univ. Hawaii, Hawaii Inst. Geophysics Rept* 70-15. 226 pp.

Russo, A.R., Dollar, S.J. and Kay, E.A. (1977). Inventory of benthic organisms and plankton at Mokapu, Oahu. *Univ. Hawaii Water Resources Research Center Tech. Rept* 101. 30 pp.

Schlanger, S. and Gillett, G. (1976). A geological perspective of the upland biota of Laysan Atoll (Hawaiian Islands). *Biol. J. Linn. Soc.* 8: 205-216.

Schreiber, R.W. and Kridler, E. (1969). Occurrence of a Hawaiian Monk Seal (*Monachus schauinslandi*) on Johnston Atoll, Pacific Ocean. *J. Mammalogy* 50: 841-842.

Schroeder, R.E. (1985). Recruitment rate patterns of coral reef fishes at Midway lagoon (Northwestern Hawaiian Islands). *Proc. 5th Int. Coral Reef Cong., Tahiti* 5: 379-384.

Scott, G.A.J. and Rotondo, G.M. (1983). A model to explain the differences between Pacific plate island-atoll types. *Coral Reefs* 1: 139-150.

Smith, S.V. (1977). Kaneohe Bay: A preliminary report on the response of a coral reef/estuary ecosystem to relaxation of sewage stress. *Proc. 3rd Int. Coral Reef Symp., Miami* 2: 577-584.

Smith, S.V., Chave, K.E. and Kam, D.T.O. (Eds) (1973). Atlas of Kaneohe Bay: A reef ecosystem under stress. UNIHI-SEAGRANT-TR-72-01. 128 pp.

Smith, S.V., Kimmerer, W.J., Laws, E.A., Brock, R.E. and Walsh, T.W. (1981). Kaneohe Bay Sewage Diversion Experiment: Perspectives on ecosystem responses to nutrient perturbation. *Pac. Sci.* 35: 279-395.

Soegiarto, A. (1972). The role of benthic algae in the carbonate budget of the modern reef complex, Kaneohe Bay. Ph.D. thesis, Univ. Hawaii.

Stanton, F.G. (1985). Temporal patterns of spawning in the demersal brooding blackspot sergeant *Abudefduf sordidus. Proc. 5th Int. Coral Reef Cong., Tahiti* 5: 361-365.

State Department of Health (1978). Non-point source pollution in Hawaii: Assessments and recommendations. *Rept of the Tech. Comm. on Nonpoint Source Pollution Control.* Tech. Rept. No. 2. 117 pp.

State Department of Health (1980a). Water quality management plan for the State of Hawaii: Summary.

State Department of Health (1980b). Water quality management plan for the City and County of Honolulu.

State Department of Health (1980c). Water quality management plan for the County of Maui.

State Department of Health (1980d). Water quality management plan for the County of Kauai.

State Department of Health (1980e). Water quality management plan for the County of Hawaii.

State Department of Health (1987). Public health regulations. Chap. 54, Water quality standards.

State of Hawaii, Div. Fish and Game (1970). Marine preserves survey, Island of Maui. 6 pp.

State of Hawaii, Div. Fish and Game (1972). Natural areas reserve survey, La Perouse - Cape Kinau - Ahihi Bay, Maui. 3 pp.

State of Hawaii, Div. Fish and Game (1976). Marine survey of Honolua, Mokuleia and Napili Bays, Island of Maui. 15 pp.

State of Hawaii, Div. Fish and Game (1977). Marine survey of Molokini Islet, County of Maui. 3 pp.

Stearns, H.T. (1940). Geology and groundwater resources of the islands of Lanai and Kahoolawe, Hawaii. *Hawaii Division of Hydrography, Bull.* 6.

Stearns, H.T. (1966). *Geology of the State of Hawaii.* Pacific Books, Palo Alto, California. 266 pp.

Stearns, H.T. (1967). Geology and groundwater resources of the island of Niihau, Hawaii. *Hawaii Division of Hydrography, Bull.* 2.

Stearns, H.T. (1974). Submerged shorelines and shelves in the Hawaiian Islands and a revision of some of the eustatic emerged shorelines. *Geol. Soc. Am. Bull.* 85: 795-804.

Stearns, H.T. and MacDonald, G.A. (1942). Geology and groundwater resources of the island of Molokai,

Hawaii. *Hawaii Division of Hydrography, Bull.* 11.

Stearns, H.T. and MacDonald, G.A. (1946). Geology and groundwater resources of the island of Hawaii. *Hawaii Division of Hydrography Bull.* 9. 363 pp.

Summers, C.C. (1964). Hawaiian fishponds. *B.P. Bishop Mus. Spec. Pub.* 52: 1-26.

Sunn, Low, Tom and Hara, Inc. (1976). Kaneohe Bay data evaluation study. Prep. for U.S. Army Corps of Engineers, Pacific Ocean Division.

Technical Committee on Water Quality Standards (1977). An ecosystem approach to water quality standards. Prepared for the State of Hawaii, Dept of Health. *208 Technical Report* 1. 214 pp.

U.S. Air Force (1974). Environmental Impact Statement of disposition of Orange Herbicide by incineration, Nov. 1974. U.S. Air Force nvironmental Health Laboratory, Kelly Air Force Base, Texas.

USACE (U.S. Army Corps of Engineers, Pacific Ocean Division) (1984). Final environmental impact statement: Johnston Atoll chemical agent disposal system. U.S. Army Corps of Engineers, Pacific Ocean Division, Fort Shafter, Hawaii.

USACE (U.S. Army Corps of Engineers, Pacific Ocean Division) (1985). Draft environmental impact statement for the designation of a deep ocean disposal site near Johnston Atoll for brine and solid waste. Ft Shafter, Honolulu.

USACE (U.S. Army Corps of Engineers, Pacific Ocean Division) (1987a). Defense Restoration Account Inventory Project Report: Howland Island National Wildlife Refuge, Howland Island. Project No. H09HI000400, Ft Shafter, Honolulu.

USACE (U.S. Army Corps of Engineers, Pacific Ocean Division) (1987b). Defense Restoration Account Inventory Project Report: Baker Island National Wildlife Refuge, Baker Island. Project No. H09HI000700, Ft Shafter, Honolulu.

USACE (U.S. Army Corps of Engineers, Pacific Ocean Division) (1987c). Johnston Atoll Chemical Agent Disposal System (JACADS): Draft supplemental environmental impact statement. Ft Shafter, Honolulu. 82 pp. + 10 appendices.

Vaughan, T.W. (1907). Recent madreporaria of the Hawaiian Islands and Laysan. *U.S. Nat. Mus. Bull.* 59. 427 pp.

Vaughan, T.W. (1910). Summary of the results obtained from a study of the recent madreporarea of the Hawaiian Islands and Laysan. *Proc. 7th Int. Zool. Cong.* BPBM QL Pan. 227.

Wass, R.C. (1976). Removal and repopulation of fishes on an isolated patch coral reef in Kaneohe Bay, Oahu, Hawaii. M.S. Thesis, Univ. Hawaii.

Wilson, J. (1963). A possible origin of the Hawaiian islands. *Can. J. Phy.* 41: 863-870.

Wimberly and Cook, Architects Ltd. (1961). Preliminary planning report for a resort development at Palmyra. Unpub. ms.

Woodward, P.W. (1972). The natural history of Kure Atoll, Northwestern Hawaiian Islands. *Atoll Res. Bull.* 164. 318 pp.

HAWAII ISLAND (SEVERAL MARINE LIFE CONSERVATION DISTRICTS)

Geographical Location South-east end of the archipelago, at approx. 19°35'N, 155°30'W.

Area, Depth, Altitude Hawaii is the largest island in the chain, with an area of 1 041 446 ha. The highest elevations are Mauna Kea (4205 m) and Mauna Loa (4168 m), which are the highest peaks in the State.

Physical Features Hawaii (the "Big Island") is the youngest island in the Hawaiian Archipelago. It is roughly triangular in shape, with five volcanoes, two of which - Mauna Loa and Kilauea - are still active, the latter being the most active volcano in the world. Two of the others, Mauna Kea, and Kohala, are virtually extinct and the third, Hualalai, last erupted in 1859. Except for the windward slope of Kohala, the island as a whole is little eroded. The only perennial streams are on the north-eastern slopes of Mauna Kea and Kohala; the latter has been deeply eroded into a number of heavily vegetated valleys such as Waipio. The high permeability of the fresh lavas forming the surface of Kilauea, Mauna Loa and Hualalai inhibits the development of permanent streams (Stearns, 1966). Large areas on these mountains are relatively bare of vegetation, and the south-western side of Kilauea is a desert (Kau Desert).

Rainfall varies greatly in different localities and is heaviest on the windward side of the north-east coast where it is about 350 cm/yr year. In the Kau Desert rainfall is generally less than 100 cm/yr. Only four significant stream mouth estuaries are found on the island: Hilo, including Wailoa and Wailuku Rivers, Waipio Stream, Waimanu Stream, and Pololu Stream (Grace, 1974). Much of the shoreline is cliffed or bordered by jagged lava fields. Several black sand beaches exist on the windward coast near Kalapana whereas white sand beaches are restricted to the Kona and western coasts. The geology and groundwater resources are described by Stearns and MacDonald (1946).

Reef Structure and Corals Hawaii has virtually no fringing reefs, but some well developed subtidal reefs occur off the Kona Coast and western coasts (Nolan and Cheney, 1981). These are characterized by a weak spur-and-groove system. The largest reef extends about 1 km from shore at Kawaihae Bay. Most of the reefs off Hawaii are Holocene in age, no more than 6000 to 9000 years old (Grigg, 1983). In areas where the best developed reefs are found, coral cover averages about 67%. Dominant species are *Porites lobata* and *P. compressa*, and to a much lesser extent *Pavona varians*, *Montipora verrucosa* and *Pocillopora meandrina*. Some of the best conditions for diving (calm, clear water and well developed reefs) in the State exist off the Kona Coast. Reef zonation in this area has been studied by Dollar (1975). Several recent underwater lava flows support reef assemblages at various stages of succession (Grigg and Maragos, 1974).

Noteworthy Fauna and Flora The marine fauna of the northern Kona Coast is described by Brock and Brock (1974).

Scientific Importance and Research Because much of the island is of recent origin, processes associated with the development of reef morphologies and community structure can be easily studied. Since weather conditions on the Kona Coast are calm much of the year this area is an ideal site for long-term reef studies. Hobson (1974) discusses feeding relationships of teleost fishes in this region.

Economic Value and Social Benefits The main industries on Hawaii are sugarcane, cattle ranching, tourism and diversified agriculture. The island has the only commercial coffee crop in the United States and the largest orchid production in the world; cultivation of marijuana, although illegal, is probably the single most economically important agricultural crop. An outstanding deep-sea fishing area is found off the Kona Coast and is a popular attraction for tourists. SCUBA diving and snorkelling is also excellent off the Kona Coast, where extremely clear water exists nearly year-round. Subsistence fishing for reef and offshore species is important to the local population. A deep-sea handline fishery for tuna (kashibi) occurs seasonally off Hilo and Kona.

Disturbance or Deficiencies The most common source of disturbance to reefs is abrasion, scour and breakage from high waves (Grigg and Maragos, 1974; Dollar, 1982). In places along the Kona Coast, submerged freshwater springs inhibit the growth of coral. Maragos *et al.* (1977) studied the impact of offshore sand mining at Keauhou Bay on the Kona coast. Siltation from stream discharge and sugar mills is a problem in some areas along the Hamakua Coast, but there are few pollution problems. The discharge of bagasse from the Hamakua Sugar Mills ceased altogether in 1975 and sediment discharges have been greatly reduced. Impacts at the present time (1984) have been judged to be minimal and highly localized (Grigg, 1985b). Overfishing of reef resources is not as serious a problem on Hawaii as it is on the other major islands, although this could change rapidly.

Legal Protection Several areas around the island of Hawaii are designated as Marine Life Conservation Districts, where no living or geological specimens may be removed. They include the submerged lands and overlying waters at the following locations: Kealakekua Bay, where permitted activities are restricted to fishing Opelu *Decapterus pinnulatus*, Akule *Triachurops crumenophthalmus* and crustaceans by any legal method except traps, and other finfish only by hook-and-line or thrownet; Lapakahi State Historical Park, where Opelu fishing is permitted with nets, and other finfish or crustaceans can be taken by hook-and-line or thrownet; and Wailea Bay, where fishing with pole or hook-and-line for finfish is permitted. Other areas designated as Marine Fisheries Management Areas include Hilo Bay (plus part of Wailoa and Wailuku Rivers), Puako Bay and Reef, and Kailua Bay. Various fishing restrictions apply to each area.

The waters off Kawaihae Bay at the north-west end of the island, and around the north-east and south-east coasts are all designated Class "A" by the Department of Health. The waters from Ka Lae (South Cape) up to the west coast (Kona Coast) are classified "AA". Adjacent to boat piers and wharves, water is class "B" (State Department of Health, 1980e and 1987). All historic era (dated) submerged lava flows are also classified "AA" (the highest level of protection).

Management Regulations for the Marine Life Conservation Districts cited above are listed under chapters 29, 33 and 35 respectively, of Subtitle 4, part I, Title 13 of the Division of Aquatic Resources, Department of Land and Natural Resources, State of Hawaii. Regulations for the Marine Fisheries

Management Areas in Hawaii Island are listed under chapters 47, 52 and 54 respectively, of subtitle 4 part II, Title 13 of the same Division. Water quality standards and regulations are found in Chapter 54, Hawaii Revised Statutes.

Recommendations An assessment of all nearshore fishery pressures (commercial, subsistence and recreation) with the ultimate goal of controlling future overfishing is required. A comprehensive tourism development plan for Kona - North Kohala to control future levels of impact of tourism on reefs should be produced. Early recommendations for marine conservation districts are given in Kimmerer and Durbin (1975).

KAHOOLAWE ISLAND

Geographical Location 11 km west of the south-western extremity of Maui across the Alalakeiki Channel; 20°32'N, 156°38'W.

Area, Depth, Altitude The island is 11 656 ha in area and projects 450 m above sea level. Reefs extend to about 30 m depth with isolated corals occurring somewhat deeper.

Land Tenure Federal government, under indefinite "lease" or jurisdiction from the State of Hawaii.

Physical Features Kahoolawe Island was formed from a single volcano and is now largely barren, dry and lacking in significant vegetation. The coastline has steep pali cliffs in the east while the western end is characterized by numerous pocket beaches and low rocky cliffs. The island is dry, with an average rainfall of about 63 cm a year. The surface is deeply weathered by wind and rain although there are no perrenial streams. Huge clouds of red dust are often observed blowing seaward during strong winds (Grace, 1974). The geology and groundwater resources are described by Stearns (1940).

Reef Structure and Corals In general, reefs are poorly developed due to the steep insular shelves and periodic stress from heavy siltation during storms. The best reefs are found off the west end at Kuia Shoal; elsewhere they are patchy and often only thin veneers on underlying basalt foundations.

Scientific Importance and Research Kahoolawe, having been a bomb site for the Navy, presents an opportunity to assess impact caused by this activity, although this has not yet been done.

Economic Value and Social Benefits The island once supported a small population, but is now uninhabited and is used as an aerial bombing target by the U.S. military. Access to the island is restricted although fishing around the island up to the shoreline is periodically allowed on weekends or other short periods. Fishing and diving are good at Kuia Shoal.

Disturbance or Deficiencies Introduced goats, sheep and cattle have destroyed virtually all of the native vegetation and have stripped the upper soil levels, preventing regeneration. Bombing has almost certainly caused damage but the extent is unknown.

Legal Protection   None.

Management   Restricted access but otherwise none. The island is currently under naval jurisdiction. Danger zones consisting of aerial bombing targets and naval shore bombardment areas extend across the island and seaward about 2 miles from all sides of the island.

Recommendations   The State of Hawaii should reclaim jurisdiction over Kahoolawe. Bombing should be prohibited and a programme to restore and manage both the terrestrial and marine resources would be desirable and beneficial to the long-term interests of the state. A study to assess the damage caused by bombing should be carried out.

KAUAI ISLAND (SEVERAL FISHERIES MANAGEMENT AREAS)

Geographical Location   113 km north-west of Oahu; 22°05'N, 159°30'W.

Area, Depth, Altitude   Kauai ranks fourth in size of the Hawaiian Islands, with an area of 143 710 ha. The highest peak, Kawaikini, is 1600 m, followed by Waialeale, 1540 m high.

Land Tenure   Over 50% of the land belongs to the private sector, 43% to the state, and the rest to the federal government.

Physical Features   Kauai is geologically the most complex and oldest of the major Hawaiian islands. It is roughly circular in shape and consists of a single shield volcano, the flanks of which have been greatly reduced by erosion, mainly by streams, except along the dry west coast. The mountains on the west and north sides descend in steep jagged ridges, while the gentle east and south slopes are cut by shallow gulches. A narrow coastal plain borders much of the shoreline to the base of the dissected upland.

Kauai is the wettest island in the Hawaiian Archipelago. At Waialeale near the centre of the island, the world's highest mean annual rainfall (over 1140 cm/yr) has been recorded. Rainfall decreases towards the coasts to a minimum of about 50 cm/yr on the leeward coast. The winter months (December to March) have the strongest winds, sometimes accompanied by heavy rains.

The major streams of Kauai have their headwaters in the massive rocks of the large central caldera of the volcano. Their stability is enhanced by low slopes, deep soil, high rainfall, and tall vegetation, and on the east side of the island by springs from the dike complex within the volcano. Kauai has 11 distinctive bays formed by the drowning of river valleys and 15 streams feed into significant estuaries (Grace, 1974). The geology and ground-water resources are described by MacDonald *et al.* (1960). Kauai has the largest and best beaches in the Hawaiian islands, although some of them at Kapaa, Hanapepe and Kekaha have undergone serious

erosion. In many places the sand is cemented into beach rock. AECOS, Inc. (1982) provide an inventory of the coastal resources of Kauai, and Manoa Mapworks (1983) provide a coastline resource atlas.

Reef Structure and Corals   Most of Kauai's coastline is bordered by fringing reef. In many areas, the reef flats are very wide and in some places extend more than 1 km from shore. Off shore, at depths between 5 and 30 m, significant reefs are found on virtually all sides of the island. An ancient, drowned "barrier reef" exists off Polihale Beach on the western shore at a depth of about 20 m and there are numerous drowned sea level terraces at 20-60 m around the island. The reefs are generally holocene veneers dominated by *Porites lobata* (Grigg, 1983); other common species are *Pocillopora meandrina*, *Montipora patula*, *Leptastrea purpurea* and *M. verrucosa*. Small apron reefs dominated by crustose coralline algae jut out from the shoreline at Milolii and Nualolo-Kai along Kauai's sea cliff coastline.

Noteworthy Fauna and Flora   The lack of heavy fishing pressure combined with excellent habitat has resulted in a diverse and abundant fish fauna. At depths of 30-50 m black corals are abundant along drop-offs and undercut areas.

Scientific Importance and Research   Little research has been done on the reefs off Kauai and abundant opportunities exist for the study of reef fish and coral populations. Kohn and Helfrich (1957) studied reef productivity at Kapaa on the east coast. The apron reefs at Milolii and Nauolo-Kai, and the composition of algae and corals (which are quite abundant on the reef flats) are unique in Hawaii.

Economic Value and Social Benefits   Kauai's major industry is the growing and processing of sugarcane, and to a lesser extent pineapple cultivation and diversified agriculture. Tourism is becoming an increasingly significant activity, particularly along the south and east shores. The most popular aquatic activities are swimming, surfing, snorkelling and SCUBA diving. Some reefs are productive for commercial and sport fishing. Excellent tuna grounds exist off black rock (Haupu) on the southern coast, and Kanai is one of the best fishing grounds in the state. Black coral is a commercially valuable resource, but stocks are somewhat depleted by past harvesting.

Disturbance or Deficiencies   The many streams which discharge into nearshore waters lead to a high degree of freshwater input and siltation, which exert a much greater influence than elsewhere in the Hawaiian islands. Agricultural wastewater and sewage discharge is reducing water quality in some places such as Hanapepe Bay. Along the south-west coast, ground-water seepage has caused nutrient enrichment and stimulated algal growth in some areas, locally inhibiting coral growth.

Legal Protection   Both Waimea Bay and Waimea Recreational Pier on the south coast, and Hanamaulu Bay and Ahukini Recreational Pier on the east coast, are regulated as Marine Fisheries Management Areas. Within 46 m of Waimea Recreational Pier it is prohibited to use spear, trap, thrownet, draw, drag, seine or any other type of net except a crab net. At Hanamaulu Bay, within 46 m of the Ahukini Recreational Pier, the use of a thrownet or spear is prohibited, while within the

Ahukini Pier portion of the bay, draw, drag, seine or other nets except for thrownets and crab nets are prohibited.

The apron reefs at Miloili and Nualolo-Kai are State Parks. Several spots on Kauai are Natural Wildlife Refuges which provide habitat for species of waterbirds or seabirds, such as the wetlands along lower reaches of Huleia River, Hanalie Valley and Kilauea Point.

The waters extending clockwise from Hikimoe Valley (between Polihale and Kaawiki Ridges) north-east to the east end of Hanalei Bay, are classified as "AA" by the Department of Health. The remaining waters are class "A" except near boat-docking facilities where they are "B" (State Department of Health, 1980d).

Management The Marine Fisheries Management Areas and the Natural Wildlife Refuges are administered by the Division of Aquatic Resources, Department of Land and Natural Resources (DLNR), 1151 Punchbowl St., Honolulu, HI 96813. Regulations pertaining to Waimea Bay and Hanamaulu Bay Marine Fisheries Management Areas are listed under Chapters 49 and 50, of Part II, Subtitle 4, Title 13 of the DLNR.

Recommendations Draft regulations regarding the monitoring of black coral harvesting should be implemented and enforced by the state.

LANAI ISLAND (TWO MARINE LIFE CONSERVATION DISTRICTS)

Geographical Location Approximately 15 km south of Molokai and west of Maui; 20°50'N, 156°57'W.

Area, Depth, Altitude Lanai ranks sixth in size of the Hawaiian Islands with an area of 36 520 ha. The highest point is Lanaihale Peak at 1020 m.

Land Tenure 98% of Lanai is owned by the Dole Company for pineapple agriculture; a small percentage of the land belongs to other private owners and 0.1% are federal lands.

Physical Features Lanai was formed from a single volcano. The slopes on the easterly side are steep and cut by gulches. Those on the westerly side are more gentle, terminating in a rolling plain between 300 and 600 m known as Palawai Basin, located in the south-central part of the island and covered with pineapple fields. The island has a generally barren appearance and rainfall is low; maximum precipitation averages 89 cm a year on the highlands and only 25 cm a year on the coasts (Grace, 1974). No perennial streams reach the coast and there are no estuaries; although there are some inland valleys there are no significant coastal bays. The north and east coasts have a narrow coastal plain bounded by sand beaches with offshore fringing reefs. The rest of the coast consists of volcanic cliffs. The geology and groundwater resources are described by Stearns (1940).

Reef Structure and Corals The best developed reefs are off the north and north-east coasts. Both are partially sheltered from winter swell by the islands of Molokai and Maui. Reef thickets dominated by *Porites compressa* are found at depths of 5-20 m and in many places approach 100% cover. The south and west sides of the island are more sparsely populated by corals. Near Manele Bay, a reef area known as "Cathedrals" offers spectacular diving by virtue of its deeply eroded caves and steep drop-offs.

Noteworthy Fauna and Flora The crest of Lanaihale Peak contains a residual forest with many endemic and indigenous plant species.

Scientific Importance and Research Reef and fish populations on Lanai are relatively undisturbed providing opportunities for baseline research. An emerged fossil coral facies has been described from an elevation of 300 m. The University of Hawaii Sea Grant Program is conducting a coral reef inventory along the east and south coasts.

Economic Value and Social Benefits Approximately one-fifth of Lanai Island is devoted to pineapple agriculture, making it the largest single pineapple plantation in the world. However the owners of the island are beginning to phase out agriculture and develop plans for resorts. Tourism at present is virtually non-existent but diving and fishing charters from Maui are common, as some of the best diving in the Hawaiian Islands is found here. Large beds of black coral *Antipathes dichotoma* and *A. grandis* exist at depths of 40-110 m off the north coast, used by small fisheries based out of Lahaina, Maui.

Disturbance or Deficiencies Lanai, being sparsely populated and relatively remote, receives little diving and fishing pressure. Most damage to corals is caused by sediment run-off or breakage during Kona storms. On the east coast, waters around Maunalei Gulch south to Kamaiki Point are often affected by erosion from the goat-denuded flanks of the island.

Legal Protection The portion of submerged lands and overlying waters of Hulopoe Bay and the offshore waters of Palawai, covering an area of 68 ha (subzone A), and Manele Bay and offshore waters of Kamao, covering an area of 56 ha, are regulated as a Marine Life Conservation District. Fishing or removal of any living or geological specimen is prohibited here except catching fish or crustaceans by hook-and-line from the shoreline within subzone A, and fishing by any legal method except spear, trap and net (other than thrownet) within subzone B. Special permits may be procured for scientific or other purposes.

Manele Boat Harbor has been designated as a Marine Fisheries Management Area where no draw, drag, seine or other type of net can be used except for thrownets of legal mesh, and crab and hand nets for shrimp. Commercial fishermen with a bait license may use nets to take baitfish only when an open season has been declared.

Most of the waters around Lanai have been designated class "AA" by the State Department of Health, except for the east coast from about the mouth of Maunalei Gulch southward to Kamaiki Point which is classified "A". Waters near wharf areas at Kaumalapau and Manele Bay are designated class "B" (State Department of Health, 1980a).

Management Manele-Hulopoe Marine Life Conservation District and Manele Harbor Marine Fisheries Management Areas are Aquatic Resources listed under Chapter 30 and 53 respectively, of Subtitle 4, Title 13 of the Division of Aquatic Resources, Department of Land and Natural Resources, 1151 Punchbowl St, Honolulu, HI 96813.

Recommendations Periodic monitoring of Marine Life Conservation Districts and Fisheries Management Areas is required to determine long term natural fluctuations in coral and reef fish populations as well as to assess the effects of management policies. A coastal resources inventory and management plan should be conducted in conjunction with long term resort development plans for the island.

MAUI ISLAND (SEVERAL PROTECTED MARINE AREAS)

Geographical Location 46 km north-west of Hawaii; 20°48'N, 156°20'W.

Area, Depth, Altitude Maui is the second largest Hawaiian island, with an area of 188 554 ha; the highest peak is Haleakala, at 3056 m.

Land Tenure Most of the land is owned by the state or by large private land owners; less than 6% belongs to the federal government.

Physical Features Maui is formed from two distinct volcanic masses joined by a low, flat isthmus. The easternmost and highest of the two peaks is Haleakala volcano, last active in 1790 although this activity was post-erosional in nature. The western part of the island consists of a single volcanic shield which has been deeply eroded into valleys and gulches. At the center stands Puu Kukui, 1760 m in elevation. A coastal plain has formed along most of the west and south-west coasts.

Reefs are sparsely developed along much of Maui's north coast. Off Kahului, near the isthmus, there are two shallow fringing reefs, on either side of Kahului Bay, known as Spartan and Waihee Reefs. At the head of the bay is a dredged harbour protected by breakwaters, which is the most important deepwater and commercial port in Maui. Fringing reefs are scattered along the rest of the eastern, southern and western shores.

Precipitation is heavier on the eastern and windward side of the island, and on East Maui average rainfall is about 760 cm a year near the summit. Perennial streams are restricted to the northern slopes. On West Maui maximum rainfall at the summit is 1016 cm/yr, decreasing to about 100 cm/yr on the windward coast and about 38 cm/yr on the leeward coast. Perennial streams drain radially from the summit area of West Maui. The waters around Maui do not receive a significant input of pollutants, and mixing conditions are good.

Honolua Bay is described by Environmental Consultants, Inc. (1974a) and State of Hawaii, Div. Fish and Game (1976). In the latter, Mokuleia and Napili Bays are also

described. The Makena-Ahihi region is described by Environmental Consultants, Inc. (1977) and Makena-La Perouse State Park by Mogi *et al.* (1977). The La Perouse-Cape Kinau-Ahihi Bay area is described in State of Hawaii, Div. Fish and Game (1972). A marine preserves survey is reported on in State of Hawaii, Div. Fish and Game (1970).

Reef Structure and Corals The major foundations on which Maui reefs have formed are about 0.8 to 1.3 million years old (Schlanger and Gillett, 1976). However, because of eustatic changes in sea level, most nearshore reefs in Hawaii are Holocene in age, ranging between 6000 and 9000 years old (Grigg, 1983) and are generally not well developed. The northern coasts are characterized by patchy and poorly formed reefs which are no more than thin veneers on outcrops of pleistocene limestone or volcanics of various origin. The best reefs are found on the southern coast of West Maui between Maalaea and Lahaina, where there are moderately developed fringing reefs, and at depths of 10-20 m rich thickets dominated by *Porites compressa* are common. Coral cover averages 49% (Grigg, 1983). Other common species on Maui are *Montipora verrucosa*, *Porites lobata*, *M. patula* and *Pocillopora meandrina*. AECOS, Inc. (1979b and 1981b) provide an inventory of the reefs of Maui. Shallow-water biota of Maalaea Bay is described by Kinzie (1972). A new coral species and genus record for Hawaii, *Gardineroseris planulata*, was reported from La Perouse Bay in 1986.

Noteworthy Fauna and Flora Some bays, such as Maalaea Bay on the southern part of the isthmus and Ahihi Bay on the south-western coast of East Maui, have very rich and diverse invertebrate faunas and algal floras. Off Maalaea, about 59 species of algae and 20 species of corals have been recorded, including the normally deepwater form *Cycloseris vaughani*. A rich fauna of molluscs, sponges, bryozoans, and sea urchins can be found. The gastropods *Pusinus novaehollandiae*, *Cypraea gaskoini*, *Conus quercinus*, *C. leopardus*, *C. publicarius* and the bivalve *Pinna semicostata* are abundant. The small mussel, *Brachiodontes crebristriatus* is conspicuous on rock outcrops. Just east of Maalaea two inland ponds (Kealia and Kanahe) make up the main habitat in Hawaii for the endangered Hawaii Stilt *Himatopus mexicanus knudseni*. From late November to early May the Humpback Whale *Megaptera noveangliae* breeds mainly in the Auau Channel between Maui and its neighbouring island, Lanai. The Auau Channel is also the area in which the best-known and richest black coral, *Antipathes dichotoma* and *A. grandis*, beds in Hawaii are found.

Scientific Importance and Research A number of drowned limestone notched terraces exist at various depths in the Auau Channel (Stearns, 1974) which have not been adequately sampled or dated and therefore represent an important scientific opportunity for future research on sea level history. In addition to the black corals in the Auau Channel, large beds of the Pen Shell *Pinna semicostata* are present and have never been studied. Opportunities for research on whales also exist in this area.

Economic Value and Social Benefits Maui has the fastest growing population in the state and is second to Oahu in the number of visitors received each year. The most desirable beaches are on the southern shores. A variety of water sports are practised, mainly swimming, snorkelling, diving, surfing, windsurfing and sailing. Sport fishing is popular on charter boats out of Lahaina, as are glass bottom boat tours, sunset dinner tours and whale watching excursions.

Disturbance or Deficiencies Perhaps the main source of pollution to the reefs off Maui is periodic silt-laden run-off from streams, particularly during Kona storms. Several small sewer outfalls off Mala Wharf in Lahaina and adjacent to Kahului Harbor, discharge untreated sewage effluents directly into the sea. Although ammonium and phosphate levels and coliform bacteria counts are sometimes high, the impact is small in both regions. Maui Electric Company discharges heated water into the sea east of Kahului Harbor. The increasing resident population and number of visitors is causing competition for limited resources. Dunlap (1987) reports dying reefs off Lahaina, Olowalu, in Maalaea Bay and off Kihei.

Legal Protection The submerged lands and overlying waters of Honolua and Mokuleia Bays are designated as a Marine Life Conservation District where no fishing or removal of any living or geological specimen is allowed without special permits. Within Kahului Harbor three small areas, two on the eastern side of the harbour, and a third on the western side, are designated Marine Fisheries Management Areas, where no thrownet, draw, drag, seine or any other type of net except for crab nets and hand nets for shrimp are permitted.

Cape Kinau and offshore waters, together with portions of Ahihi and La Perouse bays and a portion of their hinterlands have been designated as a State Natural Area Reserve where only non-consumptive human activities are permitted. Water quality designations are included in State Department of Health (1980c).

Management Regulations pertaining to the Honolua-Mokuleia Bay Marine Life Conservation District are listed under Chapter 32, Part I, Subtitle 4, Title 13 of the Division of Aquatic Resources, Department of Land and Natural Resources (DLNR), 1151 Punchbowl St., Honolulu, HI 96813. Regulations pertaining to Kahului Harbor Marine Fisheries Management Areas are listed under Chapter 51, Part II, Subtitle 4, Title 13 of the DLNR.

Recommendations Regulations to manage space for multiple uses are needed off Kaanapali, where conflicts often arise between bathers, surfers, jetskiers, divers and sailors. There is also a need for better enforcement of existing fish and game regulations pertaining to fishing and the harvest of black corals.

MOLOKAI ISLAND

Geographical Location Between Oahu and Maui; 21°08'N, 157°00'W.

Area, Depth, Altitude Molokai is the fifth largest Hawaiian island, with an area of 67 082 ha (61 km x 12 km). The highest peak is Kamakou at 1515 m. The depth limit of most reefs around the island is 40 m.

Land Tenure About two-thirds of the island is privately owned, 31% is owned by the state and only 0.5% are federal lands. All waters within three miles (4.8 km) of the shore are under state jurisdiction.

Physical Features Molokai is oriented east-west and is formed from a deeply eroded volcanic doublet. East Molokai is an extinct volcano, the top of which is Kamakou. Mountain slopes are deeply eroded and cut by numerous gorges, which terminate in a narrow strip of rolling land near the coast. The gentle west slope is cut by smaller gulches and contains several small, extinct craters. West Molokai is a bare tableland about 400 m in elevation and is cut by numerous small ravines. This region supports pineapple plantations, ranches and small farms.

Rainfall is very heavy on the north-east side of the island, averaging 508 cm/yr at the summit. In contrast, West Molokai is quite dry and averages only 76 cm/yr. Perennial streams are restricted to the central and eastern parts of the island. Many of the valleys on the north coast are deeply eroded and provide some of the most spectacular scenery in Hawaii. There are no perennial streams on West Molokai (Grace, 1974). Manoa Mapworks (1984) provide a coastal resource atlas. The geology and ground water resources are described by Stearns and MacDonald (1942).

Reef Structure and Corals A well-developed fringing reef exists along the entire south coast of the island, over 1 km in width in many places. West of Pukoo, along the inner portion of the reef flat, there are a number of old Hawaiian fishponds built between 1500-1800. Along the west and north coasts of Molokai reefs are poorly developed, either because of frequent disturbance from large north-westerly ocean swells, or perhaps from slumping of volcanic blocks due to earthquakes.

Noteworthy Fauna and Flora Deepwater macroalgae on Penguin Bank, a broad carbonate platform extending from the island, were described by Agegian and Abbott (1985).

Economic Value and Social Benefits The principal products of Molokai are pineapples and cattle. The island is still a quiet, rural place, with population centres focused around plantations, ranches, fishing villages and small tourist resorts. The tourist industry is now increasing and construction of new hotels is in progress. The reefs are used primarily for subsistence fishing. No large commercial fishing occurs off Molokai. Recent efforts to raise penaeid shrimp in pond culture have produced promising results.

Disturbance or Deficiencies In the late 1960s a large aggregation of Crown-of-thorns Starfish *Acanthaster planci* was reported off the south-east side of Molokai (Branham *et al.*, 1971). Efforts by the state to eradicate the starfish were unsuccessful, but short-term damage to the reefs was not considered significant. No follow-up surveys have been conducted since 1975. Many coral reefs surrounding Molokai have been exposed to a variety of other stresses, including heavy siltation and abrasion from high waves, and are in various stages of recovery. Off the south-west coast siltation caused by heavy run-off from barren lands is a source of periodic stress to the reefs, especially the inner third or half of the reef flats. Mangroves, introduced to Hawaii at the turn of the century, have formed thick coastal forests along the south-west coast west of Kauna Kakai.

Legal Protection   No Marine Life Conservation Districts are located on Molokai.   The waters around the island have been designated as "AA" quality by the Department of Health, from Kolo Wharf on the south shore clockwise, including the west and north shores up to Lamaloa Head on the north-west side of Halawa Bay. This bay, and waters around to Kolo Wharf, including the east and south shores, are classified "A".   Waters of "B" quality are limited to boat-docking facilities (State Department of Health, 1980a).

Management   Jurisdiction and authority for managing the marine resources of Molokai belong to the Division of Aquatic Resources, Department of Land and Natural Resources, 1151 Punchbowl St., Honolulu, HI 96813.   All laws pertaining to the management of state marine resources apply.

Recommendations   Monitoring of reefs subject to significant and chronic long-term stress should be carried out, particularly those affected by the 1970 *Acanthaster* outbreak.

MOLOKINI ISLET MARINE LIFE CONSERVATION DISTRICT

Geographical Location   Alalakeiki Channel, about 4.5 km off the south-west coast of Maui; 20°37'N, 156°35'W.

Area, Depth, Altitude   The total area of Molokini is 1.5 ha; maximum elevation is 49 m.   Lagoon depth is 30 m and the depth of the surrounding water is about 150 m.

Land Tenure   State lands.

Physical Features   Molokini is a cone formed by an undersea explosion following an eruption on the south-west rift of Haleakala (East Maui).   The crescent-shaped island is the remnant of the southern rim of a volcanic cone which is now heavily eroded.   The northern portion of the cone and much of the crater floor is submerged, forming a large lagoon on the north side of the islet. Within the lagoon the crater floor slopes gently to a depth of 30 m. The bottom of the lagoon consists of patch reefs, coarse sand, volcanic rock and boulders. Outside the lagoon the bottom drops sharply to about 150 m, the depth of the surrounding channel.

Alalakeiki Channel is exposed to strong tradewinds and seasonal Kona storms.   Currents are occasionally strong off the exposed sides of the island, but the cove enclosed by the north face of the islet is well protected.   Waters within the lagoon are generally quite clear (AECOS, 1979b). A survey of the marine environment is reported on in State of Hawaii, Div. Fish and Game (1977).

Reef Structure and Corals   In the lagoon corals are patchy but more than eleven species have been recorded. Dominant species include *Pocillopora meandrina* at depths less than 5 m and *Porites lobata* between 5 and 10 m.   Within patches, coral cover of up to 70% is common.   Several rare corals which are present include *Porites brighami* and *P. pukoensis*, which may be an unusual form of *Porites lobata*. *Pocillopora eydouxi* occurs in deeper water, and the reef coral *Pavona maldivensis*, rare in Hawaii, is locally abundant inside Molikini Crater (Maragos, 1977).

Noteworthy Fauna and Flora   The lagoon is rich in invertebrates such as sea urchins, cucumbers and asteroids.   More than 89 species of fish have been recorded and the White-tip Shark *Triaenodon obesus* is common in the cove. The island is important for seabirds.

Scientific Importance and Research   Because Molokini is protected from reef fishing and coral and shell collecting, it represents an undisturbed natural ecosystem ideal for a variety of marine biological studies.

Economic Value and Social Benefits   The cove is an attractive site for diving charters from Maui.

Disturbance or Deficiencies   *Acanthaster planci* is present in the lagoon but damage has not been reported. Anchor damage in the lagoon has become considerable. There are also bomb and ammunition craters within Molokini Crater, with evident damage to the corals. These are evidence of errant or cavalier bombing practice by the users of the nearby Kahoolawe bombing range.

Legal Protection   The submerged lands and overlying waters surrounding Molokini Islet to a depth of 55 m are a Marine Life Conservation District. Fishing or removing any living or geological specimens are prohibited, except with special permits for scientific or other purposes. Fishing of finfish is permitted by trolling with artificial lures.   Molokini Islet is a State Seabird Sanctuary with strict limitations on human activities.   Only non-consumptive activities such as snorkelling, observation and photography are permitted in the cove.

Management   Regulations pertaining to the Molokini Shoal Marine Life Conservation District are listed under Chapter 31, Part I, Subtitle 4, Title 13 of the Division of Aquatic Resources, Department of Land and Natural Resources, 1151 Punchbowl St., Honolulu, HI 96813.

The U.S. Army Corps of Engineers in 1986 investigated the extent to which unexploded bombs were present and how they could be removed.   They concluded that there were only a few bombs left and that detonating them in place, the only safe disposal method and the practice followed by the Navy, would be too damaging to marine life. Moving the bombs is dangerous but dive clubs have done this for several of them as they objected to the Navy detonating in place those brought to their attention.

Recommendations   Continued use as a Marine Life Conservation District.   The Navy should be formally approached to take measures to prevent any further bomb damage to the area.

NIIHAU ISLAND

Geographical Location   The westernmost of the eight major Hawaiian Islands, 27 km north-west of Kauai; 21°55'N, 160°10'W.

Area, Depth, Altitude   Niihau is the seventh largest Hawaiian Island, with an area of 18 650 ha. The highest elevation, Paniau, is 390 m.

Land Tenure   Since 1864 the island has been privately owned by the Robinson family, which operates it as a cattle ranch.

Physical Features   Niihau is the remnant of an old shield volcano. A high tableland exists near the middle of the island where the maximum elevation is 390 m (Paniau). The northerly and easterly sides of the tableland are precipitous, varying between 180 and 300 m in height, while the southerly and westerly slopes are gradual. There are a dozen lakes on the coastal plain some of which are isolated from the sea by beach ridges or dunes. There are no perennial streams as rainfall is low, and there are no deep valleys or true estuaries except for a small blind estuary at the mouth of the intermittent Keanauhi Stream on the west coast. Rather prominent beaches are present along much of the north-west coast. Reefs are found mainly along the west coast. Two small islands lie close to Niihau, Lehua Rock, 1 km off the north end, and Kaula Rock, 34 km to the south-west. Both are the remnants of conic volcanic caldera. Geology and groundwater resources are described by Stearns (1967).

Reef Structure and Corals   Reefs are not well developed around Niihau presumably because the island is exposed to large sea and swell on all coasts. Where corals are found, they form thin veneers on limestone and basalt outcrops, but there is little information. Black corals *Antipathes* have been found to be common off Lehua Rock, and a reef coral rare in Hawaii, *Cosinaraea wellsi*, has been reported there at a depth of 120 ft (37 m) (Maragos 1977).

Scientific Importance and Research   A general survey of the corals and reefs off Niihau is necessary before other scientific research is planned.

Economic Value and Social Benefits   As the island is privately owned, it is inaccessible to the public. A population of about 300 Hawaiians resides on the island and is perhaps the only community left in the Hawaiian Islands which uses Hawaiian as its language.

Disturbance or Deficiencies   No information.

Legal Protection   None.

Management   Not known.

Recommendations   It is recommended that a general survey of marine and reef related resources be conducted by DLNR or the Army Corps of Engineers.

NWHI - HAWAIIAN ISLANDS NATIONAL WILDLIFE REFUGE

Geographical Location   The refuge covers the North-western Hawaiian Islands (NWHI), excluding Kure and Midway (*see separate account*) and includes eight major islands: Nihoa (23°04'N, 161°55'W), Necker, French Frigate Shoals, Gardner Pinnacles (25°00'N, 168°W), Maro Reef, Laysan, Lisianski, and Pearl and Hermes Reef (27°50'N, 175°50'W).

Area, Depth, Altitude   Total area of refuge is 123 159 ha; the area of emerged land is 733 ha; marine area within refuge is 122 433 ha. The largest island is Laysan (403 ha). The altitude of the islands ranges from 277 m at Nihoa to about 3 m at Pearl and Hermes Reef. Lagoon depths are variable; maximum depths in lagoons of Pearl and Hermes Reef and French Frigate Shoals are 30 m.

Land Tenure   Federal government, although jurisdiction over the refuge is held jointly by federal and state agencies.

Physical Features   The NWHI constitute the north-west two-thirds of the Hawaiian archipelago and are a chain of small rocky islands, atolls, coral islands and reefs which become progressively older and generally smaller towards the north-west. The north-westernmost of these, Midway and Kure, are described in a separate account. The north-westernmost remnant of basalt present above sea level in the chain is Gardner Pinnacles. All the islands further to the north-west are either atolls, coral islands, or reefs and shoals of limestone construction. Estimates for the ages of the islands given by Schlanger and Gillett (1976) are as follows: Nihoa (7.0±0.3 ma), Necker (10.0±0.4 ma), French Frigate Shoals (11.7±0.4 ma), Pearl and Hermes Reef (20.0±0.5 ma).

At the south-east end of the chain is Nihoa, the largest of the volcanic islands (77 ha) which is a remnant of a former volcanic cone. The northern edge is a steep cliff made up of successive layers of lava through which numerous volcanic extrusions (dikes) are visible. Further west, French Frigate Shoals (FFS), is nearly an atoll. A small vestige of basalt (La Perouse Pinnacle) extends above sea level on the western edge of the island but otherwise FFS is a classic atoll with a well-formed barrier reef and lagoon. Ten islets are found within the atoll, including Tern Island. The last and probably oldest volcanic island in the chain is Gardner Pinnacles, situated about 210 km north-west of FFS and only 52 m in altitude.

To the north-west the next major feature in the chain is Maro Reef, which is a system of highly reticulated reefs lacking a barrier reef and emergent lands and therefore not considered to be an atoll or a coral island. The next two islands (Laysan and Lisianski) are coral islands. Laysan is the largest of the NWHI but is only 15 m above sea level, and has a salt-water lake in the centre. Lisianski also has a central depression but it is dry. The last island in the National Wildlife Refuge is Pearl and Hermes Reef or Atoll, which has six coral islets, the largest of which is South-east Island.

Waters around the NWHI are slightly more productive than along the rest of the Hawaiian Archipelago (Hirota *et al.*, 1980). Several physical gradients exist, moving to the north-west; solar radiation gradually declines with increasing latitude and the depth of the thermocline decreases, especially during summer months. Sea surface temperature differences are strongly seasonal; during summer there are almost no differences across the chain while in winter differences can be as large as 6°C. The NWHI are more disturbed than the high islands by long-period swell due to their small size (Grigg, 1983).

Reef Structure and Corals The community structure and species composition of corals in the NWHI have been documented by Grigg (1983). In general the fauna is impoverished (42 species) relative to the Indo-West Pacific but most species are found off all islands. A major exception to this trend are *Acropora cytherea*, *A. valida* and *A. humilis*, which are most abundant in the centre of the chain and totally absent at both ends. Differences in community structure of coral reef assemblages between the NWHI are primarily differences in species organization (dominance) and coral cover. Reefs characterized by high dominance are generally reefs which are protected from disturbance. Coral cover is also a function of disturbance but gradually declines to the north-west due to decreases in growth rate and increased frequency of disturbance. Pearl and Hermes Reef is described by Galtsoff (1933).

Noteworthy Fauna and Flora The NWHI are important for a number of endemic and rare species. Except for the Hawaiian Monk Seal *Monachus schauinslandi*, the endemics are primarily terrestrial and include a variety of birds, plants and insects. Many of the rare species are classified as endangered or threatened by the U.S. Department of Interior and are protected by virtue of refuge regulations. A listing of the endemic plants (12), land molluscs (7), arthropods (64) and land birds (4) from the NWHI can be found in Conant *et al.* (1983). Green Turtles *Chelonia mydas* nest and the refuge is a major breeding area for tropical North Pacific seabirds (17 species) including Laysan Albatross *Diomedea immutabilis* and Black-footed Albatross *D. nigripes*, three boobies, three terns, three noddies, shearwaters, petrels, and frigatebirds. At least 600 fish species have been recorded.

Scientific Importance and Research The first major scientific study of the NWHI was undertaken by the U.S. Fish Commission in 1902. After two decades of almost no scientific activity, the Tanager Expedition was launched in 1923 primarily to compile an inventory of the terrestrial fauna of the islands. From 1963 to 1968 the Pacific Ocean Biological Survey (POBSP) of the Smithsonian Institution, Washington D.C., in conjunction with the Bureau of Sports Fisheries and Wildlife and the Hawaii Division of Fish and Game, extensively surveyed the NWHI, concentrating on terrestrial biota. Islands visited include French Frigate Shoals (Amerson, 1971), Gardner Pinnacles (Clapp, 1972), Laysan (Ely and Clapp, 1973), Lisianski (Clapp and Wirtz, 1975), Pearl and Hermes Reef (Amerson *et al.*, 1974), Necker (Clapp and Kridler, 1977), and Nihoa (Clapp *et al.*, 1977). Little research was conducted on any marine species until the early 1970s. In 1973, the National Marine Fisheries Service (NMFS) of NOAA began a series of expeditions to the NWHI to assess the marine resources. In 1975 a formal agreement between NMFS, the U.S. Fish and Wildlife Service (FWS), and the Division of Fish and Game (now Division of Aquatic Resources) of the Hawaii Department of Land and Natural Resources was established to conduct a 5-year survey and assessment of marine resources. In 1977 the University of Hawaii through the Sea Grant College Program joined the study. The major objective of these studies was resource assessment and ecology for the purpose of protecting unique wildlife and managing potential fishery resources. Since then two major symposia on Resource Investigations in the North-western Hawaiian Islands have taken place, in 1980 and 1983 (Grigg and Pfund,

1980; Grigg and Tanoue, 1983). Several of these research projects are still underway. Information on fish studies carried out in the area is given in Parrish *et al.* (1985). Studies on production and ecosystems have been carried out by Atkinson and Grigg (1984), Grigg *et al.* (1984) and Polovina (1984).

Economic Value and Social Benefits With the exception of the small field station at French Frigate Shoals, the islands are uninhabited but are visited occasionally by scientists. The rational development of selected fisheries for lobsters, bottom fish, shrimp and precious corals is being planned.

Disturbance or Deficiencies Currently, stringent management by FWS is maintaining the NWHI in an almost undisturbed state. This, however, was not always the case. For example, in the early 1900s guano mining on Laysan led to the introduction of tobacco plants and rabbits which devastated the terrestrial ecosystems of the island. This caused the extinction of the Laysan Honey-eater *Himatione sanguinea freethii*, the Laysan Rail *Porzanula palmeri*, the Laysan Millerbird *Acrocephalus familiaris familiaris*, the Loulu Palm *Pritchardia* and the Sandlewood *Santalum*. Lesser but similar kinds of impacts have occurred on some of the other islands. Given the research efforts in recent years in the NWHI and continued management by FWS, future prospects for the National Wildlife Refuge would appear good. Grigg and Dollar (1981) describe the effect of the spill of 2200 tonnes of kaolin when a freighter grounded on French Frigate Shoals in April 1980. Impact on the reefs was very minor and highly localized, with a 2-3 m deep channel ploughed through the reef and some coral smothering and bleaching within 50 m of this.

Legal Protection In 1909 the NWHI were designated as the "Hawaiian Islands Bird Reservation". In 1940 the area was re-designated the "Hawaiian Islands National Wildlife Refuge" and in 1966 it became part of the "National Wildlife Refuge System". The NWHI are also part of a State Wildlife Refuge. Strict Natural Areas have been zoned at Pearl and Hermes Reef, Laysan Island, Lisianski Island, Necker Island and Nihoa Island. The Division of Aquatic Resources (DLNR) designates the waters around the NWHI as a Marine Fisheries Management Area, where permits are required for all fishing vessels subject to specific regulations regarding species and area quotas, season and gear. The National Marine Fisheries Service (NMFS) designated critical habitat for the endangered Hawaiian Monk Seal which includes all waters from the 10 fathom (18.3 m) depth contour to the vegetation line on all islands and reefs in the NWHI except for Maro Reef. NMFS is now investigating the possibility of extending the critical habitat to the 20 fathom (36.6 m) depth contour around the same reefs and islands.

Management Management responsibility over the Hawaiian Islands National Wildlife Refuge belongs to the U.S. Fish and Wildlife Service, P.O. Box 50167, Honolulu, Hawaii 96850, and DLNR in the State of Hawaii, 1151 Punchbowl St., Honolulu, Hawaii 96813. Regulations pertaining to the North-western Hawaiian Islands Marine Fisheries Management Area are listed under Chapter 46, Part II, Subtitle 4, Title 13 of the DLNR. Tern Island on FFS has been converted into a rectangular landing strip and serves as a field station for the U.S. Fish and Wildlife Service.

Recommendations   Continuation of the current management regime and continued basic research related to resource management.

NWHI - KURE MARINE FISHERY MANAGEMENT AREA AND MIDWAY ATOLL

Geographical Location   Midway and Kure are the north-westernmost atolls in the Hawaiian chain, and are separated by 90 km; 28°15'N, 177°20'W, and 28°25'N, 178°20'W respectively.

Area, Depth, Altitude   Midway Islands (Sand and Eastern) have a combined area of 518 ha, the largest of all North-western Hawaiian Islands. The area of Kure is 96 ha. Both atolls consist of two sandy islets with a maximum elevation of about 6 m.

Land Tenure   Midway Islands are not part of the State of Hawaii but belong to the federal government of the United States. Kure Atoll is part of the State of Hawaii.

Physical Features   Midway and Kure Atolls are remarkably similar, although Kure is slightly smaller. Both atolls have two coral islets lying inside the southern part of the reef, separated by a small break forming a pass into the lagoon of each island. On the western side of each atoll is a large open pass. General information on Midway is given in Maragos (1986) and geological studies on this atoll are described by Ladd *et al.* (1967). Midway is estimated to be 27.0 million years old (Dalrymple *et al.*, 1977).

Reef Structure and Corals   Perhaps the most remarkable thing about the species composition of corals on Midway and Kure is their similarity to the other islands in the chain. With the exception of the acroporid corals, most species are present, but coral abundance is much lower on Midway and Kure than on the other islands, except in the lagoons. Outside the barrier reefs, living corals consist of patchy veneers averaging only about 10% living cover at optimal depths (10 m) (Grigg, 1983). Corals of Kure are described by Dana (1971).

Noteworthy Fauna and Flora   Midway and Kure do not have any endemic species or subspecies of animals or plants, but support large populations of the Laysan Albatross or Gooney Birds *Diomedea immutabilis*. The Short-tailed Albatross *D. albatrus* is a constant visitor to Midway. The Laysan Rail *Porzanula palmeri* and the Laysan Finch *Psittirostrata cantans* were introduced to Midway many years ago, but both are now extinct there (the Laysan Rail is completely extinct). Hawaiian Monk Seals *Monachus schauinslandi* and Green Turtles *Chelonia mydas* are abundant on each island. A moderately rich flora is found on both atolls including the introduced tree *Casuarina* on Midway and the shrub *Scaevola* (Naupaka) on Kure where it is extremely abundant. Terrestrial fauna and flora of Kure are described in detail in Woodward (1972).

Scientific Importance and Research   The physical location of Midway and Kure as the most northerly atolls in the Pacific, if not the world (Bermuda is further north but is not an atoll in the strict sense), make these islands very important for the study of reef processes near the latitudinal limits of coral growth. Excellent facilities (housing, boats, etc.) exist on both islands. Kure was the subject of surveys by the Pacific Ocean Biological Survey Program (POBSP) of the Smithsonian Institution from 1963 to 1969 (Woodward, 1972). Schroeder (1985) studied recruitment rates of reef fish in Midway lagoon.

Economic Value and Social Benefits   The most important value of Midway Atoll to the United States is its strategic position as a military base and naval defense area. During World War II, it served as the Pacific DEW Line (Distant Early Warning) and is now operated by the Navy as a small military installation. Kure is important as a Coast Guard Station and its Loran Station is a valuable navigational aid. Both atolls have large landing strips. There is limited recreational fishing around the islands.

Disturbance or Deficiencies   The first human disturbance on Midway was dredging in 1870. In 1903 a cable station was established and in 1935 Pan American Airlines used the lagoon as a seaplane or clipper landing station. This ultimately led to Midway being developed as a military base. During World War II, the battle of Midway devastated the island, but the damage was quickly repaired and no wildlife was permanently destroyed (Carlquist, 1980). Recently the Navy has carried out extensive dredging at Midway to enlarge and/or deepen the entrance channel and turning basin for the deep-draft harbour and this is expected to have affected the reefs. The reefs on Kure are reportedly little impacted by man at the present time. A limited amount of fishing takes place on each island, but populations of reef fish and corals are reasonably pristine. However there have been unconfirmed reports of ciguatera fish poisoning on Midway during the past five years. Most impacts are natural, associated with either storm waves or sedimentation in the lagoons caused by strong winds re-suspending fine bottom sediments.

Construction on both Midway and Kure has reduced the nesting habitat for albatross and conflicts remain between these birds and aviation. Nevertheless, the birds are abundant on both islands. Apfelbaum *et al.* (1983) describe the effects of the introduced *Casuarina* on dune vegetation.

Legal Protection   For research of any kind, permits from the state (Kure only) and federal government (Midway only) are required. Kure Atoll is part of the State Seabird Sanctuary and a Marine Fisheries Management Area.

Management   Midway is under jurisdiction of the U.S. Navy, and the U.S. Fish and Wildlife Service assists in the management of wildlife under a memorandum of agreement with the Navy. Programs are underway to control avian disease and to reduce the mortality of seabirds and the hazard of aircraft strike for birds.

Recommendations   Continuation of existing management policies including support for marine research is recommended. The future development of a fisheries landing and processing port on Midway Island would enhance the development of albacore and precious coral fisheries in the north Pacific Ocean.

OAHU ISLAND

Geographical Location   Oahu is 216 km north-west of Hawaii; its approximate midpoint is 20°29'N, 157°59'W.

Area, Depth, Altitude   Oahu is the third largest of the Hawaiian islands, with a total area of 154 000 ha. Its highest peak, Mount Kaala, is 1233 m in altitude.

Land Tenure   Almost 50% of the land is privately owned, with the balance divided between the state and federal government.

Physical Features   The island consists of the tops of two shield volcanoes: Koolau, the youngest, to the north-east and Waianae to the west. Koolau has been extensively eroded, especially at its east end. Between the two mountain ranges lies a large elevated plain, which extends from Pearl Harbor in the south to Haleiwa in the north and rises to about 300 m at Wahiawa. Most of the plain is under sugar cane and pineapple cultivation.

Precipitation varies widely, ranging from 630 cm/yr on the higher elevations of the Koolaus to 50 cm/yr on the coastal plains of Waianae in the shadow of the Waianae Mountains. The major streams of the Koolaus are perennial and those of the deeper valleys which cut into dike-confined groundwater bodies are very stable. In contrast, most of the streams draining the Waianae Range are intermittent. Several of the perennial streams discharge into large estuaries, such as Pearl Harbor, a coastal plain estuary, and Kaneohe Bay (*see separate account*), an estuary-lagoon, both fed by multiple streams and springs (Grace, 1974).

The beaches of Oahu are described by Clark (1977) and the coastline by Richmond and Muller-Dubois (1972). The reefs are almost all of the fringing type. Fringing and fossil reefs are described by Pollock (1928). Other than Kaneohe Bay, the best developed reefs are in Hanauma Bay (*see separate accounts*), off west Oahu, and just north of Barbers Point. Ala Wai, Waikiki and Diamond Head are described in a separate account. Water circulation around Oahu is described by Bathen (1978). A survey of Pearl Harbor is described in Evans (1974).

Reef Structure and Corals   The north and north-west coasts are exposed to large north-westerly swells in winter and have the least well-developed reefs. The best developed (thickest) reefs lie off the south coast and windward north-east coast, including Kaneohe Bay (*see separate account*). Reefs in Hanauma Bay are described in a separate account. Surveys conducted prior to Hurricane Iwa (see below) off the south-west coast established the following rank order abundance of coral species: *Porites compressa*, *P. lobata*, *Montipora patula*, *Pocillopora meandrina*, and *M. verrucosa*. Coral cover at stations on the south-west coast averaged 54% (Grigg, 1983). An inventory of the reefs of Oahu has been compiled by AECOS, Inc. (1979a and 1981a). The coral *Pavona maldivensis*, rare in Hawaii, has been reported from Kahe Reef.

Noteworthy Fauna and Flora   Russo *et al.* (1977) provide an inventory of benthic organisms and plankton at Mokapu.

Scientific Importance and Research   Numerous opportunities for coral reef research exist on Oahu, facilitated by the presence of the main campus of the University of Hawaii, its faculty and library being in close proximity to abundant and diverse coral reef habitats. The Hawaii Institute of Marine Biology of the University of Hawaii is situated in Kaneohe Bay (*see separate account*). The Bishop Museum and Waikiki Aquarium are also situated on Oahu.

Economic Value and Social Benefits   About 82% of the population of Hawaii lives on Oahu, which is the social, political and economic centre of the state. Tourism and military expenditure are the main economic activities. Reef-related activities are a major part of the tourist industry which offers snorkelling, diving, surfing, canoeing, swimming and sailing. The reefs also serve as a source of food for recreational and native fishermen. The primary fishing area for the aquarium fish industry is Ewa Beach on the south coast (Pfeffer and Tribble, 1985).

Disturbance or Deficiencies   The reefs of Oahu have suffered widespread disturbance in recent years. Disturbance from long period and hurricane waves have impacted both northern and southern coasts (Pfeffer and Tribble, 1985). In 1982 Hurricane Iwa caused large-scale damage to most reefs between Diamond Head and Barber's Point, many of which had been dominated by the finger coral *Porites compressa*, which is particularly vulnerable to wave destruction (Dollar, 1982). Non-point source sediment-laden run-off has damaged numerous reefs, including Kaneohe Bay (*see separate account*) and leeward reefs off Hawaii Kai, but improved land management practices should remedy this. In the 1970s pollution from sewer outfalls severely impacted Kaneohe Bay (*see separate account*) and the reefs off Sand Island (Grigg, 1975). These problems have now been solved by diversion of sewage into deeper waters and improvement in sewage treatment (Dollar, pers. comm. to Grigg). The impact of the construction of runways at Honolulu International Airport on adjacent reefs was studied by Chapman (1979) and AECOS, Inc. (1979c). The effect of heated effluent on corals at Kahe Point is described by Jokiel and Coles (1974) and Coles (1984 and 1985). The environment in the vicinity of the Waianae sewage ocean outfall was studied by Reed *et al.* (1977). The impact of the Honolulu power plant is described by McCain *et al.* (1975). Many thousands of recreational fishermen, and commercial fishing, have caused severe depletion of nearshore (reef) fish populations along most coasts. The great abundance and diversity of fish in Hanauma Bay (*see separate account*), in which fishing is not allowed, provide clear indications of the effect of fishing on the rest of Oahu's reefs. Johannes (1978) describes the impact of coral collecting in the 1970s.

Legal Protection   The submerged lands and overlying waters off Pupukea Beach Park and adjoining areas are designated as a Marine Life Conservation District. Only finfish fishing is permitted using fishing poles and a hook-and-line from shoreline, or using spears without SCUBA. The use of legal nets for finfish and crustaceans in the northern portion of Pupukea District and the removal of algae and limu are also permitted. The Hanauma Bay Marine Life Conservation District and Waikiki-Diamond Head Marine Fisheries Management Area are described in separate accounts.

Most waters around Oahu are classified "A" by the D.O.H. (State Department of Health, 1980b). Waters classified "AA" are mainly the following: part of West Loch in Pearl Harbor, Hanauma Bay, South Waimanalo Bay, Kaneohe Bay in the central portion, Kahana Bay, Waialua Bay to Kaena Point, and Makua Bay (south-western coast). Waters designated as class "B" are most of Pearl Harbor, Keehi Lagoon, Honolulu Harbor, Ala Wai Yacht Harbor, Kewalo Basin, state and KMCAS boat harbors at the south-east end of Kaneohe Bay (north-eastern coast) and Haleiwa Boat Harbor (north-western coast).

Management The collection of reef fish or corals comes under the jurisdiction of the Division of Aquatic Resources, Department of Land and Natural Resources (DLNR), 1151 Punchbowl St., Honolulu, HI 96813. Regulations pertaining to Pupukea Marine Life Conservation District are listed under Chapter 34, Part I, Subtitle 4, Title 13 of the DLNR. Improved land management practices should decrease siltation in the future.

Recommendations Lack of enforcement of fish and game regulations is a major problem. More wardens and stricter enforcement are both strongly recommended; in particular greater control of fishing is urgently needed. Early recommendations for marine conservation districts are given in Kimmerer and Durbin (1975).

OAHU - ALA WAI AND WAIKIKI-DIAMOND HEAD FISHERIES MANAGEMENT AREA

Geographical Location Eastern part of Mamala Bay, south coast of Oahu.

Area, Depth, Altitude The shoreline in this section is about 7.5 km long. Reefs extend to depths of 25 m and generally are poorly developed. Diamond Head, the highest point, is 230 m in altitude.

Land Tenure Private and state lands (Kapiolani Park, Ala Moana Beach Park).

Physical Features The land behind Waikiki is a broad low plain extending inland about 2 km to the highly dissected southern slope of Koolau Range. Most of the Ala Wai and Waikiki area, except for the beach ridge along the shore, was originally swampy, with taro patches, rice paddies, duck ponds and fishponds. The swamp was fed by the perennial streams of Manoa and Palolo and several small streams and springs which had their common outlet in Waikiki. In the 1920s, the Ala Wai Canal was dredged to intercept run-off from the streams and the marshy lowlands were filled, transforming the area into the most valuable piece of real estate in Hawaii. Much of Honolulu, including Waikiki, rests on an emergent Pleistocene coral reef.

Ala Wai Canal intercepts the run-off of a large urban drainage area. Waters are highly turbid and nutrient enriched, often containing relatively high concentrations of faecal bacteria, and the sediment and biota have high concentrations of heavy metals. The canal empties into the Ala Wai Yacht Harbor, the entrance to which is through a 7 m deep channel dredged through the reef. Waters in the harbor are often discoloured with silt following heavy rains, and oil, floating trash and debris are common. Magic Island and Ala Moana Park lie to the west of the harbour. A beach about 50 m wide consisting mainly of imported sand fronts Ala Moana Park.

Directly off the Ala Wai Yacht Harbor is a shallow reef flat consisting largely of sand and rubble. The area to a depth of about 5 m off Ala Moana and Ala Wai consists of a fringing reef dissected by many surge channels. Below 5 m, low outcrops of coral alternate with patches of sand. The bottom at 20 m off Ala Moana is generally a sand/rubble zone with scattered colonies of *Porites lobata* and *P. compressa*.

East of Ala Wai Harbor is Waikiki Beach, perhaps the most famous beach in the world. It varies from less than 8 m to 60 m in width and some parts are almost covered during high tide. Most of the beach is now imported sand held in place by groins and seawalls. In the past, tons of imported sand moved offshore, filling low areas on the reef flat and altering the form of Waikiki's famous surfing breaks. A shallow discontinuous fringing reef protects most of the beach but many shallow areas have been largely cleared of coral heads. Limestone outcrops and rubble dominate the inner platform, and large sand pockets occur offshore. The reef front is usually limestone ridges separated by scattered sand channels. The reef slope is normally gentle and quite sandy, with scattered limestone outcrops. Live coral cover is generally low (less than 5%). Several channels run seaward across the reef, most of them being the remains of ancient streams (AECOS, 1979a).

Diamond Head, Hawaii's most famous volcanic cone, is believed to have been built by a simple volcanic event about 150 000 years ago which took place during the Honolulu volcanic series, the secondary eruptive phase for the Koolau Volcano, after it had been dormant for more than one million years. The seaward portion of the volcano has been greatly eroded by run-off and waves (Stearns, 1966). To the west, the shore consists of nearly continuous seawalls. West of the lighthouse there is a narrow terrace or bench formed of beach rock or basalt or tuff deposits and covered in places by sand and talus. Waves reach the foot of the cliff during high tides. A reef platform 0.5-2 m below sea level lies off the south-west face of Diamond Head. It is broken and discontinuous, and is characterized by sand, rubble and limestone base rock extending 460 m seaward. The reef margin is consolidated reef rock, interrupted by shallow surge channels cutting 1-3 m deep which broaden seawards into extensive sand patches (AECOS, 1979a).

Reef Structure and Corals The reefs are poorly developed. On the reef flats off Ala Moana Park, fleshy and calcareous algae make up the dominant cover. Coral cover ranges from 10 to 20%. *Pocillopora meandrina*, *Porites lobata* and *Montipora verrucosa* are the commonest species but at least eight species are present. Below 10 m, algal cover decreases and coral cover (mostly *P. lobata*) averages 20% or less. On the shallow reef flat off Waikiki Beach, living corals are very sparse. Coral cover increases seawards but only in areas of limestone outcrops where patches of *P. lobata* and *P. compressa* vary in abundance from 15 to 50% cover.

The reef platform off the shoreline west of Diamond Head lighthouse is rather barren and corals are sparse inside the reef margin. Beyond the margin of the shallow reef platform (3-4 m deep) the surf-swept reef front is dominated by algae (particularly *Dictyopteris australis*) covering up to 85% of the bottom. Coral cover is about 5%, dominated by *Pocillopora meandrina* and secondly *Porites lobata*. In water deeper than 10 m off the south-east face of Diamond Head, areas of low coral cover, mostly *Pocillopora meandrina*, alternate with flat limestone outcrops dominated by the alga *D. australis* (AECOS, 1979a).

Noteworthy Fauna and Flora In the Ala Wai Canal the crustaceans *Thalamita crenata* and *Podophthalmus vigil* are abundant and at least 19 species of fish have been recorded. Off Ala Moana, invertebrates include the sea urchins *Diadema paucispinum* and *Tripneustes gratilla*, and worms and the gastropod *Hipponix* sp. which are abundant on upper reef slopes. Sabellids and sponges occur on deeper slope areas. More than 92 species of fish have been recorded off this section of the Oahu coast.

Several species of fleshy and coralline algae are also present, some very abundant such as *Sargassum* sp., *Dictyosphaeria* sp., *Halimeda discoidea* and *Dictyopteris australis* on the shallow reefs off Waikiki and the surf-swept reef front off Diamond Head (Doty, 1971).

The rim of Diamond Head provides a habitat for several species of rare or endangered coastal strand plants, such as *Santalum ellipticum* var. *littorale, Schiedea verticillata, Sicyos* sp., and *Bidens cuneata* (AECOS, 1979a).

Scientific Importance and Research Reefs off Waikiki and Diamond Head have been subjected to a variety of discontinuous but long term anthropogenic disturbances. Species composition and community structure could be used to quantify environmental impact in this area.

Economic Value and Social Benefits This part of the coast is the most popular tourist area in all the Hawaiian Islands. The Ala Wai Canal is used for boating and shoreline fishing but the latter activity has decreased with increasing siltation and coliform bacteria in the water. The Ala Wai Yacht Harbor is the largest harbor for small crafts on Oahu. The beaches offer excellent swimming, snorkelling, board and canoe surfing, canoe paddling, catamaran riding, motorized jet skis riding, etc. Offshore there are some good sites for SCUBA diving. Concessions offer rental of surfboards and beach equipment and operate cruises and diving activities.

Fishing takes place all along the coast (spearfishing, net fishing, pole fishing, etc.) in open seasons (see below). There is occasional shell and tropical fish collecting, and stony corals, mainly *Pocillopora meandrina*, have occasionally been commercially collected off Ala Moana and Diamond Head. The beaches, parks and the Waikiki Aquarium are excellent attractions for both locals and tourists.

Disturbance or Deficiencies As stated above, the Ala Wai Canal periodically contains water with high concentrations of sediment and coliform bacteria which sometimes contaminate the harbour as well as nearshore waters. There are no other important discharges along this section of the coast.

The whole coast from Ala Moana Beach Park to Diamond Head is subject to tsunami flooding. Sand imported into the area is periodically transported offshore by waves and has filled low areas on the reef flat and altered the form of many surfing breaks. Hurricane Iwa recently caused large-scale damage to the reefs. Most shallow areas in the region have been cleared of coral heads and fishing pressure is sometimes very high during the open "seasons".

Legal Protection The section of shore and reef off Waikiki Beach east of the Kapahulu groin to the Diamond Head lighthouse and within 457 m of shore (or to the reef edge if further) is managed under the Waikiki-Diamond Head Shoreline Fisheries Management Area regulations. This involves alternation of "closed to fishing" and "open to fishing" periods every two years. Removal of living corals is prohibited even during the "open to fishing" periods. Special permits may be issued for scientific, educational or other purposes.

The water quality of Ala Wai Canal and Yacht Harbor do not meet the Department of Health standard for microbiological contamination. Nearshore waters off of Waikiki and Diamond Head are classified "A" by the Department of Health (DOH). The DOH monitors coliform bacteria weekly or monthly at different stations within this area (State Department of Health, 1980b).

Management Regulations pertaining to the Waikiki-Diamond Head Marine Fisheries Mangement Area are listed under Chapter 48, Part II, Subtitle 4, Title 13 of the Department of Land and Natural Resources (DLNR), 1151 Punchbowl St., Honolulu, HI 96813.

Recommendations Continued monitoring of water quality by DOH and fish abundance patterns in the Fisheries Management Area off Waikiki and Diamond Head by DLNR.

OAHU - HANAUMA BAY MARINE LIFE CONSERVATION DISTRICT

Geographical Location South-eastern coast of Oahu.

Area, Depth, Altitude Hanauma Bay is about 500 m wide and 900 m in length, covering an area of about 40 ha. Depths in the middle of the Bay range from 5 to 20 m. The surrounding cliffs are about 180 m above sea level.

Land Tenure State lands.

Physical Features Hanauma Bay is a drowned multiple crater formed as a result of several violent eruptions in the recent past. There is evidence that at least one of these was less than 32 000 years ago (Stearns, 1966). The Bay has been formed by an infilling of the sea into the craters, and is almost completely surrounded by vertical walls which are 30 m or more in height. Around the rim of the bay is an elevated wave cut terrace or bench about 5 ft (1.5 m) above sea level, although portions of the bench are somewhat higher, suggesting at least two stands of the sea above the present level.

At the head of the bay is a narrow beach formed largely of imported sand. Adjacent to the beach is a fringing reef which extends about 200 m seaward. Numerous channels and wide sand holes reticulate the reef and incidentally provide swimming areas for tourists. Outside the fringing reef in the middle of the bay, well developed reefs are found at depths of 5-15 m.

Reef Structure and Corals Hanauma Bay has some of the best reefs on Oahu. On the reef flat, corals are sparse and only occupy about one percent of living cover, and the area is dominated by coralline and other fleshy algae. The outer edge of the fringing reef edge consists of elongate grooves and ridges of limestone which support a variety of coral species and invertebrates to a depth of about 20 m. *Porites lobata* and *P. compressa* are dominant but at least 12 other species are well represented in the bay. On the surfaces of the elongate ridges, coral cover in many places is 80% or more. Easton and Olson (1976) provide a radiocarbon profile of Hanauma Reef and Anderson (1978) describes coral community structure. The coral *Montipora studeri*, rare in Hawaii, is locally common on the upper reef slope.

Noteworthy Fauna and Flora The fish fauna of Hanauma Bay is remarkably abundant and diverse, and more than 80 species have been recorded. Surgeonfish are very well represented and comprise over half of the most common species.

Scientific Importance and Research Since designation as a Conservation District, a remarkable increase in abundance and diversity of fishes in the Bay has been documented and is exceptional compared to other areas around Oahu. For example, large parrotfish and jacks, almost decimated in other waters, are abundant.

Economic Value and Social Benefits In addition to the scenic value of Hanauma Bay and the beach, which are major attractions for tourists, the Bay is one of the most popular places on Oahu for picnicking, swimming, snorkelling, and SCUBA diving. The diversity and abundance of fish have added to its popularity. Dive tours to Hanauma Bay are available for tourists on a daily basis.

Disturbance or Deficiencies Hanauma Bay is normally free of pollution and there is no fishing pressure. During heavy rains, small amounts of suspended sediment may enter the bay but are usually flushed out to sea before settling out on the bottom. Marine life has noticeably improved since the area has been managed.

Legal Protection The submerged lands and overlying waters of Hanauma Bay were designated as a State Marine Life Conservation District in 1967, and fishing or removal of any living or geological specimen is prohibited, except with special permits.

Management Hanauma Bay Marine Life Conservation District is regulated by the Division of Aquatic Resources, Department of Land and Natural Resources, State of Hawaii.

Recommendations Hanauma Bay is perhaps the best example in the State of a well managed coral reef habitat. It is also very accessible to many people. The recovery of marine life, particularly reef fish, strengthens the argument for similar management efforts (in particular complete restriction of fishing) in other reef areas in Hawaii and elsewhere.

OAHU - KANEOHE BAY AND COCONUT ISLAND MARINE REFUGE

Geographical Location Eastern coast of Oahu, off the north-east flank of the Koolau Range, between Mokapu Peninsula and Kualoa Point; 21°13'N, 157°47'W.

Area, Depth, Altitude This is the largest embayment in the Hawaiian islands, and is about 13 km long and 4 km wide, surrounded by 129 km of shoreline. The average depth in the lagoon is 12 m.

Land Tenure Private, state and federal lands (mainly the Kaneohe Marine Corps Air Station at Mokapu Peninsula) surround the Bay.

Physical Features Kaneohe Bay occupies a drowned stream-valley system, fed by nine perennial and several intermittent streams, and contains the largest estuarine habitat in the state, but morphologically it is more a reef-sheltered lagoon. Most of the shoreline consists of artificial sea walls and landfills, and natural muddy sands. Beach deposits are rare, but gravel, mud and silt form a conspicuous component of the sediments. There are several islands in the Bay, including Coconut Island (Moku o Loe), the marine environment of which is described by Environmental Consultants, Inc. (1975).

A fringing reef less than 1 m deep extends 300-750 m off most of the shoreline. It is interrupted where streams enter the Bay, and a number of cuts and channels have been dredged, particularly in the south-east basin. Originally 79 patch reefs ranging from 10 to 900 m in diameter rose to within one meter or so of the water surface. About 25 of these have been partially dredged. A wide (almost 2 km) barrier reef extends about 6 km across the mouth of the Bay and is cut by two sand-bottom channels, Mokolii and Kaneohe. The northern channel has been dredged to about 11 m and a ship channel running through the Bay lagoon connects it with the south-eastern basin which has been dredged for ship passage and for a seaplane landing area.

The mean tidal range between mean lower low water and mean high high water is 0.6 m. The extreme tidal range under normal conditions is 1.3 m. The major currents in the Bay are tidal in nature and in the south-east part there is a fairly persistent counter-clockwise eddy. Bathen (1968) describes the physical oceanography of the Bay. Smith *et al.* (1973) provide an atlas of the Bay.

Reef Structure and Corals On fringing and patch reefs in Kaneohe Bay *Porites compressa* accounts for 85% of the total coral cover followed by *Montipora verrucosa* (10%). Both corals are most common on the outer margins of the reefs. *Fungia (Pleuractis) scutaria* is also very common in the middle and northern sectors of the

Bay. About 29 other species of coral have been identified, including rare forms such as *Psammocora verrilli*, *Porites duerdeni*, *Montipora dilatata*, *Cycloseris hexagonalis*, *Pocillopora molokensis* and *Porites (Synaraea) irregularis*. In the south-east basin, the soft corals *Zoanthus pacificus* and *Palythoa psammophilia* are also abundant (Maragos, 1972). To date Kaneohe Bay is the only known locality of *Montipora dilatata* (Maragos, 1977; Vaughan, 1907). Brock (1976) describes marine communities of shoreline structures.

Noteworthy Fauna and Flora The reef flats and reef slopes harbor a wide variety of invertebrates. In the southern and mid-Bay sectors, filter-feeding and detrital-feeding forms are abundant. The green alga *Dictyosphaeria cavernosa* is abundant on reef slopes in the central basins, often excluding other algae and corals. About 170 species of fish have been recorded from the Bay. The mullet *Mugil cephalus* and Nehu *Stolephorus purpureus* are commercially important. The soft mud floor of the lagoon is inhabited mainly by crustaceans and infaunal polychaetes.

The Red Mangrove *Rhizophora mangle* was introduced in the early 1900s and has now spread throughout the Bay, and two other species of mangrove, *Bruguiera sesangula* and *B. gymnorhiza* are also present. The freshwater marshes provide habitat for many seabirds and water birds, especially the threatened Hawaiian Stilt *Himatopus mexicanus knudseni* and the gallinule *Gallinula chloropus sandvicensis*. Mokoli'i Island has native strand vegetation including the uncommon *Panicum carteri*, proposed as a threatened species, and is frequented by seabirds, Wedge-tailed Shearwaters *Puffinus pacificus chlororynchus* and Bulwer's Petrel *Bulweria bulweri*. Seabirds are also abundant on Kakapa Island (AECOS, 1979a and 1981a).

Scientific Importance and Research Kaneohe Bay is one of the most intensively studied bodies of water in the Hawaiian Islands, if not in the world. The Hawaii Institute of Marine Biology (HIMB) of the University of Hawaii is situated on Coconut Island and is perhaps the best equipped marine laboratory in the U.S. for reef research. Laboratory and housing facilities are available for visiting scientists. In 1978, two sewage outfalls that had previously discharged into the Bay were diverted to the outer coast off Kailua. Between 1976 and 1980 HIMB scientists conducted and completed a total ecosystem experiment in the Bay (Smith *et al.*, 1981) aimed at assessing and quantifying post-diversion recovery of infaunal, reef and water column communities. Reef coral recovery in the bay was documented six years after termination of sewage discharge (Maragos *et al.*, 1985). Publications up to 1970 are listed in Gordon and Helfrich (1970). Studies include reef productivity (Gordon and Kelly, 1962), plankton ecology (Clutter, 1969), sedimentation (Fan and Burnett, 1969), bacterial pollution (Gunderson and Stroup, 1967), pollution (Quan, 1969), marine fouling (Rastetter and Cooke, 1978), phytoplankton productivity (Krassick, 1973), reef fish (Losey, 1976; Wass, 1976; Stanton, 1985), coral populations (Maragos, 1972, 1974a and 1975), bathymetric configuration (Roy, 1970), and benthic algae (Soegiarto, 1972).

Economic Value and Social Benefits Both sport and commercial fishing are important in the Bay. Many methods are used, including netting, pole or handline,

trapping, spearfishing and torch fishing. Bait collection for the commercial tuna (aku) fishery is particularly important. The collection of tropical fish is carried out on a small scale. Twelve of the original 30 fishponds dating from before the 20th century are found around the Bay and two are used for aquaculture; the others were filled or dredged for urban development (Summers, 1964). The Bay is excellent for sailing, waterskiing, windsurfing and snorkelling.

Disturbance or Deficiencies For many years Kaneohe Bay was exposed to heavy sedimentation caused by the run-off of numerous silt-laden streams during periods of heavy rainfall. This problem was exacerbated by increases in population and urbanization and poor land management, particularly in the 1960s and 1970s. During this period, ecological damage associated with heavy rainfall was occasionally severe and included infilling of the south-eastern basin with land-derived sediments (Banner and Bailey, 1970; Banner, 1968 and 1974; Chave and Smith, 1974). Cutterhead dredging operations for the channels and seaplane areas generated additional sedimentation.

Up until 1978, two sewage outfalls discharged directly into the Bay and were the cause of eutrophic conditions, particularly in the southern sector (Maragos, 1972, 1977; Smith *et al.*, 1973). Since about 1980 the diversion of the sewage outfalls to the outer coast off Kailua, and improvement in land grading practices, have resulted in a dramatic recovery of benthic and coral reef communities. The clarity of the water has also improved markedly and is related to decreases in suspended sediment, nutrients, and standing crops of phytoplankton and zooplankton. Continuing research at the HIMB is monitoring these large-scale ecological changes (Maragos *et al.*, 1985; Smith *et al.*, 1981). Historical changes in the area have been documented by Devaney *et al.* (1976) and the site has been evaluated by Sunn *et al.* (1976).

Legal Protection A marine life refuge exists around Coconut Island. All collecting, including fishing, is prohibited without a scientific permit from DLNR. Mokoli'i Island is the only island off Oahu not designated as a bird sanctuary.

The waters of the bay are designated Class "AA" by the DOH. The margins are designated Class "A", except for three small areas close to the small-boat harbors which are Class "B".

Management The agency responsible for the management of marine resources is the Division of Aquatic Resources, Department of Land and Natural Resources (DLNR), 1151 Punchbowl St., Honolulu, HI 96813. Water quality management is conducted by the D.O.H. Enforcement of existing fish and game regulations could be improved by increasing the number of wardens on Oahu and by stricter observance of the laws.

Recommendations Stricter enforcement of Fish and Game regulations and land grading ordinances and continued monitoring of the recovery of reef populations in the lagoon of Kaneohe Bay are required.

BAKER ISLAND NATIONAL WILDLIFE REFUGE

Geographical Location 2657 km south-west of Honolulu and 500 km north-west of the Phoenix Islands; 0°13'N, 176°29'W.

Area, Depth, Altitude 12 849 ha total; 138 ha land, 12 711 ha marine; max. alt. 5-8 m.

Land Tenure Federal government.

Physical Features The island is oval in shape and has no fresh water. Winds are fairly steady and rainfall is very low.

Reef Structure and Corals The island is surrounded by fringing reef.

Noteworthy Fauna and Flora The island is covered with low grasses and sparse and low bush. Eleven migratory seabirds and four migratory shorebirds have been identified. In 1973 there were an estimated 7000 nesting birds including Brown Noddies *Anous stolidus*, Sooty Terns *Sterna fuscata*, Blue-faced, Brown and Red-footed Boobies *Sula dactylatra, S. leucogaster, S. sula*, and Lesser Frigatebirds *Fregata ariel*. Green Turtles *Chelonia mydas* have been recorded.

Scientific Importance and Research Occasional wildlife monitoring trips are made by the U.S. Fish and Wildlife Service.

Economic Value and Social Benefits The island is currently uninhabited but was occupied during World War II.

Disturbance or Deficiencies The bird populations were nearly eradicated during the war but are now recovering. Drums with oil and fuel were abandoned after World War II and until clearance posed some risk of pollution and hazard to nesting seabirds.

Legal Protection Declared a National Wildlife Refuge in July 1974.

Management Administered by the Hawaiian Islands and Pacific Islands National Wildlife Refuge Complex, Honolulu. In 1986 the U.S. Army Corps of Engineers sponsored a survey and complete cleanup of the oil and fuel from the island. Some 400 of the 3400 drums contained oil or fuel, which was burned *in situ* and the drums turned on their side to prevent seabirds falling into them. Observations also indicated that recent cat eradication programmes had been successful (USACE, 1987b).

HOWLAND ISLAND NATIONAL WILDLIFE REFUGE

Geographical Location 2657 km south-west of Honolulu and 500 km north-west of the Phoenix Islands; 0°48'S, 176°38'W.

Area, Depth, Altitude Total 13 178 ha; 162 ha land (1.5 x 0.5 mi. (2.4 x 0.8 km)), 13 016 ha marine; max. alt. 6-7 m.

Land Tenure Federal government.

Physical Features The island is low, flat and elongated in a north-south direction. The surface is sand, coral rubble and flat sheets of guana rock. The climate is windy and sunny with low rainfall and there is no fresh water.

Reef Structure and Corals The island is surrounded by a 100 m wide fringing reef.

Noteworthy Fauna and Flora The island is covered with grasses, prostrate vines and low shrubs. There are an estimated 200 000 Sooty Terns *Sterna fuscata*, 6000 Lesser Frigatebirds *Fregata ariel*, 1000 Blue-faced Boobies *Sula dactylatra* and a few Red-Footed *S. sula* and Brown Boobies *S. leucogaster* and White Terns *Gygis alba*, all of which nest. Blue-grey *Procelsterna cerulea* and Black Noddies *Anous minutus*, Wedge-tailed Shearwaters *Puffinus pacificus*, Phoenix Petrels *Pterodroma alba* and White-throated Storm Petrels *Nesofregetta fuliginosa* also occur. Green Turtles *Chelonia mydas* are found in the surrounding waters.

Scientific Importance and Research Some research carried out.

Economic Value and Social Benefits An airfield was built in 1935 for Amelia Erhardt's transpacific flight and a memorial plaque was erected. The island is now uninhabited; formerly there was a resident Coast Guard.

Disturbance or Deficiencies Feral cats were formerly present and preyed on seabirds; fuel and oil drums also posed some hazard to seabirds.

Legal Protection Declared a National Wildlife Refuge in July 1974.

Management Administered by the Hawaiian and Pacific Island National Wildlife Complex, Honolulu. A recent cat eradication programme appears to have been successful. In 1986 the U.S. Army Corps of Engineers turned the 12 or so fuel drums on their side to prevent seabirds falling into them; no fuel or oil was found in any of the drums (USACE, 1987a).

Recommendations Proposed as an Island for Science under the International Biological Programme.

JARVIS ISLAND NATIONAL WILDLIFE REFUGE

Geographical Location 2093 km due south of Honolulu; 0°22'S, 160°01'W.

Area, Depth, Altitude Total 15 189 ha; 445 ha land (2 x 1 mi. (3.2 x 1.6 km)), 14 744 ha marine; max. alt. 6-8 m.

Land Tenure Federal government (uninhabited dependency).

Physical Features The island is roughly trapezoidal in shape and is sandy and flat with a coral-reef rock base. There is no fresh water.

Reef Structure and Corals   A 100 m wide shallow fringing reef encircles the island.

Noteworthy Fauna and Flora   The main vegetation is low vines, succulents, grasses and sparse shrubs. Red-Footed *Sula sula* and Blue-faced Boobies *S. dactylatra*, Greater *Fregata minor* and Lesser Frigatebirds *F. ariel* and Sooty Terns *Sterna fuscata* are among the commonest of the nesting seabirds. Six other seabird species visit. The Green Turtle *Chelonia mydas* is found.

Scientific Importance and Research   An annual survey of birds, and other wildlife is made. Some studies were carried out during International Geophysical Year.

Economic Value and Social Benefits   Uninhabited now although colonized during World War II.

Disturbance or Deficiencies   Some rubbish left since World War II occupation, but there is no abandoned oil or fuel.

Legal Protection   Established as a National Wildlife Refuge in July 1974.

Management   Administered by the Hawaiian Islands and Pacific Islands National Wildlife Refuge Complex, Honolulu.

JOHNSTON ISLAND NATIONAL WILDLIFE REFUGE

Geographical Location   1328 km south-west of Honolulu; 16°44'N, 169°30'W.

Area, Depth, Altitude   Total of 13 252 ha; 252 ha land, 13 000 ha marine; below sea level to 4-5 m; lagoon 3-9 m deep.

Land Tenure   Federal government (unincorporated territory administered by Defense Nuclear Agency).

Physical Features   Johnston Atoll is a true atoll (*contra* Keating, 1985), as it has a lagoon, thriving coral formations and a shallow reef platform at depths of 0-30 m, clearly within the photic zone. It is unusual in that it is tilted about a SW/NE axis, so that the south-east quadrant is open and deeper, and the north-west quadrant is raised slightly; it thus forms an open "C" shape, rather than a closed ring. Four islands, two of which, Akau, or North (7 ha) and Hikina, or East (5 ha) were made by dredging and two, Johnston (231 ha) and Sand (9 ha) that were enlarged by dredging, lie within the atoll. The mean annual temperature is 26°C with a range from 16.5°C to 31.5°C; mean annual rainfall is 663 mm although extremes of 1085 mm and 435 mm have been recorded. There is no natural fresh water; 10 mph (16 km per hr) trade winds are fairly constant. The geomorphology is described by Ashmore (1973) and Keating (1985). The atoll was also described by Kinsey (1983).

Reef Structure and Corals   Flourishing reef corals and healthy reef formations including pinnacles, mounds and platforms are found over the entire upper platform of the

atoll, although development is shallowest and greatest in the leeward north-west sector, where there is a barrier or perimeter reef about 12 km long (Keating, 1985). Live corals are found on fringing reefs surrounding the islands of the atoll, and on the submerged south-east sector of the barrier or perimeter reef. The structure of the reef was described by Emery (1956). Five distinct zones were observed: seaward slope, lithothamnium ridge, coralline algal reef flat, coral slope and a lagoonward (or leeward) debris slope. A more detailed description is given in Maragos and Jokiel (1986). A total of 30 species of scleractinians and three hydrozoans have been described. Although the fauna has the strongest affinity with that of Hawaii, the general appearance of the reef is quite different because of the dominance of *Acropora* (which is not found off the high islands of Hawaii) and the total lack of the common Hawaiian coral *Porites compressa*. Despite low species diversity, coral coverage is extremely high in most places.

Noteworthy Fauna and Flora   Vegetation on Johnston is quite dense in some areas but very sparse on the other three islands and is highly modified by man. There were originally only two species (a low grass and a vine) but over 100 plants have been introduced by man. Fifty-three bird species have been recorded including 22 migratory seabirds of which 13 nest on Johnston. Common species are Red-footed *Sula sula*, Brown *S. leucogaster* and Blue-faced Boobies *S. dactylatra*, Great Frigatebirds *Fregata minor*, Wedge-tailed Shearwaters *Puffinus pacificus* and Brown Noddies *Anous stolidus*. One hundred and ninety-four species of fish have been identified (Fowler and Ball, 1925; Gosline, 1965; Randall *et al.*, 1986). The butterfly fish *Centropyge nigriocellus* and *C. flammeus* may be endemic. The Green Turtle *Chelonia mydas* is common. The Hawaiian Monk Seal *Monachus schauinslandi* occasionally visits the atoll (Schreiber and Kridler, 1969). The National Marine Fisheries Service recently released a dozen bachelor male Monk Seals at Johnston to decrease harassment on a limited number of female seals in the NWHI. A few of the males may still reside at Johnston. Algae are described by Agegian and Abbott (1985) and Bugglen and Tsuda (1966 and 1969). Checklists of the fauna and flora are given in Amerson and Shelton (1976).

Scientific Importance and Research   Johnson Island is extremely isolated, with one of the most remote reefs in the world, some 800 km from the nearest in the Hawaiian Islands (Maragos and Jokiel, 1986). Military installations provide research facilities. A number of studies have been carried out, reported in Amerson and Shelton (1976) and early reef work is summarized in Maragos and Jokiel (1986). Extensive biological surveys were carried out between 1975 and 1984. Thirty five dives were carried out in the course of an environmental assessment of the atoll (Keating, 1985; Lobel, 1985; Maragos and Jokiel, 1986; USACE, 1987c).

Economic Value and Social Benefits   The fishing potential of the area is described in Ralston *et al.* (1986).

Disturbance or Deficiencies   All terrestrial and the immediate surrounding reef areas have been seriously disturbed by dredging or man-made structures. About 28 million sq. m of reef and lagoon were seriously affected by sediment-laden waters and 4 million sq. m of reef were totally destroyed by dredging. There was also a reduction in numbers of echinoderms and fish species present (Brock *et al..*, 1965 and 1966). Johnston Island was levelled (from its previous height of 15 m at Summit Peak) and enlarged to 85 ha between 1939 and 1942. By 1944, the atoll had been developed as an airforce base. Further major military developments took place in the early 1960s, with the enlargement of the islands and runway and the formation of the two artificial islands. At least three seabirds no longer nest on the islands. The atoll played a central role in the U.S. nuclear testing programme of the late 1940s to 1963. Some contamination of land and water occurred due to leaking of plutonium (USACE, 1984). After the Vietnam war, drums of Herbicide Orange, a chemical defoliant, were stored on Johnston Island. Leakage from these also caused considerable contamination, but in 1976 the defoliant was incinerated at sea using the specially designed Dutch ship *Vulcanus* (U.S. Airforce, 1974). Currently thousands of tons of chemical munitions (explosively configured nerve and mustard agents) are stored on the island, but a plant is being built by the U.S. Army to incinerate and destroy all munitions in the next few years (USACE, 1984). However these plans involve a doubling of the population on the atoll which could cause increased recreational impacts to fish and wildlife. The impact of the proposed deep ocean disposal of brine waste has been studied by Lobel (1984 and 1985) and USACE (1985 and 1987c).

Legal Protection   Declared a bird refuge in 1926 under Presidential Executive Order and later upgraded to National Wildlife Refuge.

Management   Administered by the Hawaiian Islands and Pacific Islands National Wildlife Refuge Complex. There are three zones: 1) natural areas, 2) a Coast Guard LORAN Station and 3) a U.S. military toxic chemical storage area. During the nuclear tests of the 1950s, measures were taken to safeguard the seabird populations. The U.S. Army is now working with the U.S. FWS to avoid damage during the disposal of chemical munitions, and to inventory and monitor the ecological resources and establish conservation programmes (Maragos, 1986).

Recommendations   Maragos (1986) states that the U.S. Army should continue their plans to rid the atoll of chemical amunitions, and that the chemical incinerator for munition destruction should not be expanded. Resource conservation programmes should be implemented. Greater access to the atoll for scientific purposes is required.

PALMYRA ATOLL

Geographical Location   Central Pacific, 1000 mi. (1600 km) south of Hawaii near western end of the Line Islands; 5°52'N, 162°05'W.

Area, Depth, Altitude   Max. alt. 10 ft (3 m); max. lagoon depth over 50 ft (15 m); land area ca 1400 acres (567 ha).

Land Tenure   Privately owned and a sovereign territory of the U.S. The owners put the atoll up for sale in January 1987.

Physical Features   Small elliptical atoll with 50+ islets around a lagoon complex on a platform reef. There are no natural passes but the U.S. Navy has dredged a channel 200 ft (61 m) wide and 30 ft (9 m) deep along the western open side of the atoll. The lagoon consists of three sub-lagoons which are progressively more closed-off to ocean circulation to the east. Navy construction of causeways and landfills completely closed off the eastern lagoon (Dawson, 1959) and nearly closed off the central lagoon but subsequent erosion has breached the causeways and reopened or improved lagoon circulation. Palmyra has a wet climate (rainfall ca 200 in. (5100 mm) per year) and is the only "wet" atoll under U.S. jurisdiction. Storms are infrequent and unrecorded from the atoll and no storm berms are present.

Reef Structure and Corals   Only brief observations of the corals of Palmyra have been made (Maragos, 1979b). The atoll probably has a coral fauna comparable to other nearby atolls (Tabuaeran, Kiritimati) in the Line Islands which have been studied (Maragos, 1974b), probably 50-100 species and of higher diversity than those of Hawaii.

Noteworthy Fauna and Flora   Vegetation is very lush, with coconut palms, several species of ferns, and the trees *Pisonia*, and *Neiosperma* (*Ochrosia*) common. Coconut *Birgus latro* and land crabs are abundant but so are rats, brought recently by man or ship to the islands. Seabird nesting is confined to tree nesting species (Red-footed Boobies *Sula sula*, Black Noddies *Anous minutus*) except for ground-nesting on the open areas of the runway by Sooty Terns *Sterna fuscata* and Common Noddies *Anous stolidus*. There is a rich fish population in the lagoon (Dawson *et al.*, 1955; Maragos, 1979b and 1986).

Scientific Importance and Research   Palmyra's geographic isolation in the central Pacific and proximity to the intertropical convergence make it of value to reef scientists, physical oceanographers, pelagic fishery biologists, and meteorologists. Except for outbreaks of ciguatera during the occupation from 1940 to 1960, the marine biota have only been partially studied. As with Kiritimati Island and other nearby atolls of the Line Islands, Palmyra may have served as a stepping stone or source of much of the coral reef biota in the Eastern Pacific (Glynn and Wellington, 1983). Archaeologists might be interested in Palmyra to document whether Polynesians or Micronesians ever visited or inhabited the atoll, as has occurred at nearby Fanning Atoll.

Economic Value and Social Benefits   Currently uninhabited. From a military standpoint, the atoll is strategically located in a part of the Pacific with few islands, and the U.S. Navy attempted to retain control of the atoll after World War II. The landowners brought suit and eventually the U.S. Supreme Court returned the atoll to the private owners. It has obvious potential for aquaculture (because of the enclosed nature of the lagoons and large mullet populations), a tuna fishing base, a refuelling stop for aircraft, and possible resort development (Wimberley and Cook, 1961).

Disturbance or Deficiencies   Previous extensive dredging and filling by the Navy has had long term impact, modifying shorelines and water circulation, and destroying reef and fish in the central and eastern lagoons. However the last forty years has seen considerable recovery by fish and vegetation (Maragos, 1986). A proposal to use Palmyra for nuclear waste storage in 1979 was rejected.

Legal Protection None.

Management None.

Recommendations Reef research on biogeography and recovery from dredging, filling and other human disturbances should be carried out. Ecological reserve status may be appropriate, on account of the wet climate, unique biota and uninhabited status of the atoll.

WAKE ATOLL (ENEN KIO)

Geographical Location Central Pacific, several hundred km north of the Marshall Islands; 19°17'N, 167°E.

Area, Depth, Altitude Max. alt. 10 ft (3 m); max. lagoon depth 15 ft (4.6 m); land area 2.85 sq. mi. (7.4 sq. km) and lagoon area 3.75 sq. mi. (9.7 sq. km).

Land Tenure Federal Government (unincorporated territory administered by U.S. Air Force).

Physical Features Wake is a small triangular atoll with three islands and a shallow reef enclosing a shallow lagoon. There are no passes. Small coral heads, mounds and pinnacles occur scattered on sand in the lagoon. Periodic storms shape the geomorphography of the atoll: storm berms of coral rubble and shingle are found along seaward island shorelines and robust upper reef slopes.

Reef Structure and Corals Coral development and diversity appear to be low and are probably limited by storms, geographic isolation, shallowness of lagoon, and perhaps suboptimal light and temperature conditions. Corals achieve moderate to high abundance on the leeward (western) upper reef slopes.

Noteworthy Fauna and Flora The vegetation, described by Fosberg (1959), consists of indigenous and introduced species characteristic of arid atolls. Green *Chelonia mydas* and Hawksbill *Eretmochelys imbricata* Turtles forage but do not nest. An endemic flightless bird, the

Wake Rail *Rallus wakensis*, was extinct by 1945, probably due to the wartime Japanese occupation which also decimated nesting sea bird populations. Several species of sea birds have re-established large nesting populations. Fishes of Wake are described by Fowler and Ball (1925).

Scientific Importance and Research There have been no quantitative surveys of the reefs. Wake's ancient age and geographic isolation make it of potential interest to reef scientists. The palaeobiology or biology of the extinct rail may also be of interest.

Economic Value and Social Benefits The Marshall Islanders are said to have visited Wake before the advent of recent western discovery of the atoll and have a Marshallese name for it (Enek Kio) (Anon., n.d.). Pan American operated a seaplane base as part of the first trans-Pacific commercial airline route between San Francisco and Manila in the mid-1930s. Wake is strategically important due to its location for military missions and refuelling of aircraft. It was one of the first U.S. possessions to fall to the Japanese at the beginning of World War II and there are monuments, bunkers and other reminders of the tragic battles and occupation during the war, of historic significance.

Disturbance or Deficiencies The bombing and occupation of Wake by the Japanese and subsequent constant bombing by U.S. Forces devastated the terrestrial ecology of the atoll. Bombing, dredging, and filling also affected reefs although the extent of damage was never documented. The birds were particularly affected although the seabirds have now recovered (Bryan, 1959; Krauss, 1969; Maragos, 1986). In recent years, eroded shorelines and docks have been subject to repair. A tanker ran aground, sank and caused a major oil spill and reef damage in 1970 (Gooding, 1971). Partial recovery had occurred by 1979 (Maragos, 1979a and 1986). There are now plans to build facilities at Wake as part of the Strategic Defense Initiative of the U.S.

Legal Protection Air Force regulations have designated important sea bird nesting areas as refuges.

Management Air Force management of sea bird nesting areas has been adequate.

Recommendations Zoogeographic and reef evolutionary studies should be carried out and environmental assessment is required before new military missions.

HONG KONG

INTRODUCTION

General Description

Hong Kong is located on the eastern bank of the Pearl River which drains a vast area of China (estimated at 228 000 sq. km). The river is some 100 km wide at the mouth, with Macau on the western bank. The annual flow is estimated to be 308.26 billion cu.m and some 85.5 million tonnes of sediment are deposited each year. It imposes an estuarine environment on Hong Kong's western shores. The north-western shores are most strongly affected and support no hermatypic corals. The impact of the river to the east is minimal and any effect it might have cannot be differentiated hydrographically from the effects of local runoff. Eastern shores are thus washed by oceanic waters which are characteristically highly saline. The central region is a zone of compromised hydrography between western (estuarine) and eastern (oceanic) zones. However, because of the extensive pollution of this area (see below), today only the eastern waters are capable of sustaining hermatypic coral growth.

The hydrography of the South China Sea is complex but is essentially controlled by climatic changes. Hong Kong can best be described as sub-tropical. It is influenced in winter by the continental anticyclone system (the north-east monsoon) and the weather is thus cold and dry. At this time, the north-east monsoon pushes the cold Taiwan Current into the South China Sea. However, the Kuroshio enters the South China Sea via the Luzon Straits at the same time, and serves to keep Hong Kong's waters warm and possibly helps to sustain the coral community over the winter. In summer, the tropical cyclonic system (the south-east monsoon) results in weather which is hot and wet. The tropical, south-east monsoon pushes the warm Hainan Current into the South China Sea and sea temperatures rise. Hong Kong's eastern and south-eastern coasts are washed by these water masses and mean normal sea temperatures range between 16 and 27°C. Inshore, however, where sea temperature more closely follows that of the air, temperatures may range between 15 and 30°C (Cope, 1984 and 1986; Morton and Wu, 1975). Spring and autumn are times of transition. The hydrography of the inshore waters of Hong Kong have been studied by Morton (1982 and 1985), Morton and Wu (1976) and Cope (1986).

Although the total land area of Hong Kong is small (1040 sq. km), the deeply incised "drowned" mainland coastline, with 235 offshore islands representing previous mountain tops, is almost 800 km long. In the east and south-east there are high mountains, so that shores are steep, exposed, and often inaccessible, except by boat. In the north-west, sedimentation from the Pearl River has created flat alluvial plains bordered by extensive mud flats. Thus, geologically too, coral growth is restricted to the south and east.

Hong Kong is located on the borders of the temperate Palaearctic and tropical Indo-Pacific regions. In the sea, however, the major component is Indo-Pacific and the hermatypic coral community comprises some 49 species

(Scott, 1984; Veron, 1982; Zou, 1982). There is a tendency for turbinarian corals to increase in diversity from south to north (Zou and Chen, 1983). There are however no true reefs and coral communities grow on any hard substrate, in the narrow coastal fringe, on the eastern and south-eastern hard shores and in oceanic waters. Because of exposure to prevailing wind and wave action, branching, foliose and erect forms are few in exposed situations, and the community is dominated by low, submassive, encrusting faviids and poritids. In suitably sheltered bays, with clear water, relatively rich coral communities with 100% cover develop and a wider range of species and forms are represented (Cope 1981 and 1984; Thompson and Cope, 1982). For example, at Hoi Ha Wan on the north-east coast, 36 species have been recorded with a mean diversity (H') of 1.4 (Cope, 1984). Cope (1984) and Thompson and Cope (1982) have shown that greatest abundance and diversity occur at 1.3 to 2.3 m depth. Species diversity has also been shown to decrease towards more estuarine, enclosed sites in Tolo Harbour (Scott and Cope, 1982). Interspecific coral interactions at Hoi Ha Wan have been studied by Cope (1981), and hydrographic patterns at the same site by Cope (1986).

Corals are generally subtidal as cold temperatures in winter are likely to kill exposed colonies. Probably also because of inshore light attenuation following local freshwater runoff, with commensurate high turbidity, local hermatypic corals typically do not extend below 10 m depth. In deeper water, the hard benthos is characterized by a gorgonian and antipatharian community (Zou and Scott, 1982; Zou and Zhou, 1984). The coral community is held in delicate balance and many species are probably existing at the extremes of their hydrographic range.

Reef Resources

Information has not been obtained on fishery resources which are important but not directly related to reefs. Marine algal resources are described in Win Shin-Sun (1987). SCUBA diving is an increasingly popular activity.

Disturbances and Deficiencies

In winter, and because lowest tides are in the early morning, air temperatures below 10°C are often recorded for short periods, with the associated possibility of wind chill. Following such a cold spell in January 1981, those tidally exposed corals, notably *Porites lobata*, had expelled zooxanthellae. Instances of coral death, following similar climatic events have been recorded (Cope, 1984).

The central region of Hong Kong is the most densely occupied with the majority of the territories' 6 million inhabitants. Into these waters are discharged most of the urban, industrial and agricultural effluents (Morton, 1982). Coral growth is therefore virtually impossible, although it is known from the evidence of skeletons, as in Tolo Harbour (Scott and Cope, 1982) that corals did once grow here.

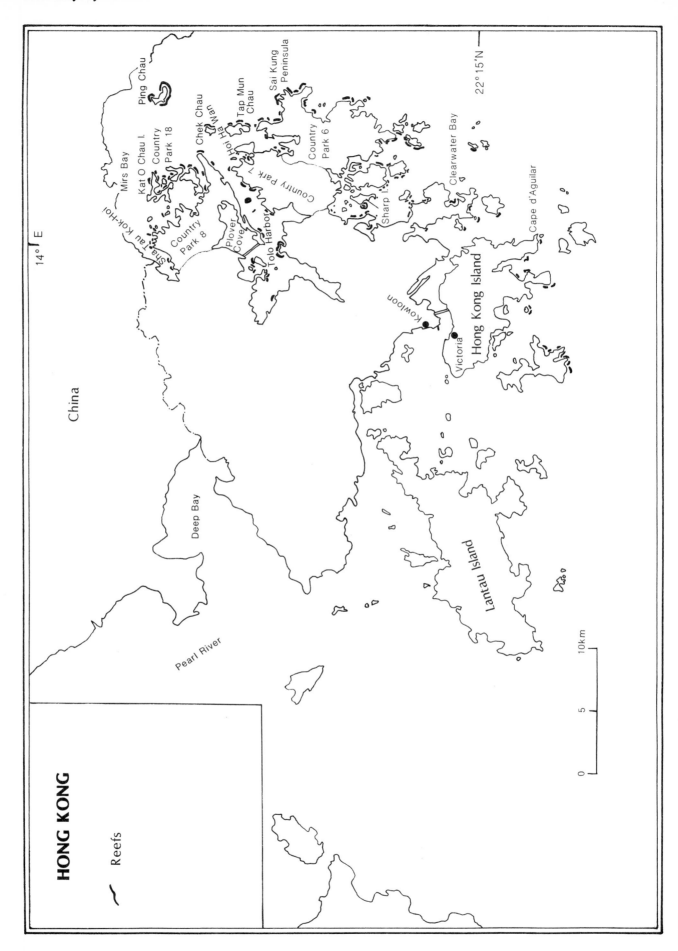

HONG KONG

Reefs

China

14° E

22°15'N

Mirs Bay

Ping Chau

Kat O Chau I.

Country Park 18

Chek Chau

Hoi Ha Wan

Tap Mun Chau

Sai Kung Peninsula

Country Park 6

Clearwater Bay

Cape d'Aguilar

Sha Tau Kok-Hoi

Country Park 8

Plover Cove

Country Park 7

Tolo Harbor

Sharp I.

Hong Kong Island

Kowloon

Victoria

Deep Bay

Pearl River

Lantau Island

10km

5

0

Pollution and especially land reclamation are now major threats to the coral communities of the north-east, although the local tropical mangrove community, which here also approaches the northern limit of its range, is under greatest threat, with intertidal alluvial flats most amenable to reclamation,(Morton, 1976a). The coral communities however, are particularly under threat from the combined effects of over-fishing (by many methods including dynamiting), coral collection for lime, increased SCUBA diving and shell collecting. The greatest threat however is from pollution on eastern shores with concomitant reclamation, dumping of dredged materials at sea, effluent discharge and a dramatic overall decline in inshore water quality (Cope, 1984; Morton, 1976a; 1982; Morton and Morton, 1983; Scott and Cope, 1982; Thompson and Cope, 1982). Scott (1984) believes that two corals (*Pocillopora* sp. and *Fungia* sp.) have already disappeared and Cope (1984) includes *Galaxea* cf. *astreata* in this category. While it is impossible to be absolutely certain that these species are locally extinct, it is known that the Giant Clam *Tridacna maxima* once occurred in Hong Kong (Rosewater, 1965), but is no longer present (Morton and Morton, 1983). Scott and Cope (1982) have found a progressive decline in the number of corals along six sites ranging from the open sea to the inner regions of Tolo Harbour. Dudgeon and Morton (1982) recorded a concomitant greater intensity of coral borers at ecologically stressed sites and believe that this is pollution-induced. Further information on marine pollution in Hong Kong is given in Morton (1976b, 1976c and 1982) and Morton and Morton (1983).

Legislation and Management

Legislation relating to the sea has been discussed by Faulkner (1976). The Country Parks Ordinance, Chapter 208 (plus amendments) (1976), defines a Country Park as "Any area that is designated as a Country Park" and a Special Area means "Any area designated as such". In Hong Kong there are (at present) 21 Country Parks covering an area of 41 296 ha or 40% of the total land area. Of these, parks such as 10 and 11 (Lantau Island), 19 (Cape D'Aguilar), 16 (Clearwater Bay), and 17 (Sharp Island) but most importantly 6 and 7 (Sai Kung Peninsula), 8 (Plover Cove) and 18 (islands of eastern Mirs Bay), have a sea coast. Parts of the island of Kat O Chau in north-eastern Mirs Bay (with a rich surrounding coral community) is the only one of 13 Special Areas that has a coastal component. In addition there are a number of sites of "Special Scientific Interest" (SSSIs) but, unlike Country Parks or Special Areas, these have no legal protection or provision for controlling their use and management. Of 45 local SSSIs, 19 are coastal and include some offshore islands and parts of the mainland coastline fringed by a coral community.

However, the Country Parks Ordinance does not specifically encompass the sea bed so that even those Country Parks and Special Areas with a coastline do not afford protection to any marine element within their boundaries. Although activities in coastal waters may be regulated by a variety of other Ordinances (Faulkner, 1976), in essence no coastal marine community in Hong Kong is given specific protection and there are no Marine Country Parks or Special Areas.

The Fisheries Protection Ordinance Chapter 171 (1964) provides for the conservation of fish and other aquatic organisms. To date, the only Regulations promulgated prohibit the use of explosives and poison in fishing.

Recommendations

Morton (1976a) and Morton and Morton (1983) recommend that the whole of the coastline of the north and eastern regions, i.e. Country Parks 6, 7, 8 and 18, should be afforded Marine Reserve status.

References

Cope, M. (1981). Interspecific coral interactions in Hong Kong. *Proc. 4th Int. Coral Reef Symp., Manila* 2: 357-362.

Cope, M. (1982). A *Lithophyllon* dominated coral community at Hoi Ha Wan, Hong Kong. In: Morton, B. and Tseng, C.K. (Eds), *Proceedings of the First International Marine Biological Workshop: The marine flora and fauna of Hong Kong and Southern China.* Hong Kong University Press, Hong Kong. Pp. 587-594.

Cope, M. (1984). *An ecological survey of the scleractinian coral community at Hoi Ha Wan, Hong Kong.* Unpub. M.Phil. Thesis, University of Hong Kong.

Cope, M. (1986). Seasonal, diel and tidal hydrographic patterns, with particular reference to dissolved oxygen, above a coral community at Hoi Ha Wan, Hong Kong. *Asian Marine Biology* 3: 59-74.

Dudgeon, D. and Morton, B. (1982). The coral-associated Mollusca of Tolo Harbour and Channel, Hong Kong. In: Morton, B. and Tseng, C.K. (Eds), *Proceedings of the First International Marine Biological Workshop: The marine flora and fauna of Hong Kong and Southern China.* Hong Kong University Press, Hong Kong. Pp. 627-650.

Faulkner, R. (1976). The Law of the Seashore. Chap. 1. In: Morton, B. (Ed.), *The Future of the Hong Kong Seashore.* Oxford University Press, Hong Kong. Pp. 3-12.

Morton, B. (1976a). Future plans for the Hong Kong Seashore. Chap. 9. In: Morton, B. (Ed.). *The Future of the Hong Kong Seashore.* Oxford University Press, Hong Kong. Pp. 176-189.

Morton, B.S. (1976b). The Hong Kong Seashore - an environment in crisis. *Env. Cons.* 3(4): 243-254.

Morton, B.S. (1976c). *The Future of the Hong Kong Seashore.* Oxford University Press.

Morton, B. (1982). An introduction to Hong Kong's marine environment with special reference to the North-eastern New Territories. In: Morton, B. and Tseng, C.K. (Eds), *Proceedings of the First International Marine Biological Workshop: The marine flora and fauna of Hong Kong and Southern China.* Hong Kong University Press, Hong Kong. Pp. 25-53.

Morton, B. (1985). The sub-tropical hydrographic environment of Hong Kong - a challenge to pollution impact assessment and monitoring. *Proceedings of the Asia and Pacific Regional Conference: Pollution in the urban environment.* Elsevier Press.

Morton, B. and Morton, J. (1983). *The Seashore Ecology of Hong Kong.* Hong Kong University Press, Hong Kong.

Morton, B. and Wu, R.S.S. (1975). The hydrology of the coastal waters of Hong Kong. *Env. Res.* 10: 319-347.

Rosewater, J. (1965). The family Tridacnidae in the Indo-Pacific. *Indo-Pacific Mollusca* 1(6): 347-396.

Scott, P.J.B. (1984). *Corals of Hong Kong.* Hong Kong University Press, Hong Kong.

Scott, P.J.B. and Cope, M. (1982). The distribution of scleractinian corals at six sites within Tolo Harbour and

Channel. In: Morton, B. and Tseng, C.K. (Eds.), *Proceedings of the First International Marine Biological Workshop: The marine flora and fauna of Hong Kong and Southern China*. Hong Kong University Press, Hong Kong. Pp. 575-586.

Thompson, G.B. and Cope, M. (1982). Estimation of coral abundance by underwater photography. In: Morton, B.S. and Tseng, C.K. (Eds.), *Proceedings of the First International Marine Biological Workshop: The marine flora and fauna of Hong Kong and Southern China*. Hong Kong University Press, Hong Kong. Pp. 557-574.

Veron, J.E.N. (1982). Hermatypic Scleractinia of Hong Kong, an annotated list of species. In: Morton, B.S. and Tseng, C.K. (Eds), *Proceedings of the First International Marine Biological Workshop: The marine flora and fauna of Hong Kong and Southern China*. Hong Kong University Press, Hong Kong. Pp. 111-125.

Win Shin-Sun, R.S.S. (1987). Marine algal resources of Hong Kong. In: Furtado, J.I. and Wereko-Brobby, C.Y. (Eds), *Tropical Marine Algal Resources of the Asia-Pacific Region: A Status Report*. Commonwealth Science Council Publication Series 181. Pp. 31-40.

Zou, R.L. (1982). A numerical taxonomic study of *Turbinaria* (Scleractinia) from Hong Kong. In: Morton, B.S. and Tseng, C.K. (Eds), The Marine Flora and Fauna of Hong Kong and Southern China. *Proceedings of the First International Marine Biological Workshop: The marine flora and fauna of Hong Kong and Southern China*. Hong Kong University Press, Hong Kong. Pp. 127-134.

Zou, R.L., and Chen, Y.Z. (1983). Preliminary study on the geographical distribution of shallow-water Scleractinia corals from China. *Nanhai Studia Marina Sinica* 4: 89-95. (In Chinese with English abstract).

Zou, R.L. and Scott, P.J.B. (1982). The Gorgonacea of Hong Kong. In: Morton, B. and Tseng, C.K. (Eds.), *Proceedings of the First International Marine Biological Workshop: the marine flora and fauna of Hong Kong and Southern China*. Hong Kong University Press, Hong Kong. Pp. 135-160.

Zou, R.L. and Zhou, J. (1984). Antipatharians from Hong Kong waters with a description of a new species. *Asian Marine Biology* 1: 101-105.

JAPAN

INTRODUCTION

General Description

Japan has a 27 000 km long coastline and consists of a chain of submerged continental islands separated from the Asian continent by the Sea of Japan. Hokkaido and the northern part of the main island, Honshu, facing the Pacific Ocean are affected by the cold Chishima Current from the north, but the other islands are washed by the warm Kuroshio Current and its branch the Tsushima current. The former flows north along the margin of the continental shelf of mainland China in the East China Sea, about 100 km to the west of the Ryukyu-retto or Ryukyu Archipelago (the southern part of the Nansei-shoto or Nansei Archipelago), and then veers east into the Pacific through the Tokara Channel south of Kyushu. The climate is therefore mild, and subtropical marine life is found along the coasts of the southern islands (Tamura, 1972).

Extensive work has been carried out in recent years on Japanese reefs, mainly at the Sesoko (University of the Ryukyus), Seto (Kyoto University), Sabiura and Kuroshima (Marine Parks Center), Iriomote (Tokai University) and Yaeyama Marine Laboratories and the reefs of the Ryukyu-retto are becoming particularly well known scientifically. The Sabiura Marine Laboratory carries out its reef-related work at Kuro-shima off southern Kyushu (Muzik *in litt.*, 5.8.87), and extensive work has also been carried out on Miyake-jima by the Tatsuo Tanaka Memorial Biological Station. WWF Japan has recently carried out a survey of reefs in Nansei-shoto, especially Ishigakishima (*see separate account*), in the course of a broad survey of conservation aspects (Tokuyama, 1985). Hori (1980) gives an overview of reef distribution and Fujiwara (1979) describes the distribution of reef-building corals. Tsuya (1977) gives an overview of the geology of the three island arcs (Honshu, Satsunan-Ryukyu or Nansei and Izu-Ogasawara) and briefly mentions the distribution of coral reefs. The Environment Agency undertakes a national survey of the natural environment every five years, in accordance with the Nature Conservation Law. In 1978, coral reefs were included in this survey, using aerial photographs and a rather limited number of ground-truthing field trips; the following figures for reef and coral assemblage distribution at that time are taken from Okinawa Prefectural Government (1979b), amended by Yamaguchi (*in litt.*, 11.5.87).

The northern part of Japan is rather marginal for coral growth. Forty coral genera have been recorded from around Honshu, Shikoku and Kyushu islands (Eguchi, 1977), and several of the temperate zone marine parks have corals (e.g. in Tottori Prefecture on Honshu (Eguchi, 1970b), and see table). Yabe and Sugiyama (1935a and b) reported the northernmost localities for reef-building corals on Honshu as Tateyama Bay (tip of Boso Peninsula, 35°00'N, 139°50'E) and Enoura Bay (north-east Suruga Bay, 35°01'N, 138°12'E), seven species having been recorded at the former and 34 at the latter, although higher figures are now reported from these areas (Tribble and Randall, 1986). Reef-building corals have also been described from Tokyo Bay (Eguchi,

1935) and around the Boso Peninsula (Eguchi, 1971b) including in the Katsuura Marine Park, and on the west coast of Miura Peninsula or east Sagami Bay, also at about 35°N (Eguchi, 1968a; Yamaguchi *in litt.*, 11.5.87). Corals of the southern Izu Peninsula are described in Eguchi (1972a). Tokioka (1966) and Utinomi (1966) describe coral communities on the west coast of the Kii Peninsula; those in Kushimoto Marine Park are described in a separate account. In the Korea Strait, reef corals are found on the island of Tsushima (129°14'E, 34°25'N), in Nagasaki Prefecture (Eguchi, 1973); the coral fauna of Nagasaki Prefecture is described by Eguchi (1975a). Other studies on marine fauna in Nagasaki Prefecture include Eguchi and Takuda (1972). Takahashi *et al.* (1985) reported the northern limit of the coral reefs to be the island of Mageshima (31°45'N), 10 km west of the island of Tanega-shima in the Osumi-shoto (Osumi Islands) off the southern tip of Kyushu, but fringing and patch reefs are found further north around Iki and Goto-retto (for example in the Genkai, Wakamatsu and Fukue Marine Parks) (Eguchi, 1971a and 1977). Ito (1984) and Utinomi (1965) describe corals from the coastal waters of Shikoku; Tada (1983) provides information for the southern part of Shikoku. Eguchi (1938) describes reefs around Aoshima and corals of Miyazaki Prefecture on Kyushu. Information on the Amakusa-shoto, off the west coast of Kyushu, is given in Kikuchi and Araga (1968). Fukuda (1984) described coral assemblages and coral fauna (33 species in 22 genera of hermatypic corals, with extensive stands of tabular *Acropora*) in the Danjo-gunto (Danjo Islands), south of the Goto-retto (32°00'N, 128°24'E).

The Satsunan-Ryukyu arc or Nansei-shoto, consisting of a volcanic arc and a frontal arc, includes 100-200 islands. The Satsunan-retto include the volcanic arc, consisting of the Tokara-retto and Tori-shima, and the northern part of the frontal arc. Hirata (1967) describes reef-building activity in Takara-jima and Kodakara-jima in the Tokara-retto. Tokioka (1953) and Utinomi (1956) described the invertebrate fauna of the intertidal zone of the Tokara-retto. The Ryukyu-retto, part of the volcanic arc, stretch from Taiwan north to Okinawa and include the southern islands of the frontal arc north to the Amami-shoto. Their geology is described by Konishi *et al.* (1974) and Kawana (1985). Shallow water octocorals of the area are described by Utinomi (1976 and 1977) and other studies of the reefs include Mezaki and Toguchi (1985), Takahashi *et al.* (1985) and Yamazato (1971). There are abundant fringing reefs, particularly around the islands south of Amami O-shima. In general reefs are 150 m wide, with lagoons of a depth of 1-3 m. The morphology of the reef crest is influenced by northward waves generated by the winter monsoon, although in the south, the southward waves of the summer monsoon have an influence. The islands in the central and southern Ryukyu-retto have mainly fringing reefs although small barrier reefs are found off the Yaeyama-retto in the south and the island of Kume in the centre.

Reefs around Amami O-shima have been described by Hirata (1968a and b) and the invertebrates by Yabe and Sugiyama (1937) and Ooishi and Yagi (1964) (see also

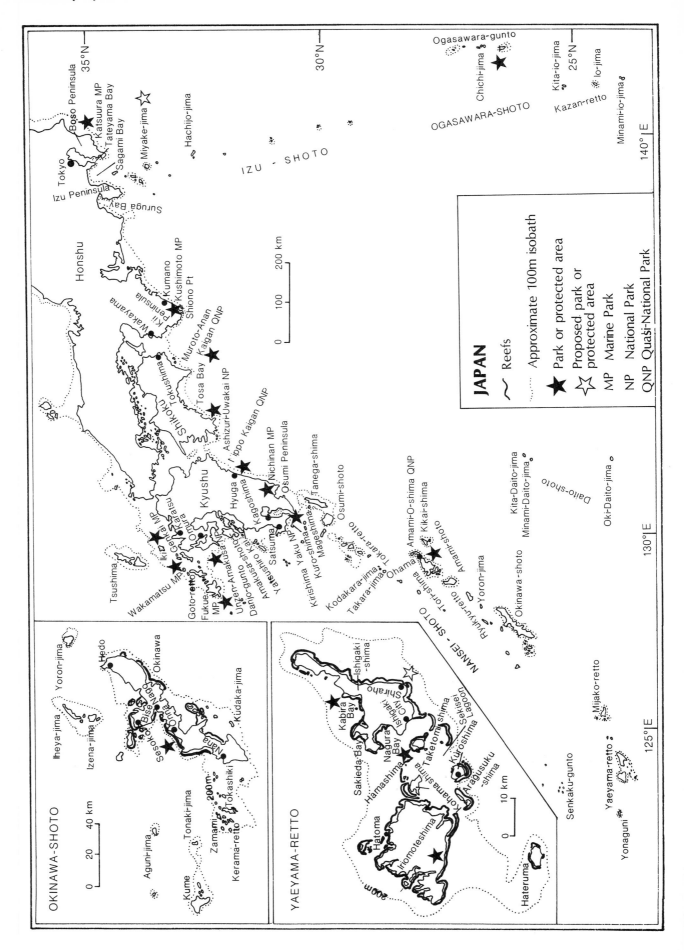

Sea area	Prefecture	Ha living 1978	% "extinct" 1973-1978	Notes
Noto Peninsula	Ishikawa	8	0	cold water *Rhizopsammia* community
Izu-Shichitou Minami	Tokyo	190	0	excl. Miyake-jima
Ogasawara gunto	Tokyo	233	0.4	dredging, 1976
Kumano Sea	Mie	21	0.1	
	Wakayama	188	1.1	road building, 1970s
Wakayama	Wakayama	756	0	
Tokushima	Tokushima	23	0	
Tosa Bay	Ehime	66	0	*A. planci* 1987
	Kohchi	14	55.7	*A. planci* 1976-77
Hyuga	Ohita	139	0	
	Miyazaki	36	0	
Karatsu Imari	Nagasaki	2	0	
Tsushima	Nagasaki	8	0	
Omura Bay	Nagasaki	1	0	
Goto	Nagasaki	50	0	inc. Danjo-gunto
Amakusa Bay	Kumamoto	62	0	
Yatsushiro	Kumamoto	1	0	
Osumi	Kagoshima	50	7.4	*A. planci* 1976
Kagoshima Bay	Kagoshima	14	19.1	*A. planci* 1977
Satsuma	Kagoshima	27	15.2	*A. planci* 1978
Osumi-shoto	Kagoshima	202	6.6	*A. planci* 1977-78
Tokara-retto	Kagoshima	228	8.8	*A. planci* 1975
Amami-shoto	Kagoshima	3534	26.8	*A. planci* 1973-78
Total coral assemblages		**5853**	**18.9 (1362 ha)**	
Okinawa	Okinawa	36 535	0.4	dredging, reclamation 1970s
Miyako-retto	Okinawa	12 193	0.03	reclamation 1973
Yaeyama-retto	Okinawa	30 974	0	
Total coral reef*		**79 702**	**0.21 (168 ha)**	

* Yaebishi Reef and some others not included, due to lack of aerial photographs; area refers only to reefs to about 20 m depth. Further work to complete and correct the figures obtained in this survey is being carried out (Yamaguchi *in litt.*, 11.5.87).

marine parks table list). Sixty-nine coral genera have been recorded including *Porites*, *Plesiastrea*, *Acropora*, *Favia* and *Caulastrea* (Eguchi, 1977). Further south, the reefs at Yoron-jima, Kagoshima Prefecture, 23 km north of Okinawa, are described by Hirata (1955, 1956, 1958, 1968c and 1977) (see also table of marine parks). This island is terraced and surrounded by "fringing barrier reefs" with a lagoon 2-3 m deep. Hirata (1977) lists 45 species of coral from Tomori Reef off the island; *Acropora pectinata* was dominant, followed by *A. cf. macrostoma*, but most of these reefs were destroyed by *Acanthaster* in the 1970s (Muzik, 1985).

The reefs of Okinawa Prefecture, which covers the island groups of Okinawa-shoto, Miyako-retto and Yaeyama-retto, were recently assessed by the Environment Agency and figures for the existing reef area are given in the above table (Yamaguchi *in litt.*, 20.3.85). Reefs of Ikema-jima, in the Miyako-retto, are described by Ohba and Aruga (1978). Corals of Okinawa are described by Yamazato (1975a) and Sakai and Nishihira (1986). Reefs are found around most of the islands in this area, including the Kerama-retto, west of Okinawa, such as Tokashiki, Zamami, Geruma, Aka, Yakabi and Fukaji. Corals from Kudaka-jima, south-east of Okinawa, are described by Nishihira *et al.* (1987). Reefs on the west coast of Okinawa, in part of the Kerama-retto, around Ishigakishima and in the vicinity of the Yaeyama-retto are described in separate accounts.

The Izu-Ogasawara arc forms a link between Japan and the Marianas, and also consists of a volcanic arc (seven Izu-shoto and the Kazan-retto or Volcano Islands) and a frontal arc (Ogasawara-shoto or Bonin Islands) (Tsuya, 1977). Reefs around Miyake-jima in the Izu-shoto are described in a separate account. Shallow water scleractinian corals are also found around Hachijo-jima (33°N), the southernmost of the Izu-shoto; 184 species in

46 genera (of which 28 species in 8 genera were ahermatypic) were recorded (Takahashi, 1983). To the south lie the Volcano Islands (Kita-io-jima, Io-jima, Minami-io-jima) which have no true reefs due to lack of suitable substrate and frequent volcanic activity although there are some coral communities. The outer arc includes some thirty islands. Douglas (1969) lists 20 islands in Ogasawara-shoto. Tsuya (1977) reports rich fringing areas around some of the islands and 41 coral genera have been recorded from this area (Eguchi, 1977). Research is described by Eldredge (1975). The Ogasawara-shoto and Kazan-retto are also described by Tuyama (1953) and Ooishi (1970).

Reefs around the islet of Parece Vela (Oki-no-Tori Shima) (20°25'N, 130°01'E) in the Philippine Sea south-west of Io-jima are described by Tayama (1952) and Kurata (1986). The three islands of the Daito-shoto (Kita-Daito-jima and Minami- Daito-jima in the north, and Oki-Daito-jima in the south) lie in the north central Philippine Sea to the east of Okinawa. Tsuya (1977) describes their geological history. Kita-Daito-jima and Minami Daito-jima are raised atolls; the third is a raised table reef. There is little coral because of the steep slopes (Muzik *in litt.*, 15.4.85). They were once mined for phosphate and two of the islets are important breeding grounds for sea birds. Oki-Daito-jima is now used for U.S. military target practise. Marcus (Minami-Tori-shima) is an atoll of 750 acres situated to the east of the main body of Japanese islands. It has a coral island and fringing reef and is important for seabirds, and is briefly described by Douglas (1969), Bryan (1903) and Kuroda (1954). Tsuda (1968) describes the benthic algae.

Other publications on reefs in Japan include Sugiyama (1934), Yabe (1933), Yabe and Sugiyama (1932a and b, 1933a and b, 1936, 1938 and 1941), and Yamazato (1959). Reef fish are described by Masuda *et al.* (1975) and Randall *et al.* (1981). Nakazono *et al.* (1985) discuss the reproductive ecology of the reef fish *Paracerpis snyderi*. Seabirds are described in Hasegawa (1984); many of the islands with important reefs also have major seabird nesting colonies.

Reef Resources

The reefs of Japan play an important role for local people as a source of food. In northern Japan there is a long tradition of professional diver-fishermen called "ama" who dive for a variety of marine resources (Muzik *in litt.*, 5.8.87; Yamaguchi *in litt.*, 8.4.85). Extensive research has been carried out on the fishery industry by the Fisheries Experimental Station of Okinawa Prefecture, but this is largely published in Japanese. The pelagic fisheries are of greater economic importance but with increasing fuel costs and the introduction of EEZs in waters formerly used by the Japanese, reef fisheries are becoming more significant. Important reef resources include edible seaweeds, Giant Clams, the gastropods *Turbo argyrostomus*, *T. marmoratus*, *Trochus niloticus* and *Strombus luhuanus*, squids, octopi, the sea urchin *Tripneustes gratilla* and many reef fish (Yamaguchi *in litt.*, 11.5.87). Reef fisheries of the Ryukyu-retto are described by Fujimori (1964). The Boring Clam *Tridacna crocea* is an important food resource in Okinawa (Murakoshi, 1986).

Since most people in Japan live on the narrow coastal plains, much outdoor recreation is focused on coastal environments (Marsh, 1985). SCUBA diving is growing in popularity, although most tourists are still content to snorkel in shallow water. The reefs around Sekisei Lagoon, Ishigakishima and Iriomoteshima are particularly popular. At least 60 dive-shops are located in Okinawa and adjacent islands alone, and there are numerous diving clubs (Yamaguchi *in litt.*, 11.5.87). An increasing amount of accommodation is being built for SCUBA divers (Muzik *in litt.*, 15.4.85).

Disturbances and Deficiencies

The table at the beginning of this section gives figures supplied by the Environment Agency for reef decline in Japan. Japanese reefs have suffered considerably from outbreaks of *Acanthaster planci* and there are numerous publications on the problem including Environment Agency, Japanese Government (1973 and 1974), Fukuda (1976), Fukuda and Okamoto (1976), Fukuda and Miyawaki (1982), Hayashi (1975), Hayashi and Tatsuki (1975), Kurata (1984), Matsusita and Misaki (1983), Misaki (1974 and 1979), Nishihira and Yamazato (1972 and 1973), Okinawa Tourism Development Corporation (1976), Suzuki (1975), The Japan Science Society (1979), Uchida and Nomura (in press), Ui (1985), Utinomi (1962) and Yamaguchi (1986a, b and c, 1987). Recent research on *Acanthaster* in Japan is outlined by Yamaguchi (1987). A number of projects are under way at the Kuroshima Laboratory of the Marine Parks Center, the Iriomote Marine Research Station, Sesoko Marine Science Center and the Dept. of Marine Sciences, University of the Ryukyus. Kagoshima Prefectural Fisheries Experimental Station has been trying, as yet unsuccessfully, to raise larval *Charonia tritonis*, a known predator of *Acanthaster*, for release on the reefs.

Nishihira and Yamazato (1974) document outbreaks in the Okinawa-shoto (*see also account for* Okinawa Marine Park and Iriomoteshima). The first outbreak was in 1957 on Miyako-shima, an island to the north of Ishigakishima (Yamazato, 1969) but almost all the reefs in the Ryukyus are now considered affected (Yamaguchi, 1986a). Outbreaks on the Ohama coast of Amami O-shima were described by Nakamura (1986), who noted that live coral was now very rare in depths of ca 3 m or more on the reef flat. Recovery of corals is thus only noticeable on the reef crests, in waters 1-2 m deep (Muzik *in litt.*, 5.8.87). The Tonaki reefs have a lower density of *Acanthaster* and corals were still present. Control methods are described by Yamaguchi (1986a), many of these having failed probably due to lack of continuous or regular application.

At the time of the massive outbreaks of *Acanthaster* in the Ryukyu- retto, the Pacific coasts of Kyushu, Shikoku and Honshu were also experiencing unusual populations of this species (Yamaguchi, 1986a and 1987). It is thought that these outbreaks might be related to changes in the course of the warm Kuroshio current. *Acanthaster* was never recorded around the Izu Peninsula in the north but permanent populations were found in the Amami-shoto; between these two areas, populations of variable persistency were found along the temperature gradient (Yamaguchi, 1987). An outbreak has occurred at Miyake-jima (*see separate account*). *Acanthaster* occurs in the Ogasawara-shoto but infestations have not been reported (Kurata, 1984).

There have been recent reports of coral bleaching in the Yaeyama-retto, Okinawa (Makiminatao Reef) and Iriomoteshima (Glynn, 1984; Kamezaki and Ui, 1984; Yamazato, 1981 and *see separate accounts*). Sea surface temperatures were abnormally high in 1983, suggesting a correlation with the abnormal El Niño year, and there was extensive mortality of corals and alcyonaceans to 10 m depth. In 1986, similar bleaching was found in shallow waters around Kudaka-jima, off south-east Okinawa (Nishihira *et al.*, 1987) and on the southern coast of Okinawa itself (Nishihira, 1987). Mass mortality of the sea urchin *Echinometra mathaei* was also observed (Nishihira, 1987). Reefs around Miyake-jima and Kushimoto Marine Park have been affected by outbreaks of the gastropod *Drupella* (*see separate accounts*). Epidemics of the encrusting sponge *Terpios* have been recorded as causing temporary damage to some reefs, for example at Tokuno-shima in the northern part of the Ryukyu-retto, and Tonaki-jima west of Okinawa (Yamaguchi, 1986d).

Japanese reefs are coming under increasing pressure from human activity. In Okinawa, reef damage has occurred since the Battle of Okinawa in 1945 when numerous ships were sunk and extensive damage was caused during salvaging activities. Since then, coastal construction, land reclamation and particularly intensive agriculture have caused increasing sedimentation (Okinawa Prefectural Government, 1979a; Sano *et al.*, 1984; Yamazato, 1987; Muzik, 1985; Kühlmann, 1985; Nishihira, 1987). A survey of the reefs of the Ryukyu-retto, from Kikai-shima in the Amami-shoto to Yonaguni in the south-west, in the early 1980s showed that of 224 sites, 63% had less than 10% living coral cover, 30% had 10-50% and only 7% had over 50% (Muzik, 1985). A survey was carried out in more detail around Okinawa and a dramatic decline in live coral cover was found (Muzik, 1985) compared with the results of Nishihara and Yamazato (1974). Reefs at Kume west of Okinawa were reported by Yamazato (1974) to have 50% coral cover but now only have 1-3% coral cover (Muzik, 1985). Tomori Reef, east of Yoron-jima, had 45 coral species in 1975 (Hirata, 1977) but was dead in 1981, and even reefs on the remote island of Yonaguni showed signs of deterioration (Muzik, 1985). At Bise, in northern Okinawa, coral diversity as well as living cover had declined, 78 species having been recorded in 1976 (Yamazato, 1976) and only 18 species being found recently (Yamazato *et al.*, in prep.).

Omija (1985) recorded significant terrestrial run-off in 174 out of 223 rivers in Okinawa, most of this a result of agricultural practices although construction was a contributing factor. A more detailed survey of discharge from the Ginoza-Fukuchi River on the east coast of Okinawa showed that living coral in the vicinity declined between 1977 and 1978 (Okinawa General Affairs Bureau, 1978 and 1979; Yamazato, 1987). Surveys of the east coast of Okinawa in 1972 (Nishihira and Yamazato, 1974), 1974 (Yamazato, 1975b) and 1984 (Coral Reef Survey Committee, 1985) showed that there had been increased siltation of the reefs over this period.

Pollution has become a major problem since the 1960s (Numata, 1977; Ogura and Sakamoto, 1984; Tamura, 1972; Uda *et al.*, 1977;). Closed bays are now seriously polluted; for example the Seto Inland Sea between Kyushu, Honshu and Shikoku (although not a reef area) has been polluted by crude oil from a refinery which has had damaging effects on the fisheries, and the Ryukyu-retto mangroves and beaches have been affected by oil pollution and siltation (Numata, 1977). Problems associated with the increasing appearance of red tides are described by Okaichi (1984).

There is little information on the impact of exploitation of reef resources, although there are numerous incidences of illegal collecting. However, reef gleaners are considered to cause a significant disturbance, and the 11th Regional Maritime Safety Agency, who police the sea, have arrested many poachers and gleaners, particularly around Okinawa and Amami O-shima. Taiwanese fishing boats are also known to exploit the Ryukyu reefs (Yamaguchi *in litt.*, 11.5.87). The Boring Clam *Tridacna crocea* has been overexploited in Okinawa, production having declined from 481 mt in 1973 to 92 mt in 1981 (Murakoshi, 1986).

Threats to Miyake-jima are described in a separate account. The proposed expansion of the military base on uninhabited Io-jima in Kazan-retto could pose similar threats to the marine as well as the terrestrial environment (Suzuki, 1986).

Legislation and Management

Moves to establish Marine Parks in Japan started as early as 1962. In 1967, the Marine Parks Center was established in Tokyo, and 50 potential sites were investigated (Marsh, 1985; Tamura, 1972). In 1969, the Ecological Society of Japan drew up a system of marine reserves which covered 10 areas from Hokkaido to Kyushu, excluding the Ogasawara-shoto and Ryukyu-retto (Tamura, 1972). In July 1970, Kushimoto and nine other areas were designated as marine parks (equivalent to Strict Nature Reserves) and in January 1971 a further 12 were designated. There are a total of 57 marine parks, those containing reef communities being listed in the table below. In addition there are marine sanctuaries at Nagura Bay and Kabira Bay on Ishigakishima, created with the co-operation of local fishermen and under the responsibility of the fishery agencies, and a nature conservation area or strict nature reserve at Sakiyama Bay on Iriomoteshima (*see separate account*). The Okinawa Prefectural Government issued regulations for conservation of the natural environment in 1973 which included provisions for the establishment of "conservation areas", but none have yet been established (Nishihira, 1987).

Most areas outside the Ryukyu-retto with good coral are within National Parks. Some of the Kazan-retto are protected for their seabird colonies (Tsuya, 1977) and Douglas (1969) reports that Minami-io-jima is possibly one of the least disturbed islands in the world. National Parks and reserves were reported to be under consideration for the Ogasawara-shoto in the 1960s (Douglas, 1969). A marine park is situated on the south-west promontory, Minamisaki, of Chichijima (Tsuya, 1977).

Japanese marine parks containing reefs and coral formations
(taken from Marine Parks Centre of Japan, 1975)

Name of Marine Park, adjacent land park, location, no. of areas, acreage, date designated	Facilities and main characteristics
MINAMIBOSO Q.N.P. (Honshu) Katsuura Katsuura, Chiba Pref. 1 (14.5 ha), 7.6.74	*Ecklonia cava* and *Eisenia bicyclis*, *Sargassum giganteifolium*, *S. fulvellum*, *Cystophyllum sisymbrioides* and *Joculator maximus*; fish include *Chaetodon collare*, *Chromis notatus*, *Prinurus microlepidotus*, *Leptoscarus japonicus*, *Girella punctata* and Labridae; invertebrates include: *Allopora boschmai*, *Solanderia secunda*, *S. misakiensis*, *Melithaea flabellifera* and *Echinogorgia rigida*; underwater observation tower and visitor centre.
YOSHINO-KUMANO N.P. (Honshu) Kushimoto	*(See separate account).*
MUROTO-ANAN KAIGAN Q.N.P. (Shikoku) Awaoshima Mugi, Tokushima Pref. 3 (15.5 ha), 22.1.71	Seabed of sandstone is covered with benthic organisms; seaweed *Padina arborescens*, corals *Acropora studeri*, *Porites tenuis*, *Leptoria phrygia*, *Favia speciosa* and *Turbinaria rugosa*; Actiniaria is characteristic; tropical fish e.g. *Amphiprion clarkii*, *Pomacentrus coelestis* and *Chaetodon auripes* abundant.
Awatakegashima Shishikui, Tokushima Pref. 2 (9.9 ha), 1972	Rather shallow seabed of sandstone; *Acropora studeri*, *Pavona decussata* and *Favia speciosa*; tropical fish e.g. *Pomacentrus coelestis*, *Abudefduf vaigiensis* and *Chaetodon auripes*.
ASHIZURI-UWAKAI N.P. (Shikoku) Tatsukushi Tosashimizu, Kochi Pref. 4 (24.9 ha), 10.11.72 (was designated as a part of Ashizuri Q.N.P. on 1.7.70 and 22.1.71)	Information given in Shirai (1971). 5 glass-boats; underwater observatory; shell museum; in Minokoshi Bay, large *Pavona decussata* of aesthetic and scientific interest which was designated a natural monument in 1964 (Tokioka, 1968); tropical fish e.g. *Pomacentrus coelestis* and *Chaetodon auripes* abundant.
Okinoshima Sukumo, Kochi Pref. 5 (36.3 ha), 10.11.72	*Acropora leptocyathus* and Alcyonacea; tropical, temperate and sometimes large migratory fish.
Kashinishi Otsuki, Kochi Pref. 2 (16.8 ha), 10.11.72	Environment similar to inland bay or lagoon; corals of every kind abundant.
Uwakai Nishiumi, Ehime Pref. 6 (32.3 ha), 10.11.72 (was designated on 1.7.70 as a part of Ashizuri (Q.N.P.)	5 glass-boats; visitor centre; Alcyonacea, corals and fish abundant; *Tubastrea aurea*, *Melithaea flabellifera*, *Anthoplexaura dimorpha* in caves at Kashima-no-hora.
GENKAI Q.N.P. (Kyushu) Genkai Karatsu, Yobuko, Chinzei, Saga Pref. 5 (45.5 ha), 1.7.70	2 glass-boats; visitor centre; underwater observation tower; caves and cliffs of basalt; tropical and temperate fauna and flora; abundant seaweeds and corals at northern limit of range (Eguchi, 1970a).

Japanese marine parks containing reefs and coral formations (contd)

Name of Marine Park, adjacent land park, location, no. of areas, acreage, date designated	Facilities and main characteristics
SAIKAI N.P. (Kyushu) Wakamatsu Wakamatsu, Nagasaki Pref. 3 (19.2 ha), 16.10.72	Volcanic sedimentary coast; inland-sea with many islands; thick growths of Alcyonacea particularly at Katashioseto; at Gotejima: *Porites tenuis* and *Acropora studeri*; at Harunomendo: abundant Gorgonacea on rockfaces.
Fukue Fukue, Nagasaki Pref. 2 (11.2 ha), 16.10.72	North-western marine area of Takenokojima and Yaneojima off Fukue Port; due to warm current, many Scleractinia e.g. *Acropora squarrosa* and *Podabacio elegans lobata*, soft corals e.g. *Melithaea flabellifera* and Alcyonacea; *Sargassum* and *Zostera marina*; tropical fish abundant; corals on Goto Islands described by Eguchi (1971a).
UNZEN-AMAKUSA N.P. (Kyushu) Tomioka Reihoku, Kumamoto Pref. 2 (16.2 ha), 1.7.70	Seaweeds, *Martensia denticulata, Eckloniopsis radicosa, Corallina pilulifera, Amphiroa dilatata* and stony corals abundant; in shallow water, Alcyonacea and tropical fish abundant.
Amakusa Amakusa, Kumamoto Pref. 1 (5.1 ha), 1.7.70	2 glass-boats; Ogase is composed of 10 large and small groups of reefs; each reef has complicated topography; abundant Sceleractinia, tropical fish, Alcyonacea and *Anthoplexaura dimorpha*.
Ushibuka Ushibuka, Kumamoto Pref. 4 (30.4 ha), 1.7.70	2 glass boats; Alcyonacea is characteristic, abundant *Antipathes japonica*, Scleractinia, Actiniaria, *Melithaea flabellifera* and *Lobophytum batarum*, and tropical fish.
NIPPO KAIGAN Q.N.P. (Kyushu) Nanpokuura Nobeoka, Kitaura, Miyazaki Pref. 5 (28.7 ha), 15.2.74	1 glass-boat; large *Acropora leptocyathus, A. studeri, Favia speciosa, Hydroides norvegicus* and *Dendronephthya*; tropical fish e.g. *Chaetodon auripes*, Labridae, *Chromis notatus, Amphiprion clarkii* are abundant.
Kamae Kamae, Oita Pref. 4 (33.5 ha), 15.2.74	*Acropora leptocyathus, Pavona decussata, A. studeri* and Alcyonacea widespread; tropical fish such as *Chaetodon auripes*, Labridae, *Chromis notatus* and *Amphiprion clarkii*.
NICHINAN KAIGAN Q.N.P. (Kyushu) Nichinan Nango, Kushima, Nichinan, Miyazaki Pref. 6 (55.9 ha), 1.7.71	3 glass-boats; Alcyonacea, *Anthoplexaura dimorpha* and *Melithaea flabellifera* on reefs; at Komenotoura Bay, dense coverage of large stratiform *Acropora leptocyathus*; tropical fish such as *Chaetodon auriga* and Labridae.
KIRISHIMA-YAKU N.P. (Kyushu) Sakurajima Nishisakurajima, Kagoshima Pref. 2 (14.7 ha); 1.7.70	Aquarium; Hakamagoshi has rugged seabed topography, formed by Taisho lava current; abundant soft and hard corals, seaweed, *Padina aroborescens*; at Okinokojima diverse coral and Actiniaria.
Sata Misaki Sata, Kagoshima Pref. 2 (11.8 ha); 1.7.70	1 glass-boat; abundant and diverse corals; also *Melithaea flabellifera, Anthoplexaura dimorpha, Sarcophyton* and Alcyonacea.

Japanese marine parks containing reefs and coral formations (contd)

Name of Marine Park, adjacent land park, location, no. of areas, acreage, date designated	Facilities and main characteristics
AMAMI O-SHIMA Q.N.P. (Ryukyu-retto)	
Kasari Hanto-Higashi Kaigan Kasari, Kagoshima Pref. 1 (93 ha); 15.2.74	1 glass-boat; lagoon surrounded by outer reefs; branching corals e.g. *Acropora squarrosa*, *Stylophora pistillata* and massive *Porites tenuis* are abundant with *Pocillopora damicornis* and foliose *Montipora catus*; fish *Chaetodon auripes*, *Zebrasoma flavescens*, *Naso unicornis* and Labridae.
Surikozaki Naze, Kagoshima Pref. 1 (70 ha), 15.2.74	*Chaetodon auripes*; *Acropora*; *Sargassum*.
Kametoku Tokunoshima, Kagoshima Pref. 1 (70 ha), 15.2.74	3 glass-boats; *Acropora studeri*, *Favia speciosa*, *Goniopora planulata*, *Platygyra lamellina* and *Echinophyllia aspera* abundant; fish such as *Holocanthus semicirculatus*, *Chaetodon auriga* and *Amphiprion clarkii* abundant.
Setouchi Setouchi, Kagoshima Pref. 3 (58 ha), 15.2.74	1 glass-boat; shore reef fronting sand bottom; transition in distribution of fauna from open-sea kind to inland sea; corals: *Pocillopora damicornis*, plate-shape *Acropora studeri*, *Platygyra lamellina*, *Favia speciosa*, *Montipora foliosa*, massive *Porites tenuis*, and branch-shape *Montipora catus*; fish: *Chaetodon auripes*, *Chromis notatus*, *Labroides dimidiatus* and *A. lineatus* abundant.
Yoronto Yoron, Kagoshima Pref. 27°02'N; 128°26'E; 23 km north of Okinawa 3 (155 ha), 15.2.74	5.5 x 5 km, 97 m alt., 2 glass-boats, 16 glass-boats; corals in lagoon including branch- and plate-shape *Acropora*, *Porites tenuis*, *Stylophora pistillata*, *Montipora catus* and *Favia speciosa*; fish: *Chaetodon auripes*, *Heniochus acuminatus*, *Holocanthus semicirculatus*, *Amphiprion clarkii*, *Abudefduf sexfaxciatus*, *Prinurus microlepidotus*, *Balistes niger*, Labridae and *Leptoscarus japonicus* are abundant; reefs described by Hirata (1977): popular tourist resort but not for diving; reefs reportedly destroyed (Muzik *in litt.*, 25.3.85).
OKINAWA KAIGAN Q.N.P. (Ryukyu-retto)	(*see separate account*).
IRIOMOTE N.P. (Ryukyu-retto)	(*see account for* Yaeyama Marine Park).
OGASAWARA N.P. (Ogasawara-shoto) Ogasawara Hyotanjima, Hitomarujima, Anijimaseto (2), Minamishima, Miyukinohama, Hirashima, Ogasawara, Tokyo Metropolis 7 (463 ha), 16.10.72	Two areas: 1. Hyotanjima, Hitomarujima, Anijimaseto and Hahajima with topography of complicated volcanic rock, agglomerate and tuff seabed; 2. Minamishima with dipped caldera topography of limestone (unique in Japan); mainly *Acropora leptocyathus*; large tropical fish e.g. *Paracanthurus hepatus*, *Leptoscaruscus japonicus*; coral e.g. *Euphyllia fimbriata* abundant; park described by Fujiwara (1985).

N.P. = National Park
Q.N.P. = Quasi-National Park

There are numerous publications on the marine park system including Habe *et al.,* (1966), Marine Parks Center (1975), Okada (1966), Sabiura Marine Park Research Station (1977), Saito (1975), Tamura (1968, 1969, 1972 and 1973), Tamura and Ino (1985), Tokioka (1966) and Marsh (1985). The National Parks Law of 1931 permitted the creation of national parks for scenery protection and recreation, and a 1957 revision provided for three categories: national, quasi-national and prefectural nature parks, which vary in degree of scenic quality, protection and type of administration. The Environment Preservation Law of 1972 designated the Environmental Agency responsible for national parks (Marsh, 1985).

Marine parks include the adjacent land which is generally an extension of the national park system. They are usually small (10-200 ha, unless zoned), in water less than 20 m deep, extend 1-2 km from shore, and sport, recreation and education are considered high priorities in their design (Tamura, 1972). According to criteria drawn up by the Environment Agency, marine parks must:

a)   be within a National or Quasi-National Park;
b)   have space for lodges, visitor centres, and other ground facilities;
c)   be in water with visibility to the sea bottom;
d)   in areas with gentle tides and and wave action;
e)   in areas with a topographically interesting sea bottom and abundant and colourful marine life;
f)   be located away from sources of pollution;
g)   be established in collaboration with local fishermen.

The local (or Prefectural) Government wishing to establish a marine park, collates the available information and provisionally identifies suitable areas. The Marine Parks Center surveys the sites and makes recommendations to the Environment Agency but is not involved in the management of the parks (Hunnam, 1977). The parks have good recreational and educational facilities; visitor centres are well equipped with slide shows and museums, and recreational activities such as SCUBA diving are popular (Yamaguchi *in litt.*, 11.5.87) (although Moyer (*in litt.*, 6.5.87 reports that, apart from Ogasawara and Yaeyama National Parks, SCUBA diving is restricted to park staff). The parks are divided into separate units, each usually less than 12 ha in area on the basis of traditional fishery rights, although the more recently designated parks generally cover a greater area. According to Marsh (1985), marine parks consist of three zones: an Ordinary Area which is that part of the park lying within 1 km of the coast; a Marine Park Area which covers the specific features warranting protection; and an Ordinary Area or Buffer Zone which covers 1 km surrounding the park. Fishermen may retain their fishing rights within the parks which can create problems (Hunnam, 1977; Ino, 1966) although it is reported that they often avoid the area (Marsh, 1985).

Tamura (1972) describes the legislation relevant to marine parks. The following activities may not be carried out within a marine park without the permission of the Director-General of the Environment Agency in the case of National Parks and the permission of the prefectural governor in the case of Quasi-National Parks:

a)   Erection, rebuilding or extension of structures;
b)   Mining of minerals or gathering of soils or stones (also prohibited within 1 km of the Park);

c)   Putting up or setting up of advertisements on the structures and the like;
d)   Collection or capture of tropical fish, coral, seaweed, or plant and animal life similar to them that has been designated for each National or Quasi-National Park by the Directory-General of the Environment Agency with the consent of the Minister of Agriculture and Forestry;
e)   Reclamation of foreshore or land reclamation by drainage;
f)   Changing the feature of the seabed;
g)   Mooring;
h)   Discharge of polluted matter or waste through facilities set up for that purpose.

Those wishing to mine minerals, gather soils and stones or change features of the seabed within 1 km of the marine park area, i.e. within the Ordinary Area, must notify the prefectural governor.

Enforcement within the parks is described by Marsh (1985). Day to day management is a co-operative effort involving Environment Agency staff, Prefectural Government staff, tourist facility operators, Marine Parks Center scientists and volunteers. Patrols can usually only be made a few times each month, so enforcement largely depends on public responsibility and contributions by people working in the park and vicinity. The parks are heavily used for recreational purposes, and receive an estimated 18 million visitors a year. Most parks have extensive interpretive facilities and activities, although occasionally the commercial aspects of these enterprises may dominate the educational (Marsh, 1985).

Japanese small-scale fishermen have legally-guaranteed equitable access to and "ownership" of the living aquatic resources in nearshore waters. All coastal waters, with the exception of ports, their adjacent areas and tract reclaimed for industrial zones, are divided up among Fisheries Cooperative Associations (FCAs). Under Fisheries Law (1949), administered by the Ministry of Agriculture, Forestry and Fisheries, fisheries rights in the sea area under the jurisdiction of an FCA are the *bona fide* personal property of the individual members of the Association. Each Association establishes regulations for the control and operation of various types of fishery in an equitable, efficient and sustained manner, as local conditions dictate. The complexity of this system, which involves time-honoured customary procedures that have been gradually incorporated into modern legislation, is described in Ruddle (1985 and 1987a and b).

In the interest of resource conservation, the Ministry places limits and sometimes bans on some fisheries. Turtle eggs, and corals in the orders Scleractinia, Gorgonacea and Stolonifera, may not be collected in Okinawa Prefecture without permission and may not be owned, processed or sold if they originate from prefectural waters. There are size limits, closed areas and closed seasons for certain species. The Green Turtle *Chelonia mydas* and the Hawksbill *Eretmochelys imbricata* may not be taken from June 1 to July 31, and there is a size limit (ventral shell length 25 cm) for Hawksbills. Spiny lobster fisheries are closed from April 1 to June 30. There are restrictions on mesh sizes for nets, types of fishing methods used and in some cases on numbers of boats. Licenses are required for collecting corals, aquarium fish and a number of other fish (Ruddle, 1985). Recreational spearfishing with SCUBA equipment has been banned by the Okinawan Prefectural Government but this is not enforced (Nishihira, 1987). In Tokyo Prefecture, the collection of either living or dead organisms by SCUBA divers is prohibited except under permit for scientific purposes; this legislation applies throughout the Izu-shoto and Ogasawara-shoto (Moyer *in litt.*, 6.5.87). Additional discussion of the problems of fisheries and coastal zone management is given in Sakiyama (1982).

Major efforts have been made to control the *Acanthaster* outbreaks, in the Ryukyu-retto and also in Ashizuri-Uwakai (south-west Shikoku), Kushimoto and Miyake-jima in the Izu-shoto; these have involved local diving associations using SCUBA (Yamaguchi, 1986a), professional fishermen (in the Ryukyus) (Yamaguchi *in litt.*, 11.5.87) and local government agencies in Miyake-jima (*see separate account*). To date they have met with little success. Boring Clams are being farmed at the Yaeyama Branch of Okinawa Prefectural Fisheries Experimental Station (Murakoshi, 1986). There have been a number of artificial reef projects to encourage growth of lobsters and sea urchins; these are carried out by the fisheries agencies and publications are in Japanese (Yamaguchi *in litt.*, 11.5.87).

Recommendations

There is an urgent need to consider the terrestrial environment and catchment area and land use practices in the context of reef conservation and sedimentation (Nishihira, 1987). Hunnam (1977) made a number of recommendations for improvement of the marine park system, including the establishment of wardens for the parks, the provision of advice on the planning and construction of park buildings to avoid damage to the reefs, and discouragement of the sale of marine curios within the park. The Danjo-gunto, south of the Goto-retto, were proposed as a strict nature reserve and surveys were carried out by Nagasaki Prefecture and the Environment Agency (Fukuda, 1984). However it was not designated because of land tenure problems (Yamaguchi *in litt.*, 11.5.87). There is clearly a need for improvement in environmental education (Yamaguchi *in litt.*, 8.4.85).

References

(N.B. Many of these references are in Japanese and have not been consulted by the compiler directly; we are grateful to Dr. M. Yamaguchi for translating many of the titles.)

Anon. (n.d. a). *Kushimoto Marine Park*. Brochure.
Anon. (n.d. b). *Sea in Okinawa*. Brochure.
Bell, L.J., Moyer, J.T. and Numachi, K. (1982). Morpohological and genetic variation in Japanese populations of the anemonefish *Amphiprion clarkii*. *Mar. Biol.* 72: 99-108.
Bryan, W.A. (1903). A monograph of Marcus Island. *Occ. Pap. B.P. Bishop Mus.* 2(1): 77-139.
Coral Reef Survey Committee (1985). A report on the coral reef survey of Okinawa Island to the Association of Local Administrative Offices of the Northern Okinawa. 114 pp. (In Japanese).
Cousteau Society (1984). Reef survey at proposed airport site on coral reef, Shiraho, Ishigaki, Okinawa,

Japan. Report May 1984. The Cousteau Society, Los Angeles.
Douglas, G. (1969). Checklist of Pacific Oceanic Islands. *Micronesica* 5(2): 327-463.
Eguchi, M. (1935). Madreporarian coral of the Tokyo Bay and its environs. *Plant and Animal* 4: 66-74. (In Japanese).
Eguchi, M. (1938). On the reef coral fauna of Aoshima and its environs and Madreporarian corals of Miyazaki Prefecture. *Plant and Animal* 6(2): 43-52. (In Japanese).
Eguchi, M. (1968). The hydrocorals and scleractinian corals of Sagami Bay collected by H.M. The Emperor of Japan. Ed. by Biological Laboratory, Imperial Household, Maruzen Co. Ltd, Tokyo.
Eguchi, M. (1970a). On geology and corals of marine park area of Saga Prefecture. *Rep. Marine Park Center, Saga Prefecture*: 25-34. (In Japanese).
Eguchi, M. (1970b). On geology and corals of marine park site of Tottori Prefecture. *Rep. Marine Park Center, Tottori Prefecture*: 11-16. (In Japanese).
Eguchi, M. (1971a). Scleractinia of Goto Island. *Rep. Marine Park Center, Nagasaki Prefecture*: 19-31, 2 pls. (In Japanese).
Eguchi, M. (1971b). Invertebrate fauna in offshore of south of Boso Peninsula. *Rep. Marine Park Center, Chiba Prefecture*: 15-31. (In Japanese).
Eguchi, M. (1972a). Corals of the coast of South Izu, Shizuoka Prefecture. *Rep. Marine Park Centre, Shizuoka Prefecture*: 19-25. (In Japanese).
Eguchi, M. (1972b). On coral fauna of Kushimoto and nearby. *Marine Parks J.* 21: 7-9.
Eguchi, M. (1973b). Coral fauna of the Tsushima Islands, Nagasaki Prefecture. *Rep. Marine Park Center, Nagasaki Prefecture*: 45-56. (In Japanese).
Eguchi, M. (1975a). Coral fauna of Nagasaki Prefecture. *Rep. Marine Park Center, Nagasaki Prefecture*: 56: 39-46. (In Japanese).
Eguchi, M. (1975b). Notes on coral genera of the Yaeyama Island Group with description of a new species, *Cladocora kabiraensis* n.sp. *Proc. Jap. Soc. Syst. Zool.* 11: 1-4.
Eguchi, M. (1977). Distribution of corals in marine parks in Japan. In: Sabiura Marine Park Research Station (1977): 21-24. (Abst.).
Eguchi, M. and Fakuda, T. (1972). Invertebrata (chiefly coral fauna) of Marine Park Iki Island, Nagasaki Prefecture. *Rep. Marine Park Center, Nagasaki Prefecture*: 45-58. (In Japanese).
Eguchi, M. and Miyawaki, T. (1975). Systematic study of the scleractinian corals of Kushimoto and its vicinity. *Bull. Marine Park Res. Stations* 1: 47-62.
Eldredge, L.G. (1975). Biological research in the Bonin Islands. *Atoll Res. Bull.* 185: 34-37.
Environment Agency, Japanese Government (1973). Studies on the population explosions of the Crown-of-thorns Starfish and its control. 74 pp. (In Japanese).
Environment Agency, Japanese Government (1974). Studies on the population explosions of the Crown-of-thorns Starfish and its control; a follow-up. 65 pp. (In Japanese).
Fosberg, F.R. (1986). Coral reef degradation in Ryukyu Islands. *Env. Cons.* 13(3): 270.
Friends of the Earth, Japan (1984). Help save the Shiraho Coral Reef of Ishigaki Island, Okinawa, Japan. Brochure.
Fujimoiri, S. (1964). Plan for promotion of fisheries industry in coastal waters of Ryukyu Islands through

agriculture. Bureau of Economy, Govt. of the Ryukyus. 138 pp. (In Japanese).

Fujiwara, K. (1979). Distribution of reef-building corals in Japan. Thesis, University of the Ryukyus. (In Japanese).

Fujiwara, S. (1985). Marine Park in Ogasawara Islands. *Marine Parks J.* 68: 11-14.

Fukuda, T. (1976). Mass migration of the Crown-of-thorns Starfish at Hakoma Island, Okinawa. *Marine Parks J.* 38: 7-10. (In Japanese).

Fukuda, T. (1984). Hermatypic corals in Danjo Islands. *Marine Parks J.* 6: 3-6.

Fukuda, T. and Miyawaki, I. (1982). Population explosions of the Crown-of-thorns Starfish in the Sekisei Lagoon, Yaeyama Islands, Okinawa. *Marine Parks J.* 56: 10-13. (In Japanese).

Fukuda, T. and Okamoto, K. (1976). Observations of the *Acanthaster planci* population in the Yaeyama Islands, Okinawa. *Sesoko Mar. Sci. Lab. Tech. Rep.* 4: 7-17.

Glynn, P.W. (1984). Widespread coral mortality and the 1982-1983 El Niño warming event. *Env. Cons.* 11 (2): 133-146.

Habe, T., Ino, T. and Horikoshi, M. (1966). Marine parks and animals in the Japanese waters. *Marine Parks in Japan.* Nature Conservation Society, Tokyo. Pp. 21-29.

Hamada, R. (1963). Corals. *Geological Figures of Chiba-Ken* 4: 1-30, pl. 1-45.

Hasegawa. H. (1984). Status and conservation of seabirds in Japan, with special attention to the Short-tailed Albatross. In: Croxall, J.P., Evans, P.G.H. and Schreiber, R.W. (Eds), *Status and Conservation of the World's Seabirds.* ICBP Technical Publication 2. Pp. 487-500.

Hayashi, K. (1975). Occurrences of the Crown-of-thorns Starfish. *Acanthaster planci* (L.), along the southern coast of Kii Peninsula, Japan. *Bull. Mar. Park Res. Stations* 1(1): 1-9.

Hayashi, K. and Tatsuki, T. (1975). *Acanthaster planci* plague in Japan. *Bull. Mar. Park Res. Stations* 1(1): 11-17.

Hirata, K. (1955). The existing and raised coral reefs in Yoron Island and their shell bearing molluscs. *Bull. Mar. Biol. Station Asamushi, Tohuku Univ., Jap.* 7(2-4): 89-99.

Hirata, K. (1956). Ecological studies on the recent and raised coral reefs in Yoron Island. *Sci. Rep. Kagoshima Univ.* 5: 97-112.

Hirata, K. (1958). Ecological studies on the recent and raised coral reefs in Yoron Island. 2. *Sci. Rep. Kagoshima Univ.* 7: 69-84.

Hirata, K. (1967). Distribution of raised reef limestone and present reef-building activity in Takara-jima and Kodakara-jima, Nansei Islands Japan. *Sci. Rep. Kagoshima Univ.* (16): 75-89, pl. 1-18.

Hirata, K. (1968a). Coral reefs in north area of Amami-osima. *Rep. Marine Park Center, Kagoshima Prefecture*: 223-224. (In Japanese).

Hirata, K. (1968b). Coral reefs in Yamato-son, Amami-osima. *Rep. Marine Park Center, Kagoshima Prefecture*: 223-224. (In Japanese).

Hirata, K. (1968c). Coral reefs in Yoron-jima. *Rep. Marine Park Center, Kagoshima Prefecture*: 333-340, pl. 1-24. (In Japanese).

Hirata, K. (1977). Coral population in Tomori Reef, Yoron-jima, Nansei Islands, Japan. In: Sabiura Marine Park Research Station (1977): 24-26. (Abst.).

*Hori, N. (1980). Coral reefs in Japan. *Nature of Japan.* (In Japanese).

Horikoshi, M. (1981). Kabira Cove: Interdisciplinary study of a physiographic unit in tropical coastal waters of Japan. *Proc. 4th Int. Coral Reef Symp., Manila* 1: 699-706.

Hunnam, P. (1977). Japan's Marine Parks. In: Sabiura Marine Park Research Station (1977): 276-279.

Ida, H. and Moyer, J.T. (1974). Apogonid fishes of Miyake-jima and Ishigaki-jima, Japan with description of a new species. *Jap. J. Ichthyol.* 21(3): 113-128.

Ino, T. (1966). Marine Parks and fisheries. *Marine Parks in Japan.* Nature Conservation Society, Tokyo. Pp. 30-34.

International Marinelife Alliance Canada (1988). Shiraho Coral Reef and the proposed new Ishigaki Island Airport, Japan. Preliminary Report. Part I. SSC/IUCN, Gland, Switzerland. 231 pp.

Ito, T. (1984). Tropical invertebrates, particularly reef-building and soft corals in the coastal waters of Shikoku. *Animals and Nature* 14: 19-24. (In Japanese).

Kamezaki, N. (1986). Notes on the nesting of the sea turtles in the Yaeyama group, Ryukyu Archipelago. *Jap. J. Herpet.* 11: 152-155.

Kamezaki, N. and Ui, S. (1984). Bleaching of hermatypic corals in Yaeyama Islands. *Marine Parks J.* 61: 10-13. (In Japanese).

Kato, M. (1987). Mucus-sheet formation and discolouration in the reef-building coral *Porites cylindrica*: Effects of altered salinity and temperature. *Galaxea* 6: 1-16.

*Kawaguti (1983). Survey report on the marine sanctuaries and their management, 1982. Yaeyama Branch of Okinawa Prefectural Fisheries Experimental Station. 51 pp. (In Japanese).

Kawana, T. (1985). Holocene sea-level changes and seismic crustal movements in a marginal coral reef area; the Central and South Ryukyu Islands, Japan. *Proc. 5th Int. Coral Reef Cong., Tahiti* 3: 205-210.

Kikuchi, S. and Araga, T. (1968). Underwater views in offshore of Amakusa Islands. *Rep. Marine Park Center, Kuamoto Prefecture*: 43-47. (In Japanese).

Kobayashi, A. (1984). Regeneration and regrowth of fragmented colonies of the hermatypic corals *Acropora formosa* and *Acropora nasuta. Galaxea* 3: 13-23.

Konishi, K., Omura, A. and Nakamichi, O. (1974). Radiometric coral ages and sea level records from the late quaternary reef complexes of the Ryukyu Islands. *Proc. 2nd Int. Coral Reef Symp., Brisbane* 2: 595-613.

Kühlmann, D.H. (1985). Wälder schützen Korallenriffe. *Biol. Rdsch.* 23: 367-370.

Kurata, Y. (1984). *Acanthaster planci* in Ogasawara islands. *Marine Parks J.* 61: 7-9. (In Japanese).

Kurata, Y. (1986). Tropical island Okino Shima. *Marine Parks J.* 70/71: 9-13.

Kuroda, N. (1954). Report on a trip to Marcus Island with notes on the birds. *Pac. Sci.* 8: 84-93.

Kushimoto Marine Park Center (1977). Madeporarian corals in Kushimoto. 54 pp. (In Japanese).

McAllister, D.E. (1987). Bibliography related to the coral reef environment at Shiraho, Ishigaki Island, Japan. International Marine Life Alliance Canada, Ottowa. 26 pp.

Marine Parks Center (1975). *Marine Parks in Japan.* Tokyo, Japan. May. Pp. 1-23.

Marine Parks Center (1984). Report on a survey of the marine nature reserve at Sakiyama Bay, Iriomote Island, Okinawa. 134 pp. (In Japanese).

Marine Parks Center (n.d.). Marine Parks Research Stations.

Marsh, J.S. (1985). Japan's Marine Parks. In: Lien, J. and Graham, R. (Eds), *Marine Parks and Conservation - Challenge and Promise*, NPPAC Henderson Park Book

Series No. 10. Vol. 2. Pp. 29-44. (National and Provincial Parks Association of Canada).

Masuda, H., Amaoka, K., Araga, C., Uyeno, T. and Yoshino, T. (Eds) (1985). *The fishes of the Japanese Archipelago. Text.* Tokai Univ. Press, Tokyo. 445 pp.

Matsuba, S. (1986). Reef-building corals, seaweeds and sediments of the reef flat and moat off Shiraho. In: *Conservation of the Nansei Shoto.* Separate Volume. WWF Japan Scientific Committee. Pp. 15-28. (In Japanese with English summary).

Matsusita, K. and Misaki, H. (1983). On the distribution of *Acanthaster planci* L. in Yaeyama Archipelago, Okinawa. *Marine Parks J.* 59: 14-16. (In Japanese).

Mezaki, S. and Toguchi, K. (1985). Reef morphology and physiography of the Ryukyu Islands, southwestern Japan. *Proc. 5th Int. Coral Reef Cong., Tahiti* 2: 245. (Abst.).

Misaki, H. (1974). Spawning of the Crown-of-thorns Starfish. *Marine Pavilion* 3: 52. (In Japanese).

Misaki, H. (1979). Spawning of the Crown-of-thorns Starfish. *Marine Pavilion* 8: 59. (In Japanese).

Misaki, H. (1985). On the survival of corals against low temperatures in winter. *Marine Parks J.* 68: 15-19. (In Japanese).

Moyer, J. (1965). Bounty of the Kuroshio: A study of an Izu Island fishing settlement. In: Beardsley, R.K. (Ed.), *Studies in Japanese Culture: 1.* University of Michigan Press, Ann Arbor. Pp. 106-138.

Moyer, J.T. (1976). Geographical variations and social dominance in Japanese populations of the anemonefish *Amphiprion clarkii. Jap. J. Ichthyol.* 23: 12-22.

Moyer, J.T. (1978). Crown of thorns starfish invades Miyake-jima. *Marine Parks J.* 44(1): 17. (In Japanese).

Moyer, J.T. (1980). Influence of temperate waters on the behavior of the tropical anemonefish *Amphiprion clarkii* at Miyake-jima, Japan. *Bull. Mar. Sci.* 30: 261-272.

Moyer, J.T. and Bell, L.J. (1976). Reproductive behavior of the anemonefish *Amphiprion clarkii* at Miyake-jima, Japan. *Jap. J. Ichthyol.* 23: 23-32.

Moyer, J.T., Emerson, W.K. and Ross, M. (1982). Massive destruction of scleractinian corals by the muricid gastropod *Drupella* in Japan and the Philippines. *The Nautilus* 96(2): 69-82.

Moyer, J.T., Higuchi, H., Matsuda, K. and Hasegawa, M. (1985). Threat to unique terrestrial and marine environments and biota in a Japanese National Park. *Env. Cons.* 12(4): 293-301.

Murakoshi, M. (1985). Fishery of the giant clam at Ishigaki Island in Okinawa. Yaeyama Branch of Okinawa Pref. Fish. Exp. Stn. Pp. 13-28. (In Japanese).

Murakoshi, M. (1986). Farming of the boring clam *Tridacna crocea* Lamarck. *Galaxea* 5: 239-254.

*Muzik, K. (1983). The runway on the reef. *Earthscan.*

Muzik, K. (1984a). Ishigaki Lagoon Coral Survey. *Research on Environmental Disruption. Toward Interdisciplinary Cooperation* 13(4): 68-70.

Muzik, K. (1984b). Ishigaki Reef Survey II. *Research on Environmental Disruption. Toward Interdisciplinary Cooperation* 14(1): 70-72.

Muzik, K. (1985). Dying coral reefs of the Ryukyu Archipelago. *Proc. 5th Int. Coral Reef Cong., Tahiti* 6: 483-489.

Nakamura, R. (1986). A morphometric study of *Acanthaster planci* (L.) populations in the Ryukyu Islands. *Galaxea* 5(2): 223-237.

Nakazono, A., Nakatani, H. and Tsukahara, H. (1985). Reproductive ecology of the Japanese reef fish *Paracerpis snyderi. Proc. 5th Int. Coral Reef Cong., Tahiti* 5: 355-360.

Nishihira, M. (1987). Natural and human interference with the coral reef and coastal environments in Okinawa. *Galaxea* 6: 311-321.

Nishihira, M. and Yamazato, K. (1972). Brief survey of *Acanthaster planci* in Sesoko Island and its vicinity, Okinawa. *Sesoko Mar. Sci. Lab. Tech. Rep.* 1: 1-20.

Nishihira, M. and Yamazato, K. (1973). Resurvey of the *Acanthaster planci* population on the reefs around Sesoko Island, Okinawa, 1973. *Sesoko Mar. Sci. Lab. Tech. Rep.* 2: 17-33.

Nishihira, M. and Yamazato, K. (1974). Human interference with the coral community and *Acanthaster* infestation of Okinawa. *Proc. 2nd Int. Coral Reef Symp. Brisbane* 1: 577-590.

Nishihira, M., Yanagiya, K. and Sakai, K. (1987). A preliminary list of hermatypic corals collected around Kudaka Island, Okinawa. *Galaxea* 6: 53-60.

Nomura, K. and Kamezaki, N. (1987). The present condition of crown-of-thorns starfish and corals in Hateruma Island, Yaeyama Group. *Marine Parks J.* 73: 16-19. (In Japanese).

Numata, M. (1977). Ideas and facts on coastal parks and reserves. In: Sabiura Marine Park Research Station (1977): 80-82.

Ogura, M., Yokochi, H., Ishimaru, A., Kohno, H., Kishi, K., Ninomiya, K. and Mitsuhashi, H. (1985). Spawning period and the discovery of juveniles of the Crown-of-Thorns Starfish *Acanthaster planci* (L.) in the northwestern part of Iriomote Island. *Bull. Inst. Oceanic Res. and Develop., Tokai Univ.* 7: 25-31.

Ogura, N. and Sakamoto, M. (1984). Environmental conditions in Japanese coastal waters. In: *Man's Impact on Coastal and Estuarine Ecosystems.* Proc. MAB/COMAR Regional Seminar, Tokyo, November 1984. MAB Co-ordinating Committee of Japan. Pp. 99-102.

Ohba, H. and Aruga, Y. (1978). Coral reefs of Ikema-jima, Miyako Islands, Japan. *La Mer* 16(4): 198-210. (In Japanese).

Okada, Y. (1966). The marine fishes and marine parks in Japan. *Marine Parks.* Special Symp. on Marine Parks, 11th Pacific Science Congress, Committee on Marine Parks, The Nature Conservation Society of Japan, Tokyo. Pp. 30-33.

Okaichi, T. (1984). Red tides in Japanese coastal waters. In: *Man's Impact on Coastal and Estuarine Ecosystems.* Proc. MAB/COMAR Regional Seminar, Tokyo, November 1984. MAB Co-ordinating Committee of Japan. Pp. 95-98.

Okinawa General Affairs Bureau (1978). Coastal Survey report for 1977. 305 pp. (In Japanese).

Okinawa General Affairs Bureau (1979). Coastal Survey report for 1978. 245 pp. (In Japanese).

Okinawa General Affairs Bureau (1983). Report on the environmental survey for harbor improvement plans. Digest edition. 78 pp. (In Japanese).

Okinawa Prefectural Government (1979a). Survey report on the effect of reddish clay on the fishing ground. 62 pp. (In Japanese).

Okinawa Prefectural Government (1979b). Survey report of the tidal flats, seagrass and seaweed beds, and coral reefs. 54 pp. (In Japanese).

Okinawa Tourism Development Corporation (1976). Effects of the Crown-of-thorns Starfish on the coral reef communities. 110 pp. (In Japanese).

Omija, T. (1985). Survey on the water pollution by reddish soil in Okinawa. The present situation of water pollution in northern Okinawa Island. *Ann. Rep. Okinawa Prefectural Inst. Public Health* 8: 71-86. (In Japanese).

Ooishi, S. (1970). Marine invertebrate fauna of the Ogasawara and Volcano Islands. Asahi Shimbun Publ. Co., Tokyo, 75-104, pl. 1-25.

Ooishi, S. and Yagi, S. (1964). Invertebrate fauna of Amami-osima. *Rep. Mar. Biol. Exp. Amami-osima*. Asahi Shinbun Publ Co., Tokyo, 43-46, pl. 1-6. (In Japanese).

Plowden, C. (1984). The battle to save the Shiraho Coral Reef. Waterlife Association and Greenpeace.

Randall, J.E., Ida, H. and Moyer, J.T. (1981). A review of the damselfishes of the genus *Chromis* from Japan and Taiwan, with a description of a new species. *Jap. J. Ichthyol.* 28: 203-242.

Ruddle, K. (1985). The continuity of traditional management practices: The case of Japanese coastal fisheries. In: Ruddle, K. and Johannes, R.E. (1985). *The Traditional Knowledge and Management of Coastal Systems in Asia and the Pacific*. Unesco/ROSTEA, Jakarta. Pp. 157-179.

Ruddle, K. (1987). Administration and conflict management in Japanese coastal fisheries. *FAO Fish. Tech. Pap.* 273. 93 pp.

Ruddle, K. (1987a). The management of coral reef fish resources in the Yaeyama Archipelago, south-western Okinawa. *Galaxea* 6: 209-235.

Sabiura Marine Park Research Station (1977). *Collected Abstracts and Papers on the International Conference on Marine Parks and Reserves*, Tokyo, Japan. May 1975. Kushimoto, Japan.

Saito, K. (1975). The present aspects of utilization of marine parks in Japan and its problems. Marine Parks Center of Japan, Tokyo. 21 pp.

Sakai, K. (1985). Brief observations on the population of the *Acanthaster planci* (L.) and coral assemblages around Sesoko Island, Okinawa, in 1983. *Galaxea* 4: 23-31.

Sakai, K. and Nishihira, M. (1986). Ecology of hermatypic corals. In: Nishihira, M. (Ed.), *Coral Reefs of Okinawa*. University Extension Program Committee, Univ. Ryukyus. Pp. 71-85. (In Japanese).

Sakiyama, T. (1982). Japan: Problems of fisheries in coastal zone management. In: Soysa, C.H., Chia, L.S. and Collier, W.L. (Eds), *Man, Land and Sea: Coastal Resource Use and Management in Asia and the Pacific*. Agricultural Development Council, Bangkok, Thailand. Pp. 135-145.

Sano, M., Shimizu, M. and Nose, Y. (1984). Changes in structure of coral reef fish communities by destruction of hermatypic corals: Observational and experimental views. *Pac. Sci.* 38: 51-79.

Sato, M. (1985). Mortality and growth of juvenile coral *Pocillopora damicornis* (Linnaeus). *Coral Reefs* 4: 27-33.

Senou, H. (1986). Preliminary report on the fish fauna of the *Heliopora* zone of Shiraho fringing reef, Ishigaki Island. In: *Conservation of the Nansei Shoto*. Separate Volume. WWF Japan Scientific Committee. Pp. 29-36. (In Japanese with English summary).

Shepard, J.W. and Moyer, J.T. (1980). Annotated checklist of the fishes of Miyake-jima, Japan. Pt. 1. Pomacentridae, Chaetodontidae and Pomacanthidae. *Publ. Seto Mar. Biol. Lab.* 15(1-4): 227-241.

Shirai, S. (1971). Underwater views in Inan area of Kochi Prefecture. *Rep. Marine Park Center, Kochi Prefecture*, 6-23.

Shirayama, Y. (1979). The ecology of coral-associated animals at some biotopes in Kabira Cove. M.Sc. Thesis, Univ. Tokyo. 42 pp. (In Japanese).

Shirayama, Y. and Horikoshi, M. (1982). A new method of classifying the growth form of corals and its application to a field survey of coral-associated animals in Kabira Cove, Ishigaki Island. *J. Oceanogr. Soc. Japan* 38: 193-207.

Sugiyama, T. (1934). Comparison of reef building corals between Nanyo-Islands and near Japan. *J. Geol. Soc. Japan* 41: 404-406.

Sugiyama, T. (1937). On recent reef building corals in offshore of Japan. *Cont. Inst. Geol. Palaeont. Tohoku Imp. Univ.* (6): 1-60. (In Japanese).

Suzuki, M. (1986). Battle for Shiraho coral reef. *Japan Environment Review* Spring 1986: 10-18.

Suzuki, T. (1975). Report on an eradication of the Crown-of-thorns Starfish. *Marine Pavilion* 4: 2. (In Japanese).

Swinbanks, D. (1986). Okinawa runway under attack. *Nature* 324: 100.

Tada, M. (1983). The drama beneath the sea of Ashizuri. Hoikusha. 133 pp. (In Japanese).

Takahashi, K. (1983). Shallow water scleractinian corals around Hachizyo Island. *Marine Parks J.* 60: 7-10. (In Japanese).

Takahashi, T. and Koba, M (1978). A preliminary investigation of the coral reef at the southern coast of Ishigaki Island, Ryukyus. *Sci. Rep. Tohoku Univ. 7th Ser. (Geog.)* 28: 49-60.

Takahashi, T., Koba, M. and Nakamori, T. (1985). Coral reefs of the Ryukyu Islands. Reef morphology and reef zonation. *Proc. 5th Int. Coral Reef Cong., Tahiti* 3: 211-216.

Tamura, T. (1968). The Marine Parks Centre of Japan. *Biol. Cons.* 1(1): 89-90.

Tamura, T. (1969). Japan's initial project for a Marine Parks Research Institute. *Biol. Cons.* 2(1): 66-68.

Tamura, T. (1972). Marine parks in Japan in the past ten years. Marine Parks Center of Japan, Tokyo. 8 pp.

Tamura, T. (1973). A new marine parks research station for Japan. *Biol. Cons.* 5(1): 64-66.

Tamura, T. and Ino, T. (1985). Some observations on marine parks in Japan with particular reference to fishing activities. In: *I Parchi Costieri Mediterranei*. Atti del Convegno Internazionale, Salerno, Catellabate 18-22 June 1973. Regione Campania Assessorato per il Turismo. Pp. 747-753.

Tayama, R. (1952). Coral reefs in the South Seas. *Bull. Hydrogr. Office* 2. (In Japanese).

The Japan Science Society (1979). Basic research on controlling measures against the infestations of the Crown-of-thorns Starfish *Acanthaster planci*. 73 pp. (In Japanese).

Tokioka, T. (1953). Invertebrate fauna of the intertidal zone of the Tokara Islands. *Publ. Seto. Mar. Biol. Lab.* 3(2): 123-149.

Tokioka, T. (1966). Reef coral communities and problems of their conservation on the west coast of Kii Peninsula. *Marine Parks*. Special Symp. on Marine Parks, 11th Pacific Science Congress, Committee on Marine Parks, The Nature Conservation Society of Japan, Tokyo. Pp. 41-46.

Tokioka, T. (1968). Preliminary observations made by Mr. S. Hamahira on the growth of a giant colony of the madreporarian coral *Pavona frondifera* Lamarck, found in a core on the southwestern coast of Shikoku Island. *Publ. Seto Mar. Biol. Lab.* 16: 55-59.

Tokuyama, A. (1985). The field investigation and the study concerning the conservation of coral reef around

Nansei Shoto, especially Ishigaki Island. In: *Conservation of the Nansei Shoto*, Part 2. WWF Japan Scientific Committee. Pp. 265-268.

Tribble, G.W., Bell, L. and Moyer, J. (1982). Subtidal effects of a large typhoon on Miyake-jima, Japan. *Publ. Seto Mar. Biol. Lab.* 27: 1-10.

Tokuyama, A. and Ichinose, H. (1986). The chemical study of water around Shiraho, Ishigaki Island. In: *Conservation of the Nansei Shoto*. Separate Volume. WWF Japan Scientific Committee. (In Japanese with English summary).

Tribble, G.W. and Randall, R.H. (1981). Description of the coral environments and community structure of the shallow-water corals of Miyake-jima, Japan. *Proc. 4th Int. Coral Reef Symp., Manila* 1: 721. (Abs.).

Tribble, G.W. and Randall, R.H. (1986). A description of the high-latitude shallow water coral communities of Miyake-jima, Japan. *Coral Reefs* 4(3): 151-159.

Tsuda, R.T. (1968). Some marine benthic algae from Marcus Island, Bonin Islands. *Micronesica* 4(2): 207-212.

Tsuya, H. (1977). The Izu-Ogasawara and Satsunan-Ryukyu Island-arcs in the north-west Pacific. In: Sabiura Marine Park Research Station (1977): 225-242.

Tuyama, T. (1953). On the phytogeographical status of the Bonin and Volcano Islands. *Proc. 7th Pac. Sci. Cong.* 5: 208-212.

Uchida, H. and Nomura, K. (in press). On a method of rearing for the pelagic larvae of *Acanthaster planci* (L.). Proceedings of a Symposium on Recent Findings in *Acanthaster* Biology and Implications for Reef Management, Guam, 1986. *Bull. Mar. Sci.*

Uda, M., Kishi, A. and Nakao, T. (1975). Marine environment and its indicators for marine parks and reserves - particularly water transparency, its distribution and changes in the Western Pacific Ocean and in waters adjacent to Japan. In: Sabiura Marine Parks Research Station (1977): 27-28. (Abst.).

Ueno, A., Tanaka, M., Teruya, K., Nagano, T., Fujimaki, U. and Shibamoto, Y. (1986). A study of crown-of-thorns starfish in the northwestern part of Iriomote Island. Bachelor Thesis, Fac. Mar. Sci. and Tech., Tokai Univ. 182 pp.

Ui, S. (1985). Past and current distribution of Crown-of-thorns Starfish and hermatypic corals in Sekisei Lagoon, Yaeyama Islands, Okinawa. *Marine Parks J.* 64: 13-17. (In Japanese).

Utinomi, H. (1956). Invertebrate fauna of the intertidal zone of the Tokara Islands. 14. Stony corals and Hydrocorals. *Publ. Seto Mar. Biol. Lab.* 5(3): 37-44.

Utinomi, H. (1962). Recent evidence for the northward extension of the range of some tropical echinoderms in Japanese waters. *Zool. Magazine, Tokyo* 71: 102-108. (In Japanese).

Utinomi, H. (1965). A revised catalogue of scleractian corals from the south-west coast of Shikoku in the collections of the Ehime University and the Ehime Prefectural Museum, Matsuyama. *Publ. Seto Mar. Biol. Lab.* 13(3): 243-261.

Utinomi, H. (1966). Shallow-water scleractinian fauna along the coast of Kii Peninsula. *Sci. Rep. Marine Park of Wakayama Prefecture, Surv. Rep. Jap. Ass. Nat. Protect.* 27: 97-102. (In Japanese).

Utinomi, H. (1968). A study of Marine Park Planning. Kumano Coast Marine Park. *Rep. Marine Park Center, Wakayama Prefecture* 97-102.

Utinomi, H. (1972). On coral fauna of Kushimoto and nearby. *Marine Parks J.* 21: 10. (In Japanese).

Utinomi, H. (1976). Shallow-water octocorals of the Ryukyu Archipelago (Part 1). *Sesoko Mar. Sci. Lab.*

Tech. Rep. 4: 1-5.

Utinomi, H. (1977). Shallow-water octocorals of the Ryukyu Archipelago (Part 2 and 3). *Sesoko Mar. Sci. Lab. Tech. Rep.* 5: 1-34.

Yabe, H. (1933). Distribution of reef corals in Japan, past and present. *Rep. Japan Soc. Sci.* 8(3): 335-341. (In Japanese).

Yabe, H. and Sugiyama, T. (1931). Reef-building coral fauna of Japan. *Proc. Imp. Acad.* 7(9): 35-36.

Yabe, H. and Sugiyama, T. (1932a). Living species of Stylocoenia recently found in Japan. *Jap. J. Geol. Geogr.* 9(3-4): 153-154.

Yabe, H. and Sugiyama, T. (1932b). Reef corals found in the Japanese sea. *Sci. Rep. Tohoku Imp. Univ. (Geol)* 15(2): 143-168.

Yabe, H. and Sugiyama, T. (1933a). Notes on three new corals from Japan. *Jap. J. Geol. Geogr.* 11(1-2): 11-18.

Yabe, H. and Sugiyama, T. (1933b). Geographical distribution of reef corals in Japan, Past and Present. *Proc. 5th Pac. Sci. Cong.*, 3115-3116.

Yabe, H. and Sugiyama, T. (1935a). Geological and geographical distribution of reef corals in Japan. *J. Paleont.* 9(3): 183-217.

Yabe, H. and Sugiyama, T. (1935b). Revised lists of the reef corals of the Japanese seas and of the fossil reef corals of the raised reefs and the Riukiu Limestone of Japan. *J. Geol. Soc. Japan* 42(502): 379-403.

Yabe, H. and Sugiyama, T. (1936). Recent reef-building corals from Japan and the South Sea Islands under the Japanese Mandate. 1. *Sci. Rep. Tohoku Imp. Univ. (Geol.)* Spec. Vol. 1: 1-66.

Yabe, H. and Sugiyama, T. (1937). Two new species of reef-building corals from Yoron-zima and Amami-O-sima. *Proc. Imp. Acad.* 13(10): 425-429.

Yabe, H. and Sugiyama, T. (1938). *Stylocoeniella*, a new coral genus allied to *Stylocoenia* and *Astrocoenia*. *Jap. J. Geol. Geogr.* 12(3-4): 103-105.

Yabe, H. and Sugiyama, T. (1941). Recent reef-building corals from Japan and the South Sea Islands under the Japanese Mandate II. *Sci. Rep. Tohoku Imp. Univ. (Geol.)* Spec. Vol. 2: 67-91.

Yamaguchi, M. (1986a). *Acanthaster planci* infestations of reefs and coral assemblages in Japan: A retrospective analysis of control efforts. *Coral Reefs* 5: 23-30.

Yamaguchi, M. (1986b). Population fluctuations of the tropical benthic invertebrates in the subtropical coasts of Japan. Chap. 3. In: *Introduction to the Study of Coral Reefs* Vol. 8(1): 2-7.

Yamaguchi, M. (1986c). The Crown-of-thorns Starfish problem. Chap. 8. Introduction to the Study of Coral Reefs. *Aquabiology* 47: 408-412.

Yamaguchi, M. (1986d). Coral reef Sponges. Chaps 4,5,6,7. In: *Introduction to the Study of Coral Reefs* Vol. 8(2): 88-92, 168-171, 248-252, 330-334.

Yamaguchi, M. (1987). Occurrences and persistency of *Acanthaster planci* pseudo-population in relation to oceanographic conditions along the Pacific coast of Japan. *Galaxea* 6: 277-288.

Yamaguchi, M. (1987). Recent research on *Acanthaster planci* problem in Japan. Unpub. rept.

Yamamoto, T. (1976). Seasonal variations in abundance, size composition and distributional patterns of damselfishes residing in Sesoko Island, Okinawa. *Sesoko Mar. Sci. Lab. Tech. Rep.* 4: 19-42.

Yamazato, K. (1969). *Acanthaster planci*, a coral predator. *Kon-nichi no Ryuku* 13: 7-9. (In Japanese).

Yamazato, K. (1971). Bathymetric distribution of corals in the Ryukyu Islands. *Proc. Symp. Corals and Coral Reefs, 1969*: 121-133.

Yamazato, K. (1974). Studies on the abnormal appearance of Crown of Thorns starfish and its control at Kume Island. Kankyocho (Environmental Agency of Japan). March 1974. Pp. 25-31. (In Japanese).

Yamazato, K. (1975a). Coral species. In: Kizaki, K. (Ed.). *Nature in Okinawa*. Heibonsha, Tokyo. Pp. 23-28. (In Japanese).

Yamazato, K. (1975b). Coral reef communities in the red clay polluted coasts: A report on the central east coast of Okinawa Island. In: Fujikawa, T. (Ed.), *Studies on the Environmental Changes and Disaster Protection in Association with the Development Activities of Okinawa*. Pp. 98-105. (In Japanese).

Yamazato, K. (1976). Ecological distribution of the reef associated organisms in the Bise-Shinzato coast of Okinawa. *Ecol. Stud. Nat. Cons. Ryukyu Islands* 2: 1-30. (In Japanese).

Yamazato, K. (1981). A note on the expulsion of zooxanthellae during summer, 1980, by Okinawan reef-building corals. *Sesoko Mar. Sci. Lab. Tech. Rep.* 8: 9-18.

Yamazato, K. (1987). Effects of deposition and suspension of inorganic particulate matter on the reef building corals in Okinawa, Japan. *Galaxea* 6: 289-309.

Yamazato, K., Kamura, S., Nakasone, Y., Aramoto, Y. and Nishihira, M. (1976). Ecological distribution of the reef associated organisms in the Bise-Shinzato coast of Okinawa. *Ecol. Stud. Nat. Cons. Ryukyu Isl.* 2: 1-30.

Yamazato, K., Nishihira, M., Nakasone, Y., Kamura, S. and Aramoto, Y. (1974). Biogeomorphological notes on the Sesoko Island reefs, Okinawa. *Ecol. Stud. Nat. Cons. Ryukyu Islands* 1: 201-212 (In Japanese with English summary).

Yamazato, K., Sakai, K. and Muzik, K. (in prep.). Status of the coral reefs around the mainland of Okinawa, 1984 survey.

Yamazato, M. (1959). Ecological studies on the coral reefs. 1. Distribution of the coral reefs. *Bull. Arts and Sci. Div. Ryukyu Univ.* 3: 53-64.

Yasumoto, M. (1986). Coral community of Shiraho coral reef areas correlating ecological problems with geographic advantages. In: *Conservation of the Nansei Shoto*. Separate Volume. WWF Japan Scientific Committee. Pp. 1-14. (In Japanese with English summary).

Yokochi, H. and Ogura, M. (in press). Spawning period and discovery of juveniles of *Acanthaster planci* (L.). (Asteroidea) in the northwestern part of Iriomote-jima, Ryukyu Islands. Proceedings of a Symposium on Recent Findings in *Acanthaster* Biology and Implications for Reef Management, Guam, 1986. *Bull. Mar. Sci.*

Zenji, Y. (1987). Petition to request the designation of blue coral (*Helipora corulea* (sic)) as a natural monument. Su-Moguri Diving Association. 9 pp.

IRIOMOTE NATIONAL PARK, YAEYAMA MARINE PARK AND SAKIYAMA BAY NATURE CONSERVATION AREA

Geographical Location Southern Japan, Sekisei Lagoon, between the islands of Ishigakishima and Iriomoteshima, Yaeyama-retto. Iriomote National Park is at 24°10'-24°30'N, 123°40'-124°10'E; the location of Sakiyama Bay has not been obtained.

Area, Depth, Altitude Iriomote National Park is reported to cover 12 506 ha including land and sea (Nature Conservation Bureau, 1985 *per* Moyer *in litt.*, 6.5.87), although Tamura (1972) gives the area of the park as 12 506 ha land and 32 100 ha sea; this may have been the proposed area of the park (Moyer *in litt.*, 6.5.87). Yaeyama Marine Park, within the National Park, is 213.5 ha and consists of four separate areas: Taketomishima Takidonguchi (36.7 ha), Taketomishima Shimobishi (83.1 ha), Kuroshima Kyanguchi (45.5 ha) and Aragusukushima Maibishi (48.2 ha). Reefs in Sekisei Lagoon cover an area of 20 x 15 km. Sakiyama Bay Nature Conservation Area covers 128 ha.

Physical Features The climate is oceanic; average temperature 23.3°C, average annual rainfall 2630 mm. On the southern shore of Iriomoteshima, a fault-line scarp faces the sea. The shores of adjacent small islands are covered with limestone formed from elevated coral reefs. Ishigakishima (*see separate account*) and the adjacent islands are surrounded by a well-developed reef, stretching 15 km N-S and 20 km E-W (Tamura, 1972).

Reef Structure and Corals Some 250 species of coral have been described from the Sekisei Lagoon (Eguchi, 1975b) which is reported to be the largest reef area in Japan. Coral distribution is described by Ui (1985). Fringing reefs are found around most of the islands and the lagoon has double and triple barrier reefs. A number of studies have been carried out around Iriomote including a survey of Sakiyama Bay, funded by the Environment Agency (Yamaguchi *in litt.*, 20.3.85) and a survey of Sakiyama Bay by the Marine Parks Center (1984). Sakiyama Bay has coral reefs and seagrass beds, with large colonies of *Galaxea fascicularis* (Ikenouye *in litt.*, 2.4.85). The Marine Park Centre has also surveyed the reefs around Kuroshima and the reefs of Sekisei Lagoon using colour aerial photographs (Ikenouye *in litt.*, 2.4.85). The island of Hamashima in Sekisei Lagoon is surrounded by reef, 1-2 m deep, with patchy coral assemblages (Nakamura, 1986).

Yaeyama Marine Park is centred on the islands of Kohamashima, Taketomishima, Kuroshima and Aragusukushima which lie in the channel between Iriomoteshima and Ishigakishima. The Taketomishima Takidonguchi area of the marine park includes outer reef flat, patch reefs, and fringing reefs from north-west Taketomishima to Takidonguchi channel on the west. Spur and groove formation is found at the reef margins and there is dense coral cover on the shallow reef-flats. The Taketomishima Shimobishi area includes the reef margin and reef-flat of Shimobishi Reef and extends to south-west Taketomishima; spur and groove formation; abundant corals on shallow reef-flat and has characteristics of an outer reef. The Kuroshima Kyanguchi area includes lagoon and reef-flat of barrier reef extending off Kyan, on east Kuroshima; lagoon 2-7 m deep, with large "forest" of *Acropora* and many fish. The Aragusukushima Maibishi area includes the Maibishi Reef off north-west Kamiji with small patch reefs; lagoon 1-3 m deep; and has abundant corals and fish.

Noteworthy Fauna and Flora Mangroves are found around some of the shores. Dugong *Dugong dugon* are occasionally sighted in the area (Yamaguchi *in litt.*, 11.5.87). Turtle nesting in the Yaeyama-retto is described by Kamezaki (1986).

Scientific Importance and Research The Yaeyama Marine Parks Research Station is situated on Kuroshima (Tamura, 1972). The Iriomote Research Station of Tokai University is situated on the west coast of Iriomoteshima, and work here is concentrated on *Acanthaster* (Yamaguchi *in litt.*, 11.5.87). Extensive research work has been carried out by the Marine Parks Centre of Japan, particularly on *Acanthaster* and on the reefs of Iriomoteshima, especially at Sakiyama Bay (Nishihira, 1987; Uchida and Nomura, in press). The effects of sedimentation on corals has been studied experimentally at the Taketomi reef complex (Okinawa General Affairs Bureau, 1983).

Economic Value and Social Benefits There are extensive reef fisheries in this area. Tourism is important in the area. Sekisei Lagoon and the large colony of *Galaxea* in Sakiyama Bay are popular with SCUBA divers (Ikenouye *in litt.*, 2.4.85; Yamaguchi *in litt.*, 11.5.87).

Disturbance or Deficiencies The reefs of Sekisei Lagoon have been damaged by fishing boat anchors, dredging for ship channels, walking on the reef and siltation as a result of construction work along the shore and riverbanks (Ikenouye, *in litt.*, 1986).

Coral bleaching has been reported in the Yaeyama-retto (Kamezaki and Ui, 1984). There were population explosions of *Acanthaster planci* in the Sekisei Lagoon and Yaeyama-retto in the 1970s and early 1980s (Fukuda and Okamoto, 1976; Fukuda and Miyawaki, 1982; Matsusita and Misaki, 1983; Ui, 1985; Nishihira, 1987; Yokochi and Ogura, in press), and the infestation is continuing. Nakamura (1986) describes a high density at Hamashima, and Ogura *et al.* (1985) and Ueno *et al.* (1986) describe an outbreak on the north-west coast of Iriomote. The Environment Agency has recently financed a survey of outbreaks in Sakiyama Bay. Although most coral cover in this area has been damaged by *Acanthaster*, the larger colonies of *Galaxea* have survived (Yamaguchi *in litt.*, 11.5.87). A 1986 survey by the Kuroshima laboratory for the management office of Iriomote National Park indicated a steady decrease in *A. planci* numbers and some recovery of coral communities since the peak outbreak in 1981 (Yamaguchi, 1987). An outbreak in 1985 at Hateruma to the south is described by Nomura and Kamezaki (1987).

Legal Protection Yaeyama Marine Park was established 1.7.77 within the Iriomote National Park. The Sakiyama Bay Nature Conservation Area is the first strict marine reserve in Japan and was designated in 1983 by the Environment Agency under the Nature Conservation Law of 1982 (Ikenouye *in litt.*, 2.4.85). Regulations are similar to those for marine parks (Nishihira, 1987).

Management Within the marine park, there are visitor centres set up by the Environment Agency in Taketomishima Takidonguchi, Taketomishima Shimobishi, Kuroshima Kyanguchi. There are also booklets on coral ecology and snorkelling available, and staff offer talks to visitors and schools. Kuroshima also has an underwater nature trail (Marsh, 1985). Efforts are still being made to reduce the number of *Acanthaster* on the reefs.

Recommendations In 1981 the Environment Agency published a survey report on conservation and utilization of Sekisei Lagoon. A model plan to use the recreational resources of the area was proposed by the Marine Parks Center; this was to include underwater trails for divers (Yamaguchi *in litt.*, 11.5.87). It is not known if this plan was implemented, although at least one underwater trail exists (see above). Subsequently, a master plan for the management of Iriomote National Park was published in 1985. This is largely concerned with the terrestrial environment but also includes coral reefs. Emphasis is laid on the need for regular reef monitoring to detect disturbances such as coral bleaching and *Acanthaster* outbreaks. Such work is to be carried out as a collaboration by the University of the Ryukyus, Tokai University, Okinawa Prefectural Fisheries Experimental Station, Ishigaki Meteorological Observatory, Fisheries Cooperative of Yaeyama and Kuroshima Marine Laboratory (the Yaeyama Marine Park Research Institute) (Yamaguchi *in litt.*, 11.5.87).

ISHIGAKISHIMA (INCLUDING SHIRAHO LAGOON, AND NAGURA BAY AND KABIRA BAY MARINE SANCTUARIES)

Geographical Location Southern Ryukyu-retto in the Yaeyama-retto, 441 km south-west of mainland Okinawa; ca 124°25'N, 124°10'E.

Area, Depth, Altitude Ishigakishima is 33 x 19.2 km; Kabira Bay Marine Sanctuary covers 275 ha; Nagura Bay Marine Sanctuary covers 68 ha.

Physical Features Winter temperatures in the area may drop as low as 16-18°C in bays and inlets but generally do not fall below 20°C outside these due to the influence of the warm Kuroshio Current. Light intensity in the winter also decreases significantly.

Kabira Bay consists of an inlet, an outer reef with moat and several raised reef islands which almost close the mouth of the bay (Horikoshi, 1981). The main body of the inlet is connected to the open sea by two channels, and has depths of about 15 m and is divided into two sections by a constriction in the middle of the bay. There are strong tidal currents in the channel which are described in more detail in Horikoshi (1981). There is a wide tidal flat along Nagura Bay and seagrass beds are extensive (Yamaguchi *in litt.*, 11.5.87).

At Shiraho, visibility may be greater than 20 m. There is relatively deep water off shore, with strong currents. A shallow reef separates the shore from the fringing reef and consists of a mosaic of patch reefs, sand and algae beds (Cousteau Society, 1984; Yasumoto, 1986).

Reef Structure and Corals Takahashi and Koba (1978) carried out a preliminary investigation of reefs along the southern coast of Ishigakishima. Reefs around Ishigakishima, including Shiraho Lagoon, were briefly surveyed in 1983 and 1984 (Muzik, 1983 and 1984a and b; Kühlmann, 1985). Most work has been carried out at Shiraho in the south and Kabira Bay on the north-west coast.

Initial surveys of the Shiraho reef found high (80-90%) coral cover in some areas (Cousteau Society, 1984) and diverse corals (Muzik, 1984a and b). In 1985 a survey was carried out, funded by WWF Japan (Tokuyama, 1985; Matsuba, 1986; Senou, 1986; Tokuyama and Ichinose, 1986; Yasumoto, 1986), and a more detailed survey was carried out in 1987 (International Marinelife Alliance Canada, 1988). The reef flat and moat off Shiraho were briefly studied by Matsuba (1986). The reef flat along the transect site was divided into six zones: rocky intertidal flat; sea grass zone; sandy flat zone; *Heliopora* zone; *Acropora* zone; reef crest zone. The *Heliopora* zone is of special interest, with *Heliopora* microatolls, mostly 1-3 m in diameter and 1.5 m high, being densely aggregated and sometimes forming huge colonies in a belt over 300 m wide. Twenty-two genera of reef building corals and 40 species of algae were recorded during the survey (Matsuba, 1986). As well as *Heliopora* and *Acropora*, *Porites* is also dominant in some areas and *Montipora foliosa* is present (Yasumoto, 1986). Sixty six species of coral were recorded by Muzik (1984a and b).

The reefs at Kabira Bay are described by Shirayama (1979), Horikoshi (1981), Shirayama and Horikoshi (1982) and Takahashi *et al.* (1985). A total of 231 species in 65 genera of corals (of which five genera and seven species are ahermatypic) have been recorded in the bay (Kawaguti, 1983). In the main channel, coral assemblages are characterized by abundant acroporids such as *Acropora hebes*. The shallow-water assemblage at the head of the bay is characterized by *Porites lobata*, *P. lutea*, *Cyphastrea serailia* and *C. chalcidicum*. Large balls of crustose coralline algae, about 20 cm in diameter, are found in the main channel where tidal currents are strong (Horikoshi, 1981).

Noteworthy Fauna and Flora   Loggerheads *Caretta caretta* have been recorded nesting on Shiraho Beach (Yamaguchi *in litt.*, 11.5.87). Poorly developed mangroves are found in several bays including Kabira and the southern part of the Shiraho area (Horikoshi, 1981; Yasumoto, 1986). A total of 102 fish species in 27 families, including a new species of *Pleurosicya* (Gobiidae), were recorded in a 3-day survey of the *Heliopora* zone of the Shiraho fringing reef in 1985 (Senou, 1986).

Scientific Importance and Research   Considerable research has been carried out on the Shiraho reefs and in Kabira Bay (see above). Kabira Bay was the site of an interdisciplinary survey of a tropical coastal ecosystem carried out by the Ocean Research Institute of the University of Tokyo (Horikoshi, 1981). The Yaeyama Branch of the Okinawa Prefectural Fisheries Experimental Station is situated near the mouth of the bay. Shiraho Reef is considered to be a particularly rich reef, with exceptional water clarity, few *Acanthaster planci*, and much living coral, in particular large stands of *Heliopora coerulea*. The reason for this is not known but is possibly due to the currents and topography in this area (Muzik, 1984a and b); this is discussed further in Yasumoto (1986). A bibliography of publications and research related to the Ishigakishima reefs has been compiled by McAllister (1987).

Economic Value and Social Benefits   Subsistence fishermen derive their living from fish, shellfish including lobsters, octopus, squid, cuttlefish *Sepia latimanus*, sea urchins and seaweed collected from the reef (Muzik, 1983). Reefs around Ishigakishima are some of the most popular for recreational activities such as diving, and the

tourist industry is increasing. The giant clam fishery is important (Murakoshi, 1985).

Disturbance or Deficiencies   The reefs around Ishigakishima have been devastated in the past by the Crown-of-thorns Starfish *Acanthaster planci*. Although it has been claimed that Shiraho had escaped, only two specimens of *Acanthaster* having been found in the course of a survey, and the reefs being regarded as still healthy in 1985 (Tokuyama, 1985), the reef slope here has been damaged (Yamaguchi *in litt.*, 11.5.87). There is also a noticeably high incidence of white- and black-band disease around the island which may be associated with the turbid conditions. *Acanthaster* has been observed to be more abundant in areas with dying coral (Kühlmann, 1985).

Additional threats are now posed by the intensification of agriculture and coastal development (Fosberg, 1986; Muzik, 1984a, b and 1985; Nishihira and Yamazato, 1974). Hills in Ishigakishima are being cleared for new pineapple plantations and for a new dam, large drainage channels are being constructed in Nagura Valley and local rivers are being straightened and lined with cement. All these activities could have long-term effects on the reefs through increased siltation. The current most serious threat is a proposal to construct a new airport near the village of Shiraho. This would involve the levelling of Karadake Hill which would be used to fill Shiraho Lagoon. About 3.6 km of reef would be destroyed in the creation of a 2.5 km runway. The purpose of the new airport is to deal with the increasing numbers of tourists visiting Ishigakishima; however, it will destroy one of the best reefs that visitors come to see (Muzik, 1983; Suzuki, 1986; Swinbanks, 1986).

Kühlmann (1985) points out the correlation between good coral growth and the presence of a coastal fringe of forest or vegetation which prevents excessive siltation. Siltation from construction activities near Ishigaki Town, the laying of submarine cables and the use of agricultural chemicals are cited as damaging factors. Sakieda Bay and Kabira Bay have considerable coral coverage, but much of this is dead and the water is turbid.

Legal Protection   Marine sanctuaries were established at Kabira Bay and Nagura Bay in 1975. Collection of boring clams within Kabira Bay is prohibited (Murakoshi, 1985 and 1986).

Management   The marine sanctuaries are under the control of fisheries agencies, not environmental agencies. The Kabira Bay Sanctuary was established on account of an aquaculture farm situated there which produces cultivated black pearls. Enforcement of the sanctuaries depends on understanding between local residents and visitors (Yamaguchi *in litt.*, 11.5.87). Enforcement of the Kabira Bay Marine Sanctuary was difficult in the early days but has improved recently particularly in the area around the Yaeyama Branch of the Okinawa Prefecture Fishery Experimental Station where the management office for the Sanctuary is located. Studies have shown that the giant clam population has increased since the establishment of the reserve and there is some evidence of an increase in areas outside the sanctuary (Murakoshi, 1986).

Recommendations   Opposition to the Shiraho airport proposal has been considerable and is being organized by a coalition of Ishigakishima fishermen and farmers, with Shiraho-born professors at the University of the Ryukyus in Okinawa (Friends of the Earth, Japan, 1984; Muzik, 1983; Plowden, 1984). The campaign has generated international interest through the efforts of a number of NGOs, but further pressure is required to halt the proposed development. It has been suggested that the area should be designated a marine park (Muzik, 1984a and b) and there has been a proposal to designate the blue coral stands at Shiraho as a "ten-nen kinenbutsu" (natural monument) (Nishihira, 1987; Zenji, 1987). The Cousteau Society (1984) recommended that the area be reserved for the use of local people only. A resolution was passed at the 17th Session of the IUCN General Assembly in Costa Rica, February, 1988, urging the Japanese Government to reconsider the construction of the airport, requesting protection for Shiraho Reef and suggesting the implementation of a research programme on the reefs. A detailed report with recommendations has been produced by the International Marinelife Alliance Canada (1988).

Kabira Bay has been recommended for Marine Park status on the basis of its terrestrial and marine scenery but there is opposition from local inhabitants. The Cousteau Society (1984) recommended the creation of an underwater trail at an appropriate site on Ishigakishima.

KUSHIMOTO MARINE PARK

Geographical Location   West of Shionomisaki, Wakayama Prefecture, the southernmost point of the Kii Peninsula, Honshu; 31°30'N, 135°30'E.

Area, Depth, Altitude   39.2 ha.

Physical Features   The warm Kuroshio Current comes close to shore in this area, permitting the establishment of tropical and subtropical species. Visibility was about 20-30 m, but has deteriorated to about 5 m (Moyer *in litt.*, 6.5.87).

Reef Structure and Corals   There are no true coral reefs in this area. Scleractinian corals of the Kii Peninsula are described by Utinomi (1966) and those of the Kushimoto area by Eguchi (1972b), Eguchi and Miyawaki (1975), Kushimoto Marine Park Center (1977) and Utinomi (1972). About 60 scleractinian species in 30 genera have been described. There are large coral communities at Meikizaki. Around Fudeshima, *Acropora leptocyathus*, *A. studeri*, *Hydnophora exesa*, *Platygyra lamellina* and *Pavona decussata* are abundant and tabular corals are particularly common. The nearby marine park of Kimanonada-Nikijima No. 1. has submarine cliffs with numerous *Melithaea flabellifera*, and coral communities of *Solanderia secunda*, table coral, *Dendronephthya habereri*, *Platygyra lamellina*, *Favia speciosa* and seaweeds (Marine Parks Center, n.d.).

Noteworthy Fauna and Flora   Butterfly fish, wrasse, damselfish and tangs are common from late spring to autumn but disappear in the winter. In the nearby marine park of Kimanonada-Nikijima No. 2 there are abundant fish including *Cheilodactylus* (*Goniistius*)

zonatus, *Prionurus microlepidotus*, *Chaetodon auripes*, *Apogon semilineatus*, *Ditrematemminki* and Labridae.

Scientific Importance and Research   The Sabiura Marine Park Research Station of the Marine Parks Center is located in the park and carries out work in this area. It is funded by income from the recreational facilities at the Park (Tamura, 1972). Studies are under way on the effect of winter temperature on corals (Misaki, 1985).

Economic Value and Social Benefits   The Park plays an important recreational role, and had up to 1000 visitors a day and 500 000 a year in the early 1970s (Tamura, 1972). Visitors averaged 359 000 a year from 1975 to 1983, but declined to 258 000 in 1983 (Marsh, 1985). The aquarium and observatory are major educational aids (Tamura, 1972). In the 1970s there were plans for building a hotel, lodges and a harbour, and two species of commercially important algae were being harvested within the Park (Hunnam, 1977).

Disturbance or Deficiencies   The Kushimoto area was affected by an outbreak of *Acanthaster planci* in 1973 (Hayashi, 1975; Yamaguchi, 1986a) but no starfish have been seen since 1980 (Yamaguchi, 1986a). Serious damage from the gastropod *Drupella* has also been recorded (Moyer *in litt.*, 6.5.87).

Legal Protection   Designated a Marine Park 1.7.70 as part of the Yoshino-Kumano Quasi-National Park. Two other marine parks lie in the same area. Kimanonada-Nikijima No. 1 (Nikijima, Kumano, Mie Prefecture) consists of two areas covering 7.8 ha and was established 19.12.75. Kimanonada-Nikijima No. 2 (Suno, Kumano, Mie Prefecture) consists of one area covering 6.6 ha.

Management   There is an aquarium (privately operated), underwater observatory, two glass-bottomed boats, restaurant, parking and gardens. Brochures are available for visitors (Anon., n.d. a). The park consists of four areas. An underwater trail has been established. Attempts were made to control the *Acanthaster* outbreak by the staff of the Sabiura Marine Park Research Station under contract to the Environment Agency (Yamaguchi, 1986a).

MIYAKE-JIMA

Geographical Location   One of the Izu-shoto, about 160 km south of Tokyo; 34°05'N, 139°31'E.

Area, Depth, Altitude   55.1 sq. km; max. alt. 518 m.

Physical Features   Miyake-jima is an active volcano and fourteen eruptions have been recorded on the island since AD 1085, the most recent being in 1962 and 1983. The island is nearly circular in shape, roughly 8 km in diameter and is dominated by the dissected slopes of Mount Oyama (Tribble and Randall, 1986). Annual rainfall is about 3000 mm. The warm Kuroshio Current strikes the island in the south-west and creates a gyre which flows along the coast to the west and then up the west coast. (Moyer *et al.*, 1985). Average monthly surface seawater temperatures range from 14.2°C in February and March to 27°C in July and August, although extremes of 13°C and 29.5°C have been recorded (Shepard and Moyer, 1980). Winds are generally strong and have a strong westerly component during most months except October and November. Twenty typhoons were recorded between 1955 and 1959 (Moyer, 1965).

Reef Structure and Corals   The richest coral communities are found on the south and west coasts which are influenced by the Kuroshio Current (Moyer *et al.*, 1985). The corals are widely scattered in small outcroppings, rather than true coral reefs, with coral cover rarely exceeding 10%. However, elements of reef-building have been found at Igaya Bay on the north-west side of the island, which has a more diverse coral fauna and greater live coral substrate cover than other localities (Tribble and Randall, 1986). Some relatively large *Acropora* patches are present in scattered locations around the island, the largest of which (1200 sq. m) is at Toga Bay in the south-west (Moyer *et al.*, 1982). A total of 91 species from 44 genera of shallow water (surface to 30 m depth) scleractinian corals have been identified, of which 80 species in 36 genera are hermatypic and 11 species in 8 genera are ahermatypic. The dominant families are Faviidae (23 species), Acroporidae (17 species) and Poritidae (10 species). The Oculinidae and Merulinidae are absent and *Heliopora*, *Tubipora* and hydrozoan corals have not been identified (Tribble and Randall, 1981 and 1986). Descriptions of nine transect sites in Igayama Bay are given in Tribble and Randall (1986). Live coral cover exceeded 30% in some parts of the bay and *Favia speciosa* constituted 26% of total coral cover in the bay as a whole (Tribble *in litt.*, 11.6.87).

Noteworthy Fauna and Flora   The island is largely forest-covered and has an important bird fauna with several species endemic to the Izu Islands, and an interesting terrestrial flora. More than 100 species of tropical marine algae have been found, despite the island's northerly location (Moyer *et al.*, 1985). Ida and Moyer (1974) describe cardinalfishes (Apogonidae) and an annotated checklist of fish is given by Shepard and Moyer (1980). More recent work has revealed additional species, including at least six endemics and over 70 species of Labridae (Moyer *et al.*, 1985). This is the world's highest latitude annual breeding location of the anemonefish *Amphiprion clarkii* (Moyer, 1976; Moyer and Bell, 1976). Moyer (1980) discusses the effects of high latitude on the species's social structure and behaviour; further information on the species is given in Bell *et al.* (1982).

Scientific Importance and Research   These are some of the highest latitude coral communities in the Indo-Pacific province, and coral diversity compares favourably with reefs in the Caribbean, Hawaii and Gulf of Aqaba. The warm Kuroshio Current has a major influence on the marine environment permitting the establishment of tropical species; fluctuations in water temperature through geological time have also contributed to speciation. The diversity of fish found around the island is considered to be particularly high for this latitude (Moyer *et al.*, 1985). Studies of the marine environment are being carried out by the Tatsuo Tanaka Memorial Biological Station situated on the island (Moyer *et al.*, 1985), and now concern exclusively the following: zoogeography of the marine fauna of Miyake-jima and

other Izu-shoto; family by family study of reef fishes for an eventual checklist; ecology of Miyake-jima's reef community, with strong emphasis on fishes; reproductive biology of Miyake-jima's fishes. An important theme is the study of short-term fluctuations in water temperature and their effect on local fauna (including mating systems). A project, beginning in 1987 and to continue for several years, is concerned with discovering whether or not corals spawn in the region (Moyer *in litt.*, 6.5.87). Most work on the reefs has been carried out in Igaya and Toga Bays (Tribble and Randall, 1986).

Economic Value and Social Benefits   Fishing is of economic importance and is described in Moyer (1965). Tourism is the main economic activity of Miyake-jima, the island being a popular summer destination owing to its proximity to Tokyo. It is now recognized as one of the main SCUBA diving sites in Japan, with about 5000 SCUBA divers visiting the island annually (Moyer *in litt.*, 6.5.87; Tribble *in litt.*, 11.6.87). Sport fishing is also popular (Moyer *et al.*, 1985).

Disturbance or Deficiencies   The northern locality of these coral communities means that they are subjected to a variety of natural stresses. The generally cool water temperatures probably act to slow growth and exceptionally cold water in winter can result in substantial mortality (Tribble and Randall, 1986). In 1987, the area was still under the influence of the most recent El Niño, with the strengthened, cold water Oyashio current having overwhelmed the warm water Kuroshio current, resulting in a dense, cold water mass occupying the area (Moyer, *in litt.*, 6.5.87). This caused a decrease in coral diversity compared with that found by Tribble and Randall (1986) in 1979, at the end of the warm period that began in the early 1970s (Shepard and Moyer, 1980; Moyer *in litt.*, 6.5.87). A major typhoon damaged large numbers of *Acropora* corals but the coral fauna was not in general severely damaged (Tribble *et al.*, 1982). The relative frequency of volcanic eruptions and the youth of the island suggest that the geological environment may be too unstable to allow rich coral communities to become established and reefs to develop (Tribble and Randall, 1986).

Since 1976, population explosions of *Drupella fragum*, a muricid gastropod, have been observed annually at Miyake-jima (Moyer *et al.*, 1982). The reefs have also been affected by *Acanthaster planci* (Moyer, 1978), and as a result of these two predators, there has been extensive coral mortality. Details of the *Drupella* outbreaks are given by Moyer *et al.* (1982), and a correlation is noted at the study sites between heavy silting from construction programmes and the beginnings of the outbreaks 2-4 years later. It is speculated that the *Drupella* outbreak may have been triggered by the siltation. *Acanthaster* was subject to a control programme in 1980, and is believed to have been finally eliminated mainly as a result of the El Niño current. There have been no further outbreaks although a small cluster survived around a volcanic hot spring at a depth of ca 15 m in 1984; it was no longer there in 1987 (Moyer pers. comm. to Yamaguchi, 1986 and *in litt.*, 6.5.87).

Deforestation for logging is an increasing problem on the island (Moyer *et al.*, 1985) and could lead to increased sediment loading of coastal waters. The construction of new coastal roads also poses a threat to inshore waters and reefs (Moyer *in litt.*, 6.5.87). The large number of visitors is causing environmental damage. SCUBA divers, attempting to gain easier access to the *Acropora* patch at Toga Bay have destroyed much of the grassland and other vegetation directly above the bay by driving over it, greatly increasing terrestrial runoff and siltation. Rubbish left by sport fishermen both in and out of the water is also a problem (Moyer *in litt.*, 6.5.87).

The Defense Agency of the Japanese Government and the U.S. Dept of Defense have designated the southern portion of Miyake-jima as a proposed site for a military airport to be used by U.S. carrier-based aircraft as a practice landing site. This proposal is considered to represent a serious threat to both the terrestrial and marine environment (Moyer *et al.*, 1985). Since 1985 the local government has been strongly resisting this and many Japanese and international organizations have filed formal protests with the governments of Japan and the U.S.A.

Legal Protection Miyake-jima lies within the limits of the Fuji-Hakone-Izu National Park (Moyer *et al.*, 1985). There are several areas on the island designated for special nature protection by the Environmental Agency of Japan and the Ministry of Agriculture, Forestry and Fisheries. The marine environment is not protected (Tribble *in litt.*, 11.6.87). However, the Tokyo Prefecture, which has jurisdiction over Miyake-jima and all other islands of the Izu-shoto and Ogasawara- shoto, prohibits the collection of organisms (living or dead) by SCUBA divers without permits which are issued only for scientific research (Moyer *in litt.*, 6.5.87).

Management At Miyake-jima, efforts are made to enforce the law concerning SCUBA diving; diving sites are patrolled by government workers during the working week (i.e. excluding Saturday afternoon and Sunday) in the diving season (Moyer *in litt.*, 6.5.87). In 1979 Igayama Bay was reportedly policed against spear-fishing by local fishermen (Tribble *in litt.*, 11.6.87). In 1980 a programme to control *Acanthaster* was initiated by the cooperative efforts of the Tokyo Prefectural and Miyake-jima local governments assisted by the Fishermen's Cooperative. The programme lasted one year and the outbreak was controlled largely by divers collecting the *Acanthaster* (Moyer *in litt.*, 6.5.87).

Recommendations Prior to the plan to build the airport in southern Miyake-jima, the Environmental Agency, in cooperation with the Tokyo Prefectural and Miyake-jima local governments, had planned to construct a Nature Park, including a Marine Park, at Toga Bay - Toga Shrine (Moyer *et al.*, 1985; Moyer *in litt.*, 6.5.87). The Environmental Agency and Tokyo Prefecture have since withdrawn support for this plan (Moyer *in litt.*, 6.5.87).

Moyer (*in litt.*, 6.5.87) recommends that the local government make efforts to protect Miyake-jima from the adverse effects of the large number of visitors, attracted by the international interest in the natural environment of the island. He makes the following specific recommendations: termination of the construction of new coastal roads (see Kühlmann, 1985); rapid implementation of reforestation programmes using natural cliffside and seaside vegetation; expansion of the government patrols of diving sites to cover weekends; elimination of rubbish-dumping by sport fishermen; construction of a parking lot at Toga Bay. It is of great importance that attempts be made to coordinate the conservation and research efforts by both terrestrial and marine specialists and a qualified independent team of scientists covering both these fields should further assess the potential damage from the construction of the proposed airport.

OKINAWA KAIGAN QUASI-NATIONAL PARK AND SESOKO-JIMA

Geographical Location Okinawa Kaigan Quasi-National Park includes three marine areas, Onna Kaigan, Tokashiki and Zamami Marine Parks, all in Okinawa Prefecture. The Quasi-National Park is discontinuous, extending from Hedo, at the northern tip of Okinawa (26°55'N, 128°15'E), along the north-west coast to Haneji, and then from Nago, south of the Motobu Peninsula (which is not included in the park), southwards to the Kerama-retto (ca 26°10'N, 127°20E) (Muzik *in litt.*, 5.8.87). Onna Kaigan Marine Park is located on the north-west coast of Okinawa in Nago at Buzena Saki (at the southern limit of Nago City, near Onna Village); Tokashiki Marine Park and Zamami Marine Park are located in the Kerama Islands.

Area, Depth, Altitude Onna Kaigan Marine Park covers 140 ha, Tokashiki Marine Park 120 ha, and Zamami Marine Park 233 ha (Anon., n.d. b).

Physical Features The waters are generally shallow and calm.

Reef Structure and Corals There are no publications relating specifically to the reefs of the park. Onna Kaigan Marine Park has a barrier reef and lagoon and the shore has a sandy beach (Anon., n.d. b). Reefs around Okinawa were surveyed in 1972 (Nishihara and Yamazato, 1974) in the course of a survey of *Acanthaster*, and in 1984 (Yamazato *et al.*, in prep.). The Environment Agency produced a survey report of the Onna Kaigan area in 1982 (Yamaguchi *in litt.*, 11.5.87). The reef flat west of Cape Zampa, on the west coast of Okinawa, has many large patches of *Porites cylindrica* at 0.3-3 m depth (Kato, 1987). Tokashiki has good reefs around the south-west coast. Reefs around Sesoko-jima are not included in the Quasi-National Park but are better known scientifically and are described by Yamazato *et al.*, (1984); the reef adjacent to Sekoko Marine Science Center, on the south-east coast of Sesoko-jima, is shallow and extends about 100 m offshore (Sato, 1985).

Noteworthy Fauna and Flora Some 90 species of fish and 21 species of non-coral invertebrates, including giant clams, have been recorded and turtles may be sighted (Anon., n.d. b). Damselfish on reefs around Sesoko-jima have been described by Yamamoto (1976). Dugongs *Dugong dugon* are occasionally sighted in the area (Yamaguchi *in litt.*, 11.5.87).

Scientific Importance and Research The Sesoko Marine Science Centre of the University of the Ryukyus carries out extensive work on the reefs of the area, publishing many of the results in the journal *Galaxea*. Coral growth and regeneration studies have been carried out by Sato (1985) and Kobayashi (1984). Mucus sheet formation has been studied by Kato (1987). Staff of the Ocean

Exposition Memorial Park Aquarium, located at Motobu, have carried out research on a variety of reef associated species (Yamaguchi *in litt.*, 11.5.87). Ecology of reef organisms on the Bise-Shinzato coast along the Motobu Peninsula, which is not within the Quasi-National Park, is described by Yamazato *et al.* (1976).

Economic Value and Social Benefits   Onna Kaigan Marine Park has an underwater observatory, aquarium, shell museum and recreational amusement park. There is an aquarium (the Ocean Exposition Memorial Park Aquarium) and an aquaculture laboratory at Motobu, the latter established in 1983 by the Okinawa Prefectural Government to produce finfish and sea urchins *Tripneustes gratilla*. The Kerama-retto are extremely popular with divers (Yamaguchi *in litt.*, 11.5.87).

Disturbance or Deficiencies   Nishihari and Yamazato (1972, 1973 and 1974) document outbreaks of *Acanthaster planci* on the west coast of Okinawa. The first outbreak was at Seragaki on the central western coast in 1969 (Yamazato, 1969). Following that, *Acanthaster* expanded north and south on this island, badly damaging the reefs. Despite attempts to control the population, it was still abundant in 1973 on many parts of the Okinawa coast and large areas of reef were killed (Environment Agency, 1973; Okinawa Tourism Development Corporation, 1976). More recent observations on *Acanthaster* around Sesoko-jima have been made by Sakai (1985). By 1984, some coral recovery was evident on the Okinawan reefs but *Acanthaster* was still numerous and there had been some reinfestation (Nishihira, 1987). Bleaching of corals was recorded on Okinawan reefs in 1980 (Yamazato, 1981).

Nishihira and Yamazato (1974) and Muzik (1985) provide details of the status of reefs around Okinawa. Many have been disturbed by a variety of types of human activity. Reclamation of reef flats and dredging of reef slopes has destroyed some reefs. Land development projects have led to erosion and siltation. Increasing agriculture, tourism and residential development have increased the pressures. It is reported that the construction of the submarine observatory in Onna Kaigan Marine Park damaged the adjacent coral communities. The Kokuba Developing Company is creating a major resort on the south-west coast of Tokashiki (Muzik *in litt.*, 5.8.87).

Legal Protection   Onna Kaigan Marine Park was established 15.5.72; Tokashiki and Zamami Marine Parks were established 19.12.78. The collection of any marine organism is prohibited (Anon., n.d. b); further details of legislation are given in the introduction.

Management   Okinawa Kaigan Quasi-National Park is administered by the Okinawa Prefectural Government, and Onna Kaigan Marine Park by the Okinawa Tourism Development Corporation, a non-profit organization created by the government. Attempts have been made to control populations of *Acanthaster* with the removal of some 240 000 starfish from the area between Cape Maeda-misaki and Point Bise-zaki on the Motobu Peninsula.

Recommendations   The 1982 report by the Environment Agency recommended that the regenerating coral communities should be preserved by careful land development and pollution controls. Underwater trails and a visitor centre are also planned (Yamaguchi *in litt.*, 11.5.87).

KIRIBATI

INTRODUCTION

General Description

The Republic of Kiribati comprises all the islands of the Gilbert and Phoenix groups, eight of the eleven Line Islands (the other three - Jarvis, Palmyra and Kingman Reef - being dependencies of the U.S.A.) and Banaba or Ocean Island. The total land area is ca 684 sq. km (of which Kiritimati in the Line Islands makes up nearly half), scattered over more than 5 000 000 sq. km of ocean and straddling both the equator and the international date line. With the exception of Banaba (a raised reef), all are low islands related to atolls. The climate is oceanic-equatorial, with a mean daytime air temperature of 32°; for much of the year (March - October) prevailing light east to south-east trades bring settled weather while the stormy season (November - February) is characterized by westerlies. Annual rainfall is variable with a mean of 1520 mm (range 380-3000 mm) in the Gilberts which are subject to droughts. Ocean surface temperatures in the Gilberts range from 27°C to 30°C, with surface salinities in the range 35.10 to 35.55 ppt. Prevailing ocean currents are south equatorial, running east to west with an east set around Tarawa which lies in an equatorial upwelling region (Anon., 1985b; Zann, n.d.). Terrestrial aspects of the Line and Phoenix Group islands are described in detail by Garnett (1983). Other publications on terrestial aspects of the islands include Catala (1957), Christopherson (1927), Fosberg (1953), Smith and Henderson (1978) and Stoddart (1976).

Table of Islands

Banaba (Ocean) (X) 2.5 sq. mi. (6.5 sq. km); raised limestone (elevated reef), 265 ft (81 m); very deep water off shore; fringing reef.

Gilbert Islands

Makin (Little Making Making Meang) (X) 2.8 sq. mi. (5.4 sq. km); one large island, 3 islets; no lagoon; reef.

Butaritari (X) 4.5 sq. mi. (11.7 sq. km); atoll with 11+ islets; reef; turtles nest on Katangateman sandbank; wave impact described by Groves (1982).

Marakei (Marake) (X) 3.9 sq. mi. (10 sq. km); atoll with semi-circular islets around pear-shaped lagoon; reef; marine algae described (Tsuda, 1964).

Abaiang (X) 11 sq. mi. (28 sq. km); atoll with many islets, 6 main ones, around large elongated lagoon; surveyed for clams (Munro, 1986).

Tarawa (X) (see separate account).

Maiana (X) 10.4 sq. mi. (27 sq. km); atoll with one main island and several small islets forming 2 sides of rectangular lagoon; lagoon shallow and turbid with few corals except near southern passage; western shelf has many large dead coral blocks; poorly developed western

barrier reef; relatively wide coral-studded shallow shelf to west of barrier; surveyed for Giant Clams (Munro, 1986).

Kuria (X) 4.9 sq. mi. (12.7 sq. km); 2 islets with fringing reef.

Aranuka (X) 6 sq. mi. (15.5 sq. km); atoll with 2 islets, small, shallow lagoon; triangular reef; mangrove.

Abemama (X) 9 sq. mi. (23.3 sq. km); atoll with 6+ islets forming continuous land rim to north-east of rectangular lagoon; south-east lagoon very turbid and little coral; lagoon near western pass between Abatiba and Bike islands has good coral; very narrow shelf between reef and sea; surveyed for Giant Clams (Munro, 1986); marine algae described (Tsuda, 1964); some mangroves.

Nonouti (X) 9.8 sq. mi. (25.4 sq. km); atoll with 8+ islets along north-east side of reef; lagoon with access for small boats only; ocean reef flat subject to continual heavy surf; marine algae described (Tsuda, 1964); turtles nest on nearby sandbank (Dahl, 1980).

Tabiteuea (X) 19 sq. mi. (49 sq. km); atoll with string of islets along north-east; largest at north and south ends, with many small ones between; elongated reef.

Beru (X) 8.1 sq. mi. (21 sq. km); atoll with one main islet; small reef.

Nikunau (Nukunau) (X) 7 sq. mi. (18 sq. km); atoll with one island and enclosed, landlocked lagoon, very small; fringing reef.

Onotoa (X) (see separate account).

Tamana (X) 2.0 sq. mi. (5.2 sq. km); island with fringing reef; no lagoon; marine algae described (Tsuda, 1964).

Arorae (X) 10 sq. mi. (26 sq. km); island with fringing reef; no lagoon.

Phoenix Islands

Kanton (Canton, Abariringa) (X) 910 ha; atoll, 5 m; elongated lagoon surrounded by broken rim of land; passages blasted into lagoon; 2 m deep with coral patches; fringing reef; algae described by Dawson (1956); Green Turtle breeding area (Garnett, 1983); island described by Degener and Gillespie (1955) and Lawrence (1979); vegetation described by Hatheway (1955); previously U.S. air base.

Enderbury 510 ha; island, 7 m; no lagoon except for shallow remnants in centre; no satisfactory anchorage; fringing reef 50-200 m wide cut by several channels; most important Green Turtle breeding site in Phoenix group (Garnett, 1983).

Birnie 20 ha; atoll with one island, 4 m, and shallow brackish lagoon; fringing reef; no anchorage; least spoiled island (Garnett, 1983).

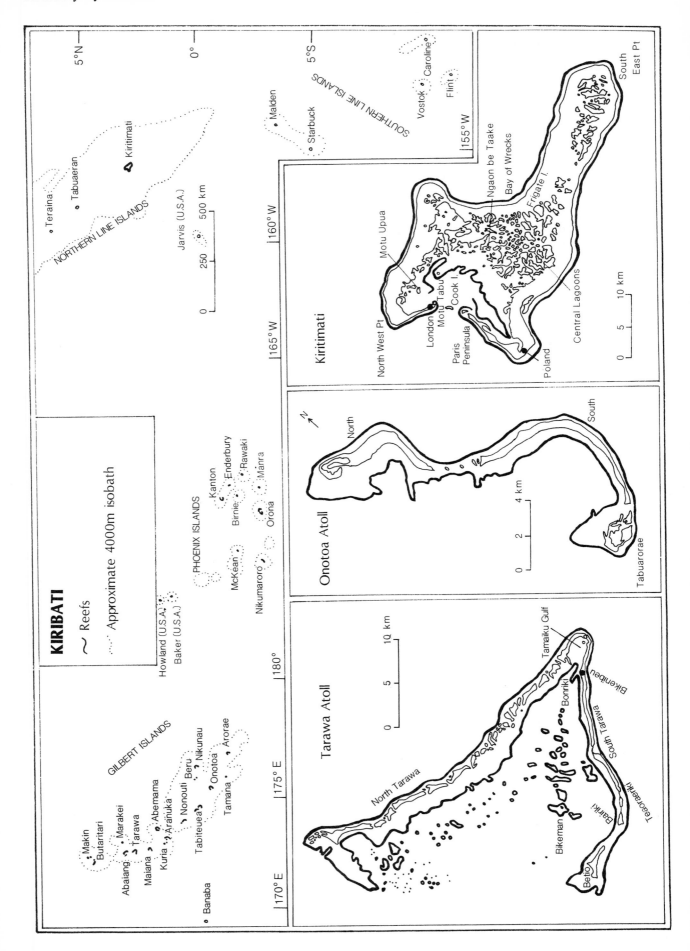

Rawaki (Phoenix) 65 ha; atoll, 6 m, with one island and shallow lagoon in centre; narrow fringing reef, 30-100 m wide; small numbers of breeding Green Turtles (Garnett, 1983).

Manra (Sydney) 436 ha; atoll, 15-20 ft (4.6-6.1 m), enclosed saline lagoon completely landlocked, one island; good anchorage but difficult landing; narrow fringing reef 50 m wide; breeding Green Turtles (Garnett, 1983).

Orona (Hull) 391 ha; atoll with rectangular lagoon surrounded by 24+ low islets; good landing and anchorage on west; fringing reef 80-250 m wide, emergent at low tide; few emergent coral heads in lagoon; Green Turtles breed and may be numerous (Garnett, 1983).

McKean 57 ha; atoll with 1 islet, 5 m, and central land-locked lagoon (Garnett, 1983); fringing reef, 100-200 m wide, emergent close to shore; corals described by Dana (1979).

Nikumaroro (Gardner) 414 ha; atoll with wedge-shaped lagoon and two low elongated islets almost enclosing it (Garnett, 1983); narrow fringing reef, 100-300 m wide; emergent near shore; numerous coral heads and shallow patch reefs in lagoon; abundant fish, birds and Coconut Crabs; Green Turtles breed (Dahl, 1980).

Southern Line Islands

Caroline 380 ha; atoll with 24+ islets around shallow, elongated lagoon; 5+ m; fringing reef off south end extends 800 m seawards; otherwise 400 m wide and steep; shallow central lagoon has much living coral; fish abundant; Green Turtle *Chelonia mydas* breeding area; Coconut Crabs *Birgus latro* abundant (Garnett, 1983).

Flint 324 ha; elongated island, 7 m; no lagoon; difficult landing; emergent fringing reef, mainly 100 m wide, but on north extends to 1500 m, and on south extends 500 m ESE; Green Turtle breeding area; Coconut Crabs abundant (Garnett, 1983); vegetation described by St John and Fosberg (1937).

Vostok 24 ha; small flat triangular island, 4-5 m; no lagoon; fringing reef extends 500 m off shore on north, south-west, and south-east; difficult landing; abundant Coconut Crabs (Garnett, 1983); vegetation described in Fosberg (1937).

Starbuck 16.2 sq. mi. (42 sq. km); atoll with one island, 5 m; enclosed shallow lagoons; no safe anchorage; fringing reef, ca 100 m wide, steep, except on south and east where extends 400-1400 m; Green Turtles nest (Garnett, 1983).

Malden 39.3 sq. mi. (102 sq. km); atoll with one triangular island 8 m; in north-west and south-west; saline lagoon connected to sea by underground channel; fringing reef, 100 m wide and steep; spur and groove formations extend 300-400 m off shore on north-west and south-east; Green Turtles nest (Garnett, 1983).

Northern Line Islands

Kiritimati (Christmas) (*see separate account*).

Tabuaeran (Fanning) 33.7 sq. mi (87.3 sq. km); atoll with enclosed lagoon encircled by almost continuous islands; 10-12 ft (3-3.6 m) (Chave and Kay, 1974; Garnett, 1983); flora described by St John (1974); narrow fringing reef; reef development and sedimentation described by Roy and Smith (1971) and community structure by Maragos (1974b); 70 species of hermatypic coral (Maragos, 1974a); reefs comparatively well known as result of Univ. of Hawaii expeditions in 1970 and 1972; comparison of these reefs with other Pacific reefs given in Glynn and Wellington (1983).

Teraina (Washington) 14.2 ha; atoll with one island, 5 m; reefs surveyed by Holthus (in prep.), generally narrow fringing reef, variable in width on north; extensive, gradually submerging reef extends out from western end; Green Turtles nest; privately owned; described by Teebaki (1979/1980).

(X) = Inhabited

There is relatively little information on the marine environment, although reefs at Tabuaeran, Teraina (Holthus in prep.), Tarawa and Onotoa (*see separate accounts*) have been described. Many of the characteristics of the Tarawa and Onotoa reefs are typical also of Butaritari. With the exception of Butaritari and Marakei, the atolls of the Gilbert Islands show continuous wave-breaking reef and almost continuous land on their windward (east) sides. The leeward sides are characterized by irregular outer reefs and few or no islands. All passes into the central lagoons are to leeward (Cloud, 1952). The atolls rise from deep water (4000-6000 m) with the angle of the outer reef slope probably ranging from 30° to 60°. Biota between 200 m and 500 m depth are thought likely to resemble Fiji's with carid shrimps, sponges, ahermatypic corals, gorgonians and snappers (Zann, n.d.). A coral species list for Kiribati is given in Zann (n.d.).

The Pacific Equatorial Research Station of the University of Hawaii is located on Tabuaeran (Fanning) in the Line group, which was visited by expeditions from the University in 1970 and 1972 (Chave and Kay, 1974). An Atoll Research Unit has been established on Tarawa, supported by the University of the South Pacific Institute of Marine Resources (*see separate account*).

Four species of giant clam occur in the central Gilbert Islands (*Tridacna gigas, T. squamosa, T. maxima* and *Hippopus hippopus*) (Munro, 1986). Holothurians of Tabuaeran are described by Townsley and Townsley (1974). Tsuda (1964) provides an annotated species list of marine algae in the Gilbert Islands. Marine turtles are described in Anon. (1979b) and Balazs (1975). The Green Turtle *Chelonia mydas* appears to be widespread, although in low numbers, in the Line and Gilbert Islands; it is likely to be more abundant in the Phoenix group. The Hawksbill *Eretmochelys imbricata* has been recorded but has not been found to nest. The Phoenix and Line Islands contain the principal breeding areas for seabirds for the Central Pacific, with rookeries containing many thousands and sometimes millions of birds (see e.g. Child, 1960; Garnett, 1983 and 1984; King, 1973).

Reef Resources

Most of the Kiribati islands are dry and infertile; in consequence the sea provides virtually all the protein in the diets of the I-Kiribati, with per capita fish consumption being one of the highest in the world (average consumption of whole fish is 565 g/head/day on rural atolls and 320 g/head/day on urban South Tarawa). Reefs are very heavily fished, although virtually all (outside South Tarawa) on a subsistence basis, and provide yields of ca 6-25 tonnes of finfish per sq. km per year. Fishing is carried out with a wide variety of techniques, involving different types of nets, hook and line, spears, nooses, traps and fences, poisoning and gleaning. Many hundreds of taxa of marine organisms are eaten, including algae, jellyfish, sipunculid worms, molluscs, crustaceans, echinoderms, virtually all non-toxic species of fish over a few cm in length, turtles, seabirds and cetaceans. However, finfish have generally been by far the most preferred food, with invertebrates and algae comprising only a small part of the diet on most islands, often only being eaten during times of hardship. An exception to this is in urban South Tarawa where lagoon bivalves (mainly *Anadara maculata, Gafrarium tumidum* and *Asaphis violascens*) have become staples in the diet, with landings exceeding that of all finfish combined. Elsewhere, on atolls with well-developed lagoons, lagoon fish generally form the major part of the diet while on reef islands flyingfish and tuna are more important (Zann, 1985). The Giant Clam fishery is described by Munro (1986). With the termination of phosphate mining on Banaba, previously the principal source of income for the country, the fishing industry is now thought to offer one of the best prospects for economic development (Oyowe, 1982). Most emphasis is placed on open sea fisheries, especially of tuna, but there are also plans to export reef and lagoon species (a small amount is exported at present from Kiritimati - *see separate account*) (Carleton, 1982). Kleiber and Kearney (1983) provide an assessment of the skipjack and baitfish resources. CCOP/SOPAC was contracted to undertake a survey of black coral resources in 1985 (Carleton and Philipson, 1987). Marine algal resources are described in Why and Reiti (1987).

Tourism at present is at a low level; such that occurs, especially on Kiritimati, is largely by naturalists and sports fishermen. Government policy was reported in 1982 as being to encourage controlled increase, but not mass tourism (Groves, 1983; Oyowe, 1982).

Disturbances and Deficiencies

Acanthaster planci has been reported in the past on reefs in Tarawa, Abaiang and Kiritimati (Edmondson, 1946).

Virtually the whole population of Kiribati (estimated at 60 302 in 1982) lives in the Gilbert Islands, with only Banaba (pop. 300) and Teraina (pop. 416), Tabuaeran (pop. 434) and Kiritimati (pop. 1800) in the Line Islands elsewhere being permanently inhabited. A large proportion of the population lives in the southern (urban) part of Tarawa. Environmental problems here are more acute than in the rest of Kiribati (*see separate account*). Coastal erosion in Kiribati is described by Howarth (1982).

Elsewhere in the Gilberts, although population densities are high, population increase is limited by shortage of land, and in particular by limits to the amount of terrestrial food which can be grown. Overfishing appears to be the principal potential form of disturbance (Lawrence, 1985; Zann, 1985). However, although waters adjacent to villages in the rural atolls may be overfished, those of more distant reefs, deeper waters and reefs exposed to strong wave action (the eastern and southern aspects of the atolls) are often underfished and may act as refuges (Zann, 1985). Ciguatera outbreaks, which are a problem in some areas (Cooper, 1964), often lead to reefs being left unfished for several years.

Nevertheless some species have evidently been overexploited, including turtles, giant clams, and coconut crabs, the latter having probably already been extirpated from Teraina (Garnett, 1983). Munro (1986) found stocks of all giant clams on Abaiang, Abemama, Maiana and Tarawa, except *T. maxima*, to be relatively low (with densities at Tarawa very low), and evidence of fairly intensive harvest for local consumption. Attention was shifting to *T. maxima*, especially on Tarawa, and there was a danger that this might become overfished.

The high urban demand for fresh fish, changes in fishing technology, the development of mechanised artisanal and small scale commercial fisheries, cold storage and improved communications are increasing fishing pressure. Plans to export reef and lagoon fish from Tarawa and the rural atolls would also greatly increase fishing pressure and could lead to increased conflict between subsistence and artisanal fisheries over fishing grounds, and between the subsistence sector and the national tuna corporations over baitfishing (Zann, 1985).

Legislation and Management

In pre-contact times land owners in Kiribati held tenure of reefs and lagoons adjacent to their lands and had exclusive rights to fisheries and passage. Most land, particularly in the southern Gilberts, was owned by groups of extended families ("utu") who lived in small, scattered hamlets ("kaainga"), although in the northern atolls the ruling king had control of a large area of land, reef and lagoons and dispensed fishing rights to the various clans in the area. In the late nineteenth century, this system began to break down: under British colonial rule, sea tenure *per se* was not recognized, although there was an attempt to codify traditional fishing rights and the government did recognize tenure of fish weirs, reclaimed areas, fish ponds and other accretions (Zann, 1985).

However, there are still a number of laws and customs regulating different aspects of fishing activities on many of the atolls. Many of these are formulated and applied by individual Island Councils. Controls can include limitations on the time and duration of fishing, restrictions on gear used and imposition of catch limits for resources susceptible to being overfished. Abemama has prohibited the taking of *Tridacna* clams by visitors, Nikunau has banned some fishing techniques in areas close to villages and Tamana has banned visitors on inter-island boats from fishing while they are in its waters; many islands also ban monofilament gill nets because they are too effective (Zann, 1985).

Zann (1985) argues that dietary preferences and local food taboos in the Gilberts often provide an effective mechanism for preventing overexploitation of vulnerable resources. Thus the abundant, pelagic flying fish and tunas, and, on atolls with large shallow lagoons, the lagoonal bonefish (*Albula*), mullet (Mugilidae), milkfish (*Chanos*), silver biddies (*Gerres*) and goatfish (Mullidae) are generally fished in preference to reef fish. Fishing on reefs on islands with limited reef areas is generally limited to periods of bad weather and to those people who are incapable of going to sea. Dietary taboos are believed to have been important in conserving some groups, such as turtles (formerly prohibited to commoners because of their rarity); this taboo has apparently largely broken down in recent years, and this has been implicated in the decline in turtle numbers.

Environmental legislation is described in SPREP (1980). The Fisheries Ordinance 1957 provides some blanket regulations, such as prohibition of fishing without a licence, and with explosives, gas or poison (Zann, 1985). The Wildlife Conservation Ordinance (1975, amended 1979) protects all birds throughout the year and Green Turtles (when on land), and their eggs and nests in the Line and Phoenix Islands, except on Kanton, Enderbury, Tabuaeran and Teraina. It is not clear whether Green Turtles are protected at sea (Anon., 1985b). The Ordinance also provides for the establishment of Wildlife Sanctuaries, within which killing or capturing any animal, other than fish, and disturbing of breeding sites is prohibited. Habitats or vegetation are not specifically protected. The following islands are Wildlife Sanctuaries, principally to protect seabird rookeries:

- Kiritimati: 32 100 ha (29.5.75)
- Malden: 3930 ha (29.5.75)
- Starbuck: 1620 ha (29.5.75)
- Rawaki: 65 ha (21.6.38)
- McKean: 57 ha (21.6.38)
- Vostock: 24 ha (19.6.79)
- Birnie: 20 ha (21.6.38)

Any area within a sanctuary can be declared a closed area, with a permit required for access. Seven closed areas have been declared so far: the entire islands of Malden and Starbuck and five within the Kiritimati Wildlife Sanctuary (*see separate account*).

All the protected areas except one of the closed areas on Kiritimati have been included in the International Ramsar Convention on Wetlands of International Importance (Anon., 1985b). At present there are no marine protected areas, and no protected areas of any type in the Gilbert Group.

The Wildlife Conservation Unit, based on Kiritimati, is the only government division responsible for day to day work in conservation management in the Line and Phoenix Islands. It is responsible for the enforcement of the Wildlife Conservation Ordinance, education/public awareness, survey and research, advice to the Government, control of introduced species, and tourism, although it is not the only agency involved in all these activities. At present only the closed areas and sanctuary on Kiritimati are fully enforced.

Recommendations

In 1980, the Government of Kiribati published a statement of its policy concerning nature conservation in the Line and Phoenix Islands, which recognises the need to integrate conservation with development of the islands' natural resources. This includes a commitment to, among other things, review and where necessary survey the entire wildlife resource, including marine flora and fauna, to implement the terms of the Wildlife Conservation Ordinance and to welcome scientific expeditions to the Line and Phoenix Islands.

A Management Plan for the Line and Phoenix Islands produced in 1983 identifies the areas of concern where action is required to fulfil the policy, and analyses the importance of each island in relation to a variety of parameters, largely non-marine (vegetation, flora, etc.) (Garnett, 1983). Existing and proposed developments on the islands are discussed, but their impact on the reefs is not touched upon. The main areas of concern are the development of tourism, fisheries, sand dredging for minerals and aquaculture. A number of islands would benefit from control programmes to eliminate introduced predators, especially rats and feral cats. Turtles should be given full protection. Substantial amendments to the Wildlife Ordinance are required.

The management plan also recommends the creation of further protected areas (Garnett, 1983):

1. Enderbury as a Wildlife Sanctuary and closed area (the most important turtle nesting site in the Phoenix group);
2. Birnie, McKean, Rawaki, Vostok as closed areas;
3. Turtle nesting areas on Caroline, Flint and Kanton.

A national park has been proposed for the Phoenix group, possibly excluding Nikumaroro, Manra and Orona. Dahl (1980) suggested a national or international reserve for the group with Kanton as communications link and surveillance centre and including Enderbury, Birnie, McKean, Rawaki and Orona and possibly Manra for its saline lagoon. Regular enforcement visits could be undertaken by government ships going to and from Kiritimati (Dahl, 1980). Birnie, Rawaki, Vostok and Malden were proposed as Islands for Science in 1971 (Elliot, 1973).

Dahl (1980) stresses the need for protection of representative marine areas, as well as for seabirds, turtles and land crabs, in the Gilbert group. In particular he recommends reserves on Butaritari at Kotabu and Nabini Islets for atoll forest and seabird rookeries, and on Teirio Islet on Abaiang for nesting turtles (Dahl, 1980).

It is considered that none of the areas investigated in the Gilbert group would support sustained exploitation of Giant Clams for export markets. Licensed pulse fishing, every 5-6 years, by foreign vessels paying a fee would be a possibility, but is not recommended as it would compete with domestic consumption. Giant clams are, however, recommended for mariculture, with the suggestion that a hatchery facility be developed at Tanea on Tarawa (Munro, 1986).

References

* = cited but not consulted

*Anon. (1979a). Wildlife Conservation Unit, Kiritimati. Report No. 4. Rep. Kiribati, Central Pacific.

Anon. (1979b). Notes on marine turtles of Republic of Kiribati. Paper presented at joint SPC-NMFS Workshop on Marine Turtles in the Tropical Pacific Islands. Noumea, New Caledonia, 11-14 December 1979. SPC-NMFS/Turtles/WP.15.

Anon. (1985a). Feral animal eradication programme, Kiritimati. *Report of the 3rd South Pacific National Parks and Reserves Conference, Apia* 2: 276-277.

Anon. (1985b). Country report - Kiribati. *Report of the 3rd South Pacific National Parks and Reserves Conference, Apia* 3: 115-124.

*Bailey, E. (1977). *The Christmas Island Story*. London.

Baines, G.B.K. (1982). Pacific Islands: Development of coastal marine resources of selected islands. In: Soysa, C., Chia, L.S., and Coulter, W.L. (Eds), *Man, Land and Sea: Coastal Resource Use and Management in Asia and the Pacific*. The Agricultural Development Council, Bangkok, Thailand. Pp. 189-196.

*Balazs, G.H. (1975). Marine turtles in the Phoenix Islands. *Atoll Res. Bull.* 184: 1-7.

Banner, A.H. and Randall, J.E. (1952). Preliminary report on marine biology study of Onotoa Atoll, Gilbert Islands. *Atoll Res. Bull.* 13. 62 pp.

Bolton L.A. (1982). The intertidal fauna of southern Tarawa Atoll Lagoon, Republic of Kiribati. Institute of Marine Resources, Univ. S. Pacific. 54 pp.

Carleton, C. (1982). Kiribati: Development of an export oriented reef fish fishing industry. Report to the Export Market Development Division of the Commonwealth Secretariat. Nautilus Consultants, Marlborough, U.K.

Carleton, C.C. and Philipson, P.W. (1987). Report on a study of the marketing and processing of precious coral products in Taiwan, Japan and Hawaii. South Pacific Forum Fisheries Agency, Honiara, Solomon Islands.

*Catala, R.L.A. (1957). Report on the Gilbert Islands: Some aspects of human ecology. *Atoll Res. Bull.* 59: 1-187.

*Chave, K.E. and Kay, E.A. (Eds) (1974). Fanning Island Expedition July-Aug 1972. Hawaii Inst. of Geophysics Rep. HIG-73-13.

*Child, P. (1960). Birds of the Gilbert and Ellis Islands Colony. *Atoll Res. Bull.* 74: 1-38.

*Chock, A.K. and Hamilton, D.C. (1962). Plants of Christmas Island. *Atoll Res. Bull.* 90: 1-7.

*Christopherson, E. (1927). Vegetation of Pacific equatorial islands. *B.P. Bishop Mus. Bull.* 44. 79 pp.

Cloud, P.E. (1952). Preliminary report on geology and marine environments of Onotoa Atoll, Gilbert Islands. *Atoll Res. Bull.* 12. 73 pp.

*Cooper, M.J. (1964). Ciguatera and other marine poisoning in the Gilbert Islands. *Pac. Sci.* 18: 411-440.

Dahl, A.L. (1980). Regional ecosystems survey of the South Pacific Area. *SPC/IUCN Technical Paper* 179. South Pacific Commission, Noumea, New Caledonia.

*Dana, T.F. (1979). Species-numbers relationships in an assemblage of reef-building corals: McKean Island, Phoenix Islands. *Atoll Res. Bull.* 228: 1-27.

*Dawson, E.Y. (1956). Some algae from Canton Atoll. *Atoll Res. Bull.* 65. 6 pp.

*Degener, O. and Gillaspie, E. (1955). Canton Island, South Pacific. *Atoll Res. Bull.* 41: 1-50.

*Doran, E. (1960). Report on Tarawa Atoll, Gilbert Islands. *Atoll Res. Bull.* 72. 54 pp.

Edmondson, C.H. (1946). *Spec. Publ. B.P. Bishop Mus.* 22.

Elliot, H. (1973). Pacific Oceanic Islands recommended for designation as Islands for Science. *Proceedings and Papers, Regional Symposium on Conservation of Nature - Reefs and Lagoons*. South Pacific Commission, Noumea, New Caledonia. Pp. 297-305.

*Floyd, C.H. (1978). Christmas Island (Gilbert Islands): Physical Planning Report. UNDAT Report.

*Fosberg, F.R. (1937). Vegetation of Vostok Atoll, Central Pacific. *B.P. Bishop Mus. Spec. Pub.* 30: 1-19.

*Fosberg, F.R. (1953). Vegetation of central Pacific atolls. *Atoll Res. Bull.* 23: 1-26.

Garnett, M.C. (1981). Project 1401. Wildlife Conservation Programme, Republic of Kiribati. *WWF Yearbook, 1980-81*. Pp. 115-117.

Garnett, M.I. (1983). A Management Plan for Nature Conservation in the Line and Phoenix Islands. Part 1: Description; Parts 2 and 3: Policy and Recommendations. Rept prepared for the Government of Kiribati. 449 pp.

Garnett, M.C. (1984). Conservation of seabirds in the South Pacific Region: A review. In: Croxall, J.P., Evans, P.G.H. and Schreiber, R.W. (Eds), *Status and Conservation of the World's Seabirds*. ICBP Technical Publication No. 2. Pp. 547-558.

Glynn, P.W. and Wellington, G.M. (1983). *Coral and Coral Reefs of the Galapagos Islands*. Univ. California Press, Berkeley/Los Angeles/London. 330 pp.

*Goodenough, W.H. (1963). Ecological and social change in the Gilbert Islands. *Proc. 11th Pacific Science Congress, Vol 3. Anthropology and Social Sciences*. Pp. 167-169.

*Groves, G.W. (1982). High waves at M'akin in December 1979. Atoll Research Unit Tech. Rept, Tarawa, Kiribati.

Groves, G.W. (1983). Opportunities for scientific research at Kiritimati Island. Atoll Research Unit Tech. Rept, Tarawa, Kiribati.

*Groves, G.W. and Ting, B.Y. (1982). Flow through Tarawa channels. Atoll Research Unit Tech. Rept, Tarawa, Kiribati.

*Hatheway, W.H. (1955). The natural vegetation of Canton Island, an equatorial Pacific island. *Atoll Res. Bull.* 43: 1-9.

Holthus, P. (in prep). Washington Island. Draft ms.

*Howarth, R. (1982). Coastal erosion in Kiribati. Tech. Rept No. 1. 13 pp.

*Johannes, R., Kimmerer, W., Kinzie, R., Shiroma, E. and Walsh, T. (1979). The impacts of human activities on Tarawa lagoon. Unpub. report to South Pacific Commission, Noumea, New Caledonia.

King, W.B. (1973). Conservation status of birds of central Pacific Islands. *Wilson Bull.* 85: 89-103.

Kleiber, P. and Kearney, R.E. (1983). An assessment of the skipjack and baitfish resources of Kiribati. *Skipjack Survey and Assessment Programme Final Country Report* 5. South Pacific Commission, Noumea, New Caledonia.

*Lawrence, P.K. (1979). Report of a visit to Canton Island. Unpub. rept.

Lawrence, R.J. (1985). The transformation of traditional agriculture in the atolls of Kiribati. *Proc. 5th Int. Coral Reef Cong., Tahiti* 5: 595-601.

*Maragos, J.E. (1974a). Reef corals of Fanning Island. *Pac. Sci.* 128(3): 247-255.

*Maragos, J.E. (1974b). Coral communities on a seaward reef slope, Fanning Island. *Pac. Sci.* 28(3): 257-278.

Marshall, J.F. and Jacobsen, G. (1985). Holocene growth of a mid-Pacific atoll: Tarawa, Kiribati. *Coral Reefs* 4(1): 11-17.

*Moul, E.T. (1957). Preliminary report on the flora on Onotoa Atoll, Gilbert Islands. *Atoll Res. Bull.* 57. 48 pp.

Munro, J.L. (1986). Status of giant clam stocks and prospects for clam mariculture in the Central Gilbert Islands group, Republic of Kiribati. Report to Fisheries Division, Ministry of Natural Resources Development, Kiribati and South Pacific Regional Fisheries Development Programme, Suva, Fiji.

*Northrup, J. (1962). Geophysical observations on Christmas Island. *Atoll Res. Bull.* 89. 2 pp.

Oyowe, A. (1982). Kiribati - living without phosphates. *The Courier* 76: 28-32.

*Perry, R. (1980). Wildlife conservation in the Line Islands, Republic of Kiribati (formerly Gilbert Islands). *Env. Cons.* 7: 311-318.

*Roy, K.J. and Smith, S.V. (1971). Sedimentation and coral reef development in turbid water: Fanning Lagoon. *Pac. Sci.* 25: 234-248.

Ryland, J.S. (1981). Natural resources of Christmas Island (Line Islands). Abst. *Future Trends in Reef Research*.

St John, H. (1974). The vascular flora of Fanning Island, Line Islands, Pacific Ocean. *Pac. Sci.* 28(3): 247-256.

*St John, H. and Fosberg, F.R. (1937). Vegetation of Flint Island, Central Pacific. *Occ Pap. B.P. Bishop Mus.* 12(24): 1-4.

Schoonmaker, J., Tribble, G.W., Smith, S.V. and Mackenzie, F.T. (1985). Geochemistry of saline ponds, Kiritimati. *Proc. 5th Int. Coral Reef Cong., Tahiti* 3: 439-444.

*Silver, H. (1982). Bêche-de-mer in the deeper waters of Tarawa Atoll. Atoll Research Unit Tech. Rept.

*Smith, S.V. and Henderson, R.S. (Eds) (1978). Phoenix Islands Report 1. *Atoll Res. Bull.* 221. 183 pp.

SPREP (1980). Kiribati. *Country Report* 7. South Pacific Commission, Noumea, New Caledonia.

Stanley, D. (1984). *South Pacific Handbook*. Moon Publications, Chico, California. 581 pp.

Stevens, C. and Burell, B.G. (in press). Coral as aircraft pavement material - case histories. *Proc. Inst. Civil Engineers*.

Stoddart, D.R. (1976). Scientific importance and conservation of central Pacific atolls. Unpub. rept.

*Teebaki, K. (1979/1980). Reports of visits to Washington Island. Unpub.

*Townsley, S.J. and Townsley, M.P. (1974). A preliminary investigation of the biology and ecology of the holothurians at Fanning Island. In: Chave and Key (1974).

Tsuda, R.T. (1964). Floristic report on the marine algae of selected islands in the Gilbert group. *Atoll Res. Bull.* 105. 13 pp.

*Weber, J.N. and Woodhead, P.M.J. (1972). Carbonate lagoon and beach sediments of Tarawa Atoll, Gilbert Islands. *Atoll Res. Bull.* 221.

*Wentworth, C.K. (1931). Geology of the Pacific equatorial islands. *Occ. Pap. B.P. Bishop Mus.* 9: 1-25.

Why, S. and Reiti, T. (1987). The marine algae of Kiribati. In: Furtado, J.I. and Wereko-Brobby, C.Y. (Eds), *Tropical Marine Algal Resources of the Asia-Pacific Region: a Status Report*. Commonwealth Science Council Publication Series 181. Pp. 53-56.

Zann, L. (1985). Traditional management and conservation of fisheries in Kiribati and Tuvalu atolls. In: Ruddle, K. and Johannes, R.E. (Eds), *The Traditional Knowledge and Management of Coastal Systems in Asia*

and the Pacific. Unesco/ROSTEA, Jakarta. Pp. 53-77.

Zann, L.P. (n.d.). The marine ecology of Betio Island, Tarawa Atoll, Republic of Kiribati. Institute of Marine Resources, Univ. S. Pacific. 27 pp.

Zann, L.P. and Bolton, L. (1985). The distribution, abundance and ecology of the blue coral *Heliopora coerulea* (Pallas) in the Pacific. *Coral Reefs* 4: 125-134.

KIRITIMATI (CHRISTMAS ISLAND) WILDLIFE SANCTUARY

Geographical Location   Northern Line Islands, about 285 km south-west of Tabueran; 2°00'N, 157°20'W.

Area, Depth, Altitude   Land area 321.37 sq. km; Cook Islet 23 ha; Motu Tabu 4 ha; Ngaon te Taake 25 ha; Motu Upua 16 ha; large tidal lagoon of 160 sq. km in west; complex of land-locked lagoons cover 168 sq. km in east; maximum altitude 13 m.

Land Tenure   Government-owned.

Physical Features   The island, a modified atoll, is part of a line of submarine volcanoes, the Christmas Island Ridge, and consists of coral and other biogenic rocks overlying volcanic rocks. The single, large, flat island is D-shaped with a stem oriented south-east and has a large tidal lagoon opening to the north-west. The eastern end of the lagoon consists of a series of smaller land-locked lagoons, separated by causeways and larger tracts of land. There are substantial areas of fossil reef and rubble scree, as well as beach-rock. The island reaches 7-13 m in height along the north coast of the south-east peninsula where there are dunes, and is semi-arid, covered with low scrub and vegetation in the centre and west. It is surrounded by a beach of coral sand. The main lagoon is bordered to the north, east and south by extensive low-lying lagoon flats; such flats are also found on the south-east peninsula. The lagoons contain hundreds of islets, the three principal ones being Cook, Motu Tabu and Motu Upua.

The island lies at the edge of the doldrums zone and has a sparse and erratic rainfall (Ryland, 1981), heaviest from January to June. Easterly trade winds prevail throughout the year and temperature is very stable, varying diurnally from 24° to 30°C. Geology of the island is described by Wentworth (1931) and geophysical observations were made by Northrop (1962). Garnett (1983) gives information on climate, hydrology and geology. Geochemistry of saline ponds on the island is discussed in Schoonmaker *et al.* (1985). A comprehensive description of the island is found in Bailey (1977). Glynn and Wellington (1983) discuss the possible evolution of the reefs.

Reef Structure and Corals   A reef platform, ranging from 30-120 m wide extends from the shoreline around all parts of the island, being widest along northern coasts. Along its seaward margin, there is a spur and groove formation which is best developed on the lee side of the island. Along the windward coast and especially in the Bay of Wrecks, there are ridges covered with reddish-purplish algae which are emergent at low tide

(Garnett, 1983). The narrow windward fringing reef has been visited at low tide and 25 species of coral and hydrocoral were collected (Ryland, 1981). On the sheltered eastern side of the main lagoon, the lagoon reef is moderately well developed. Elsewhere, the high turbidity and rate of sedimentation inhibits coral growth. Patch reefs and coral heads are found in the western part of the main lagoon wherever conditions are more favourable (Garnett, 1983).

Noteworthy Fauna and Flora   A detailed description of the flora and fauna is given in Garnett (1983). The island is most important for its seabirds, a 1981 estimate giving six million birds using or breeding on the island. A total of 22 species have been recorded of which 18 breed. Notable are Sooty Terns *Sterna fuscata* (several million), Phoenix Petrel *Pterodroma alba* and Christmas Shearwater *Puffinus nativitatis*. There is one resident land-bird, the Christmas Island Warbler *Acrocephalus aequinoctialis*. There is a small population of Green Turtles *Chelonia mydas*, which nest sporadically in the Bay of Wrecks and between Poland and Paris. Coconut Crabs *Birgus latro* are found occasionally (Garnett 1983). Exploitable semi-terrestrial crustacea present in great numbers are the crab *Cardisoma carnifex* and the hermit crab *Coenobita perlatus*. The most valuable resources are spiny lobsters *Panulirus* sp. collected from the reef (Ryland, 1981).

Scientific Importance and Research   Garnett (1983) provides a comprehensive bibliography. The island is reported to have the largest land area of any coral atoll and has been the subject of numerous studies and surveys. It was visited in 1978 for a preliminary assessment of its natural resources and development potential (Floyd, 1978). The University of Hawaii has several projects in the Line Islands which use Kiritimati as a transhipment and marshalling point; the University also has projects measuring sea level and water temperature, and making meteorological observations on Kiritimati. The New Zealand meteorological office maintains a weather station at London (Groves, 1983).

Economic Value and Social Benefits   The history of the island is described by Garnett (1983). Large scale development of Kiritimati, which is by far the largest island in Kiribati, is hampered by the shortage of fresh water and the distance from the Gilbert group, the administrative and population centre of the country. There is a resident population of ca 1800 Gilbertese who prepare copra. There are proposals for the development of salt production facilities which could become a major industry (Garnett, 1983). Export of brineshrimp and their eggs, cultured in the salt-ponds, is also planned; however a similar project to culture brineshrimp for export was abandoned in 1978 after substantial financial investment had been made.

The construction of a 20-room hotel and improved air connections with Honolulu have led to the development of small scale tourism in the form of visits by groups of sports fishermen and naturalists; the hotel organizes fishing and SCUBA diving (Garnett, 1983; Groves, 1983). Stanley (1984) notes that diving can be hazardous because of the presence of large numbers of grey sharks; the safest area is reportedly that between North-west Point and Paris. Groves (1983) mentions tentative plans for a tourist complex near the town of Poland.

The export of live crayfish and chilled ocean and lagoon fish has recently been undertaken, especially of milkfish, of which in 1982 around one ton per week was exported to Honolulu and Nauru. A local enterprise also exports small, live reef fish, collected by divers using SCUBA gear, to Honolulu for the aquarist trade (Groves, 1983).

Disturbance or Deficiencies   *Acanthaster planci* was reported to be abundant in the 1940s (Edmondson, 1946). Unusual weather conditions were experienced during the abnormal El Niño of 1982-83, with a high sea level and very high rainfall (Groves, 1983) but it is not known if the reefs were affected. The island was used in the U.K. nuclear test programme, 1956-58, and the U.S.A. programme in 1962 but it is not known if damage to the reefs occurred. The terrestrial environment and fauna have suffered damage from feral cats and pigs.

Legal Protection   Kiritimati was proclaimed a bird sanctuary under old legislation on 20.12.60 and the islets of Cook, Motu Tabu and Motu Upua, were declared reserves with restricted access. On 29.5.75, Kiritimati was declared a Wildlife Sanctuary under the Wildlife Conservation Ordinance and five closed areas were designated: North-west Point Reserve, Motu Tabu Reserve, Motu Upua Reserve, Cook Island Reserve and Ngaon te Taake Islet Reserve. All of these are seabird rookeries (Anon., 1985a and b; Garnett, 1983).

Management   The Wildlife Conservation Unit was established in 1977 with the direct aid of the British Ministry for Overseas Development and the support of WWF/IUCN (project 1401) (Garnett, 1981 and 1983; Perry, 1980). It comprises a warden and two assistants and has as its primary aim the protection of the seabird colonies. Its achievements are described by Garnett (1983), and include development of an education programme, improvement of law enforcement, a preliminary survey of the natural history of the island (excluding marine habitats), advice to the Government and controlling numbers of exotic animals.

Recommendations   Garnett (1983) draws up detailed recommendations for the island. It is considered that it should no longer, as a whole, retain its wildlife sanctuary status, but instead sanctuaries should be established in the following areas: North-west Point, Central Lagoons, South-east Peninsula, Paris Peninsula and on all off-shore islets. Closed areas should be retained for Cook Island, Motu Upua, Ngaon le Taake and Motu Tabu and additional closed areas have been proposed for Frigate Island and all Sooty Tern colonies in the breeding season. A proposal for a five-year project to eradicate feral cats and pigs was prepared in 1983 by the Wildlife Conservation Unit in conjuction with the New Zealand Wildlife Service (Anon., 1985a) but has not yet been started (Garnett *in litt.*, 22.10.87).

There are considered to be good propects for economic development of the island, centring on marine products, tourism, salt, and possibly resettlement of land-poor I-Kiribati from islands in the Gilbert group (Garnett, 1983; Groves, 1983). Exploitation of reef fish and lobsters by individual fishermen or private companies for export could be increased markedly, as long as care is taken to avoid potential ciguatera contamination, and protected fisheries areas have been proposed in the Central Lagoons and on the South-east Peninsula. There are also plans to extract sand from the lagoon. It is important that all such development plans be assessed for their impact on the reefs.

ONOTOA ATOLL

Geographical Location The most southerly atoll of the Gilbert Islands (Tamana and Arorae Islands lie further south); 1°47'S, 175°29'E.

Area, Depth, Altitude Land area 5.2 sq. mi. (13.5 sq. km); lagoon area 21 sq. mi. (54 sq. km).

Physical Features The atoll lies between the west-flowing South Equatorial Current and the east-flowing Equatorial Countercurrent, at about the northern limit of the south-east trade winds. The hydrology is described by Cloud (1952). For most of the year there is a steady easterly trade wind although from October to March occasional west and north-west gales occur. Hurricanes have been recorded. Rainfall averages about 40 in (1016 mm), with January the wettest month. Onotoa is considered a "dry" atoll. This accounts for the sparse ground cover vegetation. Cloud (1952) provides a map of depth contours of the lagoon which is very shallow (maximum depth 8 fathoms (14.6 m)). The general bottom topography, excluding numerous small patch reefs, consists of three shallow basins. Between the patch reefs, the bottom consists of calcium carbonate sand, silt or gravel.

The islands (Anteuma, Abenecnec, North (Tanyah), South (Otoae) and Tabuarorae) consist mostly of unconsolidated sand and gravel with little solid rock.

Reef Structure and Corals Banner and Randall (1952) and Cloud (1952) provide a description of the reefs. The exposed, windward reef, along the northern, eastern and southern shores of the atoll, is almost unbroken and varies in width from 300 ft (91 m) to over quarter of a mile (366 m), being more extensively developed in the south. The shore consists of either consolidated or eroded coral rock or of moderately fine sand. The reef flat, emergent but with frequent pools at low tide, may be 650 ft (198 m) wide; seaward of the flat is a depression (the back-ridge trough), 50-100 ft (15-30 m) wide and of variable depth. The reef flat consists largely of consolidated coralline algae and dead *Heliopora*. The exposed surface of the coral and, in some areas, the tidal pools are covered with a more or less dense growth of algae. Seaward of the trough is a coralline ridge, 50-100 ft (15-30 m) wide, its shoreward edge presenting an almost continuous front of reddish coralline algae and its seaward side with deep surge channels perpendicular to the shore. This spur and groove formation is largely limited to the surf zone. Surge channels range in length from 50-120 ft (15-37 m) and are about 6 ft (1.8 m) deep at midlength, deepening to 8 ft (2.4 m) at the reef front. Living coral and algae are abundant only at the crest and upper sides of intervening buttresses.

The outermost reef shelf is relatively narrow, about 300 ft (91 m) wide, and slopes rather rapidly from about 10 ft (3 m) depth on the shoreward side to over 35 ft (10.7 m) deep on the seaward side. It consists of irregular mounds of living coral interspersed with areas strewn with dead coral fragments. Beyond the shelf, the reef drops away suddenly. The reef shelf shows a rough zonation. *Pocillopora meandrina* is dominant in shallow water near the coralline ridge, *Acropora* dominates the deeper water of the middle and outer shelf and massive heads of *Porites lobata* are also conspicuous. Large areas of the bottom are covered with dead, loose branches of *Acropora*.

The leeward reefs, on the ends of the islands, west and north-west of Tabuarorae and its northern reef are similar to the windward reef but do not have a well-developed reef flat, and the back ridge trough and coralline ridge are completely absent. The reef flat which is usually submerged at low tide instead changes quite abruptly into conditions similar to the windward reef shelf. In water of 2-5 ft (0.6-1.5 m), the major elements of the fauna are the same as the back-ridge trough on the windward reef. The major exception to these generalities lies in the region northward of the *Heliopora* flats off Antenma, where conditions are similar to the area within the reef to the west of Abenecnec Island.

Heliopora flats are found in protected areas behind the windward reef at the south end and north-west tip of the north island. The southern *Heliopora* flat consists of an extensive tide pool with a sandy bottom scattered with *Heliopora* heads. The northern *Heliopora* flat is similar but shows a transition from a typical *Heliopora* flat to a consolidated condition as found on the windward reef.

A variety of habitats are found in the lagoon, including sand flats scattered with patch reefs which are dominated by *Heliopora*. Patch reefs are almost continuous toward the leeward reef and passes, and gradually decrease in number towards the windward reef platforms. In many areas the sand flats grade gradually into a region of dead coral reefs, as off the south-east of Anteuma, the southern portion of North Island, the northern portion of South Island and the passage between, and they are extensively developed off Tabuarorae and in the south-western portions of the lagoon. These areas provide a variety of habitats and are faunistically the richest in the lagoon.

Cloud (1952) provides a list of corals found at Onotoa which includes 50-60 species in 26 genera. Generalized descriptions of the green alga zone, the red alga zone (including back ridge trough), the coralline ridge and the benched reef slope of the windward reef are also given.

Noteworthy Fauna and Flora There are extensive seagrass beds of *Thalassia* particularly in the northern part of the lagoon. The lagoon has an abundant fish population and the ichthyofauna of the atoll is described by Banner and Randall (1952). A total of 352 species have been recorded. Economically important invertebrates are also listed (*see below*). The flora of the atoll has been described by Moul (1957). Ten species of seabird have been recorded.

Scientific Importance and Research Studied by a team from the Pacific Science Board in June-August 1951. Changes in sea-levels are discussed by Cloud (1952), work on the reefs at Onotoa providing evidence for a recent 6 ft (1.8 m) fall in sea-level.

Economic Value and Social Benefits The only suitable anchorage for larger vessels is on the leeward shelf outside the gap in the outer reef. Banner and Randall (1952) describe the islanders utilization of marine invertebrates, which are considered a supplement to the main diet of fish. A list of species which are eaten is given and includes the spiny lobster *Panulirus pencillatus*, the Coconut Crab *Birgus latro*, the molluscs *Trochus*, *Turbo*, *Pinctada margaritifera*, *Tridacna maxima* and *T. squamosa*. *Charonia tritonis* is collected for use as a trumpet. Fishing methods are also described by Banner and Randall (1952).

Disturbance or Deficiencies The lagoon reef is not considered to be very productive for edible molluscs. Banner and Randall (1952) thought that this might be a result of overfishing (the atoll is heavily populated) combined with a natural low productivity. There was no evidence of depletion of fish stocks although fish were more abundant in the less populated areas.

Legal Protection None.

Management Sections of the reef flat and lagoon were traditionally owned by different people who had exclusive right to fish there, but this system of reef tenure began to break down with the arrival of the British in 1892; in 1951 the inhabitants agreed by referendum to declare the lagoon public domain (Banner and Randall, 1952; Goodenough, 1963). The only remaining traditional law is that fishing is prohibited in the vicinity of an individual stone fish trap at or near low tide (Banner and Randall, 1952).

TARAWA ATOLL

Geographical Location In the Gilbert group; 1°20'N, 173°00'E.

Area, Depth, Altitude 21 sq. km land area; Betio is 1.3 sq. km; 375 sq. km lagoon; mean depth 25 m.

Physical Features Tarawa Atoll has an L-shaped chain of over 15 islets enclosing a wide shallow lagoon open to the west. Rural North Tarawa comprises twelve major islets with a land area of 14.7 sq. km. Urban South Tarawa comprises five elongated islets (most linked by causeways) with an area of 7.2 sq. km. Betio lies at the extreme western point (Bolton, 1982; Zann, n.d.). The atoll is described in Doran (1960). Little is known of its geology but it is probably similar to Bikini and Eniwetok in the Marshalls. Mean daily air temperature range is 25-32°C. From March to October, light east to south-east trade winds prevail; the stormy season is characterized by westerly winds (Zann, n.d.). The lagoon is fairly turbid (Munro, 1986), and sediments are described in Weber and Woodhead (1972). Mean tidal range in the lagoon is 1.5 m (Bolton, 1982) and currents have been investigated by Groves and Ting (1982). Annual sea temperature range is 28-29°C and salinity ranges from 35.4 ppt to 36.0 ppt (Zann and Bolton, 1985).

Reef Structure and Corals A 22 mile (35 km) reef forms the south and north-east sides of a triangle, a sunken reef forming the north-west side (Doran, 1960).

To the south-west, there is a diffuse non-continuous reef north of the main ship channel. This consists mainly of massive spurs of luxuriant coral growth running east-west. The lagoon has deep patch reefs of *Porites* and other massive corals (Munro, 1986). Immediately west of Betio is a shallow area of dense coral growth. Holocene coral growth is discussed in Marshall and Jacobson (1985).

In South Tarawa, the seaward upper reef slope (15-40 m depth) is relatively steep (up to ca 45°) and covered by a skree of calcareous sediments with living and dead coral outcrops; living coral (*Heliopora* dominant) reached 20% cover at 15 m depth. This gives way at 10-15 m to a gently sloping terrace up to 380 m width with 10-50 m long coral buttresses at least 50 m wide and separated by sandy channels; living coral cover on the buttresses is ca 20-30% with *Heliopora* and *Acropora* dominant, and coral reef fish abound. Above this is a spur-and-groove system, the spurs with ca 5% live coverage of small, mainly encrusting corals (*Porites*, *Montipora* and *Pocillopora*), and the grooves with 5-10% cover, dominated by branching *Pocillopora*, *Acropora* and *Heliopora*. This gives on to a raised algal ridge at the edge of a wide, flat reef platform, which completely dries out at most low tides, and consists of a limestone basement, partially covered by sand, coral rubble and boulder tracts, and notably lacking in living corals. Along the inner (lagoonal) margin of the reef flat is a discontinuous line of small patch reefs and micro-atolls, in depths of up to 3 m, mostly consisting of dead, algal encrusted *Porites*, although living coral cover (*Pocillopora*, *Porites*, *Montipora* and *Acropora*) reached 40% at one site. The lagoon itself is shallow (ca 10-15 m on average) with many sand shoals and numerous coral patch reefs, both living and dead, towards the open, western side of the atoll. Generally the percentage of living coral cover increases with distance from the shore. Further details of the ecology of the reefs, including a detailed description of the area around Bairiki and Betio, are given in Zann (n.d.) and Zann and Bolton (1985). The coral fauna is impoverished Western Pacific in character, and 88 species in 44 genera have been recorded (Zann and Bolton, 1985). The ecology of *Heliopora coerulea* on the reefs is described by Zann and Bolton (1985).

Noteworthy Fauna and Flora The intertidal lagoon fauna is described by Bolton (1982). Tsuda (1964) described marine algae. Mangroves occur along part of the coast (Doran, 1960).

Scientific Importance and Research Johannes *et al.* (1979) carried out a baseline study on the impact of human activities on the lagoon at the request of the Government as a result of an outbreak of cholera in 1977. Reefs in both North and South Tarawa have been surveyed for Giant Clams (Munro, 1986). The Atoll Research Unit is based on Tarawa and is supported by the University of the South Pacific Institute of Marine Resources. It carries out research on the marine environment, particularly environmental problems affecting Tarawa such as the studies by Bolton (1982) and Zann (n.d.).

Economic Value and Social Benefits The I-Kiribati are traditionally subsistence fishermen and many urban dwellers fish around Betio, commercially, for subsistence and for recreation. The reef front and terrace in the vicinity of Bairiki and Betio are important fishing grounds but are apparently depleted. The large North Betio reef is more productive (Zann, n.d.). Bêche-de-mer resources in deep waters have been investigated by Silver (1982).

Disturbance or Deficiencies The population of South Tarawa numbers nearly 20 000, over half living on Betio. As a result, environmental problems here are more acute than in the rest of Kiribati (Anon., 1985b; Zann, n.d.). Several activities in South Tarawa may adversely affect the reefs, including foreshore reclamation, construction of solid-fill causeways between islets, dumping of rubbish on the lagoon shore, and sewage discharge (Bolton, 1982; Zann, n.d.; Johannes et al., 1979). The acute shortage of building land has led to extensive reclamation of the lagoon foreshore, particularly off Bonriki islet, where Tamaiku Gulf has been reclaimed to provide airstrip approaches and other features. Coral boulders (mainly living or dead *Porites*) are collected from the lagoon reefs and rafted to shore for the erection of retaining walls, with fill being gleaned from beaches, the lagoon or adjacent land. This has resulted in local destruction of reef areas. The siltation resulting from dredging activity has probably also adversely affected adjacent coral reefs. Several channels have been constructed (by blasting, excavation or dredging) through the lagoonal reef platforms to allow small boats access to land; the largest of these is at Betio, where a long channel and a basin have been excavated in the reef and island, with smaller channels or harbours at Bairiki, Teaoraeriki and Bikenibeu.

Most of the islands in South Tarawa have been joined by causeways, which has resulted in shoreline erosion, increased sedimentation, interference with migration and spawning runs of fish such as milkfish and destruction of fishing grounds. Construction of the final, 3 km leg of the causeway, joining Bairiki and Betio, was due to recommence in 1982, having been begun in 1977 but then abandoned because of engineering problems. The platform on which the causeway was to be constructed was identified as an area of low diversity and productivity. However the potential long-term effects of construction on more productive and diverse parts of the lagoon, particularly nearby coral communities, was thought likely to be severe, both in increased sedimentation from dredge spoil and disruption of water flow across the flat.

Bolton (1982) found that sewage has not yet led to significant pollution in the lagoon, but there have been problems in the past, including an outbreak of cholera in 1977. A sewage system for most of South Tarawa has been constructed with an outfall into the ocean. The impact of sewage on the lagoon environment has thus been minimised, although Zann (n.d.) comments that possible eutrophication on the ocean-facing reef should be monitored, particularly as an enormous aggregation of the sea urchin *Tripneustes gratilla* was recorded adjacent to the outfall.

Abnormally large amounts of recently dead coral, mainly *Porites lobata* heads and micro-atolls, were recorded by Zann (n.d.) around the edge of the lagoon between Ambo and Betio and on patch reefs between the islets and Bikeman. Living corals were common only on a few lagoonal patch reefs off the Betio-Bairiki platform which were regularly flushed by oceanic water. Large areas of dead coral were also seen on the lagoonal patch reefs off North Betio, mid-way towards Bikeman, off Bikeman and on the terrace buttress reefs of west Betio. Dead corals in the lagoon are probably due to siltation, and evidence suggests that mortality is a continuing process and not the result of a single event such as a storm. No evidence of siltation was seen at the site 7 nautical miles (13 km) north of Betio or off north and west Betio in oceanic areas. *Acanthaster* outbreaks may be implicated in these areas; *A. planci* outbreaks occurred in the early 1970s (Weber and Woodhead, 1972).

There is intensive exploitation of the lagoon resources (Baines, 1982). Nearly 700 turtles were being taken annually on Tarawa in the 1970s (Anon., 1979b). Stocks of all species of Giant Clams, apart for *Tridacna maxima* are very low and there is evidence of intensive harvest for local consumption. Attention is now shifting to *T. maxima* and there is the danger that this may become overfished (Munro, 1986). However, ciguatera outbreaks have led to reefs being left unfished for several years, for example after World War II and in the early 1980s.

Legal Protection Dredging and blasting for construction materials on reefs is reportedly no longer permitted (Newill *in litt.*, 5.8.87; Stevens and Berrell, in press).

Recommendations Zann (n.d.) recommended the insertion of a bridge or culvert in the proposed causeway between Betio and Bairiki, to lessen its effects on the lagoon environment, and the monitoring of the area during and after the construction of the causeway. It is also recommended that regular monitoring of the composition of water in the lagoon and of its benthic fauna be carried out to ensure that pollution, particularly from human wastes, does not become a problem (Bolton, 1982; Zann, n.d.).

It has been suggested that a hatchery facility for Giant Clams should be developed at Tanea (Munro, 1986).

MARSHALL ISLANDS

INTRODUCTION

General Description

The Republic of the Marshall Islands (REPMAR) became a freely associated state in close association with the U.S.A. in October 1986. It consists of 29 coral atolls and five coral islands of very low elevation (5-20 ft (1.5-6 m)), a few reaching 25 ft (7.6 m)), totalling some 70 sq. mi. (180 sq. km) land area and over about 750 000 sq. km marine area, lying between 4° and 15°N. They form two chains, the Ratak Islands to the east and the Ralik Islands to the west. They are mostly irregular in shape, the land area of each atoll or island generally increasing from north to south. While some are very small, the group includes Kwajalein which has the largest lagoon in the world. The emergent parts (islets or motus) are mainly composed of coral debris and sand with some compact coral rock outcroppings. In general, the westward or lee sides of the islets rise steeply while the eastward sides slope gently. Bryan (1972) provides the most concise description of the Marshalls and information on many of the islands is given in Fosberg (1966) and Wiens (1957 and 1962).

Climate is tropical with mean annual temperature 82°F (27.8°C) and mean monthly temperature varying by 2°F (1°C) although diurnal range is considerably greater. Rainfall increases from north to south (from around 37 ins (940 mm) on Wake (see section on Hawaii), just north of the Marshalls, to 159 ins (4040 mm) on Jaluit), while seasonality decreases; in the north heaviest rainfall occurs from September to November whereas in the south it is heavy throughout the year (Amerson, 1969). In the northern Marshalls, the north-east trade winds predominate throughout the year; in the south these predominate from December to April while during the rest of the year east to south-east winds increase in frequency, becoming predominant in autumn. Typhoons are rare, with only four reported since 1900 (in 1905, 1951, 1958 and 1967) (Amerson, 1969), although tropical storms and large waves from distant storms strike the islands (Wells, 1954 and 1957a and b).

Ocean currents consist primarily of westward moving North and South Equatorial Currents and the Equatorial Countercurrent moving eastward between them. The North Equatorial Current is usually above ca 9°N, is very broad and averages 0.4 knots (with flow up to 2 knots), the South Equatorial Current below around 4°N is stronger, also broad, reaching 3-4 knots, while the Countercurrent (usually between 5° and 8°N) ranges between 0.4 and 2 knots (Amerson, 1969) and is narrower (Sverdrup et al., 1942).

Table of Islands

Ralik Islands

Enewetak (Eniwetok) (see separate account).

Ujelang (X) 0.67 sq. mi. (1.7 sq km); lagoon 25.5 sq. mi. (66.0 sq. km); atoll with 32 islets on elongated, east-west narrow reef.

Bikini (see separate account).

Ailinginae 1.1 sq. mi. (2.8 sq km); lagoon 40.9 sq. mi. (105.9 sq. km); atoll low and dry with 25 islets.

Rongrik (Rongerik) 0.65 sq. mi. (1.7 sq. km); lagoon 55.6 sq. mi. (144.0 sq. km); atoll with 17 islets; N-S elongated reef; turtles formerly abundant.

Rongelap 3.1 sq. mi. (8.0 sq km); lagoon 387.8 sq. mi. (1004.4 sq. km); atoll with 61 islets, arid, poor sandy soil; sea cucumbers studied (Bonham and Held, 1963).

Wotho (X) 1.7 sq. mi. (4.4 sq. km); lagoon 36.7 sq. mi. (95.1 sq. km); atoll with 13 islets, low and fairly wet; Coconut Crabs common.

Ujae (X) 0.7 sq. mi. (1.8 sq. km); lagoon 71.8 sq. mi. (186.0 sq. km); atoll with 14 islets on elongated E-W reef rim.

Lae (X) 0.6 sq. mi. (1.6 sq. km); lagoon 6.8 sq. mi. (17.6 sq. km); atoll with 17 islets.

Lib (?) 0.4 sq. mi. (1.0 sq. km); one island, low and sandy with central depression forming freshwater pond; fringing reef; no atoll-like lagoon (Maragos in litt., 10.8.87).

Kwajalein (X) (see separate account).

Namu (X) 2.4 sq. mi. (6.2 sq. km); lagoon 153.5 sq. mi. (397.6 sq. km); atoll with 51 islets on north-east rim of lagoon.

Jabwot (Jabat) 0.2 sq. mi. (0.5 sq. km); one island, no lagoon; surrounded by reefs; close to Ailinglaplap.

Ailinglaplap (Ailinglapalap) (X) 5.7 sq. mi. (14.7 sq. km); lagoon 289.7 sq. mi. (750.3 sq. km); large atoll with 52 islets on almost continuous reef rim; mangroves.

Jaluit (X) 4.4 sq. mi. (11.4 sq. km); lagoon 266.3 sq. mi. (689.7 sq. km); atoll with 84 islets on large diamond shaped reef, almost continuous land rim; mangroves; reef; fishing.

Kili (X) 0.4 sq. mi. (1.0 sq km); one low island, no lagoon; surrounded by fringing reef.

Namorik (X) 1.1 sq. mi. (2.8 sq. km); lagoon 3.2 sq. mi. (8.3 sq. km); atoll with closed lagoon and 2 main islands.

Ebon (X) 2.2 sq. mi. (5.7 sq. km); lagoon 40.1 sq. mi. (103.9 sq. km); atoll with 22 islets.

Ratak Islands

Bokaak (Taongi) 1.25 sq. mi. (3.2 sq. km); lagoon 30.1 sq. mi. (78.0 sq. km); atoll, crescent-shaped with 11 islets; low, dry, stony; no passage through reef; landing difficult; important seabird colony; nesting sea turtles (Maragos in litt., 10.8.87).

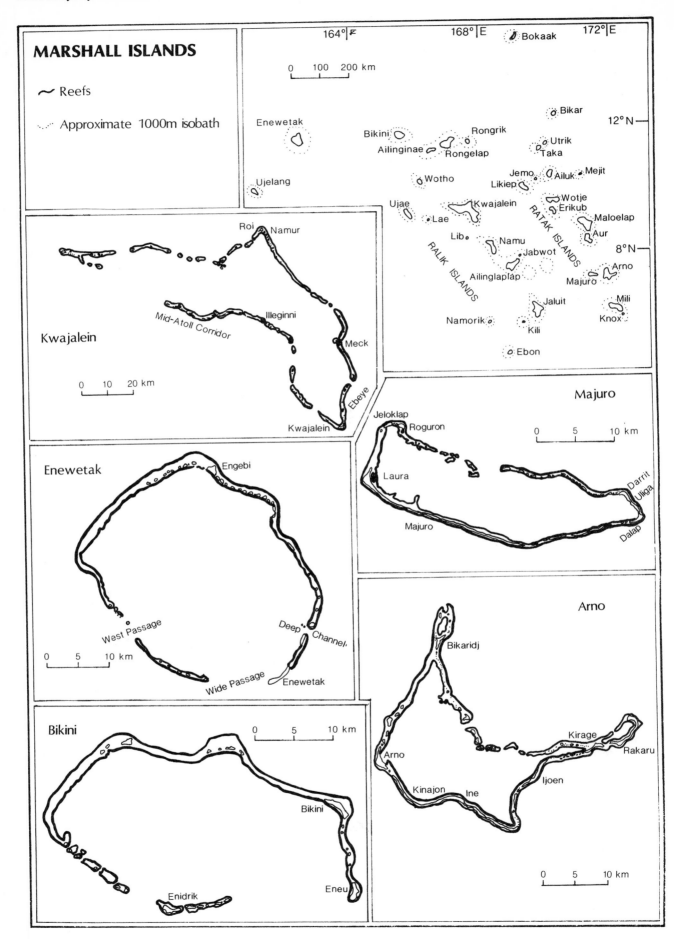

MARSHALL ISLANDS

～ Reefs

Approximate 1000m isobath

164°E 168°E 172°E

0 100 200 km

Enewetak

Ujelang

Bokaak

Bikar

Bikini Rongrik
Ailinginae Rongelap
Utrik
Taka

Wotho Jemo Ailuk Mejit
Likiep
Ujae Kwajalein Wotje
Erikub
Lae Maloelap
Lib Aur
Namu
Jabwot
RALIK ISLANDS RATAK ISLANDS
Ailinglaplap Arno
Majuro
Jaluit Mili
Namorik Knox
Kili
Ebon

12°N

8°N

Kwajalein

Roi Namur

Illeginni
Mid-Atoll Corridor
Meck
Ebeye
Kwajalein

0 10 20 km

Majuro

Jeloklap Roguron
0 5 10 km
Laura
Darrit
Uliga
Majuro
Dalap

Enewetak

Engebi

West Passage
Deep Channel
Wide Passage Enewetak

0 5 10 km

Arno

Bikaridj

Kirage
Arno Rakaru
Kinajon Ine Ijoen

0 5 10 km

Bikini

0 5 10 km

Bikini
Enidrik Eneu

Bikar 0.2 sq. mi. (0.5 sq. km); lagoon 14.4 sq. mi. (37.3 sq. km); diamond-shaped atoll with 6 islets, low and dry; reef; visited for fishing; important breeding ground for *Chelonia mydas* and seabirds.

Utrik (Utirik) (X) 0.9 sq. mi. (2.3 sq. km); lagoon 22.3 sq. mi. (57.8 sq. km); atoll with 6 islets.

Taka 0.2 sq. mi. (0.5 sq. km); lagoon 36.0 sq. mi. (93.2 sq. km); atoll with 5 islets, low and dry; important Sooty Tern *Sterna fuscata* colony.

Mejit (X) 0.7 sq. mi. (1.8 sq. km); 1 small islet, no lagoon but central small pond, mangroves; surrounded by fringing reef.

Ailuk (X) 2.1 sq. mi. (5.4 sq. km); lagoon 68.5 sq. mi. (177.4 sq. km); atoll with 35 islets on elongated N-S reef; some typhoon damage.

Jemo 0.06 sq. mi. (0.16 sq. km); 1 egg-shaped islet on linear reef; no lagoon; moderate rainfall; sea turtle nesting area.

Likiep (X) 4.0 sq. mi. (10.4 sq. km); lagoon 163.7 sq. mi. (424.0 sq. km); atoll with 64 islets.

Wotje (X) 3.2 sq. mi. (8.2 sq. km); lagoon 242.1 sq. mi. (627.0 sq. km); atoll with 72 islets.

Erikub 0.6 sq. mi. (1.6 sq. km); lagoon 88.9 sq. mi. (230.3 sq. km); atoll with 14 small islets on reef rim; close to Wotje.

Maloelap (X) 3.8 sq. mi. (9.9 sq. km); lagoon 375.6 sq. mi. (972.8 sq. km); atoll with 71 islets, triangular reef.

Aur (X) 2.2 sq. mi. (5.6 sq. km); lagoon 92.6 sq. mi. (239.8 sq. km); atoll with 42 islets; square reef; vegetation described by St John (1951).

Majuro (X) (*see separate account*).

Arno (X) (*see separate account*).

Mili (X) 5.8 sq. mi. (15.0 sq. km); lagoon 293.4 sq. mi. (759.9 sq. km); atoll with 84 islets, poor sandy soil; connected to Knox by deep submerged reef:

- *Knox* 0.4 sq. mi. (1.0 sq. km); lagoon 1.32 sq. mi. (3.4 sq. km); elongate "atoll" with 10 islets.

NB: Land area and number of islands for many of the atolls varies in different sources; this is partly the result of constant changes in the atolls owing to storms and coastal processes. Figures quoted here were provided by Maragos (*in litt.*, 10.8.87). Orthography follows Motteler (1986).

(X) = Inhabited

There have been few general descriptions of the reefs. Information on Bikini, Kwajalein, Enewetak, Majuro and Arno is given in separate accounts. In general, windward ocean reef slopes have submarine terraces and often descend gradually in contrast to leeward ocean reef slopes which plummet nearly vertically (Maragos *in litt.*, 10.7.87). Corals are described by Wells (1954), plankton by Johnson (1954), foraminifera by Cushman *et al.* (1954) and marine annelids by Hartmann (1954). *Heliopora coerulea* dominates the seaward reef flats (Emery *et al.*, 1954; Wells, 1954). The ecological relationships of coral reef fish are described by Hiatt and Strasbourg (1960) and Schultz and collaborators (1953, 1960 and 1966) describe the fish fauna in detail. The marine and terrestrial plants of the islands are described by Taylor (1950).

The Green Turtle *Chelonia mydas* and the Hawksbill *Eretmochelys imbricata* occur in the area; in the 1950s the former was reportedly very numerous in the north (Amerson, 1969). In the 1970s, Bikar was identified as the most important turtle nesting site with Bikini and Bokaak (Taongi) possibly second (Pritchard, 1981). Both Bikar and Bokaak are uninhabited, and Bikini was temporarily uninhabited; its importance for turtles is expected to decrease when it is resettled (Maragos *in litt.*, 10.8.87). The avifauna of the islands is discussed in detail by Amerson (1969). Thirty-one species of seabird have been recorded with 15 definitely breeding and a further two (Bulwer's Petrel *Bulweria bulwerii* and the Grey-backed Tern *Sterna lunata*) probably doing so.

Reef Resources

The human population was estimated at 27 000 in 1977. Subsistence fishing is an important activity on many atolls but the commercial fishing potential is undeveloped. Smith (1947) described traditional fishing methods in Micronesia, including the Marshall Islands. Government employment and the U.S. military missile testing range at Kwajalein (*see separate account*) provide the main source of income (Maragos 1986 and *in litt.*, 10.8.87). The potential for tourism has been considered but is not high at present, given the deteriorating state of the natural resources of the islands (Maragos, 1986).

Disturbances and Deficiencies

Tropical storms and large waves from distant storms greatly affect island reef geomorphology (Wells, 1954 and 1957a and b) and typhoons have caused damage on some atolls such as Utrik and Jaluit, but it is not known to what extent they have affected the reefs. Damage on the islands of Jaluit is now being modified by regrowth of vegetation (Blumenstock 1958 and 1961; Blumenstock *et al.*, 1961; McKee, 1959). Infestations of *Acanthaster planci* have been recorded (Chesher, 1969).

Many of the islands have been extensively modified by man, particularly for copra production and through military activities. Bikini and Enewetak were used as atomic test sites by the U.S.A. from 1946 to the 1960s (Telegadas, 1961). Environmental disturbance on these, and on Kwajalein and Majuro, the two most heavily populated atolls, is discussed in separate accounts. The atolls of Utrik and Rongelap were, and Rongrik and Taka may have been, exposed to fallout contamination from the 1954 nuclear test "Bravo" on Bikini. Rongrik, where the Bikinians briefly settled after their evacuation from Bikini in 1946, has a reputation for poisonous fish (Maragos *in litt.*, 10.8.87). There was much destructive bombing and tree felling on Wotje. Erikub has suffered some disturbance by visitors. Seabirds and turtles and their eggs are taken throughout the atolls but the long term effects of this are not known. The absence of environmental legislation, since U.S. Trust Territory regulations ceased to apply in October 1986 (see below), is a cause of serious concern, particularly given the small size and fragile nature of most of the atolls (Maragos *in litt.*, 10.8.87).

Legislation and Management

Traditionally property rights extend out to sea. Some unpopulated islands and reefs used to be set aside as reserves for the protection of seabirds, eggs, turtles, Coconut Crabs and crayfish, and harvesting was only permitted on certain occasions. However, these customs were broken during the Japanese occupation (Johannes, in press).

U.S. environmental laws no longer apply to the Republic of the Marshall Islands, except at the Kwajalein Missile Range and other sites where activities are carried out by the U.S. There have been no counterpart environmental regulations established to date to govern Marshallese Government and private actions, although these were promised as part of the agreement to terminate the Trust Territory (Maragos *in litt.*, 10.8.87).

Previously Trust Territory-wide law provided complete protection for Hawksbill Turtles and all turtle eggs, had two closed seasons for all turtles (1 June - 31 August and 1 December - 31 January) and forbad taking of Green Turtles under 34 ins (86 cm). Public Law 4C-65 of 1972 prohibited fishing with explosives, poisons or chemicals. These regulations were poorly enforced. Pollution legislation under Trust Territory Law is described in SPREP (1980). Bokaak and Bikar were proposed as Islands for Science under the International Biological Programme and were previously protected by order of the District Administrator for their seabird and turtle colonies; their reefs were never protected.

There are no marine reserves. Plans are under way to clear up Bikini atoll of its radiological contamination and an environmental impact statement (EIS) is to be prepared for the U.S. army missile range at Kwajalein (*see separate accounts*).

In 1939-40 the Japanese, then in control of the islands, introduced the Mediterranean sponge *Spongia officinalis* ssp. *mollissima* into the lagoon of Ailinglaplap where they were cultured on wires. These were still present in 1947 but were never apparently commercially harvested, despite being of very high quality. Between 1930 and 1937 Trochus *Trochus niloticus* were transplanted to Jaluit and Ailinglaplap from Belau (Smith, 1947).

Recommendations

The Republic of the Marshall Islands needs to establish new environmental laws and regulations, giving greater control over pollution, construction and development, to protect the natural resources of its atolls and islands. Environmental assessments of Majuro, Arno and Kwajalein are considered the highest priority needs at present (Maragos *in litt.*, 10.8.87 and 23.11.87); a coastal

resources management plan is urgently required for the entire atoll of Kwajalein (Maragos, 1986). Some degree of protection is recommended for Jemo. The creation of seabird and turtle reserves on Bikini has been recommended. Dahl (1980 and 1986) recommends protection of samples of undisturbed windward and leeward atoll reefs and lagoon habitat.

References

* = cited but not consulted

*Agegian, C.R., Chave, K.E., Lauritzen, R., Ratigan, E., Suzumoto, A. and Tribble, G. (1987). Marine environmental assessment for the rehabilitation and resettlement of Bikini Atoll. Status report prepared under Bikini Atoll Rehabilitation Committee (BARC) contract. Appendix C, BARC Suppl. Doc. No. 2, Part 2. 100 pp.

Amerson, A.B. (1969). Ornithology of the Marshall and Gilbert Islands. *Atoll Res. Bull.* 127. 348 pp.

Amesbury, S.S., Tsuda, R.T., Zolan, W.J. and Tansy, T.L. (1975a). Limited current and underwater biological surveys of proposed sewer outfall sites in the Marshall Island District: Ebeye, Kwajalein Atoll. *Univ. Guam Mar. Lab. Tech. Rept* 22. 30 pp.

*Amesbury, S.S., Tsuda, R.T., Zolan, W.J. and Tansy, T.L. (1975b). Limited current and underwater biological surveys of proposed sewer outfall sites in the Marshall Island District: Darrit-Uliga-Dalap area, Majuro Atoll. *Univ. Guam Mar. Lab. Tech. Rept* 23. 30 pp.

Atkinson, M., Smith, S.V. and Stroup, E.D. (1981). Circulation in Enewetak Atoll Lagoon. *Proc. 4th Int. Coral Reef Symp., Manila* 1: 335-338.

*Blumenstock, D.I. (1958). Typhoon effects at Jaluit Atoll in the Marshall Islands. *Nature (Lond.)* 182: 1267-1269.

*Blumenstock, D.I. (Ed.) (1961). A report on typhoon effects on Jaluit atoll. *Atoll Res. Bull.* 75: 1-105.

*Blumenstock, D.I., Fosberg, F.R. and Johnson, C.G. (1961). The re-survey of typhoon effects on Jaluit Atoll in the Marshall Islands. *Nature (Lond.)* 189: 618-620.

*Blumenstock, D.I. and Rex, D.F. (1960). Microclimate observations at Enewetak. *Atoll Res. Bull.* 71. 158 pp.

*Bonham, K. and Held, E.E. (1963). Ecological observations on the sea cucumbers *Holothuria atra* and *H. leucospilata* at Rongelap Atoll. *Pac. Sci.* 17(3): 305-314.

*Boucher, L. (1986). Coral predation by muricid gastropods of the genus *Drupella* at Enewetak, Marshall Islands. *Bull. Mar. Sci.* 38(1): 9-11.

*Brost, R.B. and Coale, R.C. (1971). *A Guide to Shell Collecting in the Kwajalein Atoll.* Charles E. Tuttle Co. 157 pp.

*Bryan, E.H., Jr. (1972). *Life in the Marshall Islands.* Pacific Scientific Information Center, Honolulu. 237 pp.

Buddemeier, R.W. (1981). The geohydrology of Enewetak Atoll islands and reefs. *Proc. 4th Int. Coral Reef Symp., Manila* 1: 339-345.

*Buddemeier, R.W., Maragos, J.E. and Knutson, D.W. (1974). Radiographic studies of reef coral exoskeletons: rates and patterns of coral growth. *J. Exp. Mar. Biol. Ecol.* 14: 179-200.

*Buddemeier, R.W., Smith, S.V. and Kinzie, R.A. (1975). Holocene windward reef-flat history, Enewetak Atoll. *Bull. Geol. Soc. Am.* 86: 1581-1584.

*Bussing, W.A. (1972). Recolonization of a population of supratidal fishes at Enewetak Atoll, Marshall Islands. *Atoll Res. Bull.* 154: 1-7.

Chesher, R.H. (1969). Destruction of Pacific corals by the sea star *Acanthaster planci*. *Science* 165: 280-283.

*Colin, P.L. (1986). Benthic community distribution in the Enewetak Atoll lagoon, Marshall Islands. *Bull. Mar. Sci.* 38(1): 129-143.

*Colin, P.L., Devaney, D.M., Hillis-Colinvaux, L., Suchanek, T.H. and Harrison, J.T. (1986). Geology and biological zonation of the reef slopes, 50-360 m depth at Enewetak Atoll, Marshall Islands. *Bull. Mar. Sci.* 38(1): 111-128.

*Cushman, J.A., Todd, R. and Post, R.J. (1954). Recent foraminifera of the Marshall Islands. *U.S. Geol. Surv. Prof. Paper* 260-H: 319-384.

Dahl, A.L. (1980). Regional ecosystems survey of the South Pacific Area. *SPC/IUCN Technical Paper* 179. South Pacific Commission, Noumea, New Caledonia.

Dahl, A.L. (1986). *Review of the Protected Areas System in Oceania*. IUCN/UNEP. 239 pp.

Dahl, A.L., Macintyre, I.G. and Antonius, A. (1974). A comparative survey of coral reef research sites. In: Sachet, M.H. and Dahl, A.L. (Eds). Comparative investigations of Tropical Reef Ecosystems: background for an integrated coral reef program. *Atoll Res. Bull.* 172: 37-77.

*Dawson, E.Y. (1955). An annotated list of marine algae from Enewetak Atoll, Marshall Islands. *Pac. Sci.* 11(1): 92-132.

*Devaney, D.M., Reese, E.S., Burch, B. and Helfrich, P. (Eds) (1988). *Natural History of Enewetak Atoll*. U.S. Dept of Energy, Oak Ridge, Tennessee.

Douglas, G. (1969). Check List of Pacific Oceanic Islands. *Micronesica* 5(2): 327-463.

*Ebert, T.A. and Ford, R.F. (1986). Population ecology and fishery potential of the spiny lobster *Panulirus penicillatus* at Enewetak Atoll, Marshall Islands. *Bull. Mar. Sci.* 38(1): 56-67.

*Eldredge, L.G. (in press). Protochordates of Enewetak. In: Devaney, D.M., Reese, E.S., Burch, B. and Helfrich, P. (Eds), *Natural History of Enewetak Atoll*. U.S. Dept of Energy, Oak Ridge, Tennessee.

*Emery, K.O., Tracey, J.I. and Ladd, H.S. (1954). Geology of Bikini and nearby atolls. Part 1. *U.S. Geol. Survey Prof. Paper* 260 A. 265 pp.

*Environmental Consultants, Inc. (1977). Reconnaissance survey of terrestrial and marine environments on Illeginni, Kwajalein Atoll. Prep. for U.S. Army Engineer Division, Pacific Ocean, Honolulu.

*Fosberg, F.R. (1966). Northern Marshall Islands land biota: Birds. *Atoll Res. Bull.* 114: 1-35.

*Gerber, R.P. and Marshall, N. (1982). Characterization of the suspended particulate organic matter and feeding by the lagoon zooplankton at Enewetak Atoll. *Bull. Mar. Sci.* 32(1): 290-300.

Gladfelter, W.B., Ogden, J.C. and Gladfelter, E.H. (1980). Similarity and diversity among coral reef fish communities: A comparison between tropical western Atlantic (Virgin Islands) and tropical central Pacific (Marshall Islands) patch reefs. *Ecology* 61(5): 1156-1168.

*Harrison, J. (1986). Recent marine studies at Enewetak Atoll, Marshall Islands. *Bull. Mar. Sci.* 38(1): 1-3.

*Hartmann, O. (1954). Marine annelids from the northern Marshall Islands. *U.S. Geol. Surv. Prof. Paper* 260-Q: 619-644.

*Hatheway, W. (1953). The land vegetation of Arno Atoll, Marshall Islands. *Atoll Res. Bull.* 16: 1-68.

*Hiatt, R.W. (1951). Factors influencing the distribution of corals on the reef of Arno Atoll, Marshall Islands. *Proc. 8th Pacific Sci. Cong.* 3A: 929-970.

*Hiatt, R.W. and Strasbourg, D.W. (1960). Ecological relationships of the fish fauna on coral reefs of the Marshall Islands. *Ecol. monogr.* 30(1): 65-127.

*Highsmith, R.C. (1979). Coral growth rates and environmental control of density banding. *J. Exp. Mar. Biol. Ecol.* 37: 105-125.

Hillis-Colinvaux, L. (1977). *Halimeda* and *Tydemania*: Distribution, diversity and productivity at Enewetak. *Proc. 3rd Int. Coral Reef Symp., Miami* 1: 365. (Abs.).

Hillis-Colinvaux, L. (1985). *Halimeda* and other deep fore-reef algae at Enewatak atoll. *Proc. 5th Int. Coral Reef Cong., Tahiti* 5: 9-14.

Hillis-Colinvaux, L. (1986). *Halimeda* growth and diversity on the deep fore-reef of Enewetak Atoll. *Coral Reefs* 5: 19-24.

*Hines, N.O. (1962). *Proving Ground: An account of the radiobiological studies in the Pacific, 1946-1961*. University of Washington Press.

Hudson, J.H. (1985). Long-term growth rates of *Porites lutea* before and after nuclear testing: Enewetak Atoll, Marshall Islands. *Proc. 5th Int. Coral Reef Cong., Tahiti* 6: 179-185.

Johannes, R.E. (in press). The role of Marine Resource Tenure Systems (TURFs) in sustainable nearshore marine resource development and management in U.S.-affiliated tropical Pacific islands. In: Smith, B.D. (Ed.), Topic Reviews in Insular Resource Development and Management in the Pacific U.S.-affiliated islands. *Univ. Guam Mar. Lab. Tech. Rept* 88.

*Johnson, M.W. (1954). Plankton of Northern Marshall Islands. *U.S. Geol. Survey Prof. Paper* 260-F: 301-314.

*Knutson, D.W., Buddemeier, R.W. and Smith, S.V. (1972). Coral chronometers: Seasonal growth bands in reef corals. *Science* 177: 270-272.

*Ladd, H.S. (1973). Bikini and Enewetak Atolls, Marshall Islands. In: Jones, O.A. and Endean, R. (Eds). *Biology and Geology of Coral Reefs*. Vol. 1. Geology 1. Academic Press, N.Y. Pp. 93-112.

*Ladd, H.S., Ingerson, E., Townsend, R.C., Russell, M. and Stephenson, K.H. (1953). Drilling on Enewetak Atoll, Marshall Islands. *Bull. American Association of Petroleum Geologists* 17(10): 2257-2280.

*Ladd, H.S. and Schlanger, S.O. (1960). Drilling operations on Enewetak Atoll. *U.S. Geol. Survey Prof. Paper* 260-Y.

*Ladd, H.S., Tracey, J.I., Lili, G., Wells, J.W. and Cole, W.S. (1948). Drilling on Bikini Atoll, Marshall Islands. *International Geological Congress, Report of the 18th Session U.K.* Part 8.

Lamberson, J.O. (1982). *A guide to the terrestrial plants of Enewetak Atoll*. Pac. Sci. Inf. Centre, Honolulu. 78 pp.

*Losey, G.S., Jr (1973). Study of Environmental Impact for Kwajalein's Missile Range: A Contract Report Prepared by the University of Hawaii for the U.S. Army Engineer Division, Pacific Ocean, Honolulu, Hawaii, as per contract DACA 84-73-C-0008.

*McKee, E.D. (1959). Storm sediments on a Pacific atoll. *J. Sed. Petr.* 29: 354-364.

*McMurtry, G.M., Schneider, R.C., Colin, P.O., Buddemeier, R.W. and Suchanek, T.H. (1986). Vertical distribution of fallout radionuclides in Enewetak lagoon sediments: Effects of burial and bioturbation on the radionuclide inventory. *Bull. Mar. Sci.* 38(1): 35-55.

Maragos, J.E. (1986). Coastal resource development and management in the U.S. Pacific Islands: 1. Island-by-island analysis. Office of Technology Assessment, U.S. Congress. Draft.

Maragos, J.E. (in prep). Recent reef coral collections at Bikini Atoll with an updated checklist and notes on geomorphology.

*Marsh, J.A. Jr, (in press). Reef processes: Energy and materials flux. In: Devaney, D.M., Reese, E.S., Burch, B. and Helfrich, P. (Eds), *Natural History of Enewetak Atoll*. U.S. Dept of Energy, Oak Ridge, Tennessee.

Motteler, L.S. (1986). Pacific Island Names. *B.P. Bishop Mus. Misc. Publ.* 34. 91 pp.

*Odum, H.T. and Odum, E.P. (1955). Trophic structure and productivity of a windward coral reef community on Enewetak Atoll. *Ecol. Monogr.* 25: 291-320.

Pritchard, P.C.H. (1981). Marine turtles of Micronesia. In: Bjorndal, K.A. (Ed.), *Biology and Conservation of Sea Turtles*. Smithsonian Institution Press, Washington D.C. Pp. 263-274.

*Randall, J.E. (1986). 106 new records of fishes from the Marshall Islands. *Bull. Mar. Sci.* 38(1): 170-252.

*Ristvet, B.L., Tremba, E.L., Couch, R.F., Fetzer, J.A., Goter, E.R., Walter, D.R. and Wendland, V.P. (1978). Geologic and geophysical investigations of the Enewetak nuclear craters. U.S. Air Force Weapons Laboratory Report AFWL-TR-77-242. Kirtland Air Force Base, NM.

*Schlanger, S.D., Goldsmith, J.R., Graf, D.L., MacDonald, G.A., Potratz, H.A. and Sackett, W.M. (1963). Subsurface geology of Enewetak Atoll. *U.S. Geol. Survey Prof. Paper* 260-BB.

*Schultz, L. and collaborators (Chapman, W.M., Herald, E.S., Lachner, E.A., Welander, A.D. and Woods, L.P.) (1953, 1960 and 1966). Fishes of the Marshall and Mariana Islands. *Bull. U.S. Nat. Mus.* 202, 3 volumes. 1299 pp.

Smith, R.O. (1947). Fishery resources of Micronesia. *Fishery leaflet* 239. Fish and Wildlife Service, U.S. Dept of the Interior. 46 pp.

SPREP (1980). Trust Territory of the Pacific Islands. *Country Report* 14. South Pacific Commission, Noumea, New Caledonia.

*St John, H. (1951). Plant records from Aur Atoll and Majuro Atoll, Marshall Islands. Pacific Plant Studies 9. *Pac. Sci.* 5: 279-286.

*Streck, C.F., Jr (1987). Archaelogical reconnaissance of Bikini Atoll. Environmental Resources Section, U.S. Army Corps of Engineers, Pacific Ocean Division. Appendix B, BARC Suppl. Doc. No. 2, Part 3. 211 pp.

*Suchanek, T.H. and Colin, P.L. (1986). Rates and effects of bioturbation by invertebrates and fishes at Enewetak and Bikini Atolls. *Bull. Mar. Sci.* 38(1): 25-34.

*Sverdrup, H.U., Johnson, M.W. and Fleming, R.H. (1942). *The Oceans, Their Physics, Chemistry and General Biology*. Prentice Hall, Englewood Cliffs. (Second Edition 1946).

*Taylor, W.R. (1950). Plants of Bikini and other northern Marshall Islands. *University of Michigan Studies, Sci. Ser.* 18: 1-227.

*Telegadas, K. (1961). Announced nuclear detonations. *USAEC HASL Report* 3: 176-182.

*Thresher, R.E. and Colin, P.L. (1986). Trophic structure, diversity and abundance of fishes in the deep reef (30-300 m) at Enewetak, Marshall Islands. *Bull. Mar. Sci.* 38(1): 253-272.

*Tracey, J.I. and Ladd, H.S. (1974). Quaternary history of Enewetak and Bikini atolls. *Proc. 2nd Int. Coral Reef Symp.* 2: 537-550.

*Tsuda, R.T. (in press). Marine benthic algae of Enewetak atoll. In: Devaney, D.M., Reese, E.S., Burch, B. and Helfrich, P. (Eds), *Natural History of Enewetak Atoll*. U.S. Dept of Energy, Oak Ridge, Tennessee.

USACE (U.S. Army Corps of Engineers, Pacific Ocean Division) (1986). Environmental assessment for initial resettlement of Eneu Island, Bikini Atoll. *Bikini Atoll Rehabilitation Committee Supplementary Report* 1.

USACE (U.S. Army Corps of Engineers, Pacific Ocean Division) (1987). Interim Environmental Impact Statement for the rehabilitation of soil at Bikini Atoll. *Bikini Atoll Rehabilitation Committee Supplementary Report* 2 and 3.

*Wells, J.W. (1954). Recent corals of the Marshall Islands. *U.S. Geol. Survey Prof. Paper* 260-I: 285-486.

*Wells, J.W. (1957a). Corals. In: Hedgpeth, J. (Ed.), Treatise on Marine Ecology and Paleoecology. *Geol. Soc. America Memoir* 67(2): 773-782.

*Wells, J.W. (1957b). Coral Reefs. In: Hedgpeth, J. (Ed.), Treatise on Marine Ecology and Paleoecology. *Geol. Soc. America Memoir* 67(2): 1087-1104.

*Wiens, H.J. (1957). Field notes on atolls visited in the Marshalls (1956). *Atoll Res. Bull.* 54. 23 pp.

*Wiens, H.J. (1962). *Atoll Environment and Ecology*. Yale University Press, New Haven.

ARNO ATOLL

Geographical Location Southern Marshall Islands, very close to Majuro Atoll; 7°05'N, 171°45'E.

Area, Depth, Altitude Atoll 21 x 6-15 mi. (34 x 10-24 km); 5 sq. mi. (13 sq. km) dry land; 131 sq. mi. (339 sq. km) of lagoon; max. alt. 6-8 ft (1.8-2.4 m).

Land Tenure Land tenure is vested in the "Wato" system (*see account for* Majuro).

Physical Features The atoll is an irregular rectangle in shape and has about 130 islets, most of which are coconut covered, although some have scrub vegetation. The water is warm and exceptionally clear, with no change in temperature (Dahl *et al.*, 1974). The beaches are of rubble on the seaward side but sandy on the lagoon side. Annual rainfall is 100-120 in. (2540-3048 mm); mean air temperature is 82°F (27.8°C) and the prevailing wind is from the east.

Reef Structure and Corals Hiatt (1951) describes some aspects of the reefs and Wells (1954) reports on the stony corals of the atoll. The reef at Kinajon is described by Dahl *et al.* (1974). The outer reef is on the sheltered south side. The rocky fore-reef flat, 100 m wide, develops good coral coverage and surge channels towards its edge at 10 m depth. A steep slope with *Porites* heads dominant drops to 30 m where the angle of slope lessens and sand patches appear. *Pachyseris* then becomes dominant. At 60 m the coral cover is still 20%. The reef near the deep entrance is considered to be even better developed. The lagoon of Arno is very deep (over 200 ft (61 m)) and contains two small enclosed sub-lagoons, one at each end of the atoll, and a large central open lagoon (Maragos *in litt.*, 10.8.87).

Noteworthy Fauna and Flora Arno Atoll is the type locality for the endemic pigeon *Ducula oceanica* (Amerson, 1969). There are five species of seabirds. The vegetation is described by Hatheway (1953). There is some mangrove.

Scientific Importance and Research   The atoll and reefs were considered to be relatively untouched in the early 1970s (Dahl *et al.*, 1974). There have been a number of scientific expeditions (Amerson, 1969) and its proximity to Majuro makes it accessible to researchers.

Economic Value and Social Benefits   About 1300 inhabitants in 1964 (Amerson, 1969) and "over 1000" in 1971, in a largely subsistence society with copra, and possibly handicrafts, the only major exports (Maragos *in litt.*, 10.8.87).

Disturbance or Deficiencies   The atoll is considered relatively undisturbed (Dahl *et al.*, 1974). However, it is potentially vulnerable to urban and tourism pressures from the nearby, overcrowded atoll of Majuro (Maragos *in litt.*, 10.8.87).

Legal Protection   None.

Management   None, although traditional systems are reported still to be in effect (Maragos *in litt.*, 10.8.87).

Recommendations   It is recommended that Arno be surveyed as soon as possible with the goal of identifying management strategies in the future should it become subject to increased environmental pressures owing to its proximity to Majuro. The U.S. Army Corps of Engineers has suggested carrying out Coastal Resource Management and atlas/inventory studies of the two atolls at the same time (Maragos *in litt.*, 10.8.87).

BIKINI ATOLL

Geographical Location   Central Pacific. north-west Marshall Islands; 11°35'N, 165°23'E°.

Area, Depth, Altitude   Land area 2.82 sq. mi. (7.3 sq. km); lagoon area 229 sq. mi. (593 sq. km). Max. alt. is on Bikini Island (16 ft (4.9 m)); max. depth of lagoon about 165 ft (50 m).

Land Tenure   The atoll is owned by the people of Bikini.

Physical Features   Bikini is a large rectangular atoll with the long axis in an east-west direction. About 23 islets currently surround a deep ocean lagoon along all sides except the west. Two islands and portions of two others were destroyed during the nuclear bomb tests. There are several deep passes through the reef on the south side, one of which is very large (over 8 mi. (13 km) wide). The long axis (25 mi. (40 km)), the open nature of the lagoon and the strong north-east tradewinds make navigation between islands, even within the lagoon, difficult to hazardous. Islands and reefs are vulnerable to storm waves approaching from the south. The largest islands, Bikini and Eneu, occur at the eastern corners of the atoll.

Reef Structure and Corals   Bikini was extensively studied by geologists and marine biologists before the testing era (Ladd *et al.*, 1948; Emery *et al.*, 1954; Ladd, 1973; Wells, 1954, 1957a and b) and more recently as part

of an evaluation to clean up the atoll (USACE, 1986 and 1987; Agegian *et al.*, 1987; Maragos, in prep.).   Wells's studies are the early classic descriptions of coral ecology and shallow atoll reef structure.   Terraces and spur-and-groove systems occur off windward ocean reef slopes, steep drop-offs along leeward ocean reef slopes and wide reef flats occur along most perimeter reefs. Numerous reef pinnacles and mounds occur in the lagoon.   Together Wells (1954) and Maragos (in prep.) have reported nearly 250 stony coral species, taking into consideration recent taxonomic revisions.   Coral diversity appears comparable to Enewetak, the only other Marshall atoll extensively studied. Wells (1954) described many new species and several species new to science from specimens collected at Bikini.   Several common genera appear missing, suggesting geographic isolation.

Noteworthy Fauna and Flora   *Pisonia* forests have re-established on the south-west islands and turtle nesting occurs extensively on Enidrik Island.

Scientific Importance and Research   Baseline reef and coral ecology, the movement of radionuclides in the marine ecosystems, and the impact of nuclear testing have been and will continue to be important research subjects at Bikini. Aspects of the natural history of the atoll are discussed in Colin *et al.* (1986) and Suchanek and Colin (1986). Recently the oldest archaeological dates from Micronesia were reported from Bikini (1960 B.C.) and included an intact village site dating from 1200 B.C. These archaeological results and other studies now planned are very significant scientifically (Streck, 1987).

Economic Value and Social Benefits   Bikini and Eneu islands supported large permanent communities prior to the 1946 nuclear testing era. At present the Bikinians do not live on the island because of contamination during the nuclear testing program between 1946 and 1958, when they were removed by the U.S. Government.   Some returned to the atoll in 1969 but became contaminated with radioactivity and had to leave in the same year. Billions of dollars of nuclear research has been conducted at Bikini but the atoll has not been cleaned up adequately to permit the permanent resettlement of the Bikini islanders. The clean-up of Bikini, which is now being planned, will cost millions of dollars while the clean-up of Rongelap and Utrik, if any, has yet to be contemplated.   The Bikinians, once resettled to their home atoll, hope to promote tourism, scientific study, fisheries, aquaculture and other commercial activities.

Disturbance or Deficiencies   Bikini was part of the U.S. Pacific Proving Grounds between 1948 and 1958 and was the site of numerous nuclear tests. The full extent of the disruption of the atoll ecosystem from the nuclear testing program and associated construction has not been fully documented or evaluated.   At least 15 historic era shipments sunk during the 1946 Baker test rest on the floor of the eastern lagoon. The 1954 Bravo test, the first and largest thermonuclear explosion by the U.S., spread fallout to several other northern Marshall atolls (Rongelap, Utrik, possibly Rongerik) and caused inestimable damage and social disruption to many reef and island communities. Reef and coral recovery in the craters has now begun but may never fully achieve pre-test levels of ecological development. Vegetation on most islands is heavily disturbed but is recovering in many places. The lack of habitation has spared turtles, seabirds and reef fish populations from fishing and harvesting pressures, but this may change in the near future should the islanders return.

Legal Protection   U.S. actions in the Marshall Islands (including the cleanup of Bikini atoll) will be subject to the substantive provisions of many U.S. environmental laws and to the National Environmental Policy Act.   An environmental impact statement (EIS) is about to be released which describes progress on planning for the cleanup (USACE, 1987).

Management   None.

Recommendations   Several south-east islands should be established as forest and sea bird preserves, Enidrik Island to the south should be established as a nesting marine turtle preserve and nesting seabirds on some northern islands should be protected.   Further dredging and filling of valuable reef areas should be discouraged and monitoring should be continued to document the status of radioactivity in the ecosystem and the recovery of ecosystems from past physical and radiological disturbances.

ENEWETAK (ENIWETOK) ATOLL

Geographical Location   One of north-westernmost atolls of the Marshall Islands, in the Ralik chain; 11°31'N, 162°15'E.

Area, Depth, Altitude   Atoll is 40 x 32 km; lagoon area 388 sq. mi. (1005 sq. km), max. depth 60 m.   40 islands with total dry land area 6.7 sq. km; largest islands about 1 sq. km in area; max. alt. 13 ft (4 m).

Land Tenure   Two major tribes have land rights to the atoll: the Dri-Engebi over the northern half and the Dri-Enewetak over the southern half.   Land ownership is vested as the "Wato" system (*see account for* Majuro) (Maragos *in litt.*, 10.8.87).

Physical Features   The atoll is roughly elliptical in shape, and the lagoon has three main openings to the sea: Deep Channel in the south-east is 1.5 km wide with a maximum depth of 60 m; Wide Passage in the south is 10 km across with a maximum depth of 30 m; and West Passage is little more than an irregular series of interruptions in the superficial reef structure with average depths of 2-3 m.   There are also many shallow connections between the lagoon and ocean over the reef flats between the islands.   The reef supports more than 30 small, low-relief islands of carbonate sand and gravel, the smaller of which are vegetated.   The north-east trade winds predominate.   There is an average annual rainfall of 1470 mm, mostly during the August-December period. Rainfall is highly variable, annual totals ranging from 605 to 2422 mm.   Tides are mixed semi-diurnal with a maximum range of about 1.8 m.

The geology of Enewetak Atoll has been extensively studied (Buddemeier *et al.*, 1975; Emery *et al.*, 1954; Ladd, 1973; Ladd and Schlanger, 1960; Schlanger *et al.*, 1963; Tracey and Ladd, 1974).   Recent geological and geophysical investigations are reported by Ristvet *et al.* (1978).   Buddemeier (1981) describes the geohydrology.

Atkinson *et al.* (1981) describe the water circulation in the lagoon; most of the outflow takes place through Wide Passage in the south.   Other studies include microclimate (Blumenstock and Rex, 1960) and the results of drilling (Ladd *et al.*, 1953).   An extensive "Natural History of Enewetak" has been compiled by the U.S. Department of Energy (Devaney *et al.*, in press).

Reef Structure and Corals   The deep reef slope is described by Colin *et al.* (1986) and the lagoonal benthic community by Colin (1986).   Odum and Odum (1955) describe reef productivity and Marsh (in press) describes energy and materials flux within the reef.   As a result of the high yield nuclear tests between 1948 and 1958 (Telegadas, 1961), corals incorporated fall-out induced radio-nuclides into their skeletons as discrete growth line horizons and numerous growth rate studies have been carried out (Buddemeier *et al.*, 1974; Highsmith, 1979; Knutson *et al.*, 1972).

Noteworthy Fauna and Flora   Marine algae are described by Dawson (1955), Hillis-Colinvaux (1977, 1985 and 1986) and Tsuda (in press), and lagoon zooplankton by Gerber and Marshall (1982).   Protochordates are described by Eldredge (in press).   Boucher (1986) describes coral predation by the muricid gastropod *Drupella*. Bussing (1972) describes some of the fish and Gladfelter *et al.* (1980) studied the fish community at the south-east end of the atoll.   Recent research on the fish includes that by Randall (1986) and Thresher and Colin (1986).   Seventeen seabird species have been recorded from the atoll (Amerson, 1969). Lamberson (1982) provides a guide to the terrestrial plants, and checklists of all flora and fauna known from the atoll are found in the "Natural History of Enewetak" including numerous newly recorded species (Devaney *et al.*, in press).

Scientific Importance and Research   The atoll has been the site of a large number of scientific investigations.   The University of Hawaii's Mid-Pacific Marine Research Laboratory (EMBL) operated for nearly 30 years on the atoll with the support of the U.S. Atomic Energy Commission and Department of Energy (DoE).   With the decrease in funding however, the laboratory is no longer available for outside researchers (Maragos, 1986) and DoE support ended with the termination of the jurisdiction of the Trust Territory of the Pacific Islands in the Marshall Islands in October 1986 (Maragos *in litt.*, 10.8.87).   Recent marine research is described by Harrison (1986).   A four-volume compilation of work published under the auspices of EMBL and the Mid-Pacific Marine Laboratory includes 223 articles published up to 1977 (Devaney *et al.*, 1988).   The U.S. Dept of Energy is funding a program to determine the physical, chemical and biological mechanisms controlling the distribution and transport of fallout radionuclides in the atoll environment (Buddemeier, 1981; McMurtry *et al.*, 1986); this programme extends to Bikini, Rongelap and Utrik Atolls (Maragos *in litt.*, 10.8.87).

Economic Value and Social Benefits   In the 1960s, only U.S. civilian and military personnel were resident, all native inhabitants having been moved off the atoll to Ujelang in 1947.   Some of the islanders were resettled in the southern part of the atoll in 1978 but many have since left or resettled elsewhere (Maragos, 1986).   The fishery potential of the spiny lobster *Panulirus penicillatus* is described by Ebert and Ford (1986).

Disturbance or Deficiencies   The atoll was a major World War II battle site and was subject to saturation bombing before the 1944 invasion by U.S. armed forces (Maragos *in litt.*, 10.8.87). Between 1948 and 1958 it was part of the U.S. Pacific Proving Grounds and was the site of numerous nuclear tests, mainly in the north-east quadrant (Buddemeier, 1981; Hines, 1962; Maragos, 1986). Hudson (1985), studying *Porites lutea*, found that average vertical growth rate had remained relatively constant, the nuclear testing having had no apparent effect. However there are suggestions that corals in close proximity to nuclear detonations may have been sterilised by radiation. Four of the larger islets are barren shot sites (Douglas, 1969). The tests and associated dredging and filling caused residual damage to many reefs and islands (Maragos, 1986).

Legal Protection   None.

Management   A radiological clean-up programme was carried out between 1976 and 1978, but some radiological hazards remain in the lagoon and some parts of the soil (Maragos, 1986).

Recommendations   The need for new environmenal laws and regulations in the Republic of the Marshall Islands is discussed in the introduction.

KWAJALEIN ATOLL

Geographical Location   Central Marshall Islands; 09°05'N, 167°20'E.

Area, Depth, Altitude   Land area 6.3 sq. mi. (16.3 sq. km); lagoon area is 839 sq. mi. (2173 sq. km), the largest atoll lagoon in the world. Max. alt. about 40 ft (12.2 m) at some man-made hills, average, natural max. alt. 10 ft (3 m). Max. lagoon depth about 200 ft (61 m).

Land Tenure   Land owned by private individuals or extended families; portions leased to U.S. Government, including all of Kwajalein Island, Roi-Namur Island and a series of about 15 islands along an east to west zone (Mid-Atoll Corridor) in the southern half of the atoll. The leases run for 15 years from 1986 with an option to extend another 15 years or more.

Physical Features   Kwajalein is located in the strong north-east tradewind belt. It is a very large boomerang-shaped atoll with 92 islets and many passes encircling the lagoon. The largest islands tend to occur at the corners and bends of wide atoll reefs. Kwajalein has more land area, lagoon area and reef length than any other Marshall atoll. The lagoon is wide, particularly along the north-south axis, and the long fetch generates large wind waves. Water circulation in the lagoon is facilitated by the many passes and open reefs without islands. The atoll is typical of the Marshalls in having very steep slopes with submerged wave cut terraces and well developed spur-and-groove systems on the windward (east) side, and pronounced leeward reefs.

Reef Structure and Corals   Reefs and corals are typical of other well studied Marshall atolls. Many pinnacles and patch reefs are scattered in the lagoon and extensive sand terraces fringe particularly the windward lagoon margin and deep lagoon. Reefs off Ebeye are described by Amesbury *et al.* (1975a), off Illeginni by Environmental Consultants Inc. (1977) and the Mid Atoll Corridor Islands by Losey (1973).

Noteworthy Fauna and Flora   Although insufficiently documented, Green Turtles *Chelonia mydas*, Hawksbills *Eretmochelys imbricata* and many seabirds probably nest on many of the sparsely to uninhabited islands and beaches. Molluscs have been described by Brost and Coale (1971).

Scientific Importance and Research   The atoll as a whole has never been adequately studied although the U.S. controlled areas and activities have generated a number of studies near Kwajalein, Meck, Roi-Namur and the Mid-Atoll Corridor Islands. Many other environmental studies have been accomplished during the past 15 years at U.S. controlled and developed sites (Maragos, 1986).

Economic Value and Social Benefits   After Majuro, Kwajalein is the most populous atoll in the Marshalls. For three decades it has served primarily as a down-range missile tracking and testing facility, mainly for U.S. defensive (anti-ballistic) missiles and many millions of dollars have been invested in ports and channels, antenna tracking facilities, housing and recreational facilities, power, sewage, water, health care, air fields, roads, etc. to support the several thousand defence workers and contractors stationed largely at Kwajalein Island and to a lesser extent at Roi-Namur. The promise and reality of jobs have also attracted many Micronesian islanders to Ebeye Island, three miles (4.8 km) north of Kwajalein but not within U.S. control or jurisdiction. Up to 10 000 people, living in slum-like squalid conditions have lived on 70-acre (28 ha) Ebeye in the past. Although the population is now down to 8 000 and living conditions are better, there is much room for improvement in health care, housing, sanitation, recreation, access, and diet. Other problems have included lack of islander access to Mid-Atoll Corridor fishing grounds, Kwajalein Island, and inadequate lease rent payments by the U.S.

Disturbance or Deficiencies   The atoll was a major World War II battle site and was subject to saturation bombing before the 1944 invasion by U.S. armed forces. Most of the larger islets on the atoll are almost completely altered by military activities, dredging and filling (Maragos, 1986). The natural resources of Ebeye have been devastated. An environmental survey was carried out prior to the establishment of a new sewer outfall site on Ebeye and revealed a very poor coral community, typical of silt-stressed environments (Amesbury *et al.*, 1975a). The cumulative impact of all activities and development has never been established.

Legal Protection   Under the terms of the Compact of Free Association, U.S. actions (including missile range and Strategic Defense Initiative actions) are required to comply with all provisions of the National Environmental Policy Act (the U.S. law for EIS) and the substantive provisions of a host of other environmental laws on endangered species, historic preservation, solid waste, hazardous waste, air quality and water quality. There are no protected areas.

Management Management responsibilities for Kwajalein Atoll are divided between the U.S. (about one third of the atoll, devoted to the missile programs) and REPMAR (the Republic of the Marshall Islands) which controls the remaining two thirds. Enactment and application to Kwajalein of many U.S. environmental laws since 1970 has substantially reduced unnecessary reef damage from dredging, filling, quarrying and construction of individual military projects. The Kwajalein Atoll Development Authority, a REPMAR government-sponsored corporation, is attempting to plan projects to redistribute the crowded Ebeye population. Kwajalein is now ear-marked for major research and development as part of the U.S. Strategic Defense Initiative.

Recommendations The U.S. Army's Strategic Defense Command has agreed to the preparation of an Environmental Impact Statement for the Army's missile range. The U.S. Army Corps of Engineers will be carrying out the EIS and contracting out studies including: atlas, marine biology, water quality, archaeology, social impact assessment. A comprehensive environmental survey of the reefs previously disturbed by dredging, quarrying, filling and other construction should be accomplished and compared to an earlier effort by Losey (1973). Basic biology and ecology of reefs near inhabited islands or those proposed for military development should be studied. The respective responsibilities are not clearly defined and there are gaps in some management needs. So far no REPMAR counterpart environmental laws and regulations have been enacted.

MAJURO ATOLL

Geographical Location Southern Marshall Islands; 7°05'N, 171°10'E.

Area, Depth, Altitude 3.5 sq. mi. (9 sq. km) (dry land); the atoll is 21 x 3-10 miles (33.8 x 4.8-16.1 km); total lagoon area 113.9 sq. mi. (295 sq. km).

Land Tenure Ownership is vested in "Watos" or land subdivisions established for each resident extended family group. Each "Wato" usually constitutes a strip of land across the islet from the lagoon to the ocean side. Outsiders must seek permission to use land or must lease it from the owners. It is not clear who owns lands created by dredging or filling, but a fee simple system may operate for these (Maragos *in litt.*, 10.8.87).

Physical Features The atoll is roughly rectangular and has about 57 islets, many thickly wooded. The average annual temperature is 81°F (27°C) and rainfall measures 140 in. (3556 mm) (Amesbury *et al.*, 1975b). Prevailing winds are from the east and south-east.

Reef Structure and Corals Amesbury *et al.* (1975b) conducted a marine survey in the Darrit-Uliga-Dalap area and found a rich coral community dominated by *Acropora* and *Pocillopora* on the reef flat at the north end of the runway. The inner reef flats are devoid of corals. Corals at the Elementary School were limited to individual colonies of *Leptastrea purpurea*, *Pocillopora danae*, and *Favia pallida*. Thirty-three species of benthic algae were noted. The reef off Laura, a sheltered location on the west, was surveyed by Dahl *et al.* (1974). The 60 m wide fore-reef consisted of 20 m of bare surface, 20 m with a dense algal cover and 20 m with corals, mainly *Acropora*, leading to the rugged reef edge at 5-8 m depth. Deep surge channels cut into the reef. There is no real drop-off but a gently rounded slope with valleys and ridges perpendicular to the shore and many shore-parallel steps. Coral cover decreases from 90% at the edge (5-8 m depth) to 60% at 12 m depth, and 50% at 20 m depth with increasing algal cover. The slope gets steeper with depth, becoming vertical at 40 m where coral cover is 20%.

Noteworthy Fauna and Flora Amesbury *et al.* (1975b) and Dahl *et al.* (1974) recorded an abundant fish fauna, although Amesbury *et al.* (1975b) found a low diversity, only 33 species being recorded. Thirty-three species of benthic algae were also recorded. The vegetation of the islets is described by St John (1951). Seven seabird species are known from the atoll but are not known to breed (Amerson, 1969).

Scientific Importance and Research A survey was carried out at the proposed sewer outfall site in the Darrit-Uliga-Dalap area (Amesbury *et al.*, 1975b).

Economic Value and Social Benefits The population is estimated at 11 000-14 000 (Maragos *in litt.*, 10.8.87), compared with ca 4500 in 1964. Majuro is the capital of the Republic of the Marshall Islands, and it and Kwajalein are the only atolls with large populations.

Disturbance or Deficiencies The reefs have been reported to be disturbed although Dahl *et al.* (1974) found them in good condition. Dredging and filling for roads and the runway have caused pollution and stagnation of the lagoon in the Darrit-Uliga-Dalap area, which has seriously deteriorated (Maragos, 1986). The absence of breeding seabirds may be because of human predation (Maragos *in litt.*, 10.8.87).

Legal Protection None.

Management Some attempts are being made to decrease pollution in the Darrit-Uliga-Dalap area (Maragos, 1986).

Recommendations A coastal resources inventory is considered a high priority by the government of the Marshall Islands and funds for an atlas, report and management plan have been requested. The U.S. Army Corps of Engineers may be asked to carry out the studies (Maragos *in litt.*, 10.8.87).

NAURU

INTRODUCTION

General Description

The Republic of Nauru is an oval-shaped volcanic island of ca 21.2 sq. km (6 x 4 km) with a circumference of 16 km, situated less than 60 km south of the equator at 0°32'S, 166°56'E. It rises 4800 m above the sea floor to form a plateau with an average height of about 50 m above sea level (max. alt. 70 m). Six-sevenths of the island is phosphate bearing. The plateau is largely composed of phosphate rock in the interstices of the pinnacles and is encircled by cliffs which give way to a flat, fertile coastal belt, 90-270 m wide. The soil is an admixture of sand and fine corals, and with an irregular rainfall, restricts cultivation of the coastal belt.

The climate is tropical but tempered by sea breezes. Average shade temperature ranges from 24.4° to 33.0°C, and humidity is normally between 70 and 80%. Annual rainfall is variable (recorded range 300-4600 mm) and averages 1500 mm. Usually, the heavy rains occur during the westerly monsoon from November to February. Easterly winds prevail throughout the rest of the year. Tidal range is around 2.0 m.

There is very little published material on the marine environment of Nauru but general information is available in De la Porte (1907), Ellis (1936), Grimble (1952), Kayser (1917-18, 1921-24 and n.d.), Maude (1971) and Petit-Skinner (1981). There is no true reef and no lagoon; the island is surrounded by an almost consistent 150-200 m wide intertidal platform, cut into the original limestone of the island and typified by the presence of numerous emergent coral pinnacles. The platform is dominated by large yellow-brown algae and little or no coral growth occurs on the reef flat. However, a rich fauna is evident in deeper water, although species diversity has not been documented. The benthic fauna is quite well represented with many common Indo-Pacific species. Reef crabs and other crustaceans are found as well as molluscs, urchins, sea cucumbers and other invertebrates. There are around 80 commonly caught fish species (Petit Skinner, 1981 amended by Lili *in litt.*, 29.12.87).

Reef Resources

In 1983, the population of Nauru was just over 8000 people with 62% of the total population being Nauruans and 38% non-Nauruans comprising mainly Kiribatese, Tuvaluans and Solomon Islanders. Most of the population lives in houses among the coconut palms near the main coastal road and/or the shoreline. There are no natural harbours. Subsistence fishing is carried out mainly by other Pacific islanders working on Nauru and reef fish and benthic organisms are subjected to fairly intensive harvesting for food. Outboard powered aluminium dinghies are used mainly by Nauruans for trolling and small outrigger canoes by Pacific islanders for hand lining for pelagic and bottom species. Some spear fishing with SCUBA is carried out in deeper waters and black corals, shells and other marine invertebrates are occasionally collected for ornamental purposes.

Milkfish ('ibia') are reared in small quantities in ponds, most of which are bomb craters from World War II. Fry are collected from the reef at low tide, acclimatized for 2-3 weeks and then released into the ponds. Although milkfish production could provide a major source of food during periods of unfavourable weather when sea fishing is impracticable, production is limited because of introduced *Tilapia* which could outcompete the milkfish. This problem is currently of major concern to the pond owners and to the Government of Nauru.

There is reported to be an active underwater diving club (Eldredge *pers. comm.*, June, 1987), although only a handful of people now dive.

Disturbances and Deficiencies

The constant pounding of waves and ocean currents confronting the almost vertical coastline results in corals being dislodged or broken and being washed up the shore.

There is no evidence of over-exploitation of marine resources. Coastal waters are relatively clean although there may have been one or two instances of silt accumulating on some parts of the reef flat. To date there is no recorded damage to the reef fauna and flora. The vegetation has been greatly modified in the interior due to phosphate mining. Land with little or no top soil has been denuded to allow mining to proceed but there has generally been no run-off as the bedrock is very porous (Manner *et al.*, 1984). However, continuing denudation may cause long-term micro climatic changes and there has been no long-term environmental monitoring of the impact of phosphate mining on the island.

Phosphate dust causes few problems during loading at the cantilevers because of the constant oceanic current upwelling around the island. Oil spillage occurs only in a limited part of the boat harbour during ship unloading. Oil washed from the power station and machinery workshop gutters nearby may be the cause of a 200 m stretch of shoreline being polluted.

Legislation and Management

In general, customary rights over the reefs restrict overharvesting, and allow the recovery of exploited resources, especially on the reef slopes. The Marine Resources Act 1978 makes provisions for the exploitation, conservation and management of fish and aquatic resources in territorial waters and the exclusive fisheries zone. A Fisheries Officer has recently been appointed. There are no marine protected areas.

Recommendations

A Commission of Inquiry into the Rehabilitation of the Worked-Out Phosphate Lands in Nauru has been established to look at the issue of rehabilitation of the island and its cost and feasibility (Anon., 1987). It will look at all forms of alternative land usage, including

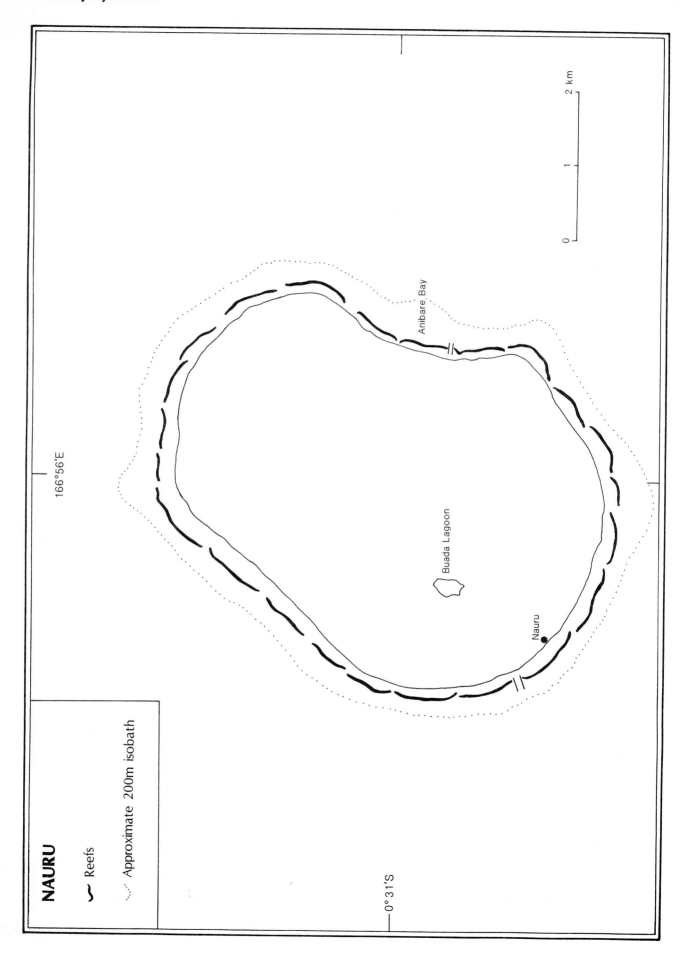

NAURU

⌇ Reefs

⋰ Approximate 200m isobath

166°56'E

0° 31'S

Anibare Bay

Buada Lagoon

Nauru

0   1   2 km

agriculture, and at the impact of phosphate mining on fisheries and marine resources. Financial compensation is being sought from Australia, Great Britain and New Zealand, the main consumers of phosphate. A study should be started immediately to provide baseline data for future environmental impact monitoring. Efforts should be made to establish protected areas, including reefs and important cultural sites. The Local Government Council should be responsible for such areas.

References

Anon. (1987). Commission of Inquiry into the Rehabilitation of the Worked-out Phosphate Lands in Nauru. Republic of Nauru.

De La Porte, P. (1907). Nauru as it was and is now. *The Friend* (Honolulu) June, July, August.

Ellis, A. (1936). *Ocean Island and Nauru*. Angus and Robertson, Sydney.

Grimble, A. (1952). *A Pattern of Islands*. John Murray, London.

Kayser, A. (1917-18). Die Eingebornen von Nauru (Sudsee). *Anthropos* (Vienna) 12/13 (1/2): 313-337.

Kayser, A. (1921-24). Spiel und Sport auf Naoero. *Anthropos* (Vienna) 16/17 (4,5,6): 681-716; 18/19 (1,2,3): 297-328.

Kayser, A. (n.d.). Die Fischerei auf Nauru. (Reference unknown).

Manner, H.I., Thaman, R.R. and Hassall, D.C. (1984). Phosphate mining induced changes on Nauru Island. *Ecology* 65: 1454-1465.

Maude, H. (1971). *The String Figures of Nauru Island*. Libraries Board of South Australia, Adelaide.

Petit-Skinner, S. (1981). *The Nauruans*. MacDuff Press, San Francisco.

NEW CALEDONIA

INTRODUCTION

General Description

New Caledonia, an overseas territory of France, consists of a large elongated island, Grande Terre, and an additional 4000 sq. km of dependencies. The Loyalty Islands form a chain of elevated reefs running parallel to Grande Terre, outside the main reef system. The larger islands are elevated and complex whereas those to the north-west consist of irregular atolls. Conway Reef (21°46'S, 174°31'E) is claimed by both Fiji and New Caledonia. It is described in the Fiji section. "Sandy Island" ("Ile de Sable") marked on many maps in the central part of the Coral Sea does not exist (Bouchet *in litt.*, 8.8.87).

There are four seasons: April-June (short, dry), July-August (short, wet), September-November (long, dry) and December-March (long, wet). The warmest months are from December to February and the coolest from July to September. The southern extremity has cool spells in winter as it is close to the temperate zone. Rainfall is very variable, from 1000 mm a year on the west coast to 3000 mm a year on the east coast, the north-east being noticeably wetter than the south-west. Cyclones may occur from December to March. Tides are semi-diurnal with a range of 0.1 to 1.8 m (Taisne, 1965). The hydrography and a general description of the lagoon is given in Taisne (1965) and the hydrology of the sector from Prony Bay, 40 km south-east of Noumea, to St Vincent Bay, 50 km north-west of Noumea, is described by Rougerie (1985). Additional general information is given in ORSTOM (1981).

Table of Islands

Hunter   100 acres (40 ha); volcanic, 974 ft (297 m); intermittent activity; steep cliffs; several coral species found on stones but colonies very small; fish abundant; seabird colony; described by Rancurel (1973b).

Matthew   30 acres (12 ha); volcanic; 465 ft (142 m); corals not recorded and probably no reefs because of recent volcanic activity; habitat similar to Hunter; seabird colony; described by Priam (1964) and Rancurel (1973b).

Isle of Pines (Kunié, Ile des Pins)   (X) 50 sq. mi. (130 sq. km); serpentine and coral; max. alt. 873 ft (266 m) with 350 ft (107 m) plateau; undercut coral cliffs; coral to 100 ft (30 m) depth; sand banks; coral patches in Oupie Bay; adjacent islets and numerous reefs include: Nokanhui (Nokankoui) Reef, Ana, Ami, Kutomo, Brosse, Du-Ami, Du-Ana, Koumo, Moneoro, Gié, Noéno, Ouatomo, Koungouati; islands between Isle of Pines and the mainland include N'Da, Kouaré, Koko, N'Do, Nge, Uatio, Gi, Uaterembi, Nerembi, Ieroue, Ndie, Nukue, Vua, Mato, Puemba, Piimbo, Noe, Totea, Uie, Mambae, Mbore, Manodi, Nouere, Uo, Ua, Ugo, Nouare, Kié and Amere (*see account for* Yves Merlet Reserve); some of these reefs described by Guilcher (1965a); plants and seabirds of many of the islands are listed by Rancurel (1974a); strandings of seals and giant squids described by Rancurel (1973c).

Ouen (Uen)   (X) alt. 332 m; separated from Grande Terre by Woodin Canal; Niagi and U reefs are nearby (Guilcher, 1965a).

Grande Terre (New Caledonia)   (X) 16 890 sq. km (400 x 50 km); 1628 m; two parallel ridges run the length of the island; west coast is large hilly plain with low and marshy coastline; on east, hills drop steeply to sea; general description in Pope (1962); geology described by Coudray (1976); barrier reef complex with diverse marine habitats; numerous inshore islets; *Caretta caretta* nesting (Pritchard, 1987); (*see separate accounts for reserves*).

Yandé   (X) 301 m.

Belep Islands

Art   (X) 283 m, steep; *Chelonia mydas* nesting and some *Eretmochelys imbricata* nesting (Pritchard, 1987);

Pott   157 m;

Also sometimes included:

North Daos Is   group of rocks and small islands between Art and Grande Terre; largest Nienane;

South Daos Is   group of rocks and small islands between Art and Grande Terre;

Daouinth   island just north of North Daos Is.

D'Entrecasteaux Reefs   (*see separate account*).

Loyalty Islands

Walpole   310 acres (126 ha; 3 km x 400 m); raised limestone; 70-90 m cliffs; flat-topped; fringing reefs; varied coral diversity and abundant fish; seabird colony; guano extraction until 1936; described by Rancurel (1973b) and Condamin (1978).

Maré   (X) 240 sq. mi. (642 sq. km); elevated atoll, 60-129 m; 2 small volcanic buttes in "lagoon" area; apron reef; geomorphology and reefs described in Chevalier (1968); geology described in Bourrouilh-Le Jan and Gaven (1980).

Dudune (Ndoundure)   steep shores.

Léliogat   steep shores, low and bare.

Oua (Uoa)   steep shores, low.

Tiga   (X) raised coral, 250 ft (76 m), perpendicular cliffs.

Vauvilliers   steep shores.

Lifou (Lifu)   (X) 1196 sq. km; coral limestone, 90 m, cliffs; fringing reefs; some *Chelonia mydas* and *Eretmochelys imbricata* nesting (Pritchard, 1987).

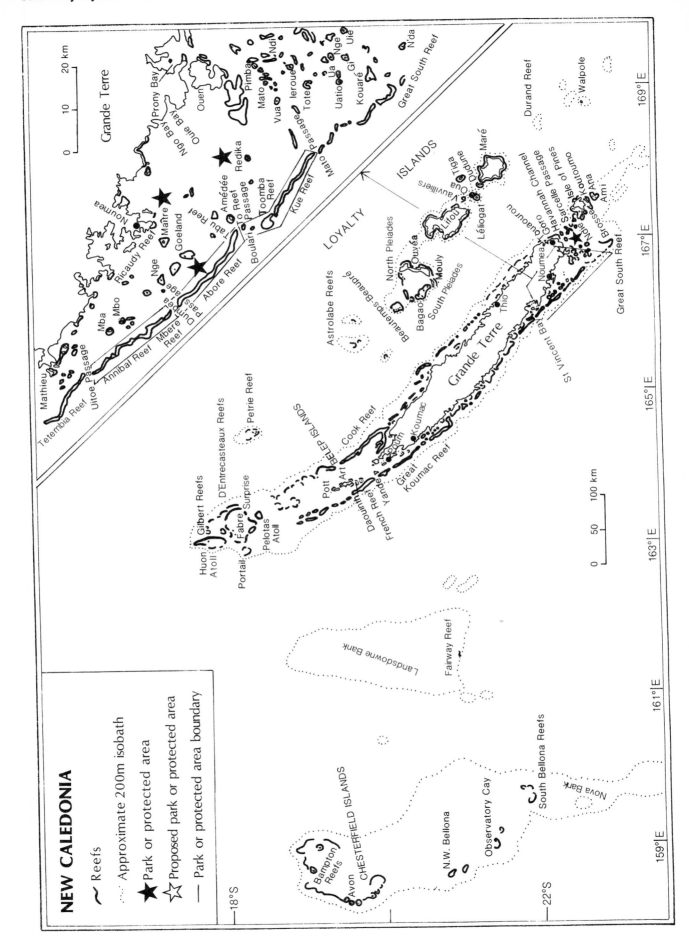

NEW CALEDONIA

〜 Reefs

··· Approximate 200m isobath

★ Park or protected area

☆ Proposed park or protected area

— Park or protected area boundary

Ouvéa (Uvea) (X) tilted atoll, 39 m; raised coral on one rim; enclosing large lagoon; possibly *Chelonia mydas* nesting (Pritchard, 1987):

- *S. Pleiades* 7 islets on north-west rim of Ouvéa atoll, including Bagao, Gué, Guetché;

- *N. Pleiades* 8 islets, on south-west rim of Ouvéa atoll. including Haute, Djekoutène, de la Table, Wégnec, de la Tortue, de la Baleine.

Beautemps-Beaupré atoll, 3 low islets, sand dunes; main island

(Eo, Heo) 1.5 km long; *Chelonia mydas* nesting (Pritchard, 1987); sand bar of Motou-Tapou 4-5 mi. (6.4-8 km) north of Beautemps-Beaupré has important seabird colony; private fishing ground belonging to people now living on Ouvea.

Chesterfield Is (*see separate account*).

Bellona Reefs coral barrier only just submerged, oriented to south-east; lagoon 67 m deep; Observatory Cay (Caye de l'observatoire) has vegetation (Richer de Forges and Pianet, 1984); endemic mollusc *Lyria grangei*; seabirds (Rancurel, 1973a); visited 1979 by New Zealand Oceanographic Institute.

(X) = Inhabited

The reefs and lagoons of New Caledonia cover an estimated 24 000 sq. km (Conand and Chardy, 1985) and have been described by Davis (1926), Avias (1959), Guilcher (1965b) and Chevalier (1964, 1968 and 1973). In the early 1960s, a major French expedition carried out studies on the reefs and these are published in the Singer-Polignac Foundation reports. Currently, reef-related research is carried out by the Centre ORSTOM de Nouméa including projects on *Acanthaster*, trochus, coral exploitation and fish (ORSTOM, 1986). High resolution satellite imagery (SPOT-LANDSAT) is now being used to map reefs in the lagoon (Bour et al., 1985a and c). Recent reef-related research is discussed by Coudray (1982). Coudray et al. (1985) and Guilcher (1965b) compare geomorphology and other characteristics of the New Caledonia and Mayotte barrier reefs.

Grande Terre is surrounded by a barrier reef, more than 1600 km in length, which borders a lagoon of clear and rather shallow water covering 16 000 sq. km, second only in size to the Australian Great Barrier Reef. The reef is continuous except in the south-east where it is partly submerged or absent. The lagoon averages 10 km in width and is widest (30 km) in the south-west; depth averages 25 m in the west and 40 m in the east with a maximum of 80 m (Coudray et al., 1985). There are few cays on the reef itself but numerous islets are found in the lagoon.

The northern section extends to the Belep Islands, French Reef (Récif des Français), and the Great Koumac Reef (Grand Récif de Koumac), a barrier reef with an enclosed lagoon which has been mapped (Bour et al., 1985a). The lagoon to the north of Grande Terre,

enclosed by Cook and French Reefs, is 30 miles (48 km) wide and 80 (128 km) miles long and encloses the Belep Islands. At the northern extremity, the lagoon is open over a 20 mile (32 km) wide, 55 m deep passage. The D'Entrecasteaux Reefs are separated from the lagoon by the 40 mile (64 km) wide Grand Passage. Some of the marine invertebrates recorded in this area are listed in Richer de Forges and Bargibant (1985). Great Koumac Reef, Great Poum Reef and adjoining areas are described in detail by Thomassin (1986). Tuo Reef and the adjacent lagoon lie along the north-east coast and form a double barrier with cays on the inner barrier. Coral knolls are numerous in the outer reaches of the lagoon and on the submarine ridge connecting the reefs of the inner barrier. Sedimentation from erosion on the mainland has a greater impact on reef development here than in the south-west described by Guilcher (1985).

The west coast is bordered by two main coral reef systems that show well developed barrier reef lines and numerous lagoonal reefs. The southern section is separated from the northern section by a narrow middle zone with poorly developed fringing reefs and a boat channel. It extends down to the Great Southern Reef (Grand Récif Sud) and Isle of Pines. The south-west reef system, bounded by the deep Prony submarine trench, is well known. Thomassin and Vasseur (1981) have described the geomorphology. The barrier reef is almost continuous, surrounding a deep lagoon which becomes wider to the south, where the coastal bays become more open. Thomassin and Coudray (1981) describe lagoonal hard bottoms in the area behind the Aboré barrier reef, including Ngé Reef (a wooded cay reef flat), Laregnere Reef (a patch reef seaward of Ngé), Goelands Reef, Maître Reef (*see separate account*) and Canards Reef. Faure et al. (1981) describe reef coral assemblages on the windward slopes of Goelands Reef and Maitre Reef. On Goelands Reef there is an intertidal reef flat at 0.5 m, characterized by high coral coverage (more than 80%) and low species diversity, with typical species such as *Acropora digitifera*, *A. humilis* and some *Pocillopora* species. Calcareous algal cover is less than 20%. The ratios of *Acropora/Pocillopora* and *A. digitifera/A. humilis* vary according to the strength of the surf. A gentle slope (with 60-80% coral coverage) descends to 2 m depth, with increasing species diversity. Massive faviids are common, interspersed with branched or tabular *Acropora*. The zone from 2 to 8 m depth is dominated by branched *Acropora* (80-100% coral cover); from 6 to 8 m depth there may be a drop-off with a small cliff overhanging the hard grounds of the lagoon bottom. This is covered by large massive colonies of *Porites lutea*, with occasional colonies of *Platygyra daedalea* and *Leptoria phrygia*, and faviids. From 9 m to 12 m depth, coral coverage decreases to less than 5% and the bottom is covered by detrital material, with scattered faviid colonies. From 12 to 15 m depth, brown algae dominates, and beyond 15 m depth there is a zone of deep sediment with scattered solitary scleractinians and a few colonial forms.

Tétembia Reef, north of the pass of Uitoé, has been studied in detail as a pilot site for investigating the use of SPOT satellite imagery for mapping reefs in relation to trochus distribution (Bour et al., 1985) and in the course of a coral exploitation study (Jouannot, 1985). The reefs around Maître and Amedée islets, and the three barrier reefs off Noumea are described in separate accounts. Ricaudy Reef, a fringing reef off the end of the Noumea peninsula has been mapped (Bour *et al.*, 1985a). Bigot *et al.* (1985) describe the M'Bé lagoonal patch reef which has a reef flat with a cental saline shallow "pond" enclosed within a ring of emergent shingle and rubble. The southern reefs are described by Guilcher (1965a). On the south-east coast at Ounia, in the Mamié region, there is a submerged barrier reef line and fringing reef system including an open lagoon 7 km wide, described briefly in Cabioch *et al.* (1985).

At the end of Grande Terre, there is a reef complex around the Isle of Pines. Rancurel (1974a) briefly describes a visit to the islands between the Great Southern Reef, which has numerous sandy cays, and the Isle of Pines (see table), all of which are surrounded by reefs. Further details of these reefs are given in Guilcher (1965a), particularly for Ua, Ndito, Kuta, Ndo, Fer à Cheval, Cimenia Reef, and further north-east Kouaré and the faroes of Puakue and Tootira. Coral communities and some reefs are found around Hunter, Walpole and Matthew (see table).

Unusual coral banks have been discovered in muddy environments in the inner lagoon and enclosed bays, of which Gail Bank, in the south-west lagoon between Mont-Dore and Ouen Island, is the most developed complex, extending over 100 sq. km at 30-35 m depth. There is a high level of sedimentation. The epibiotic fauna is dominated by hermatypic corals, octocorals, sponges and attached bivalves. The corals are mainly solitary species or are found in isolated colonies; acroporids and pocilloporids are almost entirely lacking (Joannot *et al.*, 1983).

Reefs of the Loyalty Islands have been described by Haeberlé (1952b). To the north of Grande Terre, barrier reefs are replaced by extensive meandering atoll formations in the D'Entrecasteaux system (*see separate account*). Reefs are also found around many of the more distant islands.

The reefs of the Chesterfield Islands are described in a separate account and the Bellona reefs are another important area. There are also a number of submerged banks in this area. Fairway Bank lies at 50 m depth and is largely covered with *Halimeda* sand and alcyonarians. Adjacent to this is Fairway Reef (21°00'S, 161°49'E) which is two miles (3.2 km) long and submerged at low tide. Lansdowne Bank extends north-west from Fairway and this plateau at a depth of 70-80 m covers an area of 21 000 sq. km and is largely sandy; Nereus Reef lies on the northern side. Nova Bank is a guyot 40 miles (64 km) south of Bellona (Richer de Forges and Pianet, 1984).

Scleractinian corals of New Caledonia are described by Chevalier (1971 and 1975) and Wijsman-Best (1972), about 108 species having been described. Octocorals are described by Tixier-Durivault (1970).

New Caledonia is a popular place for shell collectors, particularly along the coast from Saint-Vincent to the Isle of Pines. Several species of volute are endemic to New Caledonia: *Cymbiolacca thatcheri* (*see account for Chesterfield Islands*), *Cymbiola deshayesi* (from the north) and *C. rossiniana* (found only in the south-west between the Isle of Pines and Boulari Passage). Several cone shells are endemic (Estival, 1981) and the area is well known for its melanistic cowries (Chatenay, 1977; Pierson and Pierson, 1975). Echinoderms are described

by Guille *et al.* (1985). The Green Turtle *Chelonia mydas*, the Hawksbill *Eretmochelys imbricata* and the Loggerhead *Caretta caretta* nest (Pritchard, 1987). Additional information on marine fauna is given in Salvat (1964 and 1965) and Laboute and Magnier (1979). Garnett (1984) briefly mentions the distribution of major seabird colonies.

Reef Resources

Fish, crustaceans and shellfish are intensely exploited in certain parts of the lagoon particularly around Noumea. The commercial fishery has been extensively documented and includes publications by Loubens (1978a and b, 1980a and b), Kulbicki (1987 and in prep.), Kulbicki and Moutham (in prep.), Baillon (1986) and Thollot (1987). *Trochus niloticus* has been exploited commercially for seventy years and is exported in large quantities. It is collected from most areas in the lagoon apart from the reserves (Bour and Hoffschir, 1985) and 296 tonnes were collected in 1984 (Anon., 1985b). Growth studies on trochus have been carried out at Amédée Islet, the Yves Merlet Reserve, Kouaré, Touaourou (south-west New Caledonia) and Nienane and Pott in the Belep Islands. Small scale aquarium fish collecting has been started recently (Jourde, 1985). There is also a small-scale coral trade. Worked handicraft items, such as lampshades and ornaments, are produced for sale locally and for export to France and Italy. Raw coral is also exported to the U.S.A and Europe (Wells, 1985). In 1984, 75 t of raw coral were collected (Anon., 1985b). The U.S.A. imported 62 tonnes of raw coral from New Caledonia in 1983 and 15 tonnes in 1984 (Wells, 1985). In 1986, 192 tonnes of stony coral were collected, of which 33 t were exported, consisting primarily of worked Favidae (Bour *in litt.*, 6.10.87). Baitfishing in the lagoon was studied by Conand (1985). Traditional Kanak fishing is being studied in the north (Teulieres, 1985). The potential for commercial exploitation of holothurians is described in Conand (1981) and Conand and Chardy (1985).

Tourism is currently only of minor importance (Anon., 1985b). SCUBA diving is a popular recreational activity in the Noumea region. There is some tourism on the Isle of Pines and a dive shope operates at Gadji.

Disturbances and Deficiencies

Some of the lagoon reefs have been affected by *Acanthaster* since 1980, although the effect has not been disastrous (Dahl, 1985b). The great majority of reefs are probably in good condition. New Caledonia is not heavily populated for its size and about 50% of the population is concentrated in Noumea in the south-west. However, large areas of the interior have been destroyed through mining and bush fires which has led to erosion and siltation of rivers and lagoons. New Caledonia has some of the largest deposits of nickel in the world and the development of the mining industry is a major issue. This could pose a threat to the reefs through sedimentation and pollution. Sedimentation in the lagoon and its possible relation to erosion caused by mining is discussed by Guilcher (1985) and Bird *et al.* (1984). The major mining area is at Thio on the east coast; the river and shallow inshore waters have become heavily silted up (Dupon, 1986). The reefs of the north-west are potentially threatened by the NORCAL project, which involves mining at Dôme de Tiebaghi and the establishment of an industrial complex at Paagoumène. An environmental impact study has been carried out and if the recommendations are implemented, some of the damage could be averted (Thomassin, 1986). Industrial and domestic pollution may become a problem but at present is only significant in the vicinity of Noumea.

Exploitation of lagoon resources has not caused major problems to date because of the comparatively low population density, apart from around Noumea. Artisanal fisheries are more active here and their impact is compounded by other activities such as recreational fishing, spearfishing and tourist related activities, which have caused damage to reefs in this area (Dahl, 1985; Jourde, 1985). There has been some over-exploitation of trochus, a major decline in the harvest occurring after a peak in 1978 when 2000 tonnes were exported (Bour and Hoffschir, 1985). Melanistic cowries are found in the region from Saint Vincent to Goro and are much sought after, and have become rare. They were collected in the past using destructive methods including crowbars (Bouchet, 1979), but this has largely ceased. They are now collected at night with lights, although this is prohibited. The volutes *Cymbiola rossimiana* and *C. deshayesi* are also heavily collected but are not considered threatened (Bouchet *in litt.*, 8.8.87). Turtles are taken illegally around the Belep Islands and Loyalty Islands (Pritchard, 1987).

Legislation and Management

Traditional fishing methods are poorly documented but coastal people used to claim property rights over fishing areas. Details of customary regulations are no longer known, but, for example, there used to be a ban on fishing for the first six months of the year on Ouvéa (Dahl, 1985a). On the Isle of Pines, certain "clans" were designated fishing clans and would fish for other clans and for customary feasts (Leblic, 1985). With the recent major changes in fishing technology and the introduction of the commercial fishery, these customs have been lost (Dahl, 1985a; Leblic, 1985).

The Territory has full legislative and judicial powers in environmental matters. Eude (1973) describes legislation relevant to reefs in existence in the early 1970s. Environmental legislation is embodied in Law 76-1222 of 28 December 1976. Law 64-1331 (26 December 1964), Law 73-477 (16 May 1973) and Law 79-5 (2 January 1979) prohibit marine pollution by hydrocarbons. Law 76-599 (7 July 1976) concerns prevention and control of pollution by dumping from ships and aircraft. Regulations concerning the exploitation of living resources have proved difficult to enforce. Délibération 245 of 2.7.81, modified by Délibération 510 of 16.12.82 which covers trochus fishing, regulates fishing including prohibiting dynamite fishing (previously prohibited under Délibération 9, 2.8.67). Trochus collection is also controlled under Arrête 83-002/CG of 4.1.83; there is a size limit of 9-12 cm and collectors require a permit. Night fishing is prohibited but is poorly enforced (Bouchet, 1979). The collection of turtles is also controlled.

Specific measures have been introduced for the exploitation of coral, aquarium fish, bryozoa and sponges

in protected areas so that handicraft industries can continue. Coral collection used to be banned but is now authorised under Délibération 509 of 16.12.82 and Arrêté 85-321/CM of 19.6.85 on a trial basis in the Tetembia section of the barrier reef just north of the rotating reserve. Collection of 18 genera of corals is permitted. Coral collectors require a permit and must co-operate with the ORSTOM scientists carrying out the stock assessment and monitoring study. Collectors must comply with the regulations as these are introduced and must submit monthly reports on their activities. A coral stock assessment study is underway (Joannot, 1985) to monitor take, in view of the increased interest in collection and export. After a period it is hoped that there will be a precise definition of coral exploitation procedures and that only worked coral will be exported (Joannot, 1985; Jourde, 1985).

Aquarium fish collecting is permitted only by day and without SCUBA equipment, as for all other fishing. However, considerations are being given to permitting the use of SCUBA to avoid damage to the habitat, in which case, collecting with SCUBA would be restricted to one area only, would be permitted only during the day and would be licenced; the collecting area would probably be one of the two open areas of the rotating reserve and would change every three years. Quarterly monitoring of catch statistics would be carried out (Jourde, 1985).

Similarly, consideration is being given to permitting the take of Bryozoa and sponges with SCUBA equipment. Two well defined areas would be opened on a trial basis for this purpose; one to the north of Noumea between Nou and Mathieu Islands, at the southern entry of St Vincent Bay, for Bryozoa and one to the south, to the east of Amédée surrounding Redika islet, for sponges. A permit will be required and collectors will have to submit quarterly reports.

New Caledonia has ratified the Convention on Conservation of Nature in the South Pacific (SPREP, 1980). The Department of Waters and Forests is responsible for the enforcement of fishing regulations and the management of reserves and sanctuaries. A Marine Fisheries and Industries Commission is consulted for all measures aimed at safeguarding the marine resources of the Territory and ensuring the preservation of marine species (SPREP, 1980).

Law 56-1106 (3 November 1956) and Resolution 225 (17 June 1965) concern protected sites and the establishment of various forms of reserve. Revised legislation, Délibération 108 (9 May 1980), enforced by Decree 1504 (21 May 1980) concerns nature reserves and wildlife sanctuaries and introduces the terms Réserves Naturelles Intégrales (Strict Nature Reserves), Parcs Territoriaux and Réserves Spéciales (Special Reserves). Protected zones can also be established under the Water Resources and Pollution Law, Délibération 105 of 26.8.68, where activities likely to endanger water quality can be prohibited or controlled. The following areas receive some form of protection (Anon., 1985a):

1. Délibération 111 of 27 June 1974, modified by Délibération 229 of 2 July 1981, declared a protected zone, extending 1000 m from the high water mark, along the coasts of Grande Terre, the

Isle of Pines, Ouen, Maré, Lifou and Ouvéa Islands, the islands in the Belep Archipelago, Tiga and Yandé Islands and all permanently inhabited islands less than 12 miles (19.2 km) from Grande Terre. Within the zone there are controls on fishing methods and recreational fishing is limited to 50 kg fish per boat except for pelagic fish. Coral collection is also prohibited although some limited exploitation may be permitted. In Ouvéa, the use of spearguns is reportedly prohibited by local people.

2.   Yves Merlet Marine Reserve (*see separate account*)
3.   Great Reef Rotating Reserves (Réserves tournantes sur le Grand Récif) (*see separate account*)
4.   Maître and Amédée Islet Nature Reserves (Réserves Speciales Faune et Flore d'Ilot Maître et l'Ilot Amédée) (*see separate account*)

Benezit (1981) and Dupon (1986) report on mining pollution and the measures that are being taken to prevent it. Before any mine is opened, a commission for the Protection of the Environment evaluates pollution potential and other environmental problems, as is currently being carried out for the NORCAL project (Thomassin, 1986).

As a result of the decline in trochus production, a trochus assessment study has been carried out by ORSTOM and a two stage stock management strategy developed. Initially, 100 tonnes a year will be taken, gradually rising over five years to 400 tonnes a year, when the fishery will enter a sustained production stage (Bour and Hoffschir, 1985).

Recommendations

SPREP (1980) makes a number of recommendations, particularly with respect to the research required for improving reef management. These include studies on pollution around Noumea, the potential impact of mining on lagoon ecosystems, a study of fisheries stocks of the lagoon and a survey of vulnerable marine species. Thomassin (1986) recommends that a baseline study of the Great Koumac Reef/Great Poum Reef area be carried out prior to the implementation of the NORCAL mining project.

Recommendations have also been made for additional reserve areas:

1.   The area around the Isle of Pines was suggested for marine park status, to include the Yves Merlet Marine Reserve but this is not considered appropriate at present.
2.   Bouchet (1979) recommended the protection of the bays of Ngo and Uie, and the islets Mba and Mbo for melanistic cowries.
3.   Dahl (1980) recommended the establishment of additional reserves on:
     - the barrier reef
     - Chesterfield Islands (*see separate account*)
     - the north, east and west coasts of Grande Terre (e.g. Terrain Bas and La Foa)
     - Hunter
     - D'Entrecasteaux Reefs (*see separate account*)
     - Beautemps-Beaupré

References

* = cited but not consulted

Anon. (1985a). Country review - New Caledonia. *Report of the 3rd South Pacific National Parks and Reserves Conference, Apia* 3: 125-133.

Anon. (1985b). Secteur Maritime. D.T.S.E.E. Bull. de Conjoncture 9. New Caledonia.

*Avias, J. (1959). Les récifs coralliens de la Nouvelle Calédonie et quelques-uns de leurs problèmes. *Bull. Soc. géol. Fr.* 7(1): 424-430.

*Baillon, N. (1986). Croissance de deux espèces de poissons tropicaux à partir de la lecture des otoliths. Rapport de DEA - Université d'Aix-Marseilles, Centre d'Océanologie de Marseilles. 46 pp.

Benezit, M. (1981). Report on the mining pollution in New Caledonia. *SPREP Topic Review* 1. South Pacific Commission, Noumea, New Caledonia.

Bigot, L., Picaud, J., Roman, M.-L. and Thomassin, B.A. (1985). Example of a compressed bionomical zonation upon a coral reef flat: The salted inner pond on M'be lagoonal reef (SW New Caledonian coral reef complex). *Proc. 5th Int. Coral Reef Cong., Tahiti* 2: 32. (Abst.).

Bird, E.C.F., Dubois, J.P. and Iltis, J.A. (1984). The impacts of opencast mining on the rivers and coasts of New Caledonia. United Nations University, NRTS-25/UNUP-505, Tokyo. 53 pp.

Bouchet, P. (1979). Coquillages de collection et protection des récifs. ORSTOM, Centre de Noumea. Unpub.

Bour, W., Chaume, C., Conand, C., Loubersac, L. and Rual, P. (1985a). Use of high resolution satellite-imagery (SPOT-LANDSAT) in the thematic mapping of three coral reefs of New Caledonia. *Proc. 5th Int. Coral Reef Cong., Tahiti* 2: 42. (Abst.).

*Bour, W., Chaume, R., Conand, C., Loubersac, L. and Rual, P. (1985c). Cartographie thématique récifale par traitement d'images satellitaires: exemple d'un récif d'îlot du lagon de Nouvelle-Calédonie. *Rés. Comm. Coll. Fr. Japon Océanogr. Marseille* 1985: 113-114.

Bour, W., Gohin, F. and Bouchet, P. (1982). Croissance et mortalité naturelle des Trocas (*Trochus niloticus* L.) de Nouvelle-Calédonie. *Haliotis* 12: 71-90.

Bour, W. and Hoffschir, C. (1985). Evaluation et gestion de la ressource en trocas de Nouvelle-Calédonie. Rapport Final. Centre ORSTOM de Noumea.

Bour, W., Loubersac, L. and Rual, P. (1985b). Reef thematic maps viewed through simulated data from the future SPOT satellite. Application to the biotope of topshell (*Trochus niloticus*) on the Tetembia reef (New Caledonia). *Proc. 5th Int. Coral Reef Cong., Tahiti* 4: 225-230.

Bourrouilh Le-Jan, F.G. and Gaven, C. (1980). Géochronologie (230Th-234U-238U), sédimentologie et néotectonique des faciès récifaux pléistocènes à Maré, Archipel des Loyauté; S.W. Pacifique. *Oceanis* 7(4): 327-487.

Cabioch, G., Philip, J., Montaggioni, L., Thomassin, B.A. and Lecolle, J. (1985). First sedimentological and palaeoecological results from a drill-hole through a fringing coral reef, south-east of New Caledonia: evidence of the Holocene-Pleistocene discontinuity. *Proc. 5th Int. Coral Reef Cong., Tahiti* 6: 569-574.

*Chatenay, J.M. (1977). *Porcelaines Niger et Rostrées de Nouvelle Calédonie.* 109 pp.

*Chevalier, J.P. (1964). Compte rendu des missions effectuées dans le Pacifique en 1960 et 1962 (Mission d'études des récifs coralliens de Nouvelle-Calédonie). *Cah. Pacif.* 6: 171-175.

*Chevalier, J.P. (1968). Récifs coralliens. Expéd. française récifs coralliens Nouvelle-Calédonies. Ed. Fond. Singer-Polignac, 3.

*Chevalier, J.P. (1971). Les Scléractiniaires de la Mélanésie française (Nouvelle-Calédonie, Iles Chesterfield, Iles Loyauté, Nouvelles Hebrides). 1ère partie. Expéd. française récifs coralliens Nouvelle-Calédonie. Ed. Fond. Singer-Polignac, Paris 5: 5-307.

*Chevalier, J.P. (1973). Coral reefs of New Caledonia. In: Jones, O.A. and Endean, R. (Eds), *Biology and Geology of Coral Reefs* Vol. 1. Academic Press. Pp. 143-167.

*Chevalier, J.P. (1975). Les Scléractiniaires de la Mélanésie française (Nouvelle-Calédonie, Iles Chesterfield, Iles Loyauté, Nouvelles Hebrides). 2ème partie. Expéd. française récifs coralliens Nouvelle-Calédonie. Ed. Fond. Singer-Polignac, Paris 7: 5-407.

*Conand, C. (1981). Sexual cycle of three commercially important holothurian species (Echinodermata) from the lagoon of New Caledonia. *Bull. Mar. Sci.* 31(3): 523-544.

Conand, C. and Chardy, P. (1985). Are the Aspidochirote holothurians of the New Caledonian lagoon good indicators of the reefal features. *Proc. 5th Int. Coral Reef Cong., Tahiti* 5: 291-296.

Conand, F. (1985). Biology of the small pelagics of the lagoon of New Caledonia used as bait fish for tuna fishing. *Proc. 5th Int. Coral Reef Cong., Tahiti* 5: 463-467.

Condamin, M. (1978). Compte rendu de mission aux Iles Walpole, Hunter et Matthew (6 au 8.12.1977; 4.1.78). Rapport. Zoologie Appliquée. ORSTOM, Noumea, New Caledonia.

*Coudray, J. (1976). Recherches sur le Néogène et le Quaternaire marins de la Nouvelle-Calédonie. Contribution de l'étude sédimentologique à la connaissance de l'histoire géologique post-éocène. Expéd. française récifs coralliens Nouvelle-Calédonie. Ed. Fond. Singer-Polignac, Paris 8: 1-276.

*Coudray, J. (1982). Les récifs coralliens de la Nouvelle Calédonie: état des connaissances et perspectives de recherche. *Mém. géol. Univ. Dijon* 7: 63-72.

Coudray, J., Thomassin, B.A. and Vasseur, P. (1985). Comparative geomorphology of New Caledonia and Mayotte barrier reefs (Indo-Pacific Province). *Proc. 5th Int. Coral Reef Cong., Tahiti* 6: 427-432.

Dahl, A.L. (1980). Regional ecosystems survey of the South Pacific Area. *SPC/IUCN Technical Paper* 179. South Pacific Commission, Noumea, New Caledonia.

Dahl, A.L. (1985a). Traditional environmental management in New Caledonia: A review of existing knowledge. *SPREP Topic Review* 18. South Pacific Commission, Noumea, New Caledonia. 17 pp.

Dahl, A.L. (1985b). Status and conservation of South Pacific coral reefs. *Proc. 5th Int. Coral Reef Cong., Tahiti* 6: 509-513.

Davis, W.M. (1926). *Les côtes et les récifs coralliens de la Nouvelle-Calédonie.* Librairie Armand Colin, Paris. 120 pp.

Dupon, J.F. (1986). The effects of mining on the environment of high islands: a case study of nickel mining in New Caledonia. Environmental Case Study 1. SPREP, South Pacific Commission, Noumea, New Caledonia.

Estival, J.C. (1981). *Cônes de Nouvelle-Calédonie et du Vanuatu.* Les Editions du Cagou, Noumea, New Caledonia.

Eude, J.J. (1973). Problèmes posés par la protection des récifs et du lagon en Nouvelle Calédonie et dependances. *Proceedings and Papers, Regional Symposium on Conservation of Nature - Reefs and Lagoons.* South Pacific Commission, Noumea, New Caledonia.

Faure, G., Thomassin, B.A. and Vasseur, P. (1981). Reef coral assemblages on windward slopes in the Noumea Lagoon (New Caledonia). *Proc. 4th Int. Coral Reef Symp., Manila.* 2: 293-301.

Garnett, M.C. (1984). Conservation of seabirds in the South Pacific Region: A review. In: Croxall, J.P., Evans, P.G.H. and Schreiber, R.W. (Eds), *Status and Conservation of the World's Seabirds.* ICBP Technical Publication 2. Pp. 547-558.

*Guilcher, M.A. (1965a). Récifs du Sud, Récifs de Tuo. Expéd. française récifs coralliens Nouvelle-Calédonie. Ed. Fond. Singer-Polignac, Paris 1: 133-240.

*Guilcher, A. (1965b). Coral reefs and lagoons of Mayotte Island, Comoro Archipelago, Indian Ocean, and of New Caledonia, Pacific Ocean. *Proc. 17th Symp. Coston Res. Soc.* Bristol Univ.: 21-45.

Guilcher, A. (1985). Nature and human change of sedimentation in lagoons behind barrier reefs in the humid tropics. *Proc. 5th Int. Coral Reef Cong., Tahiti* 4: 207-212.

*Guille, A., Laboute, P. and Menou, J.-L. (1985). *Guide des étoiles de mer, oursins et autres echinoderms du lagon de Nouvelle Calédonie.* Editions de l'ORSTOM. 240 pp.

*Haeberlé, F.R. (1952a). The d'Entrecasteaux Reef Group. *Am. J. Sci.* 250: 28-34.

*Haeberlé, F.R. (1952b). Coral reefs of the Loyalty Islands. *Am. J. Sci.* 250: 656-666.

Joannot, P. (1985). Suivi de l'exploitation des coraux du récif Tétembia. 1er rapport d'activité. Centre ORSTOM de Noumea.

Joannot, P., Thomassin, B.A. and Magnier, Y. (1983). Coral banks in muddy environments in the New Caledonia South West Lagoon. *Biologie et Geologie des Récifs Coralliens.* Colloque annuel, International Society for Reef Studies, Nice, 8-9 December 1983.

Jourde, J. (1985). Marine reserves in New Caledonia. *Report of the 3rd South Pacific National Parks and Reserves Conference, Apia* 2: 74-78.

*Kulbicki, M. (1987). Experimental survey of coralline fishes by bottom longline in the lagoon of New Caledonia. Paper presented at 16th Pacific Science Congress, Seoul.

*Kulbicki, M, (in prep). Bottom longlining in the south west lagoon of New Caledonia. *Aus. Fisheries.*

*Kulbicki, M and Moutham, G. (in prep.). Essais de pèche au casier à poissones dans le lagon de Nouvelle-Calédonie. Rapport Scientifique et Technique No. 47. ORSTOM-Nouméa. 23 pp.

Laboute, P. and Magnier, Y. (1979). *Underwater Guide to New Caledonia.* Les Editions du Pacifique. Tahiti, French Polynesia.

Leblic, I. (1985). Technological, economical and sociological changes in a kanak fishing society: The Pines Island fishing clans face the development of commercialization in the fishery (New Caledonia). *Proc. 5th Int. Coral Reef Cong., Tahiti* 2: 218. (Abst.).

*Loubens, G. (1978a). Biologie de quelques espèces de poissons du lagon néo-calédonien. 1. Détermination de l'âge (otolithométrie). *Cah. ORSTOM. sér. Océanogr.* 16(3-4): 263-285.

*Loubens, G. (1978b). La pêche dans le lagon néo-calédonien. Rapport scientifique et technique No. 1. ORSTOM-Nouméa. 52 pp.

*Loubens, G. (1980a). Biologie de quelques espèces de poissons du lagon néo-calédonien. 2. Sexualité et reproduction. *Cah. Indo-Pacif.* 2(1): 41-72.

*Loubens, G. (1980b). Biologie de quelques espèces de poissons du lagon néo-calédonien. 3. Croissance. *Cah. Indo-Pacif.* 2(2): 101-153.

*ORSTOM (1981). Atlas de la Nouvelle Calédonie et dépendances. ORSTOM, Paris.

ORSTOM (1986). Resumés des Travaux. No. 11. Centre ORSTOM de Noumea, Oceanographie. 91 pp.

*Pierson, R. and Pierson, G. (1975). *Porcelaines mystérieuses de Nouvelle Calédonie*. 120 pp.

*Pope, E.C. (1962). New Caledonia, the coral-ringed island. *Australian Nat. Hist.* 14(1): 3-11.

*Priam, R. (1964). Contribution à la connaissance du volcan de l'îlot Matthew (sud des Nouvelles Hébrides). *Bull. Volcanologique* 27: 331-339.

Pritchard, P.C.H. (1987). Sea turtles in New Caledonia. Report of a literature survey and field investigation. Unpub. rep. to IUCN Conservation Monitoring Centre, Cambridge, U.K.

Rancurel, P. (1973a). Compte rendu de mission aux Iles Chesterfield du 21 au 28 juin. Rapport, ORSTOM.

Rancurel, P. (1973b). Compte rendu d'une visite aux Iles Hunter, Matthew, Walpole. Rapport. ORSTOM, Centre de Noumea.

Rancurel, P. (1973c). Compte rendu de mission à l'île des Pins (3 octobre 1973). (Echouage otarie et calmars). ORSTOM, Noumea, New Caledonia.

Rancurel, P. (1974a). Compte rendu de mission à bord de la Dunkerquoise dans les îlots du sud de la Nouvelle Calédonie (5-8 Juin 1974). Rapport, Centre ORSTOM de Noumea, New Caledonia.

Rancurel, P. (1974b). Compte rendu d'une visite à l'île Surprise le 31 janvier 1974. Rapport ORSTOM. 5 pp.

Richer de Forges, B. and Pianet, R. (1984). Résultats préliminaires de la campagne Chalcal à bord du N.O. Coriolis (12-31 juillet 1984). *Rapp. Sci. Tech.* 32. Centre ORSTOM de Noumea, New Caledonia.

Richer de Forges, B. and Bargibant, G. (1985). Le lagon nord de la Nouvelle-Calédonie et les atolls de Huon et Surprise. *Rapp. Sci. Tech.* 37. Centre ORSTOM de Noumea, New Caledonia. 23 pp.

Rougerie, F. (1985). The New Caledonian south-west lagoon: Circulation, hydrological specificity and productivity. *Proc. 5th Int. Coral Reef Cong., Tahiti* 6: 17-22.

SPREP (1980). New Caledonia. *Country Report* 8. South Pacific Commission, Noumea, New Caledonia.

*Salvat, B. (1964). Prospections faunistiques en Nouvelle-Calédonie dans le Cadre de la Mission d'études des récifs coralliens. *Cah. Pacif.* 6: 77-120.

*Salvat, B. (1965). Etude préliminaire de quelques fonds meubles du lagon calédonien. (Additif). *Cah. Pacif.* 7: 101-106.

*Taisne, B. (1965). Organisation et hydrographie. Expéd. française récifs coralliens Nouvelle Calédonies. Ed. Fond. Singer-Polignac, Paris.

Teulieres, M.-H (1985). The social, technical and ecological changes brought about by the progressive introduction of commercial fishing methods in the traditional coastal exploitation of Kanak fishermen, in the north of New Caledonia. *Proc. 5th Int. Coral Reef Cong., Tahiti* 2: 377. (Abstract).

*Thollot, P. (1987). Importance de la mangrove pour l'ichtyofaune du lagon de Nouvelle-Calédonie. Rapport de DEA - Université d'Aix-Marseilles, Centre d'Océanologie de Marseilles. 43 pp.

Thomassin, B.A. (1986). Etude de l'impact du projet "NORCAL" sur l'environnement marin de Nouvelle-Calédonie. COFREMMI, Station Marine D'Endoume, Marseille, France.

Thomassin, B.A. and Coudray, J. (1981). Presence of wide hardground areas on lagoonal bottoms of the coral reef complex of Noumea (south-west New Caledonia). *Proc. 4th Int. Coral Reef Symp., Manila* 1: 512-522.

Thomassin, B.A. and Vasseur, P. (1981). The coral reef complexes of the south-west coast of New Caledonia: Building and geomorphology. *Proc. 4th Int. Coral Reef Symp., Manila* 1: 596. (Abs.).

Tixier-Durivault, A. (1970). Les Octocoralliaires de Nouvelle Calédonie. Expéd. française récifs coralliens Nouvelle Calédonies. Ed. Fond. Singer-Polignac, Paris 4: 169-350.

Wells, S.M. (1985). Stony Corals: A case for CITES? *TRAFFIC Bulletin* 8(1): 9-11.

Wijsman-Best, M. (1972). Systematics and ecology of New Caledonian Faviinae (Coelenterata - Scleractinia). *Bijdr. Dierkunde* 42: 1-90.

CHESTERFIELD ISLANDS

Geographical Location   450 naut. mi. (834 km) north-east of Noumea in the Coral Sea; 300 mi. (480 km) east of the Belep Islands, adjoining the Bellona Reef lagoon to the south; 158°20', 19°50'S.

Area, Depth, Altitude   Lagoon about 10 m deep.

Physical Features   The Chesterfield plateau covers 34 000 sq. km at a depth of 1000 m and is delimited by the reef rim of an atoll with several emergent sandy cays and a more major vegetated motu (Richer de Forges and Pianet, 1984). The lagoon has a predominantly sandy bottom with some coral outcrops 1.5-2.0 m high. The main islands are Renard Islet and Skeleton Cay to the east, Bampton Islet to the north, Loop and Longue Islands and the Mouillage Islets to the south, and Avon Islet to the west. The west and north of the atoll is a barrier reef with an abrupt outer slope falling to more than 1000 m. The east and south-east is more open with a gentle slope. There is no fresh water (Rancurel, 1973a; Richer de Forges and Pianet, 1984).

Reef Structure and Corals   The corals have been described by Chevalier (1971 and 1975). There is some coral in the lagoon which has coral pinnacles reaching within 10 m of the surface in the central and western part. Small patch reefs are found off the Mouillage Islets. A large patch reef, 80 m in diameter, has been found about one mile (1.6 km) north-west of these islets on a sandy bottom at 35 m (Rancurel, 1973a; Richard de Forges and Pianet, 1984).

Noteworthy Fauna and Flora   The volutes *Cymbiolacca thatcheri* and *Lyria grangei* is known only from these islands, Bampton and Bellona reefs. The lagoon fauna is considered rather poor but fish life is abundant, both in the lagoon and without. Loop Island and some of the other islets have important seabird colonies. The islands are major turtle nesting sites. Sea snakes are abundant in the lagoon. The marine fauna is considered very rich, particularly fish. Molluscs and algae are listed in Richer de Forges and Pianet (1984). Dolphins are very tame (Rancurel, 1973a). The islands are important for seabirds (Garnett, 1984). Plants from Loop and Longue Islands and the Mouillage Islets are listed in Rancurel (1973a).

Scientific Importance and Research   The islands were visited in 1957, 1968 and 1973 but otherwise there have been very few scientific visits (Rancurel, 1973a).

Economic Value and Social Benefits   There is an automatic weather station on Loop Island. The islands are occasionally visited by Japanese fishermen and adventurous amateur shell collectors (Rancurel, 1973a).

Disturbance or Deficiencies   In the past the islands were exploited for guano (Rancurel, 1973a). The volute *C. thatcheri* is heavily collected, particularly by commercial collectors from Queensland. Otherwise the reefs and islands are relatively undisturbed because of their inaccessibility (Bouchet *in litt.*, 8.8.87).

Legal Protection   None.

Management   None.

Recommendations   The islands were recommended for protection by Dahl (1980), and the creation of a bird reserve is considered a priority in the Action Strategy for Protected Areas in the South Pacific Region drawn up at the Third South Pacific National Parks and Reserves Conference in 1985.

D'ENTRECASTEAUX REEFS

Geographical Location   North-west of Grand Terre and Belep Islands. Includes the atolls of Huon (18°03'S, 162°58'E), Portail, Surprise (18°09'S, 163°07'E) and Pelotas and the Guilbert and Merite Reefs.

Area, Depth, Altitude   160 acres (65 ha).

Physical Features   The atoll of Huon has a large pass in the north-west at which point the lagoon is not clearly delimited by a coralline rim and the sea bottom drops very rapidly from 50 to 500 m within several hundreds of metres. There are numerous coral pinnacles in the lagoon. The western side of the lagoon has a hard coralline bottom with calcareous algae between 40 and 70 m depth. To the east and south, the windward side of the atoll, the lagoon is shallower (30-40 m deep) and has a sandy bottom. Huon Island is a sandy cay about 1 km long, 150 m wide, and 4 m in altitude, covered with vegetation.

Between Huon and Surprise, the sea bed descends rapidly to about 800 m, although the slope up to Fabre Island on Surprise is more gentle. The atoll of Surprise has several passes and an almost totally submerged rim. The lagoon has a large pass on the west side, more than 4 mi. wide and is about 50 m deep. Most of the lagoon has a sandy bottom, although coral communities are found in the

north-west and immediately to the north of the atoll (Richer de Forges and Bargibant, 1985). Fabre and Le Leizour Islands lie on the northern rim of the atoll. Fabre has an altitude of 10 ft (3 m) and Le Leizour of 12 ft (3.6 m); both are covered with vegetation. In the south-west of the atoll is Surprise Island, 600 m x 400 m, which is the largest and most vegetated of the islands. Information on the atoll is also given in Taisne (1965), although some of the data have been found to be incorrect following the work by Richer de Forges and Bargibant (1985). Surprise Island has also been described by Rancurel (1974b).

Reef Structure and Corals   The reefs were described by Haeberlé (1952a). Rich invertebrate communities with gorgonians and algae were found in the west part of the lagoon of Huon atoll. Similar communities were found at Surprise (Richer de Forges and Bargibant, 1985).

Noteworthy Fauna and Flora   Huon Island and to a lesser extent, Surprise and the other islets, are among the most important nesting sites for the Green Turtle *Chelonia mydas* in the Pacific. There is also some nesting of the Hawksbill *Eretmochelys imbricata* (Pritchard, 1987). The islands have important seabird colonies (Garnett, 1984; Rancurel, 1974b; Richer de Forges and Bargibant, 1985).

Scientific Importance and Research   The atolls were visited by an expedition from the Centre ORSTOM de Nouméa in 1985 (Richer de Forges and Bargibant, 1985).

Disturbance or Deficiencies   Surprise is occasionally visited by Belep islanders to collect turtles (Pritchard, 1987).

Legal Protection   None.

Recommendations   The atolls have been recommended for protection by Dahl (1980).

GREAT REEF ROTATING RESERVES (RESERVES TOURNANTES SUR LE GRAND RECIF)

Geographical Location   West of Noumea, on the Great Reef, including the sectors from Annibal Reef to Kué Reef; 22°15'-22°35'S, 166°10'-166°30'E.

Area, Depth, Altitude   35 000 ha; each sector covers about 10 000 ha.

Land Tenure   Public.

Physical Features   The reserves consist of three sections of reef: Annibal, Abore and Kué, separated by channels (Jourde, 1985).

Reef Structure and Corals   (See introduction for general description of the Great Reef).

Scientific Importance and Research   Extensive research has been carried out in this area by ORSTOM scientists, but there are few if any publications relating specifically to these areas.

Economic Value and Social Benefits The area is very popular for recreational activities, on account of its accessibility from Noumea.

Disturbance or Deficiencies This area used to be heavily exploited by both commercial and recreational fishermen and by the end of the 1970s, divers from ORSTOM and the Noumea Aquarium, as well as local fishermen noticed that the fauna was declining, particularly fish and crustaceans (Jourde, 1985).

Legal Protection The Réserves Tournantes sur le Grand Récif were established on 2 July 1981 by Délibération 230. The three sectors are closed in turn for the capture or destruction of marine animals and for coral collection. Closed seasons are: Sector B, 27 August 1981 - 26 August 1984; Sector C, 27 August 1984 - 26 August 1987; Sector A, 27 August 1987 - 27 August 1989. In those areas where the closed season is not in force, collecting may take place according to normal fisheries legislation. Sector A is the area between the passage of Uitoé and that of Dumbéa; Sector B runs from Dumbéa to Boulari and Sector C from Boulari to Mato. Sector A includes the reef of l'Annibal and the M'bere Reef. Sector B includes the Abore and Tabu reefs; Sector C includes the To, Sournois, Toombo and Kué reefs.

Recommendations Fish catches within the reserve should be monitored when funding and personnel are available. Although there has been no formal monitoring to date, divers and fishermen have reported an improvement in stocks in the first area to be closed (Jourde, 1985).

MAITRE AND AMEDEE ISLET NATURE RESERVES (RESERVES SPECIALES DE FAUNE ET FLORE L'ILOT MAITRE ET L'ILOT AMEDEE)

Geographical Location West of Noumea within the lagoon of the Great Reef; Maître is about 4 km from the coast at 22°20'S, 166°25'E; Amédée lies just inside Boulari Passage at 22°29'S, 166°29'E.

Area, Depth, Altitude Maître Islet Reserve is 620 ha; Amedée Islet is 154 ha; 10 m depth to 2 m altitude.

Land Tenure Public.

Physical Features Lagoon islets with some beach strand vegetation and with fringing reefs.

Reef Structure and Corals Faure *et al.* (1981) and Thomassin and Coudray (1981) describe reef coral assemblages on the windward slopes of Maître Reef. The intertidal reef flat has less than 20% coral coverage, as a result of its sheltered position, and has abundant algae *Sargassum* and *Turbinaria*. Between 1 and 2 m depth, coral cover is still low and is predominantly short branched species such as *Acropora digitifera, Pocillopora damicornis* and *P. verrucosa*, massive faviids and some branched and tabular *Acropora*. The deeper zones are similar to Goelands Reef (see introduction). The islet has been mapped using SPOT-LANDSAT satellite imagery (Bour *et al.*, 1985a and b).

Scientific Importance and Research Extensive research has been carried out in this area by ORSTOM scientists. Growth studies on trochus have been carred out at Amedée (Bour *et al.*, 1982).

Economic Value and Social Benefits The islets are very popular for recreational activities, on account of their accessibility from Noumea. There is a hotel on Maître and launch trips run from Noumea to both islands, which are also visited by numerous weekend boating and fishing enthusiasts (Jourde, 1985).

Disturbance or Deficiencies Both islands are very heavily used and are not actively protected. Corals on the reef are broken and collection of corals and shells and fishing take place (Bouchet *in litt.*, 8.8.87). In the past there was heavy recreational and commercial fishing around both islets (Jourde, 1985). There has been some damage to the reefs around Maître Islet as a result of tourism development.

Legal Protection The Réserves Speciales de Faune et Flore of Ilot Maître et Ilot Amedée were established 2 July 1981 by Délibération 231. Both sectors include the islands and their adjacent waters to a depth of about 10 m. The Amédée sector includes the island and surrounding waters to a depth of 10 m. Collection of terrestrial or marine organisms is prohibited, except by the lighthouse keepers on Amédée; it is forbidden to disturb nesting birds. Tourist development, recreation and management to maintain conservation purposes are permitted.

Management No active management.

YVES MERLET MARINE RESERVE

Geographical Location Between Havannah Canal and Sarcelle Passage, south-east of New Caledonia between the mainland and the Isle of Pines; 22°25'S, 167°08'E.

Area, Depth, Altitude 22 925 ha (?16 700 ha); 75 m depth to 2 m altitude.

Land Tenure Public.

Physical Features Includes two islets, Amere and Kié, covered with stands of *Araucaria columnaris*, and a complex of reefs.

Reef Structure and Corals Includes a major section of barrier reef, with complex topography (Bouchet *in litt.*, 8.8.87).

Noteworthy Fauna and Flora The volute *Cymbiola rossiniana* is found only in the area between the Isle of Pines and the Boulari Passage, in sediments at 15 m depth (Bouchet, 1979). There are several seabird colonies and the osprey *Pandion haliaetus* nests.

Scientific Importance and Research A study on growth rates of trochus has been carried out within the reserve (Bour *et al.*, 1982).

Economic Value and Social Benefits The reserve is fished from time to time by local people from Touaourou and Goro (Bouchet *in litt.*, 8.8.87). Trochus collection is carried out.

Disturbance or Deficiencies The area is relatively little disturbed. It is too far from Noumea for weekend boat excursions, and the complex reef topography within the reserve means that the few small boats which reach it generally keep to its edges (Bouchet *in litt.*, 8.8.87).

Legal Protection Established 17 July 1970 as a Strict Nature Reserve (Réserve Spéciale Marine) by Délibération 244, confirmed by Délibération 108, 9 May 1980. Fishing, capture, and collection of all animals, plants and minerals is prohibited. The area is out of bounds to boats either for passage or for anchoring near islands or emergent reefs but these regulations may be waived for traditional canoes and customary fishing activities (Jourde, 1985). Local people may visit the reserve once a year to collect trochus.

Management A caretaker is paid to warden the area, from the local village.

NEW ZEALAND

INTRODUCTION

General Description

The only reef-forming corals in New Zealand occur in the Kermadec Islands which lie ca 750-1000 km north-east of North Cape, New Zealand (about half way to Tonga) at 29°14'-31'S, 177°50'-178°53'W, and consist of the summits of the volcanic pinnacles of the Kermadec Ridge. There are two large and 15 smaller islands, divided into four main groups, separated from each other by depths greater than 900 m: Raoul Island and the Herald Islets (29 sq. km in land area); Macauley Island (3 sq. km); Curtis and Cheeseman Islands (less than 1 sq. km); and L'Esperance Rock (less than 50 m long). There is considerable thermal activity on Raoul and Curtis, the most recent eruption having taken place on Raoul in 1964, and underwater vents have recently been discovered around Curtis Island and off Havre Rock (near L'Esperance Rock) (Francis, 1986). There is a mild subtropical climate, rainfall averaging 1473 mm (Dept Lands and Survey, 1984).

Mean surface temperatures of the subtropical waters around the islands range from 16.8°C to 23.7°C. Coral colonies may grow to two metres in diameter but do not form true reefs. Colony size, abundance and diversity decrease southwards through the islands, correlated with winter water temperatures. The distribution of *Acanthaster* shows a similar trend; it is found occasionally at Raoul Island, rarely at Macauley Island and is unrecorded from the two southern island groups. The marine flora and fauna comprise a mixture of warm temperate New Zealand species and tropical species not found around the mainland and some taxa show a fairly high level of endemism. The fish fauna shows strongest affinity with that of Lord Howe Island (64% of species in common), but also shows links with that of Rapa and Easter Islands (both 22% of species in common). Two fish species, *Parma kermadecensis* and *Ocosia apia*, are endemic and a third, *Gymnothorax griffini*, possibly is (Francis *et al.*, 1987). The depth distribution and relative abundance of benthic organisms and fish are discussed by Schiel *et al.* 1986) and marine algae are described by Nelson and Adams (1984). The Hawksbill Turtle *Eretmochelys imbricata* occurs (Francis, 1986) as probably do other species which reach the New Zealand mainland (Francis *in litt.*, 1987). The terrestrial fauna and flora are described in Dept Lands and Survey (1984); the islands have important seabird colonies (Robertson and Bell, 1984).

Reef Resources

Because of its remoteness, fishing in the area has so far been restricted to a few longlining vessels, but interest is escalating because of over-exploitation in areas nearer the mainland and a change in legislation which now permits smaller vessels to fish in the area. The islands are becoming increasingly popular with SCUBA divers, 50-60 visiting the Kermadecs annually (Francis, 1986).

Disturbances and Deficiencies

Conservation problems on the islands are described in Dept Lands and Survey (1984). There are particular fears that the population of the Spotted Black Grouper *Epinephelus daemelii* could be over-fished (Francis, 1986).

Legislation and Management

The New Zealand Underwater Association banned the spearing of Spotted Black Grouper at the Kermadecs in 1982 (Francis, 1986). The islands were declared a nature reserve in 1934 (Dept Lands and Survey, 1984).

Recommendations

The Department of Conservation (formerly the Department of Lands and Survey) has made a formal application to the Ministry of Agriculture and Fisheries (MAF) to create a marine reserve within the 12-mile (19.2 km) zone around each Kermadec Island. MAF have proposed interim regulations until marine reserve status is obtained. These would prohibit the taking of Spotted Black Grouper; ban commercial fishing (except tuna) within 12 mi. (19.2 km) of the islands and restrict recreational fishing to pelagic species; and control the hapuku and bass fishery around the islands (Francis, 1986).

References

Dept Lands and Survey (1984). *Register of Protected Natural Areas in New Zealand.* Wellington, New Zealand.
Francis, M. (1986). A Kermadec Islands Marine Reserve? *Forest and Bird* 17(3): 16-18.
Francis, M.P., Grace, R.V. and Paulin, C.D. (1987). Coastal fishes of the Kermadec Islands. *N.Z. J. Marine and Freshwater Res.* 21: 1-13.
*Nelson, W.A. and Adams, N.M. (1984). Marine algae of the Kermadec Islands. *Nat. Mus. N.Z. Misc. Ser.* 10. 29 pp.
Roberston, C.J.R. and Bell, B.D. (1984). Seabird status and conservation in the New Zealand region. In: Croxall, J.P., Evans, P.G.H. and Schreiber, R.W. (Eds), *Status and Conservation of the World's Seabirds.* ICBP Technical Publication 2. Pp. 547-558.
*Schiel, D.R., Kingsford, M.J. and Choat, J.H. (1986). The depth distribution and abundance of benthic organisms and fish at the subtropical Kermadec Islands. *N.Z. J. Marine and Freshwater Res.* 20: 521-535.

NIUE

INTRODUCTION

General Description

Niue is an isolated island approximately 480 km east of Tonga and 560 km south-east of Samoa and is a self-governing territory of New Zealand. It is roughly circular, ca 259 sq. km in area (21 x 18 km), and is a raised atoll of coralline limestone about 62 m high with a prominent coastal terrace about 20-28 m above sea-level. The island has a slightly depressed upper surface representing the "lagoon" of the original atoll. Both the upper surface and coastal terrace are heavily vegetated except where cleared by man. The soils are shallow, reasonably rich, partly derived from outside sources (assumed to be either airfall volcanic dust or tidal wave volcanic debris swept into the original lagoon) and have an unusually high natural radioactivity of an obscure origin, discussed in Fieldes *et al.* (1960) and Schofield (1967). Details of the geology are given in Rodgers *et al.* (1982), Schofield (1959) and Schofield and Nelson (1978). Bathymetry is discussed in Summerhayes (1967).

There is no surface or running fresh water and no good harbour. The mean annual temperature is 24.7°C, the mean annual rainfall is a little over 2000 mm and the mean humidity is 79.7%. The island is on the edge of the hurricane belt and severe destructive hurricanes have occurred, most recently in 1958, 1960, 1968 and 1979.

There is no true reef and no lagoon. The island is in part surrounded by a platform, varying from a few metres to several hundred metres in width, cut in the original limestone of the island. Large parts of this are subtidal, the remainder intertidal. Much of the south and east sides of the island are entirely devoid of reef flat; some parts have 1-8 m wide pools about 1.5-2.5 m above sea level. The flat has a thin discontinuous veneer of living corals on its upper (intertidal) surface and rich coral growth over the edge in sub-tidal waters (Paulay *in litt.*, 5.10.87; Yaldwyn, 1973). Work on the coral fauna is in progress; there are at least 43 coral genera.

There is a rich, though largely undocumented invertebrate fauna (Paulay *in litt.*, 5.10.87). Reef crabs and other crustaceans, such as crayfish, are well represented with many common Indo-Pacific species. A mollusc checklist is available. The fish fauna is also fairly rich (Yaldwyn, 1973), and there is an unusual abundance of sea snakes (Paulay *in litt.*, 5.10.87). Details of the terrestrial fauna and vegetation are given in Yaldwyn (1973), and of terrestrial vegetation and flora in Sykes (1970). Given (1968) discusses the insect fauna and Myers (1986) describes the amphipods.

Reef Resources

The population in 1986 was fewer than 3000, consisting almost entirely of Niueans, with a small Tongan minority and some New Zealand contract workers; it is steadily declining due to emigration to New Zealand (Paulay *in litt.*, 5.10.87). Fishing is one of the main activities and was described by Hinds (1970). Controlled tourism is being gradually expanded (SPREP, 1980),

although in 1984 was reported as still at a very low level (Stanley, 1984).

Disturbances and Deficiencies

Coastal waters except in Alofi "urban" area, where there is some coral silting caused by run-off, are very clean. Siltation was made worse by Cyclone Ofa in 1979. Reef animals are subjected to fairly intensive harvesting for food but there is no evidence of over-exploitation. Use of poisons in fishing may still occur but there is no dynamiting (SPREP, 1980). Shell collecting is increasing with the expansion of tourism. The vegetation has been greatly modified by man in the interior and some logging is still carried out. However there is some pristine rain forest left in several large "Tapu" (traditionally protected) areas, and in the early 1970s the low coastal forest of the lower terrace was still largely intact (Paulay *in litt.*, 5.10.87; Yaldwyn, 1973).

Legislation and Management

Fishing is controlled by each village for their own territories, and there are many traditional conservation measures in effect for different times and species (Paulay *in litt.*, 5.10.87); thus customary restrictions or "fono" are applied in certain areas from time to time and allow rapid recovery of exploited resources in these temporary reef "reserves" (Yaldwyn, 1973). The Wildlife Ordinance 1972 protects listed birds and fruit bats but hunting has recently been allowed. There is a Fish Protection Ordinance 1965. There are no formally designated protected areas (except the "Tapu" forest areas).

Recommendations

An Environmental Protection Ordinance with conservation priorities was officially proposed in 1975 (Dahl, 1980 and *in litt.*, 15.9.87; Gare, 1977). This was intended to establish Environmental Protection Areas, including reef areas, subject to existing Niuean customs and traditional usage. No aquatic shell, mollusc or marine organisms were to be allowed to be taken from the area, no litter deposited, etc., and the export of shells and other marine organisms was to be closely restricted. Village Councils were to be responsible for the protected areas (Gare, 1977). There appears to have been no progress with this in recent years (Dahl *in litt.*, 15.9.87), although the National Development Plan 1980-85 included as an objective the protection and preservation of the natural environment (SPREP, 1980).

The following marine areas were proposed as reserves by Dahl (1980):

- Fatiau Tuai: deserted village near distinctive coral reef formation
- Makapu: coastal cave
- Hio: cave and beach
- Limu: complex of caves and marine pools

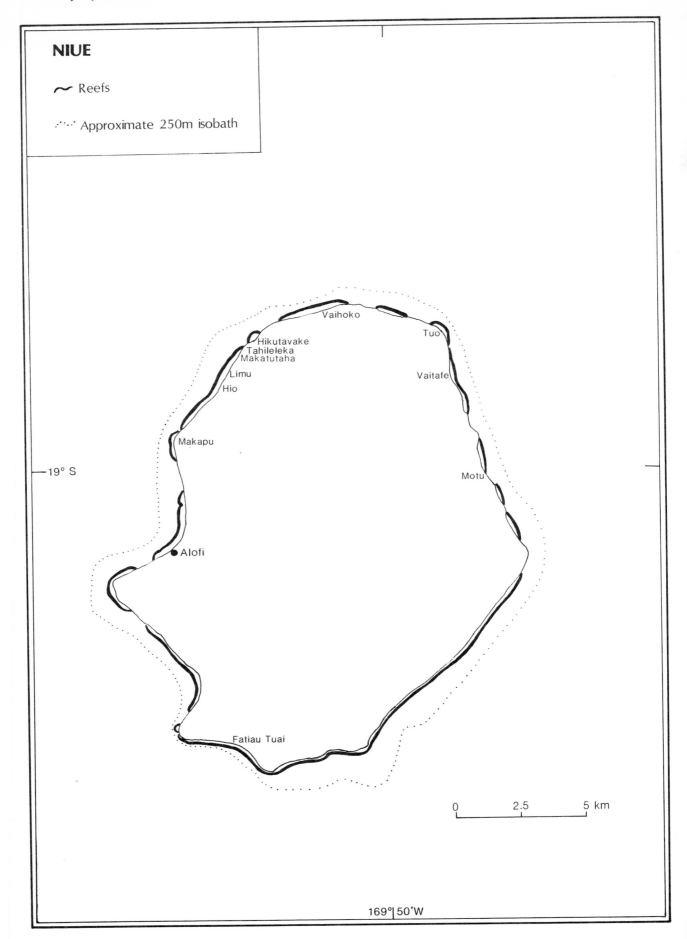

NIUE

~ Reefs

⋯⋯ Approximate 250m isobath

Vaihoko

Tuo

Hikutavake
Tahileleka
Makatutaha

Vaitafe

Limu

Hio

Makapu

—19° S

Motu

● Alofi

Fatiau Tuai

0   2.5   5 km

169° 50'W

- Makatutaha: swimming hole and cave used for storing canoes
- Tahileleka: sink-hole with underground connection to sea
- Hikutavake Reef: reef with large pools
- Vaihoko: caves and reef channels
- Tuo: reef and cave area of traditional importance
- Vaitafe: broad reef with pool and freshwater springs
- Motu: reef and caves used as canoe landing

References

* = cited but not consulted

Dahl, A.L. (1980). Regional ecosystems survey of the South Pacific Area. *SPC/IUCN Technical Paper* 179. South Pacific Commission, Noumea, New Caledonia.

*Fieldes, M., Bealing, G., Claridge, G.G., Wells, N. and Taylor, N.H. (1960). Mineralogy and radioactivity of Niue Island soils. *N.Z. J. Sci.* 3: 658-675.

Gare, N.C. (1977). Review of progress in the creation of marine parks and reserves. *Collected Abstracts and Papers of the International Conference on Marine Parks and Reserves, Tokyo, Japan, May 1975.* The Sabiura Marine Park Research Station, Japan. Pp. 139-144.

*Given, B.B. (1968). List of insects collected on Niue Island during February and March, 1959. *N.Z. Ent.* 4: 40-42.

*Hinds, V.T. (1970). *A Fisheries Report Prepared for the Government of Niue.* Cyclostyled report issued by the South Pacific Commission, Noumea. 60 pp.

*Myers, A.A. (1986). Amphipoda from the South Pacific: Niue Island. *J. Nat. Hist.* 20: 1381-1392.

*Rodgers, K.A., Easton, A.J. and Downes, C.J. (1982). The chemistry of carbonate rocks of Niue Island, South Pacific. *J. Geol.* 90: 645-662.

SPREP (1980). Niue. *Country Report* 9. South Pacific Commission, Noumea, New Caledonia.

*Schofield, J.C. (1959). The geology and hydrology of Niue Island, South Pacific. *N.Z. Geol. Survey Bull.* 62. 28 pp.

*Schofield, J.C. (1967). Origin of radioactivity of Niue Island. *N.Z. J. Geol. Geophys.* 10: 1362-1371.

*Schofield, J.C. and Nelson, C.S. (1978). Dolomitisation and quaternary climate of Niue Island, Pacific Ocean. *Pacific Geol.* 13: 37-48.

Stanley, D. (1984). *South Pacific Handbook.* Moon Publications, Chico, California.

Summerhayes, C.P. (1967). Bathymetry and topographic lineation in the Cook Islands. *N.Z. J. Geol. Geophys.* 10(6): 1382-1399.

*Sykes, W.R. (1970). Contributions to the flora of Niue. *N.Z. DSIR Bull.* 200.

Yaldwyn, J.C. (1973). The environment, natural history and special conservation problems of Niue Island. Paper 8, Section 3. *Proceedings and Papers, Regional Symposium on Conservation of Nature - Reefs and Lagoons.* South Pacific Commission, Noumea, New Caledonia. Pp. 49-55.

NORTHERN MARIANA ISLANDS

INTRODUCTION

General Description

The Commonwealth of the Northern Mariana Islands (CNMI) was officially granted the status of a sovereign territory of the U.S.A. in October 1986, having previously been a part of the Trust Territory of the Pacific Islands; it had however been functioning as a commonwealth of the U.S.A. since 1978. It comprises all the Marianas except Guam (which is described in a separate country account) and lies in the region 14°-20°5'N, 145°-146°E. The sixteen islands of the Mariana Archipelago (three of which make up Maug) are separated into two almost parallel arcs (Eldredge, 1983a). The outer or frontal arc, or southern islands, extends from Guam to Farallon de Medinilla and is composed of old volcanic islands, capped or surrounded by limestone terraces. The inner or active arc, the northern islands, runs from Anatahan to Farallon de Pajaros (Uracas) in the north. The islands are the peaks of a submerged chain of volcanic mountains, the highest in Micronesia, and cover an area of 510 sq. km. The climate is warm throughout the year with tropical conditions in the south and a subtropical climate toward the north. Rainfall totals 2000-2500 mm a year and is strongly seasonal, about 60% falling between July and November. The dry season is from January to May. Typhoons can cause extensive damage. Each of the islands is described and the climate and oceanography reviewed in Eldredge (1983b) and Maragos (1986a).

The vegetation of the islands is varied (Fosberg, 1960); their floras are described in Fosberg et al. (1975). Volcanic substrates of both the northern and southern islands are dominated by swordgrass; trees grow in associated ravines. Forests are found on some areas of limestone substrate but have been extensively altered by man.

Table of Islands

Farallon de Pajaros (Uracas)   202 ha; high (1047 ft (319 m)); active volcanic cone; entire shoreline altered during 1943 lava flow; few marine invertebrates at present (recolonization of the shoreline is taking place); no anchorages or good landing places; undisturbed and seldom visited.

Maug   205 ha; volcanic sunken cone with 3 small islets; 218 m; North and West Islands have steep cliffs; coral communities described below; sheltered anchorage; described in Anon. (1978) and Eldredge et al. (1977).

Asuncion I.   722 ha; active volcanic cone, 2900 ft (884 m).

Agrihan   (X) 18.3 sq. mi. (47.4 sq. km); dormant volcanic cone, 3165 ft (965 m), steep.

Pagan   18.6 sq. mi. (48 sq. km); cluster of active volcanoes; 1869 ft (570 m); some active coral reef growth; eruption in 1981 and still active; impact on reefs studied (Eldredge, 1982; Eldredge and Kropp, 1985); anchorage and dock.

Alamagan   (X) 4.3 sq. mi. (11 sq. km); active volcano, 2440 ft (744 m).

Guguan   412 ha; 2 volcanic cones, one active, 813 ft (248 m), one dormant 987 ft (300 m); cliffbound; difficult landing; important seabird colonies.

Sarigan   (X) 500 ha; extinct volcanic cone, 1800 ft (549 m).

Anatahan   12.5 sq. mi. (32.4 sq. km); dissected cone, 2585 ft (788 m), with active steam vents (Eldredge in litt., 1987).

Farallon de Medinilla   0.3 sq. mi. (0.8 sq. km); raised limestone islet with flat-topped limestone ridge; 265 ft (81 m); cliffs rocky shoreline; caves; difficult landing; no reefs.

Saipan   (X) 47 sq. mi. (122 sq. km); high volcanic, 1554 ft (474 m), with raised limestone terraces; geology by (Cloud, 1959); reefs described below; marine plants described by Fitzgerald and Tobias (1974); wetland vegetation mapped in Moore et al. (1977); coralline algae by Gordon (1974); Green Turtle Chelonia mydas nesting on Tanapag beach.

Tinian   (X) 39.25 sq. mi. (102 sq. km); raised limestone; 557 ft (170 m); wetland vegetation mapped in Moore et al. (1977); reefs described below.

Aguijan   2.8 sq. mi. (7.3 sq. km); small raised limestone island with steep cliffs to north; recent review by Eldredge (1985); difficult landing; shoreline surveyed; no coral reefs but diverse coral communities and active reef growth evident at many places (Maragos, 1986a).

Rota (Luta)   (X) 33 sq. mi. (85.5 sq. km); raised limestone terraces on slopes of extinct volcano; 1612 ft (491 m); studied by Sugawara (1934) and Tsuda (1969); reefs described below.

(X) = Inhabited

The shorelines and shallow reefs are briefly described by Eldredge (1983b). The entire shorelines of Aguijan, Farallon de Medinilla and most of the smaller northern islands are bordered by steep rocky slopes and sea cliffs and have no coral reefs. Rota, Tinian and Saipan have more varied coastlines including reefs. Barrier reefs and well-developed fringing reefs tend to develop along the western coasts while the eastern shorelines are generally rocky and cliff-bound. Some of the fringing reef platforms are erosional features, truncated by wave action at or near the present sea level while others are accretional due to growth of corals, calcareous algae and other reef organisms. The community structure of the corals from seven emergent islands (Pagan, Guguan, Anatahan, Saipan, Aguijan, Rota and Guam) and one shallow submerged bank (Zealandia Bank) has been outlined by Randall (1985) who designated four habitat types to assist in categorizing the reefs.

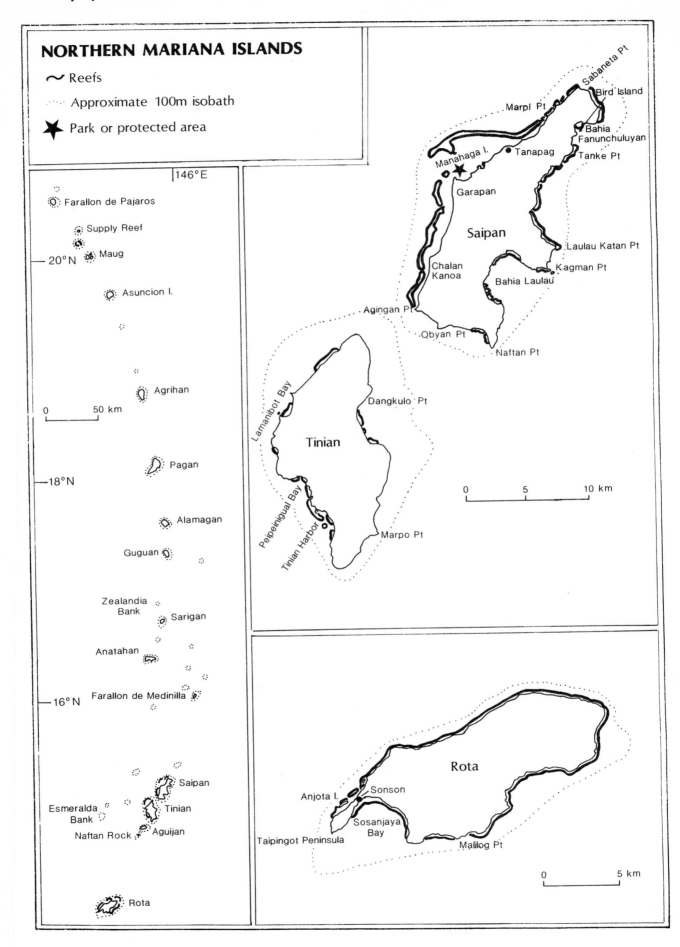

NORTHERN MARIANA ISLANDS

~ Reefs

⋯ Approximate 100m isobath

★ Park or protected area

146°E

Farallon de Pajaros

Supply Reef

Maug

—20°N

Asuncion I.

Agrihan

0   50 km

Pagan

—18°N

Alamagan

Guguan

Zealandia
Bank

Sarigan

Anatahan

—16°N   Farallon de Medinilla

Saipan

Esmeralda
Bank

Tinian

Naftan Rock

Aguijan

Rota

Sabaneta Pt

Bird Island

Marpi Pt

Bahia
Fanunchuluyan

Manahaga I.

Tanapag

Tanke Pt

Garapan

Saipan

Laulau Katan Pt

Chalan
Kanoa

Kagman Pt

Bahia Laulau

Agingan Pt

Obyan Pt

Naftan Pt

Lamanibot Bay

Dangkulo Pt

Tinian

0   5   10 km

Peipeinigual Bay

Tinian Harbor

Marpo Pt

Rota

Anjota I.

Sonson

Taipingot Peninsula

Sosanjaya
Bay

Malilog Pt

0   5 km

Saipan and Tinian have offshore barrier reefs with shallow lagoons along parts of their western coasts; the reefs and beaches of both islands, as well as Rota, have been mapped by Eldredge and Randall (1980). Saipan has smaller fringing and apron reefs at Bahia Fanunchuluyan, Unai Laulau Katan (Laulau Katan Pt) and Bahia Laulau on the east coast and Unai Obyan and Unai Dangkulo Agingan (Agingan Pt) on the south coast. Seagrass beds are well developed only on the lagoonal reefs of the west coast, off urban centres at Chalan Kanoa, Garapan and at Tanapag. Several pinnacle and patch reefs, unique in the CNMI, are found in the Garapan lagoon area, which has been recently surveyed (Maragos, 1986a and b). Corals from Saipan are described by Gawel (1974) and additional information on the reefs is given in Goreau *et al.* (1972).

Reefs on Tinian are less well developed and are confined to narrow fringing reefs at Unai Dangkulo (Dangkulo Pt) on the east coast and smaller lagoon fringing and apron reefs at Lamanibot, Peipeinigual and Tinian Harbor on the west coast. There is a single patch reef in the harbour. Descriptions of reefs and coral communities at San Jose (Tinian) Harbour, Lamanibot Bay, Unai Dangkulo and Peipeinigual Bay are given in Jones *et al.* (1974). There is a well developed narrow fringing reef along most of the north-west coast of Rota, around Anjota Island and off West Dock and the village of Songsong. Reefs straddle both sides of the Taipingot Peninsula, and narrower fringing reefs occur off Sosanjaya and both sides of Malilog Peninsula along the south coast (Maragos, 1986a). Sugawara (1934) provides an early description of the reefs of Rota.

Marine biota of Maug are described in Eldredge *et al.* (1977) and Eldredge (1983b). The central lagoon is dominated by benthic red and brown algae and only a few localized coral communities have been identified, there being no true coral reef development. Predominant corals include *Millepora platyphylla, Acropora, Porites lutea, Goniastrea retiformis, Pocillopora* and *Favia*. A narrow but well developed fringing reef extends along the southern, seaward shore of West Island, characterized by massive growths of many of the species found within the lagoon, complemented by *Lobophyllia, Diplastrea heliopora* and *Heliopora coerulea*. A total of 73 coral species has been recorded.

Detailed information on the relationships between the mollusc faunas of the northern and southern islands is provided by Vermeij *et al.* (1983). Of the 300 mollusc species known to occur in the northern islands, 18 are unknown or are very rare in the southern islands. At least 22 gastropods common in the southern islands are absent from the northern islands. Two gastropods, *Echininus viviparis* and *Nerita guamensis*, and the recently described limpet *Pateloida chamorrorum* (Lindberg and Vermeij, 1985), are known to be endemic to the southern Mariana islands, including Guam. Several other differences in the marine biota occur between the northern and southern islands. Among those known are two corals and at least three fish, as well as at least ten species of algae (Vermeij *et al.*, 1983). The marine invertebrate fauna of Maug is typical of other Indo-West Pacific coastal areas, with indications of a particularly rich opisthobranch fauna. A checklist of fishes from Maug is given in Eldredge *et al.* (1977); 232 species have been collected in five expedition surveys.

Marine benthic algae of the Northern Marianas are described in Tsuda and Tobias (1977a and b). Biological studies of the Coconut Crab *Birgus latro* in the Northern Marianas are discussed in Amesbury (1980). Few turtles appear to nest in the Northern Marianas, probably because of the scarcity of suitable nesting beaches; however Green Turtles *Chelonia mydas*, Hawksbills *Eretmochelys imbricata* and Olive Ridleys *Lepidochelys olivacea* are reported to occur in coastal waters (Pritchard, 1981).

Reef Resources

There are permanent human populations on only the three largest islands to the south, Saipan, Tinian and Rota; two other islands (Alamagan and Agrihan) are regularly inhabited. The 1980 population was about 17 000 and is predicted to more than double by 1990. Over 87% of the population live on Saipan, where fishing and tourism are the main sources of income. Local fishing techniques in use in the 1940s are described by Smith (1947). Lagoonal fisheries on Saipan have been described by Johannes (1978) and Amesbury *et al.* (1979), and deepwater snapper and grouper fisheries in the Marianas by Polovina (1985). Expansion of the fishing fleet and catch is a main objective of economic development and improvements to the shallow water harbours and deep water commercial ports at Saipan are in the planning stages. Commercial harbour facilities have been constructed on Rota and there are plans to increase harbour facilities for fishing and other purposes on Tinian, where there is one commercial port at present (Maragos, 1986a). The commercial potential of precious corals in the Northern Marianas is discussed in Grigg and Eldredge (1975). Tourist facilities in the Marianas are being expanded, with four hotel expansion or construction projects under way in 1985; 163 000 tourists visited Saipan in 1986, the majority Japanese (Eldredge *in litt.*, 1987). Two hotels have recently been built on Rota (Maragos, 1986a) and construction of a large Japanese-run hotel at Marpi on Saipan began in 1985 (Anon., 1985b).

Disturbances and Deficiencies

The impact of the volcanic eruption on Pagan is described by Eldredge and Kropp (1985). Many reefs on Saipan and elsewhere are now non-growing. The reasons for this are not known but *Acanthaster* predation or fishing with explosives have been suggested (Dahl, 1980). Alternatively, it may be owing to physical control from wave action and other disturbances from the frequent typhoons which strike this region, or long-term pollution effects of soil erosion and sedimentation from centuries of short cycle slash-and-burn cultivation on the islands, or, conceivably, to earthquakes, slumping, subsidence and volcanism through active plate tectonics in the region (Maragos *in litt.*, 10.8.87). A major typhoon hit Saipan in December 1986 (Rudolph, 1987) but its impact on the reefs is not known. During 1968 and 1971, the activities of *Acanthaster* had a devastating effect (Cheney, 1974; Chesher, 1969a; Marsh and Tsuda, 1973; Tsuda, 1971 and 1972). Its impact around Tinian is described by Jones *et al.* (1974) and around Saipan by Goreau and Isidro (1969) and most recently (1981) by Birkeland (1982).

The destructive effects of dynamite and poison as a means of collecting fish were described by Johannes (1979). World War II battles and associated military dredging and filling had major impacts on coastal environments but these have been little documented. Military activities could have an impact in the long term. There are currently threats to dredge and fill selected reefs for resort development, by overseas investors (Maragos, 1986a).

At one time the sheltered area of Maug was proposed as a transhipment or port for oil tankers but the idea met with considerable opposition and was dropped (Maragos *in litt.*, 10.8.87).

Farallon de Medinilla has been occasionally used for military target range since 1 Oct. 1978 under lease agreement with the Government of the Northern Marianas (Maragos, 1986a).

The impact of thermal effluent at Tanapag Harbour on Saipan has been studied by Doty and Marsh (1977). Concern has been expressed that excessive silt and sediment could leak into the lagoon during construction activities for a 320-room hotel development in the Marpi area of Saipan. This development was approved in 1985, after considerable modifications to the original plans following surveys by the Coastal Resource Management Office of the adjacent lagoon to help evaluate possible impacts. The Division of Environmental Quality was empowered to order changes in construction activities if damage to the lagoon occurred (Anon., 1985b). The marine environment is in general under threat from the rapidly growing, largely Japanese, tourist industry (Rudolph, 1987).

A marine environmental impact survey for a number of development projects was carried out on Tinian by Jones *et al.* (1974) and on Rota by USACE (1980). The U.S. military has an option, which is likely to be exercised soon, to develop most of the rural interior of Tinian as a military installation (Maragos *in litt.*, 10.8.87).

Legislation and Management

Traditional rights and customs relating to the marine environment appear to have been lost (Johannes, 1978 and in press). Public Law 1-8, Chapter 13, empowers the Department of Natural Resources to protect and enhance natural resources, including the marine environment. Water quality standards, including regulations controlling erosion and siltation, are the responsibility of the Division of Environmental Quality of the Northern Marianas, but enforcement is weak. There is a well organized coastal resources management program and regulatory processes, but comprehensive land use planning and controls are lacking. The Coastal Resources Management Office is funded under the U.S. Coastal Zone Management Act and its jurisdiction extends over all land areas and out to the limits of the territorial sea (Rudolph, 1987). Inventories of the coastal zone have been produced for Tinian, Saipan and Rota (Eldredge and Randall, 1980) and a natural resources study has been carried out (Coastal Zone Management Office, 1974). The Trust Territory Endangered Species Act (1975) previously protected listed species, but no longer applies (Maragos *in litt.*, 10.8.87).

Important ecosystems including coral reefs on Saipan have been afforded protective status under the Coastal Resources Management Program (CRMP) but lack enforcement (Maragos, 1986a). Guguan, Maug, Farallon de Pajaros (Uracas) and Asuncion Island are protected as uninhabited islands to be "used only for the preservation of birds, fish and wildlife and plant species" by the Constitution of the Commonwealth of the Northern Mariana Islands (Article XIV, Natural Resources), marine resources (Section 1) and uninhabited islands (Section 2) as amended in 1985. Sarigan was previously protected but has suffered serious terrestrial degradation; it is no longer protected. Under the same legislation, Managaha off Saipan is to be maintained uninhabited and used only for recreational and cultural purposes (Anon., 1985b and c). It is not clear to what extent this legislation extends to the marine environment but in any case these islands have little in the way of reefs. These areas are managed by the Division of Fish and Wildlife, Department of Natural Resources.

Public information and education are major components of the Coastal Resource Management Program. A variety of publications and audio/visual materials are available which explain the programme and offer information about issues relating to coastal management. The Coastal Resource Management Office (CRMO) has also funded preparation of a science curriculum for schools which deals with coastal and ocean resources with specific reference to local conditions in the Commonwealth. In 1985 a "Coast Week", from 7th to 14th October, was declared by the Governor of the Commonwealth, as part of the fourth annual U.S. Coastal Week, which served as a public focus for coastal management activities, including the announcement of a new citizen programme for a continuing clean-up of beaches and other private and public property in the Commonwealth (Anon., 1985b and c). Coastal resource management training programmes have also been held, organized by the CRMO in conjunction with the South Pacific Regional Environment Programme (SPREP), IUCN and the University of Guam (Anon., 1985b; Chesher, 1985).

Recommendations

There are a number of proposals for protected coastal and marine areas:

1. A 700 sq. mi. (1813 sq. km) site has been proposed for National Marine Sanctuary status by NOAA (1983) to include the waters out to 20 km from Farallon de Pajaros, Maug, Asuncion, Pagan, Guguan and Sarigan islands.
2. An area consisting of a variety of habitats in selected sites off the islands of Saipan, Rota and Tinian, as well as the waters around Aguijan and Naftan Rock has been proposed for National Marine Sanctuary status. All sites extend from high water line to the 46 m depth contour. On Saipan, the sites include the fringing reefs around Managaha Island; the barrier reef down to 46 m around the northern tip (Point Sabaneta) and south to Point Tanke; Tanapag Lagoon, the northern portion of which (Wing Beach) is a Green Turtle nesting site; and other areas. A patch reef south of the harbour on Tinian is included. On Rota, sites include the fringing reefs and submarine terrace from West

Dock south around Taipingot Peninsula to East Dock and the south-eastern portion of Sosanjaya Bay. Bird Island off Saipan and Naftan Rock off Aguijan are included for their nesting colonies of seabirds.

3. Protection was proposed for a number of coastal areas and reefs on Saipan by the District Planning Office (Dahl, 1980).

4. Protection of certain areas on the islands of Asuncion, Pagan, Alamagan, Sarigan, Anatahan and Rota has been recommended.

5. Several of the above mentioned areas are listed as separate proposed protected areas in Anon. (1985a): Naftan Pt; Kagman Peninsula and Forbidden Islands; Bird Island Reserve; Tinian; Rota; Managaha Underwater Marine Park; Rota Underwater Marine Park; and Tinian Underwater Marine Park.

Farallon de Pajaros, Maug, Guguan and Farallon de Medinilla were originally proposed as Islands for Science under the International Biological Programme (IBP). The Islands for Science programme was never pursued, although Farallon de Pajaros, Maug and Guguan are now protected, and Farallon de Medinilla is no longer considered a candidate for protection (Anon, 1985b).

Improved controls over pollution, particularly sewage, rubbish disposal, urban growth and military activities are required and explosives should be removed to prevent their use in fishing (Maragos, 1986a). Recommendations for Tinian are given in Jones *et al.* (1984).

References

* = cited but not consulted

*Amesbury, S.S. (1980). Biological studies of the coconut crab (*Birgus latro*) in the Mariana Islands. *AES College of Agriculture and Life Sciences Tech. Rept* 17. 39 pp.

*Amesbury, S.S., Lassuy, D.R., Myers, R.F. and Tyndzik, V. (1979). A survey of the fish resources of Saipan Lagoon. *Univ. Guam Mar. Lab. Tech. Rept* 52. 85 pp.

*Anon. (1978). Maug, a wildlife preserve in the Northern Marianas Commonwealth. *Pac. Sci. Assoc. Inform. Bull.* 30(2): 17-18.

Anon. (1985a). Country review - Northern Mariana Islands. *Report of the 3rd South Pacific National Parks and Reserves Conference, Apia* 3: 159-161.

Anon. (1985b). *Coastal Views* 7(2). 12 pp.

Anon. (1985c). *Coastal Views* 7(3). 12 pp.

Birkeland, C. (1982). Terrestrial runoff as a cause of outbreaks of *Acanthaster planci* (Echinodermata: Asteroidea). *Mar. Biol.* 69: 175-185.

Cheney, D.P. (1974). Spawning and aggregation of *Acanthaster planci* in Micronesia. *Proc. 2nd Int. Coral Reef Symp., Brisbane* 1: 591-594.

Chesher, R.H. (1969a). Destruction of Pacific corals by the sea star *Acanthaster planci*. *Science* 165: 280-283.

Chesher, R.H. (1969b). Divers wage war on the killer star. *Skin Diver Mag.* 18(3): 34-35, 84-85.

*Chesher, R.H. (1985). Project evaluation for the development of coastal resources in the Commonwealth of the Northern Mariana Islands: An in-country training course. South Pacific Regional Environment Programme, South Pacific Commission, Noumea, New Caledonia.

*Cloud, P.E. (1959). Submarine topography and shoal-water ecology. Geology of Saipan, Mariana Islands. Part 4. *Geol. Surv. Prof. Pap.* 280K: 361-445.

*Coastal Zone Management Office (1974). Preliminary Draft. Natural Resources Study, Commonwealth of the Northern Mariana Islands. Office of Planning and Budget, CNMI, Saipan.

Dahl, A.L. (1980). Regional ecosystems survey of the South Pacific Area. *SPC/IUCN Technical Report* 179. South Pacific Commission, Noumea, New Caledonia.

*Doty, J.E. and Marsh, J.A. (Eds) (1977). Marine survey of Tanapag Harbour, Saipan: The power barge "Impedance". *Univ. Guam Mar. Lab. Tech. Rept* 33. 147 pp.

Eldredge, L.G. (1982). Impact of a volcanic eruption on an island environment. *Pac. Sci. Assoc. Inform. Bull.* 34(3): 26-29.

Eldredge, L.G. (1983a). Mariana's active arc: A bibliography. *Univ. Guam Mar. Lab. Tech. Rept* 82. 19 pp.

Eldredge, L.G. (1983b). Summary of environmental and fishing information on Guam and the Commonwealth of the Northern Marianas: Historical background, description of the islands, and review of climate, oceanography, and submarine topography around Guam and the Northern Mariana Islands. NOAA-TM-NMFS-SWFC-40. 181 pp.

Eldredge, L.G. (1985). Aguijan revisited. Unpub. Rept. 11 pp.

Eldredge, L.G. and Kropp, R.K. (1985). Volcanic ashfall effects on intertidal and shallow-water coral reef zones at Pagan, Mariana Islands. *Proc. 5th Int. Coral Reef Cong., Tahiti* 4: 195-200.

*Eldredge, L.G. and Randall, R.H. (1980). *Atlas of the reefs and beaches of Saipan, Tinian and Rota*. Office of Coastal Resources Management of the Commonwealth of the Northern Marianas. 159 pp.

Eldredge, L.G., Tsuda, R.T., Moore, P., Chernin, M. and Neudecker, S. (1977). A natural history of Maug, northern Mariana Islands. *Univ. Guam Mar. Lab. Tech. Rept* 43. 86 pp.

*Fitzgerald, W. and Tobias, W. (1974). A preliminary survey of the marine plants of Saipan lagoon. *Univ. Guam Mar. Lab. Env. Survey Rept* 17. 20 pp.

Fosberg, F.R. (1960). The vegetation of Micronesia. 1. General description, the vegetation of the Marianas Islands, and a detailed consideration of the vegetation of Guam. *Bull. Am. Mus. Nat. Hist.* 119(1): 1-75.

Fosberg, F.R., Falanruw, M.V.C. and Sachet, M.H. (1975). Vascular flora of the Northern Mariana Islands. *Smithsonian Contr. Bot.* 22: 1-44.

*Gawel, M.J. (1974). Marine survey of Saipan Lagoon. A preliminary coral survey of Saipan Lagoon. *Univ. Guam Mar. Lab. Env. Survey Rept* 11. 13 pp.

Goreau, R.F., Lang, J.C., Graham, E.A. and Goreau, P.D. (1972). Structure and ecology of the Saipan reefs in relation to predation by *Acanthaster planci* (Linnaeus). *Bull. Mar. Sci.* 22(1): 113-152.

*Gordon, G.D. (1974). Marine survey of Saipan Lagoon. A preliminary survey of the calcareous algae of Saipan Lagoon. *Univ. Guam Mar. Lab. Env. Survey Rept* 12. 9 pp.

*Goreau, T.F. and Isidro, P. (1969). The *Acanthaster* survey of Saipan, Mariana Islands 1967. State Univ. New York, Albany Res. Foundation. AD-726-476.

*Grigg, R.W. and Eldredge, L.G. (1975). The commercial potential of precious corals in Micronesia. Part 1. The Mariana Islands. Sea Grant Publ. UGSG-75-01. 16 pp.

*Johannes, R.E. (1978). Traditional marine conservation methods in Oceania and their demise. *Ann. Res. Ecol. Syst.* 9: 349-364.

*Johannes, R.E. (1979). Improving shallow water fisheries in the Northern Mariana Islands. Northern Marianas, Coastal Zone Management. Unpub. rept.

Johannes, R.E. (in press). The role of Marine Resource Tenure Systems (TURFs) in sustainable nearshore marine resource development and management in U.S.-affiliated tropical Pacific islands. In: Smith, B.D. (Ed.), Topic Reviews in Insular Resource Development and Management in the Pacific U.S.-affiliated islands. *Univ. Guam Mar. Lab. Tech. Rept* 88.

Jones, R.S., Randall, R.H. and Tsuda, R.T. (1974). A candidate marine environmental impact survey for potential U.S. military projects on Tinian Island, Mariana Islands. *Univ. Guam Mar. Lab. Tech. Rept* 9. 143 pp.

Lindberg, D.R. and Vermeij, G.J. (1985). *Patelloida chamorrorum* spec. nov: A new member of the Tethyan *Patelloida profunda* group (Gastropoda: Acmaeidae). *The Veliger* 27(4): 411-417.

Maragos, J.E. (1986a). Coastal resource development and management in the U.S. Pacific Islands: 1. Island-by-island analysis. Office of Technology Assessment, U.S. Congress. Draft.

Maragos, J.E. (1986b). Unpublished notes on the reef corals of Garapan and nearby lagoon areas, western Saipan, northern Mariana Islands, December 1986.

*Marsh, J.A. and Tsuda, R.T. (1973). Population levels of *Acanthaster planci* in the Mariana and Caroline Islands, 1969-1972. *Atoll Res. Bull.* 170: 1-16.

*Moore, P., Raulerson, L., Chernin, M. and McMakin, P. (1977). Inventory and mapping of wetland vegetation in Guam, Tinian and Saipan, Mariana Islands. Univ. Guam Biosci. 253 pp.

NOAA (1983). Announcement of National Marine Sanctuary Program Final Site Evaluation List. *Federal Register* 48(151): 35568-35577.

Polovina, J. (1985). Variation of catch rates and species composition in handline catches of deepwater snappers and groupers in the Mariana Archipelago. *Proc. 5th Int. Coral Reef Cong., Tahiti* 5: 515-520.

Pritchard, P.C.H. (1981). Marine turtles of Micronesia. In: Bjorndal, K.A. (Ed), *Biology and Conservation of Sea Turtles*. Smithsonian Institution Press, Washington D.C. Pp. 263-274.

Randall, R.H. (1985). Habitat geomorphology and community structure of corals in the Mariana Islands. *Proc. 5th Int. Coral Reef Cong, Tahiti* 6: 261-266.

Rudolph, B. (1987). CAMP in the Northern Marianas. *CAMP Network Newsletter* Aug. 1987. P. 3.

Smith, R.O. (1947). Fishery resources of Micronesia. *Fishery leaflet* 239. Fish and Wildlife Service, U.S. Dept of the Interior. 46 pp.

*Sugawara, S. (1934). *Topography, geology and coral reefs of Rota Island*. Inst. geol. Paleontol., Tohoku Imp. Univ. (In Japanese).

*Tsuda, R.T. (Ed.) (1969). Biological results of an expedition to Rota, Mariana Islands. *Univ. Guam Mar. Lab. Misc. Rept* 4. 37 pp.

*Tsuda, R.T. (Compiler) (1971). Status of *Acanthaster planci* and coral reefs in the Mariana and Caroline Islands, June 1970-May 1971. *Univ. Guam Mar. Lab. Tech. Rept* 2. 127 pp.

*Tsuda, R.T. (1972). Proceedings of the University of Guam - Trust Territory *Acanthaster planci* (crown-of-thorns starfish) workshop. *Univ. Guam Mar. Lab. Tech. Rept* 3. 36 pp.

*Tsuda, R.T. and Tobias, W.J. (1977a). Marine benthic algae from the northern Mariana Islands, Chlorophyta and Phaeophyta. *Bull. Jap. Soc. Phycol.* 25(2): 67-72.

*Tsuda, R.T. and Tobias, W.J. (1977b). Marine benthic algae from the northern Mariana Islands, Cyanophyta and Rhodophyta. *Bull. Jap. Soc. Phycol.* 25(3): 155-158.

*USACE (U.S. Army Corps of Engineers, Pacific Ocean Division) (1980). Final Detailed Project Report and Environmental Statement: Rota Harbor, Rota, Northern Mariana Islands. Ft Shafter, Honolulu.

Vermeij, G.J., Kay, E.A. and Eldredge, L.G. (1983). Mollusks of the Northern Mariana Islands, with special reference to the selectivity of oceanic dispersal barriers. *Micronesica* 19(1-2): 27-55.

PAPUA NEW GUINEA

INTRODUCTION

General Description

Papua New Guinea comprises the eastern half of the island of New Guinea and includes all the islands making up Milne Bay Province, the Bismarck Archipelago (New Britain, New Ireland, Manus Provinces and other smaller islands) and the northern part of the Solomon Island archipelago (Bougainville and Buka). It has a total land area of 462 841 sq. km, a coastline of over 10 000 km and a marine area within the 200 mi. (320 km) Declared Fisheries Zone of 3 120 000 sq. km. The geomorphology has been described by Loffler (1977). Average temperatures are between 22°C and 31°C and annual rainfall is generally over 2500 mm (Anon., 1985b). The climate of Eastern Papua is described in Hopley and Hamilton (1973).

Within the Declared Fisheries Zone there is estimated to be a total of ca 170 000 sq. km of coralline shelf in depths of less than 200 m (Munro, 1976), and 40 000 sq. km of reef and associated shallow water in depths of 30 m or less (Wright and Kurtama, 1987; Wright and Richards, 1985). The greatest concentration of reefs is in Milne Bay Province which has an estimated 12 870 sq. km or 30% of the total (Dalzell and Wright, 1986). There is little detailed information on reef distribution and island dimensions for many areas. In the table below, details for the island groups off south-east Papua New Guinea are taken from Manser (1973a); dimensions (area/altitude) for the islands vary from chapter to chapter in this publication. Many of the smaller islands listed in the table may be seasonally or spasmodically inhabited (Lindgren *in litt.*, 22.10.87).

Table of Islands

North Solomon Is   fringing reefs but no details.

Bougainville   10 619 sq. km; volcanic high island, 3123 m; Bakawari (Pok Pok) off east coast and Taiof off northern end.

Buka   (X) 829 sq. km; 500 m.

Green (Nissan) Is   (X) Nissan and Pinipel

Takuu   atoll.

Nukumanu   (X) atoll east of Takuu.

Tulun (Carteret Gp)   (X) atoll.

Nuguria   (X) atoll.

Malum   atoll.

Louisiade Archipelago

Rossel (Yela)   (X) high island (29 x 11 km); max. alt. 920 or 839 m; surrounded by barrier reef 201 km in circumference, 80 km long and 15 km wide with few

passes, which extends in loops 32 km to west and 4 km to east to form separate lagoons; east lagoon (6 x 12 km) extends to Adele Islet, is partially choked with coral with many scattered reef patches except in west; west lagoon (8-12 x 30 km) has 3 passes, and shoal patches and rocks and is almost free of coral; on north and south coasts, barrier reef merges with fringing reef; inshore reefs mainly sediment covered; patch reefs in lagoon (Fairbridge, 1973a and b; Cahill *et al.*, 1973); some mangroves; geology described by Manser (1973b) and Hopley and Hamilton (1973); Adele (Boloba) = small limestone and coralline sand islet on barrier reef to east; Ngea (Tree Islet) = sand cay on south barrier reef.

Sudest (Tagula)   (X) high island; 70 x 15-20 km; max. alt. 800 m; geology described by Manser (1973b); early description by Davis (1922b); straight north coast; indented south coast with fringing reef twice width of north coast reef; barrier reef has relatively few passes, segments of ribbon-type reef continuous for distances of up to 40 km; lagoon depths average 40-50 m and widths vary from one to 33 km; numerous patch reefs; narrow belt of mangrove (Fairbridge, 1973a).

Quessant (Tariwerwi)   large sand cay on south barrier reef; vegetated.

Calvados Chain   (X) line of about 40 drowned mountain peaks, may represent submerged part of Owen Stanley Range; many over 200-300 m alt., with steep slopes and much landsliding; fringing reefs almost continuous except in muddy bays etc., especially well developed on exposed south coasts; mangroves on many islands; geology described by Manser (1973b) and Hopley and Hamilton (1973):

- *Utian (Brooker)*   (X) remains of crater rim to 140 m; steep slopes; fringing reefs small and limited to headlands; muddy habitat around village but some corals; surrounded by secondary barrier reef and lagoon 10-20 m deep (Fairbridge, 1973a; Cahill *et al.*, 1973);

- *Pana Rora*   (X) vertical cliffs; extensive fringing reef at west end; channels and outer reef face have wide variety of living corals (Cahill *et al.*, 1973);

- *Panatinane (Joannet, Pana tinani)*   (X) 18 x 5 km; 370 m max. alt.; fringing reef over 1000 m wide, but all inner reef sediment covered; Pananuti Pt sandy with scattered boulders with corals (Cahill *et al.*, 1973);

- *Nimoa (Pig)*   (X) 3 km long, high island, mangroves;

- *Busunlawe*   (X) high island connected by sand tombola to Nimoa; birds studied;

- *Panahubo*   high island almost connected to Nimoa at low tide;

- *Panawina*   linked to Hemanahei by extensive fringing reef; both surrounded by mangroves;

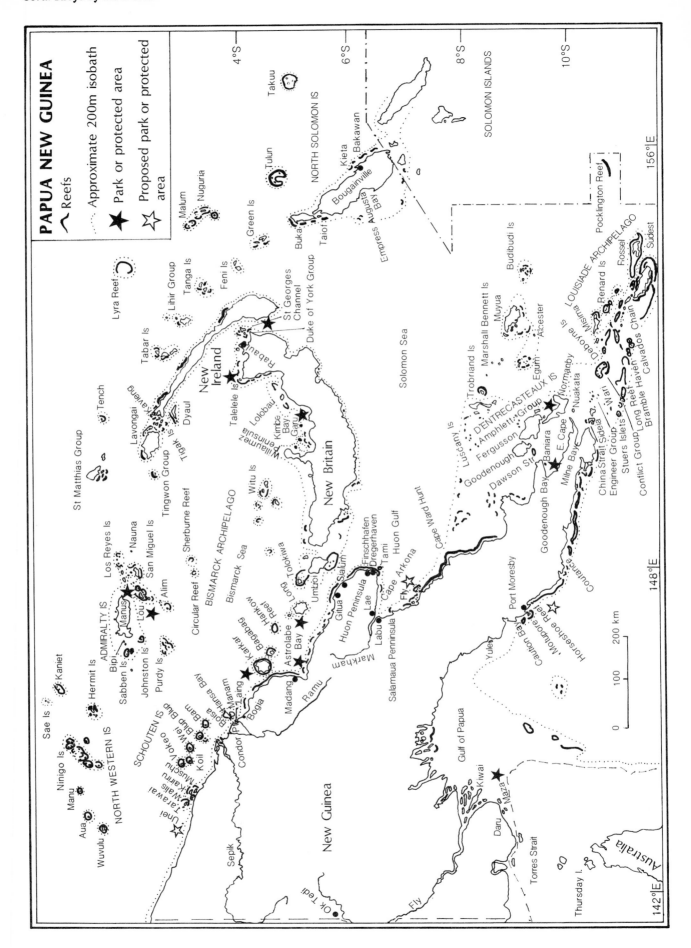

- *Yeina (Piron)* 8 x 6 km; 90 m max. alt.; mangrove fringed; fringing reef patchy inshore but better towards lagoon perhaps due to strong tidal currents; north coast sea grass sloping to sandy area with scattered corals (Cahill *et al.*, 1973);

- *Panasia* vertical cliffs to 180 m;

- *Sabari (Owen Stanley)* (X) uplifted reef limestone islet on Tawa Tawa Mal reef on north wall of Calvados barrier; smooth north shore, indented south shore, 6 x 1; fringing barrier ca 1 km offshore on north coast, 85-90% dead corals in shallow areas with scattered living corals; narrow fringing reef on south shore; mangroves (Fairbridge, 1973a; Cahill *et al.*, 1973).

  Also: Kuanak; Bagaman; Motorina; Venariwa; Ululina; Wanim.

Siwai Wa Islet low small sand cay on north Calvados barrier reef rim.

Basses Is low coral sand cays on Calvados north barrier rim including Gumai Au and Kamatal; sandy bottom at Gumai Au with scattered corals (Cahill *et al.*, 1973).

Panaroba, Pana Sagu Sagu, Pana Udi Udi low sand cays on north-west Calvados barrier reef.

Redlick Is atoll with islets on north rim; lagoon 20 m deep with no pass; separated from Deboyne Is to north-west by 1000 m wide channel.

Deboyne Is (X) complex of reefs and high islands; geology described by Manser (1973b), reefs by Fairbridge (1973a and b):

- *Panaeati (Panniet, Panniette)* elevated coral platform with extensive fringing reef shoals off south coast which merges with barrier;

- *Panapompom* (X) high island, coastal plain, coral platform, minimal fringing reef;

- *Losai* cay;

- *Nivani* (X) high island, minimal fringing reef, mainly on south;

- *Rara* cay; good example of transition from fringing to barrier reef; fringing reefs merge into 51 km long discontinuous barrier ring surrounding 20 km lagoon, 20-30 m deep.

Torlesse Is shelf atoll to west of Deboyne Is; 8 x 4 km; 3 vegetated sand cays (Pananui, Bonna Bonnawan and Tinolan) and several smaller islets.

Conflict Gp shelf-atoll reef islands and cays; lagoon 30 x 10-30 km wide (atoll 8 x 29 km); southern part of lagoon 40-56 m deep, northern half 30-36 m deep with small exposed reef patches and sand cays; ca 20 cays inc. Auriroa, Gabugabutau (dead coral 10 m from shore, large coral heads 60 m; on lagoon side coral dead except for scattered clumps), Irae (transects made on north side, live coral on reef face), Itamarina, Lutpelan, Muniara, Panaiiaii, Panasesa (cay and island; live coral rare in

shallow water; steep drop-off with luxuriant corals), Quesal, Tarapaniian, Tupit; also lagoon islets; shallow patch reef Emerald Reef lies to west (Fairbridge, 1973a; Cahill *et al.*, 1973).

Renard Is (X) islands with knife-edge ridges; inc. Kimata, Niva beno, Oreia; continuous fringing reef for 19 km.

Misima (St Aignan) (X) high island; 40 x 9 km; geology described by Manser (1973b) and Hopley and Hamilton (1973); fringing reefs narrow and fragmentary because of deep water within 1 km of coast, except east of Bwagaoia anchorage; fringing reefs off south-east point encircle Managun lagoon and create Bwagaoia harbour; no continuous barrier reef; some mangrove (Fairbridge, 1973a and b; Cahill *et al.*, 1973).

Mabui Atoll 3000 m diameter, no pass into lagoon; cay on north rim.

Jomard Is inc. Pana Rai Rai (low vegetated sand cay on east of Jomard Passage), Pana Waipona (sand cay at Jomard Entrance) on barrier reef.

Bramble Haven and Long Reef actively growing coral and algal reefs, partly exposed at low tide with several sand cays; Bramble Haven is typical barrier-rim atoll complex, ca 15 x 10 km, with lagoon (25-35 m deep) with a few coral pinnacles, and includes Duperre Islets and Panawan (all low coral sand cays); Long Reef is a large rim atoll, 39 km long (?32 x 3-11 km) with lagoon 3-10 km wide with vegetated sand cay, Lejeune, on north-east and 3 unnamed cays on north-west; Explorers Islet is a sand cay on north rim (Fairbridge, 1973a).

Laseinie Is barrier reef; inc. Dawson I.

Stuers Islets vegetated sand cays on main south barrier wall; inc. Marai and Taliwelai.

Wari (Teste) (X) high island, 3 km long; volcanic, with lagoon and barrier reef merging into a fringing reef, small but broad on south coast (Fairbridge, 1973a); four associated stacks on west = Mamamara Mama Weino (Bell Rock), Ika Ika Keino, Kera Kera and one unnamed.

D'Entrecasteaux Is

Normanby (Duau) (X) 1040 sq. km; high island, 1100 m; geology described by Manser (1973b) and Fairbridge (1973a); sedimentary and cliff shoreline; main reefs on prong extending 5 km south of Cape Chesterfield forming incipient barrier across part of entrance of Sewa Bay; leatherback turtle nesting (Quinn *et al.*, 1985b).

Dobu (X) volcanic island at entrance to Dawson Str.

Sanaroa (Welle) off south-east coast of Normanby.

Fergusson (Moratau) (X) 1340 sq. km (50 x 32 km); high active, volcanic island, 1830 m; geology described by Davis and Ives (1965) and Manser (1973b); narrow sedimentary and coral fringe; main reefs on Cape Mourilyan in south-west, headlands in south-east around Cape Doubtful and offshore islets of Kwaiope and Nekumara (Fairbridge, 1973a).

Goodenough (Morata) (X) 751 sq. km (35 x 22 km); high volcanic island (not active), 2545 m; geology described by Davis and Ives (1965) and Manser (1973b); surrounded by soft alluvial fans and shallow belt of reefs usually less than 100 m wide except in north-west where 2 km wide; short sections of fringing reefs in east; patch reefs off north and west (Bell, 1984; Fairbridge, 1973a).

Amphlett Gp (X) volcanic islands in grouping suggesting they represent peaks of remnants of crater rim; inc. Wamea (Dum Dum), Wawiwa, Tuboa, Urasi, Wata, Watota, Wawasi, Yabwaia, Yaga (high), Mumo (low), Gumatabu (cay); steep slopes to shore and only small patches of reef; larger reefs on Urasi (coral growth extensive), Wamea, Wata and Wawasi; landslides common; no offshore reefs (Fairbridge, 1973a; Cahill *et al.*, 1973).

Kwailuia low vegetated sand cay on reef platform between Lusancay and Amphlett groups.

Solomon Sea Is

Lusancay Is small, high volcanic islets and a series of low cays and reef limestones; area is thought to be an emergent shoal, exposed points of which are islands; geology described by Manser (1973b); extensive shoal areas 10 cm-1.5 m deep extend south and west from reef islets; vigorous fringing reef 100-200 m wide along many shores (Fairbridge, 1973a; Cahill *et al.*, 1973); including:

- *Boimago (Boinagi)* cay;

- *Buriwadi* limestone; leeward side has sandy sloping bottom with scattered *Porites, Seriatopora* and *Fungia*; windward side has dead coral and scattered living corals;

- *Gabwina* sand cay; sandy bottom, scattered coral growth; patch reef to south-east; seabirds;

- *Gilua* limestone;

- *Kadai* uplifted islet;

- *Kanapu* sand cay;

- *Kawa* volcanic, no reefs;

- *Kibu* limestone;

- *Konokonowana* cay;

- *Kuia* limestone;

- *Labi* cay;

- *Matagu* volcanic, minimal reef;

- *Nauria (Bare Islet)* volcanic, minimal reef;

- *Nuata (Munuwata)* cay; Green Turtle nesting (Groombridge and Luxmoore, 1987);

- *Nubiam* limestone;

- *Simlidon* Green Turtle nesting (Groombridge and Luxmoore, 1987);

- *Sim Sim* volcanic with rim of raised reef; transect made on west side;

- *Tuma* 8 km long; greater coral variety than Buriwadi; mainly rock substrate on leeward side and some massive colonies of brain coral; steep drop-off to windward with little coral; Green Turtle nesting (Groombridge and Luxmoore, 1987).

Also: Baimabu; Gwadarab; Wagalasa; Lumina.

Trobriand Is (X) low islands, four largest are coralline and algal limestone (Fairbridge, 1973a; Cahill *et al.*, 1973); includes Vakuta, Bomapau and:

- *Kiriwina* (X) large low limestone and sand island; exposed fringing reef on north-east, reef at Idateka Pt studied; shallow water, some *Porites* microatolls; Hawksbill turtle nesting (Groombridge and Luxmoore, 1987);

- *Kitava* (X) uplifted atoll, 6 km wide;

- *Muwo* low vegetated, probably limestone;

- *Kaileuna* coral reef rim islet; north and east coasts have narrow belt of fringing reef; shoal area to south in 1-5 m, reef patches, many turtles.

Marshall Bennett Is (X) uplifted atoll; inc. Kwaiawata, Gawa (174 m alt.; 5 km diam.).

Egum oval atoll, 22 km diameter; inc. Yanaba (8 km long, 65 m alt.) and 14 islets; lagoon 7-12 m deep (Fairbridge, 1973a).

Uluma Reef (Suckling) horseshoe-shaped atoll, 5 km wide, at south-east extremity of Papuan Barrier Reef; 2 passes in north; no islets; lagoon 20 m deep (Fairbridge, 1973a).

Alcester northern rim of tilted and partly uplifted atoll (Fairbridge, 1973a).

Muyua (Woodlark, Murua) (X) 56 x 11 km, max. alt. 410 m; uplifted limestone; fringing-barrier transition reef complex; fringing reefs evolve into barrier reef on west and south-west sides; several mid-lagoon reefs (Fairbridge, 1973a); Nasai off west coast, Madau off northern tip.

Budibudi Is (Laughlan, Nada) (X) horseshoe-shaped atoll, 5 km diam., 7 low cays; open to west; Cannac = 6 km wide reef a few km to the west (Fairbridge, 1973a).

Pocklington Reef 150 km east-north-east of Yela; extremely isolated ribbon reef between Milne Bay reefs and New Georgia Group of Solomon Islands (Lindgren *in litt.*, 22.10.87).

Bismarck Archipelago

New Britain (X) on west coast 40 km long barrier reef, several km offshore; scattered offshore reefs to north and south in 15-20 m depth; Kimbe Bay reefs high diversity, 50 genera corals recorded (Kojis *in litt.*, 16.12.85); reefs around Talele and Nanuk islands; leatherback turtle nesting (Quinn *et al.*, 1985b); Lolobau (X) off north-west coast; Watom (X) off northern end.

Witu Is (Vitu Is)   (X) inc. Garove, Mundua Is, Narage, Unea (Bali), Ottilean Reef, Whirlwind Reefs.

Duke of York Gp   (X) between New Britain and New Ireland; inc. Duke of York, Makada, Ulu.

New Ireland   (X) ca 8500 sq. km; volcanic high island, 2290 m; Kavieng area mapped using LANDSAT (Quinn *et al.*, 1985a); Leatherback Turtle nesting (Quinn *et al.*, 1985b).

Dyaul (Djaul)   (X) ca 100 sq. km.

Tigak Is   (X) 24 islands off west coast of New Ireland; many raised reefal limestone, others atolls; mangrove swamps; reefs (Wright and Richards, 1985; Wright *et al.*, 1983).

Lavongai (New Hanover)   (X) 1190 sq. km; volcanic, 870 m.

Tingwon Gp   (X) south-east of Lavongai.

Feni (Anir) Is   (X) inc. Ambitle, Babase, Balum.

Tanga Is   (X) inc. Boang, Malendok, Lif, Tefa.

Lyra Reef

Lihir Gp   (X) inc. Lihir, Mali, Masahet, Mahur.

Tabar Is   (X) inc. Tabar, Tatau, Mabua, Simberi.

Tench (Enus)   (X)

St Matthias Gp   (X) inc. Mussau (Green Turtle nesting (Groombridge and Luxmoore, 1987)), Eloaua, Emananus, Emirau.

Admiralty Is   many surrounded by reefs, although few details:

- *Manus and Los Negros*   (X) 1640 sq. km; volcanic, 719 m; geology and climate summarized by Kisokau (1980); Green and Sander (1979) identified the species composition of benthic, coastal, marine, intertidal and mangrove communities around Los Negros, at eastern end of Manus; south-east coast and reefs of Manus briefly described in Kisokau and Lindgren (1984); Leatherback and Hawksbill Turtle nesting (Quinn *et al.*, 1985b; Groombridge and Luxmoore, 1987);

- *Pak*   (X) uplifted limestone; south-west of Tong;

- *Tong*   (X) atoll; south-west of Los Reyes Is;

- *Los Reyes Is*   Green Turtle nesting (Groombridge and Luxmoore, 1987);

- *Nauna*   (X) uplifted limestone;

- *San Miguel Is*   (X) atoll;

- *Lou*   (X) Green Turtle nesting (Groombridge and Luxmoore, 1987); dugong (Anon., 1985a);

- *Tuluman*   0.28 sq. km; volcanic islet, emerged 1954; corals starting to colonize basalt rocks; 33 spp. recorded in initial survey; terrestrial and marine fauna and flora described in Kisokau *et al.* (1984);

- *Purdy Is*   (X) inc. Mole, Rat, Mouse, Bat; Green Turtle nesting (Groombridge and Luxmoore, 1987);

- *Sabben Is*   Green Turtle nesting (Groombridge and Luxmoore, 1987);

- *Alim*   (X) atoll;

- *Johnston Is*   uplifted limestone;

- *Bipi*   (X) uplifted limestone.

  Also: Rambutyo (X); Horno Is; Fedarb Is; St Andrew Is; Pam (Pom) Is (X); Baluan (X); Papialou Is; Circular Reef; Sherburne Reef; Mbuke Is (inc. Mbuke, Vogali) (X); Peterson Reefs; Western; Massong Is (X).

Northwestern Is

- *Kaniet (Anchorite Is)*   (X) atoll; Green Turtle nesting (Groombridge and Luxmoore, 1987);

- *Hermit Is*   (X) ca 12 islands covering 8 sq. km in 300 sq. km of sea; immature atoll with two major basaltic islands (Luf, 260 m, and Maron, 50 m) occupying the centre of the lagoon; Luf is well forested; reef front is pristine and highly spectacular with steep drop-offs; a broad channel through the reef from the west and to the north of this there are several large enclosed areas within the reef with a depth of 10-15 m which are extremely rich in corals and fish; deep water close to shore as on the north-west of Maron where a reef descends to 40 m close to shore; tridacnids abundant; well-developed perimeter reef continues around to the south and north of atoll with a broad and sparsely populated reef top; to east, reef is replaced by large areas of sand (Bell, 1984; Strauss *in litt.*, 25.2.85); also described by Gerasinov (1983); Green Turtle nesting (Groombridge and Luxmoore, 1987);

- *Ninigo Is (L'Exchequier)*   (X) a cluster of ca 20 mature atolls, including Heina, Pelelun (Pelleluhu), Ninigo, Sama, Sumasuma, Awin and Liot; formed by eruption of undersea volcano with multiple vents; surrounded by almost vertical undersea cliffs with rich fringing reefs; max. alt. about 3 m; sandy islets; about six are permanently inhabited and fishing and copra are the main sources of income; Heina has an almost land-locked lagoon, with a well-defined channel through the reef on the south-west; lagoon depth of 3 m near Kat Islet and 1 km in diameter and forms sandy bays and channels towards the east; less coral than in the Ninigo Group; mangroves are prominent, and there is a small transient population (Strauss *in litt.*, 25.2.85); Green Turtle nesting (Groombridge and Luxmoore, 1987);

- *Wuvulu (Maty)*   (X) ca 10 sq. km; circular outline; very steep cliff with numerous overhangs and underwater caves; no harbour; encircled by well developed fringing reef (Claereboudt *in litt.*, 3.8.87).

Also: Sae Is; Manu (Allison); Aua (Durour).

New Guinea Gp

New Guinea   see main text.

Islands east of China Str.   drowned mountain chain islands; narrow fringing reefs in many places usually less than 100 m wide; recent land slides on many south coasts so fewer reefs (Fairbridge, 1973a); include:

- *Logeia (Rogeia)*   steep cliffs, no reefs (Fairbridge, 1973a); between Sideia and mainland;

- *Samarai*   (X) high island between Sideia and mainland; no reefs;

- *Sariba (Hayter)*   (X) between Sideia and mainland; wider fringing reef north of Namoai Bay;

- *Sideia*   (X) wider fringing reef on north coast;

- *Engineer Gp*   (X) discontinuous barrier reef and second ring in centre; islets inc. Berri Berrije (Slade), Nara Nara wai (Skelton) and Watts with fringing reefs up to 200 m wide; Haszard (Tuatua) and Hummock at east end are lower and almost connected by fringing reef;

- *Nuakata*   small fringing reefs on headlands (Fairbridge, 1973a);

- *Nari (Mudge)*   considerable fringing reefs.

   Also: Basilaki, east of Sideia; Anagusi (Bentley).

Umboi (Rooke)   (X) ca 800 sq. km; volcanic; Sakar lies off north coast, Tolokiwa (Lokep) (X) between Umboi and Long.

Long   (X) Leatherback and Green Turtle nesting (Quinn *et al.*, 1985b; Groombridge and Luxmoore, 1987).

Karkar   (X) ca 400 sq. km; active volcanic; reefs (Dahl, 1986); Bagabag (X), Crown and Hankow Reef lie between Karkar and Long.

Manam   (X) (*see account for* Hansa Bay proposed Marine Park).

Boisa (Aris)   (*see account for* Hansa Bay proposed Marine Park).

Schouten Is (X)

- *Bam*   (X) small volcanic island with well developed fringing reef (Claereboudt *in litt.*, 3.8.87);

- *Kadover*   (X) small volcanic island;

- *Blup Blup*   (X) small volcanic island with well developed fringing reef (Claereboudt *in litt.*, 3.8.87);

- *Wei*   (X) uplifted limestone;

- *Koil*   (X) uplifted limestone;

- *Vokeo*   (X) volcanic, poorly developed fringing reef with deep water close to shore (Wright *in litt.*, 28.12.87).

Muschu   (X) Dugong *Dugong dugon* common (Dahl, 1986).

Kairiru   (X) uplifted limestone; Dugong common (Dahl, 1986).

Walis (Valif)   (X) uplifted limestone; Dugong common (Dahl, 1986).

Tarawai   Dugong common (Dahl, 1986).

(X) = Inhabited

Early work in the Papuan Islands is described in Fairbridge and Manser (1973) and Freeman *et al.* (1973) and includes Davis (1922a and b; 1928). Whitehouse (1973) summarized information on the Papua New Guinea reefs to that date; his conclusions regarding the reefs of the north-east Papua New Guinea mainland and associated islands have since been superseded (see below). Reef surveys have been carried out along the Papuan coast (Weber, 1973; Anon, 1982) and by the Fairbridge New Guinea Coral Reef Expedition, from the Geology Department of the University of Papua New Guinea. The latter study involved a reconnaissance of the reefs and lagoons of the Trobriand Islands, Lusancay Group, D'Entrecasteaux Islands, the Louisiade Group (including the Conflict Group, Deboyne Islands, Misima, Rossel and Sudest) and parts of the Papua Barrier Reef (Manser, 1973a). Reefs along the north-east mainland coast and associated islands have also been recently surveyed (Kojis *et al.*, 1984 and 1985). Some experimental mapping work has been carried out using LANDSAT satellite imagery (Quinn *et al.*, 1985a). Research on reefs is carried out by the Fisheries Department, at Port Moresby (recently moved from Lae). The Université Libre de Bruxelles has run a small research station, the King Leopold III Biological Station, on Laing Island in Hansa Bay, Madang Province, since 1974. Work here has resulted in a large number of publications on reef systematics, but until recently relatively little on reef ecology, distribution or conservation (Anon. n.d. b, and *see account for* Hansa Bay proposed Marine Park). In 1983 the University of Papua New Guinea Marine Station on Motupore Island began a long term project to map marine habitats around the island (*see separate account*).

Whitehouse (1973) distinguished three broad ecological zones along the southern coast of the mainland: to the west, in the Kiwai - Torres Strait area of Western Province, coral reef and lagoon habitats predominate, although the reefs do not appear to be prolific; the mouth of the Fly River and the Gulf Province coastline consist largely of extensive mangrove swamps bordering the Gulf of Papua which is devoid of coral growth owing to high turbidity and large freshwater inflow; and farther east the Central and Milne Bay Provinces are again dominated by reef and lagoon habitats.

The reefs of Western Province consist of numerous large patch reefs with extensive seagrass beds. There is no detailed scientific information about them. In the Port Moresby area, some 60 km of reef, including all reef environments except the fore-reef slope of large barrier reefs, was surveyed between 1962 and 1972 by Weber (1973). There is a high diversity of corals and 65 genera were described. The most profuse and diverse coral assemblages are found on the steep leeward back-reef slopes of the barrier reefs and on the nearly vertical flanks of island fringing reefs which face deep channels and passes with strong tidal currents, such as Motupore and Manunouha. The reefs in this area are also remarkable for the enormous growth forms of some corals. Coral cover is almost 100% in some areas. Weber (1973) describes the particular oceanic conditions of this area which contribute to good coral growth. The area is considered unusual in having well-developed reef structures in close proximity to a high rugged continental land mass subject to vigorous tropical rock weathering. The coastline is highly indented, the two major embayments being Port Moresby Harbour and Bootless Inlet, and the headlands are bordered by fringing reefs usually less than 100 m wide. Reefs in the vicinity of Horseshoe Reef and Motupore are described in separate accounts. A more detailed description of Bootless Inlet is given by Genolagani (1984); Watson (1983) describes the ecology of Port Moresby Harbour.

The Papuan Barrier Reef lies 5 km off the south coast and runs parallel from just west of Port Moresby eastwards to the Louisiade Barrier at about 151°E, a distance of about 563 km. There is a gap at the western end separating it from the Queensland reefs. Near Port Moresby, it encloses a lagoon which represents an ancient submerged coastal plain. This has an opening in the north-west at Caution Bay; to the south-east the lagoon narrows towards Round Point where the fringing reefs and barrier reefs meet. The barrier reef is broken by four major channels and many small ones. During low tide most of the reefs are emergent apart from a large portion of Bootless Inlet. Weber (1973) describes the sectors near Port Moresby known as Sinavi Reef and Nateava Reef. Long sectors are "sunken" forming a rim 10-20 m deep.

To the east of mainland New Guinea lie the Papuan Islands which include the Trobriand Islands, the associated Lusancay Reefs, the d'Entrecasteaux Islands, Muyua (Woodlark) and the reefs and islands of the Louisiade Archipelago (Manser, 1973a). It has been estimated that there are about 1480 km of linear reefs in this area. Manser (1973b) describes the geology. Fairbridge (1973a) describes a variety of types of reef. The Trobriand-Normanby Barrier Reefs extend approximately 400 naut. mi. (740 km) from a very poorly surveyed sector near the offshore Papua New Guinea border (8°S) around the outer Lusancays and Kiriwina and then probably to Grind River south-east of Normanby. Extensive reef-free areas are shown on navigation charts east of Fergusson but it is likely that the reef extends discontinuously south to Kiriwina and Vakuta to near Normanby Island. The East Cape-Nuakata Barrier Reefs are situated on the northern tongue of the Louisiade Shelf Reef (10°18'S, 150°53'E) and are discontinuously traceable to Blakeney Island (10°26'S, 151°13'E). Gallows Reef (around 10°17'S, 151°11'E) forms an isolated loop separating the East Cape Barrier from the Trobriand-Normanby Barrier.

The Sideia-Basilaki Barrier Reefs extend 15 naut. mi. (28 km) eastwards from the northern tip of Sideia, discontinuously to Cape Lookout on Basilaki Island.

Farther east the same trend continues through the North Louisiade Barrier Reefs. The latter include the Shortland Reefs (10°32'S, 151°08'E), the Engineer Group (10°35'S, 151°15'E), Esmerald Reef (10°37'S, 151°34'E), the Conflict Group and Torlesse Island, the Deboyne Group and Redlick Atoll. A narrow passage marks the break between this reef group and the north-west part of the Sudest Insular Barrier Reef. All these reefs are varieties of atolls. North of this reef system, and parallel to the northern Sudest Insular Barrier Reef, lie the reef-rimmed Misima and Renard Islands, the Manuga Reefs and Rossel Island with its insular barrier reef. The South Louisiade Barrier Reefs begin near Uluma or Suckling Reef (11°05'S, 150°58'E) where the Papuan Barrier Reefs are joined by a ridge that runs from near Samarai and the high islands Rogeia and Doini, through the Siriki Shoals to join the outer barriers between Wari Island and Uluma Reef. They form a barrier reef between the Coral Sea Basin to the south and the Louisiade Trough to the north which is marked by many drying reefs, numbers of islets and barrier-rim atolls (Long Reef and Bramble Haven). It is separated from the Sudest Barrier by the narrow Jomard Entrance with Jomard Reef and Island in the middle, and is about 130 mi. (209 km) long or 500 mi. (805 km) long if thought of as part of the whole Louisiade Loop.

There are two major insular barrier reefs in the area. The Calvados Barrier Reef (500 km in circumference, 180 km in length and 48 km in diameter) is part of the giant barrier loop and is one of the most varied and complex insular barrier reefs in the world. It is one of the largest atoll structures in the world and is delimited in great detail in nineteenth century Admiralty charts (Lindgren *in litt.*, 22.10.87). It is separated by only narrow channels from the rest of the New Guinea barrier system. It surrounds Sudest and the Calvados Chain from which it is separated by a lagoon 8-32 km wide, although near Rabuso Creek in the north-east of Sudest it almost merges with the fringing reef for a few km. The lagoon averages 50-60 m in depth away from reef patches and high islands, the deeper regions being almost reef-free. Around the south-east and along the south side, the barrier is essentially a simple "ribbon-reef". In the south-west there is 40 km of "sunken barrier" with depths of 10-20 m. Along the north and west sides are faroes. Passes in the barrier are mainly 10-20 m deep but a few probably exceed 50 m. The Lawik Reef in the south-east has very few passes but the Tawa Tawa Mal Reef in the west and north has passes over 1-8 km. The second insular barrier reef is the Rossel Barrier Reef (see table above) which is isolated from the main loop of the New Guinea Barrier Reefs and rises from a small separate platform, 11-16 km east and north of Sudest, separated by depths of 600-700 m.

Secondary barrier reefs are found at Utian, on the north-west side of the Calvados Barrier where the Calvados Chain intersects the barrier rim, and around a number of small isolated high islands south-east of the East Cape, such as Nuakata, the Laseinie Islands and Wari. Fringing-barrier transition reefs are found in the Deboyne Islands, Misima and Muyua. There are a number of atoll reefs in the area, including the Conflict Group (the only large shelf atoll), the Torlesse Islands,

Redlick and Mabui atolls (all shelf atolls), Uluma Reef, Long Reef, Bramble Haven and Montemont, Duchateau, Tawal and Tawa Tawa Mal atolls on the Sudest Barrier Rim Atoll (all barrier-rim atolls), Egum and Budibudi (the only oceanic atolls), and Alcester Islands, Kitava and the Marshall Bennett Islands (uplifted atolls).

Fringing reefs are found throughout this area around many of the islands including the Lusancay Group, Amphlett Group, D'Entrecasteaux Islands, the Louisiade Archipelago, Calvados Chain, Deboyne Islands, Trobriand Islands and islands off the mainland. There are also numerous patch reefs. An early description of the Louisiade Reefs is given in Davis (1922a). Corals recorded from the reefs in the Papuan Islands are listed in Cahill *et al.* (1973).

Reefs along the northern side of the Papua New Guinea mainland have never been properly surveyed and had been underestimated until recently (Kojis *et al.*, 1984 and 1985). Fringing reefs of high coral cover and species diversity are common, at least from Condor Point, in western Madang Province near the mouths of the Sepik and Ramu Rivers, eastwards to East Cape at the far eastern end of the mainland (Kojis *et al.*, 1985). Within this area, a relatively continuous fringing and barrier reef exists from Condor Point to Cape Ward Hunt in southern Morobe; Kojis *et al.* (1985) suggest that this should be called "New Guinea's Great Fringing Reef", with over one-half of the coastline of the mainland and surrounding islands in this region being fronted by reefs. Most of the reefs are of the fringing and barrier type and there is a conspicuous absence of rubble and large boulders on the crest which is not subjected to extremely high seas. Off shore reef development in this area is limited by the great depths that generally occur within a kilometer of shore (Kojis *et al.*, 1985).

Reefs fringe the shore for about 110 km of the 230 km of eastern coast south from Lae. In addition, there are 23 offshore islands with 50 km of coastline of which over 95% has reefs. The lack of large offshore shoals is probably a function of the steeply sloping nature of the off-shore sea bed. Many of the islands in the Fly (several low coral cays) and Longuerue (mainly high islands) Islands groups are surrounded by fringing reefs similar to those on the mainland. Coral cover is commonly over 70% in the top 10 m and 10% coral cover is regularly found at depths of over 30 m. Reefs around Salamaua Peninsula are described in a separate account.

The Markham River carries considerable sediment into the Huon Gulf and effectively limits reef growth as far as Cape Arkona, 45 km from its mouth, to sporadic fringing reefs, the closest of which is at Singaua, 18 km from the river mouth. The hydrology and influence of the Markham River is described in Quinn and Kojis (1982 and 1984b). From Cape Arkona to Finschhafen, fringing reefs commonly line the coast except near river mouths. Tami Atoll lies 10 km off Finschhafen, is 3 km in diameter and has three elevated limestone islets, 10 m in altitude. The lagoon is about 18 m deep with a sandy bottom and coral growth is restricted to the top 10 m depth. Strong oceanic currents sweep the atoll and the outer coral assemblages vary with exposure. Coral communities are diverse and extensive. In the Huon Gulf, 55 genera of scleractinians have been identified (Kojis *et al.*, 1985).

The geology of the upraised reefs along the north-east coast of the Huon Peninsula has been extensively studied but the living reefs have only been briefly described (Chappell, 1974). Fringing reefs are dominant from Finschhafen to Madang. The seaward reef margin is well defined and consists of a steep fore-reef slope dropping abruptly into deep water. The reefs are narrow (less than 100 m) and have no well defined lagoon, although lagoons are found in some sites such as at Gitua, Sialum and Dregerhaven. Coral communities in this region are abundant and diverse, despite the area being seismically active (Stoddart, 1972; Kojis *et al.*, 1985). Around Madang, and extending about 10 mi. (16 km) north, is a double barrier reef which delimits a large area of maximum depth 50 m with hundreds of small reef islands (Claereboudt *in litt.*, 3.8.87). Fifty-three coral genera have been identified from Madang harbour (Kojis *et al.*, 1985). From Madang north to Condor Point in Hansa Bay, a discontinuous fringing reef follows the shore. The development of this reef is very irregular and depends mainly on the nature of the shore. In areas with volcanic sand and high sediment load at river mouths there are no active reefs, while areas with rocky limestone shore often have an extensive fringing reef. At some sites the reefs show a terrace structure similar to, though on a smaller scale than, that on the Huon Peninsula described by Chappell (1974) (Claereboudt *in litt.*, 3.8.87). Reefs in Hansa Bay, including those around Laing, Boisa and Manam Islands, are described in a separate account. Octocorals of the Bismark Sea have been described by Verseveldt and Tursch (1979).

Central, Bougainville, New Ireland and Manus Provinces have extensive reefs. New Britain and East and West Sepik are characterized by extremely narrow shelves with discontinuous reef development. The reasons for this pattern of development are not clear but appear to be related to tectonic factors and the steepness and geologically unstable nature of the shorelines (Whitehouse, 1973). A U.S. National Museum expedition visited the Bismarck Archipelago in 1978 and the reefs of the Ninigo Atoll Group, Heina Atoll and Hermit Atoll were studied (Strauss *in litt.*, 1.6.84).

There are few publications on reef-associated species, apart from those resulting from work carried out at Laing Island (Anon n.d. b). A fish checklist is given in Munro (1967). Aerial surveys in the 1970s indicated that the Dugong *Dugong dugon* was widespread and locally abundant in coastal waters (Thornback and Jenkins, 1982). Leatherback *Dermochelys coriacea*, Hawksbill *Eretmochelys imbricata* and Green Turtles *Chelonia mydas* nest in several areas (Quinn *et al.*, 1985b; Groombridge and Luxmoore, 1987). Seagrasses of the Port Moresby area are described by Johnstone (1975) and mangroves by Frodin *et al.* (1975).

Reef Resources

About 13% of the total population live in rural, coastal areas (Wright *in litt.*, 28.12.87); many of these can be expected to use marine resources for some of their subsistence needs. Reefs have been traditionally exploited for food, shells and other products for use as decorations, artefacts and currency. Dalzell and Wright (1986) provide an overview of coral reef fisheries in the country. Pernetta and Hill (1981) review marine resource use on the southern coast, and mention dolphins, dugong, turtles, crocodiles, fish, crustaceans and molluscs. Traditional fishing methods are still used, although modern equipment, including monofilament nets and steel hooks, is now more common (Wright *in litt.*, 28.12.87).

Munro (1976) discussed the potential productivity of coral reef fisheries in Papua New Guinea and suggested that these could be substantially increased as stocks were virtually unexploited. The total yield in 1984 was 8344 t and the subsistence yield in coral reef areas is estimated at 7235 t/year (Dalzell and Wright, 1986). Munro (1976) estimated a sustainable yield of 45 kg/ha/yr, compared with an average yield for the whole country at present of less than 2kg/ha/yr (Wright *in litt.*, 28.12.87). Kojis *et al.* (1985) point out that in fact Munro's figures for total potential productivity are underestimated as the area of reef in Papua New Guinea seems to be considerably greater than previously thought.

Only two artisanal commercial reef fisheries in Papua New Guinea have been studied in any detail (Dalzell and Wright, 1986). One is based on the reefs around Port Moresby, where exploitation is intense (Lock, 1986a, b, c and d); the second is in the Tigak Islands, New Ireland and is described by Wright and Richards (1983 and 1985) who found that a comparatively small proportion of the available resources is taken. The tuna fishery supports a live bait fishery, principally in Ysabel Passage and Cape Lambert (Dalzell, 1984). Rapson (1973) describes fisheries in the Louisiade Archipelago and Eastern Papua. The Lusancay Reefs are particularly rich, especially for mother-of-pearl species and the Sudest and Rossel lagoons are potential good fishery sources. Chesher (1980) provides a stock assessment of commercial invertebrates of the Milne Bay reefs. The effect of coral reefs on estuarine fish communities has been studied at Labu Estuary, near the mouth of the Markham River, and Mis Inlet, Madang, by Quinn and Kojis (1984a and 1985), and the reef fisheries of the Port Moresby area are also being studied. Marine resources of Manus Island are described briefly by Kisokau (1980) and the value of the islands' reefs for food and tourism is discussed. The marine resources of New Ireland Province are reviewed by Wright *et al.* (1983).

Reef areas are important sources of the crayfish *Panulirus ornatus*, particularly around Daru, the Torres Strait and Yule Island. Commercial prawn trawlers in the Gulf of Papua have been banned, leading to increased recruitment and good catches of *P. ornatus* since 1984 for the artisanal fishery based at Daru, with 32 t of tails landed in that year, increasing to 69 t in 1985 and 74 t in 1986. Catches at Yule Island in the eastern Gulf of Papua have also increased, from a negligible catch in 1983/84 to 20 t in 1986/87 (Dalzell and Wright, 1986; Wright *in litt.*, 28.12.87).

Dalzell and Wright (1986) provide a brief overview of the bêche-de-mer fishery. Between 1960 and 1984 an average of 5.5 t was harvested per year. At the start of the century bêche-de-mer was the object of an extensive commercial fishery in the Torres Strait but this subsequently declined, in part possibly because of the disappearance of Chinese traders from the country's economy in the 1960s and 1970s. More recently production of it, along with that of trochus and green snail, has increased owing to rapid price increases for

these commodities. In the early 1980s, a considerable amount came from North Solomons Province, from Nugura, Mortlock and the Carteret Group (Dalzell and Wright, 1986; Wright *in litt.*, 28.12.87). Traditional uses of molluscs for food and shell artefacts are described by Pernetta and Hill (1981) and the shell trade is described by Wells (1982). Export of ornamental shells is a fairly recent development and is still on a small scale. A number of species (e.g. *Oliva boloui, Cypraea coxeni*) are endemic to the region (Hinton, 1979) and several are rare enough to fetch high prices. Shells are collected in the Port Moresby and Rabaul areas and from Milne Bay and Manus. Milne Bay is particularly rich, especially waters around Muyua, Rossel, Sudest, Misima and Normanby Islands. Papua New Guinea is one of the few countries where Giant Clams are still relatively abundant. They are traditionally exploited for their meat, and in some areas e.g. Manus and the Louisiade Archipelago, are reared in clam gardens (Chesher, 1980; Wells, 1982). Large quantities were being exported from Milne Bay Province in 1987, having been harvested in the Louisiade Archipelago (Anon., 1987). Papua New Guinea is a major supplier of commercial or mother-of-pearl shell, particularly *Trochus niloticus* and *Turbo marmoratus*. The trade is discussed briefly by Dalzell and Wright (1986), Glucksman and Lindholm (1982) and Wells (1982). Manus, New Ireland, New Britain and Milne Bay are important sources. Pearl farming operations existed in Port Moresby and Samarai but both ventures have been closed.

Black coral is comparatively abundant in Papua New Guinea where there has been little exploitation. It is reported still to be abundant off Central Province, Manus, New Ireland, East New Britain, the North Solomons and Milne Bay. There is a small jewellery industry using coral taken mainly from Central Province (Wells, 1982). Stony corals are collected in several areas, especially in Manus and East New Britain, for making lime. The potential for harvesting commercial sponges was investigated in Conflict Lagoon at the end of the last century but at that time was not considered very high (Rapson, 1973). Marine algal resources are described in Kudak (1987).

Tourism is comparatively poorly developed and very little of it is reef oriented. Expatriates make use of the reefs for recreational purposes and there are at least five dive tour operators, running tours out of Lae, Madang and Port Moresby on the mainland, Kimbe and Rabaul on New Britain and Kieta on Bougainville; more distant destinations include Hansa Bay (*see separate account*), Milne Bay and Wuvulu in the Bismark Archipelago (Gardiner, n.d.; Halstead, 1985; Lindgren *in litt.*, 22.10.87; Wright *in litt.*, 28.12.87).

Disturbances and Deficiencies

Cahill *et al.* (1973) suggest that storms have a considerable impact on the reefs in the Papuan Islands, the Conflict Group having been badly affected. Cyclone Annie in 1967 destroyed the western end of this lagoon (Rapson, 1973). Stoddart (1972) described the impact of an earthquake on reefs near Madang. *Acanthaster planci* has been reported periodically from the reefs, for example by Pyne (1970) and Weber and Woodhead (1970) who noted its occurrence near Port Moresby, Rabaul and Buka Island. It was not recorded in the

Papuan Islands in the early 1970s (Cahill *et al.*, 1973). Quinn and Kojis (1987) on the basis of widespread reef work in Papua New Guinea between 1980 and 1986 and a questionnaire survey concluded that *Acanthaster* was present but rare on most reefs; they saw no large aggregations and found no devastated reefs in over 800 person-hours of diving during that period. However, large numbers have been seen in Milne Bay, from Naukata Island to Samarai, for the past 13 years and some isolated reefs are reported to be completely denuded of *Acropora* although other reefs are unaffected. In 1983 a massive infestation was reported in Milne Bay which subsequently diminished; as of July 1985, a few reefs had large numbers of adult *Acanthaster* but most, including reefs which had previously been attacked, had none (Quinn and Kojis, 1987).

Although the reefs of Papua New Guinea are virtually pristine compared to those of many countries, they are coming under increasing threat. Deforestation and poor land management is leading to increased soil erosion. As yet this has had little effect on the coastal environment in terms of increased siltation loads but this problem may lie not far off in the future. Although there have been no major oil spills, there have been localized incidents, particularly in the Milne Bay area with non-tanker traffic, which have caused oil damage to reefs (SPREP, 1980). Reefs in Madang Province are potentially threatened by a possible oil field in the Ramu-Sepik basin (Claereboudt *in litt.*, 3.8.87). Tailing effluent from the Bougainville copper mine is known to have caused some problems in the Empress Augusta Bay area, but other reefs have not been affected by industrial activities (SPREP, 1980). Port Moresby Harbour, however, is already subject to pollution and physical degradation, and Bootless Bay could be affected. These areas are likely to suffer increasingly heavy and conflicting use for industrial, shipping, recreational and fishery purposes and to be variously affected by problems of effluent disposal from the growing urban and industrial areas. There is already evidence of high levels of sewage pollution on the foreshore area from Koki to Hanuabada (Munro, n.d. b). Transport of copper ore from the Ok Tedi mines into Port Moresby harbour is a further potential threat (Genolagani, 1984). There is some evidence that the reefs of Port Moresby Harbour are degraded by dredging (Gilmour, 1978).

Except in some urban areas, reefs and lagoons have remained relatively free from over-exploitation for either subsistence or commercial purposes (Lindgren *in litt.*, 22.10.87; SPREP, 1980). With the exception of Port Moresby, there is little collection of shells and reef products for the tourist trade. From time to time there has been interest in a more intensive exploitation by outside investors but none of these ventures has been approved as they are seen to be more appropriate as village-based industries (SPREP, 1980). However there are some signs of over-exploitation. Poaching of marine resources such as black coral, shells and clams by Japanese and Taiwanese fishing boats occurs in many areas (Wells, 1982) and there is potential for overcollection of the Giant Clam stocks. Depletion of shells has been inferred from results of a study at Pari Village (Swadling, 1980) and of holothurians from a study at Aroma (Genolagani, 1984). There is some evidence that the holothurian *Microthele nobilis* has been overfished in the Cartaret Islands in North Solomons Province, with a marked decline in harvest between 1982

and 1984 (Dalzell and Wright, 1986; Wright *in litt.*, 28.12.87). Deterioration of reefs caused by fishing with explosives is widespread and increasingly common along the north coast of mainland Papua New Guinea and in the northern islands. Explosives, from road construction and mining works, and 2nd World War ammunition dumps, are abundant and popularly used, despite being illegal and known to be very dangerous (Claereboudt *in litt.*, 3.8.87; Wright *in litt.*, 28.12.87). Deterioration of some of the reefs in the Rabaul area may be due to overcollection of stony corals for the lime industry (Wells, 1982). Turtle over-exploitation is of major concern. Fishery problems are briefly discussed in SPREP (1980).

Legislation and Management

Traditional restrictions on fishing and collecting shells, marine tenure and other traditional customs and rights are described in Eaton (1985), Morauta *et al.* (1980), Polunin (1984b), Bulmer (1982), Wright and Richards (1983) and Wright (1985). Customary rights, or ownership, including fishing rights, in connection with the sea, reef, seabed, rivers and lakes are specifically recognized in the Native Customs (Recognition) Act of 1963 (Chapter 19, Section 5b) (Wright *in litt.*, 28.12.87). In the Ninigo Islands, the sea and marine resources are believed to belong to all men, but many other villages have a system of marine tenure (Wright, 1985). Traditional reserves have been reported on the northern coast and islands such as the Trobriand Islands (Malinowski, 1918), Tanga Islands, off New Ireland (Bell, 1953-54), New Britain (Panoff, 1969-70) and Ponam Island, off Manus (Carrier and Carrier, 1983). In New Ireland, coastal land owners also own the adjacent marine areas and fishing is prohibited in the tenured area for a certain period of time after the death of the owner (Wright, 1985). Polunin (1984b) discusses the conservation effects of such areas on marine resources and concludes that traditional ownership often presents problems in terms of modern conservation and management. Traditional rights are now often passed to young people remaining in the village who show the most potential for using them, whether or not they are blood relations.

The Torres Strait Treaty, a bilateral treaty between Papua New Guinea and Australia, which was signed 18.12.78, amended in 1984 and ratified 15.2.85, primarily deals with the definition of the boundaries between the two countries but also provides, among other things, for the establishment of a Protected Zone in which traditional resource exploitation can continue (Antram *in litt.*, 9.11.87; Baines, 1982; Haines, 1983). Fisheries resource and management issues are discussed in Haines (1983). A Torres Strait liaison officer is based at Thursday Island on the Australian side of the strait (Antram *in litt.*, 9.11.87).

General environmental planning, conservation and legislation is discussed in SPREP (1980), Eaton (1985) and Anon. (1985b). The National Parks Act (1982) allows for the establishment of a variety of types of protected areas. The Conservation Areas Act (1978) allows for the establishment of conservation areas but has not yet been implemented; such areas would be the responsibility of local management committees. The Fauna (Protection and Control) Act (1966) provides for the establishment of Wildlife Sanctuaries and Wildlife Management Areas. The latter are declared only at the request of landowners and are designed to assist customary landowners to control wildlife resource exploitation; regulations are drawn up by the landowners themselves and may be based on local conditions and practices. Wildlife Management Areas are considered the most appropriate form of protected area in Papua New Guinea, and efforts are focused on declaring these (Lindgren *in litt.*, 22.10.87).

The Wildlife Division and the National Parks Service of the Department of Environment and Conservation are responsible for the establishment of marine nature reserves. The National Park Service has declared a number of areas as marine reserves but these have never been implemented. The Wildlife Division is responsible for the establishment of Wildlife Management Areas for Dugongs and marine turtles and other coastal resources. Genolagani (1984) and Eaton (1985) describe the following existing protected areas as including coral reefs:

Declared Marine Parks/Reserves

Management established:

E. New Britain
- Talele Island Nature Reserve: 40 ha; 8 islands; coralline; established 1.10.73; complete protection of fauna and flora including surrounding marine life; seabirds, turtles, mangroves, fringing and lagoon reefs.
- Nanuk Island Provincial Park: 14 ha (4 ha land); in St George's Channel; complete protection; established 26.11.73.
W. New Britain
- Garu Wildlife Management Area: Willuamez Peninsula; 16 997 ha; established 1976; turtles, birds; no information on reefs; general information in Liem (1976).
Manus Province
- Lou Island Wildlife Management Area: mainly for turtle and dugong; described in Anon. (1985a).
- Ndrolowa Wildlife Management Area; eastern Manus I.; covers variety of wildlife resources, including reef fisheries; described in Kisokau and Lindgren (1984).
Western Prov.
- Maza Wildlife Management Area: 184 230 ha; entirely marine; includes large area of reef to west of mouth of Fly River; established mainly for Dugong; management problems described by Eaton (1985).
Milne Bay
- Sawataetae Wildlife Management Bay: north side of Normanby; 700 ha; birds, mangroves, forest; established mainly to control hunting.
- Baniara Island Protected Area.
Madang
- Ranba (Long Island) Wildlife Man. Area: 41 992 ha; established mainly for turtles.
Karkar Island
- Bagiai Wildlife Management Area: 13 760 ha; high volcanic peak, reefs, islands; fishing controls.

No established management:

Central Prov.
- Horseshoe Reef (Tahira, Aioro) Marine Park.

E. Sepik
- Unei Island Village Reserve.
Morobe
- Fly Island Marine Park.

The following park has been recently created (Claereboudt *in litt.*, 3.8.87):

Madang
- Macclay Park, south Astrolabe Bay.

It is not known if there is any established management for this area.

The Fisheries Act (Chapter 214) defines the powers of the Minister and officers of the Department of Fisheries and Marine Resources in relation to licensing and fisheries management and applies to all fisheries resources except whales and sedentary organisms. The Continental Shelf (Living Natural Resources) Act (Chapter 210) controls the harvesting of sedentary organisms such as clams, trochus and Gold-lip Pearl shell (Lindgren *in litt.*, 22.10.87; Wright *in litt.*, 28.12.87). Harvest can only be carried out by citizens of Papua New Guinea or non-citizens normally resident in the country (Wright *in litt.*, 28.12.87). The export of shells for commercial purposes by expatriates is prohibited and shells can now only be marketed by nationally owned companies. Dynamite fishing is prohibited but still occurs in some areas (Claereboudt *in litt.*, 3.8.87; Kwapena *in litt.*, 25.3.85). Papua New Guinea has signed the Convention on Conservation of Nature in the South Pacific.

An oil spill contingency plan is in effect and some equipment is available at all major ports but a more sophisticated plan will be required in the future. The Papua New Guinea University of Technology is running a pilot monitoring project in the Huon Gulf to study the effects of mining, forestry and agricultural activities in the area, sponsored by the South Pacific Regional Environment Programme. Baseline studies of pollutants in Port Moresby Harbour and the Huon Gulf are being undertaken under the SPREP Coastal Water Quality Monitoring and Control programme.

The Department of Fisheries and Marine Resources is conducting a survey of shell and other invertebrate marine resources, initially in New Ireland but later to include areas such as Manus and Milne Bay. Resources receiving attention include bêche-de-mer, Trochus, Green Snail, Black-lip Pearl shell and Giant Clam (Wright *in litt.*, 28.12.87). A Giant Clam mariculture project is under way at Motupore Island (*see separate account*) as part of the International Giant Clam Mariculture Project, and a facility has recently been set up at the Christensen Research Institute (Anon., 1986).

Recommendations

Hill (1977) discussed proposals for marine parks in the 1970s, but these have since been superseded. The following are listed by Genolagani (1984):

Proposed Marine Parks/Reserves

Manus
- Wuvulu Is, Ninigo Is, Hermit Is, Western Is, Sabben Is, Alim Is.

New Ireland
- St Matthias Group; islands between Lavongai and Kavieng; Daul Is; Tabar Is; Lihir Group; Tanga Is; Feni Is.

N. Solomons
- Pinipel-Nissan Group; Kulu, Manus, Passau.

W. New Britain
- Hoskins Bay; Cape Anukur.

W. Sepik
- Tumleo Ali, Seleo and Angel.

E. Sepik
- Schouten Islands.

E. Sepik/Madang
- Hansa Bay (*see separate account*).

Madang
- Manam; Madang*.

Morobe
- Tami Is; Labu Lakes; Salamaua Peninsula.

Northern Prov.
- Mangrove Island; Cape Nelson.

Milne Bay
- Trobriand Is; Muyua; Goodenough; Fergusson; Normanby; Pocklington; Misima; Yela Is (Rossel); Calvados Chain; Conflict Group; Wari Is; Milne Bay Is.

Central Prov.
- Coutance Is; Papuan Barrier Reef; Motupore (*see separate account*); Idler's Bay.

National Capital
- Taurama Beach Recreational Park.

*Madang Park was to have included Madang Harbour, the double barrier reef and associated reef islands. However, Madang, the capital of the province, would have been completely contained within the park, with all the economic and industrial restrictions implied by this. Political changes in the province in 1985 led to the project being dropped (Claereboudt *in litt.*, 3.8.87).

There are plans for the Dept of Defence to assist in surveying marine and reef areas (Kisokau pers. comm., June 1985). Tuluman Island, south of Manus, a volcanic islet which emerged in the 1950s, is of considerable scientific interest and monitoring of coral and reef development would be worthwhile (Kisokau *et al.*, 1984).

Coastal zone management has not as yet received high priority and with increased interest in fishing opportunities (particularly by external powers) and harvesting of reef resources, administrative attention must be given to this issue, particularly in areas of high population density. Rapson (1973) makes some general recommendations for future research and fisheries management projects. As in most countries, improved reef management will depend on increased public awareness of the value of reefs (for example tourism), funding for manpower and acquisition of skills to assess areas of significance, and support from the government or outside funding agencies to implement the above.

References

* = cited but not consulted

*Anon. (1982). Coral reefs. In: Ok Tedi Environmental Study. Port Moresby Harbour Studies, Townships and Regional Review. 7: 13-82.

Anon. (1985a). Lou Island - Landowners Managed Area. *Report of 3rd South Pacific National Parks and Reserves Conference, Apia, 1985* 1: 253-260.

Anon. (1985b). Country Review - Papua New Guinea. *Report of 3rd South Pacific National Parks and Reserves Conference, Apia, 1985* 3: 175-194.

Anon. (1986). University of Papua New Guinea. *Clamlines* 1: 7-8.

Anon. (1987). Papua New Guinea giant clam fishery. *Clamlines* 2: 7.

Anon. (n.d. a). General information on the Motupore Island Research Station. Motupore Island Research Department, Univ. Papua New Guinea.

Anon. (n.d. b). Publications effectuées à partir de matériel récolté à Laing. Unpub. rept. 13 pp.

*Baines, G.B.K. (1982). Traditional conservation practices and environmental management: The international scene. In: Morauta *et al.* (Eds). Pp. 45-57.

*Bell, F.L.S. (1953-54). Land tenure in Tanga. *Oceania* 24: 28-57.

Bell, H.L. (1984). Data sheets on Goodenough, Hermit and Ninigo Is. Unpub. International Council for Bird Preservation, Cambridge, U.K.

*Bouillon, J., Claereboudt, M. and Seghers, G. (n.d.). Hydroméduses de la baie de Hansa. Publication of the King Leopold III Biological Station, Free University of Brussels.

Braley, R.D. (1984). Reproduction in the Giant Clams *Tridacna gigas* and *T. derasa in situ* on the North-Central Great Barrier Reef, Australia and Papua New Guinea. *Coral Reefs* 3(4): 221-227.

*Bulmer, R.N.H. (1982). Traditional conservation practices in Papua New Guinea. In: Morauta *et al.* (Eds).

*Carrier, J.G. and Carrier, A.H. (1983). Profitless property: Marine ownership and access to wealth on Ponam Island, Manus Province. *Ethnology* 2: 133-151.

Chappell, J. (1974). Geology of coral terraces, Huon Peninsula, New Guinea: A study of Quaternary tectonic movements and sea-level changes. *Bull. Geol. Soc. Am.* 85: 553-570.

Chesher, R.H. (1980). Stock assessment of commercial invertebrates of Milne Bay coral reefs. Fisheries Division, Dept Primary Industries, Papua New Guinea. 56 pp.

*Claereboudt, M.R. and Bouillon, J. (n.d.). Coral associations distribution and diversity on Laing Island reef (Papua New Guinea). Publication of the King Leopold III Biological Station, Free University of Brussels.

Coleman, N. (1980). *Post Courier*, 25th August.

Coleman, N. (1982). Tahira Marine Park. *Underwater* 2: 8-12. Sea Australia Productions Pty Ltd, Caringbah, N.S.W., Australia.

*Coppejans, E. and Meinesz, A. (n.d.). A. The Caulerpales of Hansa Bay area (Province Madang - P.N.G.). The genus *Caulerpa*. Publication of the King Leopold III Biological Station, Free University of Brussels.

Dahl, A.L. (1986). *Review of the Protected Areas System in Oceania*. IUCN/UNEP. 239 pp.

Dalzell, P.J. (1984). The population biology and management of bait fish in PNG waters. *Res. Rept, Dept Primary Ind.*, Port Moresby 83-03. 24 pp.

Dalzell, P. and Wright, A. (1986). An assessment of the exploitation of coral reef fishery resources in Papua New Guinea. In: Maclean, J.L., Dizon, L.B. and Hosillos, L.V. (Eds), *The First Asian Fisheries Forum*. Asian Fisheries Society, Manila, Philippines.

Davis, W.M. (1922a). Coral reefs of the Louisiade Archipelago. *Proc. Nat. Acad. Sci. U.S.A.* 8: 7-13.

Davis, W.M. (1922b). The barrier reef of Tagula, New Guinea. *Assoc. Am. Geogr. Ann.* 12: 97-151.

Davis, W.M. (1928). The coral reef problem. *Am. Geogr. Soc. Spec. Publ.* 9. 596 pp.

Davis, H.L. and Ives, D.J. (1965). The geology of Fergusson and Goodenough Islands, Papua. *Aust. Bur. Min. Resour. Rept.* 82: 1-65.

Eaton, P. (1985). Land Tenure and Conservation: Protected areas in the South Pacific. *SPREP Topic Review* 17. South Pacific Commission, Noumea, New Caledonia.

Fairbridge, R.W. (1973a). Morphology of the reefs. Chap. 7. In: Manser (1973a).

Fairbridge, R.W. (1973b). Geomorphology of the reef islands. Chap. 5. In: Manser (1973a).

Fairbridge, R.W. and Manser, W. (1973). Introduction and Narrative. Chap. 1. In: Manser (1973a).

Freeman, C., Fairbridge, R.W. and Manser, W. (1973). Historical Review. Chap. 2. In: Manser (1973a).

*Frodin, D., Huxley, C. and Kirina, K. (1975). Mangroves of the Port Moresby Region. *Dept. Biol. Univ. Papua New Guinea Occ. Pap.* 3. 53 pp.

Gardiner, T. (n.d.). Capital delights. *Paradise* 29: 31-32

Genolagani, J.M.G. (1984). An assessment on the development of marine parks and reserves in Papua New Guinea. In: McNeely, J.A. and Miller, K.R. (Eds), *National Parks, Conservation, and Development: The Role of Protected Areas in Sustaining Society.* Smithsonian Institution Press, Washington D.C. Pp. 322-329.

*Gerasinov, E. (1983). Volcanic islands - the Hermit Atoll in the New Guinea Sea. Paper presented at 15th Pacific Science Congress, Dunedin, New Zealand.

*Gilmour, A.J. (1978). Report on dredging environmental study: Port Moresby. Papua New Guinea Harbours Board.

Glucksman, J. and Lindholm, R. (1982). A study of the commercial shell industry in Papua New Guinea since World War Two with particular reference to village production of trochus (*Trochus* sp.) and Green Snail (*Turbo marmoratus*). *Science in New Guinea* 9(1): 1-10.

*Green, W. and Sander, H. (1979). Manus Province Tuna Cannery Environmental Study. Office of Environment and Conservation, Papua New Guinea.

Groombridge, B. and Luxmoore, R. (1987). The Green Turtle and Hawksbill (Reptilia: Cheloniidae). World Status, Exploitation and Trade. Draft Report to the CITES Secretariat. IUCN Conservation Monitoring Centre, Cambridge, U.K.

Haines, A.K. (1983). Fisheries management under the Torres Strait treaty. *Aus. Fish.* 42: 31-37.

Halstead, B. (1985). Adventures in Milne Bay. *Paradise* 50.

Hill, M.A. (1977). Marine Parks in Papua New Guinea. *Collected Abstracts and papers of the International Conference on Marine Parks and Reserves.* Tokyo, May 1975. Sabiura Marine Park Research Station, Japan.

Hinton, A. (1979). *Guide to Shells of Papua New Guinea*. Robert Brown and Assocs. Pty Ltd, Papua New Guinea.

Hopley, D. and Hamilton, D. (1973). Geomorphology of the high islands. Chap. 4. In: Manser (1973a).

Johnstone, I. (1975). The seagrasses of the Port Moresby region. *Dept. Biol. Univ. Papua New Guinea Occ. Pap.* 7. 38 pp.

Kisokau, K.M. (1980). *Manus Province - a Biophysical Resource Inventory*. Office of Environment and Conservation, Boroko, Papua New Guinea.

Kisokau, K.M. and Lindgren, E. (1984). Nrdolowa wildlife management area, a report on proposals to establish a wildlife management area for a variety of wildlife resources in Manus Province. Ofice of Environment and Conservation, Papua New Guinea.

Kisokau, K.M., Pohei, Y. and Lindgren, E. (1984). Tuluman Island after thirty years. Office of Environment and Conservation, Boroko, Papua New Guinea.

Kojis, B.L. (1986). Sexual reproduction in *Acropora (Isopora)* (Coelenterata: Scleractinia). 2. Latitudinal variation in *A. palifera* from the Great Barrier Reef and Papua New Guinea. *Marine Biology* 91: 311-318.

Kojis, B.L. and Quinn, N.J. (1984). Seasonal and depth variation in fecundity of *Acropora palifera* at two reefs in Papua New Guinea. *Coral Reefs* 3: 165-172.

Kojis, B.L., Quinn N.J. and Claereboudt, M.R. (1985). Living coral reefs of northeast New-Guinea. *Proc. 5th Int. Coral Reef Cong., Tahiti* 6: 323-328.

Kojis, B.L., Quinn N.J., Claereboudt, M.R., and Tseng, W.Y. (1984). Coral reefs of the Huon Gulf and Hansa Bay, Papua New Guinea. Paper presented at the joint meeting of The Atlantic Reef Committee and the International Society for Reef Studies, *Advances in Reef Science*, Miami, Florida, October 1984.

Kudak, M.M. (1987). Marine algal resources in Papua New Guinea. In: Furtado, J.I. and Wereko-Brobby, C.Y. (Eds), *Tropical Marine Algal Resources of the Asia-Pacific Region: A Status Report*. Commonwealth Science Council Technical Publication 181. Pp. 81-89.

*Liem, D. (1976). Report on the habitat survey and habitat assessment of Garu Wildlife Management Area, West New Britain. Dept of Natural Resources, Wildlife Branch, P.O. Box 2585, Kanedobu.

*Lock, J.M. (1986a). Economics of the Port Moresby artisanal reef fishery. *Tech. Rept Dept Primary Ind. (Papua New Guinea)* 86/4. 35 pp.

*Lock, J.M. (1986b). Effects of fishing pressure on the fish resources of the Port Moresby Barrier and fringing reefs. *Tech. Rept Dept Primary Ind. (Papua New Guinea)* 86/3. 31 pp.

*Lock, J.M. (1986c). Fish yields of the Port Moresby barrier and fringing reefs. *Tech. Rept Dept Primary Ind. (Papua New Guinea)* 86/2. 17 pp.

*Lock, J.M. (1986a). Study of the Port Moresby artisanal reef fishery. *Tech. Rept Dept Primary Ind. (Papua New Guinea)* 86/1. 56 pp.

Loffler, E. (1977). *Geomorphology of Papua New Guinea*. Australian National University Press, Canberra.

*Malinowski, B. (1918). Fishing and fishing magic in the Trobriand Islands. *Man* 18: 87-92.

Manser, W. (Ed.) (1973a). New Guinea Barrier Reefs: Preliminary results of the 1969 coral reef expedition to the Trobriand Islands and the Louisiade Archipelago, Papua New Guinea. *Dept Geol. Univ. Papua New Guinea Occ. Pap.* 1. 356 pp.

Manser, W. (1973b). Geological setting and geology of the islands. Chapter 3. In: Manser (1973a).

Morauta, L., Pernetta, J. and Heaney, W. (Eds) (1980). *Traditional Conservation in Papua New Guinea: Implications for today*. IASER Discussion Paper 35, Institute of Applied Social and Economic Research, PNG.

*Munro, I.S.R. (1967). *The Fishes of New Guinea*. Dept. Agriculture, Stock and Fisheries, Port Moresby.

Munro, J. (1976). Potential productivity of coral reef fisheries. In: Lamb, K.P. and Gressitt, J.L. (Eds), *Ecology and Conservation in Papua New Guinea*. Wau Ecology Institute Pamphlet 2. Pp. 90-97.

Munro, J.L. (n.d. a). Motupore Island Research Centre - on the Papuan Barrier Reef.

Munro, J.L. (n.d. b). A proposal for study of the Papuan coastal lagoon system in the vicinity of Port Moresby, Papua New Guinea. Proposal prepared for the Biology Dept, Univ. Papua New Guinea.

*Panoff, M. (1969-70). Land tenure among the Maenge of New Britain. *Oceania* 40: 177-194.

Pernetta, J.C.and Hill, L. (1981). A review of marine resource use in coastal Papua. *J. Soc. Océan.* 72-73: 175-191.

Polunin, N. (1984a). Annual report of the Motupore Island Research Department for 1983.

Polunin, N.V.C. (1984b). Do traditional marine "reserves" conserve? A view of Indonesian and Papua New Guinean evidence. In: Ruddle, K. and Akimichi, T. (Eds), *Maritime Institutions in the Western Pacific*. Senri Ethnological Studies 17. Pp. 15-31.

Pyne, R.R. (1970). Notes on the crown-of-thorns starfish: Its distribution in Papua and New Guinea (Echinodermata: Asteroidea: Acanthasteridae). *Papua and New Guinea Agricultural Journal* 21: 128-138.

Quinn, N.J., Dalzall, P. and Kojis, B.L. (1985a). LANDSAT as a management tool for mapping shallow water habitats in Papua New Guinea. *Proc. 5th Int. Coral Reef Cong., Tahiti* 6: 545-550.

*Quinn, N.J. and Kojis, B.L. (1982). The hydrology of the Markham River intrusion into the Huon Gulf using LANDSAT imagery and *in situ* observations. *Science in New Guinea* 9(3): 115-129.

Quinn, N.J. and Kojis, B.L. (1984a). Does the presence of coral reefs in proximity to a tropical estuary alter the estuarine fish assemblage? Paper presented at the joint meeting of the Atlantic Reef Committee and the International Society for Reef Studies, *Advances in Reef Science*, Miami, Florida, October 1984.

*Quinn, N.J. and Kojis, B.L. (1984b). Remote sensing of the Markham River intrusion into the Huon Gulf, Papua New Guinea. *Proc. 3rd Australian Remote Sensing Conference, Queensland* 1984: 740-744.

Quinn, N.J. and Kojis, B.J. (1985). Does the presence of coral reefs in proximity to a tropical fish estuary affect the estuarine fish assemblage? *Proc. 5th Int. Coral Reef Cong., Tahiti* 5: 445-450.

Quinn, N.J. and Kojis, B.L. (1987). Distribution and abundance of *Acanthaster planci* in Papua New Guinea. *Bull. Mar. Sci.* 41(2): 688-691.

Quinn, N.J., Kojis, B.L., Angaru, B., Chee, K., Keon, O. and Muller, P. (1985b). The status and conservation of a newly "discovered" Leatherback Turtle (*Dermochelys coriacea* Linneaus, 1766) chelonery at Maus Buang, Papua New Guinea. *Report of the 3rd South Pacific National Parks and Reserves Conference, Apia* 2: 90-99.

Rapson, A.M. (1973). Fisheries. Chapter 10. In: Manser (1973a).

SPREP (1980). Papua New Guinea. *Country Report* 10. South Pacific Commission, Noumea, New Caledonia.

*Stoddart, D.R. (1972). Catastrophic damage to coral reef communities by earthquake. *Nature* 239: 51-52.

*Swadling, P. (1980). Shell fishing and management in Papua New Guinea. In: Morauta *et al.* (1980).

Thornback, L.J. and Jenkins, M.D. (1982). *The IUCN Mammal Red Data Book Part 1*. IUCN, Gland and Cambridge. 516 pp.

*Tursch and Tursch (1982). Soft corals in Hansa Bay. Publication of the King Leopold III Biological Station, Free University of Brussels.

UPNG (Univ. of Papua New Guinea) (1984). Welcome to Motupore Island. (Brochure).

*Verseveldt, J and Tursch, A. (1979). Octocorallia from the Bismark Sea. Part 1. *Zool. Mededelingen* 54: 133.

*Watson, J.E. (1983). Ecology of Port Moresby Harbour, Papua New Guinea. Paper presented at 15 Pacific Science Congress, Dunedin, New Zealand.

Weber, J.N. (1973). Reef corals and coral reefs in the vicinity of Port Moresby, south coast of Papua New Guinea. *Pac. Sci.* 27: 377-390.

*Weber, J.N. and Woodhead, P.M.J. (1970). Ecological studies of the coral predator *Acanthaster planci* in the South Pacific. *Mar. Biol.* 6: 12-17.

Wells, S.M. (1982). Marine conservation in the Philippines and Papua New Guinea with special emphasis on the ornamental coral and shell trade. Report on a visit sponsored by the Winston Churchill Memorial Trust.

*Whitehouse, F.W. (1973). Coral reefs of the New Guinea Region. In: Jones, O.A. and Endean, R. (Eds), *Biology and Geology of Coral Reefs. Vol. 1*. Academic Press, N.Y. and London. Pp. 169-186.

Wright, A. (1985). Marine resource use in Papua New Guinea: Can traditional concepts and contemporary development be integrated? In: Ruddle, K. and Johannes, R.E. (Eds), *The Traditional Knowledge and Management of Coastal Systems in Asia and the Pacific*. Unesco/ROSTEA, Jakarta. Pp. 79-99.

Wright, A., Chapman, M.R., Dalzell, P.J. and Richards, A.H. (1983). The marine resources of New Ireland Province. A report on present utilization and potential for development. *Res. Rept, Dept Primary Ind., Port Moresby* 83-13. 54 pp.

*Wright, A. and Kurtama, Y.Y. (1987). Man in Papua New Guinea's Coastal Zone. *Resource Management and Optimization* 4(3-4): 261-296.

*Wright, A. and Richards, A.H. (1983). The yield from a Papua New Guinea reef fishery. Rep. No. 83-07. Dept Primary Industry, Fisheries Div., Port Moresby.

Wright, A. and Richards, A.H. (1985). A multispecies fishery associated with coral reefs in the Tigak Islands, Papua New Guinea. *Asian Marine Biology* 2: 69-84.

HANSA BAY PROPOSED MARINE PARK (NATURAL RESERVE)

Geographical Location   The western part of the Bismarck Sea from the mouth of the Sepik River to Bogia, ca 50 km south-east, in Madang and East Sepik Provinces on the northern coast of mainland PNG. Includes Manam, Boisa and Laing Islands. Manam is at 4°08'S, 145°05'E; Laing Island is in the middle of Hansa bay, 25 km south-east of the mouth of the Ramu River and 40 km south-east of the mouth of the Sepik River, at 4°10'S, 144°52'E.

Area, Depth, Altitude   The average depth of Hansa Bay is 25 m; Laing Island is 850 m long and max. 150 m wide; max. alt. 1 m.

Land Tenure   Laing Island is privately owned.

Physical Features   Laing Island is the only emergent portion of a reef string aligned along a north-south axis in Hansa Bay; this string may represent an old coastline which developed during the Holocene transgression. A small open lagoon (400 x 200 m) lies along the western shore of Laing Island. Surface sea temperatures lie in the range 27-32°C, with salinity of ca 34 ppt. Tides are semi-diurnal with a maximum range of 1.2 m. Two meteorological seasons influence sea water quality in Hansa Bay. During the rainy season from November to April, the water comes from the north and is often muddy and very rich in drift materials, such as branches, logs, aquatic ferns etc., released by the Sepik and Ramu Rivers. A strong swell breaks then on the northern part of the reef. During the dry season, the current comes from the south-east, bringing clear water, short swell and strong waves to the reef. Due to rainstorms, local rivers sometimes release large amounts of fresh water into the sea (Claereboudt *in litt.*, 3.8.87). Mean vertical transparency in the bay waters is 9 m during the wet season and 19 m in the dry season (June-October), with a range of 2-35 m (Kojis *et al.*, 1985). Manam Island is a very active volcano, and lies 10 mi. (16 km) distant from Laing Island. Boisa Island is a small volcanic island with no recent volcanic activity.

Reef Structure and Corals Laing Island is surrounded by a well developed fringing reef including a small lagoon on the western shore which has been described by Claereboudt and Bouillon (n.d.). The reef flat is emergent at most spring tides. Coral growth is luxuriant and a very wide variety of reef biotopes can be found around the island, from exposed reef crest to deep (60 m) bank reefs (Claereboudt *in litt.*, 3.8.87). Some portions of the reef are densely covered by soft corals (Tursch and Tursch, 1982). Bank reefs of different depths (5, 15, 25, 40, 60 m) and sizes can be found in the bay. In general coral cover is very high (more than 60%) in exposed areas, high on partially exposed sites and moderate in sheltered sites (30-60%). Coral cover diminishes to about 5% between 20 and 30 m. Two hundred and fifty species of hermatypic corals belonging to 73 genera have been recorded from the vicinity of Laing Island. *Acropora palifera* is only abundant in the partially exposed reef between 2 and 5 m where it is co-dominant (Kojis *et al.*, 1984 and 1985). Manam Island is surrounded by an incomplete fringing reef up to 300 m wide with reef slopes which are usually very steep. Boisa Island has an extensive fringing reef and a shallow bank reef off the north-west coast of the island (Claereboudt *in litt.*, 3.8.87).

Noteworthy Fauna and Flora Laing Island is densely forested and more than 100 species of vascular plants have been recorded. There are mangroves on Laing Island and some very large mangrove areas around the river mouths on the mainland shore. Noddy terns, megapodes, sandpipers, fish eagles, reef herons and frigates are frequently observed on Laing Island and Green Turtles *Chelonia mydas* nest there in small numbers (four in 1984). A Dugong *Dugong dugon* was recorded in Laing Island lagoon in February 1987 and there are occasional unverified reports of the species in nearby rivers. Sharks are also recorded in the bay (Claereboudt *in litt.*, 3.8.87). Hydromedusidae are described by Bouillon *et al.* (n.d.) and the alga *Caulerpa* by Coppejans and Meinesz (n.d.).

Scientific Importance and Research Hansa Bay contains the last major reefs on the north coast before the large Ramu and Sepik Rivers. A small laboratory (the King Leopold III Biological Station) has been operated by the Université Libre de Bruxelles on Laing Island since 1974. Over 150 scientific papers have been published on research carried out since then, including a large number on reef systematics. Other topics include:

coral-inhabiting gastropods; marine molluscs; polychaete systematics and ecology; hydrozoan biology, ecology and systematics; echinoderm studies; plant ecology and systematics; spider systematics; biology, behaviour and systematics of social insects; herpetology (Anon., n.d. b). Until recently, however, little work had been carried out on reef ecology, distribution or conservation. No scientific work has been carried out on the reefs around Boisa and Manam.

Economic Value and Social Benefits Fishing and shell collecting are important activities for the local people. Around 5000 people live on Manam and ca 700 on Boisa; the latter, at least, engage in traditional fishing activities. There are very attractive dives on the 2nd World War wrecks in Hansa Bay which are becoming increasingly popular; however the distance from Madang (250 km) and the lack of tourist facilities have prevented the development of mass tourism in the area. The lagoon along the western shore of Laing Island provides a safe anchorage for vessels up to 70 ft (21 m) long.

Disturbance or Deficiencies It is possible that the active Manam volcano may have impact on the reefs around Laing Island through acid rain, ashes and lava flows, but this has not been studied. The only major human disturbance to the reefs at present is fishing with explosives; this activity is illegal but is very popular with fishermen, largely owing to the ready availability of explosives, from road construction, mining activities or 2nd World War ammunition dumps. Some divers are reported to collect artefacts from the war wrecks in the bay, despite these being legally protected. A possible oil exploitation field in the Ramu-Sepik basin is a potential threat to the reefs of Madang province in general (Claereboudt *in litt.*, 3.8.87).

Legal Protection None.

Management Catching of adult turtles and consumption of turtle eggs appears to be traditionally forbidden in the villages around Hansa Bay.

Recommendations The area was recommended as a marine park or reserve (Genolagani, 1984). Plans were reportedly drawn up but, owing to political difficulties in 1984-85, the project has since stopped. The area of the proposed park or reserve is very large; the protection of a smaller area, around Laing Island, may be more practicable.

HORSESHOE REEF-TAHIRA (AIORO) PROPOSED MARINE PARK

Geographical Location Central Province, south-east Papua New Guinea, covering the portion of the barrier reef off Bootless Inlet, next to Nateava Reef; no coastal areas or islands are included; 9°36'S, 147°19'E.

Area, Depth, Altitude 395.9 ha.

Land Tenure Traditionally owned.

Physical Features The portion of the barrier reef included within the reserve is characterized by its horseshoe shape. The western part of the reef is exposed during the low-water spring tides, the eastern part remaining submerged. The north-west part borders the Padana Nahua channel, one of the four major channels in the western sector of the barrier reef. Water temperatures average 32°C at the surface and 24-27°C at 15m depth.

Reef Structure and Corals The reef is poorly studied but a brief faunal survey was carried out in 1980 (Coleman, 1980 and 1982). 70 coral species were recorded, including 37 hard corals, 43 alcyonarians and seven hydroids (Coleman, 1982). Genolagani (1984) lists some of the hard coral genera. Weber (1973) gives a general description of reefs in this area (*see* introduction). The sunken barrier reef (6 km x 200-300 m), within the proposed park, is considered to have once been part of Nateava Reef.

Noteworthy Fauna and Flora Wobbegongs or Carpet Sharks *Orectolobus dasypogon*, Catsharks *Hemiscyllium ocellatum*, and other sharks *Sphyrna lewini*, *Carcharchinus spallanzani*, *Carcharchinus* sp. and *Triaenodon obesus* were recorded in the 1980 survey but no turtles or Dugongs (Coleman, 1982). Major attractions for divers are a big grouper *Epinephelus* sp. called "Goblette", and a moray eel *Gymnothorax flavomarginatus* called "Nessie" which inhabit the *Parama* and *Pai* wrecks which now form an artificial reef (Lindgren *in litt.*, 22.10.87). Coleman (1982) lists a large number of species found in the proposed park area. Lindgren (*in litt.*, 22.10.87) has recorded over 200 species of fish from the area.

Scientific Importance and Research The University of Papua New Guinea has conducted a limited amount of research within the proposed park area. A number of boats have been scuttled within the park boundaries to provide the National Parks Service with sites for artificial reef studies (Coleman, 1982; UPNG, 1984). The largest, *Pacific Gas*, 65 m in length, lies upright at 40 m depth (Lindgren *in litt.*, 22.10.87). A basic species list has been prepared (Coleman, 1982; UPNG, 1984).

Economic Value and Social Benefits The main use of the reserve is by divers, particularly Tropical Diving Adventures Pty Ltd. There is considerable potential for the expansion of the tourist industry due to the proximity of the area to Port Moresby and its accessibility. Marine life identification courses have been held through the diving centre based in Bootless Bay. Fishermen from the villages of Pari, Tubusereia and Barakau are occasionally seen on the reef, and it is also visited by the more recent settlers in the Tahira and Mirigeda areas of Bootless Inlet.

Disturbance or Deficiencies Fishermen use spear guns or nets to take fish, and also collect shells, but their impact on populations within the area appears to be minimal (Lindgren *in litt.*, 24.10.87).

Legal Protection None at present. There have been considerable problems in the establishment of the reserve which are discussed in Genolagani (1984). Boundary descriptions for the Park were reported to be complete in 1980 and the Park was gazetted by the Lands Department in July 1981. The Park should subsequently have been declared under the National Parks Act of 1982. Problems of traditional ownership by local villagers (the area lies within a "traditional fishing zone") and the interest of the

diving company in the area has led to conflicts which the current marine park staff of the National Park Service are unable to deal with.

Management None at present.

Recommendations Genolagani (1984) outlines the basic needs for the final establishment of the area as a park:

1. Provision of biological data on the area including surveys of the reefs, data on coral species numbers, diversity and distribution; data on lobster population structure and breeding sites; data on commercial fish distribution, numbers taken; data on other species of interest such as turtles, dugongs and dolphins.
2. Provision of socio-economic data on local fishing villages, fishing grounds, location of industry and proposed developments, local agriculture and its affects of the reefs, sites of specific interest (e.g. artificial reefs).
3. Identification of the conservation potential of the area.
4. Formation of a management plan.

MOTUPORE ISLAND PROPOSED WILDLIFE MANAGEMENT AREA

Geographical Location 15 km south-east of Port Moresby, 1 km from the mainland; 9°32'S, 147°17'E.

Area, Depth, Altitude Island is 0.8 x 0.2 km; maximum altitude 60 m.

Land Tenure The island has been classed as National Cultural Property, but this designation does not extend to the reefs.

Physical Features The area lies within the lagoon of the Papuan Barrier Reef, amidst a cluster of islands which span the entrance to a reef-fringed bay. The island is small and hilly, and is a cigar-shaped emergent ridge oriented almost north-south, with a fringing reef on the shore platform around the island. Between Motupore and the outer barrier are a range of reefs, with associated seagrass, algal beds, mangroves and extensive carbonate sand and mud areas (both intertidal and sub-littoral) (Anon., n.d. a). A sheltered bay lying between a sand bar and the northern end of the island provides a safe intertidal anchorage during most of the year. Surface water temperatures range from 28° to 30°C and salinity is 30 ppt. Maximum tidal range is 2.9 m (Munro, n.d. a).

Reef Structure and Corals There is a well developed fringing reef and reef crests are emergent at most spring tides (Munro, n.d. a). Spectacular growth is found at the extreme southern end of the fringing reef where the reef surface is characterized by prominent "swell and swale" topography covered by a luxuriant and diverse coral fauna. The prolific spread of corals and relatively rapid elongation of reefs in a southward direction is almost certainly made possible by the strong north-west and north-flowing currents. The reefs around Motupore are considered to be some of the best in the area (Weber, 1973).

Noteworthy Fauna and Flora The island is covered with monsoon scrub, mangroves and palms. The marine fauna and flora is extremely diverse but very poorly known (Munro, n.d. a).

Scientific Importance and Research The island has an important archaeological site, including a midden which has been intensively studied. The University of Papua New Guinea Marine Station is situated on the island and receives many visitors; it is described in leaflets available from the university (Anon., n.d. a; UPNG, 1984). In 1983 a long-term project to map the marine habitats around the island was initiated. Reef-related research carried out in the period 1982-86 includes: reef fish ecology; reef coral assemblages and distribution; sponge taxonomy and distribution; distribution and burrowing behaviour of Ghost-shrimps; population ecology of *Strombus*; ecology and productivity of seagrass beds; nutrient cycles, including ambient levels of nutrients, supply in terrestrial runoff and fluxes across reef surfaces; oceanography and weather; sediment origins, movement and foraminifera (Polunin *in litt.*, 7.11.87). A study of the reproductive biology of Giant Clams *Tridacna* spp. has been carried out (Braley, 1984) and a project on Giant Clam farming is under way as part of an international programme sponsored by the Australian Centre for International Agricultural Research (Anon., 1986; Polunin, 1984a; UPNG, 1984).

Economic Value and Social Benefits Local people from nearby villages on the mainland fish and collect shells in the area, often employing traditional methods. The University restricts access to the island to research and training groups so there is little tourism, which is catered for on nearby Loloata Island, 1 km to the south-east (Lindgren *in litt.*, 22.10.87).

Disturbance or Deficiencies Exploitation of marine resources is not heavy and the people do not use explosives, or mine coral. There is virtually no pollution.

Management Displays and a Nature Trail have been established on the island (UPNG, 1984).

Recommendations Proposed as a marine park or reserve (Genolagani, 1984); emphasis should be placed on the importance of this area for research. The University intends to propose the island as a Wildlife Management Area, although this would include the sea only up to 100 ft (30 m) from the shore.

SALAMAUA PENINSULA

Geographical Location Morobe Province, south of the Huon Gulf on the eastern coast, 37 km south of Lae; 7°0'S, 147°E.

Area, Depth, Altitude Max. alt. about 274 m; peninsula and isthmus about 5.5 km long .

Land Tenure Some traditional ownership of peninsula; reef tenure unknown.

Physical Features The steep-sided peninsula is connected to the mainland by a sandy isthmus about 1 m above sea level. The peninsula is fringed by reefs which extend half way down the isthmus on the north side. Those on the south side are protected from the south-east swells which occur from June-September. Water currents are generally slight with only one notable tidal cycle a day; the maximum spring tidal range at Lae is 1.1 m (Kojis and Quinn, 1984). Visibility is good. Water temperature is 27.5°-32°C and is warmest from December to March and coolest from July to August. Salinity is usually 32 ppt on the south side of the peninsula.

Reef Structure and Corals The fringing reef drops off rapidly to about 40 m after which the slope diminishes and calcareous sand largely replaces the coral substrate. Coral cover is high (43-58%) at depths less than 20 m; moderate (23-28%) at 20-35 m and low (less than 10%) at depths greater than 35 m. At 45 m there is about 5% coral cover. There is a shallow reef flat dominated by massive *Porites* colonies and *Acropora palifera* which is most abundant in shallow water and was not recorded below 15 m; coral diversity is high here as well. Ninety-five species from 48 genera have been recorded from the Huon Gulf (Kojis *et al.*, 1984).

Noteworthy Fauna and Flora Black coral occurs.

Scientific Importance and Research Fecundity of *A. palifera* at Salamaua has been compared with that at a reef at Busama, 28 km south of Lae where sedimentation rates are higher (Kojis and Quinn, 1984); sexual reproduction of *A. palifera* at both these reefs has been compared with that at Lizard Island and Heron Island reefs on the Great Barrier Reef, Australia (Kojis, 1986). The area is used for biology field trips by the University of Technology in Lae and high schools.

Economic Value and Social Benefits The reefs are used for subsistence fishing by local villagers, particularly at night, and are popular for snorkelling and SCUBA diving. A resort (Naus Kibung) has been built and is operated by the Morobe Provincial Government; it provides a conference venue, accommodation and cooking facilities for tourists, and caters for about 30 people.

Disturbance or Deficiencies There is concern that dredging for the proposed port development at Lae could cause damage to the reefs. An environmental impact statement was in preparation in 1981. Slash and burn agriculture on the peninsula could also cause siltation on the reefs as the region is mountainous with high rainfall. High sedimentation rates along the coastline to the north are probably due to these factors and corals in this area have been found to have lower fecundity when water transparency is decreased by high levels of sedimentation (Kojis and Quinn, 1984). Reefs on the north of the peninsula, especially near the mainland, are affected by sediment discharged by the nearby Francisco River (Kojis *in litt.*, 1986).

Legal Protection None.

Management None.

Recommendations Recommended as a marine park or reserve (Genolagani, 1984) and of particular value for its educational and recreational importance due to its proximity to Lae, the second largest town in Papua New Guinea.

PITCAIRN ISLANDS

INTRODUCTION

The islands, a dependent territory of the United Kingdom, comprise Pitcairn (450 ha) which is inhabited, and the uninhabited islands of Henderson, Ducie and Oeno, and together have an area of 36 sq. km. The marine environments are poorly known. However information was collected in the course of the 1970-1971 National Geographic Society-Oceanic Institute Expedition, a visit made in spring 1987 by Operation Raleigh and the Smithsonian Expedition of 1987. Henderson and Ducie are described in separate accounts. A general description is given in Oldfield (1987).

Oeno is a low coral atoll (0.25 sq. miles (0.65 sq. km), maximum altitude 3.6 m) with typical atoll vegetation (Douglas, 1969). It is occasionally visited by Pitcairn islanders. It was briefly visited by the 1987 Smithsonian Expedition. The lagoon is very shallow (most or all less than 2-3 m deep) and has many small reefs. *Montipora* is the dominant coral genus, with *Acropora* also common. Other genera recorded are *Pocillopora, Psammocora, Pavona, Porites, Cyphastrea, Plesiastrea* and *Montastrea*. Fish are abundant. In 1987 no *Acanthaster* were recorded in the lagoon (Paulay, 1987). Pitcairn has a rocky, dramatically steep, cliff-lined coast with no reefs; the marine benthic algae have been described by Tsuda (1976).

The Fisheries Zone Ordinance provides empowering legislation for management of fisheries resources (SPREP, 1980), but fishing provides a small part of the Pitcairn islanders' income and is not reef-related. There is at present little threat to the reefs. Their main interest is their extreme distance from other centres of reef distribution. As a result of this, Ducie, Henderson and Oeno were proposed as "Islands for Science", although no practical steps have been taken (Dahl, 1980). The future of the islanders on Pitcairn is currently under discussion and long term plans for protection or management of the reef areas within the island group will have to await the outcome of these deliberations.

References

* = cited but not consulted

Bourne, W.R.P. and David, A.C.F. (1983). Henderson Island, Central South Pacific, and its birds. *Notornis* 30: 233-252.
Dahl, A.L. (1980). Regional ecosystems survey of the South Pacific Area. *SPC/IUCN Technical Paper* 179. South Pacific Commission, Noumea, New Caledonia.
Douglas, G. (1969). Checklist of Pacific Oceanic Islands. *Micronesica* 5(2): 327-463.
Foreign and Commonwealth Office (1988). Nomination of Henderson Island for inclusion in the World Heritage List. Submitted by The Secretary of State for Foreign and Commonwealth Affairs, United Kingdom. Prepared by S. Oldfield. Produced by the Nature Conservancy Council. 21 pp.
Fosberg, F.R. (1984). Henderson Island saved. *Env. Cons.* 11(2): 183-184.
Fosberg, F.R. and Sachet, M.-H. (1983). Henderson Island threatened. *Env. Cons.* 10(2): 171-173.
Fosberg, F.R., Sachet, M.-H. and Stoddart, D.R. (1983). Henderson Island (south-eastern Polynesia): Summary of current knowledge. *Atoll Res. Bull.* 272. 53 pp.
Oldfield, S. (1987). *Fragments of Paradise.* Pisces Publications, Oxford. 192 pp.
Paulay, G. (1987). Comments on the Pitcairn Islands. Unpub. rept. 2 pp.
Rehder, H.A. and Randall, J.E. (1975). Ducie Atoll: Its history, physiography and biota. *Atoll Res. Bull.* 183. 40 pp.
Serpell, J., Collar, N., Davis, S. and Wells, S. (1983). Submission to the Foreign and Commonwealth Office on the future conservation of Henderson Island in the Pitcairn Group. Unpub. Rept. IUCN/WWF/ICBP.
SPREP (1980). Pitcairn. *Country Report* 11. South Pacific Commission, Noumea, New Caledonia.
St John, H. and Philipson, W.R. (1962). An account of the flora of Henderson Island, South Pacific Ocean. *Trans. R. Soc. N.Z.* 1: 175-194.
***Tsuda, R.T. (1976).** Some marine benthic algae from Pitcairn Island. *Rev. Algol.* 11(3-4): 325-331.

DUCIE ATOLL

Geographical Location   293 miles (472 km) east of Pitcairn, 830 miles (1336 km) WNW of Easter Island; 24°40'S, 124°47'W.

Area, Depth, Altitude 1.3 x 1 mile (2 x 1.6 km), including four islands named by Rehder and Randall (1975) as Acadia, Edwards, Pandora and Westward.

Physical Features   Rehder and Randall (1975) give details of the physical features of the atoll. The northern and eastern sides are formed by Acadia, the largest island, which is composed largely of coral rubble, echinoid remains and dead shells, largely *Turbo argyrostomus*. Westward Island, a horseshoe-shaped ridge with the open side facing the lagoon, has a similar composition and is situated on the south-west side of the atoll. Pandora and Edwards Islands are situated on the southern side and are also similar, Edwards being the smaller island. There is no landing place on the north coast of the atoll, but a narrow (100 yds (91 m)) and shallow boat passage is found on the south-west between Westward and Pandora, which is navigable by a small boat at high tide. Small channels between the lagoon and ocean are found at the northernmost extension of Westward and the western end of Acadia but these have little influence on water exchange within the lagoon. The submarine shelf extends off the north-east and north-west points of the atoll, but the greatest seaward extension of the reef is at the south-west, where the shelf extends 300 yds (274 m) off shore to a depth of 100 ft (30 m) before dropping off more steeply. The seaward reef is broad on the north-west but becomes narrower to the north. There is a regular semi-diurnal tide. Visibility in the lagoon is about 75 ft (23 m), salinity is 38 ppt and water temperature averages 26.5°C.

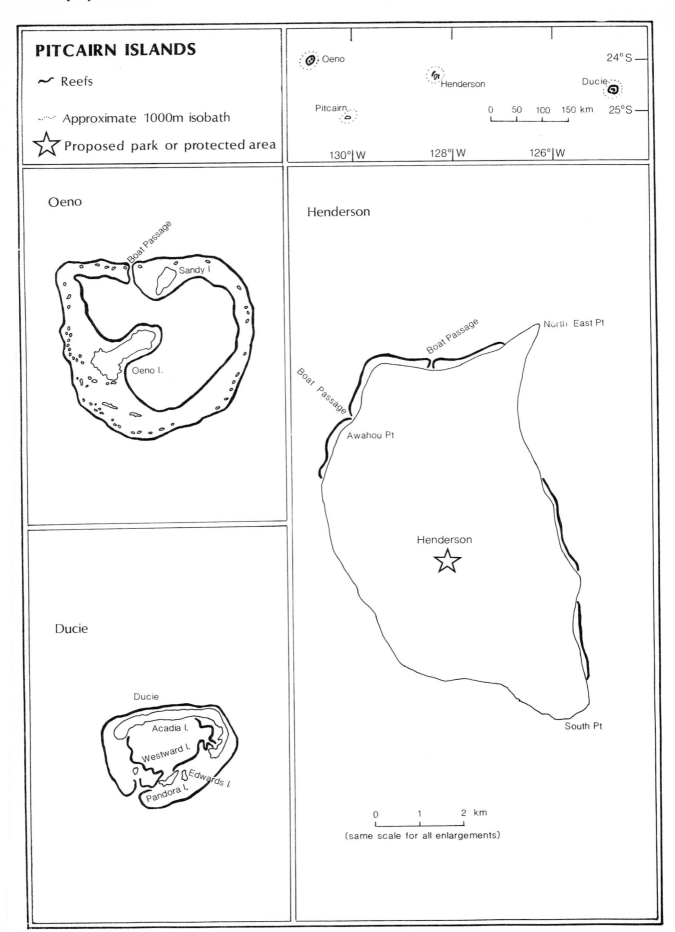

PITCAIRN ISLANDS

~ Reefs

⌁ Approximate 1000m isobath

☆ Proposed park or protected area

Oeno

Henderson

Ducie

Pitcairn

0 50 100 150 km

24°S
25°S

130°W   128°W   126°W

Oeno

Boat Passage

Sandy I

Oeno I.

Ducie

Ducie

Acadia I.

Westward I.

Edwards I.

Pandora I.

Henderson

Boat Passage

North East Pt

Boat Passage

Awahou Pt

Henderson
☆

South Pt

0   1   2 km

(same scale for all enlargements)

Reef Structure and Corals The reefs were surveyed in 1970-71, although no diving was undertaken on the seaward reefs of the northern and eastern coasts (Rehder and Randall, 1975). In 1987 the area around Acadia Island was briefly studied, mainly by snorkelling (Paulay, 1987). Acadia is largely surrounded by reef flats, the reef to the north-west consisting for the most part of a somewhat uneven reef pavement flat. It is generally covered by a thin layer of sand and fine algal growth, which may be thick in patches, and there is sparse coral cover. In 1987 *Holothuria atra* was abundant.

In 1970-71, coral on the outer reef in deeper water, mainly a short-branching *Acropora*, was largely dead and encrusted with red coralline algae (Rehder and Randall, 1975). By 1987, however, very high coral cover (over 50%) was found at depths of 50-60 ft (15-18 m), dominated by *Acropora* and *Montipora*. The fore-reef is unusual in being composed entirely of reef rock and rubble, lacking any soft sediments (Paulay, 1987).

The lagoon had fair coral cover, dominated by *Montipora* and *Astreopora* species which were situated on submerged eroded patch reef pinnacles, many interconnected at their bases. In 1970-71 these coral heads were reported to consist primarily of dead coral rock apart from a few small colonies mainly of *Montipora*. The lagoon floor between the patch reefs is made of fine white sand and there are signs of an active infauna. A total of 14 species of coral was collected in the course of the 1970-71 survey (Rehder and Randall, 1975). The genera *Psammocora, Acropora, Montipora, Astreopora, Pavona, Leptoseris, Fungia, Porites, Favia, Cyphastrea, Plesiastrea* and *Montastrea* were recorded in 1987 (Paulay, 1987).

Noteworthy Fauna and Flora Rats, probably *Rattus exulans*, have been reported. The bird life is the most striking element of the land fauna and a list is given in Rehder and Randall (1975) which includes petrels, the Red-tailed Tropicbird *Phaethon rubricauda*, boobies, the Great Frigatebird *Fregata minor*, the Sooty Tern *Sterna fuscata*, noddies, Bristle-thighed Curlew *Numenius tahitiensis*, Sanderling *Calidris alba* and Wandering Tatler *Heteroscelus incanus*. Many dead birds were found on the atoll in 1970-71; the cause of mortality was unknown but is thought to be a recent phenomenon since expeditions prior to 1922 did not report it. The fish fauna is considered to be impoverished; only 138 species were recorded in 1970-71, of which about 15 are confined to south-eastern Oceania. One species *Genicanthus* sp. is restricted to the Pitcairn group. The majority of species seen are common to much of the Indo-Pacific. Galapagos Sharks *Carcharhinus galapagoensis* and a Gray Reef Shark *C. amblyrhinchos* were recorded and the island has a reputation for its shark population (Rehder and Randall, 1975). In 1987 fish were found to be rather rare on the fore reef, the commonest being the Black Trevally *Caranx lugubris* and the Yellowtail *Seriola lalandi*. On the reef flat and in the lagoon there was a fair abundance of scarids and serranids (Paulay, 1987).

Insects, crustaceans, echinoderms and corals are discussed briefly in Rehder and Randall (1975) and an annotated marine mollusc list is given; species recorded include *Pinctada margaritifera* and *Tridacna maxima*. Fifty marine molluscs were recorded of which 34 are wide-ranging species and six are found only in Polynesia. *Heterocentrotus trigonarius*, the Purple

Slate-pencil Urchin, is a conspicuous element of the outer reef flat fauna; molluscs were abundant on the inner reef flat. *Tridacna maxima* occurs in the lagoon, but in general invertebrate fauna here was poor. Two species of land hermit crabs *Coenobita* were recorded in 1987 (Paulay, 1987). Acadia is the only vegetated island and is largely covered with *Tournefortia* which was the only plant species recorded in 1987 (Paulay, 1987).

Scientific Importance and Research Ducie is the easternmost atoll of the Indo-Pacific biogeographical region possessing a pure, though impoverished, Polynesia biota. The Whitney South Sea Expedition visited the island in 1922, it was also visited in the course of the 1970-1971 National Society-Oceanic Institute Expedition to south-east Oceania, and by Operation Raleigh in 1987. Rehder and Randall (1975) give the most complete account of the island and its history.

Economic Value and Social Benefits The reefs are not used for fishing, and the atoll is rarely visited. Ciguatera appears to be present at Ducie although not at Henderson.

Disturbance or Deficiencies In 1970, there was evidence of a relatively recent mass mortality of corals, the cause of which was not identified although a sudden drop in water temperature was postulated. The reefs had evidently recovered, at least in part, by 1987 (Paulay, 1987). Only a single specimen of *Acanthaster planci* was encountered in 1970-71 (Rehder and Randall, 1975) and none in 1987 (Paulay, 1987).

Legal Protection None.

Management None.

Recommendations Proposed for listing as an "Island for Science" in 1969 but nothing came of this recommendation (Douglas, 1969).

HENDERSON ISLAND

Geographical Location 200 km east-south-east of Pitcairn, 200 km east of Oeno and 360 km west of Ducie; 24°22'S, 128°28'W.

Area, Depth, Altitude Land area 37 sq. km (9.6 x 5.1 km); max. alt. 33 m.

Land Tenure British Government.

Physical Features Henderson is an elevated limestone island which rises as an isolated conical mound from a depth of about 3.5 km on a trend line which continues that of the Tuamotus and Gambiers eastward to Ducie; it is presumably a reef-capped volcano. There is a central depression on the top of the island, considered to be an uplifted lagoon (Paulay, 1987). The surface of the island is in large part reef rubble that is easy to walk on, with local areas of dissected limestone, especially around the periphery (early descriptions suggested that the entire surface was extremely rugged). The island is surrounded by steep cliffs of bare limestone with three main beaches, to the north, north-west and north-east. Henderson lies in the south-east trades and probably has a mean annual rainfall of 1500 mm. The tidal range at springs is probably about 1.0 m. Freshwater is almost completely absent. The geology of the island is summarized in Fosberg *et al.* (1983) and a general description of the island and its history is given in Foreign and Commonwealth Office (1988).

Reef Structure and Corals There is a fringing reef at least 200 m wide on the north, north-west and north-east sides of the island, backed by a wide beach (St John and Philipson, 1962); the south coast does not seem to have been studied. There are two narrow channels through the reef on the north and north-western coasts (Serpell *et al.*, 1983). Reefs off the north and north-east beaches are seawardly sloping reef platforms without reef crests, and are not typical fringing reefs. Coral cover is about 5%, dominated by *Pocillopora* with *Millepora* becoming dominant at depths greater than 7 m (Paulay, 1987). Submassive *Acropora* colonies are also present on the buttresses and solid substratum (Richmond *in litt.*, 16.9.87). In total, 19 genera and 29 species of coral were collected in 1987 (Paulay, 1987).

Noteworthy Fauna and Flora The top of the island and any land at the bases of the cliffs is densely vegetated with tangled scrub and scrub forest but the central part of the depression and makatea are more sparsely covered. The tallest trees are *Pandanus tectorius*. Flora and fauna recorded from the island are listed in Fosberg *et al.* (1983), Fosberg *et al.* (1985) and Foreign and Commonwealth Office (1988), and the flora is described by St John and Philipson (1962). Fifteen seabirds have been recorded, including petrels, shearwaters, noddies, three species of booby and the Greater Frigatebird *Fregata minor*. The four land birds are endemic (Bourne and David, 1983). The Green Turtle *Chelonia mydas* occasionally nests on the island (Fosberg *et al.*, 1983). A number of lizards and skinks have been recorded. The remains of a Coconut Crab *Birgus latro* were collected in 1987. There are at least two coenobite species, one, which is red in colour, was found to be the commonest crustacean on the island in 1987 (Broodbakker *in litt.*, 12.12.87 and 6.1.81). One very large *Coenobita* sp. was recorded in 1934 and 1987 (Paulay, 1987; Richmond *in litt.*, 16.9.87). The Spiny Lobster *Panulirus penicillatus* has been recorded (Paulay, 1987). Collections of marine molluscs and sponges and of as yet unidentified caridean shrimps, mostly of the family Alpheidae and probably comprising 5-8 species, were made in 1987 (Broodbakker *in litt.*, 12.12.87; Richmond *in litt.*, 16.9.87). There is a diverse echinoderm fauna. An unidentified holothurian is common on the northern reef flats (Richmond *in litt.*, 16.9.87), and the echinoid *Heterocentrotus* sp. (possibly *H. trigonarius*) is locally abundant on the sloping marginal reefs (Paulay, 1987) and shallow reef flat of the northern beach (Richmond *in litt.*, 16.9.87). Fish are sparse, with *Caranx lugubris* being the most common and obvious species (Paulay, 1987). A list of marine molluscs recorded from Henderson is given in Fosberg *et al.* (1983).

Scientific Importance and Research Fosberg *et al.* (1983) summarize published information on, and scientific expeditions to, the island to that date, of which the two most important were the Whitney South Sea Expedition in 1922 and the Mangarevan Expedition of 1934. The National Geographic Society-Oceanic Institute Expedition made two brief visits in 1971 and studied the marine fauna; an extensive molluscan collection was made. It was thought that the fauna is probably richer than that of Ducie, but the visit was too short to quantify this (Fosberg *et al.*, 1983). The island was visited by Operation Raleigh in spring 1987 when marine fauna, such as shrimps, ostracods and molluscs, were collected and a microfauna project was carried out (Broodbakker *in litt.*, 12.12.87; Parkes *in litt.*, 6.8.87). An expedition from the Smithsonian also visited the island in 1987. Henderson is the world's best remaining example of an elevated coral atoll ecosystem and is of outstanding universal value (Fosberg and Sachet, 1983).

Economic Value and Social Benefits The history of the island, which has been uninhabited apart from occasional visitors, is described in Fosberg *et al.* (1983). The island is visited by Pitcairn islanders on a fairly regular basis, chiefly to cut miro wood, *Thespesia populnea*, from which carvings are made for sale to visitors.

Disturbance or Deficiencies Goats and pigs were introduced on the island early in the century but have not survived; introduced rats are still present however (Serpell *et al.*, 1983). The terrestrial vegetation of the island is still largely pristine, with very few exotics, although there are two substantial coconut groves at the principal landing sites (Paulay, 1987). No *Acanthaster planci* were found in 1971 or 1987. In 1982/1983 the island came under severe threat as a result of the project of a wealthy American to build a house, landing facilities and airstrip on it (Fosberg and Sachet, 1983). A resolution at the 15th Pacific Science Congress in 1983 urged the British Government not to permit the proposed development before a detailed biological survey had been carried out and an assessment of the impacts made. The proposal was opposed by scientific and conservation bodies who petitioned the British Government to deny permission to carry out these plans (Serpell *et al.*, 1983), which they subsequently did (Fosberg, 1984). Had such plans gone ahead, the terrestrial fauna and flora would undoubtedly have been severely damaged and the reefs would probably have deteriorated as a result of blasting for boat channels, soil runoff and pollution.

Legal Protection Access to Henderson requires a licence issued by the Governor following approval by the Pitcairn Island Council (Foreign and Commonwealth Office, 1988).

Management None.

Recommendations Henderson has been nominated for inclusion in the World Heritage List (Foreign and Commonwealth Office, 1988) and it has been suggested that the island be declared a Biosphere Reserve under Unesco's Man and the Biosphere Programme (Oldfield, 1987). In 1969, Henderson, with Ducie and Oeno, was included in the list of Pacific Islands proposed for international scientific supervision, possibly under a proposed "Islands for Science" Convention, but no action was taken on this. Further research is required to determine the conservation status of the reefs. It is unlikely that any conservation management would be required for the island in its present condition (Foreign and Commonwealth Office, 1988).

SOLOMON ISLANDS

INTRODUCTION

General Description

The Solomon Islands have a land area of 30 000 sq. km, distributed over an area of 1 280 000 sq. km (1600 x 800 km), and consist of a double chain of elongated islands in the form of a bow, forming the western continental margin of the Pacific basin. The six larger islands are similar in having a central mountain spine with peaks rising up to 2450 m in height and only Guadalcanal has extensive coastal plains; elsewhere the mountains rise directly from the sea. Rivers and creeks are numerous on many islands. Their geology is described by Stoddart (1969a) and Thompson and Hackman (1969). The archipelago is geologically recent. Earthquakes are frequent. The islands are largely covered with tropical forest, described by Whitmore (1969). Oema, Ontong Java and Sikaiana are atolls. Rennell and Bellona are the southernmost of the Solomon Islands, and are long, narrow raised atolls with large central depressions. The climate is not well studied but has been briefly described by Brookfield (1969). South-easterly trade winds dominate and mean annual rainfall is generally well above 100 in. (2540 mm). There are occasional small tropical cyclones.

The following table includes information provided by Baines (*in litt.*, 14.3.85) (mostly for Western and Santa Isabel Provinces), Douglas (1969) and Worsnop (*in litt.*, 17.6.85); it does not list all islands.

Table of Islands

Shortland Is

- *Treasury Is* (X) Mono and Stirling; volcanic surrounded by raised coral limestone; narrow fringing reef;

- *Shortland (Alu)* (X) 413 sq. km; volcanic island with upraised reef limestone; fringing reefs; some offshore reefs with sand cays; Balalai, Poporang, Pirumeni and Magusaiai lie off coast;

- *Fauro* (X) volcanic island similar to Shortland; fringing reefs, especially to the south; Asie off west coast, Mania off southern tip;

- *Piru* reefs on the north coast;

- *Ovau* narrow fringing reef.

  Also: Oema Atoll and Oema to north.

Choiseul (Lauru) (X) ca 3400 sq. km; volcanic; narrow fringing reefs around mostof coast; a few barrier reefs along north-east coast.

Vaghena (Wagina) (X) raised reef; extensive reef-lagoon complexes forming part of south-east Choiseul complex.

Arnavon Is in Manning Str between Choiseul and Santa Isabel.

Santa Isabel (Bughotu) (X) 4014 sq. km; volcanic; some raised reef; reef complex off north-west, south-east and south-west coasts; numerous offshore islands, including:

- *Ghaghe* extensive reefs to north-west in Manning Str.;

- *Barora Fa* (X) off north-western tip of Santa Isabel; volcanic; mostly mangrove;

- *San Jorge* (X) southern end of Santa Isabel; fringing reefs on east and west coasts.

  Also: Barora Ite, south of Barora Fa; Omona, Papatura Fa, Papatura Ite off north coast; Sulei, Fera off north-east coast.

Mahige no reefs.

Ramos fringing reefs around much of coast.

Dai (Ndai) (X) fringing reefs around much of coast.

Malaita (X) east coast submergent with inlets and mangroves; west coast emergent with raised barrier reef; reefs occur in north at Lau lagoon, Manaoba I. and Leli I.; along north-east coast; fringing reef from Auki northwards; and at Langa Langa and Are Are lagoons; brief study of *Acanthaster* (Garner, 1973).

Maramasike (X) a few reefs in mouth of Maramasike Passage.

Ulawa (X) no reefs.

Vella Lavella (X) volcanic with area of uplifted reef; narrow fringing reefs with wider reefs and lagoons off crenulate NW coast and barrier reefs off north-east across Paraso Bay.

- *Liapari* (X) part of Vella Lavella reef complex.

Mbava (Baga) (X) fringing reef around island.

Ranongga (Ghanongga) (X) volcanic with upraised reef; very narrow fringing reef in north.

Simbo (X) volcanic, still active; fringing reef in west.

Ghizo (Gizo) (X) volcanic; reef at New Manra, W. Ghizo, visited (Morton, 1974); barrier reef to north; part of large reef complex including Nusatupe; fringing reef on steep south coast; at Titiana Pt reef is widest on headlands, with mean width of 250 m; reef flat divided into three zones: a) inner carbonate mud-flat with mangroves; b) moat, up to 1 m deep and 180 m wide, with scattered, small, largely dead colonies of *Acropora* and *Porites*; and c) outer rim of jaggedly eroding reef rock up to 0.2 m above high water, and seaward rim of smoother reef rock coated with calcareous algae;

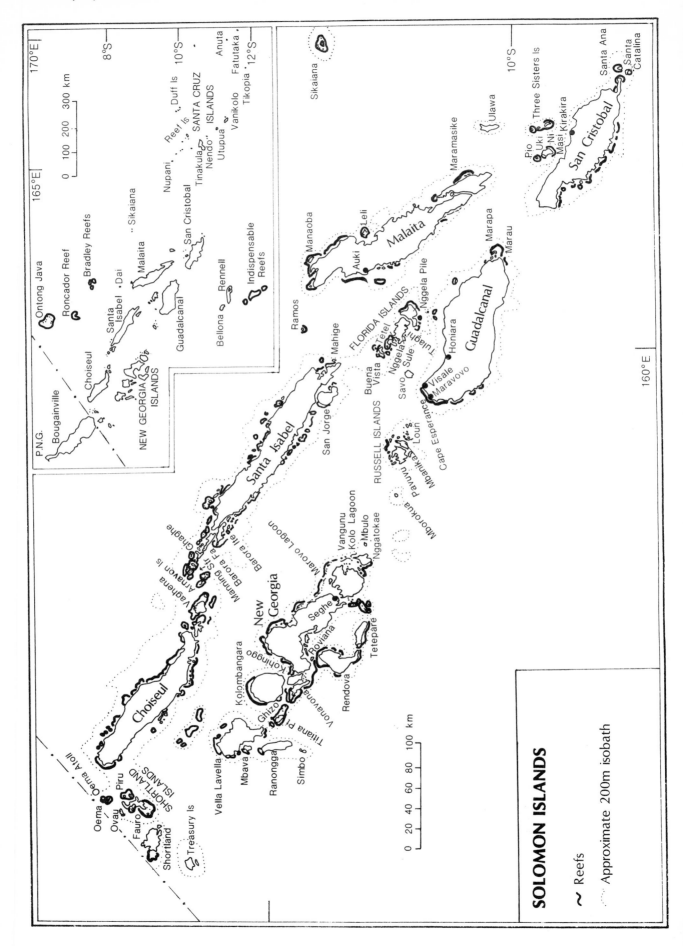

SOLOMON ISLANDS

〜 Reefs

∴ Approximate 200m isobath

living *Pocillopora, Acropora,* faviid corals and *Millepora* found in grooves in algal zone.

Kolombangara   (X) 678 sq. km; 1760 m; volcanic; very narrow fringing reef around all coasts; visited and birds surveyed by Operation Raleigh in 1986 (Taylor, 1986).

Kohinggo and Vonavona (Parara)   (X) mostly uplifted reef; a little volcanic; interesting long narrow barrier reefs on north-west and north-east coasts, also fringing; mainly mangrove.

New Georgia   (X) complex volcanic; some uplifted reef; barrier reef; great variety of reef types and lagoon complexes (*see below and account for* Marovo Lagoon); reef described at Munda (Weber, 1973).

Vangunu   (X) (*see account for* Marovo Lagoon).

Mbulo   (*see account for* Marovo Lagoon).

Nggatokae   (X) (*see account for* Marovo Lagoon).

Rendova   (X) mostly narrow fringing reef; some barrier reefs off the north coast around harbour.

Tetepare   (X) uplifted reef; fringing reef.

Mborokua   tip of forested volcanic cone surrounded by narrow fringing reef.

Russell Is   (X) south coasts abrupt and reef-less; north coasts drowned with intermittent barrier reefs; several smaller outlying islands:

-   *Pavuvu*   (X);

-   *Mbanika*   (X) reefs to north and south; Morton (1974) and Stoddart (1969a) described Sifola (south), Gvt Rest House (Yandina), Lingatu Pt, Simigan's Bay, Lever Pt (north), east shore; reefs largely dead; sea-level reef flats of east coast similar to those on Ghizo and Guadalcanal; at Sifola, shoreline is formed by 2 m shelf of elevated reef limestone, fronted by platform 60 m wide; living corals concentrated in the outermost moat where low colonies of *Goniastrea* coalesce to form reticulated pattern of microatolls with white dead crowns;

-   *Loun*   (X) reefs on north and south coasts.

Florida Islands

-   *Buena Vista (Vatilau)*   (X) possibly extensive fringing/barrier reefs;

-   *Mbokonimbeti (Sandfly I.)*   (X) between Buena Vista and Nggela Sule;

-   *Nggela Sule*   (X) separated from Nggela Pile by narrow Siota passage; areas of reef around most shores except in passage;

-   *Nggela Pile*   (X) areas of reef around most shores except in passage;

-   *Anuta*   (X) north of Nggela Sule; surrounded by reefs; reef flat dead but slope living; barrier reef may be dead;

-   *Tulaghi*   many areas of reef around adjacent islands; reef visited at Kokomtambu I. (Morton, 1974);

-   *Tetel*   in Sandfly Passage; reef visited (Morton, 1974); fringing reef on north side up to 150 m wide, extending eastwards as a narrow ridge; on south side, surface topography is steeper with narrow mangrove-fringed shelf less than 10 m wide; the reef flats are completely emergent at low tides on north side and coral growth almost entirely lacking except in crevices and holes;

-   *Switzer*   in Sandfly Passage; reef visited (Morton, 1974).

Guadalcanal   (X) north coast largely devoid of reefs, apart from narrow fringing reefs in some areas; elsewhere relatively narrow poorly developed fringing reefs but rich coral communities in some sites; fringing reef off promontory immediately south-east of Doma Cove (Ndoma), about 25 km north-west of Honiara, little more than tens of metres wide, and except for small, widely scattered colonies of encrusting corals such as *Favites, Goniastrea* and *Leptastrea*, consists of eroded, barren reef-rock awash at low tide (Weber, 1973); barrier reef here dead; reef slope at Tambaea has rich coral community; Morton (1974) described reefs at Kukum, Mamara, District Office and Rovi (near Honiara); Komimbo Bay and Cape Esperance (north-west); Naro Pt (west); Gowers Plantation, E. and S. Reef at Maraunibina I. and Jetty Reef in Paruru Bay (Marau Sound, east); Waimia (south); Marau Sound described below - reefs largely dead.

Savo

Marapa   (X)

San Cristobal   (X) many small reefs in south-west; some reef in north-east; extensive barrier at Star Harbour in east; reefs visited at Kira Kira (Morton, 1974; Stoddart 1969a); poor fringing reef; shoreline consists of a low terrace of reef limestone 100 m wide with occasional *Acropora* and *Pocillopora* on the walls of deep channels, characteristic of many Solomon Island shores, with a slightly elevated and eroding fringing reef and lack of active coral growth.

Three Sisters Is (Olu Malau)   (X) reef on east coast of Maraupaina.

Pio   no reefs.

Uki Ni Masi (Ugi, Uki)   (X) small barrier reef on north; no other reefs.

Santa Ana (Owu Rafa)   (X) fringing reef to south-east.

Santa Catalina (Owa Riki)   (X) no reefs.

Rennell   (X) 110 m; 86 km long raised atoll; cliffs, central depression; most of eastern half covered by Lake Te Nggano (largest lake in the Pacific); *Pandanus* scrub and rainforest; described by Wolff (1955-68 and 1969), described by Wolff (1955-68 and 1969), including maps and lists of marine fish (79 species), algae and mollusc (118 species); fringing reef probably all round island with only small breaks; barrier reef across Kanggava Bay and possibly other; geology described by Bourrouilh-Le Jan (1982).

Bellona (X) raised atoll 11.4 km long with honeycombed limestone surface; 79 m; fringing reef probably all round; 20 fish species recorded (Wolff, 1969); heavily cultivated; dense population.

Ontong Java (X) kidney-shaped atoll; max length 70 km, width 11-36 km; perimeter has wide reef flat broken by passes (four shipworthy), enclosing lagoon of 1420 sq. km with patch reefs; islands include: Pelau, Avaha, Keloma, Kemalu, Luaniua, Keila; climate and vegetation described in Crean (1977); Green *Chelonia mydas*, Hawksbill *Eretmochelys imbricata*, Leatherback *Dermochelys coriacea* and Loggerhead *Caretta caretta* turtles found in lagoon (Crean, 1977); additional information in McElroy (1973) and Bayliss-Smith (1973).

Roncador Reef sometimes emergent.

Sikaiana (Stewart Is) (X) atoll surrounded by reef; no ship passage into lagoon; badly damaged by low tides 1983.

Santa Cruz Is

Nupani

Duff (Wilson) Is

- *Bass Is* 200 ft (61 m); three islets (Lua, Kaa, Loreva);

- *Taumako (Disappointment, Netepa)* 1200 ft (366 m).

    Also: Ulaka; Lakao; Te Ako; Elingi; Tuleki (Anula).

Reef Is (Swallow Is) (X) include Lomlom, Fenualoa and Polynesia outliers; extensive barrier and reef systems.

Tinakula (Volcano) (X) active volcano; no reefs.

Nendö (Santa Cruz) (X) reef most of way round.

Utupua (X) barrier reef around most of island.

Vanikolo (Vanikoro) (X) barrier reef around most of island.

Tikopia (X) has a volcanic crater, 1235 ft (374 m), with lake; fringing reef; customary land held by chiefs.

Anuta (Cherry) (X) 212 ft (65 m); fringing reef; customary land held by chiefs.

Fatutaka (Fataka, Mitre) steep and rocky; seabirds.

(X) = Inhabited

Very little scientific work has been carried out on the reefs of the Solomon Islands. The 1965 Royal Society Expedition (Stoddart, 1969a), which has produced the only substantial account, found that, in general, the Solomon Islands lacked the luxuriant reefs of many parts of the Pacific due to unfavourable environmental conditions. The fringing reefs are generally associated with uplifted shores and are attached either to volcanic coastlines or grow on seaward members of successively elevated coral limestone beaches. Few flourishing coral reefs were encountered, apart from one at Haroro (Sandfly Passage, Florida Group), one north of Paruru (Marau Sound, Guadalcanal), and one at Matiu Island (New Georgia). Most reef flats were devoid of living corals and where they occurred many were found to be dead, as at Marau Sound and Honiara (Guadalcanal), in the Florida and Russell Islands, at Ghizo and Kolombangara, and on New Georgia. Live coral and calcareous red algae were largely confined, apart from pools, to the seaward zone below the dead reef. The poverty of the reefs may be partly explained by the steep and exposed shores with little suitable substrate for coral growth. Other reasons for poor coral growth are given in the section on disturbances and deficiencies. The current status of the reefs visited by the expedition over twenty years ago is not known and comparative work has not been carried out on other reefs. However, recent work in Marovo Lagoon (*see separate account*) suggests that there is a pattern of poor coral growth in shallow parts of the reefs although coral growth may be vigorous in deeper water (Baines, 1985a).

The Marine Party of the 1965 Expedition studied 36 fringing reefs in the southern Solomon Islands particularly at Tetel Island, Florida Group; Kira Kira, San Cristobal; Marau Sound, East Guadalcanal; Cape Esperance to Maravovo, north-west Guadalcanal; Honiara, Guadalcanal; Mbanika Island, Russell Group; Marovo, Gerasi and Togavai Lagoons, New Georgia; Ghizo Island and Kolombangara, New Georgia group (Stoddart, 1969a; Morton, 1974). The double barrier reefs fringing Marovo Lagoon, New Georgia (in part a triple barrier) and along the north coast of Kohinggo Island, Blackett Strait, are considered to be the best defined double barriers in the world (Stoddart, 1969a).

There are several lagoon complexes which are of particular scientific interest. In most cases these are the base of distinct cultural groupings and are therefore also of unusual cultural or anthropological interest. The name of the reef-lagoon complex is often the same as that of the associated culture and its language - for instance Langa Langa and Lau (Malaita) and Marovo and Roviana (New Georgia group). New Georgia is a volcanic complex surrounded by reef deposits which are partly elevated to form raised barrier reefs enclosing Marovo, Gerasi and Togavai lagoons on the north-east coast and Roviana Lagoon on the south-west coast, and partly drowned to form a submerged barrier south of Vangunu, enclosing the lagoon of Panga Bay. The coastline of the complex is crenulate, with numerous inlets and offshore volcanic islands. The lagoons along the north-east coast form a shelf with an area of 700 sq. km; Roviana Lagoon is 200 sq. km. The lagoon floors are irregular with depths of 37-55 m between steep-sided ridges in the southern Marovo Lagoon. The northern lagoons are narrower and shallower. Modern coral growth is limited by vertical submarine topography and lack of suitable substrates. Corals on the intertidal flat are often dead but those at deeper levels, though not luxuriant, are alive (Stoddart, 1969a). Marovo Lagoon is described in a separate account.

Other lagoon complexes include Marau Sound (Guadalcanal) and Lomlom (Reef Islands). The shores of Guadalcanal particularly on the north coast lack sea-level reef flats except for weak development at the north-west end and at Marau Sound, where the east Guadalcanal mountains plunge beneath a narrow coastal plain and re-emerge as the partially drowned high islands of Beagle, Malapa, Tawaini and Komachu. This area is described by Stoddart (1969c). A partially submerged barrier reef extends from the south coast fringing reef through Lauvie, South-east, Horohato and Round Islands to the north coast fringing reef enclosing Marau Sound, an area of 1000 sq. km. Surface sectors are linked by slightly submerged reefs lacking a surface reef flat. These submerged reefs are most extensive on the easternmost part of the barrier. Some have typical sand cays such as Niu, Maraunibina and Paipai, others have small residual nubbins of igneous rocks round which cays have accumulated (Kenra, Pelakauro). The cays are described in detail by Stoddart (1969c) and 17 islands were mapped and their sediments and flora studied. The high islands themselves have fringing reefs which in some cases are wide enough to carry cays. Outside the barrier there are further reef patches (Pari, Symons, North I. and Taunu), the three former of which have cays. Taunu Shoal is a reef patch 0.45 km in diameter 0.6 km east of the barrier. The shoal has a minimum depth of 3.6 m and is covered with growing corals in groove-spur formations. Within the Sound, channel depths increase from west to east and towards the barrier entrances, and range from 20 to more than 60 m depth. The sea level reef flats of Marau Sound cover 25 sq. km. During the earthquake of 1961, reefs and mangroves were elevated by 0.6 m in Marau Sound. The wide flats of the barrier reef may be up to 2 km broad. They are characteristically planed rock surfaces, drying at low water and are largely devoid of growing coral. Only one flourishing reef was found, just north-west of Paruru on the mainland, although the reef slopes support growing corals. Eroded, slightly elevated reefs form intertidal shelves along the south coast of Guadalcanal from Conflict Bay to beyond Kopiu Bay. In places, as at Untava, the flat is slightly above mean sea-level and has few living corals except for goniastreid microatolls on its seaward part.

Eighty-seven scleractinian species in 33 genera were recorded during the 1965 Royal Society Expedition (Pillai *et al.*, 1974). Marine algae are described by Womersley and Bailey (1969). Crustose coralline algae are dominant under strong surf conditions. *Porolithon onkodes* is dominant on the reef rim under surf conditions and *Lithophyllum moluccense* is conspicuous, with corals, under moderate wave action. *Neogoniolithon myriocarpon* is the primary cementing species. Sponges have been described by Bergquist *et al.* (1971). Mangrove woodland fringes large parts of high islands, but is only extensive on those cays with igneous outcrops or in sheltered waters; it has not been exploited on an industrial scale.

Green, Hawksbill, Olive Ridley and Leatherback Turtles (*Chelonia mydas, Eretmochelys imbricata, Lepidochelys olivacea* and *Dermochelys coriacea*) all nest in the Solomon Islands, with the Green Turtle the most abundant and widespread species. The Loggerhead *Caretta caretta* is recorded rarely and has not been demonstrated to nest (McKeown, 1977; Vaughan, 1981).

Reef Resources

The reefs are of major importance to subsistence fisheries, for the collection of shells and other marine products for decoration and ornament and for the extraction of coral sand and rock. A variety of shells and in particular the Black-lip pearl shell *Pinctada margaritifera* were used extensively for tools and implements in the past; currently shells still play an important role in jewellery and currency. Trochus *Trochus niloticus* is collected for food, and about 400 tonnes of shell are exported a year. Black-lip, Brown-lip and Gold-lip *P. maxima* pearl shell is also exported. There are at least two companies which export pearl shell (Baines *in litt.*, 2.9.87; Worsnop *in litt.*, 17.6.85). The tiny shell *Nassarius camelus* is harvested in parts of the Western Solomons such as Gerasi Lagoon and sold to the Tolai community of Papua New Guinea for which it is important as traditional currency (Baines *in litt.*, 2.9.87). Tridacnids have traditionally been an important item of diet. However it is reported that there is little trade in clam meat as many Solomon Islanders prefer finfish; most harvest is for subsistence purposes and some communities, notably Seventh Day Adventists, are forbidden to eat them (Diake *in litt.*, 18.9.87). Precious coral is being collected (Sloth *in litt.*, 9.4.87). *Acropora* coral is collected and burned for making lime to be consumed with betel nut (Baines *in litt.*, 2.9.87). All turtle species are exploited on a subsistence basis for their meat and eggs, with the Green Turtle the most highly favoured. Substantial quantities of Hawksbill tortoiseshell are exported to Japan (McKeown, 1977; Vaughan, 1981). Marine algal resources are described in Kaitira (1987). A variety of other uses of marine products is given in Anon. (1985) and development of fish marketing is described in Carleton (1981). Uses of marine resources, including fish and shells, on Ontong Java are briefly described in Crean (1977) and McElroy (1973) and a more detailed description of the bêche-de-mer export fishery, which was initiated by the Japanese earlier this century, is given. Argue and Kearney (1982) provide an assessment of the skipjack and baitfish resources.

Tourism is still in its infancy but reef-related recreational activities are being developed. There is a holiday and diving resort at Tambaea on Guadalcanal, near Visale, and diving operations exist in Honiara although these cater mainly for WWII wreck diving. There is a small diving operation on Ghizo and one at Uepi, Marovo Lagoon (*see separate account*). In the Florida Islands there is a holiday resort on Anuta (Worsnop *in litt.*, 17.6.85).

Disturbances and Deficiencies

The intact condition of the corals found by the 1965 Royal Society Expedition suggested that they were only recently dead. This could be accounted for by variations in sea-level, including extreme low tides at mid-day, and the tectonically active nature of the area (Morton and Challis, 1969; Stoddart, 1969a and b; Womersley and Bailey, 1969). At Kolombangara coral death was clearly caused by burial by fluvial sediment from an actively eroding cone. The dynamic geomorphology of reefs in the Guadalcanal area is demonstrated by the 1978 uplift of a reef formerly at several metres depth to slightly above sea-level during earthquake activity. The reef is said to be about 0.5 miles (0-8 km) long, off the south coast of Guadalcanal (Baines *in litt.*, 14.3.85). The earthquake also caused damage to some reefs, e.g. at Sikaiana (Worsnop *in litt.*, 23.4.85). More recently, a general depression of mean sea-level in the south-west Pacific during 1983, a consequence of the abnormal El Niño event, caused widespread mortality of coral reef organisms on reef flats. Exposure at low tide also caused the death of reef flat organisms at Auki on Malaita (Worsnop *in litt.*, 17.6.85). Reefs have also been extensively damaged by recent cyclones (Diake *in litt.*, 18.9.87).

The *Acanthaster* population increased noticeably on reefs at Malaita after cyclones in 1966 and 1968, but it was concluded that further research was required to determine whether these outbreaks were having deleterious effects (Garner, 1973). During the 1980s, *Acanthaster* has not been considered a threat, although localized high concentrations have sometimes been observed as at Mamara, Guadalcanal, in 1981 (Baines pers. comm., 1985).

Most of the population lives along the coast but as yet there is comparatively little evidence of human-induced damage to reefs (Dahl, 1985). Sediment disturbance from erosion of adjacent land surfaces has been reported from one or two small areas and is likely to become an increasing problem with the widespread logging activities which are underway. Unregulated, localized cutting of mangroves is under way at a small number of locations; industrial scale mangrove logging has been proposed but rejected. Liquid organic waste from the oil palm industry is discharged into the sea and in the long term sewage discharge into the sea may cause problems, and may already do so around Honiara. Light industry near Honiara is causing pollution of bays (Enekevu, 1983). Pollution has been reported from a fish cannery on Tulagi although filters are in use which prevent solid effluent being flushed into the sea (Diake *in litt.*, 18.9.87). A new port is planned at Noro (Enekevu, 1983). There was a potential threat to Rennell from bauxite mining (SPREP, 1981) but this proposal has been shelved (Baines *in litt.*, 2.9.87).

Explosive war debris has been used as a source of "dynamite" for destructive illegal fishing, causing damage in some areas, such as Langa Langa Lagoon and Marovo Lagoon. Reefs around Malaita were reported in the 1960s to have areas of dying coral, and few fish (Garner, 1973). There is some traditional exploitation of crocodiles (SPREP, 1981). Hawksbill Turtles *Eretmochelys imbricata* have declined steeply in number through intensive commercial exploitation (Vaughan, 1981). A few Dugong *Dugong dugon* remain in areas including Roviana and Marovo Lagoons (*see separate account*); traditional hunting appears to be dying out. Traditional shell money "romu shell" is reported to have become scarce near Malaita, whose societies use it, due to overcollecting (Worsnop *in litt.*, 23.4.85). However it can still be found in some other parts of the Solomons (Baines *in litt.*, 2.9.87). Giant Clams Tridacnidae and Giant Tritons *Charonia tritonis* have reportedly become locally rare (Worsnop *in litt.*, 23.4.85), although overall

there are still substantial stocks of all species of Tridacnidae (ICLARM, 1986). However, international poaching takes place. In April 1987 a Taiwanese fishing vessel, carrying Giant Clam adductor muscle representing about 10 000 harvested clams, was arrested at Roncador Reef south of Ontong Java. Most of the adductor muscles were very small (averaging 100 g) suggesting that the reef had long been stripped of its main stocks (Anon., 1987b).

"Outbreaks" of shellfish poisoning in parts of Western Province appear to have been associated with muddy river flows. Ciguatera has been reported to occur occasionally in the east but probably never in Western Solomons (Denton, 1983). Fish poisoning (type unspecified) has been known to occur in many areas, including Western Province, Ontong Java, Sikaiana, Rennell and Bellona, and Temotu Province. In Western Province it was associated with a red tide but in other areas it may be due to ciguatoxin or other related toxic agents (Diake *in litt.*, 18.9.87).

Legislation and Management

Two systems of law apply to marine areas: customary law and statute law, and there is no clear understanding yet as to the relative application of these two systems. There is a general understanding that traditional marine area rights must be respected and this is embraced under Government policy. Baines (1985b) discusses the full implications of this in relation to conservation and the inshore fisheries.

Customary rights to an area of marine resources are usually retained by the social group which has rights to the adjacent land, and there is a general understanding that traditional marine area rights must be respected. In the 1950s, interests in reefs were expressed in various ways depending on social organisation and culture, from a close subdivision on an individual basis as in the Reef Islands, through tenure based on the family group (Roviana) to a relative lack of exclusiveness in Tikopia. In Lau (Malaita) an individual holding primary interests in a reef area would exercise exclusive interests in gathering shells for commercial sale, net and trap fishing but less interest in other marine resource activities, which an outsider might therefore be permitted to carry out (Allan, 1957). With the decrease in abundance of some marine resources, such as trochus and clams, rights to certain activities are reported to be more strongly exercised (Baines, 1985b). Akimichi (1978) describes a traditional fishery in Lau, Malaita, that was subdivided into individual fishing areas, some of which were allocated to individuals with secondary rights, i.e. rights that are not acquired by inheritance but for example through marriage or by traditional purchase. The Provincial Government of Makira has reportedly made the reef near Kira Kira "tambu" (Worsnop *in litt.*, 17.6.85). In general, residents have access to subsistence and cash fisheries in shore, whilst non-residents have no access to commercial fisheries in shore but may bait fish in lagoons if the local community agrees (Enekevu, 1983). Baines (1985b) describes the system for obtaining live bait fish for the tuna industry which are harvested by employers of the tuna fishing companies in areas subject to traditional marine resource rights. This involves bilateral negotiations and the payment of royalties by the fishing companies to the traditional owners of the baitground areas (Diake *in litt.*, 18.9.87).

Regulations prescribing measures for the protection and preservation of the marine environment can be promulgated under the Delimitation of Marine Waters Act, 1978. The Fisheries Act 1972/1977, provides for the protection of fish; under the Act, a Principal Licensing Officer has the power to issue licenses and permits for commercial fishing and fish processing. The Minister in charge of fisheries has the power to set certain limits on gear and sizes of fish taken and to declare areas as not available to commercial fishing. However this power has not yet been exercised and there are no fishery regulations concerning harvestable fish sizes or closed areas. The only regulation prohibiting the use of certain gear sizes or gear types is that banning the use of nets in the vicinity of Honiara Wharf, mainly to prevent the obstruction of vessels in the area (Diake *in litt.*, 18.9.87). There are size limits of 1.5-3.5 ins (3.8-8.9 cm) for trochus collection. No turtles with carapace length of less than 75 cm may be sold.

A pilot-scale Coastal Aquaculture Centre, incorporating a Giant Clam hatchery, has recently been established as a cooperative effort by the International Centre for Living Aquatic Resources Management (ICLARM) and the Government of the Solomon Islands on a 4.8 ha site ca 25 km west of Honiara. The centre has control of 400 m of sea frontage from the inter-tidal zone to a distance of 100 m from shore (ICLARM, 1986). By mid-1987 spawning, broodstock and larval rearing tanks had been installed and a broodstock of *Tridacna gigas* had been collected (Anon., 1987a). The aims of the centre are: to test, develop and demonstrate methods for the cultivation of economically valuable aquatic organisms; investigate methods for environmental enhancement or manipulation which relate to artificial improvement of natural fisheries; investigate and develop methods for processing and marketing aquacultural products; appraise the social and economic impacts of such developments in the South Pacific Region; provide public dissemination of the results of the above activities; and serve as a regional source of fisheries and aquaculture information (ICLARM, 1986).

The National Parks Act of 1954 provides for the establishment of national parks but there is no system of marine protected areas. Five islands have been designated bird sanctuaries (Tulagi, Oema island and atoll, Mandoleana, Dalakalau and Dalakalonga) but these do not extend to the marine habitat (Anon., 1985). The Environment and Conservation Division of the Ministry of Natural Resources is now responsible for the establishment of protected areas. In 1981, a marine protected area was established, under statute law, at the Arnavon Islands in Manning Strait, between Santa Isabel and Choiseul. This move was not made in consultation with all parties who believed that they had rights to that area under customary law. An aggrieved party, in protest, destroyed a WWF-funded turtle hatchery which had been established as part of a conservation programme and it has not since been possible to re-establish the facility.

Western Province has established a system of Fisheries Management Areas to embrace areas of the province's coastal zone over which the Provincial Government Act,

1981, gives it jurisdiction, together with all areas under traditional marine tenure and some areas beyond three nautical miles but which are considered to be essential for proper management of the coastal zone (Baines, 1985a); these have not yet however been developed into practical terms (Baines *in litt.*, 2.9.87).

Recommendations

General management requirements for marine areas are outlined in Baines (1981) and SPREP (1981) and include the following recommendations:

1. commercial fisheries development in reefs and lagoons should be undertaken with care as the fisheries resource here is not as rich as it might appear to be;
2. traditional knowledge of reef and lagoon fisheries should be used as a basis on which to build modern small scale fisheries;
3. reported reef and lagoon damage in southern Roviana lagoon should be investigated;
4. coastal development, such as road, hotel and other construction activities, should be undertaken with greater care;
5. a system of protected areas should be developed, with the participation of customary landowners;
6. new legislation is required.

Some 33 proposed protected areas are listed in Anon. (1985) but none of these include reef habitat. Dahl (1980) recommended a reef reserve in Manning Strait and a selection of reef and lagoon reserves elsewhere. Rennell Island has been recommended for designation as a World Heritage Site, should the Solomon Islands become party to the World Heritage Convention.

A pilot project is being carried out within the South Pacific Coastal Zone Management Programme (SOPACOAST), at Marovo lagoon (*see separate account*) and there is interest in the applicability of this concept to other areas of the Solomons (Anon., 1986a and b). Policies of the Government of Western Province, within which Marovo Lagoon is situated, call for marine resource assessment, insist that subsistence needs be met before commercial harvest levels are determined, promote the concept of protected areas as safe breeding and nursery grounds for marine species, encourage cultural expression as a socially stabilizing element at a time of potentially disruptive change and firmly promote the idea that traditional marine resource tenure is appropriate for economic development. There is a need to reconcile differences of interpretation between National and Western Province Governments, to provide clearer administrative and legal support for the traditional marine tenure system, and to establish a better approach to baitfish harvesting in the lagoon (Baines, 1985a).

References

* = cited but not consulted

*Akimichi, T. (1978). The ecological aspect of Lau (Solomon Islands) ethnoichthyology. *Journal of the Polynesian Society* 87: 301-326.

*Allan, C.H. (1957). Customary land tenure in the British Solomon Islands Protectorate. Western Pacific High Commission.

Anon. (1985). Country Review - Solomon Islands. *Report of the 3rd South Pacific National Parks and Reserves Conference, Apia* 3: 195-209.

Anon. (1986a). Marovo Lagoon Resource Management Project. Report of the first community workshop, Seghe, Solomon Islands, 2-3 December 1985. *CSC Tech. Publ. Series* 192. 15 pp.

Anon. (1986b). Project Document. South Pacific Coastal Zone Management Programme (SOPACOAST). *CSC Tech. Publ. Series* 204. 28 pp.

Anon. (1987a). ICLARM Coastal Aquacalture Centre. *Clamlines. Newsletter of the International Giant Clam Mariculture Project* 3: 3-5.

Anon. (1987b). Taiwanese clam boat arrested in Solomon Islands. *Clamlines. Newsletter of the International Giant Clam Mariculture Project* 3: 7.

Argue, A.W. and Kearney, R.E. (1982). An assessment of the skipjack and baitfish resources of Solomon Islands. *Skipjack Survey and Assessment Programme Final Country Report* 3. South Pacific Commission, Noumea, New Caledonia.

Baines, G.B.K. (1981). Environmental management for sustainable development in the Solomons: A report on environment and resources. Prep. for Government of the Solomon Islands. Mimeo. 57 pp.

Baines, G. (1985a). Study Area One: Marovo Lagoon, Solomon Islands. Working paper on pilot project for Commonwealth Science Council.

Baines, G.B.K. (1985b). A traditional base for inshore fisheries development in the Solomon Islands. In: Ruddle, K. and Johannes, R.E. (Eds), *The Traditional Knowledge and Management of Coastal Systems in Asia and the Pacific*. Unesco/ROSTEA, Jakarta. Pp. 39-52.

*Bayliss-Smith, T.P. (1973). Ecosystem and economic system of Ontong Java atoll. Ph.D., Dept Geography, University of Cambridge.

*Bergquist, P.R., Morton, J.E. and Tizard, C.A. (1971). Some Demospongia from the Solomon Islands with descriptive notes on the major sponge habitats. *Micronesica* 7(1-2): 99-121.

Bourrouilh-Le Jan, F.G. (1982). Les étapes de la dolomitisation. Géochimie et sédimentation d'un atoll soulevé du S.W. Pacifique et de sa couverture bauxitique. Rennell, Iles Solomon. *Livre Jubilaire de Monsieur le Professeur G. Lucas. Mém. Géol. Univ. Dijon*. Pp. 3-20.

Brookfield, H.C. (1969). Some notes on the climate of the British Solomon Islands. *Phil. Trans. Roy. Soc.* B 255: 207-210.

*Carleton, C.R.C. (1981). Solomon Islands: Fish Marketing Development. FAO.

Crean, K. (1977). Some aspects of the bêche-de-mer industry in Ongtong Java, Solomon Islands. *South Pacific Commission Fisheries Newsletter* 15: 36-48.

Dahl, A.L. (1980). Regional ecosystems survey of the South Pacific Area. *SPC/IUCN Technical Paper* 179. South Pacific Commission, Noumea, New Caledonia.

Dahl, A.L. (1985). Status and conservation of South Pacific coral reefs. *Proc. 5th Int. Coral Reef. Cong., Tahiti* 6: 509-513.

*Denton, P. (1983). Ecology of ciguatera: Baseline reef studies in Vanuatu and Solomon Islands. Paper presented at 15 Pacific Science Congress.

Douglas, G. (1969). Checklist of Pacific Oceanic Islands. *Micronesica* 5(2): 327-463.

Enekevu, L. (1983). Kao Tung No. 1: Trial clam and reef fishing operation. Report, Fisheries Division, Solomon Islands. 3 pp.

Garner, D. (1973). Preliminary findings of a brief survey of the Crown-of-Thorns Starfish (*Acanthaster planci*) carried out on the island of Malaita in the British Solomon Islands Protectorate. Paper 26. *Proceeding and Papers, Regional Symposium on Conservation of Nature - Reefs and Lagoons*. South Pacific Commission, Noumea, New Caledonia.

ICLARM (International Center for Living Aquatic Resources Management) (1986). Report on progress of the International Giant Clam Mariculture Project and on the development of a pilot-scale Giant Clam Hatchery. Unpub. rept. 5 pp.

Kaitira, B. (1987). Marine algal resources of the Solomon Islands. In: Furtado, J.I. and Wereko-Brobby, C.Y. (Eds), *Tropical Marine Algal Resources of the Asia-Pacific Region: A Status Report*. Commonwealth Science Council Technical Publication 181. Pp. 101-104.

*McElroy, S. (1973). The bêche-de-mer industry: Its exploitation and conservation. Findings of an exploratory bêche-de-mer resource survey at Ongtong Java atoll. Internal Rept, Fisheries Division, Dept Agriculture, Honiara.

McKeown, A. (1977). *Marine Turtles of the Solomon Islands*. Ministry of Natural Resources, Honiara, Solomon Islands. 49 pp.

Morton, J. (1974). The coral reefs of the British Solomon Islands: A comparative study of their composition and ecology. *Proc. 2nd Int. Coral Reef Symp., Brisbane* 2: 31-53.

Morton, J.E. and Challis, D.A. (1969). The biomorphology of Solomon Islands shores with a discussion of zoning patterns and ecological terminology. *Phil. Trans. Roy. Soc.* B 255: 459-516.

Pillai, G., Stoddart, D.A. and Morton, J.E. (1974). The scleractinian corals of the British Solomon Islands. Unpub. rept.

SPREP (1981). Solomon Islands. *Country Report* 17. South Pacific Commission, Noumea, New Caledonia.

Stoddart, D.A. (1969a). Geomorphology of Solomon Islands coral reefs. *Phil. Trans. Roy. Soc.* B 255: 355-382.

Stoddart, D.R. (1969b). Geomorphology of the Marovo elevated barrier reef, New Georgia. *Phil. Trans. Roy. Soc.* B 255: 383-402.

Stoddart, D.R. (1969c). Sand cays of eastern Guadalcanal. *Phil. Trans. Roy. Soc.* B 255: 403-432.

Taylor, J.B. (1986). Kulumbangara bird report. Unpub. rept, Operation Raleigh. 10 pp.

Thompson, R.B. and Hackman, B.D. (1969). Some geological notes on areas visited by the Royal Society Expedition to the British Solomon Islands, 1965. *Phil. Trans. Roy. Soc.* B 255: 189-202.

Vaughan, P.W. (1981). *Marine Turtles: A Review of Their Status and Management in the Solomon Islands*. Ministry of Natural Resources, Honiara, Solomon Islands. 70 pp.

Weber, J.W. (1973). Generic diversity of scleractinian reef corals in the Central Solomon Islands. *Pac. Sci.* 27(4): 391-398.

Whitmore, T.C. (1969). The vegetation of the Solomon Islands. *Phil. Trans. Roy. Soc.* B 255: 259-270.

Wolff, T. (Ed.) (1955-1968). *The Natural History of Rennell Island, British Solomon Islands*. Vols 1-8. Danish Science Press Ltd., Univ. Copenhagen and British Museum (Natural History).

Wolff, T. (1969). The fauna of Rennell and Bellona, Solomon Islands. *Phils. Trans. Roy. Soc.* B 255: 321-343.

Womersley, H.B.S. and Bailey, A. (1969). The marine algae of the Solomon Islands and their place in biotic reefs. *Phil. Trans. Roy. Soc.* B 255: 433-442.

MAROVO LAGOON

Geographical Location Western Province; bounded in the north-west by the southern coast of New Georgia and extending around the west, north and east sides of Vangunu; its southern boundary on the west of Vangunu is Hele Bar, and on the east a chain of islands between Vangunu and Nggatokae; Vangunu is at 18°40'S, 158°00'E.

Area, Depth, Altitude The lagoon is 35 km long from Hele Bar in the south-west to Uipi in the north-east, and narrows at Njae Passage to less than 500 m in width. The eastern section from Tatama to Mbili is 50 km long; from Tatama to Nggatokae the average width is 8 km. West of Njae Passage, it is known as Nono Lagoon and reaches depths of 50-60 m, deepening to 80 m at the south-facing ocean entrance. Njae Passage is about 40 m deep and shallows into the northern, main body of Marovo Lagoon which is rarely more than 25 m deep except immediately inshore of passages between some eastern barrier islands where depths of 50-60 m are found. Maximum depth in the eastern part of the lagoon is 25-30 m except in the subsidiary shallow Kolo Lagoon on the east coast of Vangunu; max. alt. on Vangunu is 1000 m (Baines, 1985a).

Land Tenure The Western Province has an as yet undefined jurisdictional right over areas within three nautical miles (5.6 km) of the shores and reefs of its islands, although this cannot override traditional law as it applies to resource use and allocation (Baines, 1985a).

Physical Features The surrounding islands of Vangunu, Nggatokae and New Georgia are volcanic, the first consisting of a dramatic complex of several extinct cones, the largest of which is Mt Vangunu. The islands between Vangunu and Nggatokae are raised reef. Within the lagoon, are hundreds of small islands of varied geomorphology: sand cays, mangrove islets, raised reef islands and, inshore, occasional small islands of volcanic origin. The eastern and western outer limits of Marovo Lagoon are raised barrier reefs, some submerged at high water but with sand cays, others elevated several metres. Stoddart (1969b) described their geomorphology. Morton (1974) and Stoddart (1969a) looked at Martin, Pirikale, Paleti and Lulu Islands and Gap Reef. Mbulo is an uplifted reef with a narrow fringing reef and a series of lagoons with different physical and biological characteristics on the west coast.

Baines (1985a) describes three major habitats in Marovo Lagoon: sand cay complexes such as Mindeminde and Tinge, which are groups of numerous small vegetated sand islets on patch reefs with a thin mangrove fringe; estuarine complexes such as Ghoe River/Nono Lagoon, Kolo River and Kele Bay, which have shallow water, muddy bottoms rich in organic detritus, varying salinity and freshwater inputs rich in organic material from freshwater swamps and mangroves; and barrier islands such as Uipi which are long narrow islets with marked environmental contrast between ocean-facing fringing reefs and lagoon-side fringing reefs.

Reef Structure and Corals There are numerous reefs, at varying depths, throughout the lagoon. As is usual in the Solomons, the upper surfaces do not have rich coral growth and even less than usual is currently present owing to an extended period of mean sea-level depression in 1983. Below low water mark, reef growth is vigorous (Baines, 1985a). The reefs at Wickham Anchorage and Cheko on Vangunu are described by Morton (1974). Nggatokae has small reefs on the east and west coasts (Worsnop *in litt.*, 17.6.85).

Noteworthy Fauna and Flora Mangroves are found in estuaries, shorewards of many fringing reefs and on many islets (Baines, 1985a). The lagoon has stocks of Giant Clams, although these are smaller than those off the south-west coast of Santa Isabel (Anon., 1987a). Dugong have been recorded (Worsnop *in litt.*, 23.4.85).

Scientific Importance and Research Marovo Lagoon is the largest of the Solomons' reef-lagoon complexes and is one of the largest in the world. It is currently being studied in the course of a coastal zone management project under the auspices of the Commonwealth Science Council (Baines, 1985a).

Economic Value and Social Benefits The human population in the area numbers about 5300. Commercial fishing in the lagoon is a part-time activity and is based on small units using gill nets and lines, taking mainly reef fish; the catch is iced and shipped to Honiara. The lagoon is also an important source of baitfish for the skipjack tuna industry, and communities which have traditional rights over these fish stocks receive payment from the industry.

There is potential for tourism, particularly as the area has considerable archaeological and anthropological interest, the scenery is attractive and the reefs and sand cays in sheltered waters have great recreational value. Numerous yachts visit the lagoon, stopping at villages such as Mbili to buy handicrafts. SCUBA diving is a popular tourist activity; there is a small tourist resort on Uipi (Baines, 1985; Anon., 1986a).

Disturbance or Deficiencies Tsunamis and storm surges from the open ocean are buffered by Vangunu and barrier reefs. Hurricanes are usually of low intensity and are infrequent (only in 1951 and 1982 in recent years) and have relatively little effect. There is a potential threat from an intermittently active submarine volcano, Kavachi, located about 20 km south-west of the village of Zaira on the south coast of Vangunu. Earth tremors are common, but the most recent one to have caused noticeable geomorphological changes, such as reef and shore subsidence, was in 1939.

Although sizeable sediment loads are transported by the larger streams, the sediment is confined to mangrove-lined estuaries and does not enter the lagoon at present. However, mangroves are being systematically cleared, partly as a result of a mistaken belief that where they are adjacent to coconut plantations they will inhibit the growth of the latter. The lagoon is therefore potentially threatened by the changed amount and quality of freshwater input due to mangrove destruction, and the resultant reduction in organic matter and nutrient inputs through mangroves, as well as forest removal, mining and possibly commercial agriculture.

Activities which have directly threatened the lagoon in recent years include dynamite fishing by a small number of local individuals. There are increasing pressures for commercial fishing, due to an increasing population and growing prospects of inadequate agricultural land.

The lagoon is traversed by most trading and passenger ships travelling between Honiara and Ghizo, but there is little or no legislative provision for pollution control. This is not a problem at present but could become so in the near future. The growth of tourism and increased number of visiting yachts is another potential problem; some yachtsmen have trawled, dredged and dived for rare shells without the necessary permission from those exerting traditional jurisdiction (Baines, 1985a; Anon., 1986a). There is also considerable mineral prospecting activity, mainly for epithermal gold deposits (Baines *in litt.*, 2.9.87).

Legal Protection Most of the coastal lands and all of the submerged lands of the lagoon are subject to traditional law (Baines, 1985a).

Management Throughout the area there is a uniform system of reef-lagoon tenure which is largely traditional, although there have been some adaptations. Most areas of the lagoon are only being intermittently fished, due to the part-time nature of the industry, which results in a *de facto* rotation of fishing grounds.

Recommendations In June 1984, the Marovo Area Council called on the Western Province Government to make arrangements for a comprehensive survey of the area's resources and environment. Shortly after, a proposal for an investigation of lagoon resource management needs and preparation of appropriate management plans was prepared by the Commonwealth Science Council as part of SOPACOAST, and Marovo was accepted as a pilot project for high islands of the South Pacific region. The objectives of this project are (Anon., 1986b):

1. to define and describe the resources and environment of the area;
2. to describe the various development activities underway and the actual and potential threats;
3. to assist in the development of local community capacity to assess, monitor and use sustainably the lagoon resources;
4. to devise a coastal resource management regime, using appropriate traditional arrangements;

An important element of the project is education and training. A community awareness workshop for the project is held annually to inform the local communities of the results of project investigations and to facilitate community review and direction of the project (Anon., 1986a; Baines *in litt.*, 2.9.87). The lagoon may be used as a source of Giant Clams for the hatchery being established at the Coastal Aquaculture Centre near Honiara (Anon., 1987a).

TAIWAN

INTRODUCTION

General Description

Oceanic conditions around Taiwan are greatly influenced by the Kuroshio Current and the seasonal monsoon. During the summer, gentle south-west winds blow and the Kuroshio Current is strong, bringing warm water and calm weather. During the winter, the cold, strong north-east monsoon dominates and the Kuroshio Current is weak, creating cool water temperatures and rough sea conditions. Typhoons hit the area from June to early October every year, and damage the reefs.

Corals are found in all the waters around Taiwan except in the sandy area on the west coast. The main reef area is around the southern tip in the vicinity of Kenting (Jones et al., 1972) where coral communities with small fringing reefs are found (Su et al., 1984) (see account for Kenting National Park). These reefs are characterized by a vast and magnificent coverage of soft corals (Yang et al., 1976), but stony corals are equally well developed. The rocky north shore has flourishing solitary or patchy communities of scleractinians (Chang, 1983a) but these are much less well developed than those off Kenting (see account for Northern Reef).

Fringing reefs also flourish around offshore islands such as Lu-tao (Green Island) and Lan-yu (Orchid Island) which are located to the south-east in the pathway of the warm Kuroshio Current (Chang, 1983b) (see separate accounts). Hsiao-liu-chiu, to the south-west, has a coral fauna similar to that of Kenting Reef but better developed and dominated by reef-building species (Yang et al., 1975b) (see separate account). Reefs in the Pen-hu islands (Pescadores) are also described in a separate account. Tung-sha (Pratas) and Nan-sha (Spratly Islands) are typical tropical fringing and atoll reefs to the south of Taiwan in the South China Sea and are poorly known due to their remoteness (Chang et al., 1982; Yang et al., 1975a and see separate accounts).

The Taiwanese reefs have been studied by Cheng (1971), Dai and Yang (1981), Jones et al. (1972), Kawaguti (1953), Randall (1972), Randall and Cheng (1977 and 1979) and Yang et al. (1975a, b and 1982) and a brief overview is given in Liang (1985b). A total of 213 species and 52 genera of corals have been recorded. Marine communities have been surveyed in several areas (Chang, 1983a and b; Chang et al., 1982; Chang and Shao, 1981; Chiang, 1973; Lee, 1979 and 1980; Shen, 1984; Yang et al., 1980). Currently, a project on the basic ecology and physio-biochemistry of coral is being sponsored by the Kenting National Park and the Ministry of Interior, and is being carried out by the National Sun Yat-Sen University.

Coral reef fish communities on Taiwan are rich and diverse. A long term project is under way (Shao pers. comm.) on the southern and northern reefs to monitor community change in the coral fish fauna. There are few specific studies on invertebrates or algae of reefs but information is scattered in ecological survey reports. An invertebrate survey project is being carried out on the Kenting reefs by the Institute of Zoology, Academica Sinica.

Reef Resources

Coral reefs in Taiwan are important for both tourism and fisheries. Those within National Parks (see below) and at Pen-hu, Lan-yu and Lu-tao are major attractions for tourists and are popular for recreational fishing. Tourism is promoted by the Bureau of Tourism, the National Park Service and local government. There is little SCUBA diving at present; most takes place in the Kenting area and north and north-east coast. Fish from the waters adjacent to the reefs comprise a considerable portion of the total catch of near shore fisheries. Detailed figures are given in the following accounts.

Disturbances and Deficiencies

Heavy tourism, aquarium fish collection, siltation from unrestricted dredging and construction, explosive fishing, coral collection, as well as widely occurring pollution of different kinds are considered to be threatening many reef systems. There have been several projects on the environmental impact of nuclear power plants, in which coral and their responses to thermal effects have been studied (Su et al., 1979 and 1984; Yang et al., 1976, 1977 and 1980) and a programme on the hydrology of the waters around nuclear plants is currently under way to monitor the possible influence of outlet water on corals.

Legislation and Management

The National Park Law of 13.6.72 provides for the establishment of National Parks. These are zoned into Nature Reserves, Historical and Cultural Sites, Special Landscape Areas, Recreation Areas and Regulated Development Areas; park management plans are to be revised every five years. The National Park Department within the Construction and Planning Administration of the Ministry of the Interior is responsible for their administration (McHenry, 1984). Kenting National Park includes coral reefs (see separate account).

Coastal resources are protected under the Coastal Area Environment Protection Plan which is administered by the National Park Department. Under this, seven coastal conservation zones (CCZ) have been established:

- North-east Coast CCZ
- Kenting National Park
- Lang-Yang River Mouth CCZ
- Su-Hua CCZ
- Hua-Tung CCZ
- Chan-Yun-Chia CCZ
- Tan-Shui River Mouth CCZ

All include reefs except the last two which lie on the sandy west coast. They are managed by the Ministry of the Interior and Government of Taiwan Province. Activities permitted depend on the type of zone (e.g. lobster, abalone, wetland, mangrove). Twenty-five fishery conservation areas have also been established which include two on the north coast, two bordering Kenting National Park, one at Lu-tao, one at Hsiao-liu-chiu and one in the Pen-hu islands, all of which

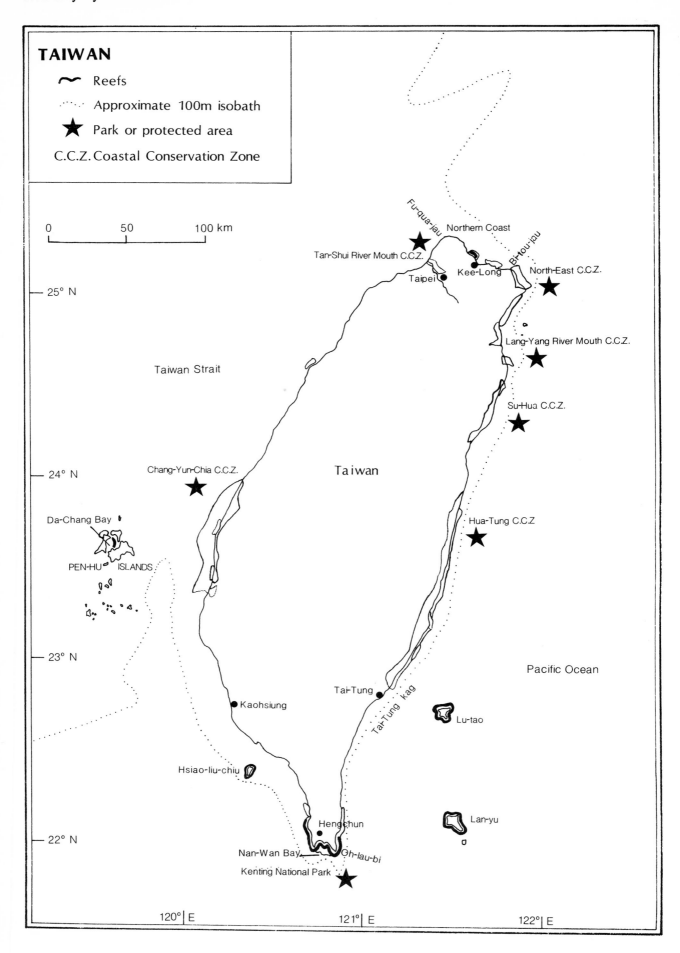

TAIWAN

~ Reefs

⋯ Approximate 100m isobath

★ Park or protected area

C.C.Z. Coastal Conservation Zone

0   50   100 km

Fu-qua-jau

Northern Coast

Tan-Shui River Mouth C.C.Z.

Kee-Long

Bi-tou-jou

Taipei

North-East C.C.Z.

— 25° N

Lang-Yang River Mouth C.C.Z.

Taiwan Strait

Su-Hua C.C.Z.

Chang-Yun-Chia C.C.Z.

— 24° N

Taiwan

Da-Chang Bay

Hua-Tung C.C.Z

PEN-HU ISLANDS

— 23° N

Pacific Ocean

Tai-Tung

Kaohsiung

Tai-Tung kag

Lu-tao

Hsiao-liu-chiu

Lan-yu

Heng-chun

— 22° N

Nan-Wan Bay

Oh-lau-bi

Kenting National Park

120° E   121° E   122° E

include or are adjacent to coral reefs. These areas are managed in the same way as the coastal conservation zones, but commercial fishing is strictly prohibited as their aim is to protect nursery areas for fisheries. Within the conservation areas, environmental impact assessments are required before approval of any development project (McHenry and Wu, 1985). Artificial reefs in Taiwan are described in Chang (1979).

On 20 September 1984, the Executive Yuan of Taiwan approved the Taiwan Nature Conservation Strategy, drawn up by the National Park Department and modelled on the World Conservation Strategy (Anon., 1984b; Construction and Planning Administration, 1985). Nine major work plans have been drawn up under this, covering ecological monitoring and inventories, land use planning, pollution, protected area planning, threatened species protection, public awareness campaigns and coastal management.

Recommendations

The National Park Department hopes to expand and devise a Coastal Zone Management Plan for the entire coastal area. However, currently a mechanism to enforce legislation is lacking. A number of recommendations for future management are given in Anon. (1984c).

References

Anon. (1984a). Director General Chang attends First World Conference on cultural parks. *Taiwan Parks Newsletter* 1(2): 6-7.
Anon. (1984b). Nature Conservation Strategy approved by the Government. *Taiwan Parks Newsletter* 1(2): 1-2.
Anon. (1984c). Coastal expert reviews Taiwan's Coastal Protection Plan. *Taiwan Parks Newsletter* 1(2): 9-10.
Anon. (1984d). Research begins at Kenting National Park. *Taiwan Parks Newsletter* 1(2): 12-13.
Bassett-Smith, P.W. (1890). Report on the corals from Tizard and Macclesfield Banks. *Ann. Mag. Nat. Hist.* 6(6): 353-374, 443-458.
Chang, K.H. (1979). Artificial reefs in Taiwan (1), (2), (3). *Monogr. Ser. Inst. Zool. Acad. Sin.* 7.
Chang, K.H. (1983a). Feasibility study on establishment of a marine science museum at northern and southern Taiwan. (In Chinese). *Inst. Zool. Acad. Sinica Monogr.* 10. 89 pp.
Chang, K.H. (1983b). Marine resources of Lanyu and Lutao islands. *Inst. Zool. Acad. Sinica Monogr.* 9. 69 pp. (In Chinese).
Chang, K.H., Jan, R.Q. and Hua, C.S. (1981). Inshore fishes at Tai-ping Island (South China Sea). *Bull. Inst. Zool. Acad. Sinica* 20(1): 87-93.
Chang, K.H., Jan, R.Q. and Hua, C.S. (1982). *Fishes of Itu Aba Island in South China Sea.* Encylopedia Sinica Inc., Taipei, Taiwan. 124 pp. (In Chinese).
Chang, K.H. and Shao, K.T. (1981). Ecological studies on distribution and habitat of coral reef fishes from the southern part of Taiwan. *Proc. 4th Int. Coral Reef Symp., Manila* 2: 588. (Abst.).
Cheng, Y.M. (1971). On some recent commensal solitary corals from Anping, Taiwan. *Taiwan Oceanogr. Sinica* 10: 1-6.
Chiang, Y.M. (1973). Studies on the marine flora of southern Taiwan. *Bull. Jap. Soc. Phycol.* 21: 97-102.
Construction and Planning Administration (1985). *The*

Taiwan Nature Conservation Strategy. Ministry of the Interior. 39 pp. (English translation).
Dai, C.F. and Yang, R.T. (1981). Alcyonaceans from Southern Taiwan. Part 1. Genus *Sarcophyton* (Alcyoniidae). *Acta Oceanogr. Taiwanica* 12: 121-131.
Fan, K.L. (1982). A study of water masses in Taiwan Strait. *Acta Oceanogr. Taiwanica* 13: 140-153.
Hayasaka, I. (1935). Reef-building corals in the bay of Su-Ao. *Geol. Bull. of Taiwan* 2: 5-8.
IUCN (in prep.). *Directory of Indo-Malayan Protected Areas.* IUCN, Gland and Cambridge.
Jones, R.S., Randall, R.H., Cheng, Y.-M., Kami, H.T. and Mark, S.-M. (1972). A marine biological survey of southern Taiwan with emphasis on corals and fishes. *Nat. Taiwan Univ., Inst. Oceanogr., Spec. Publ.* 1: 1-93.
Kawaguti, S. (1953). Coral fauna from Botel Tobago, Formosa, with a list of corals from the Formosa waters. *Biol. J. Okayama Univ.* 1(3): 185-201.
Lee, S.C. (1979). The intertidal fish of the rocky pools of Wan-Li in the northern Taiwan. *Quat. J. Taiwan Mus.* 22: 167-190.
Lee, S.C. (1980). Intertidal fishes of the rocky pools at Lanyu, Taiwan. *Bull. Inst. Zool. Acad. Sinica* 19(2): 1-13.
Liang, J.-F. (1985a). Holocene reef corals of China. In: Zeng, Z.X. (Ed.), *Coral Reefs and Geomorphological Essays of South China.* Geography Series No. 15, Institute of Geomorphology, South China Normal University, Guangzhou, China. Pp. 1-54.
Liang, J.-F. (1985b). Ecological regions of the reef corals of China. *J. Coast. Res.* 1(1): 57-70.
McHenry, T. (1984). National Parks of Taiwan. *Tigerpaper* 11(3): 24-28.
McHenry, T. and Wu, C.-A, (1985). Coastal zone protection in Taiwan. *Tigerpaper* 12(1): 13-16.
Melville, D.S. (1984). Seabirds of China and the surrounding seas. In: Croxall, J.P., Evans, P.G.H. and Schreiber, R.W. (Eds), *Status and Conservation of the World's Seabirds.* ICBP Technical Publication No. 2, Cambridge. Pp. 501-512.
Nerbonne, J.J. (1981). Coral - a treasure from the sea. *The Asia Magazine* April 26, 1981: 21-25.
Randall, H. (1972). Corala, a marine biological survey of southern Taiwan with emphasis on corals and fishes. *Inst. Ocean. Coll, Sci. Nat. Taiwan Univ.*: 39-45.
Randall, R.H. and Cheng, Y.-M. (1977). Recent corals of Taiwan. Part 1. Description of reefs and coral environments. *Acta Geologica Taiwanica* 19: 79-102.
Randall, R.H. and Cheng, Y.-M. (1979). Recent corals of Taiwan. Part 2. Description of reefs and coral environments. *Acta Geologica Taiwanica* 20: 1-32.
Shen, S.C. (Ed.) (1984). *Coastal Fishes of Taiwan.* Dept Zool. Natl Taiwan Univ., Taipei. 189 pp.
Su, J.C. (group representative) (1979). An ecological survey on the waters adjacent to the nuclear power plants in northern Taiwan. 5. *Natl. Sci. Comm. on the problem of Enviro., Acad. Sinica, Spec. Publ.* 5. 52 pp. (In Chinese).
Su, J.C. (group representative) (1984). An ecology survey on the waters adjacent to the nuclear power plant in southern Taiwan. *Natl Sci. Comm. on the problem of Enviro., Acad. Sinica, Spec. Publ.* 27. 214 pp. (In Chinese).
Yang, R.T. (group representative) (1976). A marine biological data acquisition program pertaining to the construction of a power plant in the Nan-Wan bay area. Phase 1. A preliminary reconnaissance survey. *Inst. Oceanogr. Natl Taiwan Univ., Spec. Publ.* 11. 134 pp.
Yang, R.T. (group representative) (1977). A marine biological data acquisition program pertaining to the construction of a power plant in the Nan-wan bay area.

Phase 2. Biological data acquisition. *Inst. Oceanogr. Natl Taiwan Univ., Spec. Publ.* 13. 194 pp.

Yang, R.T. (group representative) (1980). A survey of marine ecology and underwater scenery of Kentin national park. *Inst. Oceanogr. Natl Taiwan Univ., Spec. Publ.* 26. 100 pp. (In Chinese).

Yang, R.T. (group representative) (1982). Ecology and distribution of coral communities in Nan-wan bay, Taiwan. *Inst. Oceanogr. Natl Taiwan Univ., Spec. Publ.* 40. 74 pp. (In Chinese).

Yang, R.T. (1985). Coral communities in Nan-Wan Bay, Taiwan. *Proc. 5th Int. Coral Reef Cong., Tahiti* 6: 273-278.

Yang, R.T., Chiang, Y. and Huang, T. (1975a). A report of the expedition to Tung-Sha reefs. *Inst. Oceanogr. Natl Taiwan Univ., Spec. Publ.* 8. 33 pp. (In Chinese).

Yang, R.T., Chi, K.S., Hu, S.C. and Chen, H.T. (1975b). Corals, fishes and benthic biota of Hsiao-Liuchiu. *Inst. Oceanogr. Natl Taiwan Univ., Spec. Publ.* 7. 53 pp.

Yang, R., Yeh, S. and Sun, C. (1980). Effects of temperature on reef corals in the Nan-Wan Bay, Taiwan. *Inst. Oceanogr. Natl Taiwan Univ., Spec. Publ.* 23. 27 pp.

Descriptive information not cited comes from the personal observations of the author, based on field work during the past nine years.

HSIAO-LIU-CHIU REEF FISHERY CONSERVATION AREA

Geographical Location   18 mi. (29 km) south of Kaohsiung and 8 mi. (13 km) south-west of Tung-kang; 22°20'N, 120°22'E.

Area, Depth, Altitude   The island is 6.8 sq. km; coral occurs from 0 m to 40 m depth.

Land Tenure   Private and public owned; an administrative district of the local government of Pin-Tung Hsieh, Pin-Tung, Taiwan.

Physical Features   Hsiao-liu-chiu is a reef island on the edge of the Taiwan shelf which drops steeply to thousands of metres depth to the west and south. The climate is tropical to sub-tropical with a north-east monsoon in winter and a south-west wind and occasional typhoons in summer. The water is warm and clear with an average temperature of 26°C. The current is influenced by a branch of the north-flowing Kuroshio Current, although the north-east monsoon brings south-flowing cold water.

Reef Structure and Corals   Coral communities and fringing reefs on the west side of the island are best developed. Patch reefs, with high coral cover, are found in waters more than 40 m deep and are more common than fringing reefs. Hermatypic scleractinian corals dominate and are well developed in depths of 8-18 m. Dominant species are *Acropora spicifera*, *Favia speciosa*, *Goniastrea pectinata* and *Platygyra rustica*. Typical species of soft corals are *Sinularia polydactyla* and *Sarcophyton glaucum*. A total of 177 species and 48 genera of coral have been recorded (Yang *et al.*, 1975b).

Noteworthy Fauna and Flora   A total of 176 species of coral fish (87 genera, 43 families) have been recorded and the fish fauna is similar to that of southern Taiwan (Yang *et al.*, 1975b). 25 species of molluscs in 22 genera have been collected (Yang *et al.*, 1975b). Common molluscs and echinoderms are *Vasum turbinellus*, *V. ceramicum*, *Tridacna* sp., *Comanthus japonica*, *Certonardoa semiregularis*, *Diadema setosum* and *Echinometra mathaei*.

Scientific Importance and Research   A marine ecological survey project was carried out here by the Institute of Oceanography, National Taiwan University as the complexity and pristine nature of these reefs provided good research conditions; they are considered to be among the best reefs in Taiwan.

Economic Value and Social Benefits   About 16 000 people live on the island. Most are fishermen operating on the adjacent fishing ground. Tourism is an important source of income.

Disturbance or Deficiencies   Collection of coral for sale to tourists, explosive fishing, and aquarium fish collection are the major problems on the reefs. Typhoons occur from June to October.

Legal Protection   The waters around the island have been declared a fishery conservation area for lobster.

Recommendations   Controls on marine organism collection and trading should be established.

KENTING NATIONAL PARK

Geographical Location   The Park is approximately 100 km south of Kaohsiung City, at the southern tip of Taiwan, Pingtung County, and includes Nan-Wan Bay; the reef is at 21°54'-22°12'N, 120°40'-120°51'E.

Area, Depth, Altitude   The park area covers 32 631 ha, of which 14 900 ha is marine; reefs are found from 0 m to more than 40 m depth and extend 30 km along the shore; max. alt. in the park is 520 m.

Land Tenure   Nationally owned: 1.90%; Provincial: 11.65%; County: 2.78%; District: 0.56%; Ministry of Defence: 0.51%; Public Utilities: 0.26%; Forest Bureau: 32.27%; Range Bureau: 15.69%. Total Public Land: 66.62%; total private land: 33.38%.

Physical Features   The two-pronged Hengchun Peninsula juts out into the south portion of the park and provides a varied topography including hill, tablelands, lakes, open meadows, sandhills, sandfalls, sand rivers, rocky coasts, limestone coasts, cliffs and coral reefs. Particularly striking features include coastal sites such as Taiping Top, Rocky Kuanshan and Maopitou along the Taiwan Straits; Oh-lau-bi and Chialoshui on the Pacific Ocean; densely forested Banana Bay; caves in a raised coral reef, including the Stalagmite, Fairy and Silver Dragon caves; and volcanic fumeroles including a permanent natural flame burning gases emerging outside Hengchun Gate. The shore has rocky areas, sandy beaches, large blocks, solitary boulders, pebbles, and reefs on limestone bedrock. It gradually slopes west to the continental shelf and the Taiwan Strait. There are mainly fringing or patch reefs but a small barrier-like reef is found east of Ho-pi-hu.

The climate is typical sub-tropical to tropical. Mean annual precipitation for Kenting is 2030 mm, 82% falling between June and October. November to May is the dry season with monsoons that cause high winds, and typhoons occur from July to September. The average mean temperature in January is 20.5°C. and in July, 28.3°C. The water is fairly clear, with vertical visibility of 10-25 m. Water temperature ranges from 21 to 29°C, salinity from 33 to 39%, pH from 8.1 to 8.5, and oxygen saturation from 90 to 107% (5-6 mg/l) (Yang *et al.*, 1976). Current flow is mainly north or north-west; current velocity ranges from minimal to 20 cm/sec. During flood tides, water moves south-east to north-west, and the direction reverses during ebb tides (Su *et al.*, 1984).

Reef Structure and Corals Coral communities flourish but are modified by the characteristics of the coast. Coral patches start at Ho-wan, and stretch 30 km south along the shore to Oh-lau-bi. The reef is discontinuous, interrupted by sandy beaches, cliffs and Nan-wan Bay. Coral growth extends to 40 m depth. Chang and Shao (1981) describe four zones: limestone tidal pools; a subtidal reef margin with coral; a reef front zone and reef platform with 50% alcyonarian cover; and a submarine terrace of scattered coral rocks and coral mounds. Two types of coral communities are found. To the west of Nan-wan Bay, around Hou-pi-hu, Ta-lau-ku and Shiao-lau-ku, there are flourishing soft coral communities. Average coral cover is more than 75%, and up to 95% in certain areas, of which 35% is stony coral and 60% is soft coral (Su *et al.*, 1984; Yang *et al.*, 1982). East of Nan-wan Bay, stony corals predominate and coral cover averages 36-40%. The species diversity of the soft coral community is lower than that of stony corals, but both appear to be stable and to have reached climax conditions. More than 50 genera and 173 species of stony coral have been found in this area, Acroporidae, Faviidae and Poritidae being dominant (Jones *et al.*, 1972). A unique community dominated by *Junella juncea* (sea whip) has been observed between Mao-bi-tou and Hou-pi-hu in depths of 6-10 m. Yang (1985) provides a description of coral communities in Nan-Wan Bay.

Noteworthy Fauna and Flora At least 400 species of fish in more than 70 families have been recorded on these reefs (Chang and Shao, 1981). Labridae, Pomacentridae, Chaetodontidae, Apogonidae, Acanthuridae, Blenniidae, are the obvious dominant families. The only known locality in Taiwan of the cephalochordate fish, *Asymmetron lucayanum*, the Lancelet, is the sandy bottom of Nan-wan Bay (Yang *et al.*, 1977).

Approximately 250 species of benthic macroinvertebrates other than corals have been recorded: in order of abundance these are Mollusca, Arthropoda, Echinodermata and Annelida (Yang *et al.*, 1977). One-fifth of these species are associated with sandy substrates in the Nan-wan Bay system. There are ca 70 species of algae in more than 60 genera (Chlorophyta, Rhodophyta, Phaeophyta, Chrysophyta and Cyanophyta) on the reefs (Yang *et al.*, 1977). Conspicuous seasonal

variation in algal abundance which increases from late winter to early spring has been observed.

Terrestrial fauna and flora are briefly mentioned in Anon. (1984d) and are described in IUCN (in prep.).

Scientific Importance and Research The Kenting reefs have attracted the most intensive reef research work in Taiwan as a result of four factors: (1) a nuclear power plant has been built here; (2) there are unique very rich, extensively developed soft coral communities; (3) abundant fish larvae and eggs have been found, mainly in spring and summer; and (4) the area is in a national park. Baseline surveys on ecology, including hydrology, nutrients, plankton, fish, corals, algae, benthos, radio-activity and fisheries have been carried out on the waters adjacent to the nuclear power plant since the beginning of its construction in 1979 (Su *et al.*, 1984; Yang, 1985). A monitoring system has just been established. More academically oriented research, such as the physio-chemistry of corals and the recruiting mechanism of reef fish (Shao pers. comm.) is also under way. Terrestrial research projects are described in Anon. (1984d).

Economic Value and Social Benefits One major fishery port, Hou-pi-hu, and several smaller ones are located in this area. Approximately 700 fishermen operate 77 boats (724 tons total) and 225 motor rafts in the surrounding waters. Collection of milk fish larvae is also a major fishery enterprise. Annual catch may exceed five million fish, worth 250 000 U.S. dollars. About 22 100 people live within the park in several small villages. Tourism and recreational facilities are increasing rapidly due to promotion by the National Park Service. Exhibition halls for public education are being planned.

Disturbance or Deficiencies Natural events cause damage to the reefs, particularly typhoons which strike during the summer (June to October). Although there may be several a year, it seems that the reefs absorb the impact and recover quickly. Crown-of-thorns Starfish *Acanthaster planci* have been observed on the reefs from Wan-li-tung to Pai-sha. Some damage was reported (Jones *et al.*, 1972).

Coral reefs at Kenting are coming under increasing pressure. The main disturbances come from: (1) siltation due to road and housing construction along the coast; (2) harbour dredging; (3) explosives fishing; (4) collecting for the aquarium trade; (5) damage by tourists (trampling and collecting) and SCUBA divers; (6) anchor damage; and (7) repetitive specimen collection by scientists. Fish abundance has notably declined over the past nine years. In addition, the warm outlet water of the nuclear power plant is likely to have thermal and chemical (e.g. chlorine) impacts on the reefs. The plant started operating in July, 1984, and it is not yet known if there has been any negative effect on the reefs.

Legal Protection Previously a national recreation area managed by the Tourism Bureau, the boundary of Kenting National Park was designated in 1979. The Executive Yuan approved its Final Management Plan, thus officially establishing it as a National Park, on 1 September 1982, and it was finally staffed and a headquarters established on 1 January 1984 (IUCN, in prep.). The Kenting reef area was declared a National Reserve in 1982, after the establishment of the National

Park. Collection, fishing or selling of natural organisms without permission is prohibited. The area has also been declared a Coastal Conservation Zone under the Coastal Area Environment Protection Plan (McHenry and Wu, 1984).

Management The reefs are managed by Kenting National Park Service, 272 Kentin Road, Kentin, Hen-chun, Pin-ton, Taiwan, and are reasonably well protected and monitored. A long-term project is under way to monitor and prevent damage to corals from the cooling water of the nuclear power plant. Further details of management of the National Park are given in IUCN (in prep.) and McHenry (1984).

Recommendations Fishing with explosives, indirect siltation resulting from land construction and damage from tourists must be more actively controlled. The impact of the nuclear power plant must continue to be monitored. Long-term scientific research projects, perhaps with international co-operation, focusing on the physio-ecology of corals which may act as good indicator organisms of the impact of the plant, could be very rewarding.

LAN-YU REEF (ORCHID ISLAND REEF)

Geographical Location 41 naut. mi. (76 km) east of Oh-lau-bi and 49 naut. mi. (91 km) south-east of Tai-tung city; 22°N, 121°30'E.

Area, Depth, Altitude Lan-yu is 45.79 sq. km; reefs are 38.54 km long, corals grow from the surf zone to more than 50 m depth.

Land Tenure Public or freehold by local people; Lan-yu is a county run by the local government of Tai-tung Hsien, Tai-tung, Taiwan.

Physical Features Lan-yu is a volcanic island, rising from the deep bottom of the Pacific, and is situated in the main path of the warm Kuroshio Current. The weather is tropical; the water is warm and clear. A north-east seasonal wind is strong during winter which affects the development of the coral community. Fringing and patch reefs are found.

Reef Structure and Corals Stony corals are dominant. Acroporidae dominate near the surf zone, and *Porites*, Faviidae and Mussidae flourish in deeper water. Other common corals are Xenidae, *Fungia, Millepora*, Pocilloporidae, Pectiniidae, Dendrophylliidae, *Heliopora*, Stylasteridae, *Tubipora, Galaxea* and Alcyoniidae (Chang, 1983b). Near the Kai-Yan Harbour, sea fans form a special community. Average cover and diversity is fairly high.

Noteworthy Fauna and Flora A total of 347 species of fish in 49 families have been recorded. The dominant families are, in order, Labridae, Pomacentridae and Chaetodontidae; species in these families make up 39% of all fish found here (Chang, 1983b; Lee, 1980). Molluscs and crustaceans are also abundant. The Coconut Crab *Birgus latro* occurs. At least four species of sea snakes are commonly seen on the reefs, of which *Laticauda colubrina* and *L. semifasciata* are the commonest. Sea turtles often appear in the vicinity. The Brown Booby *Sula leucogaster* formerly bred on the island; the Black-naped Tern *Sterna sumatrana* may breed (Melville, 1984).

Scientific Importance and Research This is a fairly pristine area and corals and the associated community are relatively undisturbed. There is therefore good potential for research. Marine life is abundant and as the island is located in the path of the Kuroshio current, the area may be suitable for zoogeographical studies.

Economic Value and Social Benefits There is great potential for tourism and recreation, providing the culture of the Lan-yu native people is carefully preserved. About 2500 Yami people live on the island and subsist on taro farming and fishing (Anon., 1984a).

Disturbance or Deficiencies A nuclear waste disposal site was recently established on the island. Oil pollution, which may come from shipping vessels, has been observed. Collection of aquarium fish, further pollution and heavy tourism are potential problems. Explosive fishing is frequently reported. The Coconut Crab is declining due to over-exploitation. During summer, typhoons may strike the island and damage the reefs.

Legal Protection The reefs are unprotected.

Management The reefs are not managed.

Recommendations Promotion of tourism on the island should take into consideration the preservation of the native environment and culture. The island has already been considered for national park status but this was refused as it was thought that the increase in tourism that such a designation would bring would cause further disruption to the Yami people. A "cultural park", which would limit the number of visitors to the island, is therefore being considered (Anon., 1984a). Fishing with explosives should be strictly prohibited.

LU-TAO REEF (GREEN ISLAND REEF) FISHERY CONSERVATION AREA

Geographical Location 40 naut. mi. (74 km) north of Lan-yu Island, 18 naut. mi. (33 km) east of Tai-tung City; 22°40'N, 121°29'E.

Area, Depth, Altitude The area of the island is 17.32 sq. km.

Land Tenure National ownership.

Physical Features The island is volcanic and resembles Lan-yu, and the surrounding waters have similar characteristics to those surrounding Lan-yu.

Reef Structure and Corals The island is surrounded by fringing and patch reefs, the reef structure and composition resembling that of the reefs of Lan-yu. *Melithaea* and *Tubipora* are abundant around the lighthouse and Kung-kwan; Dendrophylliidae is dominant in Kung-kwan. A large sea-fan community occurs near Chung-liao.

Noteworthy Fauna and Flora A total of 310 species of fish in 52 families have been recorded. Of these, 237 species in 42 families overlap with the fish of Lan-yu (Chang, 1983b). The dominant species are similar to those of Lan-yu. A large school of *Odonus niger* has been seen in the waters of Chung-liao. Other marine life is similar to that of Lan-yu. The Black-naped Tern *Sterna sumatrana* may breed on the island (Melville, 1984).

Scientific Importance and Research The reefs have considerable research potential, similar to those of Lan-yu.

Economic Value and Social Benefits There is great potential for tourism and recreation.

Disturbance or Deficiencies Damage with the development of tourism is a potential problem.

Legal Protection This is a fishery conservation area for lobster, which are protected in the surrounding area. Certain places on the island are restricted areas.

NAN-SHA REEFS (NANSHA QUNDAO, SPRATLY ISLANDS)

Geographical Location The main island is Taiping Tao (or Itu Aba Island), 10°23'N, 114°22'E, and is located north-west of Tizard banks and reefs; the southernmost, at 4°N, is Tsan-Mou Reef; 4°-11°30'N, 109°30'-117°50'E.

Area, Depth, Altitude Taiping Tao is 1289 x 366 m; 3.8 m above sea-level.

Land Tenure Sovereignty over the reefs is disputed by a number of countries in the area, including Taiwan and China.

Physical Features The Nan-sha Reefs are a series of 104 emergent reefs and countless submerged reefs and include fringing reefs and atolls. The climate is oceanic tropical. Air temperature ranges from 26.1°C to 28.8°C. Average water temperature is 28.1°C. The north-east monsoon blows from October to March, the south-west monsoon from May to October. The current flows south-west during the former and east or north during the latter.

Reef Structure and Corals Taiping Tao is surrounded by a reef terrace and has sandy shoals, reef flats, reef barriers and reef fronts. Coral communities are flourishing and diverse, and there is high coral cover with *Acropora* the dominant species. Bassett-Smith (1890) first described corals from Tizard Bank. Liang (1985a and b) lists over 100 species of coral recorded at Tizard Atoll. There are two distinct coral zones, an upper terrace to 18 m depth and a lower terrace, from 37 to 56 m depth, with different coral composition. On the intervening slope, from 18 to 37 m, only one species was found. Some species have been recorded at considerable depth, for example *Leptoseris striata* at 65 m, *Montipora* sp. at 81 m and *Favia* sp. at 83 m. Eighteen species of

reef coral have been recorded on coral knolls in the lagoon.

Noteworthy Fauna and Flora A total of 111 species of fish in 26 families have been identified (Chang *et al.*, 1981). The fish fauna is similar to but less diverse than that of southern Taiwan. Giant clams *Tridacna* spp. have been seen. A flourishing eel grass bed was found on a sandy shoal.

Scientific Importance and Research Research expeditions, which concentrated on fish, were undertaken to Taiping Tao from the Fishery Research Institution of Taiwan and the Institute of Zoology, Academia Sinica, Taiwan, in 1976 and 1981 respectively.

Economic Value and Social Benefits Projects on the exploitation of new fishing grounds and commercial shell and lobster fisheries are now under way in the area.

Disturbance or Deficiencies Not known.

Legal Protection There is reportedly Chinese legislation stipulating a reserve in the area (Wang *in litt.*, 20.11.86) but details are lacking.

Management None known.

NORTHERN COAST AND NORTH-EAST COAST COASTAL CONSERVATION ZONE

Geographical Location Along the coast from Fu-qwa-jau to Bi-tou-jau; 40 km north of Taipei city; 25°-25°23'N, 121°30'-122°E.

Land Tenure Private or national ownership in different sections.

Physical Features This area is towards the northern edge of the range of reef-building corals. The sea bottom turns into soft substratum a few hundred metres from the surf line in most areas. The climate is moderate; average monthly air temperatures range from 14°C to 28°C. Annual precipitation is over 2700 mm and the north-east monsoon blows strongly in winter causing waves to increase from 2-3 m to as much as 6 m in height (Chang, 1983a). The Kuroshio Current exerts an influence; average water temperature in winter is 17°C and in summer, 28°C. Underwater visibility is usually less than 10 m.

Reef Structure and Corals Small dispersed coral patches occur in coastal waters. Larger communities of a few hectares are found only at Fu-qwa-jau, Yae-Liu, Shin-au and Bi-tou-jau, and there are no true reefs. Coral cover is limited, colonies growing on hard, rocky substrates, on terraces, boulders, cliffs or in small trenches. However, species diversity is fairly high and more than 82 species in 35 genera have been reported (Hayasaka, 1935). Common stony corals include *Acropora* spp., *Stylophora*, *Pectinia*, *Favia*, *Pavona*, *Turbinaria*, *Pachyseris*, *Pocillopora*, *Leptoseris*, *Hydnopora*, *Plesiastrea*, *Platygyra*, *Diploria*, and some rare ones such as *Euphyllia* spp., *Goniopora* and *Tubastrea*. *Melithaea* and *Gorgonia* occur on the walls of cliffs or trenches. Soft corals include *Sarcophyton, Lobophytum* and *Sinularia*.

Noteworthy Fauna and Flora The fish fauna is typical of rocky shores rather than of coral reefs. Dominant species include *Sebasticus marmoratus, Epinephelus diacanthus, Cephalopholis pachycentron, Choreodon azurio, Gymnothorax undulatus*, and reef families such as Acanthuridae, Blennidae, Pomacentridae, Chaetodontidae and Labridae. Sea urchins and brittle stars are abundant, particularly *Diadema setosum, D. savigni, Tripneustes gratilla* and *Ophiocoma* spp. Notable molluscs include *Ovula ovum, Ouoyula monodonta* and *Drupella cornus*. Algae vary seasonally, flourishing from late winter to spring. Species include *Ulva, Enteromorpha, Caulerpa, Endarachne, Porphyra, Bengia, Pterocladia, Padina* (Chang, 1983a; Su, 1979) and *Sargassum*.

Scientific Importance and Research There are two nuclear power plants on this coast and a joint environmental assessment project has been carried out in the adjacent waters for over ten years (Su *et al.*, 1979). In addition, a short term study was carried out in part of the area to assess an oil leaking event in 1977. Projects concerning fishery biology and artificial reefs have also been undertaken.

Economic Value and Social Benefits The international port, Kee-long, and seven small fishing ports are located in the area. Approximately 94 000 tons of fish are caught from the adjacent waters every year. Certain algae are harvested as food or for other uses. The small abalone, *Haliotis diversicolor*, and the ovaries of sea urchins are collected and are popular sea foods. Mariculture farms along the shore produce 200 tons of abalone annually.

Disturbance or Deficiencies The major disturbances come from oil pollution from large vessels and tankers. The rapid development of aquaculture farms along the coast may cause damage. Intensive spearfishing and collecting of corals, shells and other marine organisms by SCUBA divers and tourists may threaten the marine ecosystem. Fishing with explosives has been reported.

Legal Protection The North-east Coast Coastal Conservation Zone has been established along the shore east of Kee-long, primarily for its scenic attractions and as a recreational area. Within this area, the building of aquaculture installations is prohibited and existing abalone farms have been demolished so that the environment can revert to its natural state (McHenry and Wu, 1985).

Management The conservation zone is managed by the Tourism Bureau, Ministry of Transportation, Taipei, Taiwan. Sea water quality is routinely monitored by the environmental protection agency in and near Kee-long Harbour.

Recommendations Regulations for the collection of particular species, quotas, minimum sizes and closed seasons for fish and other marine invertebrates should be set. Coral collection should be prohibited.

PEN-HU REEFS (PESCADORES) AND FISHERY CONSERVATION AREA

Geographical Location A cluster of 64 islands located 30 mi. (48 km) west of Taiwan in the Taiwan Strait; 23°10'-23°50'N, 190°-190°21'E.

Area, Depth, Altitude 320 km of shoreline.

Land Tenure Private or public owned; the address of the local government is Government of Pen-hu Hsieh, Mar-Kong, Pen-Hu, Taiwan.

Physical Features The islands are located on the continental shelf in less than 200 m of water and are surrounded by cliffs, rocky shores and sandy or muddy beaches. Da-chang Bay is enclosed by several major islands and is muddy. The climate is subtropical but is influenced by the north-east monsoon. Average air temperature ranges from 16°C to 28°C. Annual precipitation is 1000 mm, 80% of which falls in the summer. Currents are dominated by a branch of the Kuroshio but during winter a cold current comes down from north (Fan, 1982). Horizontal visibility can be as low as a few metres in the bay but is more than 15 m in the open ocean.

Reef Structure and Corals Solitary or patchy coral reef communities are found and approximately 25% of the 320 km coastline has corals growing offshore. Patch reefs are most common but there are a few small fringing reefs around some islands. Common corals include *Acropora, Pocillopora, Stylophora* and brain corals. *Turbinaria* is abundant throughout. *Goniopora* spp. is abundant off Lon-man, and *Scytalium splendens* is abundant off Chi-tou and Da-Chang islands. There is a community of a few hectares of *Goniastrea* and *Favia* in the lower intertidal zone at Chi-tou which is exposed during spring tides.

Noteworthy Fauna and Flora A total of 309 species of fish in 83 families have been found in this area and the fauna is similar to that of Northern Reef. Labridae, Chaetodontidae and Pomacentridae are dominant. A total of 90 species of mollusc have been collected in Pen-hu, and Neritidae, Veneridae, Trochidae, and Cypraeidae are dominant (Chang *et al.*, 1982). The oyster *Crassostrea gigas* is commercially cultured. *Pinctada* sp. is a pearl-producing shell in the subtidal zone. Dolphins *Tursiops gilli* and *T. aduncus* come here in groups every winter. The islands are reported to have the most significant seabird breeding populations in Taiwan (Melville, 1984), including the Greater Crested Tern *Sterna bergii*, the Black-naped Tern *S. sumatrana*, the Bridled Tern *S. anaethetus* and the Brown Noddy *Anous stolidus*.

Scientific Importance and Research The oceanographic conditions and biological environment of Pen-hu have been intensively studied. At present, there is a three-year pioneer project on ocean farming under way in Da-chang Bay. A branch of the Taiwan Fishery Research Institute is based at Pen-hu and many fishery and aquaculture projects are carried out.

Economic Value and Social Benefits Pen-hu has a population of 105 141, most of whom depend on fishing; the annual catch is over 44 000 tons. Oyster culture and the squid and shrimp fishery are important. Da-chang Bay and some other areas are important nursery grounds for fish and shrimps. The islands are also the centre of the Taiwanese precious coral industry, although little is now collected in the area (Nerbonne, 1981).

Disturbance or Deficiencies There is a tendency towards over-fishing in the adjacent waters. Local people have been using poison to collect fish larvae which greatly damages coral communities. Tourism and aquaculture farm construction have put increasing pressure on the marine ecosystem. Terrestrial disturbance is reported to be moderately severe (Melville, 1984).

Legal Protection There are laws to control fish poisoning and the export of fish larvae from Pen-hu is restricted. Part of the area around Pen-hu is a fishery conservation area.

Recommendations It is recommended that the seabird breeding colonies are protected (Melville, 1984).

TUNG-SHA REEF (DONGSHA QUNDAO, PRATAS)

Geographical Location 240 mi. (386 km) south-west of Kaohsiung City, Taiwan, and 170 mi. (274 km) south-east of Hong Kong; 20°40'-43'N, 116°42'-44'E.

Area, Depth, Altitude Tung-sha (Dongsha reefs) cover ca 100 sq. km; Tung-Sha Tao (Dongsha Qundao) is 1.7 sq. km; max. alt. 5 m.

Land Tenure Sovereignty of the area is disputed by Taiwan and China.

Physical Features Tung-sha Reef is a large submerged atoll with the small island of Tung-sha Tao situated on its west arm. The island is covered with sandy coral debris and traces of guano and has a lagoon (0.6 sq. km) open on the west side. It is surrounded by a shallow terrace which drops to 40 m or more at the outer edge.

The climate is subtropical oceanic, heavily influenced by the monsoon. Average air temperature ranges from 20.2°C to 28.7°C; water temperature from 21°C to 30°C. A thermocline is found at 30-75 m in waters outside the atoll. Surface currents flow north-east with a velocity of 0.2-0.5 mi./hr (0.3-0.8 km/hr). Currents at depths of 300 m turn to the north-west with a speed below 0.2 mi./hr (0.3 km/hr). Seasonal winds also produce surface currents.

Reef Structure and Corals The reef structure is typical of atolls (Yang *et al.*, 1975a). Over 70 species of reef coral have been recorded (Liang, 1985a and b). In a three day trip, 45 coral species in 17 genera were collected: 60% of the specimens (125 pieces) were *Acropora*, 10% were *Porites* (Yang *et al.*, 1975a). The coral fauna appeared to be similar to that of Kenting and Hsiao-liu-chiu.

Noteworthy Fauna and Flora Twenty-five species of fish in 17 genera (86 specimens) have been collected. Apogonidae, Labridae, and Pomacentridae are dominant families. Twenty-six species of mollusc in 23 genera have been recorded. *Strombus, Monetaria* and *Tridacna* are abundant, while *Tutufa bubo, Oliva arnata, Terebra dimidiata* and *Arca ventricosa* are rare (Yang *et al.*, 1975a). The seagrasses *Halophila ovalis* and *Thalassia hemprichii* flourish. Eleven Chlorophycophyta, seven Phaeophycophyta and ten Rhodophyta have also been found.

Scientific Importance and Research The weather has been recorded since 1925 due to the presence of the lighthouse. An expedition sponsored by the National Science Council, Taiwan, carried out a survey in March 1975 and studied the hydrology, biology, geology, fishery and chemical composition of the reefs.

Economic Value and Social Benefits There is a permanent station established by Taiwan on the island which is an important site for navigation and ocean/weather observations. *Diegenea simplex*, an alga used pharmacologically, is collected in large quantities in this area. The adjacent fishing ground (fin and shell fish) is visited by fishermen from southern Taiwan and commercial shell and lobster fishing was under way in 1987.

Disturbance or Deficiencies There is a potential threat from overfishing of shell fish and populations of Giant Clams *Tridacna elongata* are reported to be decreasing.

Legal Protection None.

Management Reef resources are not managed.

TOKELAU

INTRODUCTION

General Description

Tokelau (9°S, 173°W), a non-self-governing territory of New Zealand lying 430 km north of Western Samoa, consists of a group of three low reef-bound atolls with a total land area of only 12.25 sq. km. Nukunonu Atoll (4.7 sq. km) has about 24 islets around a very large lagoon (98 sq. km). Fakaofo Atoll (4.0 sq. km) has about 60 islets on a rectangular reef (ca 32 km long) with a lagoon of 50 sq. km. Atafu (3.5 sq. km) has about 40 islets on a smaller triangular reef with a lagoon of 17 sq. km. Maximum altitude is 5 m and individual islets vary in length from 90 m to 6 km and in width from a few metres to 200 m. None of the lagoons has deep-water passages to the open sea but there is some water exchange across the reefs (Hooper, 1985). Stoddart (1976) gives brief descriptions of the terrestrial ecology and Hinkley (1969), Hooper (1985), Parham (1971) and Wodzicki (1968 and 1972) give extensive general information on the islands. Ecology of terrestrial arthropods is discussed by Hinkley (1969).

Average temperature is about 28°C and rainfall is heavy but irregular, with an annual average of 2500 mm. Severe storms have caused damage (see below).

There have been few studies of the reefs. Hinds (1969/1971) carried out a brief survey and reported that coral growth in the lagoons was limited to the upper portions of old coral massifs rising from the floor, the main portions of which were barren. A light capping of reddish coral was found close to the surface on massive grey coral heads on the south side of Atafu lagoon.

Laboute (in press) briefly surveyed reefs on all three atolls in 1987 to assess damage caused by Hurricane Tusi. On Fakaofo, dives were made on the outer slope about 600 m north of the village at Fenua Fala on the western tip, to the north of Fenua Fala, and in the lagoon. On Nukunonu, dives were made on the western outer slope at Ahaga-Lahi and in the south-west at Te Puka i Mua. On Atafu dives were made around the north-west tip, on the western outer reef slope of Te Alofi and in an area undamaged by the hurricane, 1.2 mi. (1.9 km) north of Edgar Island. Coral composition in this last area, which is leeward of the winds, was considered fairly representative of the outer reefs before the hurricane. The coral platform was ca 15 m wide, with the drop-off at 10-12 m depth. From 2 to 10 m depth, *Pocillopora eydouxi* and *P. verrucosa* together accounted for 50% of coral cover, the remainder being *Montipora* sp. (30% cover) and *Acropora* sp. (5-10%), with a few colonies of *Pavona* sp., *Hydnophora* sp. and *Millepora platyphylla*. The alga, *Halimeda* sp., occupied cracks and crevices between 4 and 15 m depth. At 18 m depth the main corals present were two species of *Pocillopora*, *Montipora* sp., *Pavona* sp. and *Porites lobata*. From 18 to 50 m depth, where the slope was 60°, *P. lobata* accounted for 60% of coral cover. From 55 to 60 m depth coral cover was ca 20% and consisted almost entirely of *Pachyseris speciosa* with a few small colonies of *Leptoseris* sp., *Mycedium elephantotus*, *Pavona* sp. and at least two species of *Favia*.

In Fakaofo lagoon, which suffered no damage from the hurricane, large coral pinnacles with jagged edges and steep slopes were scattered over the lagoon. Coral growth was diverse but not very luxuriant. The lagoon sediment was fine and the lagoon floor was muddy at 40-50 m, and expected to be so at greater depths. The bivalve molluscs *Tridacna maxima, Aca* sp. and *Chama* sp. were present on the coral pinnacles to 10 m depth and *Spondylus* sp. was abundant from 5 m to 50 m depth. A yellow-orange sponge (Axinellidae) was common on hard substrate. The holothurian *Holothuria atra* was present at densities of 8-12 per sq. m at the edges of the lagoon (Laboute, in press).

Tridacna spp. and thorny oysters were common on the north-east facing slopes of Atafu lagoon (Hinds, 1969/71). Good fish life was found on the seaward slope of Atafu reef down to 60-70 ft (18-21 m) but was mostly concentrated where heavy surf broke onto the reef (Hinds, 1969/71). Hinds (1969/71) reported that fish life was found to be virtually absent within the lagoons, but it is in fact fairly typical of other mid-Pacific atolls (Gillett *in litt.*, 26.7.87). The echinoderm fauna of the outer reef flat of Fakaofo included of the holothurians *Actinopyga mauritiana, Holothuria atra, Stichopus chloronotus, Bohadschia argus* and *Microthele nobilis*, the sea urchins *Heterocentrotus mamiliatus* and *Echinothrix calamaris*, and the starfish *Linckia multifora* (Laboute, in press).

The Green Turtle *Chelonia mydas*, the Hawksbill *Eretmochelys imbricata* and the Loggerhead *Caretta caretta* nest on the islands (Balazs, 1982).

Reef Resources

There is a stable human population of about 1600 (Hooper, 1985) and the marine environment is an important source of food. There is a rich and varied offshore fish fauna and fishing is a major source of protein; the fishery is subsistence only and surplus catches are distributed to kin and neighbours (Anon., 1985; Hooper, 1985). Traditional tuna fishing is described by Gillett (1986a) and Gillett and Toloa (1987). An assessment of the skipjack and bait fish resources of Tokelau is given in Tuna Programme (1983). Harvestable bait fish resources are very small, partly due to the absence of passes into the lagoons. An early description of the fisheries is given by Van Pel (1958). In 1986, *Trochus niloticus* was introduced to Fakaofo to provide an alternative source of revenue to copra and also as an additional source of protein (Gillett, 1986b; Vola, 1987). A survey was planned for November 1987 to judge the success of this (Gillett *in litt.*, 28.7.87). Trochus does not naturally occur in Tokelau but extensive consultation suggested that such an introduction would not be harmful (Gillett, 1986b). There is no tourism.

Disturbances and Deficiencies

Local reports in the late 1960s suggested that *Acanthaster planci* was present but a brief survey in 1971 revealed few

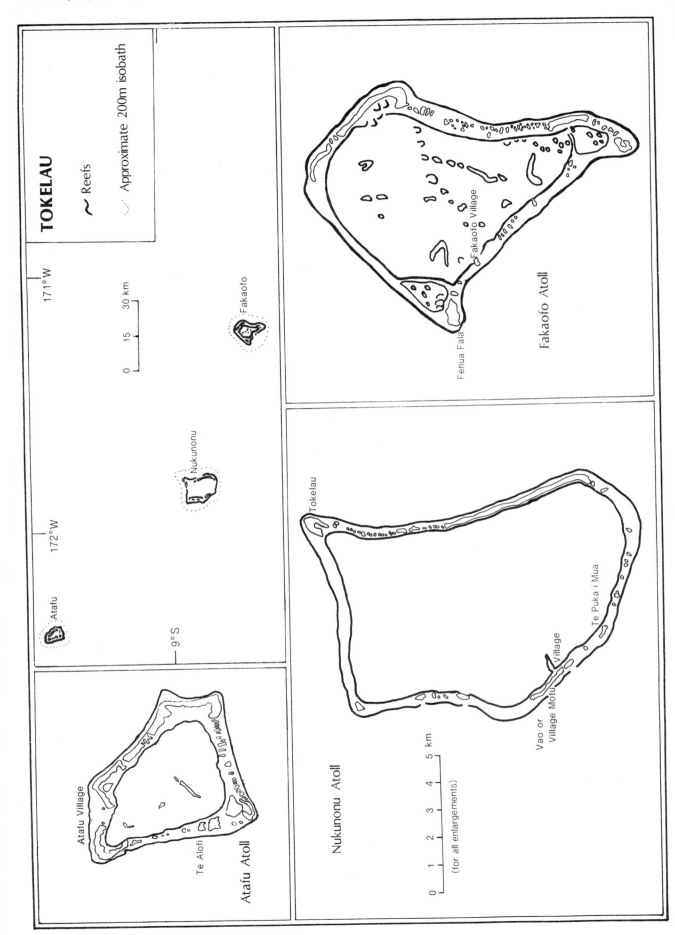

TOKELAU

〜 Reefs

·········· Approximate 200m isobath

171°W

172°W

9°S

0   15   30 km

Atafu

Nukunonu

Fakaofo

Fakaofo Village

Fenua Fala

Fakaofo Atoll

Tokelau

Vao or
Village Motu
Village

Te Puka i Mua

Nukunonu Atoll

0   1   2   3   4   5 km

(for all enlargements)

Atafu Village

Te Alofi

Atafu Atoll

specimens on any of the atolls and there appeared to be no abnormal features (Hinds, 1969/1971). In early 1983, extensive coral mortality was reported in reef shallows at Nukunonu, correlated with a sudden drop in sea-level (about 60 cm), which caused reef exposure and restricted lagoonal circulation, affecting fish and turtles as well. This may have been associated with the abnormal El Niño of that year (Glynn, 1984). In early 1987 widespread coral damage was caused on the western outer reef slopes of the atolls by a huge surf resulting from Hurricane Tusi (Gillett *in litt.*, 28.7.87; Laboute, in press). The degree of destruction of corals was variable, reaching 90% in some areas. No coral damage was evident within Fakaofo lagoon although large quantities of rubbish were deposited in the lagoon by the hurricane, near the village. Nukunonu and Atafu lagoons were not inspected (Laboute, in press).

Prior to the Hurricane, but also in 1987, corals on the north and north-western outer reef slopes of Atafu were found to be almost completely dead (mortality 90-100%) to at least a depth of 70 m, the only living corals being a few very young colonies of *Porites, Pocillopora* and *Montipora*. The cause of this die-off is not known but is thought possibly to be the result of the grounding on the reef in 1979 of a Korean ship which was blown up with dynamite in 1981. It was almost certainly not the result of storm damage, as the corals were not broken or displaced (Laboute, in press).

A significant area of Nukunonu lagoon reef was damaged by a pesticide spill in 1969; all corals in a 2 km section of the lagoon along Motu Te Kakai died with the exception of *Porites*. There was little recovery by 1975 (Marshall, 1976).

Turtles are now rare and there has been serious depletion of Giant Clams *Tridacna* spp. (Anon., 1985; Wodzicki, 1973) but there is no evidence of depletion of other marine resources (Hooper, 1985). There has been no dredging or poisoning on the reefs. Over the past twenty years a number of reef-blasting teams have visited Tokelau to blast small-boat channels across the reefs to the three village islets. The channels are small and do not facilitate exchange of sea water between the lagoon and open sea. A New Zealand Army team was deepening and widening the channels using explosives in 1986 (Hooper *in litt.*, 13.5.87).

Legislation and Management

There is no protected area system and conservation of resources has been achieved through traditional practices, which are described in Hooper (1985). Although they are tending to break down, there is still strong interest in maintaining many of these practices. An Agricultural and Fisheries Committee was set up in 1984 and representatives from the three island councils were encouraged to impose measures to reduce clam collecting and to improve turtle conservation. Fakaofo and Nukunonu have placed a ban on Giant Clam collecting, to be reviewed after one year. At Atafu the Council of Elders has placed a ban on the taking of turtle eggs, except for a few that are allowed to be removed for the purpose of hatching a pet turtle (Anon., 1985).

Recommendations

Development plans and needs for the islands are outlined in SPREP (1981) but there is little mention of marine issues. Further information on the reefs is clearly a priority. Wodzicki (1973) has recommended the further study and protection of turtle nesting beaches.

Laboute (in press) recommends monitoring for ciguatera fish poisoning for a period of 18 months following the damage caused by Hurricane Tusi, in view of the possibility that new sea bottom growth may be conducive to the development of the algae responsible for ciguatera. For the same reason, waste which has accumulated in Fakaofo lagoon south of the village, and probably also at Nukunonu, should be removed.

Fakaofo lagoon, and probably also Nukunonu and Atafu lagoons, may be suitable for culture trials of the pearl oyster *Pinctada margaritifera*, although studies of the physical conditions in the lagoons should first be carried out (Laboute, in press).

References

* cited but not consulted

Anon. (1985). Country review - Tokelau. *Report of the 3rd South Pacific National Parks and Reserves Conference, Apia* 3: 210.

Balazs, G. (1982). Sea turtles and their traditional usage in Tokelau. Unpublished project report prepared for WWF-US and Office of Tokelau Affairs.

*Gillett, R. (1986a). Traditional tuna fishing in Tokelau. *SPREP Topic Review* 27. South Pacific Commission, Noumea, New Caledonia.

Gillett, R. (1986b). The Transplantation of Trochus from Fiji to Tokelau. Report No. 86-01, UNDP/OPE Integrated Atoll Development Project. 28 pp.

Gillett, R. and Toloa, F. (1987). The importance of small-scale tuna fishing: A Tokelau case study. In: Doulman, D.J. (Ed.), *Tuna Issues and Perspectives in the Pacific Islands Region*. East-West Center, Hawaii. Pp. 177-190.

Glynn, P.W. (1984). Widespread coral mortality and the 1982-83 El Niño warning event. *Env. Cons.* 11(2): 133-136.

Hinds, V.J. (1969/71). A rapid fisheries reconnaissance in the Tokelau Islands, August 18-25, 1971. South Pacific Commission, Noumea, New Caledonia.

*Hinkley, A.D. (1969). Ecology of terrestrial arthropods on the Tokelau atolls. *Atoll Res. Bull.* 124: 1-18.

Hooper, A. (1985). Tokelau fishing in traditional and modern contexts. In: Ruddle, K. and Johannes, R.E. (Eds), *The Traditional Knowledge and Management of Coastal Systems in Asia and the Pacific*. Unesco/ROSTEA, Jakarta, Indonesia. Pp. 7-38.

Laboute, P. (in press). Mission to the Tokelau Islands to evaluate cyclone damage to coral reefs. *SPREP Topic Review* 31. South Pacific Commission, Noumea, New Caledonia.

Marshall, K.J. (1976). Critical marine habitats and insect control in the South Pacific. *Proc. SPC and IUCN Second Regional Symposium on Conservation of Nature, Apia, Western Samoa, June 1976.*

Parham, B.E.V. (1971). The vegetation of the Tokelau Islands with special reference to the plants of Nukunono Atoll. *N.Z. Jour. Bot.* 9: 576-609.

SPREP (1981). Tokelau. *Country Report* 12. South Pacific Commission, Noumea, New Caledonia.

Stoddart, D.R. (1976). Scientific importance and conservation of central Pacific islands. Unpublished report.

Tuna Programme (1983). An assessment of the skipjack and baitfish resources of Tokelau. *Skipjack Survey and Assessment Programme Final Country Report* 10. South Pacific Commission, Noumea, New Caledonia.

Van Pel, H. (1958). A survey of fisheries in the Tokelau islands. South Pacific Commission, Noumea, New Caledonia.

Vola, K. (1987). Johnny Trochusseed at work. *Hawaiian Shell News* Feb. 1987: 10.

Wodzicki, K. (1968). An ecological survey of rats and other vertebrates of the Tokelau Islands, 19 November 1966 - 25 February 1967. Government Printer, Wellington, New Zealand. 89 pp.

Wodzicki, K. (1972). The Tokelau Islands - man and introduced animals in an atoll ecosystem. *South Pacif. Bull.* 22(1): 37-41.

Wodzicki, K. (1973). The Tokelau Islands - environment, natural history and special conservation problems. Paper 10, Section 3. *Proceeding and Papers, Regional Symposium on Conservation of Nature - Reefs and Lagoons.* South Pacific Commission, Noumea, New Caledonia. Pp. 63-68.

TONGA

INTRODUCTION

General Description

The last remaining Kingdom in the South Pacific, Tonga has a total land area of 699 sq. km spread over a sea area of approximately 360 000 sq. km and lying between 15°S and 23°30'S and 173°W and 179°W. There are three major island groups (Vava'u group, Ha'apai group and Tongatapu group) with a total of 171 islands, 37 of which are inhabited. The islands are mainly elevated coral reefs which cap the peaks of two parallel submarine ridges, although some are volcanic; the region is geologically active, with earthquakes and volcanic eruptions in recent times (Anon., 1985b; Zann, 1981). The Ha'apai group is described briefly in Halapua (1981).

The climate is mild, with average temperatures varying from 23°C (range 15°-27°C) in the south to 26°C in the north. Prevailing winds are the south-east trades; during the warm months (October to March) tropical cyclones may form over the waters to the north and move southwards where they may cause considerable damage. Average rainfall in Nuku'alofa is 1733 mm (Zann, 1981).

Table of Islands

Niuafo'ou   (X) 13.41 sq. mi. (34.7 sq. km); volcanic circular crater (260 m); 70-100 ft (21-30 m) cliffs, difficult landing.

Tafahi   (X) 1.32 sq. mi. (3.4 sq. km); extinct volcanic cone; 2000 ft (610 m); difficult landing.

Niuatoputapu   (X) 6 sq. mi. (15.5 sq. km); volcanic; 350 ft (107 m); fringing and barrier reef.

Fonualei   312 ha; volcanic; 600 ft (183 m) with breached crater (abandoned because of eruptions); fringing reef.

Toku   61 ha; low, flat-topped volcano (abandoned because of eruptions); fringing reef.

Vava'u Group

'Uta Vava'u   (X) 33.16 sq. mi. (86 sq. km); raised limestone, 670 ft (204 m); very indented coastline with 300-500 ft (91-152 m) cliffs; bordered to south by maze of small islands and reefs (Chesher, 1985); many reefs.

Koloa   (X) 0.71 sq. mi. (1.8 sq. km).

Okoa   (X) 0.17 sq. mi. (0.44 sq. km).

Olou'a   (X) 49 ha.

Ofu   (X) 125 ha; swift currents; shore and reef described by Dawson (1971); algal ridge and reef; thick coral growth; microatolls of *Porites* at south end of island; diving dangerous but reef impressive.

Pangaimotu   (X) 885 ha; limestone; 290 ft (88 m); Ahunga Passage area described by Dawson (1971).

'Utungake   (X) 93 ha; raised limestone; 290 ft (88 m).

Kapa   (X) 597 ha; raised limestone; 315 ft (96 m); cliff-bound coast.

Hunga   (X) 469 ha; 245 ft (75 m) sheer cliff coastline; sandy bottomed lagoon to south.

Fofoa   83 ha; 255 ft (78 m) sheer cliff coastline; reef system linking Fofoa with Hunga, Foelifuka and Foiatu; rich coral fauna; Hunga-Fofoa reef and lagoon system undisturbed and relatively pristine as reef animals not collected (Dawson, 1971).

Lape   (X) 40 ha; channel between Lape and Langitan filled with rich coral growth (Dawson, 1971).

Nuapapu (Noapapu)   (X) 267 ha; flat-topped; 210 ft (64 m); steep cliffs; sandy beaches to south (Dawson, 1971); extensive mangrove and seagrass areas; shallow reefs on west, linking with Vaka'eitu (Chesher *in litt.*, 1.8.87).

Vaka'eitu   90 ha; 150 acres (61 ha); limestone; flat-topped; 200 ft (60 m); reefs connect with Noapapu and islet of Langitau (6.9 ha) (Chesher *in litt.*, 1.8.87).

Ovaka   (X) 130 ha; steep cliffs; reef-bound.

Taunga   (X) 36 ha; irregular shape; surrounded by reefs.

'Euakafa   (X) 52 ha; flat-topped; 270 ft (82 m).

Fonua'one'one (Lua-a-Fuleheu)   4.9 ha; good fringing reef; steep sand beach round 1/2 of island; turtle nesting (Braley, 1974).

Fangasito (Lua-a-Fuleheu)   6.7 ha; poorly developed reef; very steep beach round whole island; turtle nesting reported (Braley, 1974).

Taula   4.1 ha; fairly well-developed fringing reef; steep sand beach round 2/3 of island; turtle nesting (Braley, 1974).

Maninita   2.8 ha; well-developed fringing reef with small harbour; steep sand beach; turtle nesting (Braley, 1974).

Late   1740 ha; high volcanic, 1700 ft (518 m) cliffs; shore described by Dawson (1971); corals in pools; west of main Vava'u Group.

Other islands in the Vava'u Group: Mafana (52 ha); Faioa (26.3 ha); 'Umuna (51.8 ha); Kenutu (43.3 ha); Ota (27 ha); Sisia (12 ha); To'ungasika (6.9 ha); Luafatu (1.6 ha); Luamoko (14 ha); Kalau (4.9 ha); Fo'ilifuka (26 ha); Fo'iata (10 ha); Kulo (1.6 ha); Kitu (6.1 ha); 'Oto (26 ha); Luakapa (4.9 ha); A'a (51 ha); Mala (6.1 ha); Nuku (3.2 ha); Kiato (1.6 ha); Afo (4.0 ha); Lolo (2.4 ha); Lotuma (0.8 ha); Tapana (36 ha); Lautala (2.0 ha); Tauta (1.2 ha); Totokafonua (4.0 ha); 'Ovalau (8.1 ha); Mounu (0.8 ha); 'Euaeiki (18 ha); Ngau (3.2 ha); Pau (1.2 ha); Mu'omu'a (10 ha); Fua'amotu (24 ha); Tefito (0.8 ha); Luahiapo (0.4 ha); Tulie Is (6.2 ha); 'Alinonga Is (0.2 ha); Katafanga Is (1.2 ha); Lekaleka (0.4 ha); Kahifehifa Is

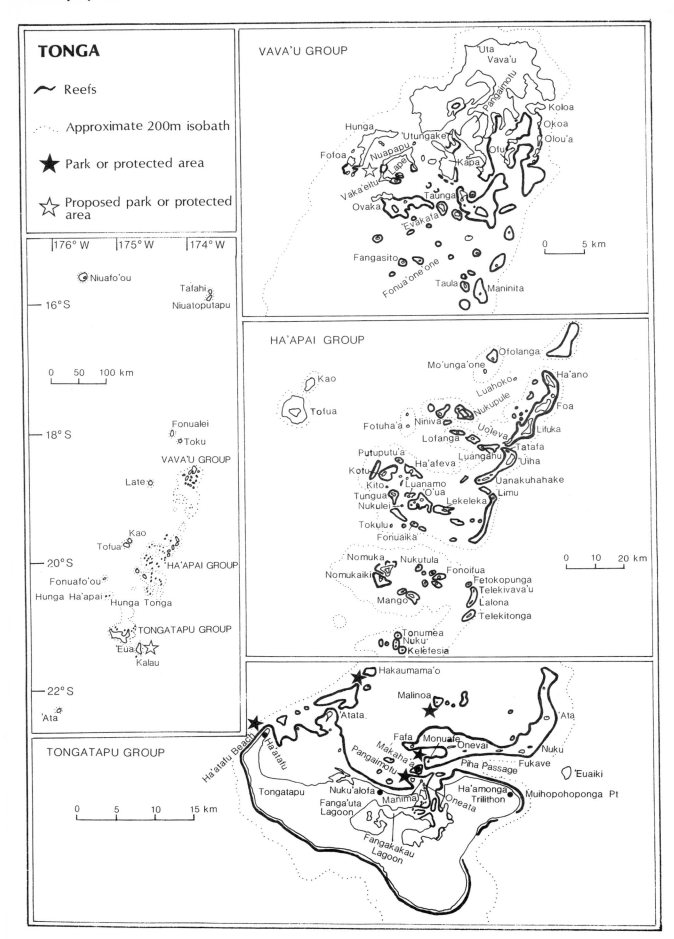

(0.2 ha); Lualoli Is (0.2 ha); Matatoa Is (0.4 ha); Kolo'uta Is (0.4 ha); Noapule Is (0.4 ha); Lateiki (3.0 ha).

Ha'apai Group

Ofolanga   89 ha; low with sandy coast; barrier reef enclosing lagoon; visited for fishing.

Mo'unga'one   (X) 115 ha; flat-topped, limestone, rocky coast; fringing reef only to south-east.

Luahoko   0.1 ha; sandy beach around most of island; fairly well-developed fringing reef; good turtle nesting (Braley, 1974).

Ha'ano   (X) 658 ha; limestone 90+ ft (27+ m); high population density.

Foa   (X) 1340 ha; low cliffs, 100+ ft (30+ m); high population density.

Lifuka   (X) 1100 ha; limestone, 100+ ft (30+ m), low cliffs to east; very high population density.

Uoleva   300 ha; limestone, (formerly inhabited).

Tatafa   32 ha; turtle nesting (Braley, 1974).

'Uiha   (X) 537 ha; 100 ft (30 m).

Luangahu   6.1 ha; has associated reefs (Halapua, 1981).

Hakauata   2.0 ha; turtle nesting (Braley, 1974); between Luangahu and Lofanga.

Lofanga   (X) 146 ha; flat-topped, low cliffs; 150 ft (46 m); narrow fringing reef with extensive lagoon (Halapua, 1981); turtle nesting (Braley, 1974).

Uanakuhihifo   10 ha; long, narrow island, connected by sand spit to Uanakuhahake; well-developed fringing reef on east side; turtle nesting on west side (Braley, 1974).

Uanakuhahake   49 ha; long, narrow island, connected by sand spit to Uanaukuhihifo; well-developed fringing reef on east side; Turtle nesting west side (Braley, 1974).

Limu   4.0 ha; turtle nesting (Braley, 1974).

Nukupule   3.2 ha; has associated reefs (Halapua, 1981).

Meama (Miama)   0.2 ha; good turtle nesting (Braley, 1974); on same reef as Nukupule.

Niniva (Miniva)   49 ha; flat-topped, has associated reefs (Halapua, 1981); good turtle nesting (Braley, 1974).

Fotuha'a   (X) 113 ha; flat-topped; 90 ft (27 m) cliffs.

Kao   (X) 1255 ha; high volcano, 3380 ft (1030 m), 2 craters; shore is black lava, described by Dawson (1971); coral patches in shallows.

Tofua   (X) 5564 ha; high active volcano; 1600 ft (488 m); shore described by Dawson (1971); small amounts of coral.

Putuputu'a   0.1 ha; turtle nesting (Braley, 1974).

Ha'afeva   (X) 202 ha; flat-topped; sandy shore; barrier reef.

Fetoa   16 ha; turtle nesting (Braley, 1974); between Ha'afeva and Kotu.

Teaupa   20 ha; between Ha'afeva and Kotu.

Matuku   (X) 34 ha; low, flat-topped; between Ha'afeva and Kotu.

Kotu   (X) 16 ha; flat-topped; 50 ft (15 m) cliffs; extensive reef (Halapua, 1981).

Kito   1.2 ha; good fringing reef; turtle nesting (Braley, 1974).

Tungua   (X) 152 ha; nearly circular, low, flat.

Nukulei (Nukulai)   1.2 ha; sandy beach around 1/4 of the island; fairly well-developed fringing reef around 3/4 of island; turtle nesting (Braley, 1974).

Luanamo (Launamu)   2.4 ha; sandy beach around 2/3 of island; good fringing reef; good turtle nesting (Braley, 1974).

'O'ua   (X) 97 ha; low cliffs; extensive reef (Halapua, 1981).

Lekeleka   12 ha; small low island.

Fonuaika   2.0 ha; small low island; fringing reef not well developed; good turtle nesting (Braley, 1974).

Tokulu   1.2 ha; small low island; turtle nesting (Braley, 1974).

Nomuka   (X) 534 ha; coral limestone, 166 ft (51 m); 180 ha saltwater lagoons 4-5 ft (1.2-1.5 m) deep; narrow fringing reef; extensive lagoon (Halapua, 1981).

Nomukaiki   (X) 66 ha; volcanic/limestone; 147 ft (45 m); many reefs.

Manoiki (Mango Iki)   1.6 ha; turtle nesting (Braley, 1974); adjacent to Mango.

Mango   (X) 64 ha; volcanic and limestone; 140 ft (43 m); extensive reefs and lagoon (Halapua, 1981).

Tonumea   12 ha; 138 ft (42 m); sandy beach around 2/3 of island; 40-50 ft (12-15 m) cliffs of sedimenary rock around rest; plantation; turtle nesting (Braley, 1974).

Nuku   1.6 ha; 2 islands; good fringing reef around large island; very poorly developed fringing reef around small island; many nesting seabirds and turtle nesting on both islands (Braley, 1974).

Kelefesia   12 ha; limestone and volcanic with 50-60 ft (15-18 m) cliffs along 1/3 of coast; 123 ft (37.5 m); plantation; turtle nesting (Braley, 1974).

Nukutula   8.0 ha; poorly developed fringing reef; turtle nesting (Braley, 1974).

Meama   2.0 ha; poor to fair fringing reef; turtle nesting (Braley, 1974); north of Fonoifua.

Fonoifua (X) 16 ha; low island, 60 ft (18 m) cliffs; maximum altitude 67 ft (20 m).

Tanoa 1.2 ha; 57 ft (17 m); south of Fonoifua.

Fetokopunga 0.8 ha; turtle nesting (Braley, 1974).

Telekivava'u 20 ha; low, flat; turtle nesting (Braley, 1974).

Lalona (Telekiha'apai) 28 ha; Low, flat; good landing places; turtle nesting (Braley, 1974).

Telekitonga 45 ha; low, flat; turtle nesting (Braley, 1974).

Nukufalau 1.2 ha; poorly developed fringing reef; good turtle nesting (Braley, 1974).

Hunga Tonga 45 ha; low volcanic; steep cliffs to 490 ft (149 m); no anchorage; adjacent to Hunga Ha'apai west of main Ha'apai Group.

Hunga Ha'apai 65 ha; low volcanic; 400 ft (122 m) ridge.

Fonuafo'ou (Falcon I.) active volcano, varying in size and form; north of Hunga Tonga.

Other islands in the Ha'apai Group include: Foua (0.4 ha); Fakahiku (1.2 ha); Fotu-amangai (Fatumanongi); Muifuiva (0.4 ha); Nukunamo (32 ha); Onoiki (1.2 ha); Teaupa (20 ha); Tofanga Is. (3.4 ha); Pepea (0.4 ha); Kolo Is. (0.4 ha).

Tongatapu Group

Tongatapu (X) 257 sq. km; raised limestone; 270 ft (82 m); generally low relief; south shore cliff-bound above cemented limestone with blowholes; fringing and patch reefs and islets off low-lying north shore; general information in Stoddart (1975); Fenn (1972) provides some information on reef and shore fauna; Fanga'uta Lagoon described by Zann *et al.* (1984) (see below); marine reserves described in separate accounts.

Islands in Fanga'uta Lagoon Nukunukumotu (132 ha); Talakite Is (1.6 ha); Mata'aho Is (6.5 ha); Mounu Is (0.2 ha); Mo'ungatapu Is. (4.0 ha); Nogofuna (0.4 ha); Kanatea (45 ha).

Islands off north coast of Tongatapu:

- *'Euaiki* (X) 105 ha; 180 ft (55 m); landing difficult;

- *Nuku* 4.0 ha; has important seabird rookery (Dahl, 1980);

- *Monuafe* (*see separate account*);

- *Pangaimotu* (?X) 22.3 ha; described by Stoddart (1975); adjacent to Pangaimotu Reef Reserve (*see separate account*);

- *Malinoa* (*see separate account*);

- *Mahaha'a (Makaha'a)* 3.9 ha; sand cay isolated from Tongatapu fringing reef by 10 m deep channel; described by Stoddart (1975);

- *Manima* 3 ha; sand cay described by Stoddart (1975);

- *'Oneata* 7.8 ha; sand cay described by Stoddart (1975).

Also: 'Atata (52 ha); Fafa (9.3 ha); Veliota Hahake (2.0 ha); Veliota Hihifo; 'Onveao (2.6 ha); 'Ata (22 ha); Motutapu (13 ha); Fukave (17 ha); Tufaka (1.8 ha); 'Alakipeau (0.8 ha); Polo'a (4.9 ha); 'Onevai (10 ha); Tau (5.7 ha); Toketoke Is (1.6 ha).

'Eua (X) 8743 ha; volcanic; 1078 ft (328 m); fringing reef; eastern side very exposed and reef has terraced pools ringed by coralline algae *Lithothamnion* sp (Anon., 1985b).

Kalau 20 ha; south-west of 'Eua.

'Ata (Pylstaart) 227 ha; extinct volcano with 2 peaks; 1165 ft (355 m); cliff bound; difficult landing; numerous seabirds.

Minerva Reefs 2 reefs lying at ca 23°S, 179°W on Lau Ridge; one ca 400 x 600 m with small sandy islet probably inundated during very high swells; some vegetation (coconuts); Giant Clams surveyed (Lewis *et al.*, 1985).

(X) = Inhabited

Coral reefs are widespread in Tonga but surveys and research have been sporadic and limited; occasional visits by marine researchers have resulted in a widely scattered literature (Chesher *in litt.*, 1.8.87). Agassiz visited the area briefly in 1899 and the Cook Bicentenary Expedition surveyed coastal features of the Vava'u group; recent general reef studies include Dawson (1971) and Chesher (1984a and b, 1985). The former provides descriptions of the area around Tofua, Kao, Late and Vava'u, the latter describes over 100 shallow water sites in the Tongatapu, Nomuka, Ha'apai and Vava'u groups investigated in the course of studies on pollution and black corals. A project using remote sensing to quantify coral reef habitats and land use began in 1986 and was still in progress in 1987 (Chesher, 1987). The goal of the project is to construct an electronic atlas (i.e. using computer- aided design programmes) of land and marine resources. Vava'u is the test site for the project which, when complete, should provide a useful tool for resource mapping and management for other small islands (Chesher, 1985 and *in litt.*, 1.8.87; Anon., 1986). The pilot study is due to involve ORSTOM SPOT satellite imagery of Vava'u. A new aerial photographic survey of Tonga was planned by Australia for 1986, although has not yet been carried out, and it is thought likely that aerial photographs may be more useful than satellite images. Ground truthing is being carried out by the Marine Research Foundation and Center for Field Research in cooperation with the Tongan Ministry of Lands, Survey and Natural Resources (Chesher, 1987).

In 1981 a study was made of Fanga'uta Lagoon in Tongatapu by the Institute of Marine Resources of the University of the South Pacific, the Hawaii Institute of Marine Biology and the Tonga Fisheries Division (Zann *et al.*, 1984). The 2830 ha lagoon is a shallow, almost enclosed embayment on the northern coast which is highly productive and the site of an important mullet fishery. Corals are virtually absent in the lagoon owing to low salinity, soft substrate and high turbidity but were more widespread in the recent past, as attested to by the presence of large beds of dead *Acropora* and dead *Porites* "microatolls". This change is ascribed to an uplift at some time between 40 and 200 years ago, probabaly in 1914, resulting in conditions within the lagoon becoming unsuitable for coral growth. Immediately outside the lagoon living coral cover on reefs reached 70% at some sites, with 20 genera of stony corals recorded in total (Zann *et al.*, 1984).

Studies have also been carried out on artisanal fisheries (Halapua, 1981; Zann, 1981) and on the biology and exploitation of tridacnids, of which four species are known to be present in Tongan waters (*Hippopus hippopus, Tridacna maxima, T. squamosa* and *T. derasa*) (McKoy, 1980). In 1973/74 a survey was carried out of islands in parts of the Vava'u and Ha'apai Groups to determine the extent of nesting by marine turtles (Braley, 1974). Green Turtles *Chelonia mydas* and Hawksbills *Eretmochelys imbricata* were both recorded; nesting levels appeared to be very low, although the Green Turtle was apparently quite widespread in the Vava'u and Ha'apai Islands. The only known nesting (species uncertain) in the Tongatapu group is on Malinoa (Braley, 1974; Wilkinson, 1979). Seabird distribution is briefly described in Garnett (1984). Reef and shore fauna are described in Fenn (1972).

Reef Resources

Over 70 000 of the estimated population of 104 000 (in 1984) live on Tongatapu (Anon., 1985b); most of the remainder live on the Ha'apai Archipelago and Vava'u, and three quarters of the islands are uninhabited. Fisheries, especially inshore, are important although demand exceeds supply and Tonga is a net importer of fish and fish products (De Backer, 1985; Halapua, 1981; Zann, 1981). A survey of the Ha'apai Islands in 1980 found that almost all fishermen there were part-time and semi-commercial; true subsistence fishing was apparently virtually unknown (Halapua, 1981). Most (70%) of the total annual catch consisted of reef-lagoon fish: the commonest finfish taken were surgeon fish (Acanthuridae), parrot fish (Scaridae), squirrel fish (Holocentridae), wrasses (Labridae), damsel fish (Pomacentridae), groupers (Serranidae), goat fish (Mullidae) and butterfly fish (Chaetodontidae). In addition the reefs provided a good harvest of octopus; this was dried and sold at Nuku'alofa market and served as an important source of income to the Ha'apai fishermen. The reefs considered particularly rich fishing grounds are listed by Halapua (1981).

Giant clams are harvested for food and their shells for decorating graves and gardens. Most clam fishing is carried out in the Tongatapu group where the 1978 catch was estimated at ca 150 000 kg, this being around a six-fold increase on the estimated harvest in 1974 (ca 24 000 kg) (McKoy, 1979 and 1980). Most is for local consumption but a small proportion was exported to American Samoa; exports in 1978 consisted mainly of *T. derasa* from around Limu Island and other species from Lofanga, Niniva, Meama and Nukupule Islands (McKoy, 1980).

Tonga has important reserves of Black Coral, *Antipathes dichotoma* and *A. grandis* (Chesher, 1984b). Small-scale commercial utilisation of this began in 1973, and the harvest has increased considerably. A firm, Imua Pacifica, based in Nuku'alofa, specializes in the production of jewellery (Anon., 1985a; Chesher, 1984b; Hamley, 1985). The eggs and meat of turtles are eaten, and carapaces are sold in tourist shops; some tortoiseshell from Hawksbill is worked into jewellery (Hirth, 1971a; Wilkinson, 1979).

Tourism is at a relatively low but growing level (92 000 visitors in 1983) with some SCUBA diving; there are dive operators in Nuku'alofa and in Vava'u (Chesher, n.d. a and *in litt.*, 1.8.87; De Backer, 1985; Hamley, 1985). Vava'u is a well known yachting centre but is off normal tourist routes (Chesher, 1987).

Disturbances and Deficiencies

Cyclone Isaac caused severe damage in 1982 (Chesher, 1985; Woodruffe, in press). *Acanthaster planci* was common on Ofu and Hunga in the Vava'u Group in 1969 (Dawson, 1971). Raj *et al.* (1984) investigated mass mortality of marine life on a reef on the eastern side of the Hihifo peninsula in north-west Tongatapu which occurred in February 1984 and affected fish (only reef species) and invertebrates, including corals, in a relatively limited area. No clear indication of the cause was evident and seawater and plankton proved to be non-toxic. It was concluded that the mortality was caused by the unusual exposure of the reef to particularly low tides, high temperature and heavy rainfall in the previous week.

From May to September of 1984, a Pollution Sources Survey and a Black Coral Survey, using interview and questionnaire techniques as well as inspection of sites, was carried out as part of the South Pacific Regional Environment Programme (Chesher, 1984a and b). Sixty five per cent of the reefs inspected showed evidence of coral destruction, with high coral mortality and infection with blue-green algae. The high *Acropora* mortality and shift in structure towards milleporids indicated long-term stress of the reef community especially in Tongatapu. This was attributed to a variety of factors including storm damage, pollution, causeway construction and destructive fishing techniques, and possibly disruption of the phosphate cycle by overfishing, unusual water level changes in Vava'u, unusual weather patterns and resultant changes in water temperatures.

Of these, destructive fishing techniques were identified as a major cause of reef degradation, and may still be the greatest threat to Tongan reefs (Chesher, 1985 and *in litt.*, 2.4.85), e.g. at Langitau in the Vava'u Group (Chesher *in litt.*, 1.8.87). These techniques principally involve smashing reefs and coral outcrops to chase fish into nets either set at the mouths of lagoons or encircling the outcrops (a technique known as "tu'afeo"), as well as turning and breaking of rocks and corals to take invertebrates. Trampling and coral breakage also occur during food gathering on reef flats (Chesher, 1984a and 1985; Wilkinson, 1977). On Vava'u, the reef flat in some areas has been reduced to rubble due to such activities, both for collecting fish and invertebrates for food and shells for sale to tourists. Dynamite and poisons, including sodium hypochlorite and pesticides, are also used, although the present extent of this is unclear. In damaged areas, coral predators such as *Acanthaster planci* and gastropods were found to be particularly common (Chesher, 1984a and 1985).

Contamination of shallow reef areas from sewage outflow, principally from hotels on Tongatapu and Vava'u, was a localized although potentially serious problem, particularly at Nuku'alofa where one hotel discharged up to a million gallons of effluent a month onto the reef flat (this sewage system has subsequently been remodelled, although the new outfall has not been monitored (Chesher *in litt.*, 1.8.87)). Siltation, mainly from construction and quarrying activities, had caused degradation of reefs in Tongatapu, particularly in the region of the Nuku'alofa waterfront, and Vava'u, especially around Neiafu Harbour. The steeper hillsides of Vava'u mean that its surrounding reefs are at even greater risk of siltation. The construction of causeways across bays and between islands has caused problems; in particular one between Lifuka and Foa islands in the Ha'apai group has halted water flow and led to degradation of reefs to the west of the causeway. Further causeways were planned in the early 1980s (Chesher, 1984a). In 1981 there were plans for dredging for building aggregate and land fill in Fanga'uta Lagoon (Zann *et al.*, 1984). The area is, however, still protected with all drilling and dredging banned, although there are plans to build a bridge and causeway across the entrance to the lagoon when funds are available (Chesher *in litt.*, 1.8.87). Mangrove cutting has theoretically been banned in the lagoon but still occurs (Chesher *in litt.*, 2.4.85).

Demand for sand for construction purposes is increasing and sand mining from beaches and the dunes behind them has become a major problem. Several beaches of particular importance to tourism and recreation have been stripped of sand, leaving only beach rock (Chesher *in litt.*, 1.8.87). Sand mining has been carried out by both the Government and private individuals; a CCOP/USGS workshop on Tongatapu in 1984 helped clarify the problem and stopped some of the destruction. Recently government sand mining of more important recreational beaches has stopped although it is not known if the cessation is permanent or not. It is possible that some reefs may be threatened in the future by the need to dredge sand.

Pressure on reefs and other ecosystems can be expected to increase in the future. For example, at present most of the 17 000 inhabitants of the Vava'u Group are still engaged in fishing and agriculture, but recent decisions by the Tongan Government have targeted Vava'u for rapid development of agricultural and tourist activities, with additional roads, causeways, resorts and harbours planned (Chesher, 1987).

Some onshore and offshore reefs (e.g. those at Lifuka, Foa, Ha'ano, Ha'afeva and Tungua in the Ha'apai Islands) have been extensively overfished (Halapua, 1981; SPREP, 1980). According to fishermen, stocks of clams, particularly *Tridacna derasa* and *T. squamosa* in shallower waters and to some extent of *T. maxima* on reef tops and edges, have declined considerably over the past 10 years, almost certainly as a result of overexploitation. By 1980, in some localities close to population centres it was very difficult to find any clams on shallow reefs; areas around Tongatapu were apparently under greatest pressure. Fishermen had to travel further; for example, clams landed at Nuku'alofa

were largely caught around Tau and 'Ata. However, because of the variable pattern of exploitation, it is difficult to make quantitative estimates of the effects of fishing on the stocks (McKoy, 1980).

Black coral has suffered from overexploitation, and large specimens have become rare especially in Nuku'alofa Harbour where siltation may have been one of the main reasons behind the death of some 49% of the colonies. The method used for searching for black coral may also be damaging since it involves dragging an anchorline along the seabed (Anon., 1985a; Chesher, 1984b).

Legislation and Management

All territorial seas and internal waters are the property of the Crown and may be subject to government restrictions and regulations. Every Tongan has the right to fish in these waters and there are no traditional fishing rights giving villages or individuals exclusive rights to fish or gather shells in certain areas, although these may have existed in the past (Eaton, 1985; Samani *in litt.*, 11.8.87). The foreshore is the property of the Crown and is defined in the Land Act as "land adjacent to the sea alternately covered and left dry by the ordinary ebb and flow of the tides and all land adjoining thereunto lying within 50 ft (15 m) of the high water mark" (Eaton, 1985).

The Parks and Reserves Act of 1976 authorizes the establishment of protected areas which are managed by the Parks and Reserves Authority. There are a number of marine and coastal protected areas:

1. Five marine reserves were established in 1979 off the north coast of Tongatapu (*see separate accounts*):
    - Hakaumama'o Reef Reserve
    - Pangaimotu Reef Reserve
    - Ha'atafu Beach Reserve
    - Malinoa Island Park and Reef Reserve
    - Monuafe Island Park and Reef Reserve

2. In 1972 the 2 km beach at Muihopohoponga at the extreme eastern end of Tongatapu, and a 23 ha terrestrial area around the Ha'amonga Trilithon, also in the east of Tongatapu, were declared reserves by Royal Proclamation (Dahl, 1978; Anon., 1979 and 1985b; Samani *in litt.*, 19.8.87). These areas have yet to be declared protected areas under the 1976 Parks and Reserves Act, although in 1987 it was reported that work was under way developing historical trails at the Ha'amonga Trilithon, indicating that gazetting was likely to occur in the near future (Samani *in litt.*, 19.8.87). Reports in Wilkinson (1977) that recreational beaches on mainland Tongatapu are protected presumably apply to these areas.

3. Fanga'uta and Fangakakau Lagoons were declared a protected area in 1974 under Act 24, amending the Birds and Fish Preservation Act of 1915; all commercial fishing, trawling and setting of fish-fences or traps, effluent discharge into the lagoon, drilling, dredging, construction of any building works, harbours, wharfs, piers or jetties, and cutting or damaging of mangrove trees was prohibited. The prohibition on fishing has apparently never been completely observed or enforced; however, a motion by the Legislative

Assembly to repeal it in September 1981 was not ratified, and it thus remains in force (Samani *in litt.*, 19.8.87 *contra* Zann *et al.*, 1984).

Reports in Eaton (1985) that restrictions against commercial fishing operate in Vaipuua Lagoon in Vava'u, are believed to be in error (Chesher *in litt.*, 1.8.87; Samani *in litt.*, 11.8.87).

The Ministry of Lands, Survey and Natural Resources is responsible for park management but is hampered by shortage of funds and personnel and there has thus been only limited development of the official protected areas, although there are signs identifying them and setting out the rules prohibiting the removal and destruction of marine life (Eaton, 1985). No biological surveys or inventories of the reserves have been carried out (Samani *in litt.*, 11.8.87). Chesher (*in litt.*, 2.4.85) reported virtually complete neglect of park boundaries and noted that destructive fishing methods were still used within the parks. In 1985 the Ministry was provided by the Australian High Commission with one 25 ft (8 m) fibreglass boat with a 25-HP outboard motor for park supervision (Samani *in litt.*, 11.8.87) and in 1987 they were in the process of appointing two park rangers (Chesher *in litt.*, 1.8.87).

The Ministry of Lands, Survey and Natural Resources, in cooperation with a diving concern, has set aside diving areas in Nuku'alofa Harbour which contain good colonies of Black Coral; a diving concern at Vava'u has also set aside areas of particularly well developed coral and Black Coral growth as protected diving sites. Protection is voluntary as none of these sites is legally protected and some damage has reportedly been done by Black Coral collection (Chesher *in litt.*, 1.8.87). The ministry and diving company has also begun a planting project by cutting Black Coral branches and relocating them; by mid-1987 some 15-20 colonies had been replanted at Ualanga Uta in Nukua'alofa Harbour. This project is planned to expand to other parts of the Tongan group (Samani *in litt.*, 11.8.87). A project organized by the Ministry of Lands, Survey and Natural Resources with the cooperation of the Fisheries Division and the Marine Research Foundation and Center for Field Research to boost Giant Clam stocks was begun during Environment Week in 1986 (see below) when a brood stock of *Tridacna derasa* was planted as a "clam circle" on a reef in Nuku'alofa harbour (Chesher, 1986 and n.d. b). In 1987 it was reported that these were doing well and a large and successful spawn was anticipated (Chesher *in litt.*, 1.8.87). The clam circle was the subject of ongoing study (Chesher, n.d. b).

The Bird and Fish Preservation Act (first enacted 1915, amended 1974) provides complete protection or at least closed seasons for some taxa including marine turtles. The capture of Leatherback Turtles *Dermochelys coriacea* is prohibited and Green Turtles *Chelonia mydas* may only be taken outside the breeding season which runs from 1 November to 31 January (Hubbard, 1979). However, the legislation concerning turtles is poorly enforced and nesting populations are already extinct on several islands and threatened through overcollecting on others. The use of dynamite and poisons for fishing is prohibited.

The Tourism Act of 1976 is designed to develop tourism for the benefit of all Tongans and to license, regulate and control the industry. Proposed Environmental Protection and Fisheries Acts were reported in 1987 as awaiting consideration for enactment by the current session of the Legislative Assembly (Samani *in litt.*, 11.8.87). Following recommendations in Chesher (1984a), Environmental Impact Statements are made prior to development activities (Chesher *in litt.*, 1.8.87); other recommendations to control siltation on reefs and in lagoons during waterfront construction have reportedly been followed and the situation has improved considerably in the past few years (Chesher *in litt.*, 1.8.87). An act, amending the 1983 Customs and Excise Act, controlling the export of unworked Black Coral, has been approved by the Privy Council and will be considered by the Legislative Assembly in 1987; anyone convicted of exporting unprocessed Black Coral without written Cabinet permission will be liable to a fine and/or up to five years' imprisonment (Samani *in litt.*, 11.8.87). Controls will also apparently extend to other unworked marine products (Chesher *in litt.*, 1.8.87).

Marine conservation education is being promoted in schools (Tongilava, 1979), although provision of educational materials is hampered by shortages of funds and staff (Chesher *in litt.*, 1.8.87). A National Environmental Awareness Week has been held each year since 1983 with considerable success (Anon., 1985; Chesher *in litt.*, 1.8.87).

Recommendations

Chesher (1984a) provided detailed recommendations for improved environmental protection; at least some, including the control of siltation and the provision of Environmental Impact Statements, have been acted on (see above). Others include recommendations for controlling sewage discharge onto reefs and into the lagoon and for the prevention of oil spills.

Further attention to the impact of subsistence oriented practices on the reef is required. Halapua (1981) provides recommendations for improved management of fisheries in the Ha'apai group and considers that reef and lagoon species do not have potential for further exploitation. Recommendations for the management of Giant Clam stocks have been made to the Government, including introducing a minimum size limit, prohibiting the use of SCUBA for collection, licensing collectors and prohibiting exports except of shells for the souvenir trade (McCoy, 1980). Chesher (1987 and n.d. b) provides details of further planned research on reintroduction of giant clams, concentrating in Vava'u. In particular, the planned follow-up to electronic atlas of Vava'u will be a survey of shallow-water marine environments with special reference to Giant Clam populations. Recommendations for the improved management of black coral stocks are given in Anon. (1985a) and include a further survey to determine distribution and stock density, legislation to control exploitation and exports (only worked products to be exported) and replanting experiments; the last two of these are being implemented (see above).

Chesher (1984a and 1985) stresses the need for educational programmes to reduce coral damage during subsistence and commercial fishing activities; he recommends the preparation of educational materials to be incorporated in the Radio Tonga Educational

Programme and into the school system and the production of an educational video aimed at the adult public. The need for educational materials in schools is recognized, but there are shortages of funds and personnel (Chesher *in litt.*, 1.8.87). Adequate enforcement of already existing legislation (such as that protecting marine turtles) has been strongly advocated (Braley, 1974).

There is an urgent need for improved management of existing protected areas. Dahl (1978) recommended that the marine reserves should extend seawards to at least 50 m depth or to the base of the reef if this is shallower, although this has not been carried out (Samani *in litt.*, 5.11.87). Additional reserves have been proposed:

1. 'Eua National Park: an area of 1400 ha to include the fringing reef, coastal region, eastern ridge and ridge summit on the eastern side of the island (Anon, 1985b; Dahl, 1980). Plans are under way for the New Zealand government to fund the development of facilities in the area; if funding is finalized, boundary surveys prior to gazetting will take place (Samani *in litt.*, 11.8.87).

2. Three protected areas have been proposed for Vava'u, with funding for development of facilities arranged through the European Community (Samani *in litt.*, 11.8.87); of these, one is a marine reserve covering the reef connecting Nuapapu with Vaka'eitu (Chesher *in litt.*, 1.8.87; Samani *in litt.*, 5.11.87).

3. 'Ata Island Biosphere Reserve (Anon, 1985b; Dahl, 1980), which would probably include reefs; there are no current developments with this proposal (Samani *in litt.*, 5.11.87).

4. Dahl (1978) recommended the establishment of a coastal reserve at Muihopohoponga (see above), to incorporate 6.4 ha of unallocated land and the foreshore, and with an adjacent marine reserve to depth of 50 m; this recommendation is being pursued (Samani *in litt.*, 5.11.87).

Chesher (*in litt.*, 1.8.87) stresses the need for a comprehensive bibliography of marine studies carried out in Tonga, using the databank being set up by the University of the South Pacific in Suva, Fiji, to provide a basis for future planning. General recommendations and conservation guidelines are discussed by Dahl (1978).

References

* = cited but not consulted

Anon. (1979). Situation report: Tonga. *Proc. 2nd South Pacific Conference on National Parks and Reserves, Sydney* 1: 89-93.

Anon. (1985a). Sustainable black coral harvesting potential in Tonga. *Report of the 3rd South Pacific National Parks and Reserves Conference, Apia* 2: 100-110.

Anon. (1985b). Country review - Tonga. *Report of the 3rd South Pacific National Parks and Reserves Conference, Apia* 3: 211-216.

Anon. (1986). Islands from the sky. *Earthwatch* 5(2): 40.

Braley, R. (1974). The present marine turtle situation in Tonga. Unpub. rept, Fisheries Section, Tongan Agriculture Dept.

Chesher, R.H. (1984a). Pollution Sources Survey of the Kingdom of Tonga. *SPREP Topic Review* 19. South Pacific Comission, Noumea, New Caledonia. 110 pp.

Chesher, R.H. (1984b). Resource Assessment Report. Black Coral of Tonga. SPREP, South Pacific Commisssion, Noumea, New Caledonia. 30 pp.

Chesher, R.H. (1985). Practical problems in coral reef utilization and management: A Tongan case study. *Proc. 5th Int. Coral Reef Cong., Tahiti* 4: 213-217.

Chesher, R.H. (1986). How to establish a clam farm for food security in future. *Tongan Chronicle* 22(3): 2-3.

Chesher, R.H. (1987). An electronic atlas for resource management and project planning in Pacific islands. Unpub. rept. 12 pp.

Chesher, R.H. (n.d. a). Satellite imagery for the improvement of island ecosystems. Project proposal, Marine Research Foundation.

Chesher, R.H. (n.d. b). The Tongan Giant Clam (vasuva) revitalization project. Unpub. rept. 6 pp.

Dahl, A.L. (1978). Environmental and ecological report on Tonga. Part 1. Tongatapu. South Pacific Commission, Noumea, New Caledonia. 48 pp.

Dahl, A.L. (1980). Regional ecosystems survey of the South Pacific Area. *SPC/IUCN Technical Paper* 179. South Pacific Commission, Noumea, New Caledonia.

Dawson, E.W. (1971). Marine Biology in Tonga - Vava'u and the western islands: An interim report. *Bull. Roy. Soc. New Zealand* 8: 107-120.

De Backer, R. (1985). Tonga, the scepter'd isle where the dawn first breaks. *The Courier* 93: 14-19.

Eaton, P. (1985). Land Tenure and Conservation: Protected areas in the South Pacific. *SPREP Topic Review* 17. South Pacific Commission, Noumea, New Caledonia. 103 pp.

Fenn, L.R. (1972). Reef and shore fauna of Tonga. Unpub. Progress Rept.

Garnett, M.C. (1984). Conservation of seabirds in the South Pacific Region: A review. In: Croxall, J.P., Evans, P.G.H. and Schreiber, R.W. (Eds), *Status and Conservation of the World's Seabirds*. ICBP Technical Publication 2. Pp. 547-558.

Halapua, S. (1981). The islands of Ha'apai: Utilisation of land and sea. Consultancy report for the Central Planning Department, Kingdom of Tonga.

*Halapua, S. (1982). Fishermen of Tonga: Their means of survival. University of the South Pacific, Suva, Fiji. 100 pp.

Hamley, P. (1985). Tonga Black Coral Kingdom of the South Pacific. *Tusitala* (Western Samoan Airlines) Autumn: 10-11.

Hirth, H.F. (1971). South Pacific Islands - Marine Turtle Resources. Report to Fisheries Development Agency Project. FAO. Report F1: SF/SOP/REG/102/2.

Hubbard, T.P. (1979). Presentation at 2nd South Pacific Conference on National Parks and Reserves, Sydney, Australia.

Lewis, A.D., Adams, T.J.H. and Ledua, E. (1985). Giant Clam Project Progress Report. Fisheries Division, Ministry of Primary Industry, Fiji.

McKoy, J.L. (1979). Giant clams in Tonga under study. *South Pacific Commission Fisheries Newsletter* 19: 1-3.

McKoy, J.L. (1980). Biology, exploitation and management of Giant Clams (Tridacnidae) in the Kingdom of Tonga. *Fisheries Bulletin* 1. Fisheries Division, Tonga.

Raj, U., Sesto, J., Puloka, S.T. and Fakahau, S. (1984). An investigation of mass mortality of marine life in the Kingdom of Tonga. Rept to Ministry of Health, Kingdom of Tonga.

SPREP (1980). Tonga. *County Report* 13. South Pacific Commission, Noumea, New Caledonia.

Stoddart, D.R. (1975). Sand cays of Tongatapu. *Atoll Res. Bull.* 181. 8 pp.

Tongilava, S.L. (1979). Development and management of marine parks and reserves in the Kingdom of Tonga. *Proc. 2nd South Pacific Conference on National Parks and Reserves, Sydney* 1: 148-152.

Wilkinson, W.A. (1977). Marine conservation in Tonga. *Parks* 2(2): 11-12.

Wilkinson, W.A. (1979). The marine turtle situation in the Kingdom of Tonga. Paper given at Joint SPC-NMFS Workshop on marine turtles in the tropical Pacific islands, Noumea, New Caledonia, 11-14 Dec. 1979.

Woodruffe, C.D. (in press). The impact of Cyclone Isaac on the coast of Tonga. *Pac. Sci.*

Zann, L.P. (1981). Tonga's artisanal fisheries. Effects of the energy crisis on artisanal fisheries in the South Pacific. Report 5. Institute of Marine Resources, University of the South Pacific, Suva, Fiji.

Zann, L.P. Kimmerer, W.J. and Brock, R.E. (1984). The Ecology of Fanga'uta Lagoon, Tongatapu, Tonga. Sea Grant Co-operative Report UNIHI-SEAGRANT-CR-84-04.

HA'ATAFU BEACH RESERVE

Geographical Location The reserve, which includes both the beach and adjacent reef, lies at the western tip of Tongatapu, 14 miles (22.5 km) north-west of Nuku'alofa, and is bounded by the following points:
1. 21°04'11.6"S, 175°20'00.3"W;
2. 21°04'05.9"S, 175°20'09.6"W;
3. 21°04'00.1"S, 175°20'04.3"W;
4. 21°04'05.2"S, 175°09'56.1"W.

Area, Depth, Altitude 8 ha; to 5 m depth.

Land Tenure Government owned.

Physical Features Spectacular beach with lagoon and fringing reef.

Reef Structure and Corals No information.

Noteworthy Fauna and Flora Fish are abundant. The Giant Clams *Tridacna maxima* and *T. squamosa* occur (McCoy, 1980).

Scientific Importance and Research Giant clams were surveyed by McCoy (1980).

Economic Value and Social Benefits The attractive beach is popular for recreation (Dahl, 1978) and the area could become one of Tonga's main tourist attractions (Tongilava, 1979).

Disturbance or Deficiencies There is reported to have been some damage to corals by *Acanthaster planci* (Tongilava, 1979). Collecting of shellfish still occurs (Eaton, 1985).

Legal Protection Established as a reserve in 1979 under the Parks and Reserves Act of 1976.

Management None at present.

Recommendations The need for improved management of all protected areas in Tonga is discussed in the introduction.

HAKAUMAMA'O REEF RESERVE

Geographical Location The reserve is 12 miles (19.3 km) north of Nuku'alofa, Tongatapu, and is bounded by the following points:
1. 20°59'30.6"S, 175°12'57.3"W;
2. 20°59'30.0"S, 175°12'04.4"W;
3. 21°00'13.2"S, 175°12'03.8"W;
4. 21°00'13.7"S, 175°12'56.8"W.

Area, Depth, Altitude 260 ha; to 5 m depth.

Land Tenure Government-owned.

Physical Features An isolated exposed barrier and algal reef, often exposed to strong currents and hurricanes, which is not associated with an island (Wilkinson, 1977; Dahl, 1978 and 1980).

Reef Structure and Corals No information.

Noteworthy Fauna and Flora The Giant Clams *Tridacna maxima* and *T. derasa* occur (McCoy, 1980).

Scientific Importance and Research An extremely diverse, almost pristine reef. Giant Clam survey carried out by McCoy (1980).

Economic Value and Social Benefits The area is used very little for fishing as it is so isolated.

Disturbance or Deficiencies Virtually undisturbed by man due to its remote location (Tongilava, 1979).

Legal Protection The area was apparently originally declared a reserve by Royal Proclamation in 1972; it was then formally established as a reserve, under the 1976 Parks and Reserves Act, in 1979.

Management None at present.

Recommendations The need for improved management of all protected areas in Tonga is discussed in the introduction.

MALINOA ISLAND PARK AND REEF RESERVE

Geographical Location The reserve is 9 miles (14.5 km) north of Nuku'alofa, Tongatapu, and is bounded by the following points:
1. 21°02'24.0"S, 175°07'59.1"W;
2. 21°01'48.7S, 175°07'59.6"W;
3. 21°01'48.2"S, 175°07'21.3"W;
4. 21°02'23.6"S, 175°07'20.8"W.

Area, Depth, Altitude The reserve is 73 ha, of which the island comprises 0.8 ha; to 5 m depth.

Land Tenure Government owned.

Physical Features A small island with fringing reef. Tides create strong currents (Tongilava, 1979).

Reef Structure and Corals Considered an excellent coral reef area (Chesher, 1984a).

Noteworthy Fauna and Flora The area is said to be rich in fish, octopus, clams and other shellfish (Eaton, 1985). The Giant Clam *Tridacna maxima* occurs (McCoy, 1980). It is the only known turtle nesting site in the Tongatapu group, although the species concerned are not known (Braley, 1974; Wilkinson, 1979).

Scientific Importance and Research The reefs were briefly surveyed by Chesher (1984a) and the Giant Clams by McCoy (1980).

Economic Value and Social Benefits The island is of historic interest (would-be assassins of prime minister shot in 1835). Snorkelling for strong swimmers only on account of tides (Tongilava, 1979).

Disturbance or Deficiencies Coral mortality is high but the cause is unknown (Chesher, 1984a).

Legal Protection Gazetted as an island park and reef reserve in 1979, under the Parks and Reserves Act of 1976. Although reported to be a National Park in Anon. (1985b), this is not the case (Samani *in litt.*, 11.8.87).

Management None at present.

Recommendations The need for improved management of all protected areas in Tonga is discussed in the introduction.

MONUAFE ISLAND PARK AND REEF RESERVE

Geographical Location The reserve is 4 miles (6.4 km) north of Nuku'alofa, Tongatapu, and is bounded by the following points:
1. 21°06'44.7"S, 175°08'37.0"W;
2. 21°06'12.7"S, 175°08'37.4"W;
3. 21°06'12.5"S, 175°08'20.0"W;
4. 21°06'44.5"S, 175°08'19.6"W.

Area, Depth, Altitude The reserve is 33 ha, of which the island comprises 2.0 ha; max. depth 5 m.

Land Tenure Government owned.

Physical Features A small sand islet with scrub, surrounded by a sheltered lagoon reef. It is situated at the confluence of two main water movements which provide good conditions for reef growth (Wilkinson, 1977).

Reef Structure and Corals The reef is a small section of a large, very diverse and vigorous reef. *Acropora* is dominant, but all species of coral known from Tonga are

found there (Tongilava, 1979). The reef to the south and west of Monuafe Island is considered to be one of the richest in the vicinity of Tongatapu and is a good example of sheltered reef development (Dahl, 1978).

Noteworthy Fauna and Flora The adjacent sand flats are rich in molluscs (Wilkinson, 1977). The Giant Clam *Tridacna maxima* occurs (McCoy, 1980).

Scientific Importance and Research The reefs were briefly surveyed by Chesher (1984a) and Giant Clams by McCoy (1980).

Economic Value and Social Benefits No information.

Disturbance or Deficiencies The reefs have suffered from a local fishing method "tu'afeo" which involves the breaking of coral to scare fish into nets (Wilkinson, 1977). Chesher (1984a) reported there had been no recent changes in reef condition.

Legal Protection The area was established as an island park and reef reserve in 1979 under the Parks and Reserves Act of 1976. Although reported to be a National Park in Anon. (1985b), this is not the case (Samani *in litt.*, 11.8.87).

Management None at present.

Recommendations The need for improved management of all protected areas in Tonga is discussed in the introduction.

PANGAIMOTU REEF RESERVE

Geographical Location The reserve lies ca 3 km north of Nuku'alofa, Tongatapu, and immediately north of Pangaimotu Island. It is bounded by the following points (marked by navigational beacons):
1. 21°07'09.8"S, 175°09'54.1"W;
2. 21°06'56.7"S, 175°09'64.2"W;
3. 21°06'56.4"S, 175°09'29.1"W;
4. 21°07'09.5"S, 175°09'29.0"W.

Area, Depth, Altitude 49 ha; to 5 m depth.

Land Tenure Government owned.

Physical Features Shallow reef, bounded on the seaward side by Piha Passage, and best developed on the northern side (Tongilava, 1979). An outer reef extends into Piha Passage. There are sand flats and extensive eelgrass beds and mangroves (Dahl, 1978; Eaton, 1985). Information on adjacent areas is given in Zann *et al.* (1981).

Reef Structure and Corals Coral is reportedly predominantly *Acropora* (Tongilava, 1979). The reef edge had at least 12 genera of corals (Zann *et al.*, 1981). There was good coral cover in shallow water (Chesher, 1984a).

Noteworthy Fauna and Flora Fish are abundant, especially *Chromis caeruleus, Holocentrus, Chaetodon* and *Amphrion* spp. (Tongilava, 1979). The Giant

Clams *Tridacna maxima* and *T. squamosa* occur (McCoy, 1980).

Scientific Importance and Research   Some of the reefs were briefly surveyed by Zann *et al.* (1981) and Chesher (1984a). Giant clams were surveyed by McCoy (1980).

Economic Value and Social Benefits   Heavily used for recreation by tourists and Tongans (Tongilava, 1979). There is some fishing around the island using arrowhead fish fences (Zann *et al.*, 1981).

Disturbance or Deficiencies   The area has suffered from over-fishing and collecting by tourists (Dahl, 1978;

Tongilava, 1979) and coral breakage and infection with blue-green algae is common (Chesher, 1984a).

Legal Protection   Established as a reserve in 1979 under the Parks and Reserves Act, 1976.

Management   None at present.

Recommendations   The area could be developed as a site for tourist and glass-bottomed boats because of its proximity to Nuku'alofa. The reef could be restored and perhaps even improved with the addition of substrates to encourage coral growth (Dahl, 1978).

TUVALU

INTRODUCTION

General Description

Tuvalu (formerly the Ellice Islands) consists of nine coral atolls and islands, rarely reaching more than 4 m in altitude, with a total land area of 26 sq. km. The five atolls (Funafuti, Nanumea, Nui, Nukufetau and Nukulaelae) generally have narrow strips of land on the east and reef with scattered islets on the west. Nanumanga, Niulakita and Niutao are reef islands consisting of single islets with brackish internal lakes and Vaitupu is intermediate in type.

Table of Islands

Niulakita (Nurakita) (X) 0.15 sq. mi. (0.4 sq. km); small reef island.

Nukulaelae (Nukulailai) (X) 0.7 sq. mi. (1.8 sq. km); atoll with 14 islets on enclosed lagoon.

Funafuti (X) (*see separate account*).

Nukufetau (X) 1.1 sq. mi. (2.8 sq. km); diamond-shaped atoll, 8+ islets; large sheltered lagoon; reef 24 mi. (39 km) in circuit.

Vaitupu (X) 2.15 sq. mi. (5.6 sq. km); atoll with one large oval island and two small lagoons; broad fringing reef.

Nui (Egg) (X) 0.8 sq. mi. (2.1 sq. km); crescent-shaped atoll with 8+ islets to east.

Niutao (X) 1.0 sq. mi. (2.6 sq. km); reef island.

Nanumanga (X) 1.1 sq. mi. (2.8 sq. km); reef island.

Nanumea 1.5 sq. mi. (3.9 sq. km); atoll with 2 islets; described in Chambers (1975).

(X) = Inhabited

The islands are of fossil coral limestone and other calcareous materials of marine origin, and appear to be rather young with changing land areas and lagoons still filling with sediment. Hurricanes have played a major role in land formation by throwing up rubble from the reef, and major hurricanes were recorded in 1891, 1958 and 1972 (*see account for* Funafuti). Royal Society expeditions visited the islands from 1896 to 1898 and provided useful baseline geomorphological information (Royal Society, 1904).

Reef platforms examined on Funafuti, Nui and Nanumea were impoverished in coral and other invertebrates. On Nanumea all the micro-atolls of *Porites* on the reef top were dead, possibly indicating a recent elevation of the land by about 0.5 m (Zann, 1980), although this could have been the result of draining a moated reef crest area

by channel blasting during World War II (Zann *in litt.*, 1987). A description of some of the shallow reef areas of Funafuti is given in a separate account. Blue Coral *Heliopora coerulea* is abundant on many reefs (Zann and Bolton, 1985).

Reef Resources

Subsistence fisheries are very important and are described by Zann (1980, 1983 and 1985). Fish are the protein staple and fisheries are mainly based on the outer reef slopes and in pelagic waters, the latter being most important. Pole-lining, gill netting, trapping and spearfishing methods are used on the reefs. Women and children forage for shellfish on the rock flats. A variety of molluscs including tridacnids are eaten. Soils, where developed, are poor and terrestrial commercial development is limited to coconut plantations for copra production for an unreliable world market. The potential of coastal and oceanic resources is therefore very important (Baines, 1982). Parkinson (1984) investigated the potential for the introduction of trochus *Trochus niloticus*. An overview of the National Fisheries Development Programme is provided by Pita (1985). Ellway *et al.* (1983) describe the skipjack and baitfish resources.

Disturbances and Deficiencies

Reefs at Funafuti were damaged by a hurricane in 1972 (*see separate account*).

Because of their small size and coral origin, the islands have very limited resources and are sensitive to environmental problems. The only urban area is on Funafuti and at present this is the only area where human activity is having a noticeable impact (*see separate account*). Although there have been recent signs of overfishing on Funafuti and Vaitupu, the resource seems largely adequate. However, giant clams and turtles are declining (SPREP, 1981). There is no commercial collecting of corals and shells but visitors occasionally come to collect rare shells (Sloth *in litt.*, 8.4.87). The changes in fishing impact from new technologies such as outboard motors are so recent that their effect is not yet noticeable, but new management controls and fisheries extension work will be needed as the impact spreads. There is a ciguatera problem in some areas and red tides toxic to fish have been reported in some lagoons (SPREP, 1981). Most seabirds are protected but they are an important traditional food source. There has been concern about channel blasting projects in the outer islands which have damaged corals (Zann, 1980).

Legislation and Management

Zann (1983 and 1985) describes the traditional utilization and conservation of marine resources, a system of "tapu" permitting regulation of natural resources. A range of penalties might be imposed for breaches of fisheries regulations. Traditionally land owners probably held tenure to the reefs and lagoons adjacent to their lands

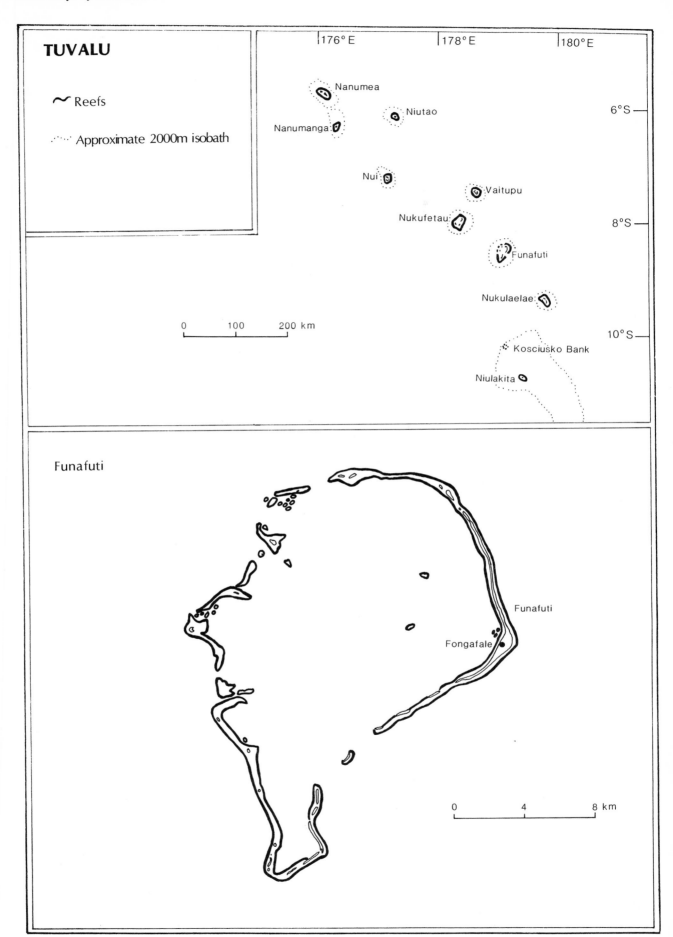

TUVALU

~ Reefs

⋯⋯ Approximate 2000m isobath

176°E 178°E 180°E

Nanumea

Niutao 6°S

Nanumanga

Nui Vaitupu

Nukufetau 8°S

Funafuti

Nukulaelae

Kosciusko Bank 10°S

Niulakita

0 100 200 km

Funafuti

Funafuti

Fongafale

0 4 8 km

and had exclusive rights to fisheries and passage. Many of these traditions have now broken down. The Fisheries Ordinance (1977) prohibits fishing with explosives and poisons. The Prohibited Areas Ordinance and the Wildlife Conservation Ordinance (1975) provide for the establishment of wildlife reserves and protection of seabirds. Modern legislation also covers pollution, regulation of sand and coral removal and waste disposal. No wildlife reserves have yet been established. However regulations are often weak or entirely lacking and implementation and enforcement is poor (SPREP, 1981).

Tuvalu is a party to the International Convention on Civil Liability from Oil Pollution Damage (1969) and the International Convention on the Establishment of an International Fund for Compensation for Oil Pollution Damage (1971). There is great concern about erosion of land areas, but these changes may be natural events and be beyond the control of any reasonable coastal engineering projects. Nearly all the islands are trying to control problems of coastal erosion with gabion sea wall construction, but there is little understanding of the currents and coastal processes involved (Baines, 1982).

Recommendations

The Government would like to see the region declared a nuclear-free zone in which all nuclear waste dumping would be prohibited, and would like better international control of oceanic fisheries. The harvest of giant clams and turtles should be controlled. Channel blasting should be monitored in case there is an associated increase in fish poisoning or coastal erosion. Wildlife sanctuaries may be necessary for seabird and turtle nesting areas (SPREP, 1981). The Kosciusko Bank, to the north of Niulakita, has been suggested as a possible reef reserve (Dahl, 1980). Dahl (1986) suggests that marine protected areas on Funafuti and Vaitupu might be considered in order to help control overfishing.

References

* = cited but not consulted

Baines, G.B.K. (1982). Pacific Islands: Development of coastal marine resources of selected islands. In: Soysa, C., Chia, L.S. and Coulter, W.L. (Eds), *Man, Land and Sea: Coastal Resource Use and Management in Asia and the Pacific*. The Agricultural Development Council, Bangkok, Thailand. Pp. 189-198.

Baines, G.B.K., Beveridge, P.J. and Maragos, J.E. (1974). Storms and island building at Funafuti, Ellice Islands. *Proc. 2nd Int. Coral Reef Symp., Brisbane* 2: 485-496.

*Baines, G.B.K. and McLean, R.F. (1976a). Sequential studies of hurricane deposit evolution at Funafuti Atoll. *Mar. Geol.* 21: M1-M8.

*Baines, G.B.K. and McLean, R.F. (1976b). Re-surveys of 1972 hurricane rampart of Funafuti Atoll, Ellice Islands. *Search* 7(1-2): 36-37.

*Buckley, R.C. (1983). Tuvalu Lagoon Bed Materials Resource Survey: Environmental Baseline and Impact Study. AMDEL Report No. 1508 to ADAB and Gibb Australia. AMDEL, Adelaide.

Buckley, R. (1985). Environmental survey of Funafuti Atoll. *Proc. 5th Int. Coral Reef Congr., Tahiti* 6: 305-310.

*Chambers, A. (1975). Nanumea Report. Victoria University, Wellington, New Zealand.

*Chapman, V.J. (1955). Algal collection from Funafuti Atoll. *Pac. Sci.* 9: 354-356.

Dahl, A.L. (1980). Regional ecosystems survey of the South Pacific Area. *SPC/IUCN Technical Paper* 179. South Pacific Commission, Noumea, New Caledonia.

Dahl, A.L. (1986). *Review of the Protected Areas System in Oceania*. UNEP/IUCN. 239 pp.

*David, T.W.E., Halligan, G.H. and Finckh, A.E. (1904). Report on dredging at Funafuti. In: Royal Society (1904). *The Atoll of Funafuti. Boring into a coral reef and the results*. London. Pp. 151-159.

*David, T.W.E. and Sweet, G. (1904). The geology of Funafuti. In: Royal Society (1904), *The Atoll of Funafuti. Boring into a coral reef and the results*. London. Pp. 61-124.

*Finckh, A.R. (1904). Biology of the reef-forming organisms at Funafuti Atoll. In: Royal Society (1904), *The Atoll of Funafuti. Boring into a coral reef and the results*. London. Pp. 125-150.

Ellway, C.P., Farman, R.S., Argue, A.W. and Kearney, R.E. (1983). An assessment of the skipjack and baitfish resources of Tuvalu. *Skipjack Survey and Assessment Programme Final Country Report* 8. South Pacific Commission, Noumea, New Caledonia.

*Hedley, C. (1899). The mollusca of Funafuti. Parts 1, 2 and Supplement. *Mem. Aust. Mus.* 3: 395-488, 489-510, 547-565.

*Johnson, J.H. (1961). Fossil algae from Eniwetok, Funafuti and Kita-Daito-Jima. *U.S. Geol. Surv. Prof. Pap.* 260-Z.

*Maragos, J.E., Baines, G.B.K. and Beveridge, P.J. (1974). Tropical cyclone creates a new land formation on Funafuti atoll. *Science* 181: 1161-1164.

*Mergner, H. (1983). Initial recolonisation of Funafuti Atoll coral reefs devastated by Hurricane "Bebe". Paper presented at 15th Pacific Science Congress, Dunedin, New Zealand.

Parkinson, B. (1984). A report on the potential for the introduction of Trochus (*Trochus niloticus*) to Tuvalu. South Pacific Commission, Noumea, New Caledonia.

Pita, E. (1985). An overview of the National Fisheries Development Programme. Fisheries Division, Government of Tuvalu, Funafuti, Tuvalu.

*Royal Society (1904). *The Atoll of Funafuti. Boring into a coral reef and the results*. London.

SPREP (1981). Tuvalu. *Country Report* 18. South Pacific Commission, Noumea, New Caledonia.

*Voronov, A.Y., Ignatiev, Y.M. and Kaplin, P.A. (1977). The trip of the Kallisto to the islands of the Pacific Ocean. *Bull. Moscow Univ. Ser. Geog.* 5: 111-118.

*Whitelegge, T. (1897a). The Crustacea. *Mem. Aust. Mus.* 3: 127-154.

*Whitelegge, T. (1897b). The sponges of Funafuti. *Mem. Aust. Mus.* 3: 323-335.

*Whitelegge, T. (1897c). The Echinodermata. *Mem. Aust. Mus.* 3: 155-164.

*Whitelegge, T. (1897d). The Alcyonaria of Funafuti, Parts 1 and 2. *Mem. Aust. Mus.* 3: 211-226, 307-322.

*Whitelegge, T. (1898). The Madreporaria of Funafuti. *Mem. Aust. Mus.* 3: 349-368.

*Whitelegge, T. and Hill, J.P. (1899). The Hydrozoa, Scyphozoa, Actinozoa and Vermes. *Mem. Aust. Mus.* 3: 369-394.

Zann, L.P. (1980). Tuvalu's subsistence fisheries. Report 4. Effects of energy crisis on small craft and fisheries in the South Pacific. Inst. Marine Resources, Univ. S. Pacific, Suva, Fiji.

*Zann, L.P. (1983). Man and atolls: Traditional utilization and conservation of marine resources and recent changes in Tuvalu and Kiribati. Paper given at 15th Pacific Science Congress, Dunedin, New Zealand.

Zann, L.P. (1985). Traditional management and conservation of fisheries in Kiribati and Tuvalu atolls. In: Ruddle, K. and Johannes, R.E. (Eds), *The Traditional Knowledge and Management of Coastal Systems in Asia and the Pacific*. Unesco-ROSTEA, Jakarta. Pp. 53-77.

Zann, L.P. and Bolton, L. (1985). The distribution, abundance and ecology of the blue coral *Heliopora coerulea* (Pallas) in the Pacific. *Coral Reefs* 4: 125-134.

FUNAFUTI ATOLL

Geographical Location Between Nukufetau and Nukulaelae in the south-central part of the Tuvalu Group; 8°30'S, 179°12'E.

Area, Depth, Altitude 1.1 sq. mi. (2.8 sq. km); 18 km diameter.

Physical Features The atoll is almost pear-shaped and has 30 islets, and a central swamp with mangroves. The deep central lagoon has three main deep channels, additional lagoon/ocean interchange taking place over the reef rim especially on the western side where the islets and small and widely separated (Buckley, 1985). Geomorphology of the Fongafale area on Funafuti Islet is described in Buckley (1983) and a detailed description of the storm beach or rubble rampart is given by Baines *et al.* (1974). David *et al.* (1904) describe early dredging studies.

Reef Structure and Corals Early descriptions of the coral fauna are given in David and Sweet (1904), Finckh (1904) and Whitelegge (1898). Large areas of the lagoon floor are covered by intermittent coral heads, particularly on the sloping margins. The coral heads found at 15-30 m depth consist almost entirely of acroporid skeletons supporting sparse living colonies of *Acropora* sp. At the lagoon rim and in isolated shallow shoals in the centre of the lagoon, the coral fauna is more diverse, although not as diverse as that of the deep channels receiving ocean water, and true reefs are formed. A total of 36 coral species was recorded from the lagoon reefs in 1983. The Blue Coral *Heliopora coerulea* is notably absent from the eastern rim reefs although dead *Heliopora* is a major component of the reef flat. Abundant live Blue Coral was found on the western rim (Buckley, 1985).

Following Hurricane Bebe in 1972, extensive studies were carried out on the newly formed rubble rampart which had a submarine origin. Its outer edge lies parallel to the reef rim leaving an outer reef platform about 30 m wide exposed at low water spring tide. A spur and groove system was well developed in the outer 15 m of the platform. The rubble rampart was situated entirely landward of this system except at the eastern end of Funafuti Islet where a few unusually long and deep fissures extended beneath the rampart into the moat.

Adjacent to the islands on the south-east coast a well developed spur and groove system occurs at the outer edge of the reef flat. Below this, the reef front drops

steeply to a sloping terrace at 7.8 m, although in places extensions of the spurs project onto the slope. A spur and groove system was virtually absent from the centre of the inter-island gaps. The reef edge is inconspicuous, merging with the seaward slope. This extended to a depth of about 10 m where there is a broad flat shelf. Below 20-30 m, the slope drops abruptly to 45-65 m, ending in a gentle slope of *Halimeda* sand.

Although not abundant, a variety of corals was found on the rocky walls below 20-30 m. Corals and algae were more common along the upper margin of the cliff. Shallower than 15-20 m, corals and other organisms recorded by Finckh (1904) were virtually absent and the slope and platform was covered with loose rubble. At shallower depths, a few scattered corals were found in the grooves but coral cover was less than 1%.

Beyond the northern end of the rubble rampart, the reef had suffered less disturbance. A variety of corals was found on the reef slope and reef buttresses, including *Millepora*, *Acanthastrea* and *Favia*, as well as genera typical of shallower water. *Porolithon* provided the dominant cover; Frinckh (1904) had also recorded the dominance of this alga on the seaward slope. The area adjacent to the north-east section of the rubble rampart had suffered intermediate storm damage, with little evidence of damage to the buttresses and spurs (Baines *et al.*, 1974).

Noteworthy Fauna and Flora There are some small areas of mangrove swamp (Buckley, 1985). Algae are described by Chapman (1955) and fossil algae by Johnson (1961). Species lists for the marine, intertidal and terrestrial zones of Fongafale are given in Buckley (1983), Whitelegge (1897a, b, c and d) and Whitelegge and Hill (1899). Over 200 demersal fish species and 400 molluscs have been described from the lagoon (Buckley, 1983; Hedley, 1899; Zann, 1980).

Scientific Importance and Research Funafuti was the site of the Royal Society Expeditions of 1896-8 (Armstrong *et al.*, 1904) and has been visited subsequently by occasional specialists. Several studies were carried out following Hurricane Bebe in 1972 (Baines *et al.*, 1974; Baines and McLean, 1976a and b; Maragos *et al.*, 1974; Mergner, 1983). Work has also been carried out by Voronov *et al.* (1977). In 1983, a one-month environmental study was carried out on Funafuti Islet (Fongafale) on behalf of the Australian Development Assistance Bureau (Buckley, 1985). As expected for an isolated central Pacific atoll, the reefs are less diverse than more extensive reef areas in Fiji and Australia, but they are nevertheless quite diverse for an atoll (Buckley, 1985).

Economic Value and Social Benefits During World War II, Funafuti was developed as a forward base, an airfield occupying much of the main island, which was built from material from the hurricane bank. The main population of Tuvalu is now situated on Funafuti Islet (Fongafale), the largest island of the atoll (Buckley, 1985). Bêche-de-mer was once successfully harvested but fishermen lacked motivation to continue (SPREP, 1981).

Disturbance or Deficiencies Hurricane Bebe in 1972 caused marked alterations to the geomorphology of the atoll. Hurricane generated waves from the south-east caused reef detritus from an offshore terrace to be elevated onto the reef flat from depths of up to 20 m to form a rubble rampart 19 km in length and up to 4 m in height along the south-east rim of the atoll. Most of the material probably came from an almost level shelf at a depth of about 10 m. In the process, the benthic communities above 10 m were destroyed. Initial recolonization of the reefs is described by Mergner (1983).

There have been recent signs of overfishing. There is a pollution risk from old American fuel dumps on some islets of Funafuti (SPREP, 1981) and the lagoon is reported to be polluted (Zann, 1980).

Legal Protection   None.

Management   None.

Recommendations   A marine protected area might be considered in order to help limit overfishing (Dahl, 1986).

VANUATU

INTRODUCTION

General Description

The Republic of Vanuatu, formerly the Anglo-French Condominium of the New Hebrides, consists of the central and southern part of an archipelago forming one of the numerous seismic arcs found in the western Pacific (the Santa Cruz Islands, politically part of the Solomon Islands, constitute the northern part of the archipelago). Vanuatu has a land area of ca 12 100 sq. km, lying between ca 13°S and 21°S, excluding Matthew and Hunter Islands; these, originally part of the New Hebrides Condominium, became part of New Caledonia about 1955 and are now claimed by Vanuatu and France (see section on New Caledonia). The archipelago essentially forms a bifurcating chain; the larger islands are found in the western part and are made up of extinct volcanoes covered with fossil or modern coral reefs, the best known being Efate, Malakula and Espiritu Santo Islands. Some volcanoes are subaerial (Ambrym, Tanna and Lopevi), others submarine as off Epi and Erromango Islands. Geological aspects of the islands are described by Karig and Mammerickx (1972). Vanuatu is fairly frequently swept by tropical cyclones, which tend to be severe. The climate is perhumid and hot, the temperature and to some extent the rainfall increasing from south (1500 mm) to north (3000 mm) (Guilcher, 1974); in some parts of the north rainfall exceeds 4000 mm p.a. Water temperatures range from 28.1°C in February to 24.5°C in September, with maximum sun in November and December (Buskirk *et al.*, 1981).

Names used in the table follow the *Official Gazetteer of Place Names* (1979). Comments on reefs in the table marked * are from Chambers (*in litt.*, 20.7.87) and are derived mainly from available maps; they refer only to the sizes of reefs as marked, not to their condition or quality.

Table of Islands

Anatom (Aneityum) (X) 40 sq. mi. (104 sq. km); volcanic; 2795 ft (852 m); high central massif; radially dissected; surrounded by fringing reefs, especially well-developed in north and south-west; Intao Reef in south-west extends into sea for 2.7 km and has a sand cay, Inyeung I., on sheltered side (Guilcher, 1974); shearwaters breed (Garnett, 1984).

Futuna (Erronan) (X) volcanic cone; 2109 ft (641 m); steep coast; *narrow fringing reef, widens to 500 m on north coast.

Aniwa (X) 42 m; raised coral overlying volcanic; *reefs up to 500 m wide.

Tanna (X) 150 sq. mi. (389 sq. km); active volcano, 1084 m, with limestone fringe; elevated fringing reefs may occur in north and west; poorly developed living fringing reefs (Guilcher, 1974); shearwaters breed (Garnett, 1984).

Erromango (Eromanga) (X) 330 sq. mi. (855 sq. km); volcanic with extensive raised coral terraces up to 800 ft (244 m); volcanoes 2906 ft (886 m); submarine volcano off island; fringing reefs, narrow in most places though some are broader (Guilcher, 1974); further data in Corner and Lee (1975) and Dahl (1980).

Efate (Vate) (X) 300 sq. mi. (777 sq. km); volcanic overlain with limestone, 2122 ft (647 m); south low plateau; raised coral terraces; geology described by Lecolle and Bernat (1985); closed lagoon; west coast of Mele Bay (Port Vila) has slightly emerged reef flat dissected by shallow pools and outer erosional spurs and grooves; fringing reefs around island up to 2 km wide on north coast, some project seawards bearing sand cay, e.g. at Mele Point (Chambers *in litt.*, 20.7.87; Guilcher, 1974); Port Vila and Erakor Lagoon surveyed (Carter, 1983); islands off coast include:

- *Eretoka* 90 m; *fringing reef; regularly dived (Chambers *in litt.*, 20.7.87);

- *Lelepa* (X) 202 m; *narrow fringing reef;

- *Moso* (X) 116 m; *narrow fringing reef on south side, more extensive on northern and western sides;

- *Nguna* (X) 1945 ft (593 m); volcanic; *narrow fringing reef;

- *Pele* (X) 649 ft (198 m); *fringing reef, generally narrow, up to 1.3 km wide on west side;

- *Emao (Mau)* (X) 448 m; volcanic; *fringing reefs up to 700 m wide.

Etarik (Monument) 155 m; *no reef; seabird colonies (Chambers *in litt.*, 20.7.87).

Mataso (X) 494 m; steep; *narrow fringing reef, up to 250 m wide.

Shepherd Group

Group of small islands south-east of Epi.

Makura (Makir) (X) 297 m volcanic islet; *narrow fringing reefs.

Emae (Mai) (X) 644 m, volcanic; *fringing reefs up to 1 km wide; Cook Reef, 5 km to east, described below.

Buninga (X) 216 m; *fringing reef, absent in west, poorly developed elsewhere.

Amora Rocks 112 ft (34 m).

Tongariki (X) 521 m; *narrow fringing reef.

Falea 100 m; *fringing reefs absent or narrow.

Ewose (Awoh) (X) 319 m; *fringing reefs absent or narrow.

Tongoa (Kuwae) (X) 487 m, volcanic, steep cliffs; *fringing reef generally narrow; high population density.

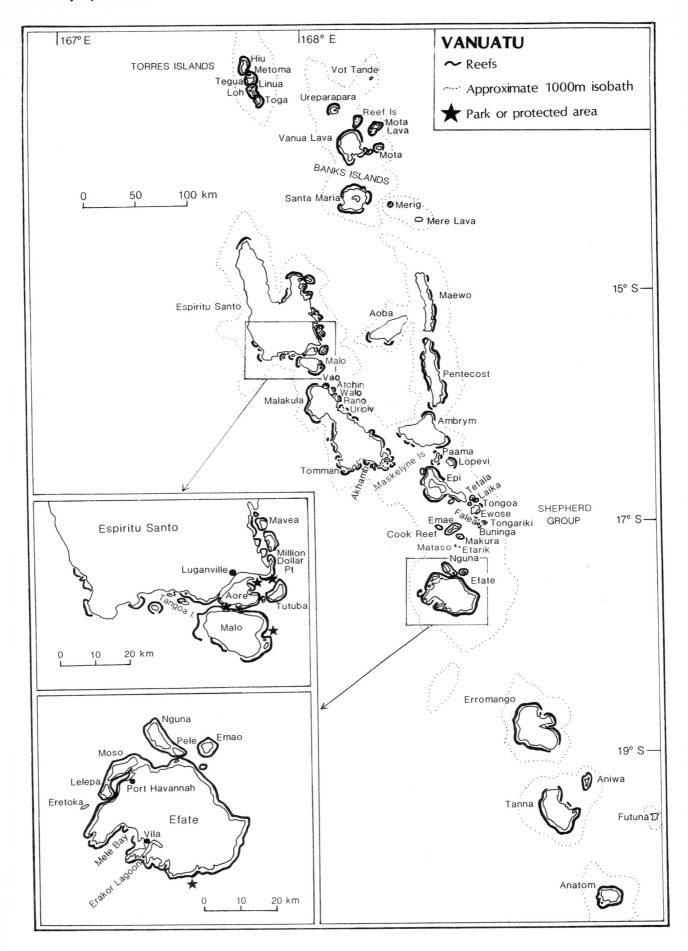

VANUATU

~ Reefs

········· Approximate 1000m isobath

★ Park or protected area

167° E
168° E

TORRES ISLANDS

Hiu
Metoma
Tegua Linua
Loh
Toga

Vot Tande

Ureparapara

Reef Is
Mota
Lava

Vanua Lava

Mota

BANKS ISLANDS

0   50   100 km

Santa Maria

Merig

Mere Lava

15° S

Espiritu Santo

Maewo

Aoba

Malo

Vao

Atchin
Walo
Rano
Uripiv

Pentecost

Malakula

Ambrym

Paama
Lopevi

Epi
Tefala
Laika
Tongoa
Ewose
Faleas
Tongariki
Emae
Buninga
Makura
Mataso
Etarik
Nguna

Tomman

Akhamb

Maskelyne Is

Cook Reef

SHEPHERD
GROUP

17° S

Efate

Espiritu Santo

Mavea

Million
Dollar
Pt

Luganville

Aore

Tangoa I.

Tutuba

Malo

0   10   20 km

Erromango

19° S

Nguna

Pele
Emao

Moso

Lelepa

Port Havannah

Eretoka

Efate

Aniwa

Mele Bay
Vila

Tanna

Futuna

Erakor Lagoon

★

0   10   20 km

Anatom

Laika   87 m; visited frequently; *narrow fringing reef.

Tefala   93 m; *narrow fringing reef.

Epi   (X) 100 sq. mi. (259 sq. km); volcanic with coral; 2732 ft (833 m); well watered; submarine volcano off island; *reefs absent in the south-east, often narrow elsewhere but up to 500 m wide, especially in the north.

Lopevi (Ulveah)   (X) volcanic; 4635 ft (1413 m); active crater; *narrow fringing reef.

Paama   (X) volcano; 1784 ft (544 m); *fringing reef generally poorly developed or absent, but up to 400 m wide in places.

Ambrym   (X) 160 sq. mi. (414 sq. km); volcanic, active cone; 4166 ft (1270 m); *fringing reef generally absent or narrow, but up to 500 m wide in north and south-east.

Malakula (Malekula, Mallicollo)   (X) 450 sq. mi. (1166 sq. km); limestone and volcanic; 863 m.   Raised coral terraces in north-west.   Reefs on north-west coast elevated over 6 m in 1965. Bordered on east by fringing reefs, sometimes wide. Isolated, often raised, reefs with wooded cays situated off coast, including Maskelynes off south-east (Guilcher, 1974):

- *Sakao (Khoti)*   (X) 102 m; *narrow fringing reef, up to 500 m wide in north-east;

- *Vulai*   83 m; *narrow fringing reef with 3 km long reef extending off north-east;

- *Awei*   88 m; *on Malakula fringing reef;

- *Lembong*   *on Malakula fringing reef;

- *Livghos*   *on Malakula fringing reef;

- *Ulendeuv (Bagatelle)*   *on Malakula fringing reef;

- *Uliveo*   (X) 30 m; *has four associated islets (Metai, Livlakhoas, Batghutong, Khuneveo); situated on patch reef of ca 4x5 km.

Other islands off Malakula include:

- *Akhamb*   (X) 30 m; *1200 m wide reef in south-east;

- *Tomman (Urur)*   (X) 84 m; *narrow fringing reef;

- *Vao*   (X) 10 m; *narrow fringing reef;

- *Atchin*   (X) 23 m; *narrow fringing reef;

- *Wala*   (X) 26 m; *narrow fringing reef;

- *Rano*   (X) 24 m; *narrow fringing reef;

- *Uripiv*   (X) 38 m; *fringing reefs up to 600 m wide.

Pentecost (Raga)   (X) 125 sq. mi. (324 sq. km); volcanic and limestone; 3103 ft (946 m); extensive elevated limestone; *fringing reefs absent or narrow in west, up to 500 m wide in east.

Maewo (Aurora)   (X) 90 sq. mi. (233 sq. km); volcanic and limestone, 2660 ft (811 m); extensive elevated limestone; *fringing reef narrow or absent.

Aoba (Ambae, Omba)   (X) 105 sq. mi. (272 sq. km); volcanic (1496 m); not surveyed; *reefs generally absent, otherwise narrow.

Espiritu Santo   (X) 1500 sq. mi. (3885 sq. km); limestone and volcanic (to 1879 m), west coast precipitous, east coast = chain of coral islands, north = hot and humid with low scrub; many raised coral terraces; reefs on north-west coast elevated over 6 m in 1965; reefs at Million Dollar Point Reserve described in separate account.

Mavea   (X) 48 m; *narrow fringing reef.

Tutuba   (X) raised coral; 118 ft (36 m);   *narrow fringing reef.

Malo   (X) coral, 300-400 ft (91-122 m) with volcanic plateau, 326 m; *generally narrow fringing reefs, but up to 1.5 km wide on east coast; submerged reef 9 km long and up to 1.5 km wide off west coast.

Aore   (X) 325 ft (99 m) raised limestone; *narrow fringing reef.

Banks Islands

Mere Lava   (X) volcanic cone; 883 m, steep coast; densely populated.

Merig   (X) volcanic, 125 m; difficult landing; *narrow fringing reef.

Santa Maria (Gaua)   (X) 797 m eroded volcano; *generally narrow fringing reef, but up to 1 km wide on north-east coast.

Mota Lava (Valua)   (X) 411 m; volcanic peak; steep cliffs; west promontory surrounded by wide fringing reef with cay (Guilcher, 1974).

Vanua Lava   (X) 946 m, volcanic; short stretch of fringing reef, especially on east (Guilcher, 1974). 2 apron reefs with large mature sand cays (Ravea, Pakea, Nowela).

Mota   (X) 411 m; volcanic; *narrow fringing reef.

Reef Islands (Iles Rowa)   (*see separate account*).

Ureparapara   (X) breached crater 764 m; fringing reef to west, very narrow except for some places in caldera (Guilcher, 1974).

Vot Tande (Vat Ganai)   64 m; two tree-covered rocks; *narrow fringing reef.

Torres Islands

Toga   (X) 240 m; raised limestone terrace, narrow central plain; *narrow fringing reef.

Loh (X) raised limestone; 155 m; mangroves; *narrow fringing reef.

Linua 8 m; *narrow fringing reef; joined to Loh by sandy passage at low tide.

Tegua (X) 254 m; raised limestone forming steep-sided central plateau; *generally narrow fringing reef.

Metoma (X) 115 m; raised limestone; *narrow fringing reef.

Hiu (X) 3 limestone terraces rising to 366 m; *generally narrow fringing reefs but up to 700 m wide on west coast.

(X) = Inhabited

Reefs are generally more extensive in the western than in the eastern part of the chain (i.e. essentially Espiritu Santo, Malakula and Efate and smaller islands asociated with these). Fringing reefs of some of the islands are briefly described in the table above; the best of these are probably those around Anatom (Guilcher, 1974). Early descriptions include Baker (1925 and 1929). The only non-fringing reefs are the Reef Islands (*see separate account*) and Cook Reef. The latter (17°04'S, 168°17'E), situated 5 km west of Emae and 25 km south of Epi, is described by Guilcher (1974) from aerial studies. It has no emergent island and consists of a roughly triangular rim on which the swell breaks, enclosing a lagoon with a depth of generally less than 10 m and in places probably less than 5 m. Coral growth has created "cells" in the lagoon which resemble those found at Mataiva Atoll in the Tuamotu Group and at Raiatea and Maupiti Islands in the Society Islands but less regular and conspicuous. Cook Reef faces south-south-east although the trade winds blow from the east or east-south-east. There is a passage through the atoll rim on the exposed side which leads to a narrow and deeper (20 m at least) lagoon. Although considered simpler and less interesting than the Reef Islands, it is thought that this reef requires further investigation. Scleractinian corals of Vanuatu are described by Chevalier (1971 and 1975).

The tectonic uplifts of 1965 and 1971 which caused living corals to become emergent have provided a natural experiment for testing factors that influence coral growth (Buskirk *et al.*, 1981) and for studying vertical tectonic movements (Taylor *et al.*, 1981). In the Espiritu Santo-Malakula area, there are at least four arc segments or tectonic blocks that are generally being tilted eastward, with uplift rates as great as 7 mm/yr in late Quaternary time (Taylor *et al.*, 1980). The association between the uplift of Malakula Island and earthquakes in 1965 is described by Benoit and Dubois (1971) and Mitchell (1968). The reefs on the north-west coasts of Malakula and Espiritu Santo are considered to be particularly interesting on account of these events.

There are estimated to be between 2500 and 3500 ha of mangroves in Vanuatu, most (ca 2000 ha) on Malakula (David, 1985). Both Green Turtles *Chelonia mydas* and Hawksbills *Eretmochelys imbricata* are reportedly common in Vanuatu, with Malakula and the adjacent Maskelynes apparently being the most important nesting areas; notable nesting also occurs in the Torres Islands

and the south-east of Epi (Pritchard, 1981). Dugong *Dugong dugon* are apparently widespread, although not abundant (Crossland, 1983). Garnett (1984) gives a brief overview of seabird distribution. The fish of Vanuatu are described by Fourmanoir and Laboute (1976) and Herre (1931). Brouard and Grandperrin (1985) describe the fish of the deep outer reef and provide estimates for maximum sustainable yields. The Blacklip Pearl Oyster *Pinctada margaritifera* is present in small numbers and is not believed to be harvested (Crossland *in litt.*, 18.8.83).

Reef Resources

Subsistence fisheries are now of major importance (although the people of Vanuatu have never made heavy use of marine resources (Dahl, 1985)), and Grandperrin (1982) provides a review of the industry. Organized reef and outer-reef slope fisheries began around 1980, following the establishment of the Fisheries Department, although prior to this Asian fishermen at a Santo-based fishing company had fished on the reefs. There are at present up to 80 village fisheries projects catching finfish, with an additional 20 catching both finfish and shellfish and a further 10 engaged in marketing. Most fish is processed in the period November to January (Bakeo *in litt.*, 13.11.87). An assessment of the skipjack and baitfish resources is given in Tuna Programme (1983). Trochus *Trochus niloticus* is widely gathered for food in many parts of Vanuatu and there is a factory ("Melanesian Shell") in Port Vila where trochus and green snail shells are made into buttons (Anon., n.d.; Bakeo *in litt.*, 13.11.87; Bour and Grandperrin, 1985; Crossland *in litt.*, 18.8.83). This operates as a monopoly enterprise and obtains all its supplies from within Vanuatu (Bakeo *in litt.*, 13.11.87). In the past it has had difficulty obtaining adequate supplies and has imported shell, and in 1983 was temporarily out of production. In the early 1980s trochus was exported (49 t in 1981, 77 t in 1982) (Crossland *in litt.*, 18.8.83), but exports of it and the green snail are no longer permitted (Bakeo *in litt.*, 13.11.87). Dugong are hunted although not apparently intensively so (Crossland, 1983). Both species of turtle are exploited for eggs and meat; the use of shell is thought to be minimal (Pritchard, 1981). Palolo worms, which only occur in the northern part of Vanuatu, are collected when spawning in October and November (Bakeo *in litt.*, 13.12.87; Horrocks, 1986) and the Coconut Crab *Birgus latro* is a popular item on restaurant menus (Anon., n.d.)

The tourist industry is developing rapidly in Vanuatu and reef-related activities are of particular importance (Anon., n.d.). Popular dive sites around Port Vila include Mele Reef in Mele Bay, Hat Island and Paul Rock. There are at least three dive operators and several glass-bottomed boats.

Disturbances and Deficiencies

Periodic hurricanes have damaged some of the reefs, including those of the *President Coolidge* and Million Dollar Point Reserve, which were extensively damaged by Cyclone Nigel in 1985 (Bakeo *in litt.*, 13.11.87; Power *in litt.*, 12.11.87 and *see separate account*) and reefs around Port Vila which were seriously damaged by Hurricane Uma in 1987 (Anon., 1987). Corals on both the north-west and south coasts of Santo have been partially or wholly killed by emergence in 1965. *Porites* and *Goniastrea* corals are known to be quite resistant to temperature extremes, which tend to occur in shallow depths in Vanuatu after tectonic uplift, until tidal and weather conditions combine to be particularly detrimental. The top halves of many *Goniastrea* colonies tend to die, but the coral polyps surrounding the bases of the colonies survive. These events have been studied on reefs around Efate, Malo, Espiritu Santo, Tongoa Island off the south coast of Espiritu Santo, and Malakula (Buskirk *et al.*, 1981; Taylor *et al.*, 1981). There is evidence of a population outbreak of *Acanthaster planci* on the reef at the end of Malapoa Point at the entrance to Vila Harbour (King, 1986).

Damage from human activities has not been documented although some types of coral have been particularly heavily collected (Crossland, 1983). Vanuatu still has a low although rapidly growing population (1979 census result of 112 304, or 7.6 per sq. km, increasing at 3.2% per annum). There are fears that increased urban growth could lead to increasing pollution in the Vila area (Dahl, 1985; SPREP, 1980). The lagoon at Efate has been disturbed by urban development (Dahl, 1980; Lam Yuen, 1980) and sand has been removed from Samoa Point at Port Havannah, Efate, damaging a site with considerable potential for recreation and tourism (Baines, 1981). Studies on ciguatera have been conducted by Denton (1983).

Legislation and Management

There is a wide variety of traditional practices which protect marine resources; these are being surveyed by students at Malapoa College (Dickinson, 1983). For example, seasonal custom "tabu" are applied in certain coastal waters where fish are known to gather for breeding (Anon., 1985).

The Fisheries Regulations (1983) of the Fisheries Act (1982) provide for the conservation and regulation of fisheries and the issue of fishing licences. Joint regulations concerned with fisheries made during the time of the Condominium have been repealed. There is a minimum size limit of 22 cm for spiny lobsters (*Panulirus pencillatus*, *P. versicolor* and *P. longipes* are found in Vanuatu) and they and slipper lobsters may not be speared. The minimum size limit for the slipper lobster *Parribacus caledonicus* is 15 cm, and for the Coconut Crab *Birgus latro* 9 cm. The regulations for the latter species are the same as those which previously applied under the Coconut Crabs (Protection) Act 1981. Minimum size limits have been established for *Trochus niloticus* (9 cm diameter), *Turbo marmoratus* (15 cm longest dimension) and *Charonia tritonis* (20 cm length). The taking of eggs of Green Turtles *Chelonia mydas* and Hawksbills *Eretmochelys imbricata* is prohibited, and the sale of Hawksbill meat and shell is not permitted. Green Turtle meat and shell may be sold, and subsistence harvesting of both species is permitted. It is prohibited to take more than three pieces of living coral in a day except with the permission of the Director of Fisheries. Written permits from the Minister responsible for fisheries are required for the export of trochus, green snail, crustaceans of any kind, aquarium fish, coral and bêche-de-mer, with quotas being set for bêche-de-mer

(Crossland, 1983). No permits are issued at present for the export of trochus or green snail (Bakeo *in litt.*, 13.11.87). Marine mammals are totally protected (Crossland, 1983).

Under Section 20 of the Fisheries Act No.37 of 1982 the Minister responsible for fisheries may, after consultation with owners of adjoining land and with the appropriate local government council, declare any area of Vanuatu waters, and the seabed underlying such waters, to be a marine reserve (Anon., 1985). Within such a reserve, fishing of any kind (including the taking of shells, crustaceans and other living organisms) is prohibited and sand, coral or wreck artefacts may not be removed (Crossland, 1983). At present Million Dollar Point is the only marine reserve (*see separate account*). There are also four small recreational reserves which include coral formations and sandy beaches:

- Whitesands, 39 ha in extent at 168°25'E, 18°12'S on Efate;
- Bucaro, 20 ha at 167°13'E, 15°33'S on Aore;
- An un-named 37 ha site at 167°7'E, 15°36'S on Aore;
- Naomebaravu, 11 ha at 167°15'E, 15°44'S on Malo.

All these were established in 1984 although there was (as of 1985) a lack of controlling legislation; complete protection of vegetation and the natural environment was intended (Anon., 1985).

An Environmental Unit currently operates in the Ministry of Lands, Energy and Water Supply.

Recommendations

Baines (1981) provided a series of recommendations for planned resource use and environmental protection. With respect to marine and coastal ecosystems, care must be taken that any commercial fisheries development based on coral reefs, lagoons and mangroves must be planned so as not to interfere with subsistence fisheries; enviromental assessment of development projects involving disturbance of soil or of submarine sediments must include provision for minimising sediment disturbance to coral reefs, to prevent declines in reef fish production; improvements in the control of waste disposal and the use of chemicals are required; and mangroves may be in need of special protection. A protected area system covering representative areas of natural ecosystems was proposed, emphasizing resource and species protection and the involvement of traditional landowners in protected area management. Traditional conservation and resource management practices should be made full use of in modern development.

Low (in Baines, 1981) proposed improvements in natural resource and enviromental administration and legislation, in particular for the following planned legislation, potentially relevant to reefs and fisheries: Marine Spaces; Town and Country Planning; Education; Tourism; Mining; National Parks; National Historic Sites; Ports and Harbours; Fisheries; Protection of Flora and Fauna. A Project Planning Act and Environmental Protection Act and the following new governmental posts were proposed: Resource and Enviromental Coordinator; Physical Resource and Land Use Planner; Town Planner; Wildlife Resource Planner.

The following reef areas have been identified as being of particular interest:

1. Anatom Island - fringing reefs;
2. Reef Islands (*see separate account*);
3. Cook Reef (Guilcher, 1974), requires further investigation;
4. North-west coasts of Malakula and Espiritu Santo (Dahl, 1980).

A joint request has been made by the Santo/Malo Local Government Council and the Santo Land Council for 9 coastal areas to be declared public land for recreational purposes, 7 on Espiritu Santo and 2 on Malo (Anon., n.d.); it is not known if these border reefs.

Funds from UNEP/SPREP have been obtained for a dugong survey to be undertaken by the Environmental Unit of the Ministry of Lands, Energy and Water Supply; a survey of fringing reefs has also been planned, which it is hoped will eventually lead to further marine protected areas (Chambers *in litt.*, 20.7.87).

Following a basic agreement passed in 1979 with the Vanuatu Government, ORSTOM (French Scientific Research Institute for Co-operation and Development) in New Caledonia has undertaken research activities in cooperation with the Vanuatu Fisheries Department. This is aimed at investigating resources along the coast, the outer reef slope and open sea, as well as studying the socio-economic aspects of the fisheries. A project has been launched to raise trochus to reseed overexploited reefs; as part of the project, trochus have been tagged on Efate (Bour and Grandperrin, 1985). Similar projects are being considered for Giant Clams and cowries such as *Cypraea aurantium*. A two year project on the biology of the Coconut Crab *Birgus latro*, funded by ACIAR, was completed in 1987. This could provide a basis for the development of conservation and management plans for the species, which is regarded as having considerable potential as an exploitable resource (Bakeo *in litt.*, 13.11.87).

References

* = cited but not consulted

Anon. (1985). Country Report - Vanuatu. *Report of the 3rd South Pacific National Parks and Reserves Conference, Apia* 3: 217-228.
Anon. (1987). Announcements. *PSA Coral Reef Newsletter* 18: 25.
Anon. (n.d.). *What to do in Vanuatu*. Port Vila.
*Baker, J.R. (1925). A coral reef in the New Hebrides. *Proc. Zool. Soc., London* (3): 1007-1019.
*Baker, J.R. (1929). *Man and Animals in the New Hebrides*. George Routledge and Sons Ltd, London. 200 pp.
Baines, G.B.K. (1981). Environmental resources and development in Vanuatu. Report to Government of Vanuatu with support of UNDAT (United Nations Development Advisory Team for the Pacific). 26 pp.
Benoit, M. and Dubois, J. (1971). The earthquake swarm in the New Hebrides archipelago, August, 1965: recent crustal movements. *Roy. Soc. N.Z. Bull.*: 141-148.
Bour, W. and Grandperrin, R. (1985). Trocas growth in Vanuatu. Mission ORSTOM Port Vila. *Not. et Doc. d'Océan.* 14. 31 pp.

Brouard, F. and Grandperrin, R. (1985). Les poissons profonds de la pente récifale externe à Vanuatu. *SPC 17th Conf. Tech. Rég. des pêches.* SPC/Fisheries 17/WP 12. 131 pp.
Buskirk, R.E., Taylor, F.W., O'Brien, W.P., Maillet, P. and Gilpin, L. (1981). Seasonal growth patterns and mortality of corals in the New Hebrides (Vanuatu). *Proc. 4th Int. Coral Reef Symp., Manila* 12: 197-200.
*Carter, R. (1983). Baseline studies of Port Vila and Erakor Lagoon, Vanuatu Cruise VA-83-1. CCOP/SOPAC Cruise Report. 53 pp.
*Chevalier, J.P. (1971). Les Scléractiniaires de la Mélanésie française (Nouvelle Calédonie, Iles Chesterfield, Iles Loyauté, Nouvelles Hebrides). 1ère partie. *Expéd. française récifs coralliens Nouvelle Calédonies.* Ed. Fond. Singer-Polignac, Paris. 5: 5-307.
*Chevalier, J.P. (1975). Les Scléractiniaires de la Mélanésie française (Nouvelle Calédonie, Iles Chesterfield, Iles Loyauté, Nouvelles Hebrides). 2ème partie. *Expéd. française récifs coralliens Nouvelle Calédonies.* Ed. Fond. Singer-Polignac, Paris. 7: 5-407.
*Corner, E.H.J. and Lee, K.E. (Eds) (1975). A discussion on the results of the 1971 Royal Society-Percy Sladen Expedition to the New Hebrides. *Phil. Trans. Royal Soc. London B* 272: 267-486.
Crossland, J. (1983). Vanuatu's new fishery conservation measures. *Naika* 12: 10-11.
Crossland, J. (1984). Vanuatu's first marine reserve. *Naika* 14: 2-3.
Dahl, A.L. (1980). Regional ecosystems survey of the South Pacific Area. *SPC/IUCN Technical Paper* 179. South Pacific Commission, Noumea, New Caledonia.
Dahl, A.L. (1985). Status and conservation of South Pacific coral reefs. *Proc. 5th Int. Coral Reef Cong., Tahiti* 6: 509-513.
David, G. (1985). Les mangroves de Vanuatu: 2ème partie, présentation générale. *Naika* 19: 13-16.
*Denton, P. (1983). Ecology of ciguatera: Baseline reef studies in Vanuatu and Solomon Islands. Paper presented at 15th Pacific Science Congress, Dunedin, New Zealand.
Dickinson, D. (1983). Conservation of coastal resources in custom 1: Tafea (Southern District); 2: Central District no. 1; 3: Northern District; 4: Central District no. 2. *Naika* 9: 9-10; 10: 16-17; 11: 10-11; 12: 15-16.
Fourmanoir, P. and Laboute, P. (1976). *Poissons de Nouvelle-Caledonie et des Nouvelles-Hebrides.* Les Editions du Pacifique, Papeete.
Garnett, M.C. (1984). Conservation of seabirds in the South Pacific Region: A review. In: Croxall, J.P., Evans, P.G.H. and Schreiber, R.W. (Eds), *Status and Conservation of the World's Seabirds.* ICBP Technical Publication 2. Pp. 547-558.
Grandperrin, R. (1982). Fisheries in Vanuatu: Situation at present and development and research prospects. ORSTOM Mission Port Vila. *Not. et Doc. d'Océan.* 1. 33 pp.
*Guilcher, A. (1972). Un banc corallien orienté: Reef Island aux îles Banks, Nouvelles Hébrides. *Intern. Geography* 1972 (abstr. 22nd Int. Geogr. Congr. Montreal): 1025-1026.
Guilcher, A. (1974). Coral reefs of the New Hebrides, Melanesia, with particular reference to open-sea, not fringing reefs. *Proc. 2nd Int. Coral Reef Symp., Brisbane* 2: 523-535.
Herre, A. (1931). A checklist of the fish recorded from the New Hebrides. *Journal of the Pan Pacific Research Institution* 6: 11-14.
Horrocks, M. (1986). Marine worm biorhythms. *Naika* 23: 16-17.

Karig, D.E. and Mammerickx, J. (1972). Tectonic framework of the New Hebrides Island Arc. *Mar. Geol.* 12: 187-205.

King, F. (1986). Observations on *Acanthaster planci*. *Naika* 23: 15.

Lam Yuen, T. (1980). Study of the coastal water quality around Port Vila. South Pacific Commission, Noumea. 33 pp.

Lecolle, J. and Bernat, M. (1985). Late Quaternary surrection history of Efate Island, New Hebrides Island arc (Vanuatu): Th/U dates from uplifted terraces. *Proc. 5th Int. Coral Reef Cong., Tahiti* 3: 179-184.

Mitchell, A.H.G. (1968). Raised reef-capped terraces and Plio-Pleistocene sea-level changes, North Malekula, New Hebrides. *J. Geol.* 76: 56-67.

Pritchard, P.C.H. (1981). Marine turtles in the South Pacific. In: Bjorndal, K. (Ed.), *The Biology and Conservation of Sea Turtles*. Smithsonian Institute Press, Washington D.C. Pp. 253-262.

SPREP (1980). Vanuatu. *Country Report* 15. South Pacific Commission, Noumea, New Caledonia.

*Taylor, F.W., Isacks, B.L., Jouannic, C., Bloom, A.L. and Dubois, J. (1980). Coseismic and Quaternary vertical tectonic movements, Santo and Malekula Islands, New Hebrides Island Arc. *J. Geophys. Res.* 85: 5367-5381.

Taylor, F.W., Jouannic, C., Gilpin, L. and Bloom, A.L. (1981). Coral colonies as monitors of change in relative level of the land and sea: Applications to vertical tectonism. *Proc. 4th Int. Coral Reef Symp., Manila* 1: 485-492.

Tuna Programme (1983). An assessment of the skipjack and baitfish resources of the Republic of Vanuatu. *Skipjack Survey and Assessment Programme Final Country Report* 9. South Pacific Commission, Noumea, New Caledonia.

PRESIDENT COOLIDGE AND MILLION DOLLAR POINT RESERVE

Geographical Location   The Reserve covers the area off the south coast of Espiritu Santo Island, approximately 6 km from Luganville (Santo) town, to seaward of the highest water mark of spring tides and bounded by a line from the "white rock", extending 180° true for 0.3 naut. mi. (556 m), then 090° true for 1 naut. mi. (1853 m) and then 000° true to the shore; 15°31'S, 167°13'E.

Area, Depth, Altitude   3 sq. mi. (ca 7.8 sq. km); the stern of the *President Coolidge* lies in ca 240 ft (73 m) of water (Anon., 1985).

Land Tenure   Part Government owned and part customary land.

Physical Features   Includes the wreck of the American wartime troopship *President Coolidge*, which sank in 1942, and the area known as Million Dollar Point (Crossland, 1984). The wreck, over 650 ft (198 m) long and 80 ft (24.4 m) wide, forms an artificial reef in an otherwise barren bottom area (Power *in litt.*, 12.11.87).

Reef Structure and Corals   There is a shallow reef within the reserve and the wreck is well covered with corals (Power *in litt.*, 12.11.87).

Noteworthy Fauna and Flora   The wreck supports a population of thousands of reef fish. The largest are two groupers *Promicrops lanceolatus*, one ca 7 ft (2.1 m) long, the other slightly smaller. The latter can be fed by hand and follows divers around the wreck; many of the smaller fish are also tame. Some individual fish, including angelfish and small cod, have been living on the wreck for at least 10 years. Grey Sharks (*Carcharhinus* sp.), ca 6 ft (1.8 m) long, are frequently seen. Females of this species and of a small (up to 4 ft (1.2 m)) hammerhead shark species (family Sphyrnidae) congregate in May for breeding. Hawksbill Turtles (*Eretmochelys imbricata*) also occur. Shallower areas of the reserve also support a very large number and variety of fish, despite the destruction of coral by Hurricane Nigel in 1985 (see below); these include a tame 6 ft (1.8 m) Moray Eel *Gymnothorax flavomarginatus* (Power *in litt.*, 12.11.87).

Scientific Importance and Research   No scientific studies to date.

Economic Value and Social Benefits   A favourite spot for local and visiting divers and the main tourist attraction of Espiritu Santo. Santo Dive Tours runs conducted dives of the reserve (Power *in litt.*, 12.11.87).

Disturbance or Deficiencies   The shallow reef within the reserve suffered extensive damage from Hurricane Nigel in 1985, with areas of living reef reduced to dead coral rubble. Regeneration of hard corals has begun but is occurring very slowly. Black corals and gorgonians in deeper water were unaffected. Some fish poaching occurs, including spearfishing and fishing with nylon lines from boats at night. The latter is believed to be responsible for the virtual disappearance of a tame population of snapper (*Lutjanus argentimaculatus*) from under the wreck's bow (Power *in litt.*, 12.11.87). Divers from visiting yachts remove artefacts from the wreck and take corals, mainly gorgonians, black coral and the red *Distichopora violacea* (Anon., 1985; Power *in litt.*, 12.11.87); prevention of this is hampered by shortage of staff. In 1975-80, prior to creation of the reserve, there was also some semi-commercial collection of black coral and gorgonians by local divers (Crossland, 1984; Power *in litt.*, 12.11.87); this has now ceased and these corals are recovering (Power *in litt.*, 12.11.87).

Legal Protection   Declared a Marine Reserve on 18th November 1983 by the Minister of Land and Natural Resources. Fishing (of all marine organisms) is prohibited; sand and gravel may not be removed and wreck souvenirs may not be taken.

Management   The Luganville branch of the Fisheries Department (P.O. Box 129, Port Vila, Vanuatu) is responsible for policing the park.

REEF ISLANDS (ILES ROWA)

Geographical Location   Northern part of Banks Group, between Mota Lava and Ureparapara Islands; 13°35'S, 167°30'E.

Area, Depth, Altitude 8.5 km x 5.2 km; area of islands is 90-100 ha, that of reef and islands combined ca 2700 ha; max. alt. 6 m at Ro Island.

Physical Features An arcuate reef, facing ESE, with the concave face on the west, this shape being related to the trade wind swell. The western bay is sheltered, the eastern side is constantly beaten by surf. The Reef Islands consist largely of dead reefs, widely exposed on the eastern, northern and north-western sides of the rim and in some places in the median part. These form a reef flat up to high spring tide level. Tidal range reaches 1.80 m at spring tides. The whole of the old ledge in the intertidal zone consists of corals in position of growth, often perfectly preserved, with large tridacnids also preserved *in situ*. The morphology of these raised fossil reefs is described by Guilcher (1972 and 1974). Apart from Ro Island, the fifteen islets are very low (less than 2 m above high tide level) and are almost continuously washed by spray. They have a low, thick halophytic vegetation consisting mostly of pemphis and ironwoods. They are separated from each other by shallow passages. Enwut (Anouit) and Wosu (Wosou) are long, narrow sandy islets in the middle of the reef and are both thickly wooded and difficult to penetrate. In the north-west corner, a sand cay is found with coconut trees and smaller trees and bushes and a shallow pond bordered by small mangroves. The centre of the Reef Islands consists of a comparatively large lagoon with depths of 0.5-4 m at low spring tide. The bottom is largely covered by turtle grass *Halodule uninervis*. The southern part is a large flat area, under 0.5-1.5 m of water at low spring tide with sand and *Halimeda*. Not far from the south-western corner is a wide hole, 4-7 m deep. Coral patches, irregular in outline and sometimes fairly large, grow in the hole, which may be a karstic feature (Guilcher, 1974).

Reef Structure and Corals No information.

Noteworthy Fauna and Flora The pools carved by coastal erosion in the old ledge of the reef have a large population of moray eels. Turtles feed in large numbers in the central lagoon (Guilcher, 1974).

Scientific Importance and Research Surveyed in 1971 (Guilcher, 1974). The Reef Islands and Cook Island are the only non-fringing reefs and atolls in Vanuatu. Guilcher (1974) considers the Reef Islands to be the most interesting reef in this area.

Economic Value and Social Benefits A village was situated on Ro Island until 1939 when the island was struck by a devastating cyclone. The people of Mota Lava and Vanua Lava catch turtles and fish for subsistence purposes (Crossland *in litt.*, 9.10.84). Enwut is used as a temporary base by fishermen (Guilcher, 1974). Otherwise uninhabited.

Disturbance or Deficiencies Hurricanes strike the islands at regular intervals, the 1972 one causing considerable damage (Guilcher, 1974). No information on damaging human activities.

Legal Protection None.

Management None.

Recommendations A joint regulation for the creation of a reserve was agreed to by the pre-independence Governments but negotiations were never completed by the owners (Dahl, 1980). The proposal subsequently awaited legislation (SPREP, 1980) but has not been followed up. In fact, the value of the islands as a subsistence fishing ground for fishermen from Mota Lava and Vanua Lava outweighs the value of declaring the area a reserve, particularly as the islands are so remote from any real threat (Crossland *in litt.*, 9.10.84).

WALLIS AND FUTUNA

INTRODUCTION

General Description

Wallis and Futuna is an Overseas Territory of France and comprises three main islands: Uvea, Futuna (Hoorn) and Alofi. Uvea (80 sq. km) is a volcanic island, with a maximum altitude of 146 m. It is surrounded by a barrier reef, about 3-4 km offshore, with about 22 reef islets; the whole complex, including Uvea, is collectively also known as the Wallis Islands. Petrography and geology are described by Aubert de la Rue (1965), Macdonald (1945) and Stearns (1945). Average annual rainfall is over 2500 mm on Futuna and over 3000 mm on Uvea (Dupon, 1986). Futuna is described in a separate account. Alofi is a small (18.5 sq. km) uninhabited, mountainous (400 m) volcanic island off Futuna with a small patch of fringing reef. A bibliography of the islands is provided by O'Reilly (1964). General information is given in Morat et al. (1983) and Dupon (1986).

A survey of the lagoons and reefs of Wallis and Futuna and of their potential resources was carried out by the Musée d'Histoire Naturelle, Paris in 1980 (Richard et al., 1982). Uvea lagoon is naturally poor in coral but is very rich in algae and is free of ciguatera (SPREP, 1982).

Terrestrial birds of Wallis and Futuna are described in Guyot and Thibault (1987) and seabirds in Thibault and Guyot (1987).

Reef Resources

Although not intimately involved with the sea, the local population obtains the greater part of its protein from marine products, particularly fish. Fishing is carried out mainly by line, net, speargun or harpoon. Bottom fishing is increasing. *Trochus* spp. from Uvea lagoon are exported in large quantities to New Caledonia (SPREP, 1982).

Disturbances and Deficiencies

In general, the reefs and lagoon are undisturbed, although exploitation of fish and molluscs should not be increased. Uvea lagoon may not be able to support the current level of harvesting of trochus. The situation on Futuna is possibly more serious and is described in a separate account (SPREP, 1982).

Oil pollution is not considered a serious problem in Wallis and Futuna and all necessary precautions have been taken in the construction of the petroleum storage depot. There is some domestic pollution of Uvea lagoon but a project to improve the sewerage system is underway. Pigs roam the reef flat in search of food and have had a destructive impact. Coral rubble and sand is dredged for road maintenance and building but has had no noticeable impact (SPREP, 1982). General information on conservation problems is given in Dupon (1984) and Anon. (1985).

Legislation and Management

Fisheries regulation Order No. 83, November 24, 1965, prohibits the use of explosives or poisons for fishing but this is poorly enforced. Traditional law is applied for the day-to-day settlement of local affairs but is no longer sufficient to prevent over-exploitation of resources. There are currently no reserves and there is no conservation legislation. However, in 1978, an environmental impact study was carried out in Uvea lagoon as a result of proposals to construct the Halalo petroleum storage depot, to improve the sanitation of the administrative centre, Mata'Utu, to establish a dump for household refuse and to dredge for coral rubble and sand (SPREP, 1982). The results are not known.

Recommendations

In 1979, the Territorial Assembly of Wallis and Futuna adopted a "Long-term Economic and Social Development Plan" which includes, among its principal aims, improvement of the use of marine resources and the protection of the national heritage of the islands. Priorities include disposal of urban effluents, prevention of sea pollution and protection of the coastal zone. It is recommended that Alofi Island be totally protected (SPREP, 1982).

Regulations to control exploitation of fish and molluscs should be introduced and the prohibition on fishing with dynamite should be enforced. A more detailed survey of the lagoon and reef resources is recommended. The establishment of a system of marine reserves is considered a priority in the Action Strategy for Protected Areas in the South Pacific Region drawn up at the Third South Pacific National Parks and Reserves Conference in 1985.

References

* = cited but not consulted

Anon. (1985). Country review - Wallis and Futuna. *Report of the 3rd South Pacific National Parks and Reserves Conference, Apia* 3: 229-231

Aubert de la Rue, E. (1965). Introduction à la géologie et à la géographie des Iles Wallis et Horn. *J. Soc. Océan.* 19: 47-56.

*Caillot, M. (1961). Un type de pêche dans le Pacifique: la pêche à Futuna. *Les Cahiers d'Outre Mer, Bordeaux* 14éme année 55, 61: 317-322.

*Doumenge, F. (1961). Observations à propos des formations coralliennes de l'île Wallis. *Bull. Assoc. Geog. française, Paris*: 186-196.

*Dupon, J.F. (1984). Les risques naturels à Wallis et Futuna: préparation - prévention - experience. ORSTOM, Noumea, Section Sciences Humaines.

Dupon, J.F. (1986). Wallis and Futuna: Man against the Forest. Environmental Case Studies 2. SPREP, South Pacific Commission, Noumea, New Caledonia. (Leaflet).

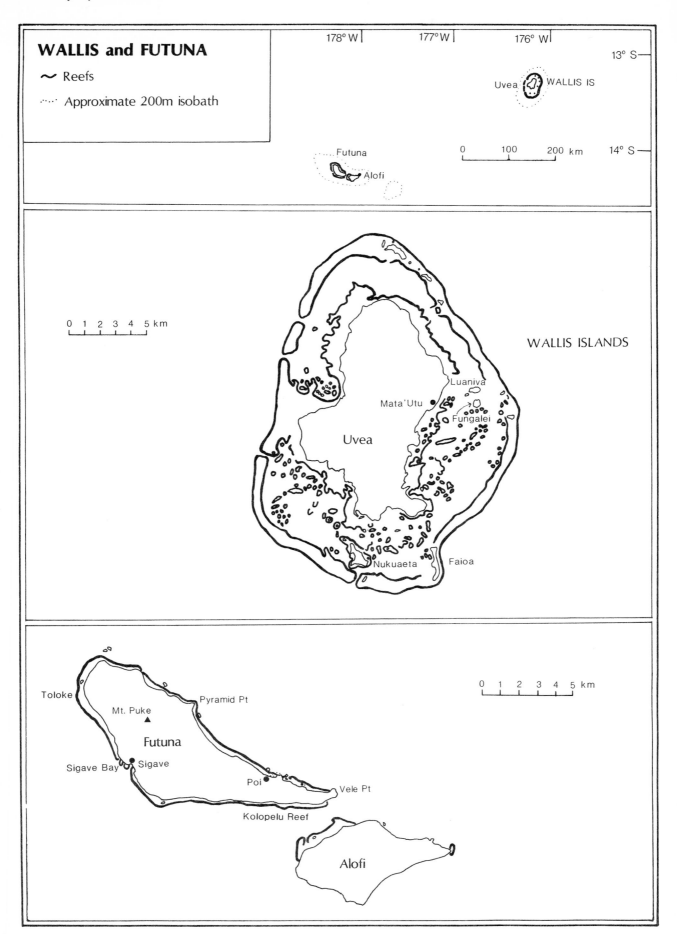

WALLIS and FUTUNA

~ Reefs

····· Approximate 200m isobath

178° W   177° W   176° W

13° S

Uvea ⬡ WALLIS IS

Futuna

Alofi

0   100   200 km

14° S

0 1 2 3 4 5 km

WALLIS ISLANDS

Luaniva

Mata'Utu ●

Fungalei

Uvea

Nukuaeta   Faioa

0 1 2 3 4 5 km

Toloke

Pyramid Pt

Mt. Puke ▲

Futuna

Sigave Bay ●

● Sigave

Poi ●

Vele Pt

Kolopelu Reef

Alofi

*Fusimalohi, T. and Grandperrin, R. (1980). Rapport sur le projet de la pêche profonde à Wallis et Futuna. *Rapport 861/80 South Pacific Commission*: 1-25.
Galzin, R. and Mauge, A. (1981). Traditional fishery in Futuna and its dangers - Horn Archipelago, Polynesia. *Proc. 4th Int. Coral Reef Symp., Manila* 1: 111. (Abs.).
Guyot, I. and Thibault, J.-C. (1987). Les oiseaux terrestres des îles Wallis-et-Futuna (Pacifique sud-ouest). *L'Oiseau et RFO* 57: 226-250.
*MacDonald, G.A. (1945). Petrography of the Wallis island. *Bull. Geol. Soc. Am.* 56: 861-872.
Morat, P., Veillon, J.M. and Hoft, M. (1983). Introduction à la végétation et à la flore du territoire de Wallis et Futuna. ORSTOM, Noumea. 24 pp.
*O'Reilly, P. (1964). Bibliographie des îles Wallis et Futuna. *Publ. de la Société des Oceanistes* 13. 68 pp.
*Richard, G., Bagnis, R., Bennett, J., Denizot, M., Galzin, R., Ricard, M. and Salvat, B. (1982). Wallis et Futuna: étude de l'environnement lagunaire et récifal des îles Wallis et Futuna (Polynesie occidentale). Rapport définitif, Muséum/E.P.H.E./Territoire des îles Wallis et Futuna/N.E.B. RL-9. 101 pp.
Richard, G., Galzin, R., Salvat, B., Bagnis, R., Bennett, J., Denizot, M. and Ricard, M. (1981). Geomorphology, ecology and socio-economy of the Futuna marine ecosystem (Horn Archipelago, Polynesia). *Proc. 4th Int. Coral Reef Symp., Manila* 1: 270-274.
SPREP (1982). Wallis and Futuna Islands. *Country Report* 19. South Pacific Commission, Noumea, New Caledonia.
*Stearns, H.T. (1945). Geology of the Wallis Island. *Bull. Geol. Soc. Am.* 56: 849-860.
*Thibault, J.-C. and Guyot, I. (1987). Notes on the seabirds of Wallis and Futuna Islands (South-west Pacific Ocean). *Bull. Br. Orn. Cl.* 107: 63-68.

FUTUNA (HOORN) ISLAND

Geographical Location 230 km south-west of Uvea; 14°15'S, 178°10'W.

Area, Depth, Altitude Island is 18 km x 6 km (44 sq. km); max. alt. 500 m.

Physical Features The island is traversed by a mountain chain, the highest point of which is Mt Puke. There is no lagoon but there is an apron reef of varying size. The island is deeply dissected with well-wooded valleys, many streams and montane forest.

Reef Structure and Corals Field studies were made at six sites during a survey in 1980 (Richard *et al.*, 1981). Sigave Bay on the west coast has a narrow reef flat with furrows containing boulders of dead coral; the flat is totally emergent at low tide and there is no living coral. The reef front has no algal ridge and *Porolithon* is only abundant down to 2 m depth. The Kolopelu Reef on the south coast has a "patterned" reef flat, more clearly defined than the one at Sigave, and is classically eroded, with much fossilized coral and many boulders. At Vele Pt, the south-eastern tip of the island, the reef flat is 100 m wide and has an area of pebbles; an area of

scattered boulders, largely dead, and a raised frontal area with more living *Madreporaria*, particularly on the outer slope. At 15 m depth, the bottom is very irregular with a strong tidal current. At Pyramid Pt on the east coast, there is a narrow reef flat.

Poi Reef, on the east coast, and Toloke Reef on the west coast are described in greater detail. Poi Reef has a reef flat dominated by algae, *Padina* and *Valonia*, and cerithiid molluscs near the shore, and ophiuroids and Foraminifera further seaward. Corals are virtually absent apart from occasional small *Porites* colonies. Molluscs are abundant. The back ridge and reef front are covered with algae, and the latter has occasional colonies of *Porites* and *Favites*. Toloke Reef is similar with ophiuroids and algae dominating the reef flat and occasional colonies of *Pavona*, *Porites* and *Montipora*. Molluscs are again well represented. Furrows in the reef flat are edged with corals, *Porites* and *Pavona*, and are filled with algae and a variety of invertebrates. The outer part of the reef flat has scattered algae, the corals *Porites* and *Montipora* and large numbers of fish. Corals, calcareous algae and zoanthids *Palythoa* are found on the reef front, with a variety of molluscs and echinoderms. Corals cover 30-50% of the outer slope. *Pocillopora*, *Acropora* and *Porites* dominate at shallower depths; *Montipora* and *Echinopora* are found in deeper waters. At 28 m alcyonarians, crinoids and algae comprise a diverse community.

Noteworthy Fauna and Flora Fish are abundant in several areas (Richard *et al.*, 1981), particularly on the outer slope of Toloke Reef. There are no mangroves or extensive seagrass beds.

Scientific Importance and Research As a result of discussions during the Colloque de la Mer at Noumea in 1979, a study of the marine ecosystem of the island was carried out in 1980 (Richard *et al.*, 1981 and 1982).

Economic Value and Social Benefits The marine resources are not fully made use of by the population of Futuna, probably as a result of a number of prohibitions and taboos which have been in effect for over a century (Fusimalohi and Grandperrin, 1980). Over the last ten years the Service de l'Economie Rurale Territorial has been encouraging improved utilization of marine resources. Fish are now widely consumed and fishing activities are intense in Sigave Bay. Crustaceans, turtles (caught at nearby Alofi Island) and echinoderms are all important foods. The fishery is described by Caillot (1961).

Disturbance or Deficiencies Serious overfishing is reported by Richard *et al.* (1981). The emergent reef flat is easily accessible, particularly at low spring tides when it is visited by large numbers of people and scavenging domestic animals. Galzin and Mauge (1981) describe two non-selective fishing methods: fishing with "Futu", a toxic substance obtained from the seeds of *Barringtonia speciosa*, and the temporary construction of artificial patch reefs to attract juvenile fish. It is thought that these fishing methods may affect the ichthyofauna as about 58% of the catch are juveniles. Sedentary species (echinoderms, etc.) have also been affected by over-exploitation and non-selective fishing methods. Soil erosion is a serious problem on the island but it is not known if this has had an impact on the reefs. Population expansion may have an indirect effect on the reefs. For example, people are now constructing houses on the previously uninhabited east coast (SPREP, 1982).

Legal Protection   None, except Fisheries Regulation 83 (see introduction).

Management   None.

Recommendations   The taking of living organisms from the reef should be controlled, perhaps by having a closed season (SPREP, 1982).

WESTERN SAMOA

INTRODUCTION

General Description

Western Samoa (13-15°S, 171-173°W), situated north of Tonga and north-east of Fiji, has a total area of ca 2930 sq. km. The two main islands are Savai'i and Upolu, the latter with the capital, Apia. There are few major rivers but numerous coastal springs are found on both islands. The climate is tropical, characterized by heavy rainfall (av. 2850 mm), high relative humidity (av. 83%) and an average temperature of 30°C. The wet season is December to March. Winds above gale force are rare (Hunter, 1977).

Table of Islands

Savai'i (X) 703 sq. mi. (1820 sq. km); active volcanic dome, 1860 m; very few mangroves; small areas of fringing reef, e.g. at Leanamoea and Cape Puava.

Apolima (X) 2 sq. mi. (5.2 sq. km); volcanic cone; 472-545 ft (144-166 m).

Upolu (X) 430 sq. mi. (1114 sq. km); high volcanic 1100 m, string of cones E-W; extensive wide barrier and fringing reefs; *see separate account for* Palolo Deep Reserve; six offshore islets of which Fanuatapu, Namu'a, Nu'utele and Nu'ulua are described in the account for Aleipata and Nu'utele Islands proposed National Park; others are:

- *Nu'usafe'e* believed to be of coral origin; has fringing reef;

- *Manono* (X) coral sand and basalt, 197 ft (60 m) no crater; linked to Upolu by fringing reef.

(X) = Inhabited

Living reefs are reportedly widely distributed around both main islands although they appear limited in extent compared with the size of the islands. Johannes (1982) estimated a total of 23 100 ha of reef and lagoon (water less than 50 m deep). Upolu is almost entirely surrounded by barrier reefs (Johannes, 1982); however reefs (both barrier and patch) are reportedly only well developed on the northern and western coastlines (Bell, 1985). Considerable areas are devoid of significant coral life (SPREP, 1980). Mangroves are of very limited extent (less than 1000 ha) (Bell, 1985).

Reef Resources

Settlements in Western Samoa are traditionally coastal and fisheries provide a major source of protein; both inshore and deep sea fishing are practised, with the former being historically of greater importance. Fisheries statistics indicate an annual inshore catch (for the period 1973-83) of 800-900 tons bottom fish, 20-25 tons shellfish and 80-100 tons of "others". Bottom fish catch increased to 1100 tons in 1984 (this is estimated to be the maximum sustainable yield) (Bell, 1985). Harvest rate in inshore waters has been estimated at ca 4 tons per sq. km which is not low by coral reef standards although the average size of the fish taken is small, indicating sustained heavy fishing (Johannes, 1982). The relatively high yield is in part explained by the very wide variety of species of finfish and invertebrates harvested, some of the more important being mullet, small snappers, scad and surgeon fish, molluscs, holothurians, jellyfish and seaweeds (Bell, 1985).

In recent years the offshore fishery (principally for tuna) has been the focus of UN-funded development projects and has increased in relative importance. The tuna catch is principally sold in the market in Apia and provides food for the urban population; reef fisheries are still of major importance to the rural population. A preliminary survey in 1984 indicated consumption of local seafoods in urban areas to be ca 100 g per head per day and in rural areas to be ca 240 g per head per day; reef fishes comprised ca 87% of total rural consumption and 60% of urban consumption (although offshore fish catch, notably of tuna, was depressed in that year) (Bell, 1985).

Principal fishing techniques used at present include hand-gleaning for invertebrates, underwater fishing using spear guns, hand-thrown spears, gill-netting and fencing using chicken-wire, handlining and trolling; dug-out canoes are still used, though apparently to a lesser extent than in the past. Some methods are destructive and overfishing and degradation of reefs have become increasing problems (see below). Gleaning of reefs is generally carried out by women and children while men are responsible for most finfishing (Bell, 1985).

Both Green Turtles *Chelonia mydas* and Hawksbills *Eretmochelys imbricata* occur in Western Samoan waters. The former is considered the commonest turtle although no definite nesting sites are known; a small population of the latter nests on the islets off the eastern tip of Upolu (*see account for* Aleipata and Nu'utele Islands proposed National Park). Human predation on turtles is reportedly severe, and they were once an important food source; however their present scarcity means that they are now mainly eaten by village chiefs on special occasions (Witzell, 1982; Witzell and Banner, 1980).

Tourism is being developed but is still at a comparatively low level compared with many other Pacific countries. There are plans for a new hotel outside Apia (Anon., 1985b). In 1985 there was one SCUBA operation on the south coast (Wells pers. obs., 1985); in 1987 this was temporarily closed but a company in Apia was offering reef tours and SCUBA diving (Bell *in litt.*, 15.7.87).

Disturbances and Deficiencies

Infestations by *Acanthaster* have occurred at least since the 1930s; a recent invasion, in the late 1960s, early 1970s and in 1977, affected several parts of the south, west and

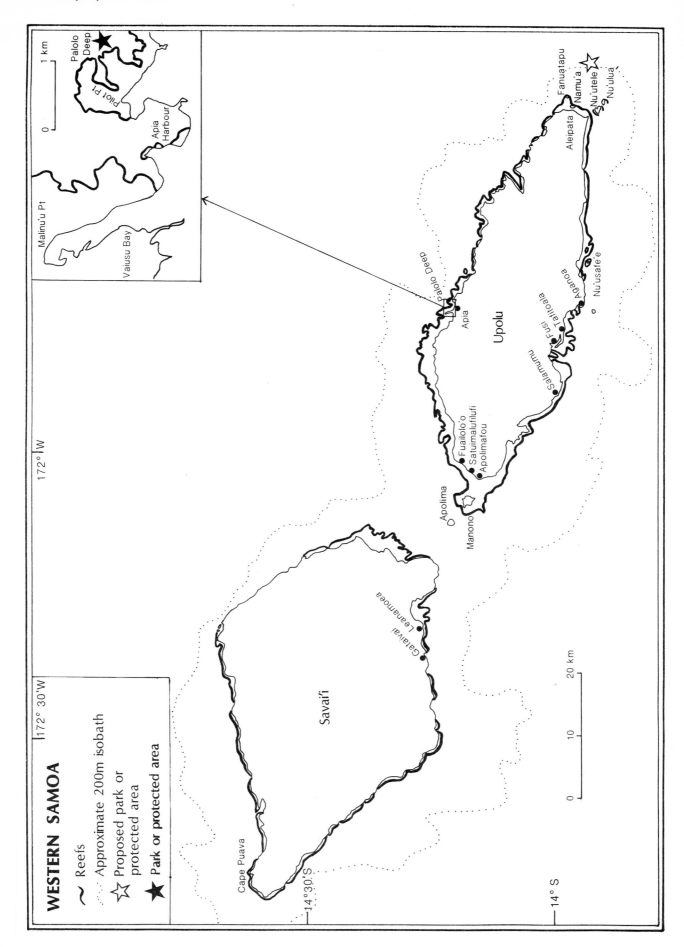

WESTERN SAMOA

~ Reefs

········· Approximate 200m isobath

☆ Proposed park or protected area

★ Park or protected area

172° 30'W

172° W

14°30'S

14° S

Savai'i

Cape Puava

Gataivai

Leone/moea

Upolu

Apia

Apolima

Manono

Fuailolo'o
Satuimalufilufi
Apolimafou

Salemumu

Fusi
Tafitoala
Aganoa

Nu'usafe'e

Aleipata

Fanuatapu
Namu'a
Nu'utele
Nu'ulua

Palolo Deep

20 km

10

0

Malinu'u Pt

Vaiusu Bay

Apia Harbour

Pilot Pt

Palolo Deep

1 km

0

east coasts of Upolu (Bell, 1985; Birkeland, 1982; Garlovsky and Bergquist, 1970).

Fishing practices have had both direct and indirect effects on reefs and their biota. Dynamiting, involving either a charge thrown at a school of fish (notably mullet and scad), or, more damagingly, a charge laid by a diver on a coral head and detonated from a fishing boat, has been a problem at least since the beginning of the century; Johannes (1982) noted that it had decreased somewhat in extent, owing to reduced availability of dynamite and to increased enforcement of laws against its use. A number of fishermen have been prosecuted for selling dynamited fish. Despite this, dynamiting continues to be a problem and was identified by Bell (1985) as one of the greatest threats to the marine environment in Western Samoa. The long-standing practice of breaking of corals to extract fish and invertebrates for food still prevails; a variant of this is the local fishing practice of "faamo'a", involving a number of people driving fish into set nets by hitting the coral substrate with wooden poles. Poisoning with root extract from *Derris elliptica* is also widespread and difficult to control; this poison kills juvenile fish, shellfish, and, at high concentrations, corals. Both bleach and the herbicide Paraquat have also been reportedly used (Bell, 1985; Johannes, 1982). Dynamiting and fishing with poison are reported to be particularly widespread on Upolu along the 37 km stretch of coast between Apia and the airport, one of the most heavily settled areas of the island (Bell, 1985).

Effects of fishing on individual species (particularly of finfish) are difficult to quantify as population data are lacking; however marine turtles, some bivalves, including Giant Clams *Tridacna*, a variety of sea cucumbers, a limcoid and some edible seaweeds are reported to have definitely become scarcer (Bell, 1985). Concern has also been expressed for shoaling species such as mullet and scad which are markedly susceptible to overfishing (Johannes, 1982).

The Palolo *Eunice viridis*, an eunicid annelid that lives in reefs and whose reproductive segments are collected and eaten during their annual rise to the water surface, has a very significant place in the culture and traditions of the Samoan people. It used to occur along the entire north coast of Upolu until the mid 1930s. It is no longer found along this coastline and its occurrence in south Upolu is spasmodic (Hunter, 1977).

Soil erosion is becoming an increasing problem, particularly as a result of cutting firewood along the banks of the river, and mangroves are being increasingly destroyed. It is estimated that from 1975 to 1981 the seabed in the eastern and central parts of Apia harbour has shallowed by up to five feet (1.5 m) owing to siltation (Bell, 1985). There is increasing use of pesticides which may pollute coastal waters (Johannes, 1982). A study of fish and shell-fish in the Apia lagoon region indicated sewage pollution (Hunter, 1977). Dredging for reef sand on Upolu (principally at Mulinu'u (Bell, 1985)), land reclamation from the sea and potentially the proposed sewage disposal system for Apia were identified as marine environmental problems in SPREP (1980). Coastal waters are affected by silting from dredging, reclamation, agriculture and road construction. There is extensive blasting of passages through the reefs for small fishing vessels.

Legislation and Management

Traditionally Samoans had elaborate customs of ownership and control of fishing rights (Bulow, 1902). The right to fish in reef, lagoon and mangroves areas was owned by adjacent villages, families or chiefs but these customs have largely disappeared as far as reefs and lagoons are concerned, in part because, following the constitution, all land lying below the line of high-water mark is now public land, and all people have the right to navigate over the foreshore and fish within the limits of the territorial waters of the state (Anon., 1985a; Bell, 1985; Eaton, 1985a and b). It is argued, however, that the maintenance of traditional fishing controls is not necessarily incompatible with this, and some controls are still maintained: for example, around Manono, only the inhabitants may catch mullet and "atule" (scad); on Savai'i, fishing for whitebait during the annual run is reportedly the exclusive right of the people of the village of Gataivai. Johannes (1982), Bell (1985) and Eaton (1985a and b) give further information.

The exploitation of marine resources is regulated through the Fish Protection Act, 1972 and the Exclusive Economic Zone Act, 1977, which controls fishing by foreign vessels. The Fish Dynamiting Act, 1972, prohibits all use of dynamite for fishing (Anon., 1985a; Bell, 1985). Section 4(f) of the Police Offences Act 1961 prohibits the use of the plants "ava niukini" (*Derris*) or "futu" (*Barringtonia*) or any derivative thereof for the purpose of capturing fish (Bell, 1985). The Water Act (1965) controls the discharge of pollutants into coastal waters. Western Samoa has signed the Convention on Conservation of Nature in the South Pacific.

The National Parks and Reserves Act of 1974 provides empowering legislation for the establishment of Marine Parks and Reserves. A distinction is drawn between national parks and reserves: the former are public lands of 600 ha or more, or islands, to which the public is guaranteed freedom of entry and access subject to any controls necessary for the preservation of the park's features; reserves (which may be nature reserves, recreation reserves, historic reserves, or "others") may include areas of territorial sea, although customary fishing rights are guaranteed, and the Minister of Agriculture may restrict access to and activities within them (Anon., 1985a). It is thus implied that national parks cannot contain areas of territorial sea, although the Aleipata and Nu'utele Islands proposed National Park has associated coral reefs (*see separate account*). A survey under the auspices of UNDAT and IUCN was carried out in 1974 to identify sites for protected areas and advize on management (Holloway and Floyd, 1975; Hunter, 1977). In 1978 a follow-up team from UNDAT provided further assistance with development of three existing and three proposed reserves and in 1979-80 a New Zealand park advisor (funded by WWF, IUCN and the New Zealand Government) became the country's first National Parks and Reserves Superintendent; a Samoan counterpart was trained during this period (Anon., 1985a). Of the existing protected areas, O le Pupu-pu'e National Park on Upolu extends to the coast but is cliff-bound and has no coral reefs, Palolo Deep Marine Reserve is largely coral reef and is described in a separate account; the remainder have no coastal or marine components.

Experimental mariculture using the Philippine Green Mussel (*Perna viridis*) was begun in 1981 using spats imported from CNEXO AQUACOP in Tahiti; these have had some success, with natural spatfalls recorded and proposals have been submitted to develop small-scale commercial mussel farming in villages (Bell, 1985). Western Samoa is taking part in the International Giant Clam Mariculture Project and will eventually be examining the possibility of re-introducing *Hippopus hippopus* (Anon., 1987a).

Recommendations

Further study and assessment of reefs and their productivity and the effects of dredging, and atmospheric pollution is required (SPREP, 1980). Johannes (1982) discusses alternative methods for managing the reef resources and suggests that at present fishery controls may not be necessary, although monitoring of some species such as mullet, is recommended. At present development in the coastal zone is not planned or managed in any integrated way. Recommendations in Johannes (1982) include entrusting dynamite only to highly responsible personnel, planting trees such as *Leucana*, which can withstand heavy cutting, along river banks to reduce soil erosion and monitoring pesticide residues in selected edible reef animals. Efforts should also be made to investigate and record remaining traditional marine tenure and resource use. Improvement in environmental education at all levels is also required. The Five Year Development Plan in the early 1980s acknowledged the need for an Environmental Unit. A proposal is under consideration for Western Samoa to accede to a number of international maritime conventions and for the implementation of a new Fisheries Act and Marine Pollution Act (Bell *in litt.*, 15.7.87).

A number of proposed national parks and reserves contain reefs; all these are taken from Holloway and Floyd (1975). Five areas (in Nos 1,3,4,7,8 below) have been recommended as coral sanctuaries (Anon., 1985a; Dahl, 1978 and 1980; Holloway and Floyd, 1975); it is unclear if the necessary legislation for the designation of most of these areas exists at present, although such sanctuaries may be regarded as forms of natural reserve.

1. Aleipata and Nu'utele Islands proposed National Park (*see separate account*).
2. Cape Puava Forest in northern Savai'i covers ca 800 acres (324 ha) and has fringing reefs off its narrow coastline (Anon., 1985a; Dahl, 1980).
3. Satuimalufilufi/Fuailolo'o reef on Upolu (ca 2 mi (3.2 km) straight-line length and in good condition) has been proposed as a coral sanctuary (Anon., 1985a); it is associated with the proposed Apolimafou reserve (120 acres (49 ha)) which is primarily a rush and reed swamp (Anon., 1985a; Dahl, 1980). The area is close to a hotel development and the airport (Anon., 1985a).
4. Fusi/Tafitaola fringing reef on Upolu (3.2 km straight-line length and in good condition) is a proposed coral sanctuary associated with a proposed 120 acre (49 ha) reserve largely of tidal mangrove (Anon., 1985a; Dahl, 1980).
5. Aganoa (200 acres (81 ha)) on Upolu is a coastal beach with associated fringing reef (Anon., 1985a; Dahl, 1980).

6. Salamumu (240 acres (97 ha)) on Upolu is also a coastal beach with fringing reef and a small rocky islet (Nu'navasa Island); it is also one of the few areas where the Palolo worm is still common (Anon., 1985a; Dahl, 1980).
7. Nu'usafe'e Island proposed reserve (230 acres (93 ha)) off Upolu is believed to be of coral origin (Anon., 1985a; Dahl, 1980). The 2 mi. (3.2 km) fringing reef has been proposed as a coral sanctuary (Anon., 1985a).
8. Leanamoea (340 acres (138 ha)) on Savai'i is a coastal area which includes a freshwater spring; the associated 3.2 km fringing reef, reportedly in moderately good condition, has been proposed as a coral sanctuary (Anon., 1985a).

Additional protection for the palolo worm may be required (Dahl, 1980).

References

* = cited but not consulted

Anon. (1985a). Country Report - Western Samoa. *Report of the 3rd South Pacific National Parks and Reserves Conference, Apia* 3: 232-269

Anon. (1985b). Western Samoa - Matai country on the threshold of recovery. *The Courier* 93: 27-36.

Anon. (1987a). Fisheries Division, Western Samoa. *Clamlines* 3: 1.

Anon. (1987b). SPREP undertakes marine park survey in Western Samoa. *Environment Newsletter* 9: 1-2.

Bell, L.A.J. (1985). Coastal zone management in Western Samoa. *Report of the 3rd South Pacific National Parks and Reserves Conference, Apia* 2: 57-73.

Birkeland, C. (1982). Terrestrial runoff as a cause of outbreaks of *Acanthaster planci* (Echinodermata, Asteroidea). *Mar. Biol.* 69: 175-185.

Bulow, W. von (1902). Fishing rights of the natives of German Samoa. *Globus* 82: 40-41. (In German).

*Chew, W.L. (1986). Aleipata Marine National Park (Western Samoa): A feasibility study. Report to Unesco. 39 pp.

Dahl, A.L. (1978). Report on assistance to Western Samoa with national parks and conservation. South Pacific Commission, Noumea, New Caledonia. 28 pp.

Dahl, A.L. (1980). Regional ecosystems survey of the South Pacific Area. *SPC/IUCN Technical Paper* 179. South Pacific Commission, Noumea, New Caledonia.

Eaton, P. (1985a). Tenure and taboo: Customary rights and conservation in the South Pacific. *Report of the 3rd South Pacific National Parks and Reserves Conference, Apia* 2: 114-134.

Eaton, P. (1985b). Land Tenure and Conservation: Protected areas in the South Pacific. *SPREP Topic Review* 17. South Pacific Commission, Noumea, New Caledonia.

Garlovsky, D.F. and Bergquist, A. (1970). *South Pacif. Bull.* 20: 47.

Garnett, M.C. (1984). Conservation of seabirds in the South Pacific Region: A review. In: Croxall, J.P., Evans, P.G.H. and Schreiber, R.W. (Eds), *Status and Conservation of the World's Seabirds*. ICBP Technical Publication 2. Pp. 547-558.

*Holloway, C.W. and Floyd, C.H. (1975). National Parks Systems for Western Samoa. UNDAT-IUCN.

Hunter, A.P. (1977). Country Report - Western Samoa. In: *Collected Abstracts and Papers of the International*

Conference on Marine Parks and Reserves, Tokyo, Japan, May 1975. The Sabiura Marine Park Research Station, Kushimoto, Japan.

Johannes, R.E. (1982). Reef and lagoon resource management in Western Samoa. Unpub. rept to South Pacific Regional Environment Programme.

SPREP (1980). Western Samoa. *Country Report* 16. South Pacific Commission, Noumea, New Caledonia.

Witzell, W.N. (1982). Observations on the Green Sea Turtle (*Chelonia mydas*) in Western Samoa. *Copeia* 1982(1): 183-185.

Witzell, W.N. and Banner, A.C. (1980). The Hawksbill Turtle, *Eretmochelys imbricata* (Linnaeus, 1766) in Western Samoa. *Bull. Mar. Sci.* 30(3): 571-579.

ALEIPATA AND NU'UTELE ISLANDS PROPOSED NATIONAL PARK

Geographical Location Eastern end of Upolu; 14°01'S, 171°25'W.

Area, Depth, Altitude Nu'utele is 108 ha; Nu'ulua is 25 ha; Fanuatapu is 15 ha; Namu'a is 20 ha.

Land Tenure Namu'a and Fanuatapu are government-owned; the other islands have complicated land ownership, now being clarified prior to negotiations for their possible acquisition by the government.

Physical Features Fanuatapu and Namu'a are connected to Upolu by the wide fringing reef, the other two are separate. Namu'a has scenically interesting cliffs and is briefly described by Dahl (1978).

Reef Structure and Corals Nu'utele has two small developing reefs, Nu'ulua has one; these reefs are in good condition. The reefs of Namu'a and Fanuatapu extend for ca 1 mi. (1.6 km). Much of the *Acropora* on the reefs is now dead (see below) but large *Porites* colonies have generally survived (Andrews and Holthus *in litt.*, 27.10.87; Anon., 1987b).

Noteworthy Fauna and Flora A small population of Hawksbills *Eretmochelys imbricata* nests on Namu'a, Nu'utele and Nu'ulua (Witzell and Banner, 1981). Seabirds, especially boobies *Sula* sp., nest (Garnett, 1984). There are some mangroves along the shore of Namu'a (Dahl, 1978). Fish are abundant.

Scientific Importance and Research The area was surveyed by SPREP in 1987 (Andrews and Holthus *in litt.*, 27.10.87; Anon., 1987b) at the request of the Samoan Government, following a Unesco feasibility study (Chew, 1986). The reefs have no particular aesthetic or biological value but are well-developed systems (Andrews and Holthus *in litt.*, 27.10.87).

Economic Value and Social Benefits The islands are uninhabited although visited by fishermen, and by villagers to collect coconuts (Anon., 1985a). Rough water and inaccessibility mean that the reefs in the area have little potential for tourism (Andrews and Holthus *in litt.*, 27.10.87). Namu'a and Fanuatapu are more accessible and has recreational potential (Dahl, 1978).

Disturbance or Deficiencies In 1978, the reef around Namu'a was heavily infested with *Acanthaster* and much of the coral in the lagoon was dead (Dahl, 1978). Most of the fringing reef slope has been heavily impacted by *Acanthaster* in the past 15 years, resulting in much large, dead, standing, algae-encrusted *Acropora* colonies. The south-facing fringing reef is exposed to greater wave activity and was apparently less badly affected, as were the reefs around the offshore islets (Anon., 1987b).

There does not appear to be any serious over-exploitation or reef degradation due to fishing, although there is occasional use of poisons or dynamite (Andrews and Holthus *in litt.*, 27.10.87).

Legal Protection None at present.

Management The Aleipata area is exclusively fished by its villages; fishermen from other areas do not venture into, and are not welcome in, the area. The few individuals who use poison or dynamite are increasingly criticised and ostracised by other villagers (Andrews and Holthus *in litt.*, 27.10.87).

Recommendations The area is being developed as a national park with the assistance of SPREP, with the primary aims of conserving the vegetation of Nu'utele and Nu'ulua islands and their nesting populations of turtles. The reefs around these two islands could be protected as fish and invertebrate seed stock areas (Andrews and Holthus *in litt.*, 27.10.87). Dahl (1978) provides recommendations for the development of Namu'a as a recreational reserve.

PALOLO DEEP RESERVE

Geographical Location Matautu, on the north coast of Upolu, one mile (1.6 km) east of Apia in Tuamasaga District; 13°49'S, 171°45'W.

Area, Depth, Altitude 22.25 ha; depth 0-10 m; altitude 0-1 m.

Land Tenure Government owned.

Physical Features Palolo Deep is a hole about 200 m in diameter and 10 m deep within a fold in the fringing reef which surrounds Pilot Point (Eaton, 1985b). There is a small sand beach on the shore extending onto the reef flat which is dotted with small basalt rocks. The beach is backed by a 10 m wide access strip of banana plants with an understorey of ornamental shrubs. On the seaward side of the deep there is a talus slope of sand and coral debris thrown over the reef. Temperatures range from 22° to 30°C, with a mean daily temperature of 27°C. Average rainfall is 2870 mm a year.

Reef Structure and Corals Lagoon with fringing reef. The fore-reef slope and extensive submarine terraces have rich coral cover (Dahl, 1978).

Noteworthy Fauna and Flora Abundant fish. Although the reserve is named after the Palolo Worm, this is now rarely found there (Eaton, 1985b).

Scientific Importance and Research   In the early 1970s, a rudimentary monitoring programme of Palolo Deep was carried out by the Regional Ecologist of the South Pacific Commission (Hunter, 1977) but this has not continued. The origin of the Deep is uncertain but it has been suggested that a freshwater spring occurred there preventing coral growth.

Economic Value and Social Benefits   Important for recreation and tourism as it is easily accessible from Apia (Dahl, 1978).

Disturbance or Deficiencies   Affected by pollution and siltation; problem with urchins and *Acanthaster*. Underwater visibility declined between 1970 and 1977 as a result of sedimentation, considered to be due to large scale reclamation for a hotel complex in the vicinity (Hunter, 1977). There is some risk from pollution. The proximity of the reserve to Apia Harbour increases its vulnerability.   In 1982, the area was being fished surreptitiously, probably at night (Johannes, 1982). There is reported to be some coral collection.

Legal Protection   Established as a reserve on 5th December 1979.

Management   An underwater nature trail was planned in 1978 (Dahl, 1978), established in the early 1980s (Dahl, 1980) and redeveloped in 1985 with assistance from New Zealand. However trained personnel for its management are lacking and although there are still plans to maintain it, lack of funds mean that this cannot be made a high priority (Uli *in litt.*, 21.7.87). Brochures for tourists were produced but not adequately distributed (Johannes, 1982). A local person has been employed as caretaker and facilities are gradually being developed. A covered platform for the use of swimmers and divers has been constructed on the reef flat at the edge of the deep (Anon, 1985a; Eaton, 1985b). The caretaker derives a small income from the hire of snorkelling equipment to visitors (Uli *in litt.*, 21.7.87). A management plan was to be produced in 1985 (Anon., 1985a).

Recommendations   Enforcement of the park regulations, particularly fishing controls, must be improved. Publicity about the reserve should also be improved (Johannes, 1982). Recommendations for the development of the reserve are given in Dahl (1978).

UNEP (1976). *Directory of Mediterranean Marine Research Centres*, 1st ed. UNEP Regional Seas Directories and Bibliographies. Geneva, UNEP. 280 pp.*

UNEP (1977). *Directory of Mediterranean Marine Research Centres*, 2nd ed. UNEP Regional Seas Directories and Bibliographies. Geneva, UNEP. 622 pp.*

NIO/UNEP (1978). *Directory of Indian Ocean Marine Research Centres*. UNEP Regional Seas Directories and Bibliographies. National Institute of Oceanography, Goa, India. 360 pp.*

UNEP/IOC (1980). *Directory of Caribbean Marine Research Centres*. UNEP Regional Seas Directories and Bibliographies. Geneva, UNEP. 500 pp.

IAEA/UNEP (1981). *Directory of Kuwait Action Plan Marine Science Centres*. UNEP Regional Seas Directories and Bibliographies. Geneva, UNEP. 100 pp.*

UNEP/CPPS (1981). *Directory of the South-East Pacific Marine Science Research Centres*. UNEP Regional Seas Directories and Bibliographies. Geneva, UNEP. 120 pp.*

UNEP/ FAO/ UNESCO/ WHO/ WMO/ IOC/ IAEA (1981). *Selected Bibliography on the Pollution of the Mediterranean Sea*. UNEP Regional Seas Directories and Bibliographies. Geneva, UNEP. 135 pp.*

UNEP/UN-ECA/UNESCO (1982). *Directory of Marine Research Centres in Africa*. UNEP Regional Seas Directories and Bibliographies. Rome, FAO. 254 pp.

UNEP (1984). *Bibliography of the Marine Environment in the Kuwait Action Plan Region, 1972-1981*. UNEP Regional Seas Directories and Bibliographies. Rome, FAO. 52 pp.

UNEP (1984). *Bibliography of the Marine Environment in the South Asian Seas*. UNEP Regional Seas Directories and Bibliographies. Rome, FAO. 39 pp.

UNEP/FAO (1984). *Bibliography of the Marine Environment in the East Asian Seas Region*. UNEP Regional Seas Directories and Bibliographies. Rome, FAO. 68 pp.

UNEP/FAO (1984). *Directory of Marine Environmental Centres in East Asian Seas*. UNEP Regional Seas Directories and Bibliographies. Rome, FAO. 138 pp.

UNEP/PSA/SPREP/UG (1984). *Directory of Coral Reef Researchers in the Pacific*. UNEP Regional Seas Directories and Bibliographies. Rome, FAO. 101 pp.

UNEP/FAO (1985). *Directory of Marine Environmental Centres in Caribbean*, 2nd ed. UNEP Regional Seas Directories and Bibliographies. Rome, FAO. 214 pp.

SPREP/FAO (1985). *Directory of Marine Environmental Centres in Mediterranean*, 3rd ed. UNEP Regional Seas Directories and Bibliographies. Rome, FAO. 302 pp.

SPREP/UNEP/FAO (1985). *Directory of Marine Environmental Centres in South Pacific*. UNEP Regional Seas Directories and Bibliographies. Rome, FAO. 147 pp.

UNEP/UNESCO/UN-DIESA (1985). *Bibliography on Coastal Erosion in West and Central Africa*. UNEP Regional Seas Directories and Bibliographies. Rome, FAO. 92 pp.

UNEP/FAO (1985). *Directory of Marine Environmental Centres in Indian Ocean and Antarctic Region*. UNEP Regional Seas Directories and Bibliographies. Rome, FAO. 226 pp.

CCA/UNEP (1985). *Directory of Environmental Education Institutions, Programmes and Resource People in the Caribbean Region*. UNEP Regional Seas Directories and Bibliographies. Rome, FAO. 89 pp.

UNEP/FAO (1985). *Bibliography of the Marine Environment in the Mediterranean, 1978-1984*, 2nd ed. UNEP Regional Seas Directories and Bibliographies. Rome, FAO. 151 pp.

Eldredge, L.G. (1987). *Directory of Coral Reef Researchers in Pacific*, 2nd ed. UNEP Regional Seas Directories and Bibliographies. Rome, FAO. 104 pp.

Eldredge, L.G. (1987). *Bibliography of Marine Ecosystems in Pacific Islands*. UNEP Regional Seas Directories and Bibliographies. Rome, FAO. 72 pp.

UNEP (1987). *Directory of Organizations and Organizational Units Co-ordinating or Contributing to the Co-ordination of the Action Plans Related to the Regional Seas Programme*. Nairobi, UNEP. 34 pp.

PNUMA/CPPS (1987). *Directorio de Centros de Investigacion Marina: Pacifico Sudeste*. UNEP Regional Seas Directories and Bibliographies. Rome, FAO. 145 pp.

*Out of print.

Titles in print are obtainable from:
Oceans and Coastal Areas Programme Activity Centre,
United Nations Environment Programme,
P.O. Box 30552, Nairobi, Kenya.